90 0929783 3

D1765280

**University of Plymouth**
**Charles Seale Hayne Library**
Subject to status this item may be renewed
via your Primo account

http:/primo.plymouth.ac.uk
Tel: (01752) 588588

# Arthropod Biology and Evolution

WITHDRAWN
FROM
UNIVERSITY OF PLYMOUTH
LIBRARY SERVICES

Alessandro Minelli · Geoffrey Boxshall
Giuseppe Fusco

Editors

# Arthropod Biology and Evolution

Molecules, Development,
Morphology

 Springer

*Editors*
Alessandro Minelli
Giuseppe Fusco
Department of Biology
University of Padova
Padova
Italy

Geoffrey Boxshall
Zoology Department
Natural History Museum
London
UK

PLYMOUTH UNIVERSITY
9009297833 .

ISBN 978-3-642-36159-3      ISBN 978-3-642-36160-9   (eBook)
DOI 10.1007/978-3-642-36160-9
Springer Heidelberg New York Dordrecht London

Library of Congress Control Number: 2013934723

© Springer-Verlag Berlin Heidelberg 2013
This work is subject to copyright. All rights are reserved by the Publisher, whether the whole or part of the material is concerned, specifically the rights of translation, reprinting, reuse of illustrations, recitation, broadcasting, reproduction on microfilms or in any other physical way, and transmission or information storage and retrieval, electronic adaptation, computer software, or by similar or dissimilar methodology now known or hereafter developed. Exempted from this legal reservation are brief excerpts in connection with reviews or scholarly analysis or material supplied specifically for the purpose of being entered and executed on a computer system, for exclusive use by the purchaser of the work. Duplication of this publication or parts thereof is permitted only under the provisions of the Copyright Law of the Publisher's location, in its current version, and permission for use must always be obtained from Springer. Permissions for use may be obtained through RightsLink at the Copyright Clearance Center. Violations are liable to prosecution under the respective Copyright Law.
The use of general descriptive names, registered names, trademarks, service marks, etc. in this publication does not imply, even in the absence of a specific statement, that such names are exempt from the relevant protective laws and regulations and therefore free for general use.
While the advice and information in this book are believed to be true and accurate at the date of publication, neither the authors nor the editors nor the publisher can accept any legal responsibility for any errors or omissions that may be made. The publisher makes no warranty, express or implied, with respect to the material contained herein.

Printed on acid-free paper

Springer is part of Springer Science+Business Media (www.springer.com)

# Contents

# Contributors

**Wasiu A. Akanni** Department of Biology, The National University of Ireland, Callan Building, Maynooth, Co. Kildare, Ireland

**Francesca Bortolin** Department of Biology, University of Padova, Via U. Bassi 58 B, I 35131 Padova, Italy

**Geoffrey A. Boxshall** The Natural History Museum, Cromwell Road, London SW7 5BD, UK

**Lahcen I. Campbell** Department of Biology, The National University of Ireland, Callan Building, Maynooth, Co. Kildare, Ireland

**Robert Carton** Department of Biology, The National University of Ireland, Callan Building, Maynooth, Co. Kildare, Ireland

**Steven R. Davis** Division of Entomology, Natural History Museum, Department of Ecology, and Evolutionary Biology, University of Kansas, 1501 Crestline Drive – Suite 140, Lawrence, KS 66045, USA

**Jason A. Dunlop** Museum für Naturkunde, Leibniz Institute for Research on Evolution and Biodiversity, Humboldt University Berlin, Invalidenstrasse 43, 10115 Berlin, Germany

**Gregory D. Edgecombe** Department of Palaeontology, Natural History Museum, Cromwell Road, London SW7 5BD, UK

**Michae S. Engel** Division of Entomology, Natural History Museum, Department of Ecology, and Evolutionary Biology, University of Kansas, 1501 Crestline Drive – Suite 140, Lawrence, KS, 66045, USA

**Thomas Frase** Allgemeine und Spezielle Zoologie, Universität Rostock, Universitätsplatz 2, 18055 Rostock, Germany

**Giuseppe Fusco** Department of Biology, University of Padova, Via U. Bassi 58 B, I 35131 Padova, Italy

**Massimo Giorgini** Istituto per la Protezione delle Piante, CNR, Via Università 133, 80055 Portici, NA, Italy

**Gonzalo Giribet** Museum of Comparative Zoology, Department of Organismic and Evolutionary Biology, Harvard University, 26 Oxford Street, Cambridge, MA 02138, USA

**Steffen Harzsch** Department of Cytology and Evolutionary Biology, Zoological Institute and Museum, Ernst Moritz Arndt University of Greifswald, Soldmansstrasse 23, 17487 Greifswald, Germany

**Matthes Kenning** Department of Cytology and Evolutionary Biology, Zoological Institute and Museum, Ernst Moritz Arndt University of Greifswald, Soldmansstrasse 23, 17487, Greifswald, Germany

**David A. Legg** Department of Earth Science and Engineering, Imperial College London, London SW7 2AZ, UK; Department of Palaeontology, Natural History Museum, London SW7 2AZ, UK

**Rudolf Loesel** Institut für Biologie, RWTH Aachen, Lukasstr. 1, 52056 Aachen, Germany

**Diego Maruzzo** Department of Biology, University of Padova, Via U. Bassi 58 B, I 35131 Padova, Italy

**Alessandro Minelli** Department of Biology, University of Padova, Via U. Bassi 58 B, I 35131, Padova, Italy

**Armin P. Moczek** Department of Biology, Indiana University, 915 E. Third Street, Bloomington, IN 47405-7107, USA

**Bernard Moussian** Interfaculty Institute for Cell Biology, Animal Genetics, Eberhard-Karls University of Tübingen, Auf der Morgenstelle 28, 72076 Tübingen, Germany

**Eoin Mulvihill** Department of Biology, The National University of Ireland, Callan Building, Maynooth, Co. Kildare, Ireland

**H. Frederik Nijhout** Department of Biology, Duke University, 125 Science Drive, Durham 27708, USA

**Günther Pass** Department of Integrative Zoology, University of Vienna, Althanstr. 14, Vienna 1090, Austria

**Francesco Pennacchio** Dipartimento di Entomologia e Zoologia Agraria "F. Silvestri", Università di Napoli "Federico II", Via Università 100, 80055 Portici, (NA), Italy

**Davide Pisani** School of Biological Sciences and School of Earth Sciences, University of Bristol, Woodland Road, Bristol BS8 1UG, UK

**Jakub Prokop** Department of Zoology, Faculty of Science, Charles University in Prague, Viničná 7, 128 44 Prague, Czech Republic

**Stefan Richter** Allgemeine und Spezielle Zoologie, Universität Rostock, Universitätsplatz 2, 18055, Rostock, Germany

**Gerhard Scholtz** Institut für Biologie, Humboldt-Universität zu Berlin, Vergleichende Zoologie Philippstr. 13, 10115 Berlin, Germany

**Paul A. Selden** Paleontological Institute, University of Kansas, 1475 Jayhawk Boulevard, KS, Lawrence 66045, USA

**Andy Sombke** Department of Cytology and Evolutionary Biology, Zoological Institute and Museum, Ernst Moritz Arndt University of Greifswald, Soldmansstrasse 23, 17487, Greifswald, Germany

**Omar Rota Stabelli** IASMA Research and Innovation Centre, Fondazione Edmund Mach, Via E. Mach 1, 38010 San Michele all'Adige (TN), Italy

**Matthew S. Stansbury** Center for Insect Science, University of Arizona, 1007 E. Lowell Street, Tucson, AZ 85721-0106, USA

**Martin Stein** Danish Museum of Natural History, University of Copenhagen, Universitetsparken 15, DK 2100 Copenhagen, Denmark

**Michael R. Strand** Department of Entomology, University of Georgia, 120 Cedar Street, 413 Biological Sciences Building, Athens, GA 30602, USA

**Nikolaus U. Szucsich** Department of Integrative Zoology, University of Vienna, Althanstr. 14, Vienna, 1090, Austria

**Markus Tögel** Department of Integrative Zoology, University of Vienna, Althanstr. 14, Vienna, 1090, Austria

**Jennifer A. White** Department of Entomology, University of Kentucky, S-225 Agricultural Science Center N, Lexington 40546-0091, USA

**Christian S. Wirkner** Allgemeine und Spezielle Zoologie, Universität Rostock, Universitätsplatz 2, 18055, Rostock, Germany

**Harald Wolf** Institute of Neurobiology, University of Ulm, Albert-Einstein-Allee 11, 89081 Ulm/Donau, Germany

**Carsten Wolff** Institut für Biologie, Humboldt-Universität zu Berlin, Vergleichende Zoologie Philippstr. 13, 10115, Berlin, Germany

# An Introduction to the Biology and Evolution of Arthropods

## Alessandro Minelli, Geoffrey Boxshall and Giuseppe Fusco

In a recent paper in the journal *Arthropod Structure and Development*, Polilov (2012) has shown that 95% of the ca. 4,600 neurons forming the brain of a tiny parasitoid wasp are anucleate. This amazing correlate of miniaturization is just one of the latest unexpected discoveries in arthropod biology, one of those whose relevance goes far beyond the limits of the largest of living phyla. Such discoveries are of the highest general interest for biology and serve to remind us that arthropods are, indeed, an unparalleled source of facts and inspiration for biologists of every brand.

In terms of sheer numbers of species, the Arthropoda is by far the largest living phylum, comprising in excess of 1.2 million extant species including just over 1 million hexapods, nearly 112,000 chelicerates, about 67,000 crustaceans, and some 12,000 myriapods. In addition, although the number is hard to estimate, considerably more than 100,000 fossil arthropod species have also been described, about half of which are ostracod microfossils. It is clear that the arthropods have been megadiverse for at least 520 million years, since the Early Cambrian.

The unique evolutionary success of arthropods deserves an up-to-date comprehensive analysis from the perspective of comparative morphology of extant as well as fossil representatives of the phylum, and developmental biology, including developmental genetics and endocrinology. Indeed, these discoveries have prompted us to extend coverage even more widely to encompass additional topics from comparative genomics to endosymbiosis. This is the ambitious target of this book. Ambitious, especially because the increasing specialization of both descriptive and experimental research has forced the vast majority of researchers to focus not only on a selected set of problems, but also on a restricted range of taxa. In this respect, even the four main arthropod groups of traditional classifications (chelicerates, crustaceans, insects and myriapods) are already too numerous and diverse to be adequately covered by one scientist's expertise. This is amply illustrated by the titles and scope of major treatises of the recent past, dealing with more or less diverse aspects of the biology of either insects or crustaceans, or more rarely of arachnids or myriapods. Contrary to this largely unavoidable trend towards increased specialization, we have tried to produce an updated overview of arthropod biology and evolution articulated in a series of

A. Minelli (✉) · G. Fusco
Department of Biology, University of Padova,
Via U. Bassi 58 B, I 35131, Padova, Italy
e-mail: alessandro.minelli@unipd.it

G. Fusco
e-mail: giuseppe.fusco@unipd.it

G. Boxshall
Department of Life Sciences, Natural History
Museum, London, SW7 5BD, UK
e-mail: G.Boxshall@nhm.ac.uk

A. Minelli et al. (eds.), *Arthropod Biology and Evolution*,
DOI: 10.1007/978-3-642-36160-9_1, © Springer-Verlag Berlin Heidelberg 2013

chapters devoted to morphology, palaeontology and developmental biology, together with selected aspects of physiology and molecular biology, from a cross-phylum comparative viewpoint. Based on this phylum-wide perspective, it will be possible to appreciate the most advanced levels of knowledge in arthropod biology and evolution thus far attained, with respect to all the main arthropod lineages, and to identify less thoroughly investigated areas to be prioritized in future research.

Arthropod phylogeny has been repeatedly and profoundly revised during the last quarter of century, eventually providing an increasingly robust, although far from definitive, background against which to read the diversity, and to reconstruct the history, of arthropod morphology, developmental biology and physiology in their different expressions. In Chap. 2, Giribet and Edgecombe set the Arthropoda in its phylogenetic context, by examining their relationships with other protostome phyla and by reviewing the competing hypotheses of the Articulata (comprising the arthropods and annelids) versus the Ecdysozoa (which unites arthropods, onychophorans and tardigrades with a group of mostly pseudocoelomate animals with which they share a cuticle that is moulted). Their conclusions on the monophyly of Ecdysozoa, Panarthropoda and an Onychophora plus Arthropoda clade provide the context for an evaluation of the internal phylogeny within the Arthropoda, which is now nearly universally accepted as monophyletic.

Giribet and Edgecombe examine relationships between the major arthropod lineages—Pycnogonida, Euchelicerata, Myriapoda and Tetraconata (or Pancrustacea) (see Table 1.1 for a summary of major taxon composition). Reviewing the broad sweep of evidence, they focus on the emerging picture of the Pycnogonida and Euchelicerata as forming a clade, the Chelicerata, with the Mandibulata as its most likely sister group. Within the Mandibulata, a monophyletic Myriapoda constitutes the sister group to the Tetraconata, which comprises a paraphyletic 'Crustacea' from within which a monophyletic Hexapoda arose. While the arthropod tree of life is considerably more stable

than twenty years ago, uncertainty remains particularly concerning the interrelationships between arachnid orders and crustacean classes. The sister group of hexapods appears to be the remipedes, but key questions remain: do the cephalocaridans group with remipedes, and are either branchiopods or malacostracans more closely related to remipedes and hexapods, or to each other?

Modern molecular estimates of the divergence events between the deep arthropod clades, such as Chelicerata versus Mandibulata, date these events to the Ediacaran Period. However, Giribet and Edgecombe (and Edgecombe and Legg in Chap. 15) note that the Ediacaran has not yet yielded credible body fossils, or trace fossils of arthropods. The dating of arthropod diversification needs further refinement based on improved clock methods and careful integration of fossil constraints.

Arthropod comparative genomics is in its infancy but is growing fast. Pisani et al. (Chap. 3) present an overview of arthropod mitochondrial and nuclear genomic resources, before exploiting the available genomic information to investigate the evolutionary origin of novel proteins (orphan gene families) in the arthropod proteome. The inclusion of the first genomic-scale data set for the Onychophora gave them an unprecedented opportunity to identify the orphan protein families that arose in the stem arthropod lineage of the Arthropoda. Pisani et al. found more than 300 complete arthropod mitochondrial genomes available, but taxonomic sampling is extremely biased towards economically relevant species, even though most major orders and classes are now represented. The majority of the currently available arthropod genomes are from closely related species (mostly insects). Overall, they conclude that the current genomic-scale information available across the Arthropoda is still too fragmentary to allow the development of a coherent view of arthropod genome evolution.

The most surprising result to emerge from Pisani et al.'s analysis was that the deepest nodes in the ecdysozoan phylogeny are not characterized by above average acquisitions of new gene families. All internodes within

**Table 1.1** A list of high-level arthropod taxa, with their composition and important synonyms

| Taxon | Composition | Synonyms |
|---|---|---|
| Cormogonida | Arthropoda excl. Pycnogonida | |
| **Paradoxopoda** | Euchelicerata + Myriapoda or Chelicerata + Myriapoda | =Myriochelata |
| **Chelicerata** | Pycnogonida + Euchelicerata | =Chelicerophora =Cheliceriformes |
| **Euchelicerata** | Xiphosura + Arachnida | =Chelicerata |
| Stomothecata | Scorpiones + Opiliones | |
| Haplocnemata | Solifugae + Pseudoscorpiones | |
| **Myriapoda** | Chilopoda + Symphyla + Pauropoda + Diplopoda | |
| Dignatha | Pauropoda + Diplopoda | |
| Schizoramia | Chelicerata + Crustacea[a] | |
| **Mandibulata** | Myriapoda + Crustacea[b] + Hexapoda | |
| **Tetraconata** | Crustacea[c] + Hexapoda | =Pancrustacea |
| 'Altocrustacea' | All Tetraconata to the exclusion of Oligostraca | |
| Atelocerata | Myriapoda + Hexapoda | =Tracheata |
| **Crustacea**[c] | Mystacocarida + Ostracoda + Ichthyostraca + Branchiopoda + Thecostraca + Tantulocarida + Malacostraca + Copepoda + Remipedia + Cephalocarida | |
| **Ichthyostraca** | Branchiura + Pentastomida | |
| Oligostraca | Ostracoda + Ichthyostraca | |
| **Branchiopoda** | Anostraca + Phyllopoda | |
| Multicrustacea | Malacostraca + Copepoda + Thecostraca | |
| **Thecostraca** | Facetotecta + Ascothoracida + Cirripedia | |
| Miracrustacea | Remipedia + Cephalocarida + Hexapoda | |
| Xenocarida | Remipedia + Cephalocarida | |
| **Hexapoda** | Entognatha + Insecta | |
| Entognatha | Collembola + Protura + Diplura | |
| Nonoculata | Protura + Diplura | |
| Ellipura | Protura + Collembola | =Parainsecta |
| **Insecta** | Archeognatha + Dicondylia | |
| **Dicondylia** | Zygentoma + Pterygota | |
| **Pterygota** | Odonata + Ephemeroptera + Neoptera | |
| Palaeoptera | Odonata + Ephemeroptera | |
| Metapterygota | Odonata + Neoptera | |
| Chiastomyaria | Ephemeroptera + Neoptera | |
| **Neoptera** | Polyneoptera + Paraneoptera + Holometabola | |

Some of them represent conflicting phylogenetic hypotheses. Names of major taxa in bold are used recurrently throughout this volume
[a] Regarded in this hypothesis as monophyletic
[b] Regarded in this hypothesis as either monophyletic or paraphyletic
[c] Regarded in this hypothesis as paraphyletic

Ecdysozoa (on the path leading to Arthropoda and within Arthropoda as well) exhibit roughly the same rate of new protein acquisition per million years. Constancy of the rate of protein family acquisition through time (from the Precambrian to the Jurassic) suggests that this rate might represent the neutral background rate of new protein family origination in Ecdysozoa.

This neutral rate is modified only at one internode, representing the stem lineage of a large group of insects. Along this lineage, the rate significantly increased, suggesting that orphan gene family acquisition was an important phenomenon in the evolution of this group.

The spectacular diversity of arthropod morphology and lifestyles is matched by an impressive variety of developmental trajectories. Ontogenetic differences may involve all embryonic stages and levels from gene expression, cleavage and gastrulation, germ band formation and growth, to segmentation and morphogenesis. In Chap. 4, Scholtz and Wolff review arthropod embryology focusing their comparative treatment on early arthropod development, encompassing the cleavage process, germ band formation and differentiation.

The two main cleavage modes, superficial and total, occur in arthropods, but the variation within arthropods spans the entire spectrum of superficial, total and mixed cleavages, as well as determinate and indeterminate cleavage modes. These are distributed across arthropod taxa in a complex pattern that does not allow for unambiguous reconstruction of the ancestral cleavage mode for the Arthropoda. Scholtz and Wolff conclude, for example, that if pycnogonids are the sister group to the remaining chelicerates, the cleavage type of the chelicerate stem species is ambiguous. In contrast, current views of tetraconate phylogeny led them to infer that the stem species of the Tetraconata underwent total cleavage. Scholtz and Wolff suggest a different perspective focusing on the pattern of arrangement of blastomeres at the four-cell stage which, they consider, might be a good starting point for a re-evaluation of arthropod cleavage patterns in general.

One of the characteristic features of arthropod development is the germ band, which is an elongate field of blastoderm cells lying at the surface on one side of the yolky egg. It is mostly formed by cell migration and aggregation and represents the embryo proper. The germ band stretches along the longitudinal axis of the embryo and marks the future ventral side, where structures such as segmental furrows and limbs

are first formed. A germ band is formed in the ontogeny of representatives of every large arthropod subgroup, and it has been considered as part of the arthropod ground pattern. Scholtz and Wolff review exceptions where a germ band is not formed and conclude that the occurrence of a germ band is related to the amount of yolk. As such, they consider that the presence or absence of a germ band might be prone to convergence.

Finally, Scholtz and Wolff examine the assumption that the posterior growth zone of arthropods buds or produces segments. They conclude that, as a general mode, arthropod segments are formed one by one in a general anteroposterior sequence. However, they regard as problematic the view that growth and patterning are initiated simultaneously by a posterior growth zone.

In Chap. 5, Minelli and Fusco review the complex and multifaceted topic of post-embryonic development in arthropods. They consider that this aspect of arthropod biology is in need of a new conceptual framework. Arthropod post-embryonic development involves two aspects of segmental organization: the production schedule of segments and the differentiation of these segments resulting in the patterning of the main body axis. Neither process is necessarily completed at the beginning of post-embryonic life. Minelli and Fusco tease apart arthropod post-embryonic development into concurrent processes and describe them based on the standard periodization provided by the succession of moults. However, they stress that the 'cuticular view' imposed by the moult-based periodization of arthropod development is not always the best framework for analysing the interactions between underlying developmental processes.

Segment production schedules are discussed in detail, moving on from the basic anamorphosis and epimorphosis modes to introduce more subtle patterns. Minelli and Fusco provide a wealth of examples from extant and fossil taxa, often selected to remind the reader of the diversity across the phylum. Their review of the relationship between segment production and segment articulation in trilobites supports the conclusion that fixation of tagmosis was independent from segment production in this taxon.

They examine the relationship between adulthood and the onset of maturity and conclude that we should distinguish between mature stages, characterized by reproductive maturity, and adult stages, characterized by a morphologically invariant final condition. The uncoupling of these two classes of developmental phenomena is a key evolutionary feature of heterochronic change. They briefly look at growth and modes of growth in arthropods, addressing topics such as size increment (including a look at Dyar's rule), growth compensation, the general concept of target ontogenetic trajectory and allometric growth.

Another major theme explored is the number of moults in the arthropod life cycle. The number of post-embryonic stages varies conspicuously across the arthropods and often also within more basal clades. Minelli and Fusco note that there are generally fewer than 15 moults and provide a tabular comparison of data from across the major taxa. They also highlight the existence of sexual dimorphism in number of moults, as well as individual variation in some taxa. Minelli and Fusco address the homology of stages between different arthropod lineages. They observe that terms used for the different post-embryonic stages of arthropods are confused and confusing and were often based on poorly supported homology. They look at the importance of larvae and metamorphosis in arthropod life histories. First asking what is a larva and what criteria can we use to define the term? Here, as throughout this chapter, numerous examples of larval types are discussed and terminological problems exposed. Minelli and Fusco end with a short account of the evolutionary patterns of arthropod post-embryonic development. They identify main trends, from the reduction in post-embryonic stages to increasing complexity and hypermetamorphosis.

Moving from understanding patterns of moults to the process of moulting itself, Nijhout (Chap. 6) reviews recent developments in our understanding of the control of growth and moulting in arthropods. In contrast to embryonic development, which appears to be largely controlled by gene regulatory cascades and networks, and gene products that move by diffusion, post-embryonic growth and differentiation in arthropods are controlled almost entirely by circulating hormones and secreted growth factors. Only a handful of developmental hormones control an extraordinarily diverse array of post-embryonic developmental processes, from growth to moulting and metamorphosis, including the development of alternative phenotypes in response to environmental signals (polyphenism).

Growth of arthropods has two components: episodic growth of the exoskeleton and more or less continuous growth of biomass. Somatic growth is controlled by hormones and secreted growth factors that regulate the onset, rate and duration of growth. The developmental regulation of growth and size can be partitioned into five questions which Nijhout addresses in turn: (1) How is moulting controlled? (2) What controls the size increment at each moult? (3) What controls the growth rate between moults? (4) What controls the timing of a moult? (5) What controls the cessation of moulting when the final size is reached?

Three classes of hormones appear to dominate in arthropods: the insulins, the ecdysteroids and the juvenoids. Some hormones, such as ecdysone, are universal across the Arthropoda, whereas others are taxon specific, such as juvenile hormone which is found in insects and androgenic hormone which is found in decapod crustaceans. Even though their function has been studied in relatively few arthropods, Nijhout considers it safe to assume that insulins play a general role in the regulation of growth while noting that if both insulin and ecdysone are required for normal growth, then variation in either could control growth.

The control process is briefly reviewed: prothoracicotropic hormone stimulates ecdysone biosynthesis and secretion, but ecdysone can have a broad diversity of effects depending on stage in the life cycle and on target tissue. In addition to inducing moulting, ecdysone stimulates context-dependent gene transcription and controls cell division, tissue growth, the switch in commitment prior to metamorphosis and the development of some seasonal polyphenisms. Nijhout concludes that growth of integument is an ecdysone-triggered episodic event, but that growth of internal organs is more or less continuous and nutrition dependent. Growth of internal organs thus requires cell growth and proliferation that is independent of the ecdysone-induced round of cell division of the epidermal cells.

Maruzzo and Bortolin consider regeneration in arthropods, and the focus of Chap. 7 is primarily limb regeneration. Only a few arthropod species are able to regenerate parts of the trunk, and even then to a limited extent; however, most arthropods are capable of regenerating organs and tissues to some degree. Limb regenerative potential fluctuates across the different arthropod groups, and many factors influence the outcome of a regeneration process, such as developmental stage, limb type and amputation level. Limb regeneration typically involves only limited cell dedifferentiation, and it is likely that slightly dedifferentiated cells produce only cells of the same type. Arthropods show good tissue regenerative potential, and regeneration can be experimentally induced in many tissues, but Maruzzo and Bortolin necessarily focus their review on the most studied models. They note that the relationship between moulting and regeneration is not yet clear, and while moulting is necessary for proper limb regeneration, at least for some insects there is no evidence of loss of regenerative potential after the final moult. In insects, it has been shown that the presence of high levels of ecdysteroids inhibits the initiation of regeneration. The observation that regenerating insects show a longer intermoult period may be correlated with a delay in the appearance of the ecdysteroid peaks, but the mechanism by which regeneration influences hormone levels is as yet unknown.

Reports of limited trunk regeneration are reviewed in this chapter and include regeneration of the telson of horseshoe crabs and malacostracan crustaceans and of the caudal appendage of a beetle larva. There are also reports of complete regeneration of one or more posterior trunk segments; however, there was high mortality in most of these studies, suggesting that both wound healing and intrinsic developmental factors might explain the limited arthropod trunk regenerative potential.

Maruzzo and Bortolin focus their discussion on the phylogenetic diversity of limb regenerative potential and on the main developmental and physiological aspects. Many arthropods have mechanisms facilitating limb loss through limb breakage at specific points along the limb. The variation in mechanisms facilitating limb loss, autotomy, is explored, and other forms of autotomy, such as autospasy and autotilly, are briefly reviewed. They conclude that our knowledge of developmental events connected to limb regeneration is based on relatively few studies and that better comparative data are needed before the variation in regenerative potential can be fully understood.

In Chap. 8, Moussian reviews the state of knowledge on the structure and function of arthropod cuticle. Cuticle is a multifunctional coat, an exoskeleton, which defines and stabilizes body shape both inside and out. It prevents dehydration and infection and protects against predators. The physical properties of cuticle may be stage specific, as in insect larvae where the body cuticle is soft and elastic serving as a hydrostatic jacket, while the head capsule comprises hard cuticle allowing for effective muscle attachment. Hard cuticle is typical of adult arthropods, and sclerites of hard cuticle are joined by soft cuticle rendering the exoskeleton pliable.

Virtually all types of cuticle are organized in three ultrastructurally distinct horizontal layers. In this chapter, Moussian uses the latest unifying nomenclature for these layers: the surface envelope, the epicuticle and the procuticle. The characteristics of each of these layers are considered, and the epithelial cells that produce cuticle plus the plasma membrane underlying this epithelium are examined in detail.

Commonly, cuticles are composed of the polysaccharide chitin, glycosylated and unglycosylated proteins, catecholamines, and lipids and waxes. The latter are mainly coating the surface and are implicated in preventing water loss. Additionally, minerals such as calcite may be incorporated. Species-, stage- and tissue-specific differences mainly rely on lipid and wax composition, differences in proteins, the amounts of chitin and the degree of covalent cross-links by catecholamines, for example. Moussian also reviews the secretion of cuticle material and emphasises the importance of processes such as cross-linking of cuticle components, sclerotization and melanization and the role of specialized components such as resilin. Finally, tracheal cuticle is described and its unique features highlighted. This provides Moussian with an opportunity to review mechanisms controlling cuticle differentiation.

In Chap. 9, Fusco and Minelli discuss arthropod segmentation and tagmosis patterns in a wide range of contexts, from developmental biology to phylogeny. Throughout, they address the disparity and inconsistencies in the use of morphological terminology related to segmentation and tagmosis across the diversity of arthropod taxa. However, the meaning and usage of the terms 'segmentation' and 'tagmosis' are analysed here with respect to the adult, and the focus is primarily on the post-cephalic section of the body, since head segmentation is discussed in Chap. 10.

Fusco and Minelli stress that an improved understanding of arthropod body organization can be obtained by dissociating the serial homology of individual periodic structures (segmentation) from the concept of the segment as a body module. They define segmental structures and segmental elements within the trunk and examine the significance of the telson. They then look at tagmosis which represents a form of higher-level modularity along the main body axis. However, they note that there is little agreement on how tagmata should be defined and their boundaries characterized, and they recognize that the concept of tagma is to a large extent arbitrary.

Fusco and Minelli survey arthropod diversity comparing various aspects of their morphological patterning including interspecific and intraspecific variation in the number of post-cephalic body segments. They review the forms of segmental mismatch, focusing on cases where the mismatch involves comparable segmental structures, for example between dorsal and ventral serial sclerites. They note other types of discordance, such as that between segmental structures of the internal anatomy and serial structures of the exoskeleton. In arthropods, there are widespread forms of periodic body pattern that are in register with the segmental organization of major structures. However, Fusco and Minelli also note cases of structures and processes that show forms of periodic pattern with a less strict connection to the more obvious external segmental organization. They end their consideration of segmentation by discussing the difficulties in answering apparently simple questions concerning the homology of trunk segments with the same ordinal post-cephalic position in series that exhibit different numbers of elements.

Turning their attention to tagmosis, Fusco and Minelli employ the same principles to examine examples of dorsoventral mismatch between tagmata and the homology of tagmata. Finally, they explore segmental pervasivity—how much of the anatomy of a given domain of the body exhibits segmental organization, irrespective of whether the different segmental structures are in register or not, as well as tagmatic pervasivity—the level of integration of the segmental elements of a given tagma. The need for greater precision in use of terms and concepts related to segmentation and tagmosis is very apparent from this chapter.

One might have assumed, wrongly it appears, that the exact composition and origin of the 'arthropod head' were well understood, but Richter et al. (Chap. 10) consider this to be an enduring problem in arthropod phylogenetic reconstruction. The 'head' is widely used for mandibulatan arthropods, and its use is linked to the presence of a dorsal cephalic shield or head capsule. Applying the concept of this head to chelicerates is particularly challenging, although the availability of gene expression data has greatly facilitated comparisons across the Arthropoda as a whole. Richter et al. also consider the importance of internal anatomical systems such as the endoskeleton and the brain, in outlining recent developments concerning the concept of the arthropod head. They then expand the discussion to take in the onychophoran head: in the mandibulatan head, there are three additional posterior segments fused with the anterior part of the head compared with the onychophoran head.

Another unresolved problem reviewed by Richter et al. is the fate of the onychophoran antenna. The segmental affinities of the labrum have been debated intensively, but the once-favoured tritocerebral origin has been more or less discounted in recent years. Evidence relevant to the potential transformation of the onychophoran antenna into the arthropod labrum is reviewed here.

Richter et al. finally focus on the fossil record and consider the possible nature of the so-called great appendage of megacheirans, as well as evidence relating to the number of appendage-bearing segments incorporated into the anterior-most unit covered by a single dorsal shield, and to the constancy of this number in different Cambrian panarthropod taxa. In their summary, they conclude that there is fossil evidence that the last common ancestor of Chelicerata and Mandibulata possessed a head comprising the ocular region and at least three, probably four, appendage-bearing segments. The anterior appendage inserts laterally to the hypostome/labrum and probably represents the deutocerebral appendage, but a smaller appendage-like structure might have been present anterior to it. Post-antennular appendages display little differentiation other than a gradual shift anteriorly towards limbs increasingly adapted to feeding.

In Chap. 11, Boxshall takes a closer look at arthropod limbs and aims to integrate the wealth of new data emerging from morphological and embryological studies, from developmental genetics and from novel fossils. The distinction between segments and annuli has been highlighted in the past and is based primarily on musculature. In limbs that possess a mix of segments and annuli, the segments tend to appear before the annuli, but Boxshall asks whether there is evidence from developmental genetics to support this distinction. The early establishment of

the proximo-distal (P-D) axis by the leg gap genes is a general feature of limb patterning during development in all arthropods and, downstream the Notch signalling pathway, plays a central role in segmentation along the P-D axis of the leg. The mechanisms controlling the formation of true segments and of annuli along the P-D axis of the limb are compared. Evidence from knockdown studies indicates that certain genes are known to affect tarsal subdivision but not basic leg segmentation, so the patterning mechanisms for leg segments and leg annuli, while similar, exhibit important differences in detail. Muscle patterns may be the key criterion for anatomists, but Boxshall points out that relatively little is known about the mechanisms governing adult leg myogenesis in the *Drosophila* leg model, but also that this is not a good model here since both segments and annuli are everted simultaneously from the imaginal disc.

Boxshall briefly reviews the two basic limb types of arthropods, the single-axis antennule originating on the deutocerebral segment and the fundamentally biramous limb present on post-antennulary segments. He focuses on the apparently profound morphological gap between an elongate sensory antennule and a short feeding chelicera, summarizing evidence supporting the hypothesized transition from the great appendage of megacheirans to the chelicera of chelicerates. The discovery of a new Silurian fossil, with long flexible antenniform chelicerae, is highly relevant to this debate. After comparing antenna and leg development in *Drosophila*, Boxshall notes that shared features indicate that despite some significant differences, the antennules and post-antennulary limbs of arthropods can be viewed as serial homologues. However, specification of the anterior-most limb as the antennule ensures that it develops as a single axis rather than biramous limb.

The morphological characteristics of the major structures of the arthropod limb are briefly examined. Comparative data from across the arthropods show that homologous patterning domains do not necessarily mark homologous morphological domains. It seems unlikely therefore that gene expression patterns will provide reference points allowing the identification of homologies, for example, between the component segments of chelicerate and mandibulatan walking limbs. However, a possible exception might be limb components with very specific functional attributes that are reflected in cellular physiology. The epipodites of the branchiopodan trunk limb and malacostracan pereopod, for example, express several genes that are not expressed elsewhere and which are presumably linked to specific cellular functions.

A close look at the early fossil record of insects was vital for Engel et al. (Chap. 12) as they reviewed the timing of the origin of insect flight. They stress that insect wings evolved only once, that is, the Pterygota is monophyletic, and in order to date this event, they consider in detail the often controversial records of pre-Carboniferous fossil hexapods. The first wings preserved in the fossil record, from the transition period between Early and Late Carboniferous (about 318 Mya), are much younger than any estimate of the age of Pterygota, and younger than the fragmentary remains of pterygotes from the Devonian. While it was only by the time of the Carboniferous coal measures that a truly diverse fauna of winged insects began to appear, Engel et al. consider that the timing of the origin of wings can be pushed back from the Carboniferous to the earliest Devonian. Engel et al. conclude that remaining uncertainty regarding the basal lineages of Pterygota renders it difficult to distinguish between competing interpretations of polarity relative to the form of the wing articulation. However, they considered that the principal lineages important for resolving basal relationships can now be characterized and that the pivotal phylogenetic uncertainties have at least been identified.

Engel et al. point to a growing body of developmental and morphological evidence in support of the inference that the wing is largely a paranotal extension which integrated appendage-patterning modules to develop a functional articulation incorporating elements of the upper pleuron. After integrating palaeontological, neontological and developmental evidence, they conclude that there is evidence for a developmental ground plan in Hexapoda that produced paranotal extensions of the thorax and that, subsequently, through the integration of appendage-patterning modules, such as those present in gills, or legs, a functional articulation developed incorporating dorsal elements of the pleuron. This provided a functional wing and the basis for further refinements of the pterygote wing, such as in wing shape, venation and the structure of the articulation of the wing to the thorax. Interestingly, they note that definitive prothoracic wing-like structures have been documented, although evidence for articulations is lacking, and that nearly a full developmental programme for wing formation has been demonstrated in the prothorax of holometabolous and hemimetabolous insects. They conclude that it appears more likely that wings were part of the ground plan for the hexapods only in the thorax and that a wing is more likely an amalgamation of tergal and pleural outgrowths which develop according to the redeployment of limb-patterning genes and portions of their pathways, as opposed to a modification of such structures as gills, epipodites, styli or other limbs that share similar developmental modules.

In Chap. 13, Loesel et al. focus on the central nervous system of arthropods and identify key common architectural principles of the arthropod ventral nerve cord and brain and highlight important evolutionary trends of these structures. They note that in arthropods, the basic segmentation of the ventral nerve cord matches body segmentation, in the form of segmental ganglia connected by paired connectives. The correspondence is closest for the more anterior regions of the body, although the fusion pattern of segmental ganglia does not always match the expressed external body segmentation. Loesel et al. compare the tract patterns of the central nervous system across the major arthropod taxa and identify the elements that are stereotypic and tend to be conserved.

The arthropod nervous system provides a wealth of information that can contribute both to our understanding of the phylogeny of arthropods and to the elucidation and description of the evolutionary transformations that have occurred within the arthropod brain. In this chapter, Loesel et al. highlight the important contribution that the rapidly expanding discipline of neurophylogeny is making to the current debate on arthropod phylogeny.

Loesel et al. conclude by attempting to reconstruct the ground pattern of the arthropod central nervous system. The three preoral neuromeres of the arthropod brain are the protocerebrum (ocular segment), deutocerebrum (antennulary/chelicera segment) and tritocerebrum. They note that the axons of bilaterally symmetrical median eyes project into a protocerebral neuropil and review variation across the arthropods. They also consider the pattern of the input of the lateral eyes into the protocerebrum and how these lateral eyes develop. The composition of the preoral frontal commissure (the stomatogastric bridge) is analysed and provides further detail of the innervation of the oesophagus and anterior part of the gut. This ground pattern can now be defined in impressive detail—to include information such as the number of serotonergic neurons present in each hemiganglion of the ventral nerve cord.

Wirkner et al. (Chap. 14) begin by describing the essential features of the arthropod circulatory system. The exoskeleton encloses a liquid-filled body cavity, the haemocoel, containing haemolymph which bathes all organs and tissues. Circulation of haemolymph is actively forced by pumping hearts, which are typically strongly muscularized sections of the vascular system. Wirkner et al. focus their review on the functional and evolutionary morphology of these organs, to provide a comprehensive picture of their diversity and evolutionary transformations undergone in the context of major environmental transitions. The arthropod vascular system exhibits clear segmental organization with individual elements reflecting an iterative configuration in a number of segments, even in unrelated lineages. Wirkner et al. explore the features of the segmental set of circulatory organ structures that might be attributable to the ground pattern of arthropods.

The vascular system of arthropods exhibits a broad spectrum of complexity. Some arthropods have a compact heart, and others have an extensive vascular system with peripheral capillarization. Fundamentally, however, it is an open system since no vessels lead directly back into the heart. In all arthropods, the haemolymph is collected in the pericardial sinus before it enters the heart via the ostia. The degree of variation in structural and functional complexity in the circulatory system is striking. The cardiovascular parts can be highly sophisticated, as in most chelicerates and malacostracan crustaceans, while in other groups, such as copepods and insects, it comprises only the dorsal vessel. The greatest variation is found in the arterial systems: reductions are apparent in many lineages, and a decrease in arterial complexity is often correlated with decreasing body size. Reduction in complexity of, and loss of, lateral cardiac arteries is common in spiders and malacostracans and is often accompanied by the loss of the posterior aorta. The anterior aorta is rarely reduced, probably due to its functional significance in supplying the cephalic region. In contrast, in some other lineages, such as the pulmonate arachnids, there is an increase in structural complexity of the vascular system.

The circulatory system fulfils an enormous range of physiological functions in arthropods, but the most important driver behind the evolution of an effective circulation system was probably the improvement of oxygen transport. The degree of concentration of the respiratory organs, together with the constraint for the shortest possible pathway to the heart, resulted in the greatest architectural transformations. In arthropods with tracheal systems, the circulatory system lost the function of oxygen transportation and such terrestrial forms are generally characterized by relatively simple vascular systems. Wirkner et al. note that the circulatory system acquired completely new tasks and features in connection with the evolution of flight in insects, such as tracheal ventilation and thermoregulation.

In their focus on fossils, Edgecombe and Legg (Chap. 15) stress that fossils provide glimpses of extinct morphologies which can contribute unique character combinations to phylogenetic analyses. In addition, the temporal information provided by fossils is vital for inferring divergence dates, fossils being the usual source of minimal divergence dates for calibrating nodes in molecular trees. Modern methods of molecular dating use relaxed clocks and probabilistic calibrations that can incorporate uncertainties in the fossil record.

Edgecombe and Legg provide brief overviews of pivotal Lagerstätte, describing the nature of the fossilization as well as highlighting some of the key taxa known from each. They review Burgess Shale-type biota (preservation of non-biomineralized fossils as more or less two-dimensional carbonaceous compressions) from the Chengjiang Lagerstätte and the Burgess Shale itself. Their taxon coverage focuses on the naraoiids, fuxianhuiids, bradoriids, various other bivalved arthropods (such as *Canadaspis* and *Isoxys*), marrellomorphs, megacheirans, anomalocaridids and *Sanctacaris*. The significance of each is briefly highlighted, and any current controversy is set into context, such as the current classification of *Anomalocaris* in the Radiodonta and the affinities of the Radiodonta with the Arthropoda. Other similar Lagerstätte, such as Sirius Passet in Greenland and the Emu Bay shale in Australia, are less familiar to zoologists, but also provide important insights into the evolutionary history of arthropods. Sirius Passet, for example, is rich in the so-called 'gilled lobopodians' which have featured prominently in the debate on character origins in arthropods and on affinities with anomalocaridids.

Orsten-type preservation refers to small fossils preserved by calcium phosphate replacement of cuticle. Edgecombe and Legg briefly mention individual taxa, such as *Agnostus* and *Rehbachiella*, but consider the most significant contribution of Swedish Orsten fossils to be the insights they have provided into the early evolution of Tetraconata, because a series of Orsten taxa can be arranged in progressively more crownward positions in the crustacean stem group. The Silurian Herefordshire Lagerstätte (525 Mya) of western England involves three-dimensional soft tissue preservation of small fossils in concretions. Reconstruction as virtual

3D fossils has allowed reconstruction of the detailed morphology of several arthropods, including phylogenetically important taxa such as *Tanazios*, *Haliestes* and *Offacolus*. Edgecombe and Legg close by considering the Early Devonian Rhynie chert, Upper Carboniferous coal deposits and fossiliferous amber from deposits ranging as far back as the Lower Cretaceous.

The fossil theme is picked up again by Dunlop et al. (in Chap. 16) who examine the water-to-land transitions of arthropods—and begin by stressing that in terms of number of extant species, terrestrial arthropod lineages massively outnumber primarily aquatic lineages. They estimate the minimum number of independent colonization events that must have taken place, but unresolved questions concerning the sister group of the hexapods and uncertainty about relationships between orders of arachnids make it difficult to infer the route taken in some of these events.

The concept of 'terrestrial' is discussed at length. Dunlop et al. support the view that for an arthropod to be considered as fully terrestrial, it should not need to return to water to complete its life cycle. They consider the time frame for the transition onto land—drawing inferences after integrating data from body fossils, from trace fossils (trackways) and from molecular clock data. By the Silurian, myriapods and arachnids were unequivocally living on land and hexapods appear soon afterwards in the Early Devonian. However, the oldest putative record of an arthropod walking across land comes from the Cambrian–Ordovician (around 488 Mya) in Canada. These trackways were interpreted as having been made in a near-shore environment and possibly by members of the Euthycarcinoidea, but in the absence of unequivocal respiratory organs in euthycarcinoids, it is unclear whether they were aquatic, amphibious or terrestrial.

Dunlop et al. note the preponderance of arachnid and myriapod fossils in the Silurian–Devonian terrestrial assemblages, as compared to the relative paucity of hexapods/insects and the complete absence of any demonstrably terrestrial crustaceans. The hexapods, in particular the winged insects, only really seem to come into their own from the Carboniferous onwards by which time land-based communities of plants and animals were already well established. Molecular clock data often suggest older dates for life on land, as compared to the direct evidence of the fossil record, but improvements in methods are beginning to generate new dates that are more consistent with the fossil record.

Finally, they explore the challenges of terrestrial life and briefly review solutions found across the various arthropod groups. They consider body size, locomotion, osmoregulation, reproductive biology, egg type, development and gaseous exchange in turn, as factors in the colonization of the land. Terrestrial arthropods, faced with options to adapt or innovate, often adapted as, for example, in the internalization of an existing system to form the book lungs of the pulmonate arachnids. Hexapods represent the first and by far the most successful colonization of the land by crustaceans. The fossil record suggests that their transition onto land may have begun slightly later than arachnids and myriapods, but they were present both as collembolans and as early jawed insects by at least 410 Mya.

In Chap. 17, White et al. provide a comprehensive overview of knowledge about the interactions between insects and endosymbiotic bacteria and viruses and highlight the impact of endosymbiosis on the evolution of arthropods. These interactions have been studied in more detail in insects, but wider comparisons are made where possible. They discuss the range of beneficial endosymbiotic associations that have evolved between insects and bacteria and the role of intracellular bacteria in manipulating the reproduction of their arthropod hosts. The role of viruses as beneficial symbionts of parasitoid wasps and other insects is also surveyed.

Bacteria, particularly $\alpha$- and $\gamma$-Proteobacteria, often establish tight interactions with arthropod tissues, either as pathogens or as mutualists. Obligate microbial symbionts are common among arthropods that have nutritionally poor or imbalanced diets. While the microbial partners are highly diverse, representing a wide array of

bacterial and fungal lineages, the majority of research has been focused on bacterial partners. Facultative endosymbionts maintain themselves in host populations through reproductive manipulation or mutualism. Bacteria that manipulate host reproduction to promote their own spread and maintenance in the host population are parasites, whereas mutualistic bacteria provide their host with fitness benefits, resulting in a selective advantage for infected hosts. Fitness benefits including defence against natural enemies, interaction with host plants and environmental tolerances are discussed. These bacteria can drive rapid evolutionary shifts in their hosts.

White et al. examine the diversity and transmission of reproductive parasites of arthropods. Most are heritable, maternally transmitted intracellular bacteria that alter the reproduction of their hosts in ways that promote their own fitness. An astonishing 66% of insect species are estimated to be infected by the endosymbiotic *Wolbachia*, and its prevalence in isopods has been estimated at 47%. Reproductive manipulators have evolved mechanisms that favour a female-biased host sex ratio and are detrimental to the non-transmitting sex (the male), including thelytokous parthenogenesis, feminization and male-killing. These are reviewed and shown to help to ensure vertical transmission to host progeny. By inducing cytoplasmic incompatibility, they inhibit the reproduction of uninfected or differently infected individuals and can spread without skewing the sex ratio of the host population. White et al. cover a wide range of topics here, including the evolution of host resistance genes, sex-determination mechanisms and gene acquisition from reproductive parasites.

The current state of knowledge on viruses as beneficial symbionts of insects is also reviewed. White et al. take a look at Polydnaviruses, Entomopoxviruses and Ascoviruses as beneficial symbionts and consider the role of Cypoviruses as modulators of Ascovirus function in parasitoids. Finally, they briefly summarize research on viruses that manipulate parasitoid behaviour, that impact aphid polyphenism, that serve as vectors of plant viruses and even that help mosquitoes take their blood meals.

In Chap. 18, Stansbury and Moczek provide an interesting and thought provoking examination of the evolvability, that is, the potential for evolutionary change and diversification, of arthropods. They explore two axes of diversification: evolvability in developmental space and in developmental time, and their contributions to facilitating evolutionary radiation within the Arthropoda. They begin by identifying anatomical and developmental qualities of arthropods that make them particularly amenable to morphological change. They consider that the potential to explore morphological space was enhanced by compartmentalization of repeating morphological units and by the redundancy inherent in such a body plan. This potential was realized in the extraordinary range of arthropod morphologies, and Stansbury and Moczek conclude that such diversification relied critically upon the degree of spatial decoupling present in the underlying genetic architecture. Thus, the modular nature of gene networks under relatively simple regulatory control enabled their transfer across a flexible regulatory scaffold by means of modest developmental genetic modifications.

Arthropods exhibit a similar potential for diversification along the axis of developmental time, through the life cycle. Immature and mature stages, with or without distinct transitional forms, have evolved to varying degrees in different groups, and this is dependent upon the developmental decoupling of different life stages. The expression of distinct life stages requires mechanisms that specify life-stage identity and their order. Endocrine mechanisms play a key role in communicating throughout the body of a developing arthropod what kind of stage in the life cycle to express and when to transition to the next stage (see also Chap. 6). Stage-specific modularity in gene expression and pathway activation facilitates niche-specific adaptation while reducing pleiotropic constraints. Stage-specific development does not require the evolution of new genes or pathways: instead, only patterns of activation, inhibition and integration must be stage specific, whereas the genes and their products themselves remain

conserved. Diversification is facilitated through changes in assembly, rather than changes in components. The authors emphasize that truly novel traits may originate when a formerly stage-restricted trait becomes expressed at a different stage.

Finally, Stansbury and Moczek look in detail at developmental plasticity—a universal property of development—and its contribution to arthropod evolvability. They explore the genetic, developmental and ecological mechanisms that may have allowed arthropods to diversify so successfully, the interactions among these mechanisms and the emergent properties of these interactions. They highlight key questions for future research and point to opportunities stemming from increased integration of evolution and ecology with developmental biology and genomics.

Our current awareness of arthropod biology and evolution, as summarized in the chapters of this volume, has expanded to a large extent due to the recent and rapidly improving use of new methods. Some of these methods are based on new sophisticated techniques applied in effectively customized way to replace the much less effective approaches used thus far. In several respects, however, these technical improvements have opened completely new dimensions in the investigation of extant and fossil arthropods.

Evidence summarized in Edgecombe and Legg's chapter on arthropod fossils rests to a large extent on new methods of extracting and studying fossils. Organic preservation in the form of cuticle fragments extracted from shales and mudstones by dissolution in hydrofluoric acid has proved to be especially informative for understanding the early history of crustaceans and has provided a wealth of data about terrestrial arthropods from the Middle Devonian at Gilboa, for example. Similarly, the extraction of Orsten fossils has revealed much about the early origins of the crustacean lineages. Small carbonaceous fossils obtained in this way are proving to be especially informative for understanding the early history of crustaceans. Fragments, such as mandibular gnathal edges, indicate that crustaceans such as Copepoda and

Ostracoda had evolved by the Cambrian. The Silurian Herefordshire Lagerstätte of western England involves three-dimensional soft tissue preservation of small fossils in concretions. The specimens are a sparry calcite fill of the void space left after decay of the animal. The sample is serially ground and then reconstructed as a virtual 3D fossil. This technique has allowed the detailed morphology of several important Palaeozoic arthropods to be reconstructed.

A whole set of new techniques is offering advanced methods to analyse developmental and anatomical data. New non-invasive, non-destructive techniques for anatomical analysis and imaging have been developed and are continually being refined. These include laser scanning confocal microscopy, micro-computed tomography and magnetic resonance imaging. Other new techniques have been developed to focus on particular organ systems, such as the application of micro-CT techniques and 3D reconstruction with corrosion casting, to the study of the arthropod circulatory system.

Non-invasive imaging by micro-computed tomography also permits three-dimensional models of fossil arthropods to be reconstructed, including body parts that are otherwise concealed in the rock, such as the distal parts of appendages. This technique has been successfully applied, for example, to otherwise much less informative fossil remains of Carboniferous arachnids.

Any comparative statement in biology requires a phylogenetic context. Almost every chapter in this volume demonstrates the need for phylogenies against which the evolutionary history responsible for generating the observed embryological, anatomical, behavioural and other patterns can be interpreted. Less visually spectacular than the applications of new techniques for reconstructing and presenting morphological evidence, but arguably more popular among researchers, are the daily improvements in molecular phylogenetics, whose application generates an unceasing production of trees, within which some important areas of consensus finally seem to be emerging.

Early molecular phylogenies relied on the target-gene approach—the direct sequencing of selected genes that were amplified with specific primers. But developments in sequencing technology and shotgun approaches ushered in a new era in the production of DNA sequence data. Next-generation sequencing uses random sequencing strategies and automated processes to collect hundreds or thousands of genes. The genes are processed automatically in phylogenomic analyses that are based on a sizeable fraction of the genome or transcriptome. High-throughput sequencing together with next-generation sequence technologies, such as Solexa *Illumina*, can produce millions of sequences per sample at a fraction of the cost of the earlier Sanger technology sequencing.

In addition to new hardware for molecular analysis, the methods in bioinformatics are constantly advancing. Analysis of arthropod mitogenomes presents particular challenges as indicated by Pisani et al. in their chapter. The problem is compositional heterogeneity, and the main source of such compositional heterogeneity in mtDNA is mutational pressure, which is correlated with a deficiency in the mtDNA repair system and with a consequent inefficiency in replacing erroneous insertions of A nucleotides. In addition, strand asymmetry also affects mtDNA, and in arthropods, most mtDNA coding genes are characterized by a negative GC-skew. Sophisticated evolutionary models which account for among site and among branch heterogeneity are useful tools for lessening the effects of mitochondrial compositional bias.

In the face of the huge number of named species of extant and extinct arthropods, the continuing description of new taxa might be perceived as simply adding minor, if abundant, detail to an already established picture. This perception, however, would be grossly off the mark. Even considering only examples from extant arthropods, the last three decades have witnessed the discovery and first description of representatives of previously unknown higher taxa, especially among the crustaceans (e.g. Remipedia, Tantulocarida, Mictacea) and even among the insects, with the totally unexpected discovery of the Mantophasmatodea, a taxon formally described with the rank of order. The continuing discovery of new fossil taxa such as the Silurian synziphosurine *Dibasterium durgae*, with its long flexible antenniform chelicerae, provides an elegant link between the typical sensory antennule and a short feeding chelicera.

In the field of molecular phylogenetics, recent progress only makes us more hungry for more extensive, but also taxonomically denser taxon sampling, together with further refining of the bioinformatics tools applicable to phylogenetic reconstructions, which had become increasingly demanding, following the exponential increase in the volume of available data. Increased taxon sampling, however, is badly needed in all aspects of descriptive and experimental biology. Too limited still, in particular, is the range of arthropods thus far investigated from the perspective of developmental genetics and endocrinology, and even for morphological evidence about critical aspects of phases of ontogeny, such as cleavage and germ band formation, or—for the holometabolous insects—the contribution of imaginal discs in giving shape to the adult are very inadequately known. Our in-depth knowledge remains too restricted to a very small number of model species.

We hope that the concise factual summaries and the questions articulated with this book, despite the obvious limitations of any attempt to summarize arthropod biology, will help increase the general appreciation of both the highlights and the darker recesses of our current knowledge on arthropod biology and evolution and stimulate younger researchers to address these problems from the vantage point of a phylum-wide comparative perspective.

Last but not least, our acknowledgments: first of all, to our authors, for their enthusiasm in accepting our invitation to contribute a chapter to this book and for their generous and successful efforts in producing this set of well-researched and thought-provoking essays; to Gerd Alberti, Carlo Brena, Derek E. G. Briggs, Jennifer A. Brisson, Maurizio Casiraghi, Karyn Johnson, Niels Peder Kristensen, Georg Mayer, André Nel, Frederick R. Schram, Lars Vilhelmsen and Paul Whitington for their precious advice on early drafts on this book's chapters; to Matteo Simonetti for his help with editing a number of illustrations; and to Sabine Schwarz and Ursula Gramm, both of Springer, for their encouragement and assistance in planning and eventual delivery of this book.

Padova-London, 30 November 2012

# Reference

Polilov AA (2012) The smallest insects evolve anucleate neurons. Arthropod Struct Dev 41:29–34

# The Arthropoda: A Phylogenetic Framework

## Gonzalo Giribet and Gregory D. Edgecombe

## Contents

G. Giribet (✉)
Museum of Comparative Zoology, Department of
Organismic and Evolutionary Biology, Harvard
University, 26 Oxford Street, Cambridge, MA
02138, USA
e-mail: ggiribet@g.harvard.edu

G. D. Edgecombe
Department of Earth Sciences, Natural History
Museum, Cromwell Road, SW7 5BD London, UK
e-mail: g.edgecombe@nhm.ac.uk

## 2.1   Introduction

Arthropoda, the best-known member of the clade Ecdysozoa, is a phylum of protostome animals, its closest relatives being Onychophora (velvet worms) and Tardigrada (water bears). Arthropods are not only the largest living phylum in terms of species diversity, with 1,214,295 extant species, including 1,023,559 Hexapoda, 111,937 Chelicerata, 66,914 Crustacea and 11,885 Myriapoda (Zhang 2011), but they have probably been so since the Cambrian. The number of fossil arthropods is even harder to estimate; the EDNA fossil insect database lists ca. 25,000 species (http://edna/palass-hosting.org/); 1,952 valid species of fossil chelicerates were reported by Dunlop et al. (2008), and the decapod crustaceans include 2,979 fossil species (De Grave et al. 2009). Trilobites (19,606 species fide Adrain 2011) and ostracods (>50,000 species) are two of the best-represented arthropod groups in the fossil record.

Arthropods are also, together with Mollusca and Annelida, among the animal phyla with the greatest body plan disparity. This astonishing diversity and disparity of extant and extinct lineages have inspired hundreds of published research articles discussing different aspects of their phylogenetic framework, first focusing on anatomy and embryology, and later being strongly influenced by functional morphology. The advent of cladistic techniques in the mid-twentieth century and the widespread use of

A. Minelli et al. (eds.), *Arthropod Biology and Evolution*,
DOI: 10.1007/978-3-642-36160-9_2, © Springer-Verlag Berlin Heidelberg 2013

molecular data in the last 25 years—the first molecular approach to arthropod phylogeny was published in 1991 by Turbeville et al. (1991)—have revolutionized our understanding of the *Arthropod Tree of Life*. Given the amount of effort revisiting and reviewing arthropod phylogenetics, this chapter will touch upon some of the most fundamental questions: (a) the relationship of arthropods with other key protostome phyla and (b) the relationships between the major arthropod lineages (often referred to as classes, superclasses or subphyla: Pycnogonida, Euchelicerata, Myriapoda and Tetraconata—Tetraconata or Pancrustacea is widely accepted as a clade of arthropods that include the traditional classes Crustacea and Hexapoda, the former often found to be paraphyletic with respect to the latter). Finally, this chapter will provide a roadmap for future focus in arthropod phylogenetic and evolutionary research.

## 2.2    Arthropods in the Animal Tree of Life

Arthropods are protostome animals, and like other protostomes, they have an apical dorsal brain with a ventral longitudinal paired nerve cord and a mouth that typically originates from the embryonic blastopore. They have been traditionally considered to have a primary body cavity, or coelom, that has been restricted to the pericardium, gonoducts and nephridial structures (coxal glands, antennal/maxillary glands) (Brusca and Brusca 2003), but the true coelomic nature of arthropods has been recently called into question. The only putative coelomic cavities in *Artemia salina*, one of the species that underpinned former ideas about arthropods having a coelom, are the nephridial sacculus in the second antennal and second maxillary segments. However, these have been shown not to be remnants of any primarily large coelomic cavity (Bartolomaeus et al. 2009). Similarly, although many authors at one time considered arthropods to have a modified spiral cleavage (Anderson 1969)—as found in annelids, molluscs, nemerteans and platyhelminths

(Maslakova et al. 2004)—this idea is now rejected (Scholtz 1998).

The systematic position of arthropods has changed radically in the past two decades as a result of refinements in numerical phylogenetic analysis and even more so by the introduction of molecular data. Traditionally, arthropods, onychophorans and tardigrades—the three collectively known as Panarthropoda or Aiolopoda—were grouped with annelids in a clade named Articulata (Cuvier 1817), in reference to the segmental body plan in these phyla (Scholtz 2002). The competing Ecdysozoa hypothesis (Schmidt-Rhaesa et al. 1998; Giribet 2003) unites arthropods, onychophorans and tardigrades with a group of mostly pseudocoelomate animals with which they share a cuticle that is moulted at least once during the life cycle and lacks epidermal ciliation. Ecdysozoa was proposed originally on the basis of 18S rRNA sequence data (Aguinaldo et al. 1997; Giribet 1997; Giribet and Ribera 1998) but has subsequently been shown to have support from diverse kinds of molecular information (Edgecombe 2009) (see examples listed below). Concurrently, support has waned for the putative clade once thought to unite arthropods with annelids, despite various morphological phylogenies that retrieved Articulata (e.g. Nielsen et al. 1996; Sørensen et al. 2000; Nielsen 2001; Brusca and Brusca 2003). Contradictory support for Articulata was also found early based on morphological data analyses that explained the similarities of annelids to molluscs and other spiral-cleaving phyla without having to force arthropods to have "lost" spiral cleavage and a trochophore larva to salvage Articulata and recovered effectively Ecdysozoa (Eernisse et al. 1992), or has been shown to depend on the interpretation of certain morphological characters (Jenner and Scholtz 2005). In some cases, authors attempted to reconcile both hypotheses by making Ecdysozoa the sister group of Annelida, nested within Spiralia (Nielsen 2003), or by making Annelida paraphyletic to the inclusion of Ecdysozoa and Enterocoela (Almeida et al. 2003). Even before the molecular support for Ecdysozoa was proposed, some visionary zoologists had already proposed a relationship of arthropods with the then known "aschelminth" phyla (Rauther 1909; Colosi 1967), and others had questioned the homology of segmentation in arthropods and annelids (Minelli and Bortoletto 1988). Kristensen (1991, p. 352), discussing the phylogenetic relationships of Loricifera, wrote

> Annulation of the flexible buccal tube, telescopic mouth cone, and the three rows of placoids are found only in Tardigrada and Loricifera (Kristensen, 1987). Because tardigrades exhibit several arthropod characters (see Kristensen, 1976, 1978, 1981), this last finding supports a theory about a relationship between some aschelminth groups and arthropods (Higgins, 1961). That theory has recently gained support derived primarily from new ultrastructural data, e.g., the fine structure of the chitinous cuticular layer, molting cycle, sense organs, and muscle attachments.

Combined parsimony or Bayesian analyses of morphology and molecules have consistently retrieved Ecdysozoa rather than Articulata (Zrzavý et al. 1998b; Giribet et al. 2000; Peterson and Eernisse 2001; Zrzavý et al. 2001; Zrzavý 2003; Glenner et al. 2004). Likewise, molecular analyses of metazoan relationships have repeatedly recovered ecdysozoan monophyly, whether using just a few genes (e.g. Aguinaldo et al. 1997; Giribet and Ribera 1998; Giribet and Wheeler 1999; Giribet et al. 2000; Mallatt and Winchell 2002; Ruiz-Trillo et al. 2002; Mallatt et al. 2004; Telford et al. 2005; Mallatt and Giribet 2006; Bourlat et al. 2008; Paps et al. 2009a, b; Mallatt et al. 2010), or large collections of genes in phylogenomic analyses (e.g. Dunn et al. 2008; Hejnol et al. 2009; Holton and Pisani 2010; Philippe et al. 2011). When Ecdysozoa was rejected in molecular analyses, as happened in some early genome-scale analyses with depauperate taxonomic sampling, the rival group was Coelomata (nematodes falling outside a group that included arthropods and vertebrates) (Blair et al. 2002; Dopazo et al. 2004; Wolf et al. 2004; Philip et al. 2005), but Articulata was never tested because no annelid was represented in those analyses. Further analyses of these initial whole eukaryotic genomes, whether using intron conservation patterns, rare genomic changes or standard sequence data, rejected Coelomata (Roy and Gilbert 2005; Irimia et al. 2007; Holton and Pisani 2010). Nowadays, even authors who once argued fervently for Articulata have accepted Ecdysozoa (e.g. Nielsen 2012).

Thus, an alliance between Panarthropoda and five moulting phyla with collar-shaped, circumesophageal brains (i.e. Nematoda, Nematomorpha, Kinorhyncha, Priapulida and Loricifera) is the strongest available hypothesis. The latter five phyla are collectively named Cycloneuralia (some authors also include Gastrotricha in this group) or Introverta. The exact position of the three panarthropod phyla within this clade has remained unsettled, often because authors questioned the monophyly of Panarthropoda. The jointed appendages of arthropods have been homologized with the lobopods of onychophorans, a view strengthened by similar genetic patterning of the proximo-distal axes of both kinds of appendages (Janssen et al. 2010), as well as with the limbs of tardigrades. The homology of these paired ventrolateral segmental appendages, which also share segmentally arranged leg nerves, provides the most conspicuous apomorphy for Panarthropoda. Earlier, the appendages were also considered possible homologues of the annelid parapodia. Although some arguments from gene expression have been made in defence of this homology (Panganiban et al. 1997), they mostly pertain to general characters of lateral outgrowths of bodies, and even authors arguing in defence of Articulata have observed that the complexity of the similarities between panarthropod legs and parapodia is not great (Scholtz 2002). Their homology is not generally accepted now.

Under the Panarthropoda hypothesis, each of the three competing resolutions for the interrelationships between the three groups has been defended in recent studies, that is, either Onychophora, or Tardigrada, or a clade composed of them both is the candidate sister group of arthropods (reviewed by Edgecombe et al. 2011; Giribet and Edgecombe 2012). Phylogenomic data have repeatedly endorsed the first option, an onychophoran–arthropod clade (Giribet and

Edgecombe 2012), but the position of tardigrades has been less clear. Two placements for tardigrades recur in broadly sampled molecular analyses, being either sister group of Onychophora + Arthropoda or Nematoda, and in fact both of these alternatives are resolved for the same EST (expressed sequence tag) datasets (Roeding et al. 2007; Dunn et al. 2008; Hejnol et al. 2009; Meusemann et al. 2010; Campbell et al. 2011; Rehm et al. 2011) or mitogenomic data (Rota-Stabelli et al. 2010) under different analytical conditions. In the latter case, conditions intended to counter certain kinds of systematic error strengthen the support for tardigrades grouping with arthropods and onychophorans rather than with nematodes, and the same pattern has also been found for EST-based analyses (Campbell et al. 2011). Tardigrades, onychophorans and arthropods have also been united as a clade based on a uniquely shared micro-RNA (non-coding regulatory genes) (Campbell et al. 2011), with another micro-RNA grouping onychophorans and arthropods to the exclusion of tardigrades.

Thus, current evidence favours panarthropod monophyly with the subgroups (Tardigrada (Onychophora + Arthropoda)), but better sampling is required within Ecdysozoa before this issue is definitely resolved, as ESTs are absent for loriciferans and scarce for kinorhynchs, nematomorphs and priapulans. A rival clade that includes Tardigrada, Nematoda and Nematomorpha, and even Loricifera, has some morphological (Kristensen 1991) and limited molecular (Sørensen et al. 2008) support. In contrast, the alliance of tardigrades with onychophorans and arthropods, along with the fossil lobopodians and anomalocaridid-like taxa ("gilled lobopodians"), is consistent with a single origin of paired, segmental ventrolateral appendages in a unique common ancestor (Liu et al. 2011; Giribet and Edgecombe 2012).

Arthropod monophyly (Lankester 1904; Snodgrass 1938) is now nearly universally accepted based on morphological, developmental and molecular evidence, but this has not always been the case. The Manton School strongly advocated for arthropod polyphyly (Tiegs and Manton 1958; Anderson 1973; Manton 1973, 1977; Willmer 1990), but this reasoning was based on differences between groups and conjectures about whether or not intermediate forms could be functionally viable; it did not provide characters that supported alternative sister group hypotheses with non-arthropod phyla. In the absence of explicit rival hypotheses, arthropod monophyly remains unchallenged and is supported by a suite of synapomorphies. These include a sclerotized exoskeleton, and legs that are composed of sclerotized podomeres separated by arthrodial membranes, two characters absent in onychophorans and tardigrades (some authors use the term Arthropoda to include Onychophora and Tardigrada, but we reject this nomenclature, as the members of those phyla have not undergone the arthropodization process). In all arthropods except pycnogonids, muscles attach at intersegmental tendons. Compound eyes across the Arthropoda share a similar developmental mode, with new eye elements being added in a peripheral proliferation zone of the eye field (Harzsch and Hafner 2006), and the presence of two optic neuropils in the inferred ancestor is apomorphic for arthropods as a whole (Harzsch 2006). Segmentation gene characters, such as a pair-rule function of the Pax protein (Angelini and Kaufman 2005; Gabriel and Goldstein 2007), and a conserved pattern of how neural precursors segregate (Eriksson and Stollewerk 2010a) map onto the tree as autapomorphies of Arthropoda compared with the states in Onychophora and Tardigrada. Under the criterion of monophyly, the parasitic Pentastomida are arthropods. This group had a long history of classification as "prot(o)arthropods" in its own phylum (Brusca and Brusca 1990), and an early divergence from the arthropod stem lineage is still endorsed by some morphologists (Castellani et al. 2011). The molecular arguments for a placement as ingroup crustaceans, grouped with branchiuran fish lice according to the Ichthyostraca hypothesis, are strong (Abele et al. 1989; Giribet et al. 2005; Møller et al. 2008; Regier et al. 2010; Sanders and Lee 2010), if in conflict with some morphological interpretations

(Waloszek et al. 2006), and are congruent with synapomorphies from sperm ultrastructure (reviewed by Giribet et al. 2005).

## 2.3   The Arthropod Tree of Life

The diversity of arthropods traditionally has included the classes (or comparatively higher-rank taxa) Chelicerata, Myriapoda, Hexapoda and Crustacea, with Pycnogonida sometimes considered part of Chelicerata (hence divided into Pycnogonida, Xiphosura and Arachnida), or their own class, due to their unique morphology and uncertain phylogenetic affinities. Recent developments have provided strong endorsement for paraphyly of Crustacea with respect to Hexapoda, and hence, we consider the extant arthropod phylogenetic conundrum as a four-taxon problem—Pycnogonida, Euchelicerata (=Xiphosura + Arachnida), Myriapoda and Tetraconata (=Pancrustacea)—with three alternative rootings (Fig. 2.1a–c).

Relationships between these groups have been debated for decades. Through much of the twentieth century, the only nearly universally accepted result was the monophyly of Atelocerata (also known as Tracheata)—a clade composed of hexapods and myriapods (e.g. Snodgrass 1938; Wheeler et al. 1993) (Fig. 2.1d). However, the addition of molecular and novel anatomical and developmental data has helped to reinterpret arthropod relationships, with the result that Atelocerata has been overturned. In most contemporary studies, hexapods are associated with crustaceans instead of with myriapods (e.g. Friedrich and Tautz 1995; Giribet et al. 1996, 2001, 2005; Regier and Shultz 1997; Giribet and Ribera 1998, 2000; Zrzavý et al. 1998a; Hwang et al. 2001; Regier et al. 2005a, 2008, 2010; Mallatt and Giribet 2006; Meusemann et al. 2010; von Reumont and Burmester 2010; Campbell et al. 2011; Regier and Zwick 2011; Rota-Stabelli et al. 2011; von Reumont et al. 2012) in a clade named Tetraconata in reference to the shared presence of four crystalline cone cells in the compound eye ommatidia in both groups (Richter 2002). A few groups of morphologists still argue in support of Atelocerata (Bitsch and Bitsch 2004; Bäcker et al. 2008), though this follows as a consequence of either examining a single character system (e.g. pleurites around the leg base in the case of Bäcker et al. 2008) or not including the rival characters for Tetraconata in the analysis. Morphologists who recognize Tetraconata have reinterpreted the putative apomorphies of Atelocerata as likely being convergences due to terrestrial habits (Harzsch 2006), and numerical cladistic analyses that incorporate the neuro-anatomical evidence for Tetraconata retrieve that group in favour of Atelocerata (Giribet et al. 2005; Rota-Stabelli et al. 2011). Perhaps, the only novel argument in support of Atelocerata in modern times is a similar expression pattern of the *Drosophila collier* gene (*col*) in the limbless intercalary segment of the head in a few studied myriapods and insects (Janssen et al. 2011). This conserved function of *col* in insects and myriapods as a putative synapomorphy is overwhelmed by a much larger body of neuro-anatomical and molecular data that speak in favour of a crustacean–hexapod clade. Thus, the *col* function could have been lost in early head development in crustaceans or may indeed have evolved convergently in insects and myriapods.

A perfectly resolved *Arthropod Tree of Life* is still elusive, but the notion that arthropod phylogeny can be depicted as "chaos" (Bäcker et al. 2008) is obsolete. Several patterns, including a basic unrooted topology, are congruent among nearly all new sources of data, and today, most authors interpret the arthropod phylogeny problem as a rooting problem (Giribet et al. 2005; Caravas and Friedrich 2010; Giribet and Edge-combe 2012) and not as alternative conflicting topologies. These three alternative rootings result in (a) Pycnogonida as sister to all other arthropods (=Cormogonida) (Zrzavý et al. 1998a; Giribet et al. 2001); (b) Chelicerata monophyletic and sister group to Mandibulata (Regier et al. 2008, 2010; Rota-Stabelli and Telford 2008; Regier and Zwick 2011; Rota-Stabelli et al. 2011), or those arthropods with true mandibles (Edgecombe et al. 2003), as opposed to chelicerates or chelifores; and (c) a

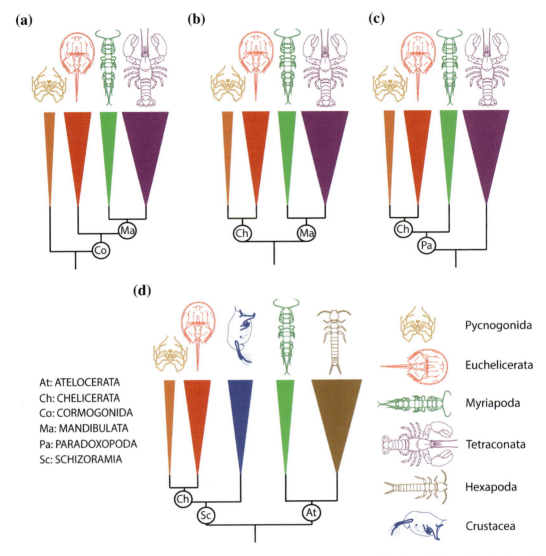

**Fig. 2.1** Alternative hypotheses of arthropod relationships, including the three currently recognized rooting options. **a** Cormogonida. **b** Chelicerata versus Mandibulata. **c** Paradoxopoda/Myriochelata. **d** A traditional view of arthropod relationships with the putative clades Schizoramia and Atelocerata/Tracheata

clade named Paradoxopoda (=Myriochelata) that joins myriapods with the chelicerate groups (Friedrich and Tautz 1995; Hwang et al. 2001; Mallatt et al. 2004; Pisani et al. 2004; Mallatt and Giribet 2006; Dunn et al. 2008; von Reumont et al. 2009; Rehm et al. 2011) (Fig. 2.1a–c). Whereas the choice between these hypotheses involves the placement of the root, a few traditional morphological hypotheses present more fundamental topological conflict. Among the conflicting hypotheses are

Atelocerata and Schizoramia (Fig. 2.1d), the latter uniting Crustaceomorpha and Arachnomorpha (Bergström 1979; Hessler 1992).

In this chapter, we focus on developments in two key areas, comparative anatomy and novel molecular approaches, each of which has advanced greatly since the publication of the first arthropod phylogenies combining morphology and multiple molecular markers (Wheeler et al. 1993; Zrzavý et al. 1998a; Giribet et al. 2001). Since then, the quantity of molecular data devoted to this problem has increased exponentially with recent genomic approaches. The techniques used to analyse developmental and anatomical data have also improved considerably as a result of new technological advances. For example, a classical technique for studying internal anatomy, histological sectioning, is now aided by computer reconstruction (e.g. Stegner and Richter 2011 for cephalocarids). Non-invasive, non-destructive techniques for anatomical imaging are continually being refined. Among these are confocal laser microscopy, micro-computed tomography and magnetic resonance imaging (Hörnschemeyer et al. 2002; Friedrich and Beutel 2010). Other new techniques have been developed to focus on particular organ systems, for example, studies on the circulatory system that apply micro-CT techniques and 3D reconstruction with corrosion casting are a source of new characters for several arthropod groups (Wirkner and Richter 2004; Wirkner and Prendini 2007; Huckstorf and Wirkner 2011). While these techniques have had an impact, they have still been applied to a limited (yet valuable) number of taxa, both fossil and extant.

## 2.3.1 Neural Cladistics

Comparative anatomy was the traditional source of data for inferring arthropod phylogeny, coupled with evidence from embryonic and post-embryonic development (Anderson 1973). Among anatomical systems that are currently receiving intensive study for their phylogenetic signal, the nervous system is perhaps prevalent, an approach that has come to be called neurophylogeny (Richter et al. 2010) or neural cladistics (Strausfeld and Andrew 2011). Nervous system characters had already played an important role in arthropod phylogenetics in the early twentieth century (Strausfeld 2012). Indeed, one of the major insights of this early neuroanatomical research was the ancestry of hexapods from crustaceans rather than from myriapods, a hypothesis that drew its support from characters that have returned to the

forefront of debate, such as eye ultrastructure and configurations of the optic neuropils (Hanström 1926). A crustacean ancestry of hexapods laid dormant through the decades in which myriapods were upheld as the closest relatives of hexapods, until the mid-1990s. Since then, neuroanatomists have provided compelling corroboration for crustacean paraphyly as well as many other key nodes in the arthropod tree by applying new staining/immunoreactivity and imaging techniques, coupled with analysis of the data by cladistic methods.

Character matrices based on the nervous system (Harzsch 2006; Strausfeld 2009; Strausfeld and Andrew 2011) consistently resolve Malacostraca and Hexapoda as more closely related to each other than either is to Branchiopoda or is to Maxillopoda, as upheld earlier by Hanström. Character support for a malacostracan–hexapod clade to the exclusion of branchiopods is provided by such shared features as optic neuropils that have a nesting of the lamina, medulla, lobula and lobula plate and their connections by crossed axons (chiasmata). To explain the distribution of character states on a tree in which cephalocarids and remipedes are positioned stemward of branchiopods within Tetraconata, branchiopod brains have been interpreted as secondarily simplified from an ancestor that shared traits seen in the brains of malacostracans and remipedes (Strausfeld and Andrew 2011). Character polarities are, however, very much dependent upon the exact pattern of relationships between these crustacean groups and Hexapoda, an area that is subject to instability between different analyses (notably for the relationship between remipedes and cephalocarids).

The mode of development of neural tissue has played a major role in recent discussion about where the root should be placed between the main extant arthropod groups, which corresponds to the controversy over Mandibulata versus Paradoxopoda. Detailed similarities in chelicerate and myriapod neurogenesis have been recognized for nearly a decade (Dove and Stollewerk 2003; Kadner and Stollewerk 2004; Mayer and Whitington 2009) and present a contrast with the stem cell–like division of neural precursors in insects and crustaceans (Ungerer et al. 2011). The question becomes one of polarity—whether the chelicerate–myriapod characters are symplesiomorphies, inherited from the ancestor of all arthropods, or are potential synapomorphies that provide anatomical support for Paradoxopoda. To resolve this matter, neurogenesis in the arthropod sister group, Onychophora, has been examined using immunohistochemistry and confocal laser microscopy (Mayer and Whitington 2009; Whitington and Mayer 2011), supplemented by new data from gene expression of *Delta*, *Notch* and *ASH* (Eriksson and Stollewerk 2010a, b). The results remain open to interpretation, the onychophorans being argued to share characters with insects and crustaceans, being thus a plesiomorphic state, which would make the condition in myriapods and chelicerates apomorphic, providing positive support for Paradoxopoda (Mayer and Whitington 2009; Whitington and Mayer 2011). Other authors instead suggest that onychophorans possess unique and divergent character states that cannot be homologized with those of insects and crustaceans and that myriapods have characters of neural precursor cells that are consistent with Mandibulata rather than with Paradoxopoda (Eriksson and Stollewerk 2010a). Knowledge on the neurogenesis of pycnogonids at this level is entirely lacking, but would constitute an obvious starting point to look into in order to possibly settle this debate.

The most recent neural cladistic analysis (Strausfeld and Andrew 2011) has retrieved Mandibulata as a monophyletic group, but it has also exposed the ongoing problem of correctly rooting Arthropoda, for example, Onychophora unite with Chelicerata as a putative clade for the same data. The latter grouping is contradicted by many other kinds of data and signals an incorrect root position, possibly resulting from a distant outgroup (annelids were used as an outgroup rather than as tardigrades and/or cycloneuralians). Though Mandibulata is depicted as the "state of play" in some recent studies (as in Regier et al. 2010; Rota-Stabelli et al. 2011), it need be cautioned that anatomical and gene expression data supporting Paradoxopoda continue to emerge. As an example, we note expression patterns along the proximo-distal axis of the limb, specifically the expression domains of homothorax (*hth*) and extradenticle (*exd*). These are comparable with chelicerates (spiders and harvestmen) and millipedes (Abzhanov and Kaufman 2000; Prpic et al. 2003; Prpic and Damen 2004; Pechmann and Prpic 2009; Sharma et al. 2012). *hth* is expressed broadly in much of the developing appendage, whereas *exd* is restricted to the proximal podomeres. Taken together with the inverse spatial relationship between *hth* and *exd* in onychophorans and pancrustaceans (Prpic et al. 2003; Prpic and Telford 2008; Janssen et al. 2010), the expression data are consistent with a sister group relationship between chelicerates and myriapods.

## 2.3.2 Novel Molecular Approaches

Understanding of arthropod relationships has been transformed by molecular data, with vast refinements in both sampling and techniques since an initial wave of analyses was conducted in the early 1990s (Abele et al. 1989; Wheeler 1989; Kim and Abele 1990; Turbeville et al. 1991; Carmean et al. 1992; Spears et al. 1992; Pashley et al. 1993; Wheeler et al. 1993). Until the past few years, molecular phylogenies relied on direct sequencing of a few selected genes that were amplified with specific primers—an approach now called a "target-gene approach". Arthropod phylogenies were often inferred from nuclear ribosomal genes (Friedrich and Tautz 1995; Giribet et al. 1996; Giribet and Ribera 2000; Mallatt and Giribet 2006; von Reumont et al. 2009), nuclear protein-encoding genes (Regier and Shultz 1997; Shultz and Regier 2000; Regier and Shultz 2001; Regier et al. 2004, 2005a), or a combination of these with mitochondrial genes (Giribet et al. 2001, 2005; Giribet and Edgecombe 2006). These studies typically used just a few genes to build trees. Other analyses instead focused on mitogenomics (Boore et al. 1995; Hwang et al. 2001; Lavrov et al. 2002; Masta and Boore 2008; Rota-Stabelli et al. 2010), the analysis of complete mito-chondrial genomes. Although the early analyses of mitochondrial genes from the 1990s some-times yielded contradictory and/or morphologi-cally anomalous results (Ballard et al. 1992), many of these problems have now been identi-fied as resulting from a deficient taxon sampling, too few molecular data, systematic error or combinations of these defects.

The target-gene approach still forms the basis for some modern work on arthropod phyloge-netics. The number of markers has substantially increased, drawing on as many as 62 nuclear protein-encoding genes (Regier et al. 2008; Re-gier and Zwick 2011), as has the taxon sampling, up to 75 taxa (Regier et al. 2010). The use of large numbers of markers obtained through standard PCR approaches has been an important advance, and in the case of the arthropod dataset, it permits a clear choice of Mandibulata over Paradoxopoda and injects new hypotheses for crustacean interrelationships (though some of these have been questioned because they do not account for serine codon usage bias and are contradicted under alternative analytical condi-tions: Rota-Stabelli et al. 2013). The downsides of this method are that it is time-consuming, it is difficult to consistently amplify large numbers of genes for many taxa, and many of the selected genes may present problems of paralogy that are difficult to detect by PCR approaches alone (Clouse et al. submitted).

Developments in sequencing technology and shotgun approaches following the sequencing of the first complete eukaryotic genomes of *Cae-norhabditis elegans*, *Drosophila melanogaster* and *Homo sapiens* ushered in a new era in the production of DNA sequence data. "Next-gen-eration sequencing" uses random sequencing strategies and automated processes to collect hundreds or thousands of genes from cDNA libraries obtained from mRNA, for a fraction of the effort required to amplify multiple markers. The genes are processed automatically in phy-logenetic analyses (Dunn et al. 2008; Edge-combe et al. 2011) that have come to be known as "phylogenomic"—based on a sizeable frac-tion of a transcriptome or a genome (Morozova

et al. 2009). The random sequencing of clones from a cDNA library generates large numbers of ESTs, and soon, studies combined the data from full genomes with novel ESTs generated for a diverse sampling of protostomes (Dunn et al. 2008; Hejnol et al. 2009) or arthropods in par-ticular (Roeding et al. 2009; Meusemann et al. 2010; Campbell et al. 2011; Rehm et al. 2011; Rota-Stabelli et al. 2011; von Reumont et al. 2012).

With respect to the basal split in Arthropoda, EST-based studies to date have come down in favour of either Paradoxopoda (Fig. 2.1b) or Mandibulata (Fig. 2.1c), generally observing the choice between the two to be sensitive to taxon sampling, but also to gene sampling. The first EST analyses supported the Paradoxopoda hypothesis (Dunn et al. 2008; Hejnol et al. 2009; Roeding et al. 2009; Meusemann et al. 2010), whereas others support a split between Chelic-erata and Mandibulata (Campbell et al. 2011; Rota-Stabelli et al. 2011). The most densely sampled analysis, which added some crustacean lineages missing from earlier studies (von Reu-mont et al. 2012), retrieved Mandibulata when their entire taxon/character sample was used, but support shifted to Paradoxopoda when the matrix was reduced according to criteria that the authors believed would lessen "noise". The two hypotheses were likewise found to be variably supported for different taxonomic samples in EST analyses by Andrew (2011).

Most EST libraries until 2010 were obtained using standard Sanger capillary sequencers. High-throughput sequencing with next-genera-tion sequence technologies such as Roche 454 (Margulies et al. 2005) and more recently Solexa *Illumina* (Illumina_Inc 2007) can produce up to hundreds of thousands or millions of sequences per sample, at a fraction of the cost of the earlier Sanger technology sequencing. These techno-logical developments will radically increase the amount of data available for analysis, especially for non-model organisms (Riesgo et al. 2012).

Molecular data have also made an important contribution towards producing reliable chron-ograms of arthropod cladogenesis and diversifi-cation (Murienne et al. 2010; Sanders and Lee

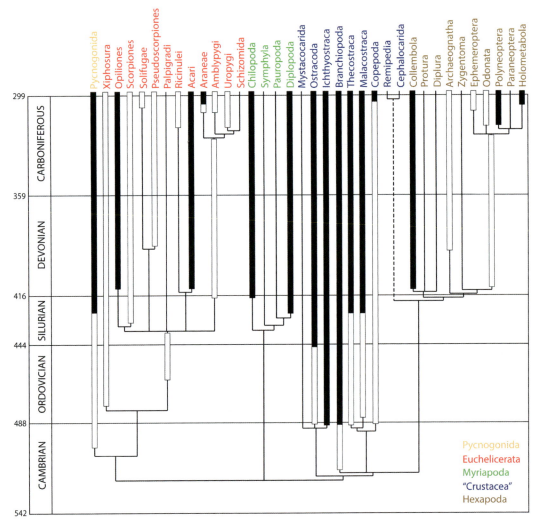

**Fig. 2.2** Relationships between living arthropod lineages with palaeontological calibration. *Solid bars* indicate the presence of unambiguous fossils assigned to the crown group, and empty bars indicate the presence of fossils assigned to the stem group. Fossil data obtained from original sources and reviews, including Dunlop 2010; Rehm et al. 2011). Palaeontology continues contributing most of the data to the age of lineages because minimum ages from fossils (Fig. 2.2) calibrate the molecular estimates for divergences. Modern molecular estimates of the splits between the deep arthropod clades such as Chelicerata versus Mandibulata (or the rival split of Paradoxopoda versus Tetraconata) date these events to the Ediacaran Period (635–542 My) (Pisani 2009; Erwin et al. 2011; Rehm et al. 2011). This is more consistent with the fossil

(2010) for Chelicerata, Edgecombe (2010) and Rehm et al. (2011). Relationships within Tetraconata mostly based on Regier et al. (2010). As a convention, divergences are depicted as shallow as warranted by fossils; deeper divergences are inferred from molecular dating; see Sanders and Lee (2010) and Rehm et al. (2011)

record than were earlier analyses that used more immature clock models, which retrieved divergences between onychophorans and arthropods and basal splits in Arthropoda dating to the Cryogenian (reviewed by Pisani 2009). Even so, the Ediacaran has not yet yielded credible body or trace fossils of arthropods, and an Ediacaran "fuse" of some tens of millions of years separates the latest molecular divergence of arthropods from the first appearance of arthropod trackways in the early Cambrian.

## 2.4 Advancing Arthropod Phylogenetics

While many of the new developments discussed above have contributed to stabilize the arthropod tree (Fig. 2.2), there are several areas in need of refinement. In this section, we navigate the main arthropod clades and suggest possible areas of inquiry.

The persistent controversy over whether the root of the arthropod tree identifies Mandibulata or Paradoxopoda as clades would best be tested by additional genomic data on Pycnogonida, the currently unsampled orders of Arachnida, and Myriapoda. Taxon sampling in those groups is sparse (e.g. only one myriapod was used in the currently best-sampled EST analyses; von Reumont et al. 2012), and the EST libraries available to date for these groups are shallow when compared to those of other arthropod groups, for which whole genomes or extensive genetic resources are at hand (Clark et al. 2007). The recent sequencing of several pancrustacean genomes, as well as the first myriapod genome for the centipede *Strigamia maritima* and the genome of the horseshoe crab *Limulus polyphemus*, should be key in resolving some of the most fundamental questions about deep arthropod phylogeny. Fossil data are also important for establishing an accurate position of the root (Edgecombe 2010), but the methodological difficulties in combining morphology with genomic-level data remain largely unexplored (Giribet 2010). New kinds of molecular characters should also be more broadly sampled to include arthropod lineages that have thus far been unexplored. For example, the hypothesis that myriapods share two novel micro-RNAs with crustaceans and hexapods that are not shared with chelicerates (Campbell et al. 2011; Rota-Stabelli et al. 2011) has been tabled as a new argument in favour of a monophyletic Mandibulata. The presence of these micro-RNAs should be determined in more myriapods (e.g. symphylans and pauropods), crustaceans and arachnids.

### 2.4.1 Chelicerata

Euchelicerata is nearly always identified as monophyletic, apart from in some mitogenomic analyses (e.g. Masta et al. 2009), which have repeatedly placed pycnogonids within Arachnida, often attracted to Acari, and in some trees that were not based on explicit data analysis (Simonetta 2004). Beyond the relatively straightforward question of euchelicerate monophyly, though, molecular datasets to date (Wheeler and Hayashi 1998; Giribet et al. 2002; Masta et al. 2009; Pepato et al. 2010; Regier et al. 2010) have mostly conflicted with morphology (Shultz 1990; Wheeler and Hayashi 1998; Giribet et al. 2002; Shultz 2007), apart from identifying the clade Tetrapulmonata (and in some cases recovering its internal phylogeny congruently with morphology; Regier et al. 2010). In many analyses, the molecules have not even recovered the basal dichotomy between Xiphosura (horseshoe crabs) and Arachnida. Possible causes for the difficulty in recovering these relationships are the old history of the group, the extinction of key lineages (arachnids include several high-ranking extinct groups such as the orders Trigonotarbida, Haptopoda and Phalangiotarbida, as well as stem-group arachnid taxa such as Eurypterida and Chasmataspidida; Dunlop 2010) or intrinsic problems of the molecular data. The monophyly and phylogenetic affinities of Acari (Dunlop and Alberti 2008; Pepato et al. 2010) and the precise position of Palpigradi and Ricinulei remain as some of the most puzzling issues. Likewise, challenging are the relationships between a set of arachnid orders that have been regarded as solidly placed from the perspective of morphology—Scorpiones, Opiliones, Pseudoscorpiones and Solifugae. The currently favoured morphological hypothesis in which scorpions and harvestmen form the clade Stomothecata (Shultz 2007) conflicts with the largest available molecular datasets for arachnids (Regier et al. 2010). The latter unite scorpions with the tetrapulmonates, but that group (Pulmonata in Regier et al. 2010) is not strongly supported. A similar situation pertains to pseudoscorpions and solifuges. Their grouping as a clade, Haplocnemata, is widely endorsed by arachnologists because of numerous shared derived morphological characters (Weygoldt and Paulus 1979; Shultz 2007). Alternative sister groups based on nuclear genes (Solifugae + Ricinulei; Pseudoscorpiones + parasitiform Acari) have weak support (Regier et al. 2010). Figure 2.2 depicts the relationships between these groups as resolved by morphology.

A sister group relationship between Pycnogonida and Euchelicerata has a long tradition among morphologists, though few strong synapomorphies have been identified (Dunlop and Arango 2005). The main alternative placement for Pycnogonida, as sister group to all other arthropods according to the Cormogonida hypothesis (Fig. 2.1a), has been based on absences of certain morphological characters shared by other arthropods, such as intersegmental tendons and a labrum or labral anlagen, being interpreted as primitively absent. Recent electron microscopic study of pycnogonid embryos in search of potential labral homologues has failed to identify a plausibly homologous structure (Machner and Scholtz 2010), which is consistent with a position of Pycnogonida outside Euchelicerata + Mandibulata. Additional characters that have been tabled as potentially plesiomorphic in pycnogonids refer to the presence of a terminal mouth at the end of a proboscis and a Y-shaped pharynx (Miyazaki 2002), both characters widely found in the introvertan ecdysozoans and in some tardigrades (Schmidt-Rhaesa et al. 1998; Giribet 2003). The choice between Chelicerata (i.e. Pycnogonida + Euchelicerata) and Cormogonida is not decisively settled with current molecular datasets, although the former seems to be preferred. The sister group relationship between pycnogonids and euchelicerates was retrieved by Regier et al. (2010) in their analyses of nuclear coding genes, though they observed a "more basal placement of Pycnogonida" (i.e. Cormogonida) to provide only a marginally poorer fit to the data.

## 2.4.2 Myriapoda

The rediscovery of myriapod monophyly has been identified as one of the successes of arthropod molecular phylogenetics (Regier et al. 2008). A long tradition of postulating that Myriapoda was non-monophyletic resulted from the Atelocerata hypothesis. In that framework, myriapods were identified as a grade from which hexapods evolved (Dohle 1980; Kraus and Kraus 1994, 1996). From the mid-1960s through the mid-1990s, myriapod paraphyly often took the form of Progoneata (symphylans, pauropods and diplopods) being sister group of Hexapoda in a putative clade called Labiophora, with Chilopoda being sister group to that assemblage. Intriguingly, key proponents of arthropod polyphyly through that era were strong defenders of the monophyly of Myriapoda (e.g. Anderson 1973). Sidnie Manton (1964) perceptively observed that myriapods share a unique structure of the tentorial endoskeleton which has come to be known as the "swinging tentorium". Throughout Myriapoda, the posterior process of the tentorium is fused to a transverse bar that extends to the lateral cranial wall (Koch 2003a); downward and outward movements of these tentorial apodemes provide the abductor force that opens the mandibles. This character system remains an autapomorphy of Myriapoda.

The rediscovery of Myriapoda is linked to the demise of Atelocerata. The unambiguous molecular and very strong neuroanatomical support for a hexapod–crustacean clade that excludes Myriapoda effectively solves the question of myriapod paraphyly; if the shared characters of Myriapoda no longer have to be seen as atelocerate symplesiomorphies, then the only parsimonious solution is to identify them as myriapod autapomorphies (Shear and Edgecombe 2010). Recent analyses that used a broad sampling of genes and taxa (Regier et al. 2010; Regier and Zwick 2011) have resolved Myriapoda as monophyletic, with strong support, corroborating previous molecular phylogenetic analyses.

A challenge to myriapod monophyly had been raised in neural cladistic analyses, specifically a possibility that Diplopoda could be basally positioned in Arthropoda, falling outside a group that united other myriapods with Tetraconata and that "partial Mandibulata" clade with Chelicerata (Loesel et al. 2002; Strausfeld et al. 2006). This hypothesis is derived from the absence of a specific midline neuropil in the brain in spirostreptid millipedes that is shared by other arthropods (as well as onychophorans). Expanded character and taxonomic sampling in neural cladistic datasets have corrected this anomalous placement of millipedes: Diplopoda and Chilopoda are sister groups in current cladograms (Strausfeld and Andrew 2011). The addition of comparable data for Symphyla and Pauropoda is an obvious target for future work.

Shifting attention from myriapod monophyly to the basal split within the group, the 75-taxon, 62-gene dataset (Regier et al. 2010; Regier and Zwick 2011) yielded a division that corresponds to the standard morphological tree, that is, Chilopoda as sister group to Progoneata. Within Progoneata, however, conflict with morphology emerges, and this presents the most pressing issue in Myriapoda as a whole. The union of diplopods and pauropods as a clade named Dignatha has not been seriously challenged from the perspectives of morphology and development (Dohle 1980; Shear and Edgecombe 2010). These putative sister groups share many detailed characters, including a limbless post-maxillary segment, the vas deferens opening on conical penes on the same trunk segment, spiracles at the bases of the walking legs that open to tracheal pouches, a motionless post-hatching ("pupoid") stage and three leg pairs in the first free-living stage. Because of the strength of support for Dignatha from these similarities, it was unexpected when sequence-based analyses instead retrieved a grouping of Pauropoda with Symphyla rather than with Diplopoda (Regier et al. 2005b; Gai et al. 2008; Regier et al. 2010; Dong et al. 2012). However, pauropods and symphylans have been seen to attract in anomalous positions (sometimes even falling outside Arthropoda) in analyses of nuclear ribosomal

genes (Giribet and Ribera 2000; von Reumont et al. 2009). Their grouping with nuclear coding genes thus needs to be critically evaluated as a possible artefact of systematic error.

### 2.4.3  Tetraconata

Monophyly of Tetraconata has long been recognized from diverse molecular datasets (see citations above) and indeed has never been challenged by Atelocerata in any sequence-based analysis. Tetraconata is in no sense a "molecular grouping", though, as explained above, it reflects a hypothesis put forward by neurobiologists in the early twentieth century, and in its contemporary form, it is reinforced by important morphological characters of eye ultrastructure (Richter 2002), brain and optic lobe anatomy (Harzsch and Hafner 2006; Strausfeld 2009; Strausfeld and Andrew 2011), serotonin reactivity in the nerve cord (Harzsch 2004) and similarities in neurogenesis (Ungerer and Scholtz 2008).

Whether crustaceans are monophyletic or paraphyletic with respect to hexapods (Schram and Koenemann 2004; Giribet et al. 2005; Richter et al. 2009) and if the latter, precisely which crustacean lineage constitutes the sister group of hexapods, remain labile (Grimaldi 2010). The case for crustacean paraphyly has mostly come from molecular datasets, but morphologists have been far from universal in endorsing the traditional hypothesis of a monophyletic Crustacea. Schram and Koenemann (2004) and Richter et al. (2009) evaluated most of the traditionally diagnostic or putatively autapomorphic characters of Crustacea and found that they are often ambiguous or likely symplesiomorphic. Cladistic analyses of neural characters, either manually computed (Harzsch 2006) or analysed using parsimony programs (Strausfeld et al. 2006; Strausfeld and Andrew 2011), resolve Crustacea as paraphyletic with respect to Hexapoda.

The alternative sister group hypotheses for each major crustacean clade have been summarized (Jenner 2010), so we focus on developments in the latest molecular analyses using large numbers of genes. Among these are some new hypotheses not anticipated based on other data sources. For example, an analysis of 62 markers suggests that a putative clade composed of Cephalocarida + Remipedia (named as Xenocarida) is sister to Hexapoda, while Branchiopoda forms a clade with Malacostraca, Thecostraca and Copepoda (Regier et al. 2010). The latter grouping, named Multicrustacea by Regier et al. (2010), has also been retrieved using different kinds of molecular data, notably the EST analyses of Meusemann et al. (2010) and Andrew (2011) and compilations of molecular and morphological data by Oakley et al. (2013). The branchiopod–malacostracan–hexapod three-taxon statement lies at the heart of current conflict between various datasets and analyses. Rather than grouping branchiopods and malacostracans together (as in Regier et al. 2010), neural cladistics instead identifies Malacostraca as the likely sister group of hexapods (Strausfeld 2009; Strausfeld and Andrew 2011). In contrast to both of these resolutions, larger gene samples in EST analyses repeatedly resolve Branchiopoda as sister group to Hexapoda (Roeding et al. 2009; Meusemann et al. 2010; Campbell et al. 2011; Rota-Stabelli et al. 2011), although Cephalocarida and Remipedia were not sampled in those studies. The first ESTs of remipedes suggest that they are indeed the sister group of Hexapoda (von Reumont et al. 2012), but an alliance with Cephalocarida has not yet been tested, and these data reflect the signal of earlier EST analyses in resolving branchiopods as more closely related to remipedes and hexapods than are malacostracans. A comparable clade composed of branchiopods + cephalocarids and remipedes + hexapods was named Allotriocarida by Oakley et al. (2013). Denser taxon sampling of key crustacean lineages (e.g. Mystacocarida) is still needed in phylogenomic analyses before a definitive solution can be proposed. In particular, the attraction of remipedes and cephalocarids warrants close scrutiny because this relationship has not been anticipated from the perspective of morphology, though it has been detected for some time in

molecular datasets (Giribet et al. 2001; Regier et al. 2005a). Reanalysis of the Regier et al. (2010) 62-gene dataset by Rota-Stabelli et al. (2013) found the remipede–cephalocarid grouping to be model dependent and sensitive to the analysis of either nucleotides or amino acids. Irrespective of the eventual placement of Cephalocarida, the congruent signal from large samples of nuclear coding genes (Regier et al. 2010) and ESTs (von Reumont et al. 2012), together with the discovery of hexapod-type haemocyanins in remipedes (Ertas et al. 2009), makes a strong case for Remipedia being closely allied to hexapods.

The issue of hexapod monophyly was for a few years disputed in some mitogenomic analyses (Carapelli et al. 2007), but has since been resolved in favour of a single origin using larger molecular datasets (Timmermans et al. 2008; Meusemann et al. 2010; Regier et al. 2010; von Reumont et al. 2012; Oakley et al. 2013). At the base of Hexapoda, the status of Entognatha as a clade or a grade remains sensitive to taxon sampling and methods of molecular data analysis (Giribet et al. 2004). Morphologists had, over the past 20 years, largely abandoned Entognatha, arguing that enthognathy in collembolans and proturans did not have a common origin with that in diplurans (Koch 1997, 2000), and the latter instead shared derived characters with Ectognatha, that is, "Entognatha" was a paraphyletic group (Bitsch and Bitsch 2004; Giribet et al. 2005, among others, from numerical cladistic analyses; Machida 2006 from embryological data; Dallai et al. 2011 from sperm ultrastructure). The resurrection of Entognatha as a possible clade is a recurring theme in molecular analyses, which also produced a novel hypothesis within that group—Nonoculata. The Nonoculata hypothesis advocates a sister group relationship between Protura and Diplura to the exclusion of Collembola. It was originally proposed based on nuclear ribosomal genes (Giribet et al. 2004; Luan et al. 2005; Gao et al. 2008; von Reumont et al. 2009), but has found further support in some phylogenomic analyses (Meusemann et al. 2010). Nonoculata was a novel solution because it conflicted with the standard morphological hypothesis of a sister group relationship between Protura and Collembola, a group named Ellipura. Morphologists have, however, observed that Nonoculata is able to accommodate some anatomical features shared by proturans and diplurans but not collembolans (Koch 2009). The situation remains contentious because denser taxon sampling in EST analyses yields trees that unite Protura and Collembola as Ellipura, rather than giving support for Nonoculata (von Reumont et al. 2012).

A few phylogenetic problems remain unresolved at the base of the insect tree. Among them is the position of the relictual silverfish *Tricholepidion* relative to remaining Zygentoma (Wygodzinsky 1961). In some analyses, *Tricholepidion* appears as sister group to Dicondylia (Zygentoma + Pterygota) (Beutel and Gorb 2001; Giribet et al. 2004), whereas other data speak in favour of it being sister group to other Zygentoma or within that group (Koch 2003b; Dallai et al. 2004).

Monophyly of the winged insects (Pterygota) has been recognized since the earliest studies of insect phylogeny, but the resolution of the basalmost lineages of winged insects, Odonata and Ephemeroptera, remains contentious to this date. Current datasets support either their grouping as a clade named Palaeoptera or that they comprise a grade leading to Neoptera in either of the two possible arrangements, which represent the Metapterygota and Chiastomyaria hypotheses (Hovmöller et al. 2002; Ogden and Whiting 2003; Whitfield and Kjer 2008; Simon et al. 2009; Trautwein et al. 2012). This conundrum has been called "the Palaeoptera problem" and qualified as presently "intractable" (Trautwein et al. 2012), although recent morphological work based on head structure adds support to Palaeoptera (Blanke et al. 2012). Neopteran monophyly is widely accepted, but two of the three putative lineages nested within it, Polyneoptera and Paraneoptera (= Acercaria), lack robust support, and the cladistic structure of the tree remains poorly understood (Trautwein et al. 2012). Exciting developments within Polyneoptera are the discovery and systematic placement of the order Mantophasmatodea (Klass et al. 2002; Terry and Whiting 2005; Cameron et al. 2006; Eberhard et al. 2011), the inclusion of Isoptera as a family of Blattodea (Terry and Whiting 2005; Inward et al. 2007) and the possible resolution of Zoraptera as the sister group to the dictyopteran orders (Ishiwata et al. 2011).

Resolution within Holometabola is now comparatively stable, including the acceptance that fleas are members of the scorpionfly order Mecoptera (Whiting 2002; Wiegmann et al. 2009; Friedrich and Beutel 2010). Recent analyses have resolved "the Strepsiptera problem" (Whiting et al. 1997) towards the Coleoptera side, placing them as the sister group of beetles (Niehuis et al. 2012). The early divergence of Hymenoptera, which comprises the sister group to all other Holometabola, has found recent support in analyses of both single-copy nuclear genes (Wiegmann et al. 2009) and morphology (Friedrich and Beutel 2010).

## 2.5  Final Remarks

New approaches to studying anatomy and molecular analyses that are increasingly becoming phylogenomic in scope have converged on many of the main issues in arthropod phylogeny. Monophyly of Ecdysozoa, Panarthropoda and an Onychophora + Arthropoda clade provides a context for evaluating the internal phylogeny of Arthropoda, which is itself unambiguously monophyletic. Pycnogonida and Euchelicerata probably form a clade, Chelicerata, and its most likely sister group is Mandibulata, though various lines of evidence still signal an alternative alliance between chelicerates and myriapods, or Paradoxopoda. Myriapoda is monophyletic and in the context of Mandibulata constitutes the sister group to Tetraconata, composed of a paraphyletic Crustacea from which a monophyletic Hexapoda arose, most probably from a shared ancestor with Remipedia (and doubtfully Cephalocarida). Key outstanding issues are the interrelationships between arachnid orders and crustacean classes, notably whether cephalocarids group with

remipedes and whether branchiopods or malacostracans are more closely related to remipedes
and hexapods or to each other. The dating of
arthropod diversification needs to be refined by
improved clock methods and careful integration
of fossil constraints. Geologically, Chelicerata
(at least Pycnogonida) have a Cambrian origin,
while Arachnida started diversifying by the
Early Silurian, probably concurrently with
Myriapoda. The deepest splits within Tetraconata demonstrably date to no younger than the
Cambrian, as shown by spectacularly preserved
Late Cambrian fossils that can be identified as
branchiopods, copepods and ostracods (Harvey
et al. 2012), and early Cambrian maxillopodan-
type metanauplius larvae (Zhang et al. 2010; see
Chap. 15). Though molecular dating and pala-
eontologically inferred ghost lineages date the
origins of Hexapoda to the Cambrian, the
clade's diversification is probably Silurian–
Devonian and has been correlated with the
origin of vascular plants (Kenrick et al. 2012).
We expect that with the current availability and
facilities for generating genomic data of a
diverse selection of arthropods, a broad consensus will be found for the most diverse group
of animals, a group with more than 500 million
years of evolutionary history.

# References

Abele LG, Kim W, Felgenhauer BE (1989) Molecular evidence for inclusion of the phylum Pentastomida in the Crustacea. Mol Biol Evol 6:685–691

Abzhanov A, Kaufman TC (2000) Crustacean (malacostracan) Hox genes and the evolution of the arthropod trunk. Development 127:2239–2249

Adrain JM (2011) Class Trilobita Walch, 1771. In: Zhang Z-Q (ed) Animal biodiversity: an outline of higher-level classification and survey of taxonomic richness. Magnolia Press, Auckland, pp 104–109

Aguinaldo AMA, Turbeville JM, Lindford LS, Rivera MC, Garey JR, Raff RA, Lake JA (1997) Evidence for a clade of nematodes, arthropods and other moulting animals. Nature 387:489–493

Almeida WO, Christoffersen ML, Amorim DS, Garrafoni ARS, Silva GS (2003) Polychaeta, Annelida, and Articulata are not monophyletic: articulating the Metameria (Metazoa: Coelomata). Rev Bras Zool 20:23–57

Anderson DT (1969) On the embryology of the cirripede crustaceans Tetraclita rosea (krauss), Tetraclita purpurascens (Wood), Chthamalus antennatus Darwin and Chamaesipho columna (Spengler) and some considerations of crustacean phylogenetic relationships. Philos Trans R Soc B 256:183–235

Anderson DT (1973) Embryology and phylogeny in annelids and arthropods. Pergamon, Oxford

Andrew DR (2011) A new view of insect–crustacean relationships II: inferences from expressed sequence tags and comparisons with neural cladistics. Arthropod Struct Dev 40:289–302

Angelini DR, Kaufman TC (2005) Comparative developmental genetics and the evolution of arthropod body plans. Annu Rev Genet 39:95–119

Bäcker H, Fanenbruck M, Wägele JW (2008) A forgotten homology supporting the monophyly of Tracheata: The subcoxa of insects and myriapods re-visited. Zool Anz 247:185–207

Ballard JWO, Ballard O, Olsen GJ, Faith DP, Odgers WA, Rowell DM, Atkinson P (1992) Evidence from 12S ribosomal RNA sequences that onychophorans are modified arthropods. Science 258:1345–1348

Bartolomaeus T, Quast B, Koch M (2009) Nephridial development and body cavity formation in Artemia salina (Crustacea: Branchiopoda): no evidence for any transitory coelom. Zoomorphology 128:247–262

Bergström J (1979) Morphology of fossil arthropods as a guide to phylogenetic relationships. In: Gupta AP (ed) Arthropod phylogeny. Van Nostrand Reinhold, New York, pp 3–56

Beutel RG, Gorb SN (2001) Ultrastructure of attachment specializations of hexapods (arthropoda): evolutionary patterns inferred from a revised ordinal phylogeny. J Zool Syst Evol Res 39:177–207

Bitsch C, Bitsch J (2004) Phylogenetic relationships of basal hexapods among the mandibulate arthropods: a cladistic analysis based on comparative morphological characters. Zool Scr 33:511–550

Blair JE, Ikeo K, Gojobori T, Hedges SB (2002) The evolutionary position of nematodes. BMC Evol Biol 2:1–7

Blanke A, Wipfler B, Letsch H, Koch M, Beckmann F, Beutel R, Misof B (2012) Revival of Palaeoptera—head characters support a monophyletic origin of Odonata and Ephemeroptera (Insecta). Cladistics 28:560–581

Boore JL, Collins TM, Stanton D, Daehler LL, Brown WM (1995) Deducing the pattern of arthropod phylogeny from mitochondrial DNA rearrangements. Nature 376:163–165

Bourlat SJ, Nielsen C, Economou AD, Telford MJ (2008) Testing the new animal phylogeny: a phylum level molecular analysis of the animal kingdom. Mol Phylogenet Evol 49:23–31

Brusca RC, Brusca GJ (1990) Invertebrates. Sinauer Associates, Sunderland

Brusca RC, Brusca GJ (2003) Invertebrates, 2nd edn. Sinauer Associates, Sunderland

Cameron SL, Barker SC, Whiting MF (2006) Mitochondrial genomics and the new insect order Mantophasmatodea. Mol Phylogenet Evol 38:274–279

Campbell LI, Rota-Stabelli O, Edgecombe GD, Marchioro T, Longhorn SJ, Telford MJ, Philippe H, Rebecchi L, Peterson KJ, Pisani D (2011) MicroRNAs and phylogenomics resolve the relationships of Tardigrada and suggest that velvet worms are the sister group of Arthropoda. Proc Natl Acad Sci USA 108:15920–15924

Carapelli A, Lió P, Nardi F, van der Wath E, Frati F (2007) Phylogenetic analysis of mitochondrial protein coding genes confirms the reciprocal paraphyly of Hexapoda and crustacea. BMC Evol Biol 7:S8

Caravas J, Friedrich M (2010) Of mites and millipedes: recent progress in resolving the base of the arthropod tree. BioEssays 32:488–495

Carmean D, Kimsey LS, Berbee ML (1992) 18S rDNA sequences and the holometabolous insects. Mol Phylogenet Evol 1:270–278

Castellani C, Maas A, Waloszek D, Haug JT (2011) New pentastomids from the late Cambrian of Sweden—deeper insight of the ontogeny of fossil tongue worms. Palaeontogr Abt A: Palaeozoology Stratigr 293:95–145

Clark AG, Eisen MB, Smith DR, Bergman CM, Oliver B, Markow TA, Kaufman TC, Kellis M, Gelbart W, Iyer VN, Pollard DA, Sackton TB, Larracuente AM, Singh ND, Abad JP, Abt DN, Adryan B, Aguade M, Akashi H, Anderson WW, Aquadro CF, Ardell DH, Arguello R, Artieri CG, Barbash DA, Barker D, Barsanti P, Batterham P, Batzoglou S, Begun D, Bhutkar A, Blanco E, Bosak SA, Bradley RK, Brand AD, Brent MR, Brooks AN, Brown RH, Butlin RK, Caggese C, Calvi BR, Bernardo de Carvalho A, Caspi A, Castrezana S, Celniker SE, Chang JL, Chapple C, Chatterji S, Chinwalla A, Civetta A, Clifton SW, Comeron JM, Costello JC, Coyne JA, Daub J, David RG, Delcher AL, Delehaunty K, Do CB, Ebling H, Edwards K, Eickbush T, Evans JD, Filipski A, Findeiss S, Freyhult E, Fulton L, Fulton R, Garcia AC, Gardiner A, Garfield DA, Garvin BE, Gibson G, Gilbert D, Gnerre S, Godfrey J, Good R, Gotea V, Gravely B, Greenberg AJ, Griffiths-Jones S, Gross S, Guigo R, Gustafson EA, Haerty W, Hahn MW, Halligan DL, Halpern AL, Halter GM, Han MV, Heger A, Hillier L, Hinrichs AS, Holmes I, Hoskins RA, Hubisz MJ, Hultmark D, Huntley MA, Jaffe DB, Jagadeeshan S, Jeck WR, Johnson J, Jones CD, Jordan WC, Karpen GH, Kataoka E, Keightley PD, Kheradpour P, Kirkness EF, Koerich LB, Kristiansen K, Kudrna D, Kulathinal RJ, Kumar S, Kwok R, Lander E, Langley CH, Lapoint R, Lazzaro BP, Lee SJ, Levesque L, Li R, Lin CF, Lin MF, Lindblad-Toh K, Llopart A, Long M, Low L, Lozovsky E, Lu J, Luo M, Machado CA, Makalowski W, Marzo M, Matsuda M, Matzkin L, McAllister B, McBride CS, McKernan B, McKernan K, Mendez-Lago M, Minx P, Mollenhauer MU, Montooth K, Mount SM, Mu X, Myers E,

Negre B, Newfeld S, Nielsen R, Noor MA, O'Grady P, Pachter L, Papaceit M, Parisi MJ, Parisi M, Parts L, Pedersen JS, Pesole G, Phillippy AM, Ponting CP, Pop M, Porcelli D, Powell JR, Prohaska S, Pruitt K, Puig M, Quesneville H, Ravi Ram K, Rand D, Rasmussen MD, Reed LK, Reenan R, Reily A, Remington KA, Rieger TT, Ritchie MG, Robin C, Rogers YH, Rohde C, Rozas J, Rubenfield MJ, Ruiz A, Russo S, Salzberg SL, Sanchez-Gracia A, Saranga DJ, Sato H, Schaeffer SW, Schatz MC, Schlenke T, Schwartz R, Segarra C, Singh RS, Sirot L, Sirota M, Sisneros NB, Smith CD, Smith TF, Spieth J, Stage DE, Stark A, Stephan W, Strausberg RL, Strempel S, Sturgill D, Sutton G, Sutton GG, Tao W, Teichmann S, Tobari YN, Tomimura Y, Tsolas JM, Valente VL, Venter E, Craig Venter J, Vicario S, Vieira FG, Vilella AJ, Villasante A, Walenz B, Wang J, Wasserman M, Watts T, Wilson D, Wilson RK, Wing RA, Wolfner MF, Wong A, Ka-Shu Wong G, Wu CI, Wu G, Yamamoto D, Yang HP, Yang SP, Yorke JA, Yoshida K, Zdobnov E, Zhang P, Zhang Y, Zimin AV, Baldwin J, Abdouelleil A, Abdulkadir J, Abebe A, Abera B, Abreu J, Christophe Acer S, Aftuck L, Alexander A, An P, Anderson E, Anderson S, Arachi H, Azer M, Bachantsang P, Barry A, Bayul T, Berlin A, Bessette D, Bloom T, Blye J, Boguslavskiy L, Bonnet C, Boukhgalter B, Bourzgui I, Brown A, Cahill P, Channer S, Cheshatsang Y, Chuda L, Citroen M, Collymore A, Cooke P, Costello M, D'Aco K, Daza R, De Haan G, Degray S, Demaso C, Dhargay N, Dooley K, Dooley E, Doricent M, Dorje P, Dorjee K, Dupes A, Elong R, Falk J, Farina A, Faro S, Ferguson D, Fisher S, Foley CD, Franke A, Friedrich D, Gadbois L, Gearin G, Gearin CR, Giannoukos G, Goode T, Graham J, Grandbois E, Grewal S, Gyaltsen K, Hafez N, Hagos B, Hall J, Henson C, Hollinger A, Honan T, Huard MD, Hughes L, Hurhula B, Erii Husby M, Kamat A, Kanga B, Kashin S, Khazanovich D, Kisner P, Lance K, Lara M, Lee W, Lennon N, Letendre F, Levine R, Lipovsky A, Liu X, Liu J, Liu S, Lokyitsang T, Lokyitsang Y, Lubonja R, Lui A, Macdonald P, Magnisalis V, Maru K, Matthews C, McCusker W, McDonough S, Mehta T, Meldrim J, Meneus L, Mihai O, Mihalev A, Mihova T, Mittelman R, Mlenga V, Montmayeur A, Mulrain L, Navidi A, Naylor J, Negash T, Nguyen T, Nguyen N, Nicol R, Norbu C, Norbu N, Novod N, O'Neill B, Osman S, Markiewicz E, Oyono OL, Patti C, Phunkhang P, Pierre F, Priest M, Raghuraman S, Rege F, Reyes R, Rise C, Rogov P, Ross K, Ryan E, Settipalli S, Shea T, Sherpa N, Shi L, Shih D, Sparrow T, Spaulding J, Stalker J, Stange-Thomann N, Stavropoulos S, Stone C, Strader C, Tesfaye S, Thomson T, Thoulutsang Y, Thoulutsang D, Topham K, Topping I, Tsamla T, Vassiliev H, Vo A, Wangchuk T, Wangdi T, Weiand M, Wilkinson J, Wilson A, Yadav S, Young G, Yu Q, Zembek L, Zhong D, Zimmer A, Zwirko Z, Jaffe DB, Alvarez P, Brockman W, Butler J, Chin C, Gnerre S,

Grabherr M, Kleber M, Mauceli E, Maccallum I (2007) Evolution of genes and genomes on the *Drosophila* phylogeny. Nature 450:203–218

Clouse RM, Sharma PP, Giribet G, Wheeler WC (submitted) Independent and isolated suites of paralogs in an arachnid elongation factor-1α, a purported single-copy nuclear gene. Mol Phylogenet Evol

Colosi G (1967) Zoologia e biologia generale. UTET, Torino

Cuvier G (1817) Le règne animal distribué d'après son organisation. A. Belin, Paris

Dallai R, Carapelli A, Nardi F, Fanciulli PP, Lupetti P, Afzelius BA, Frati F (2004) Sperm structure and spermiogenesis in *Coletinia* sp. (Nicoletiidae, Zygentoma, Insecta) with a comparative analysis of sperm structure in Zygentoma. Tissue Cell 36:233–244

Dallai R, Mercati D, Carapelli A, Nardi F, Machida R, Sekiya K, Frati F (2011) Sperm accessory microtubules suggest the placement of Diplura as the sister-group of Insecta s.s. Arthropod Struct Dev 40:77–92

De Grave S, Pentcheff ND, Ahyong ST, Chan T-Y, Crandall KA, Dworschak PC, Felder DL, Feldmann RM, Fransen CHJM, Goulding LYD, Lemaitre R, Low MEY, Martin JW, Naaaag PKL, Schweitzer CE, Tan SH, Tshudy D, Wetzer R (2009) A classification of living and fossil genera of decapod crustaceans. Raffles Bull Zool, pp 1–109

Dohle W (1980) Sind die Myriapoden eine monophyletische Gruppe? Eine Diskussion der Verwandtschaftsbeziehungen der Antennaten. Abh naturwiss Ver Hamburg NF 23:45–104

Dong Y, Sun H, Guo H, Pan D, Qian C, Hao S, Zhou K (2012) The complete mitochondrial genome of *Pauropus longriamus* (Myriapoda: Pauropoda): implications on early diversification of the myriapods revealed from comparative analysis. Gene 505:57–65

Dopazo H, Santoyo J, Dopazo J (2004) Phylogenomics and the number of characters required for obtaining an accurate phylogeny of eukaryote model species. Bioinformatics 20(Suppl 1):116–121

Dove H, Stollewerk A (2003) Comparative analysis of neurogenesis in the myriapod *Glomeris marginata* (Diplopoda) suggests more similarities to chelicerates than to insects. Development 130:2161–2171

Dunlop JA (2010) Geological history and phylogeny of Chelicerata. Arthropod Struct Dev 39:124–142

Dunlop JA, Alberti G (2008) The affinities of mites and ticks: a review. J Zool Syst Evol Res 46:1–18

Dunlop JA, Arango CP (2005) Pycnogonid affinities: a review. J Zool Syst Evol Res 43:8–21

Dunlop JA, Penney D, Tetlie OE, Anderson LI (2008) How many species of fossil arachnids are there? J Arachnol 36:267–272

Dunn CW, Hejnol A, Matus DQ, Pang K, Browne WE, Smith SA, Seaver EC, Rouse GW, Obst M, Edgecombe GD, Sørensen MV, Haddock SHD, Schmidt-Rhaesa A, Okusu A, Kristensen RM, Wheeler WC, Martindale MQ, Giribet G (2008) Broad taxon sampling improves resolution of the animal tree of life. Nature 452:745–749

Eberhard MJB, Picker MD, Klass K-D (2011) Sympatry in Mantophasmatodea, with the description of a new species and phylogenetic considerations. Org Divers Evol 11:43–59

Edgecombe GD (2009) Palaeontological and molecular evidence linking arthropods, onychophorans, and other Ecdysozoa. Evo Edu Outreach 2:178–190

Edgecombe GD (2010) Arthropod phylogeny: an overview from the perspectives of morphology, molecular data and the fossil record. Arthropod Struct Dev 39:74–87

Edgecombe GD, Giribet G, Dunn CW, Hejnol A, Kristensen RM, Neves RC, Rouse GW, Worsaae K, Sørensen MV (2011) Higher-level metazoan relationships: recent progress and remaining questions. Org Divers Evol 11:151–172

Edgecombe GD, Richter S, Wilson GDF (2003) The mandibular gnathal edges: Homologous structures throughout Mandibulata? Afr Invertebr 44:115–135

Eernisse DJ, Albert JS, Anderson FE (1992) Annelida and Arthropoda are not sister taxa: A phylogenetic analysis of spiralian metazoan morphology. Syst Biol 41:305–330

Eriksson BJ, Stollewerk A (2010a) Expression patterns of neural genes in *Euperipatoides kanangrensis* suggest divergent evolution of onychophoran and euarthropod neurogenesis. Proc Natl Acad Sci USA 107: 22576–22581

Eriksson BJ, Stollewerk A (2010b) The morphological and molecular processes of onychophoran brain development show unique features that are neither comparable to insects nor to chelicerates. Arthropod Struct Dev 39:478–490

Ertas B, von Reumont BM, Wägele JW, Misof B, Burmester T (2009) Hemocyanin suggests a close relationship of Remipedia and Hexapoda. Mol Biol Evol 26:2711–2718

Erwin DH, Laflamme M, Tweedt SM, Sperling EA, Pisani D, Peterson KJ (2011) The Cambrian conundrum: early divergence and later ecological success in the early history of animals. Science 334:1091–1097

Friedrich F, Beutel RG (2010) Goodbye Halteria? The thoracic morphology of Endopterygota (Insecta) and its phylogenetic implications. Cladistics 26:579–612

Friedrich M, Tautz D (1995) Ribosomal DNA phylogeny of the major extant arthropod classes and the evolution of myriapods. Nature 376:165–167

Gabriel WN, Goldstein B (2007) Segmental expression of Pax3/7 and Engrailed homologs in tardigrade development. Dev Genes Evol 217:421–433

Gai Y, Song D, Sun H, Yang Q, Zhou K (2008) The complete mitochondrial genome of *Symphylella* sp. (Myriapoda: Symphyla): extensive gene order rearrangement and evidence in favor of Progoneata. Mol Phylogenet Evol 49:574–585

Gao Y, Bu Y, Luan YX (2008) Phylogenetic relationships of basal hexapods reconstructed from nearly complete 18S and 28S rRNA gene sequences. Zool Sci 25:1139–1145

Giribet G (1997) Filogenia molecular de Artrópodos basada en la secuencia de genes ribosomales. Universitat de Barcelona: Departament de Biologia Animal, Barcelona

Giribet G (2003) Molecules, development and fossils in the study of metazoan evolution; Articulata versus Ecdysozoa revisited. Zoology 106:303–326

Giribet G (2010) A new dimension in combining data? The use of morphology and phylogenomic data in metazoan systematics. Acta Zool 91:11–19

Giribet G, Carranza S, Baguñà J, Riutort M, Ribera C (1996) First molecular evidence for the existence of a Tardigrada + Arthropoda clade. Mol Biol Evol 13:76–84

Giribet G, Distel DL, Polz M, Sterrer W, Wheeler WC (2000) Triploblastic relationships with emphasis on the acoelomates and the position of Gnathostomulida, Cycliophora, Plathelminthes, and Chaetognatha: a combined approach of 18S rDNA sequences and morphology. Syst Biol 49:539–562

Giribet G, Edgecombe GD (2006) Conflict between data sets and phylogeny of centipedes: an analysis based on seven genes and morphology. Proc R Soc B 273:531–538

Giribet G, Edgecombe GD (2012) Reevaluating the arthropod tree of life. Annu Rev Entomol 57:167–186

Giribet G, Edgecombe GD, Carpenter JM, D'Haese CA, Wheeler WC (2004) Is Ellipura monophyletic? A combined analysis of basal hexapod relationships with emphasis on the origin of insects. Org Divers Evol 4:319–340

Giribet G, Edgecombe GD, Wheeler WC (2001) Arthropod phylogeny based on eight molecular loci and morphology. Nature 413:157–161

Giribet G, Edgecombe GD, Wheeler WC, Babbitt C (2002) Phylogeny and systematic position of opiliones: a combined analysis of chelicerate relationships using morphological and molecular data. Cladistics 18:5–70

Giribet G, Ribera C (1998) The position of arthropods in the animal kingdom: a search for a reliable outgroup for internal arthropod phylogeny. Mol Phylogenet Evol 9:481–488

Giribet G, Ribera C (2000) A review of arthropod phylogeny: new data based on ribosomal DNA sequences and direct character optimization. Cladistics 16:204–231

Giribet G, Richter S, Edgecombe GD, Wheeler WC (2005) The position of crustaceans within the Arthropoda—evidence from nine molecular loci and morphology. In: Koenemann S, Jenner RA (eds) Crustacean issues 16: crustacea and arthropod relationships. Taylor & Francis, Boca Raton, pp 307–352

Giribet G, Wheeler WC (1999) The position of arthropods in the animal kingdom: Ecdysozoa, islands, trees, and the "parsimony ratchet". Mol Phylogenet Evol 13:619–623

Glenner H, Hansen AJ, Sørensen MV, Ronquist F, Huelsenbeck JP, Willerslev E (2004) Bayesian

inference of the metazoan phylogeny; a combined molecular and morphological approach. Curr Biol 14:1644–1649

Grimaldi DA (2010) 400 million years on six legs: on the origin and early evolution of Hexapoda. Arthropod Struct Dev 39:191–203

Hanström B (1926) Vergleichende Anatomie des Nervensystems der wirbellosen Tiere unter Berücksichtigung seiner Funktion. Springer, Berlin

Harvey TH, Velez MI, Butterfield NJ (2012) Exceptionally preserved crustaceans from western Canada reveal a cryptic Cambrian radiation. Proc Natl Acad Sci USA 109:1589–1594

Harzsch S (2004) Phylogenetic comparison of serotonin-immunoreactive neurons in representatives of the Chilopoda, Diplopoda, and Chelicerata: implications for arthropod relationships. J Morphol 259:198–213

Harzsch S (2006) Neurophylogeny: architecture of the nervous system and a fresh view on arthropod phyology. Integr Comp Biol 46:162–194

Harzsch S, Hafner G (2006) Evolution of eye development in arthropods: Phylogenetic aspects. Arthropod Struct Dev 35:319–340

Hejnol A, Obst M, Stamatakis AMO, Rouse GW, Edgecombe GD, Martinez P, Baguñà J, Bailly X, Jondelius U, Wiens M, Müller WEG, Seaver E, Wheeler WC, Martindale MQ, Giribet G, Dunn CW (2009) Assessing the root of bilaterian animals with scalable phylogenomic methods. Proc R Soc B 276:4261–4270

Hessler RR (1992) Reflections on the phylogenetic position of the Cephalocarida. Acta Zool 73:315–316

Holton TA, Pisani D (2010) Deep genomic-scale analyses of the Metazoa reject Coelomata: evidence from single- and multigene families analyzed under a supertree and supermatrix paradigm. Genome Biol Evol 2:310–324

Hörnschemeyer T, Beutel RG, Pasop F (2002) Head structures of *Priacma serrata* Leconte (Coleptera, Archostemata) inferred from X-ray tomography. J Morphol 252:298–314

Hovmöller R, Pape T, Källersjö M (2002) The Palaeoptera problem: basal pterygote phylogeny inferred from 18S and 28S rDNA sequences. Cladistics 18:313–323

Huckstorf K, Wirkner CS (2011) Comparative morphology of the hemolymph vascular system in krill (Euphausiacea; Crustacea). Arthropod Struct Dev 40:39–53

Hwang UW, Friedrich M, Tautz D, Park CJ, Kim W (2001) Mitochondrial protein phylogeny joins myriapods with chelicerates. Nature 413:154–157

Illumina_Inc (2007) DNA sequencing with Solexa® technology

Inward D, Beccaloni G, Eggleton P (2007) Death of an order: a comprehensive molecular phylogenetic study confirms that termites are eusocial cockroaches. Biol Lett 3:331–335

Irimia M, Maeso I, Penny D, Garcia-Fernàndez J, Roy SW (2007) Rare coding sequence changes are consistent with Ecdysozoa, not Coelomata. Mol Biol Evol 24:1604–1607

Ishiwata K, Sasaki G, Ogawa J, Miyata T, Su Z-H (2011) Phylogenetic relationships among insect orders based on three nuclear protein-coding gene sequences. Mol Phylogenet Evol 58:169–180

Janssen R, Damen WGM, Budd GE (2011) Expression of *collier* in the premandibular segment of myriapods: support for the traditional Atelocerata concept or a case of convergence? BMC Evol Biol 11:50

Janssen R, Eriksson JB, Budd GE, Akam M, Prpic N-M (2010) Gene expression patterns in onychophorans reveal that regionalization predates limb segmentation in pan-arthropods. Evol Dev 12:363–372

Jenner RA (2010) Higher-level crustacean phylogeny: consensus and conflicting hypotheses. Arthropod Struct Dev 39:143–153

Jenner RA, Scholtz G (2005) Playing another round of metazoan phylogenetics: historical epistemology, sensitivity analysis, and the position of Arthropoda within Metazoa on the basis of morphology. In: Koenemann S, Jenner RA (eds) Crustacean Issues 16: Crustacea and arthropod relationships. Taylor & Francis, Boca Raton, pp 355–385

Kadner D, Stollewerk A (2004) Neurogenesis in the chilopod *Lithobius forficatus* suggests more similarities to chelicerates than to insects. Dev Genes Evol 214:367–379

Kenrick P, Wellman CH, Schneider H, Edgecombe GD (2012) A timeline for terrestrialization: consequences for the carbon cycle in the Palaeozoic. Philos Trans R Soc B 367:519–536

Kim W, Abele LG (1990) Molecular phylogeny of selected decapod crustaceans based on 18S rRNA nucleotide sequences. J Crustacean Biol 10:1–13

Klass KD, Zompro O, Kristensen NP, Adis J (2002) Mantophasmatodea: a new insect order with extant members in the afrotropics. Science 296: 1456–1459

Koch M (1997) Monophyly and phylogenetic position of the Diplura (Hexapoda). Pedobiologia 41:9–12

Koch M (2000) The cuticular cephalic endoskeleton of primarily wingless hexapods: ancestral state and evolutionary changes. Pedobiologia 44:374–385

Koch M (2003a) Monophyly of Myriapoda? Reliability of current arguments. Afr Invertebr 44:137–153

Koch M (2003b) Towards a phylogenetic system of the Zygentoma. Entomol Abh 61:122–125

Koch M (2009) Protura. In: Resh VH, Carde R (eds) Encyclopedia of insects, 2nd edn. Academic Press/ Elsevier Science, San Diego, pp 855–858

Kraus O, Kraus M (1994) Phylogenetic system of the Tracheata (Mandibulata): on "Myriapoda": Insecta interrelationships, phylogenetic age and primary ecological niches. Verh naturwiss Ver Hamburg 34:5–31

Kraus O, Kraus M (1996) On myriapod/insect interrelationships. Mem Mus nat Hist Nat 169:283–290

Kristensen RM (1991) Loricifera. In: Harrison FW, Ruppert EE (eds) Microscopic anatomy of

invertebrates, vol 4., AschelminthesWiley-Liss, New York, pp 351–375

Lankester ER (1904) The structure and classification of the Arthropoda. Q J Microscop Sci 38:523–582

Lavrov DV, Boore JL, Brown WM (2002) Complete mtDNA sequences of two millipedes suggest a new model for mitochondrial gene rearrangements: duplication and nonrandom loss. Mol Biol Evol 19:163–169

Liu J, Steiner M, Dunlop JA, Keupp H, Shu D, Ou Q, Han J, Zhang Z, Zhang X (2011) An armoured Cambrian lobopodian from China with arthropod-like appendages. Nature 470:526–530

Loesel R, Nässel DR, Strausfeld NJ (2002) Common design in a unique midline neuropil in the brains of arthropods. Arthropod Struct Dev 31:77–91

Luan YX, Mallatt JM, Xie RD, Yang YM, Yin WY (2005) The phylogenetic positions of three basal-hexapod groups (Protura, Diplura, and Collembola) based on ribosomal RNA gene sequences. Mol Biol Evol 22:1579–1592

Machida R (2006) Evidence from embryology for reconstructing the relationships of hexapod basal clades. Arthropod Syst Phyl 64:95–104

Machner J, Scholtz G (2010) A scanning electron microscopy study of the embryonic development of *Pycnogonum litorale* (Arthropoda, Pycnogonida). J Morphol 271:1306–1318

Mallatt J, Craig CW, Yoder MJ (2010) Nearly complete rRNA genes assembled from across the metazoan animals: effects of more taxa, a structure-based alignment, and paired-sites evolutionary models on phylogeny reconstruction. Mol Phylogenet Evol 55:1–17

Mallatt J, Giribet G (2006) Further use of nearly complete 28S and 18S rRNA genes to classify Ecdysozoa: 37 more arthropods and a kinorhynch. Mol Phylogenet Evol 40:772–794

Mallatt J, Winchell CJ (2002) Testing the new animal phylogeny: first use of combined large-subunit and small-subunit rRNA gene sequences to classify the protostomes. Mol Biol Evol 19:289–301

Mallatt JM, Garey JR, Shultz JW (2004) Ecdysozoan phylogeny and Bayesian inference: first use of nearly complete 28S and 18S rRNA gene sequences to classify the arthropods and their kin. Mol Phylogenet Evol 31:178–191

Manton SM (1964) Mandibular mechanisms and the evolution of arthropods. Philos Trans R Soc B 247:1–183

Manton SM (1973) Arthropod phylogeny-a modern synthesis. J Zool 171:11–130

Manton SM (1977) The Arthropoda: habits, functional morphology, and evolution. Clarendon Press, Oxford

Margulies M, Egholm M, Altman WE, Attiya S, Bader JS, Bemben LA, Berka J, Braverman MS, Chen YJ, Chen Z, Dewell SB, Du L, Fierro JM, Gomes XV, Godwin BC, He W, Helgesen S, Ho CH, Irzyk GP, Jando SC, Alenquer ML, Jarvie TP, Jirage KB, Kim JB, Knight JR, Lanza JR, Leamon JH, Lefkowitz SM,

Lei M, Li J, Lohman KL, Lu H, Makhijani VB, McDade KE, McKenna MP, Myers EW, Nickerson E, Nobile JR, Plant R, Puc BP, Ronan MT, Roth GT, Sarkis GJ, Simons JF, Simpson JW, Srinivasan M, Tartaro KR, Tomasz A, Vogt KA, Volkmer GA, Wang SH, Wang Y, Weiner MP, Yu P, Begley RF, Rothberg JM (2005) Genome sequencing in microfabricated high-density picolitre reactors. Nature 437:376–380

Maslakova SA, Martindale MQ, Norenburg JL (2004) Fundamental properties of the spiralian developmental program are displayed by the basal nemertean *Carinoma tremaphoros* (Palaeonemertea, Nemertea). Dev Biol 267:342–360

Masta SE, Boore JL (2008) Parallel evolution of truncated transfer RNA genes in arachnid mitochondrial genomes. Mol Biol Evol 25:949–959

Masta SE, Longhorn SJ, Boore JL (2009) Arachnid relationships based on mitochondrial genomes: asymmetric nucleotide and amino acid bias affects phylogenetic analyses. Mol Phylogenet Evol 50:117–128

Mayer G, Whitington PM (2009) Velvet worm development links myriapods with chelicerates. Proc R Soc B 276:3571–3579

Meusemann K, von Reumont BM, Simon S, Roeding F, Strauss S, Kück P, Ebersberger I, Walzl M, Pass G, Breuers S, Achter V, von Haeseler A, Burmester T, Hadrys H, Wägele JW, Misof B (2010) A phylogenomic approach to resolve the arthropod tree of life. Mol Biol Evol 27:2451–2464

Minelli A, Bortoletto S (1988) Myriapod metamerism and arthropod segmentation. Biol J Linn Soc 33:323–343

Miyazaki K (2002) On the shape of foregut lumen in sea spiders (Arthropoda: Pycnogonida). J Mar Biol Assoc UK 82:1037–1038

Møller OS, Olesen J, Avenant-Oldewage A, Thomsen PF, Glenner H (2008) First maxillae suction discs in Branchiura (Crustacea): development and evolution in light of the first molecular phylogeny of Branchiura, Pentastomida, and other "Maxillopoda". Arthropod Struct Dev 37:333–346

Morozova O, Hirst M, Marra MA (2009) Applications of new sequencing technologies for transcriptome analysis. Annu Rev Genomics Hum Genet 10:135–151

Murienne J, Edgecombe GD, Giribet G (2010) Including secondary structure, fossils and molecular dating in the centipede tree of life. Mol Phylogenet Evol 57:301–313

Niehuis O, Hartig G, Grath S, Pohl H, Lehmann J, Tafer H, Donath A, Krauss V, Eisenhardt C, Hertel J, Petersen M, Mayer C, Meusemann K, Peters RS, Stadler PF, Beutel RG, Bornberg-Bauer E, McKenna DD, Misof B (2012) Genomic and morphological evidence converge to resolve the enigma of Strepsiptera. Current biology: CB 22:1309–1313.

Nielsen C (2001) Animal evolution: interrelationships of the living phyla, 2nd edn. Oxford University Press, Oxford

Nielsen C (2003) Proposing a solution to the articulata-ecdysozoa controversy. Zool Scr 32:475–482

Nielsen C (2012) Animal evolution: interrelationships of the living phyla, 3rd edn. Oxford University Press, Oxford

Nielsen C, Scharff N, Eibye-Jacobsen D (1996) Cladistic analyses of the animal kingdom. Biol J Linn Soc 57:385–410

Oakley TH, Wolfe JM, Lindgren AR, Zaharoff (2013) Phylogenomics to bring the understudied into the fold: monophyletic Ostracoda, fossil placement, and pancrustacean phylogeny. Mol Biol Evol. 30:215–233

Ogden TH, Whiting MF (2003) The problem with "the Palaeoptera problem": sense and sensitivity. Cladistics 19:432–442

Panganiban G, Irvine SM, Lowe C, Roehl H, Corley LS, Sherbon B, Grenier JK, Fallon JF, Kimble J, Walker M, Wray GA, Swalla BJ, Martindale MQ, Carroll SB (1997) The origin and evolution of animal appendages. Proc Natl Acad Sci USA 94:5162–5166

Paps J, Baguñà J, Riutort M (2009a) Bilaterian phylogeny: a broad sampling of 13 nuclear genes provides a new Lophotrochozoa phylogeny and supports a paraphyletic basal Acoelomorpha. Mol Biol Evol 26:2397–2406

Paps J, Baguñà J, Riutort M (2009b) Lophotrochozoa internal phylogeny: new insights from an up-to-date analysis of nuclear ribosomal genes. Proc R Soc B 276:1245–1254

Pashley DP, McPheron BA, Zimmer EA (1993) Systematics of holometabolous insect orders based on 18S ribosomal RNA. Mol Phylogenet Evol 2:132–142

Pechmann M, Prpic NM (2009) Appendage patterning in the South American bird spider *Acanthoscurria geniculata* (Araneae: Mygalomorphae). Dev Genes Evol 219:189–198

Pepato AR, da Rocha CE, Dunlop JA (2010) Phylogenetic position of the acariform mites: sensitivity to homology assessment under total evidence. BMC Evol Biol 10:235

Peterson KJ, Eernisse DJ (2001) Animal phylogeny and the ancestry of bilaterians: inferences from morphology and 18S rDNA gene sequences. Evol Dev 3:170–205

Philip GK, Creevey CJ, McInerney JO (2005) The Opisthokonta and the Ecdysozoa may not be clades: stronger support for the grouping of plant and animal than for animal and fungi and stronger support for the coelomata than ecdysozoa. Mol Biol Evol 22:1175–1184

Philippe H, Brinkmann H, Copley RR, Moroz LL, Nakano H, Poustka AJ, Wallberg A, Peterson KJ, Telford MJ (2011) Acoelomorph flatworms are deuterostomes related to *Xenoturbella*. Nature 470:255–258

Pisani D (2009) Arthropods (Arthropoda). In: Hedges SB, Kumar S (eds) The timetree of life. Oxford University Press, Oxford, pp 251–254

Pisani D, Poling LL, Lyons-Weiler M, Hedges SB (2004) The colonization of land by animals: molecular phylogeny and divergence times among arthropods. BMC Biol 2:1–10

Prpic N-M, Telford MJ (2008) Expression of *homothorax* and *extradenticle* mRNA in the legs of the crustacean *Parhyale hawaiensis*: evidence for a reversal of gene expression regulation in the pancrustacean lineage. Dev Genes Evol 218:333–339

Prpic NM, Damen WGM (2004) Expression patterns of leg genes in the mouthparts of the spider *Cupiennius salei* (Chelicerata: Arachnida). Dev Genes Evol 214:296–302

Prpic NM, Janssen R, Wigand B, Klingler M, Damen WGM (2003) Gene expression in spider appendages reveals reversal of exd/hth spatial specificity, altered leg gap gene dynamics, and suggests divergent distal morphogen signaling. Dev Biol 264:119–140

Rauther M (1909) Morphologie und Verwandtschaftsbeziehungen der Nematoden und einiger ihnen nahe gestellter Vermalien. Ergebnisse und Fortschritte der Zoologie 1:491–596

Regier JC, Shultz JW (1997) Molecular phylogeny of the major arthropod groups indicates polyphyly of crustaceans and a new hypothesis for the origin of hexapods. Mol Biol Evol 14:902–913

Regier JC, Shultz JW (2001) Elongation factor-2: a useful gene for arthropod phylogenetics. Mol Phylogenet Evol 20:136–148

Regier JC, Shultz JW, Ganley AR, Hussey A, Shi D, Ball B, Zwick A, Stajich JE, Cummings MP, Martin JW, Cunningham CW (2008) Resolving arthropod phylogeny: exploring phylogenetic signal within 41 kb of protein-coding nuclear gene sequence. Syst Biol 57:920–938

Regier JC, Shultz JW, Kambic RE (2004) Phylogeny of basal hexapod lineages and estimates of divergence times. Ann Entomol Soc Amer 97:411–419

Regier JC, Shultz JW, Kambic RE (2005a) Pancrustacean phylogeny: hexapods are terrestrial crustaceans and maxillopods are not monophyletic. Proc R Soc B 272:395–401

Regier JC, Shultz JW, Zwick A, Hussey A, Ball B, Wetzer R, Martin JW, Cunningham CW (2010) Arthropod relationships revealed by phylogenomic analysis of nuclear protein-coding sequences. Nature 463:1079–1083

Regier JC, Wilson HM, Shultz JW (2005b) Phylogenetic analysis of Myriapoda using three nuclear protein-coding genes. Mol Phylogenet Evol 34:147–158

Regier JC, Zwick A (2011) Sources of signal in 62 protein-coding nuclear genes for higher-level phylogenetics of arthropods. PLoS ONE 6:e23408

Rehm P, Borner J, Meusemann K, von Reumont BM, Simon S, Hadrys H, Misof B, Burmester T (2011) Dating the arthropod tree based on large-scale transcriptome data. Mol Phylogenet Evol 61:880–887

Richter S (2002) The Tetraconata concept: hexapod-crustacean relationships and the phylogeny of crustacea. Org Divers Evol 2:217–237

Richter S, Loesel R, Purschke G, Schmidt-Rhaesa A, Scholtz G, Stach T, Vogt L, Wanninger A, Brenneis G, Doring C, Faller S, Fritsch M, Grobe P, Heuer CM, Kaul S, Møller OS, Müller CHG, Rieger V, Rothe BH, Stegner MEJ, Harzsch S (2010) Invertebrate neurophylogeny: suggested terms and definitions for a neuroanatomical glossary. Front Zool 7:29

Richter S, Møller OS, Wirkner CS (2009) Advances in crustacean phylogenetics. Arthropod Syst Phyl 67:275–286

Riesgo A, Andrade SCS, Sharma PP, Novo M, Pérez-Porro AR, Vahtera V, González VL, Kawauchi GY, Giribet G (2012) Comparative description of ten transcriptomes of newly sequenced invertebrates and efficiency estimation of genomic sampling in non-model taxa. Front Zool 9:33

Roeding F, Borner J, Kube M, Klages S, Reinhardt R, Burmester T (2009) A 454 sequencing approach for large scale phylogenomic analysis of the common emperor scorpion (*Pandinus imperator*). Mol Phylogenet Evol 53:826–834

Roeding F, Hagner-Holler S, Ruhberg H, Ebersberger I, von Haeseler A, Kube M, Reinhardt R, Burmester T (2007) EST sequencing of Onychophora and phylogenomic analysis of Metazoa. Mol Phylogenet Evol 45:942–951

Rota-Stabelli O, Campbell L, Brinkmann H, Edgecombe GD, Longhorn SJ, Peterson KJ, Pisani D, Philippe H, Telford MJ (2011) A congruent solution to arthropod phylogeny: phylogenomics, microRNAs and morphology support monophyletic Mandibulata. Proc R Soc B 278:298–306

Rota-Stabelli O, Kayal E, Gleeson D, Daub J, Boore JL, Telford MJ, Pisani D, Blaxter M, Lavrov DV (2010) Ecdysozoan mitogenomics: evidence for a common origin of the legged invertebrates, the Panarthropoda. Genome Biol Evol 2:425–440

Rota-Stabelli O, Lartillot N, Philippe H, Pisani D (2013) Serine codon usage bias in deep phylogenomics: pancrustacean relationships as a case study. Syst Biol. 62:121–133

Rota-Stabelli O, Telford MJ (2008) A multi criterion approach for the selection of optimal outgroups in phylogeny: recovering some support for Mandibulata over Myriochelata using mitogenomics. Mol Phylogenet Evol 48:103–111

Roy SW, Gilbert W (2005) Resolution of a deep animal divergence by the pattern of intron conservation. Proc Natl Acad Sci USA 102:4403–4408

Ruiz-Trillo I, Paps J, Loukota M, Ribera C, Jondelius U, Baguñà J, Riutort M (2002) A phylogenetic analysis of myosin heavy chain type II sequences corroborates that Acoela and Nemertodermatida are basal bilaterians. Proc Natl Acad Sci USA 99:11246–11251

Sanders KL, Lee MS (2010) Arthropod molecular divergence times and the Cambrian origin of pentastomids. Syst Biodiv 8:63–74

Schmidt-Rhaesa A, Bartolomaeus T, Lemburg C, Ehlers U, Garey JR (1998) The position of the Arthropoda in the phylogenetic system. J Morphol 238:263–285

Scholtz G (1998) Cleavage, germ band formation and head segmentation: the ground pattern of the Euarthropoda. In: Fortey RA, Thomas RH (eds) Arthropod relationships. Chapman & Hall, London, pp 317–332

Scholtz G (2002) The Articulata hypothesis—or what is a segment? Org Divers Evol 2:197–215

Schram FR, Koenemann S (2004) Are the crustaceans monophyletic? In: Cracraft J, Donoghue MJ (eds) Assembling the tree of life. Oxford University Press, New York, pp 319–329

Sharma PP, Schwager EE, Extavour CG, Giribet G (2012) Evolution of the chelicera: a dachshund domain is retained in the deutocerebral appendage of Opiliones (Arthropoda, Chelicerata). Evol Dev 14:522–533

Shear WA, Edgecombe GD (2010) The geological record and phylogeny of Myriapoda. Arthropod Struct Dev 39:174–190

Shultz JW (1990) Evolutionary morphology and phylogeny of Arachnida. Cladistics 6:1–38

Shultz JW (2007) A phylogenetic analysis of the arachnid orders based on morphological characters. Zool J Linn Soc 150:221–265

Shultz JW, Regier JC (2000) Phylogenetic analysis of arthropods using two nuclear protein-encoding genes supports a crustacean + hexapod clade. Proc R Soc B 267:1011–1019

Simon S, Strauss S, von Haeseler A, Hadrys H (2009) A phylogenomic approach to resolve the basal pterygote divergence. Mol Biol Evol 26:2719–2730

Simonetta AM (2004) Are the traditional classes of arthropods natural ones?—recent advances in palaeontology and some considerations on morphology. Ital J Zool 71:247–264

Snodgrass RE (1938) Evolution of the Annelida, Onychophora and Arthropoda. Smithsonian Misc Coll 97:1–159

Sørensen M, Hebsgaard MB, Heiner I, Glenner H, Willerslev E, Kristensen RM (2008) New data from an enigmatic phylum: evidence from molecular sequence data supports a sister-group relationship between Loricifera and Nematomorpha. J Zool Syst Evol Res 46:231–239

Sørensen MV, Funch P, Willerslev E, Hansen AJ, Olesen J (2000) On the phylogeny of Metazoa in the light of Cycliophora and Micrognathozoa. Zool Anz 239:297–318

Spears T, Abele LG, Kim W (1992) The monophyly of brachyuran crabs: a phylogenetic study based on 18S rRNA. Syst Biol 41:446–461

Stegner MEJ, Richter S (2011) Morphology of the brain in *Hutchinsoniella macracantha* (Cephalocarida, Crustacea). Arthropod Struct Dev 40:221–243

Strausfeld NJ (2009) Brain organization and the origin of insects: an assessment. Proc R Soc B 276:1929–1937

Strausfeld NJ (2012) Arthropod brains: evolution, functional elegance, and historical significance. The Balknap Press of Harvard University Press, Cambridge

Strausfeld NJ, Andrew DR (2011) A new view of insect-crustacean relationships inferred from neural cladistics. Arthropod Struct Dev 40:276–280

Strausfeld NJ, Strausfeld CM, Loesel R, Rowell D, Stowe S (2006) Arthropod phylogeny: onychophoran brain organization suggests an archaic relationship with a chelicerate stem lineage. Proc R Soc B 273:1857–1866

Telford MJ, Wise MJ, Gowri-Shankar V (2005) Consideration of RNA secondary structure significantly improves likelihood-based estimates of phylogeny: examples from the Bilateria. Mol Biol Evol 22:1129–1136

Terry MD, Whiting MF (2005) Mantophasmatodea and phylogeny of the lower neopterous insects. Cladistics 21:240–257

Tiegs OW, Manton SM (1958) The evolution of the Arthropoda. Biol Rev 33:255–337

Timmermans MJTN, Roelofs D, Mariën J, van Straalen NM (2008) Revealing pancrustacean relationships: phylogenetic analysis of ribosomal protein genes places Collembola (springtails) in a monophyletic Hexapoda and reinforces the discrepancy between mitochondrial and nuclear DNA markers. BMC Evol Biol 8:83

Trautwein MD, Wiegmann BM, Beutel R, Kjer KM, Yeates DK (2012) Advances in insect phylogeny at the dawn of the postgenomic era. Annu Rev Entomol 57:449–468

Turbeville JM, Pfeifer DM, Field KG, Raff RA (1991) The phylogenetic status of arthropods, as inferred from 18S rRNA sequences. Mol Biol Evol 8:669–686

Ungerer P, Eriksson BJ, Stollewerk A (2011) Neurogenesis in the water flea *Daphnia magna* (Crustacea, Branchiopoda) suggests different mechanisms of neuroblast formation in insects and crustaceans. Dev Biol 357:42–52

Ungerer P, Scholtz G (2008) Filling the gap between identified neuroblasts and neurons in crustaceans adds new support for Tetraconata. Proc R Soc B 275:369–376

von Reumont BM, Burmester T (2010) Remipedia and the evolution of hexapods: Encyclopedia of Life Sciences. John Wiley & Sons, Chichester

von Reumont BM, Jenner RA, Wills MA, Dell'Ampio E, Pass G, Ebersberger I, Meyer B, Koenemann S, Iliffe TM, Stamatakis A, Niehuis O, Meusemann K, Misof B (2012) Pancrustacean phylogeny in the light of new phylogenomic data: support for Remipedia as the possible sister group of Hexapoda. Mol Biol Evol 29:1031–1045

von Reumont BM, Meusemann K, Szucsich NU, Dell'Ampio E, Gowri-Shankar V, Bartel D, Simon S, Letsch HO, Stocsits RR, Y-x Luan, Wägele JW, Pass G, Hadrys H, Misof B (2009) Can comprehensive background knowledge be incorporated into substitution models to improve phylogenetic analyses? A case study on major arthropod relationships. BMC Evol Biol 9:1–19

Waloszek D, Repetski JE, Maas A (2006) A new late Cambrian pentastomid and a review of the relationships of this parasitic group. Trans Roy Soc Edin Earth Sci 96:163–176

Weygoldt P, Paulus HF (1979) Untersuchungen zur Morphologie, Taxonomie und Phylogenie der Chelicerata: I Morphologische Untersuchungen. Z zool Syst Evol 17:85–116

Wheeler WC (1989) The systematics of insect ribosomal DNA. In: Fernhölm B, Bremer K, Jörnvall H (eds) The hierarchy of life: molecules and morphology in phylogenetic analysis. Elsevier Science Publishers B. V, Amsterdam, pp 307–321

Wheeler WC, Cartwright P, Hayashi CY (1993) Arthropod phylogeny: a combined approach. Cladistics 9:1–39

Wheeler WC, Hayashi CY (1998) The phylogeny of the extant chelicerate orders. Cladistics 14:173–192

Whitfield JB, Kjer KM (2008) Ancient rapid radiations of insects: challenges for phylogenetic analysis. Annu Rev Entomol 53:449–472

Whiting MF (2002) Mecoptera is paraphyletic: multiple genes and phylogeny of Mecoptera and Siphonaptera. Zool Scr 31:93–104

Whiting MF, Carpenter JM, Wheeler QD, Wheeler WC (1997) The Strepsiptera problem: phylogeny of the holometabolous insect orders inferred from 18S and 28S ribosomal DNA sequences and morphology. Syst Biol 46:1–68

Whitington PM, Mayer G (2011) The origins of the arthropod nervous system: insights from the Onychophora. Arthropod Struct Dev 40:193–209

Wiegmann BM, Trautwein MD, Kim JW, Cassel BK, Bertone MA, Winterton SL, Yeates DK (2009) Single-copy nuclear genes resolve the phylogeny of the holometabolous insects. BMC Biol 7:34

Willmer PG (1990) Invertebrate relationships: patterns in animal evolution. Cambridge University Press, Cambridge

Wirkner CS, Prendini L (2007) Comparative morphology of the hemolymph vascular system in scorpions–a survey using corrosion casting, microCT, and 3D-reconstruction. J Morphol 268:401–413

Wirkner CS, Richter S (2004) Improvement of microanatomical research by combining corrosion casts with microCT and 3D reconstruction, exemplified in the circulatory organs of the woodlouse. Microsc Res Tech 64:250–254

Wolf YI, Rogozin IB, Koonin EV (2004) Coelomata and not Ecdysozoa: evidence from genome-wide phylogenetic analysis. Genome Res 14:29–36

Wygodzinsky P (1961) On a surviving representative of the Lepidotrichidae (Thysanura). Ann Entomol Soc Amer 54:621–627

Zhang XG, Maas A, Haug JT, Siveter DJ, Waloszek D (2010) A eucrustacean metanauplius from the lower Cambrian. Curr Biol 20:1075–1079

Zhang Z-Q (ed) (2011) Animal biodiversity: an outline of higher-level classification and survey of taxonomic richness. Magnolia Press, Auckland

Zrzavý J (2003) Gastrotricha and metazoan phylogeny. Zool Scr 32:61–81

Zrzavý J, Hypša V, Tietz DF (2001) Myzostomida are not annelids: Molecular and morphological support for a clade of animals with anterior sperm flagella. Cladistics 17:170–198

Zrzavý J, Hypša V, Vlášková M (1998a) Arthropod phylogeny: taxonomic congruence, total evidence and conditional combination approaches to morphological and molecular data sets. In: Fortey RA, Thomas RH (eds) Arthropod relationships. Chapman & Hall, London, pp 97–107

Zrzavý J, Mihulka S, Kepka P, Bezdek A, Tietz D (1998b) Phylogeny of the Metazoa based on morphological and 18S ribosomal DNA evidence. Cladistics 14:249–285

# An Overview of Arthropod Genomics, Mitogenomics, and the Evolutionary Origins of the Arthropod Proteome

**3**

Davide Pisani, Robert Carton, Lahcen I. Campbell, Wasiu A. Akanni, Eoin Mulville and Omar Rota-Stabelli

## Contents

D. Pisani (✉)
School of Biological Sciences and School of Earth
Sciences, University of Bristol, Woodland Road,
Bristol, BS8 1UG, UK
e-mail: Davide.Pisani@bristol.ac.uk

R. Carton · L. I. Campbell · W. A. Akanni ·
E. Mulville
Department of Biology, The National University
of Ireland, Callan Building, Maynooth, County
Kildare, Ireland
e-mail: robcarton@gmail.com

L. I. Campbell
e-mail: lahcencampbell@yahoo.ie

W. A. Akanni
e-mail: waakanni13@gmail.com

E. Mulville
e-mail: eoin.d.mulvihill@nuim.ie

O. Rota-Stabelli
IASMA Research and Innovation Centre,
Fondazione Edmund Mach, Via E. Mach 1,
38010 San Michele all'Adige (TN), Italy
e-mail: omar.rota@fmach.it

## 3.1 Introduction

Arthropods represent the largest majority of animal biodiversity and include organisms of economic interest and key model species. It is thus unsurprising that the genome of an arthropod, the fruit fly *Drosophila melanogaster*, was among the very first to be sequenced (Adams et al. 2000) and that to date, about 21 *Drosophila*

A. Minelli et al. (eds.), *Arthropod Biology and Evolution*,
DOI: 10.1007/978-3-642-36160-9_3, © Springer-Verlag Berlin Heidelberg 2013

genomes as well as a variety of other arthropod genomes have been sequenced. Despite this promising start, current sampling is biased towards economically relevant species, and a suitable close outgroup to the arthropods, which is necessary to polarise genomic studies, is still missing. Among the suitable outgroups to the Arthropoda, the Nematoda represent one of the largest components of the extant animal biomass, and their economic importance is comparable to that of the more biodiverse arthropods. As with the Arthropoda, the importance of the nematodes is reflected in the fact that the very first animal genome to be sequenced was that of the nematode *Caenorhabditis elegans* (The C. elegans genome consortium 1998). Despite the nematodes being phylogenetically close to the arthropods (Aguinaldo et al. 1997; Copley et al. 2004; Dopazo and Dopazo 2005; Philippe et al. 2005; Irimia et al. 2007; Roy and Irimia 2008; Dunn et al. 2008; Belinky et al. 2010; Hejnol et al. 2009; Holton and Pisani 2010), this group is composed of highly derived species, both genetically and morphologically. Accordingly, their genomes are unlikely to be of great utility in understanding arthropod genome evolution. Some genomic data (mostly in the form of transcriptomes) are now available for other smaller ecdysozoan phyla, and some genomes (Priapulida and Tardigrada) are on the horizon. Nonetheless, enough genomic information is now available for the Arthropoda (Table 3.1) to justify an investigation into the evolution of their genome. Such an analysis, however, is intimately dependent on the availability of a robust phylogenetic background, and to a lesser extent, robust divergence times for the nodes in the background phylogeny.

In this chapter, we present an overview of arthropod mitochondrial genomics (Sect. 3.2) and nuclear genomics (Sect. 3.3). We then exploit the available genomic information to investigate the evolutionary origin of novel proteins (orphan gene families) in the arthropod proteome (Sect. 3.4). We notably present the first genomic-scale data set for the Onychophora and include it in our analyses to be able to consider the closest sister group of the Arthropoda (see Campbell et al. 2011) when identifying orphan gene families. Inclusion of new data for the Onychophora is key to this study as it allows the correct identification of the orphan protein families that arose in the stem arthropod lineage.

## 3.2 Arthropod Mitogenomes: Useful, but Hazardous Small Genomes

Each cell contains up to hundreds of mitochondria, and each mitochondrion possesses many copies of their own small, typically circular, genome (mitogenome or mtDNA). Therefore, mitochondrial genes largely outnumber the nuclear ones in terms of their copy number by several orders of magnitude, making mitochondrial genes easy to extract and amplify. Accordingly, there has been an exceptional amount of articles published that attempted (not always successfully) to resolve the phylogenetic relationships within Arthropoda (and more broadly Metazoa) using mtDNA. Other reasons behind the fortunes of mtDNA are as follows: a relatively conserved gene set, the unambiguous orthology of genes, the presence of rare genetic changes, and the availability of universal primers for many lineages. Other characteristics of the mitogenome, however, make it a doubled-edged sword. These are accelerated mutation rate due to uniparental inheritance, and severe biases in the composition of nucleotides that are often responsible for the dilution of the phylogenetic signal in mtDNA (Bernt et al. 2012). In this section, we review some of these aspects.

### 3.2.1 Mitogenomic Studies

Mitogenomic studies have helped throughout the 1990s and 2000s to elucidate some arthropod affinities. For example, one of the earliest studies providing robust, non-rRNA based, evidence in support of the Pancrustacea used mtDNA gene order comparisons (Boore et al. 1998) and

**Table 3.1** The most important of the available Arthropod genomes

| | Species | Genome size (Mb) | GC (%) | Chromosomes | Genes | Transcripts |
|---|---|---|---|---|---|---|
| Chelicerata Acari-Acariformes | *Tetranychus urticae* | 89.6 | 32.3 | N/A | N/A | 18,414 |
| Chelicerata Acari-Parasitiformes | *Ixodes scapularis* | 1,896.32 | 45.5 | 15 | 7,112 | 5,867 |
| Myriapoda Chilopoda | *Strigamia maritima* | 173.61 | 35.7 | N/A | N/A | N/A |
| Crustacea Branchiopoda | *Daphnia pulex* | 158.62 | 40.8 | N/A | 30,613 | 30,611 |
| Hexapoda Phthiraptera | *Pediculus humanus* | 108.37 | 27.5 | N/A | 10,993 | 10,775 |
| Hexapoda Coleoptera | *Tribolium castaneum* | 210.27 | 38.4 | 10 | 10,132 | 9,833 |
| Hexapoda Hemiptera | *Acyrthosiphon pisum* | 464 | 29.6 | 4 | N/A | 11,089 |
| Hexapoda Hymenoptera | *Apis mellifera* | 250.29 | | 16 | N/A | N/A |
| Hexapoda Lepidoptera | *Bombyx mori* | 431.75 | 37.7 | 28 | N/A | N/A |
| Hexapoda Lepidoptera | *Heliconius melpomene* | 269 | | 21 | 12,669 | N/A |
| Hexapoda Diptera | *Drosophila melanogaster* | 139.73 | 42.2 | 6 | 15,431 | 24,113 |
| Hexapoda Diptera | *Aedes aegypti* | 1,310.11 | 38.3 | 3 | 16,684 | 16,785 |
| Hexapoda Diptera | *Anopheles gambiae* | 265.03 | 44.5 | 5 | 13,240 | 14,099 |

*N/A* not available. All the values in the table were obtained either from the NCBI website or from the original genome paper

mtDNA sequence phylogeny (Hwang et al. 2001). However, in some cases, mitogenomic studies have pointed towards likely incorrect topologies, for example, suggesting a Myriapoda plus Chelicerata grouping (Hwang et al. 2001; Negrisolo et al. 2004; Pisani et al. 2004), which has also been uncovered by some analyses of nuclear coding genes (e.g. Pisani et al. 2004; Dunn et al. 2008; Roeding et al. 2009; Hejnol et al. 2009; Meusemann et al. 2010) and that most likely represent a long-branch attraction artefact (Pisani 2004; Rota-Stabelli and Telford 2008; Rota-Stabelli et al. 2010; Campbell et al. 2011; Rota-Stabelli et al. 2011). This topology was most likely the result (in the case of the mtDNA analyses) of a systematic error caused by the use of distant outgroups and compositionally biased taxa (Rota-Stabelli and Telford 2008). Such features of the mitochondrial

genomes may seriously affect phylogenetic reconstruction unless they are taken into account when inferring phylogenies (Rota-Stabelli et al. 2010).

Utility of the mitochondrial genomes is not restricted to phylogeny. The most widely used arthropod barcode is a region of approximately 650 nucleotides of the subunit I of the cytochrome oxidase complex (COX1)—a mitochondrial gene. Other mitochondrial genes (NADH4, for example) are occasionally added to COX1 to improve resolution. A possible risk with mtDNA-based barcoding is the amplification of pseudo-genes numts (nuclear copies of mitochondrial genes), which may disrupt barcoding studies. In addition, single gene barcoding has been shown to fail occasionally and the advent of NGS makes it an obsolete approach (Taylor and Harris 2012). Nevertheless, barcoding remains the method of choice for biodiversity studies (likely because its simplicity and low cost makes it appealing to founding agencies).

To date, there are more than 300 complete arthropod mitochondrial genomes, and partial sequences are in excess of a million. The taxonomic sampling is, however, extremely biased towards economically relevant species: 47 chelicerates (mostly ticks and mites), 53 crustaceans (mostly malacostracans), 198 insects (mostly beetles, dipterans, and hemipterans), and only 9 myriapods. Still, most major orders and classes are now represented, thus providing an invaluable starting point for comparative analyses.

### 3.2.2 The Structure of the Arthropod Mitochondrial Genome

Arthropod mtDNA varies in size from less than 14,000 bp in the spider *Ornithoctonus huwena* to more than 19,000 bp in *D. melanogaster*. This difference is almost entirely due to non-coding intergenic regions, particularly the major non-coding region commonly called *control region*. Due to its low structural constraints and high tendency to accumulate A and T nucleotides, this region is also called the AT-rich region. The AT-rich region is involved in both replicative and transcriptional processes and typically contains structural elements like hairpin loops and thymidine stretches (Zhang and Hewitt 1997), elements that do not seem to be conserved throughout the arthropods.

The gene content of the arthropod mtDNA is the same as in most other bilaterians; it typically consists of 13 coding genes, 2 ribosomal RNA subunits, and 20 tRNAs (Boore 1999). This gene set is highly conserved throughout the phylum, although a few exceptions can be found. Examples include a tRNA-Ser duplication in *Thrips imaginis* (Hexapoda: Thysanoptera) (Shao and Barker 2003), a tRNA-His duplication in *Speleonectes tulumensis* ('Crustacea': Remipedia) (Lavrov et al. 2004), and a tRNA-Cys triplication in *Pollicipes polymerus* ('Crustacea': Cirripedia) (Lavrov et al. 2004). Many arthropod mitochondrial coding genes lack a stop codon

(TAA or TAG) and possess a single T or TA at the 3-terminal end. The correct stop codon is then assembled by the polyadenylation of an excised, presumably polycistronic, transcript. Although most arthropod mitogenomes use the invertebrate genetic code, it has been shown that some lineages use a slightly different code (Abascal et al. 2006). Remarkably, this new genetic code is scattered throughout the arthropod tree.

Although the gene content is conserved throughout the arthropods, the gene order may vary significantly (Lavrov et al. 2004). Comparative studies have determined an arthropod ancestral gene order, which is represented (retained) by *Limulus polyphemus*, while the pancrustacean gene order differs from that of all the other arthropods by the position of one of the two leucine tRNAs. tRNAs in general are mostly responsible for variation in gene order as they are hot spots of recombination. Less often, coding genes change their position or swap strand, allowing for variation in gene-specific strand asymmetry, as detailed below.

### 3.2.3 Arthropod Mitogenomes: A Composition Nightmare

The main source of compositional heterogeneity in mtDNA is mutational pressure, which is correlated with a deficiency in the mtDNA repair system and with a consequent inefficiency at replacing erroneous insertions of A nucleotides (Reyes et al. 1998). Compared to other metazoans, arthropod lineages are typically enriched in A and T. In the absence of strong purifying selection, this mutational pressure affects also encoded proteins, which are enriched in amino acids encoded by A+T-rich codons (Foster et al. 1997; Foster and Hickey 1999; Rota-Stabelli et al. 2010). The effect of this mutational pressure depends on structural constraints acting on the genes: more conserved genes such as COX1 accumulate fewer A+T mutations than poorly constrained genes such as ATP8. In addition, not all positions of a gene are affected in a similar way: while the 1st and 2nd codon positions are

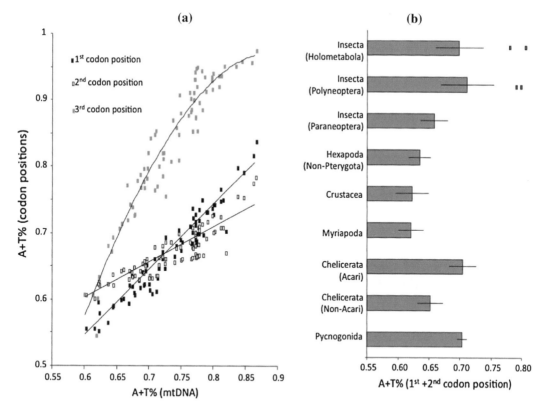

**Fig. 3.1** Compositional heterogeneity in arthropod mitogenomes. **a** A+T % content of the three codon positions plotted against that calculated on the whole mtDNA. Second codon position is the most constrained, while 3rd codon position changes so dramatically that reaches plateau in some species. **b** A+T % calculated on the whole mtDNA in different arthropod lineages. Nucleotide content varies between and within classes

more constrained by the genetic code, the 3rd codon positions are more prone to accumulate A+T mutations and experience saturation of replacement events (Fig. 3.1a). Interestingly, 1st codon positions show a different A+T replacement pattern from the 2nd. This advocates the employment of different models of evolution for the 1st and 2nd codon positions and the exclusion of the 3rd codon positions when performing phylogenetic reconstruction from nucleotide sequences. This would, at least partially, compensate for possible artefactual attraction in the case that unrelated species have a similarly increased A+T content.

The A+T content is not homogenously distributed throughout the arthropods: some groups such as Pycnogonida, Acari, and some insects are more A+T rich than other lineages (Fig. 3.1b). This uneven distribution of nucleotide content may have been responsible for the

artefactual attraction of, for example, Acari and Pycnogonida in published phylogenetic studies (Podsiadlowski and Braband 2006). In some species such as the bees *Apis mellifera* and *Melipona bicolor* and the hemipterans *Schizaphis graminum* and *Aleurodicus dugesii* (grey dots in Fig. 3.1b), the A+T content reaches extremely high values, the highest ever reported for eukaryotic coding genes.

Strand asymmetry is another type of compositional heterogeneity affecting mtDNA. This bias is related to the origin and direction of mtDNA replication (Reyes et al. 1998) and leads one strand to become enriched in G (and to a lesser extent in T), while the other strand become enriched in C (and less in A). Strand asymmetry is generally expressed in terms of GC-skew. Although all genes in a mitochondrial genome usually have a similar A+T content, homologous genes from different organisms may have extremely different, sometimes opposite, GC (and AT)-skew: this depends on the strand on which the gene is located, and on its position relative to the origin of replication (Lavrov et al. 2000). Therefore, there is a link between strand asymmetry and gene order.

In arthropods, most mtDNA coding genes are characterised by a negative GC-skew (they have more C than G), while four genes that lie on the opposite strand are characterised by a positive GC-skew. This situation is characteristic, in particular, of species characterised by the arthropod ancestral gene order (as in Fig. 3.2a). In some species, the GC-skew is opposite for all the genes, although the gene order is substantially identical to that of the ancestral arthropods (Fig. 3.2b). In such cases, it is the origin of replication (the control region) that underwent a modification, for example, a duplication or an inversion of strand. In other cases, all genes may have been translocated on the same strand, so that all the genes possess either a positive or a negative GC-skew (Fig. 3.2c).

### 3.2.4 The Hazards of Using Arthropod Mitochondrial Genomes for Phylogenetics

It has been shown that both sources of compositional heterogeneity (A+T mutational pressure and strand asymmetry) may play strong roles in generating artefactual mitogenomic phylogenies (Hassanin et al. 2005; Rota-Stabelli et al. 2010). Compositional problems are worsened by the accelerated rate of evolution of mitogenomic sequences, which is related to the uniparental inheritance characterising mitochondria. An effective approach to deal with these problems is to improve models of mitochondrial sequence evolution both at the nucleotide (Hassanin et al. 2005) and protein level (Abascal et al. 2007; Rota-Stabelli et al. 2009). However, if the biases are too strong to be accounted for using models, one might have to try to highlight potentially incorrect topologies by experimenting with character exclusion strategies targeting more affected genes or codon positions (e.g.

Rota-Stabelli et al. 2010). Sophisticated evolutionary models which account for among site and among branch heterogeneity (Foster 2004; Blanquart and Lartillot 2008) are useful to lessen the effects of these mitochondrial compositional biases. Another obvious approach is to enlarge or modify taxonomic sampling. More taxa may break problematic branches and reduce the number of homoplasies responsible for long-branch (or compositional) attractions. In some conditions, when addition of more taxa does not seem to be breaking long branches, it might be useful to carry out experiments in which taxon sampling is modified (by taxon removal) and the effect of these taxonomic reductions on the analyses is monitored (e.g. Rota-Stabelli et al. 2012; Campbell et al. 2011). More generally, it is advisable to conduct an exploratory compositional analysis of the properties of the mitochondrial genomes under consideration prior to phylogenetic inference. This is particularly true for the arthropods, which include some highly derived lineages, parasites, for example, whose particular lifestyle is responsible for bottleneck events and therefore extreme acceleration of substitution rates or divergent nucleotide compositions.

Compositional biases (and related phylogenetic artefacts) have been primarily studied using mitogenomic data sets (Foster et al. 1997). The advent of the phylogenomic-type (nuclear) data sets has been initially seen as a relief in terms of compositionally related biases. This may, however, not be the case: the community is just noticing that even large genomic data sets are not free from compositional problems that can cause serious phylogenetic artefacts (Nabholz et al. 2012; Rota-Stabelli et al. 2012). Still, the origins of such biases in nuclear genomic data are largely not known.

### 3.3 Arthropod Comparative Genomics

The study of arthropod genomics started with the sequencing of the genome of the fruit fly *D. melanogaster* (Adams et al. 2000). Currently,

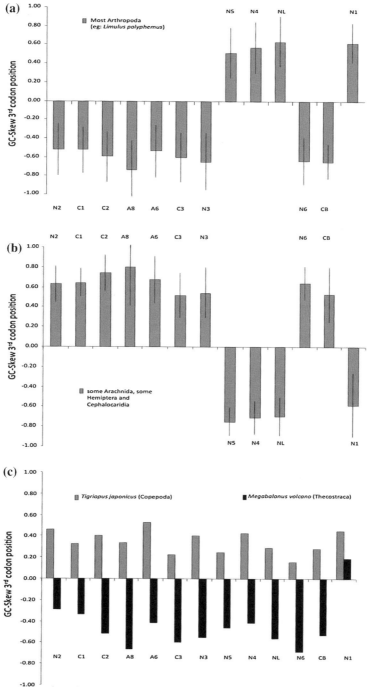

**Fig. 3.2** Strand asymmetry in arthropod mitogenomes. Each gene in the mtDNA is characterised by a different propensity to accumulate mutations towards G or C. This is because different genes lie on different strands and each strand has his own mutational pressure, described here by the GC-skew statistics. **a** In most arthropods, the majority of genes are on the same strand and possess a negative GC-skew; the ORF of NADH4, NADH5 NADH4L and NADH1 is on the opposite strand; as a consequence, these genes accumulate more G and have a positive GC-skew. **b** Some phylogenetically unrelated arthropods experienced an inversion of the replicative system, which leads to a complete inversion of GC-skew for each of the genes. **c** Some taxa underwent genomic rearrangement, so that all genes are on the same strand

genomic data are available for a relatively large number of arthropods allowing the first attempts at performing comparative genomic analyses of the Arthropoda (Vieira and Rozas 2011). However, the majority of the currently available arthropod genomes are from closely related species (mostly insects), and a coherent set of conclusions about the arthropod nuclear genomes (as presented for the mitochondrial genomes above) is still lacking.

### 3.3.1 Uneven Taxonomic Sampling

The biased taxonomic distribution of the available arthropod genomes is a persistent problem. This is because it does not allow detailed investigations into key questions in arthropod evolution, like the origin of the arthropod subphyla. Initiatives exist that aim at increasing the amount of available genomic information for the Arthropoda. Paramount among these projects are the 1KITE project—1,000 Insects Transcriptome Evolution project (http://1kite.org/), and the i5K (http://www.arthropodgenomes.org/wiki/i5K) project which plans to sequence the complete genomes of 5,000 insects and related arthropod species. Unfortunately, as commendable as these projects are, they fall short of adequately capturing the breadth of the evolutionary diversity within the Arthropoda. The 1KITE project will not even attempt to generate data for non-hexapod species, while about 87 % of the species currently nominated for sequencing as part of the i5K project are hexapods. Only 0.7 % belongs to Myriapoda and only 2.8 % to Crustacea. This is an important issue with the current initiatives, as this heterogeneous species sampling, even if reflective of species diversity, does not reflect arthropod disparity. As such, it might bias future comparative analyses and might not allow a clear understanding of the genomic factors underlying the great morphological and physiological variation observed in Arthropoda. Disparity (e.g. the morphological diversities observed between a tick and a millipede) is underlined by variation in the genomes of the considered organisms, and the way these

genomes are wired. To understand arthropod disparity, therefore, genomic data as well as protein–protein interaction networks (e.g. Giot et al. 2003) and gene regulatory networks (Davidson and Erwin 2006) would be necessary for representatives of each major lineage within each subphylum. Even though hundreds of insect genomes will be a welcomed resource, it can be expected that, while they will allow to a significant increase in our understanding of adaptations, they will not be particularly useful to explain the origin of arthropod disparity, of the arthropod subphyla and of the main lineages within these subphyla.

An important aspect to which current large-scale genome sequencing projects are not given sufficient attention is that of the arthropod outgroups. To increase the power of comparative analyses, adequate outgroups should also be sequenced, but large-scale sampling initiatives are not considering the outgroups of the Arthropoda. Indeed, to date, the only arthropod outgroups available with at the least one fully sequenced genome are the nematodes. Yet, species belonging to this phylum are too distantly related and too divergent from the Arthropoda (see also above) to be of significant utility in arthropod comparative genomics. Other more closely related genomes (those of the Onychophora and the Tardigrada) should be sequenced and used instead. As part of this chapter, to obviate the lack of genomic-scale data sets for the arthropod outgroups, we shall present a genome-wide transcriptomic data set obtained using next generation sequencing.

The 1KITE and i5K projects have not produced data yet. However, a relative abundance of arthropod genomes has been accumulating in recent years, albeit with a biased taxonomic distribution. The genomes of 21 *Drosophila* species have been sequenced and made publicly available. Transcriptomic, proteomic, and genomic data, as well as abundant functional annotations, for 12 of these species can be found in the specialised database Flybase (http://flybase.org/). Other key insects for which genomic information is available include the mosquitoes *Aedes aegypti* (Nene et al. 2007) and *Anopheles gambiae* (Holt et al. 2002), the honeybee *A. mellifera* (The honeybee genome consortium 2006), the beetle *Tribolium castaneum* (Richards et al. 2008), the body louse *Pediculus humanus* (Kirkness et al. 2010), the pea aphid *Acyrthosiphon pisum* (The pea aphid genome consortium 2010), and the silk moth *Bombyx mori* (The silkworm genome consortium 2008). A variety of other insects, for example, ants and other butterflies, have also been sequenced (Suen et al. 2011; The Heliconius genome consortium 2012). Results from these more recent studies (which generally used next generation sequencing strategies) allowed some truly surprising conclusions to be reached. For example, the *Heliconius* genome consortium was able to demonstrate the repeated exchange of large ( ~ 100-kb) adaptive regions among multiple butterfly species in a recent radiation. In this way, they were also able to uncover the pervasiveness and importance of introgressive adaptation and its role in hybrid speciation. For many of these more recently sequenced species, taxonspecific databases exist (e.g. Butterflybase— http://butterflybase.ice.mpg.de/). Differently from Flybase, which is a mature database providing, for example, a genome browser, and allowing complex searches (using Gene Ontology—GO terms and developmental stages), most of these species-specific databases are still quite immature. In any case, they represent an important resource and their utility is bound to increase with time.

While hexapod genomes are relatively abundant, the situation changes drastically when moving to other arthropod subphyla. Only one complete crustacean genome (that of the water flea *Daphnia pulex*—Colbourne et al. 2011), and one complete chelicerate genome, that of the two-spotted spider mite *Tetranychus urticae* (Grbic et al. 2011) have been released. Finally, the complete genome of one myriapod, the centipede *Strigamia maritima* (GenBank access id: GCA_000239455.1), and that of a second chelicerate *Ixodes scapularis* (GenBank access id: GCA_000208615.1) are now publicly available, although they have not yet been released.

Apart from standard genomic studies, a variety of large-scale transcriptome-wide sequencing studies have been performed, and EST data are thus available for other taxa. Even though these studies do not provide information about untranslated genomic regions, a large amount of useful data has been provided using these approaches. One of the earliest studies that employed EST generated using next generation sequencing (in that specific case it was 454 sequencing) to gain a complete snapshot of an arthropod genome was the transcriptome sequencing of the emperor scorpion *Pandinus imperator* (Roeding et al. 2009). More recently, Illumina and other sequencing techniques have been applied to other important groups for which genomic data are not available, like the harvestmen (Opiliones; Hedin et al. 2012), and the amphipod crustacean *Parhyale hawaiensis* (Zeng et al. 2011; Blythe et al. 2012). Similar approaches have started to generate extremely interesting insights into chelicerate venoms, allowing the development of the new science of venomics (Rendon-Anaya et al. 2012) and arthropod developmental biology (Ewen-Campen et al. 2011).

### 3.3.2 Heterogeneity of Genome Sizes and Shortage of microRNA

Important aspects of the key, publicly available, arthropod genomes are reported in Table 3.1. From this table, it is clear that the arthropod genomes are fairly variable. Their lengths in MB vary substantially with one of the chelicerate genomes being the smallest, while the other is the biggest overall. Similarly, GC content is quite variable with *Ixodes* having the highest GC content and the pea aphid the lowest. Also, the number of predicted protein coding genes varies substantially between genomes, with *Daphnia* having 30,613 and *Ixodes* only 7,112. A notable aspect of Table 3.1 is the difference in the number of protein coding genes and known, corresponding transcripts, for *D. melanogaster*. The fruit fly is the only species in Table 3.1 for which the number of known transcripts largely exceeds the number of predicted protein coding genes. The difference between the number of genes and the number of transcripts is most likely caused by alternative splicing. It is in fact known that approximately 40 % of the protein coding genes in *D. melanogaster* correspond to more than one transcript (Hartmann et al. 2009). The lack of knowledge of alternatively spliced genes for other taxa in Table 3.1 is likely to reflect our ignorance rather than biology. For *D. melanogaster*, deep sequencing of specimens in specific developmental stages, specific tissues, and organs allowed identification of a larger number of transcripts. It is to be expected that as knowledge of the transcriptomes of the other species in Table 3.1 will increase, the number of their known transcripts will also increase. An obvious observation emerging from an analysis of Table 3.1 is that the sequenced chelicerate taxa cannot be particularly good resources for evolutionary biologists. *Ixodes* and *Tetranychus* are highly specialised species unlikely to reflect what the analysis of more standard chelicerate genomes will uncover.

Next generation sequencing approaches have also allowed our understanding of regulatory (non-coding) microRNA to increase substantially. Genome-wide screening performed for taxa belonging to all arthropod subphyla and to the arthropod outgroups (Campbell et al. 2011; Rota-Stabelli et al. 2011) allowed identification of several arthropod-specific microRNA (miR-275 and iab-4), mandibulate-specific ones (miR-965 and miR-282), and chelicerate-specific ones (miR-3931). These studies also showed that arthropods, in contrast to other lineages (such as the mammals or annelids), have significantly less lineage specific microRNAs, suggesting that arthropod genomes, from this point of view, evolve quite differently from those of other animal lineages.

Overall, current genomic-scale information available across the Arthropoda is still too fragmentary to allow the development of a coherent view of arthropod genome evolution. However, in the last section of this chapter, we shall attempt to start obviating this problem, by presenting an evolutionary analysis of the arthropod proteomes that exploits the transcriptomic data we generated for the Onychophora.

## 3.4 A Genomic Phylostratigraphic Analysis of the Arthropod Proteomes

An interesting aspect of the arthropod genome evolution that availability of current metazoan and arthropod genomes allows us to address (given also the data we generated for the Onychophora) is that of the origin of the arthropod-specific protein coding genes (i.e. genes found only within Arthropoda). Studies of this type have been named *genomic phylostratigraphic analyses* by Domazet-Loso et al. (2007). To complete such studies (in addition to genomic information), one needs information about phylogeny and divergence times. The relationships between the arthropods and divergence times used are summarised below.

### 3.4.1 A Robust Phylogenetic Framework for Genomic Studies

Comparative genomics must be anchored on a phylogenetic tree. Significant progress in our understanding of the ecdysozoan relationships has been made (Dunn et al. 2008; Hejnol et al. 2009; Campbell et al. 2011). Similarly, some agreement on the phylogenetic relationships within the Arthropoda has recently emerged (Regier et al. 2010; Rota-Stabelli et al. 2011), but see Rota-Stabelli et al. (2012). For this study, it is important that the tree used to anchor our analyses is resolved. However, some level of incongruence still exists among the various phylogenetic studies addressing the relationships within Ecdysozoa. With reference to the current study, we shall consider the Lobopodia (Arthropoda plus Onychophora) to be the sister group of the Tardigrada within a monophyletic Panarthropoda. We shall further assume Nematoida (Nematoda plus Nematomorpha) to be the sister group of Panarthropoda, with the Scalidophora (here Priapulida and Kinorhyncha) representing the sister group of Nematoida plus Panarthropoda. That is, we shall assume the ecdysozoan relationships of Campbell et al. (2011) and Rota-Stabelli et al. (2011) to represent our working hypothesis. These relationships differ from those of Dunn et al. (2008) with reference to the placement of Nematoida that the study of Dunn and co-workers was found to be a member of Cycloneuralia, that is, more closely related to the Scalidophora than to the Arthropoda. However, because Campbell et al. (2011) only performed a Bayesian analysis of their data set and did not present bootstrap support for their results. Given that they did not find particularly strong support (low posterior probabilities) for some key contested nodes (Nematoida + Panarthropoda and Mandibulata—which are not supported in other studies, for example, Dunn et al. 2008), and given that there are few other studies (e.g. Meusemann et al. 2010) whose results contradict those of Campbell et al. (2011) and Rota-Stabelli et al. (2011) with reference to the placement of Tardigrada and the monophyly of Mandibulata, we present here a novel statistical analysis—nonparametric bootstrapping—of the data set used in Campbell et al. (2011). A detailed explanation of the methods used in this analysis is presented in the Appendix to this chapter.

Results of the bootstrap analysis that considers all the taxa in Campbell et al. (2011) are in agreement with the Bayesian analyses in that paper. This analysis shows a lack of support for many important nodes, including Nematoida (which was not recovered), Nematoda plus Panarthropoda (BP = 41), Panarthropoda (BP = 66), Lobopodia (BP = 61), and Mandibulata (BP = 64), see Fig. 3.3. We performed a leaf stability analysis (results not shown—but see Appendix) illustrating that Nematomorpha is the most unstable taxon in the data set. The nematomorph in Campbell et al. (2011) emerged as the sister group of the Nematoda in agreement with Dunn et al. (2008) and Hejnol et al. (2009). Yet, in Fig. 3.3, Nematomorpha is not the sister group of the Nematoda. Instead, it emerges as the sister of a Nematoda + Arthropoda clade. This is an artefact caused by high volume of missing data in the Nematomorpha (which is the most incomplete taxon in Campbell et al. 2011) and that is unstable in bootstrapped data sets.

Upon removal of the unstable Nematomorpha, the bootstrap support for all the other nodes increases significantly. Arthropoda plus Nematoda reaches 100 %, Panarthropoda increases to 76 %, and Lobopodia to 70 %. In conclusion, when accounting for unstable taxa, Arthropoda has a bootstrap support of 100 % and Mandibulata of 76 %. This confirms that there is a good level of support for the clades in Fig. 3.3 and those in Campbell et al. (2011).

### 3.4.2 Expanding Our Understanding of the Arthropod Comparative Genomics

Given our poor understanding of the processes through which the arthropod (nuclear) genomes evolved, we shall here present a genomic phylostratigraphic analysis (Domazet-Loso et al. 2007) of their genome. The aim of this analysis is to gain some information on the evolutionary processes responsible for the origin and evolution of the Arthropoda. Domazet-Loso et al. (2007) performed a similar analysis, but various new genomes have been published since their study, allowing for a much greater precision in the identification of orphan genes along the ecdysozoan and arthropod phylogeny. To better identify proteins that are arthropod specific, we extended our analyses to include a variety of ecdysozoans and non-ecdysozoan genomes. Particularly, we included representatives of the Lophotrochozoa, of the Deuterostomia and two non-bilaterian metazoans (a sponge, *Amphimedon queenslandica*, and a cnidarian, *Hydra magnipapillata*)—see Fig. 3.4. In addition, and most importantly, here we added data for an onychophoran transcriptome, which allowed pinpointing protein families that are specific to the Arthropoda (i.e. that originated after the Onychophora–Arthropoda split). Finally, more reliable molecular clock divergence times (Erwin et al. 2011) are now available and they have been used here to define rates of orphan gene acquisitions through time allowing for

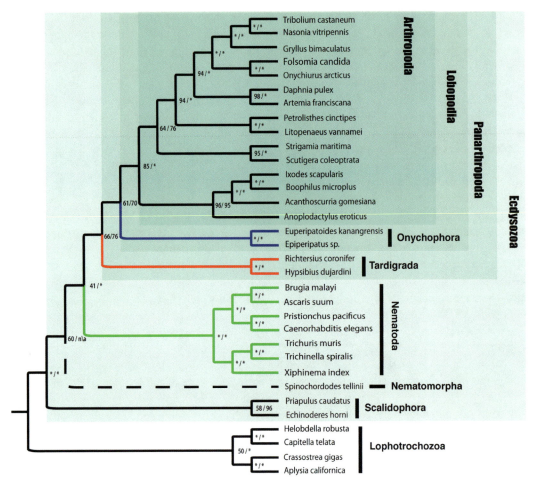

**Fig. 3.3** A phylogeny of the Ecdysozoa. The tree represents a Bayesian bootstrap analysis performed under CAT+G of the data set of Campbell et al. (2011). Values at the nodes represent bootstrap proportions. *Asterisk* = 100 % support. The leftmost value represents the bootstrap proportion obtained for a data set including all the sequences in Campbell et al. (2011). The rightmost value represents the bootstrap proportion obtained when the most unstable taxon in the data set (the nematomorph *Spinochordodes*) was excluded

better estimation of rates of new protein family acquisitions in Ecdysozoa and Arthropoda.

### 3.4.3   The Evolution of Orphan Gene Families in Arthropoda

We used the MCL algorithm (Enright et al. 2002) to identify protein families in the set of considered genomes, and identified, for each internal node in Fig. 3.4, all the proteins universally distributed in the taxa descending from each given node. These are orphan families that evolved in the branch underlying the considered

node. The average number of new families acquired across all the internodes of the considered phylogeny is 1,025. When this value is normalised (dividing by the total number of proteins in the considered set of genomes (79,052 protein coding genes)), the 1,025 protein families that are gained as novel orphan genes correspond to ∼1.2 %.

Within Arthropoda, and more broadly Panarthropoda, only the origin of the Diptera (with 2.05 % of new protein families being acquired) shows a statistically significantly higher rate of novel gene families acquisition (Figs. 3.4 and 3.5). Genomic data were not available for the

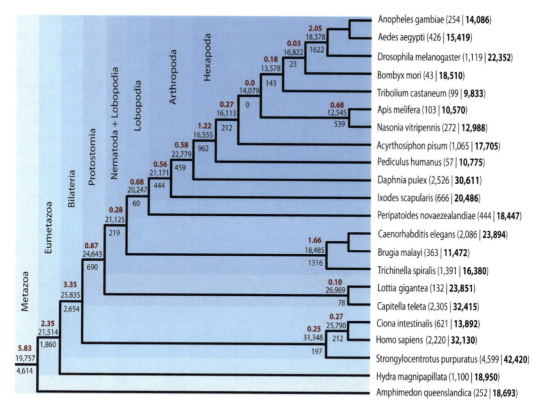

**Fig. 3.4** Orphan protein gains in Arthropoda. The number below each node quantifies the orphan families that evolved along the branch subtending the considered node. The number in black above each node represents the number of protein coding genes inferred to have existed (using squared parsimony) in the common ancestor represented by the considered node. The *red value* above the node represents the rate of orphan gene acquisition along the branch subtending the considered node. These values are normalised (calculated as the number of orphans divided by the total number of proteins in the collection of considered proteomes). The numbers reported for each terminal taxon are the number of orphan families that originated along the terminal branch and the number of genes in the genome of the corresponding organism (*in bold*). Note that the numbers of orphans for the terminal taxa are misleading and should not be considered to represent the number of new genes that emerged in the species at the tip of the tree. Instead, they represent the number of orphan in the group the species represent. For example, the number of orphans in *Hydra* represents the orphans that were acquired by the Cnidaria (to which *Hydra* belong and that *Hydra* represents) rather than by *Hydra* itself

Myriapoda when we assembled our data set, but it is clear, given the low level of proteins that originated in the branch separating Arthropoda and Pancrustacea (1.49 %) that also the origin of Mandibulata cannot be marked by a spike in the origin of new protein families (Figs. 3.4 and 3.5).

The most surprising result emerging from this analysis is that the deepest nodes in the Ecdysozoan phylogeny (origin of Nematoida plus Arthropoda, origin of Lobopodia, and origin of Arthropoda) are not characterised by above average acquisitions of new gene families (Fig. 3.5). When the number of orphan families (N-orph) acquired along a branch is divided by the length (in millions of years) of the branch along which the N-orph accumulated, the pattern in Fig. 3.5a changes quite significantly: even the mild, but somewhat continuous, increment in the rate of N-orph acquisition disappears (Fig. 3.5b). All internodes within Ecdysozoa (on the path leading to Arthropoda and within Arthropoda) roughly exhibit the same rate of new protein acquisition per million of years. Constancy of the rate of protein family

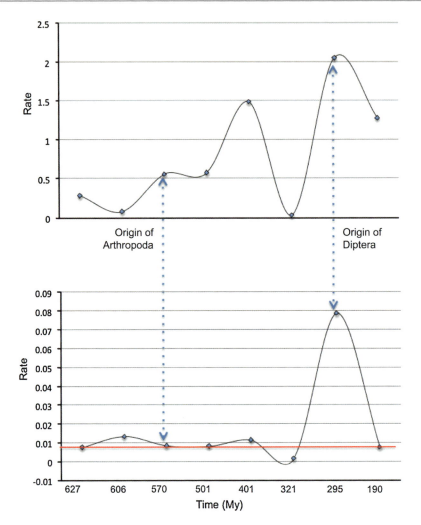

**Fig. 3.5** Protein gains through time. **a** Normalised rates of orphan acquisition (*red values* in Fig. 3.4). This panel illustrates that the normalised rates are quite variable across all the considered nodes. Note that the values were ordered from oldest to youngest to make the figure more readable. **b** Rates of orphan acquisition per millions of years. This chart was derived dividing the values in Fig. 3.5a by the length (in million years) of the branch along which the considered orphans originated. This figure clearly illustrates how the row rates and the rates per million years are substantially different, and that normalising for the time of duration of the considered internodes is key to obtain values that are biologically meaningful. The *red line* represents the average rate across the considered lineages (but excluding the Diptera). This was done to estimate the average rate of orphan protein acquisition (i.e. the neutral rate)

acquisition through time (from the Precambrian to the Jurassic—see Fig. 3.5b) suggests that this rate (identified with a red line in Fig. 3.5b) might represent the neutral background rate of new protein family origination in Ecdysozoa. The only internode where this neutral rate is modified is represented by the stem dipteran lineage. Along this lineage (Fig. 3.5b), the rate is significantly increased, suggesting that orphan

gene family acquisition was an important phenomenon in the evolution of this group.

A functional analysis of the orphan proteins that originated along the stem dipteran lineage (see Appendix for methodological details) provides a view of what kind of gene families are acquired along this branch (Fig. 3.6). When comparing the average trend estimated across all the considered stem lineages but the dipteran,

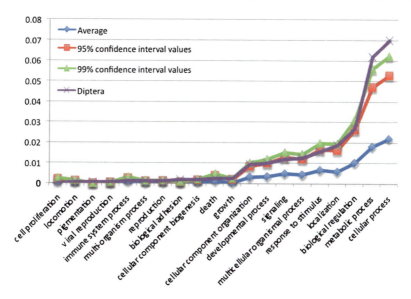

**Fig. 3.6** The function of the newly acquired families. This graph displays the average number of orphans (across all the internodes but the Diptera) for each GO (Gene Ontology) category. We also reported the values representing, respectively, the limits of the 95 and 99 % confidence intervals. Values observed for the dipteran stem lineage are reported. This figure shows that for two GO categories, the number of orphans acquired in Diptera is higher than the value bounding the 99 % confidence interval over all the other internodes, and that various other GO categories are overrepresented with reference to the other internal branches considered

with the trend observed in the dipteran, two conclusions can be reached. The first is that the trends observed are comparable in shape (i.e. there is a proportionality in the number of new genes acquired on average across the Arthropoda and specifically in Diptera). The second is that when the numbers of genes in each Gene Ontology (GO) category is analysed, it is clear that for two GO terms (metabolic processes and cellular processes), the increase observed in Diptera is significantly higher (greater than the limiting values of a 99 % confidence interval calculated across all the other internodes; Fig. 3.6). A further significantly increased category (exceeding the 95 % confidence interval calculated across all the other non-dipteran internodes) is the localisation proteins category. Finally, other GO categories for which new proteins are accumulated in Diptera to levels that are above average (but not significantly so) are as follows: biological regulation, response to stimulus, multicellular organismal processes, signalling, developmental processes, and cellular component organisation.

### 3.4.4 Conserved Rate of Gene Gain with Some Surprises

It is fairly obvious from the above results that, at least within Ecdysozoa, the origin of new protein families (orphan gene accumulation) did not play a particularly significant role in the evolution of what we recognise as high-level taxonomic groups (phyla and assemblages of phyla). In particular, we have shown here that the origin of the arthropod body plan was not characterised by an unusual rate of new protein families acquisition. One can thus argue that other processes, like the re-wiring of developmental networks (and more generally protein-protein interaction networks), might have been much more important (see also Erwin et al. 2011). Yet, these hypotheses need to be tested and will be tested in the future when more data become available.

On the other hand, the origin of the Diptera is marked by a substantial increase in the origin of orphan families. This is interesting because it suggests that (1) if increases in rate existed somewhere else in the ecdysozoan tree, we should have been able to identify them (i.e. our results do not seem to represent a methodological artefact), and (2) orphan gene acquisition is not always an unimportant process in animal evolution: hence, the need to investigate it. With reference to the Diptera, it is clear that the strong acceleration in rate of acquisition of new families observed implies that new functionalities emerged in this part of the ecdysozoan tree, and it is clear that these protein families played a role in the origin of this group. Our current GO analyses did not allow us to obtain a detailed description of what the newly acquired dipteran functions are. However, as more precise functional annotations will become available, it will become possible to pinpoint the functions of the orphan genes originating along the dipteran branch much more precisely.

One can only conjecture, given also the unimpressive amount of orphan families being fixed on the stem lineage of the Holometabola, that the origin of key innovations affecting the emergence of novel life cycles or substantially modified morphological features is generally fuelled by re-wiring of the developmental networks and by differential expressions of genes, while origin of novel protein families probably has a greater impact on adaptations to novel environmental challenges.

## 3.5 Conclusions

Here, we have tried to summarise mitogenomic and nuclear genomic information currently available for the Arthropoda. There are a large number of mitochondrial genomes available to date, but it is unclear if something that will be of any utility will be gained from the analyses of these genomes. They might have some limited utility in phylogenetics compositional bias studies, and DNA barcoding, but probably not much utility in understanding large-scale evolutionary patterns in Arthropoda.

Arthropod genomics, on the other end, is still in its infancy, very few genomes are available at this stage but within five years, we will probably have thousands of genes available (particularly

thanks to large-scale efforts like the i5k). One wonders what will be gained from having so many genomes. Perhaps a lot, but their biased taxonomic distribution might prove to be a limitation of these data sets. Data analysis will be prohibitively complex, and serious bioinformatic resources will be necessary for these data to be of any utility. In any case, the initial analysis we present in this chapter suggests that, if adequate bioinformatic resources are available, a multitude of arthropod genomes will allow us to gain detailed information on the origin and evolution of this important phylum. Yet, sequencing projects should not forget that arthropod outgroups are necessary and important to increase the power of comparative analyses.

No matter what the future will hold, it is clear that arthropod comparative genomics is still in its infancy. We are just at the dawn of what will be a laborious and complex research task which will involve the continuous effort of many research groups, from all around the world for, probably, several research cycles.

**Acknowledgments** We would like to thank Alessandro Minelli for inviting us to contribute a chapter to this book and for the patience demonstrated during the editing process. DP and RC are supported by a Science Foundation Ireland Research Frontiers Programme (SFI-RFP) grant SFI-RFP 11/RFP/EOB/3106. ORS by a Marie Curie-Trento Province COFUND Fellowship. WAA by an IRCSET PhD studentship.

## Appendix: Methods for the Analyses Presented in this Chapter

### A. Generation of the Onychophoran Transcriptome

Total RNA was extracted from three individuals of "*Peripatoides novaezealandiae* complex" (Trewick 1998), which were commercially purchased, using TriZol©. A transcriptome-wide cDNA library was generated and sequenced using two IlluminaHiseqII lanes at TrinSeq (Trinity College Dublin, Institute of Molecular Medicine, Genome Sequencing Laboratory) to an estimated coverage of <100, using 100-bp paired end reads. Row data were inspected for its quality and assembled using Abyss (Simpson et al. 2009) with k-mer of 45. This resulted in ~27,000 assembled transcripts (with lengths variable between ~70 and 1,750 base pairs). Approximately 17,000 of these transcripts had a significant blast hit against an annotated gene, while ~5,000 hit a known gene of unknown function. This set of ~22,000 genes was used to investigate the origin of orphan genes in Arthropoda. However, the 5,000 non-annotated genes were not considered for the Blast2go analysis (see below).

## B. Mitogenomic Compositional Analyses

We downloaded a set of mitochondrial genomes of 90 arthropods in order to represent the whole phylum as homogenously as possible. Coding genes were extracted and processed with DAMBE (Xia and Xie 2001) to obtain composition for each codon position.

## C. Phylogenetic Analyses

We investigated whether the low posterior probabilities observed for some nodes by Campbell et al. (2011) were caused by the presence of unstable taxa. We estimated leaf stability indices (Thorley and Wilkinson 1999) using P4 (Foster 2004) and performed Bayesian bootstrap analysis under CAT+G—the same model used by Campbell et al. (2011)—using the entire data set of Campbell et al. (2011). To perform the Bayesian bootstrap analyses, 100 bootstrapped data sets were generated starting from the alignment of Campbell et al. (2011). For each bootstrapped data set, a Bayesian analysis (2 independent runs) was performed under CAT+G (using Phylobayes; Lartillot et al. 2009). Results from each Bayesian analysis were summarised to generate a Bayesian majority rule consensus tree, and the resulting 100 trees were then summarised to generate a bootstrap majority rule consensus (results in Fig. 3.3).

## Identification of Novel Gene Families

We downloaded the entire proteomes for the taxa in Fig. 3.4 and used MCL (Enright et al. 2002) to define protein families. A Perl script written by LC was used to partition these gene families with reference to their taxon coverage. This allowed the identification of protein families that are exclusive and universally distributed within each one of the clades in Fig. 3.4. These protein families must have been present in the clade's last common ancestor (LCA) and must have been gained along the stem lineage of the considered clade. Because different genomes have different numbers of protein coding genes, the absolute numbers of newly acquired protein coding families for each internode can be misleading. We thus normalised numbers of orphan families by dividing these numbers by the total number of protein coding genes in the set of considered genomes (sum of the values in bold at the tips of Fig. 3.4). The normalised orphan counts (N-orph) can be interpreted as the fraction of some, abstract, pan-metazoan genome that was acquired at each internode of Fig. 3.4. Finally, we calculated rates of new orphan acquisition per million of years, dividing the N-orph values by the length of the internode along which the N-orph was acquired. As above, this allows the amount of orphan families gained each million year, along each internode in Fig. 3.4, to be expressed as proportions of a reference (abstract) "pan-metazoan" genome. The estimates of divergence times of Erwin et al. (2011) were used to calculate branch durations in million of years. For each internal node in our phylogeny, we also estimated (using squared parsimony—as implemented in Mesquite—http://mesquiteproject.org) the expected size of the genome of the corresponding LCA. This was done to allow evaluation of what proportion of each LCA genome was gained via new orphan family acquisition, along the corresponding stem lineage. Because squared parsimony is unlikely to be a particularly robust estimator of ancestral size, we suggest these numbers should be considered with caution, and only to represent a rough approximation of the true LCA-genomes dimensions.

Once the orphan gene families were identified for every internode of Fig. 3.4, BLAST2Go (www.blast2go.com) was used to obtain functional information for each of these families. For each protein family, the BLAST2Go analysis was performed for one protein family member only, and we assumed, by homology implication that all the other proteins in the same orphan family had the same (or similar) function.

# References

Abascal F, Posada D, Knight RD, Zardoya R (2006) Parallel evolution of the genetic code in arthropod mitochondrial genomes. Plos Biol 4:e127. doi: 10.1371/journal.pbio.0030127

Abascal F, Posada D, Zardoya R (2007) MtArt: a new model of amino acid replacement for Arthropoda. Mol Biol Evol 24:1–5

Adams MD, Celniker SE, Holt RA, Evans CA, Gocayne JD, Amanatides PG, Scherer SE, Li PW, Hoskins RA, Galle RF, George RA, Lewis SE, Richards S, Ashburner M, Henderson SN, Sutton GG, Wortman JR, Yandell MD, Zhang Q, Chen LX, Brandon RC, Rogers YH, Blazej RG, Champe M, Pfeiffer BD, Wan KH, Doyle C, Baxter EG, Helt G, Nelson CR, Gabor GL, Abril JF, Agbayani A, An HJ, rews-Pfannkoch C, Baldwin D, Ballew RM, Basu A, Baxendale J, Bayraktaroglu L, Beasley EM, Beeson KY, Benos PV, Berman BP, Bhandari D, Bolshakov S, Borkova D, Botchan MR, Bouck J, Brokstein P, Brottier P, Burtis KC, Busam DA, Butler H, Cadieu E, Center A, Chandra I, Cherry JM, Cawley S, Dahlke C, Davenport LB, Davies P, de Pablos B, Delcher A, Deng Z, Mays AD, Dew I, Dietz SM, Dodson K, Doup LE, Downes M, Dugan-Rocha S, Dunkov BC, Dunn P, Durbin KJ, Evangelista CC, Ferraz C, Ferriera S, Fleischmann W, Fosler C, Gabrielian AE, Garg NS, Gelbart WM, Glasser K, Glodek A, Gong F, Gorrell JH, Gu Z, Guan P, Harris M, Harris NL, Harvey D, Heiman TJ, Hernandez JR, Houck J, Hostin D, Houston KA, Howland TJ, Wei MH, Ibegwam C et al (2000) The genome sequence of Drosophila melanogaster. Science 287:2185–2195

Aguinaldo AMA, Turbeville JM, Linford LS, Rivera MC, Garey JR, Raff RA, Lake JA (1997) Evidence for a clade of nematodes, arthropods and other moulting animals. Nature 387:489–493

Belinky F, Cohen O, Huchon D (2010) Large-scale parsimony analysis of metazoan indels in protein-coding genes. Mol Biol Evol 27:441–451

Bernt M, Braband A, Middendorf M, Misof B, Rota-Stabelli O, Stadler PF (2012) Bioinformatics methods for the comparative analysis of metazoan mitochondrial genome sequences. Molecular Phylogenetics and Evolution published on-line ahead of press. doi: 10.1016/j.ympev.2012.09.019 [to be replaced by volume: page range on publication]

Blanquart S, Lartillot N (2008) A site- and time-heterogeneous model of amino acid replacement. Mol Biol Evol 25:842–858

Blythe MJ, Malla S, Everall R, Shih YH, Lemay V, Moreton J, Wilson R, Aboobaker AA (2012) High through-put sequencing of the Parhyale hawaiensis mRNAs and microRNAs to aid comparative developmental studies. Plos One 7(3):e33784. doi: 10.1371/journal.pone.0033784

Boore JL (1999) Animal mitochondrial genomes. Nucleic Acids Res 27:1767–1780

Boore JL, Lavrov DV, Brown WM (1998) Gene translocation links insects and crustaceans. Nature 392:667–668

Campbell LI, Rota-Stabelli O, Edgecombe GD, Marchioro T, Longhorn SJ, Telford MJ, Philippe H, Rebecchi L, Peterson KJ, Pisani D (2011) MicroRNAs and phylogenomics resolve the relationships of Tardigrada and suggest that velvet worms are the sister group of Arthropoda. Proc Natl Acad Sci USA 108:15920–15924

Colbourne JK, Pfrender ME, Gilbert D, Thomas WK, Tucker A, Oakley TH, Tokishita S, Aerts A, Arnold GJ, Basu MK, Bauer DJ, Caceres CE, Carmel L, Casola C, Choi JH, Detter JC, Dong Q, Dusheyko S, Eads BD, Frohlich T, Geiler-Samerotte KA, Gerlach D, Hatcher P, Jogdeo S, Krijgsveld J, Kriventseva EV, Kultz D, Laforsch C, Lindquist E, Lopez J, Manak JR, Muller J, Pangilinan J, Patwardhan RP, Pitluck S, Pritham EJ, Rechtsteiner A, Rho M, Rogozin IB, Sakarya O, Salamov A, Schaack S, Shapiro H, Shiga Y, Skalitzky C, Smith Z, Souvorov A, Sung W, Tang Z, Tsuchiya D, Tu H, Vos H, Wang M, Wolf YI, Yamagata H, Yamada T, Ye Y, Shaw JR, Rews J, Crease TJ, Tang H, Lucas SM, Robertson HM, Bork P, Koonin EV, Zdobnov EM, Grigoriev IV, Lynch M, Boore JL (2011) The ecoresponsive genome of Daphnia pulex. Science 331:555–561

Copley RR, Aloy P, Russell RB, Telford MJ (2004) Systematic searches for molecular synapomorphies in model metazoan genomes give some support for Ecdysozoa after accounting for the idiosyncrasies of Caenorhabditis elegans. Evol Dev 6:164–169

Davidson EH, Erwin DH (2006) Gene regulatory networks and the evolution of animal body plans. Science 311:796–800

Domazet-Loso T, Brajkovic J, Tautz D (2007) A phylostratigraphy approach to uncover the genomic history of major adaptations in metazoan lineages. Trends Genet 23:533–539

Dopazo H, Dopazo J (2005) Genome-scale evidence of the nematode-arthropod clade. Genome Biol 6:R41. doi:10.1186/gb-2005-6-5-r41

Dunn CW, Hejnol A, Matus DQ, Pang K, Browne WE, Smith SA, Seaver E, Rouse GW, Obst M, Edgecombe

GD, SØrensen MV, Haddock SHD, Schmidt-Rhaesa A, Okusu A, Kristensen RM, Wheeler WC, Martindale MQ, Giribet G (2008) Broad phylogenomic sampling improves resolution of the animal tree of life. Nature 452:745–749

Enright AJ, Van Dongen S, Ouzounis CA (2002) An efficient algorithm for large-scale detection of protein families. Nucleic Acids Res 30:1575–1584

Erwin DH, Laflamme M, Tweedt SM, Sperling EA, Pisani D, Peterson KJ (2011) The Cambrian conundrum: early divergence and later ecological success in the early history of animals. Science 334:1091–1097

Ewen-Campen B, Shaner N, Panfilio KA, Suzuki Y, Roth S, Extavour CG (2011) The maternal and early embryonic transcriptome of the milkweed bug *Oncopeltus fasciatus*. BMC Genomics 12:61. doi: 10.1186/1471-2164-12-61

Foster PG (2004) Modeling compositional heterogeneity. Syst Biol 53:485–495

Foster PG, Hickey DA (1999) Compositional bias may affect both DNA-based and protein-based phylogenetic reconstructions. J Mol Evol 48:284–290

Foster PG, Jermiin LS, Hickey DA (1997) Nucleotide composition bias affects amino acid content in proteins coded by animal mitochondria. J Mol Evol 44:282–288

Giot L, Bader JS, Brouwer C, Chaudhuri A, Kuang B, Li Y, Hao YL, Ooi CE, Godwin B, Vitols E, Vijayadamodar G, Pochart P, Machineni H, Welsh M, Kong Y, Zerhusen B, Malcolm R, Varrone Z, Collis A, Minto M, Burgess S, McDaniel L, Stimpson E, Spriggs F, Williams J, Neurath K, Ioime N, Agee M, Voss E, Furtak K, Renzulli R, Aanensen N, Carrolla S, Bickelhaupt E, Lazovatsky Y, DaSilva A, Zhong J, Stanyon CA, Finley RL Jr, White KP, Braverman M, Jarvie T, Gold S, Leach M, Knight J, Shimkets RA, McKenna MP, Chant J, Rothberg JM (2003) A protein interaction map of *Drosophila melanogaster*. Science 302:1727–1736

Grbic M, Van Leeuwen T, Clark RM, Rombauts S, Rouze P, Grbic V, Osborne EJ, Dermauw W, Ngoc PC, Ortego F, Hernandez-Crespo P, Diaz I, Martinez M, Navajas M, Sucena E, Magalhaes S, Nagy L, Pace RM, Djuranovic S, Smagghe G, Iga M, Christiaens O, Veenstra JA, Ewer J, Villalobos RM, Hutter JL, Hudson SD, Velez M, Yi SV, Zeng J, Pires-daSilva A, Roch F, Cazaux M, Navarro M, Zhurov V, Acevedo G, Bjelica A, Fawcett JA, Bonnet E, Martens C, Baele G, Wissler L, Sanchez-Rodriguez A, Tirry L, Blais C, Demeestere K, Henz SR, Gregory TR, Mathieu J, Verdon L, Farinelli L, Schmutz J, Lindquist E, Feyereisen R, Van de Peer Y (2011) The genome of *Tetranychus urticae* reveals herbivorous pest adaptations. Nature 479:487–492

Hartmann B, Castelo R, Blanchette M, Boue S, Rio DC, Valcarcel J (2009) Global analysis of alternative splicing regulation by insulin and wingless signalling in *Drosophila* cells. Genome Biol 10:R11. doi: 10.1186/gb-2009-10-1-r11

Hassanin A, Leger N, Deutsch J (2005) Evidence for multiple reversals of asymmetric mutational constraints during the evolution of the mitochondrial genome of metazoa, consequences for phylogenetic inferences. Syst Biol 54:277–298

Hedin M, Starrett J, Akhter S, Schönhofer AL, Shultz JW (2012) Phylogenomic resolution of Paleozoic divergences in harvestmen (Arachnida, Opiliones) via analysis of next-generation transcriptome data. Plos One 7(8):e42888. doi:10.1371/journal.pone.0042888

Hejnol A, Obst M, Stamatakis A, Ott M, Rouse GW, Edgecombe GD, Martinez P, Baguñà J, Bailly X, Jondelius U, Wiens M, Müller WEG, Seaver E, Wheeler WC, Martindale MQ, Giribet G, Dunn CW (2009) Assessing the root of Bilaterian animals with scalable phylogenomic methods. Proc R Soc B 276:4261–4270

Holt RA, Subramanian GM, Halpern A, Sutton GG, Charlab R, Nusskern DR, Wincker P, Clark AG, Ribeiro JM, Wides R, Salzberg SL, Loftus B, Yandell M, Majoros WH, Rusch DB, Lai Z, Kraft CL, Abril JF, Anthouard V, Arensburger P, Atkinson PW, Baden H, de Berardinis V, Baldwin D, Benes V, Biedler J, Blass C, Bolanos R, Boscus D, Barnstead M, Cai S, Center A, Chaturvedi K, Christophides GK, Chrystal MA, Clamp M, Cravchik A, Curwen V, Dana A, Delcher A, Dew I, Evans CA, Flanigan M, Grundschober-Freimoser A, Friedli L, Gu Z, Guan P, Guigo R, Hillenmeyer ME, Hladun SL, Hogan JR, Hong YS, Hoover J, Jaillon O, Ke Z, Kodira C, Kokoza E, Koutsos A, Letunic I, Levitsky A, Liang Y, Lin JJ, Lobo NF, Lopez JR, Malek JA, McIntosh TC, Meister S, Miller J, Mobarry C, Mongin E, Murphy SD, O'Brochta DA, Pfannkoch C, Qi R, Regier MA, Remington K, Shao H, Sharakhova MV, Sitter CD, Shetty J, Smith TJ, Strong R, Sun J, Thomasova D, Ton LQ, Topalis P, Tu Z, Unger MF, Walenz B, Wang A, Wang J, Wang M, Wang X, Woodford KJ, Wortman JR, Wu M, Yao A, Zdobnov EM, Zhang H, Zhao Q et al (2002) The genome sequence of the malaria mosquito *Anopheles gambiae*. Science 298:129–149

Holton TA, Pisani D (2010) Deep genomic-scale analyses of the metazoa reject Coelomata: evidence from single and multigene families analyzed under a supertree and supermatrix paradigm. Genome Biol Evol 2:310–324

Hwang UW, Friedrich M, Tautz D, Park CJ, Kim W (2001) Mitochondrial protein phylogeny joins myriapods with chelicerates. Nature 413:154–157

Irimia M, Maeso I, Penny D, Garcia-Ferna'ndez J, Roy SW (2007) Rare coding sequence changes are consistent with Ecdysozoa, not Coelomata. Mol Biol Evol 24:1604–1607

Kirkness EF, Haas BJ, Sun W, Braig HR, Perotti MA, Clark JM, Lee SH, Robertson HM, Kennedy RC, Elhaik E, Gerlach D, Kriventseva EV, Elsik CG, Graur D, Hill CA, Veenstra JA, Walenz B, Tubio JM, Ribeiro JM, Rozas J, Johnston JS, Reese JT, Popadic

A, Tojo M, Raoult D, Reed DL, Tomoyasu Y, Kraus E, Mittapalli O, Margam M, Li HM, Meyer JM, Johnson RM, Romero-Severson J, Vanzee JP, Alvarez-Ponce D, Vieira FG, Aguade M, Guirao-Rico S, Anzola JM, Yoon KS, Strycharz JP, Unger MF, Christley S, Lobo NF, Seufferheld MJ, Wang N, Dasch GA, Struchiner CJ, Madey G, Hannick LI, Bidwell S, Joardar V, Caler E, Shao R, Barker SC, Cameron S, Bruggner RV, Regier A, Johnson J, Viswanathan L, Utterback TR, Sutton GG, Lawson D, Waterhouse RM, Venter JC, Strausberg RL, Berenbaum MR, Collins FH, Zdobnov EM, Pittendrigh BR (2010) Genome sequences of the human body louse and its primary endosymbiont provide insights into the permanent parasitic lifestyle. Proc Natl Acad Sci USA 107:12168–12173

Lartillot N, Lepage T, Blanquart S (2009) PhyloBayes 3: a Bayesian software package for phylogenetic reconstruction and molecular dating. Bioinformatics 25:2286–2288

Lavrov DV, Boore JL, Brown WM (2000) The complete mitochondrial DNA sequence of the horseshoe crab Limulus polyphemus. Mol Biol Evol 17:813–824

Lavrov DV, Brown WM, Boore JL (2004) Phylogenetic position of the Pentastomida and (pan)crustacean relationships. Proc Biol Sci 271:537–544

Meusemann K, von Reumont BM, Simon S, Roeding F, Strauss S, Kuck P, Ebersberger I, Walzl M, Pass G, Breuers S, Achter V, von Haeseler A, Burmester T, Hadrys H, Wägele JW, Misof B (2010) A phylogenomic approach to resolve the arthropod tree of life. Mol Biol Evol 27:2451–2464

Nabholz B, Ellegren H, Wolf JB (2012) High levels of gene expression explain the strong evolutionary constraint of mitochondrial protein-coding genes. Molecular Biology and Evolution published on line ahead of press. doi: 10.1093/molbev/mss238 [to be replaced by volume: page range on publication]

Negrisolo E, Minelli A, Valle G (2004) The mitochondrial genome of the house centipede Scutigera and the monophyly versus paraphyly of myriapods. Mol Biol Evol 21:770–780

Nene V, Wortman JR, Lawson D, Haas B, Kodira C, Tu ZJ, Loftus B, Xi Z, Megy K, Grabherr M, Ren Q, Zdobnov EM, Lobo NF, Campbell KS, Brown SE, Bonaldo MF, Zhu J, Sinkins SP, Hogenkamp DG, Amedeo P, Arensburger P, Atkinson PW, Bidwell S, Biedler J, Birney E, Bruggner RV, Costas J, Coy MR, Crabtree J, Crawford M, Debruyn B, Decaprio D, Eiglmeier K, Eisenstadt E, El-Dorry H, Gelbart WM, Gomes SL, Hammond M, Hannick LI, Hogan JR, Holmes MH, Jaffe D, Johnston JS, Kennedy RC, Koo H, Kravitz S, Kriventseva EV, Kulp D, Labutti K, Lee E, Li S, Lovin DD, Mao C, Mauceli E, Menck CF, Miller JR, Montgomery P, Mori A, Nascimento AL, Naveira HF, Nusbaum C, O'Leary S, Orvis J, Pertea M, Quesneville H, Reidenbach KR, Rogers YH, Roth CW, Schneider JR, Schatz M, Shumway M, Stanke M, Stinson EO, Tubio JM, Vanzee JP, Verjovski-

Almeida S, Werner D, White O, Wyder S, Zeng Q, Zhao Q, Zhao Y, Hill CA, Raikhel AS, Soares MB, Knudson DL, Lee NH, Galagan J, Salzberg SL, Paulsen IT, Dimopoulos G, Collins FH, Birren B, Fraser-Liggett CM, Severson DW (2007) Genome sequence of Aedes aegypti, a major arbovirus vector. Science 316:1718–1723

Philippe H, Lartillot N, Brinkmann H (2005) Multi gene analyses of Bilaterian animals corroborate the monophyly of Ecdysozoa, Lophotrochozoa, and Protostomia. Mol Biol Evol 22:1246–1253

Pisani D (2004) Identifying and removing fast-evolving sites using compatibility analysis: an example from the Arthropoda. Syst Biol 53:978–989

Pisani D, Poling LL, Lyons-Weiler M, Hedges SB (2004) The colonization of land by animals: molecular phylogeny and divergence times among arthropods. BMC Biol 2:1. doi:10.1186/1741-7007-2-1

Podsiadlowski L, Braband A (2006) The complete mitochondrial genome of the sea spider Nymphon gracile (Arthropoda: Pycnogonida). BMC Genomics 7:284

Regier JC, Shultz JW, Zwick A, Hussey A, Ball B, Wetzer R, Martin JW, Cunningham CW (2010) Arthropod relationships revealed by phylogenomic analysis of nuclear protein-coding sequences. Nature 463:1079–1083

Rendon-Anaya M, Delaye L, Possani LD, Herrera-Estrella A (2012) Global transcriptome analysis of the scorpion Centruroides noxius: new toxin families and evolutionary insights from an ancestral scorpion species. Plos One 7:e43331. doi:10.1371/journal.pone.0043331

Reyes A, Gissi C, Pesole G, Saccone C (1998) Asymmetrical directional mutation pressure in the mitochondrial genome of mammals. Mol Biol Evol 15:957–966

Richards S, Gibbs RA, Weinstock GM, Brown SJ, Denell R, Beeman RW, Gibbs R, Bucher G, Friedrich M, Grimmelikhuijzen CJ, Klingler M, Lorenzen M, Roth S, Schroder R, Tautz D, Zdobnov EM, Muzny D, Attaway T, Bell S, Buhay CJ, Chandrabose MN, Chavez D, Clerk-Blankenburg KP, Cree A, Dao M, Davis C, Chacko J, Dinh H, Dugan-Rocha S, Fowler G, Garner TT, Garnes J, Gnirke A, Hawes A, Hernandez J, Hines S, Holder M, Hume J, Jhangiani SN, Joshi V, Khan ZM, Jackson L, Kovar C, Kowis A, Lee S, Lewis LR, Margolis J, Morgan M, Nazareth LV, Nguyen N, Okwuonu G, Parker D, Ruiz SJ, Santibanez J, Savard J, Scherer SE, Schneider B, Sodergren E, Vattahil S, Villasana D, White CS, Wright R, Park Y, Lord J, Oppert B, Brown S, Wang L, Weinstock G, Liu Y, Worley K, Elsik CG, Reese JT, Elhaik E, Landan G, Graur D, Arensburger P, Atkinson P, Beidler J, Demuth JP, Drury DW, Du YZ, Fujiwara H, Maselli V, Osanai M, Robertson HM, Tu Z, Wang JJ, Wang S, Song H, Zhang L, Werner D, Stanke M, Morgenstern B, Solovyev V, Kosarev P, Brown G, Chen HC, Ermolaeva O, Hlavina W, Kapustin Y et al (2008) The genome of the model beetle and pest *Tribolium castaneum*. Nature 452:949–955

Roeding F, Borner J, Kube M, Klages S, Reinhardt R, Burmester T (2009) A 454 sequencing approach for large scale phylogenomic analysis of the common emperor scorpion (*Pandinus imperator*). Mol Phylogenet Evol 53:826–834

Rota-Stabelli O, Telford MJ (2008) A multi criterion approach for the selection of optimal outgroups in phylogeny: recovering some support for Mandibulata over Myriochelata using mitogenomics. Mol Phylogenet Evol 48:103–111

Rota-Stabelli O, Yang Z, Telford MJ (2009) MtZoa: a general mitochondrial amino acid substitutions model for animal evolutionary studies. Mol Phylogenet Evol 52:268–272

Rota-Stabelli O, Kayal E, Gleeson D, Daub J, Boore JL, Telford MJ, Pisani D, Blaxter M, Lavrov DV (2010) Ecdysozoan mitogenomics: evidence for a common origin of the legged invertebrates, the Panarthropoda. Genome Biol Evol 2:425–440

Rota-Stabelli O, Campbell L, Brinkmann H, Edgecombe GD, Longhorn SJ, Peterson KJ, Pisani D, Philippe H, Telford MJ (2011) A congruent solution to arthropod phylogeny: phylogenomics, microRNAs and morphology support monophyletic Mandibulata. Proc R Soc B 278:298–306

Rota-Stabelli O, Lartillot N, Philippe H, Pisani D (2012) Serine codon usage bias in deep phylogenomics: pancrustacean relationships as a case study. Systematic Biology. doi: 10.1093/sysbio/sys077 [to be replaced by volume: page range on publication]

Roy SW, Irimia M (2008) Rare genomic characters do not support Coelomata: intron loss/gain. Mol Biol Evol 25:620–623

Shao R, Barker SC (2003) The highly rearranged mitochondrial genome of the plague thrips, *Thrips imaginis* (Insecta: Thysanoptera): convergence of two novel gene boundaries and an extraordinary arrangement of rRNA genes. Mol Biol Evol 20:362–370

Simpson JT, Wong K, Jackman SD, Schein JE, Jones SJ, Birol I (2009) ABySS: a parallel assembler for short read sequence data. Genome Res 19:1117–1123

Suen G, Teiling C, Li L, Holt C, Abouheif E, Bornberg-Bauer E, Bouffard P, Caldera EJ, Cash E, Cavanaugh A, Denas O, Elhaik E, Fave MJ, Gadau J, Gibson JD, Graur D, Grubbs KJ, Hagen DE, Harkins TT, Helmkampf M, Hu H, Johnson BR, Kim J, Moeller JA, Munoz-Torres MC, Murphy MC, Naughton MC, Nigam S, Overson R, Rajakumar R, Reese JT, Scott JJ, Smith CR, Tao S, Tsutsui ND, Viljakainen L, Wissler L, Yandell MD, Zimmer F, Taylor J, Slater SC, Clifton SW, Warren WC, Elsik CG, Smith CD, Weinstock GM, Gerardo NM, Currie CR (2011) The genome sequence of the leaf-cutter ant *Atta cephalotes* reveals insights into its obligate symbiotic lifestyle. Plos Genet 7:e1002007. doi:10.1371/journal.pgen.1002007

Taylor HR, Harris WE (2012) An emergent science on the brink of irrelevance: a review of the past 8 years of DNA barcoding. Mol Ecol Resour 2:377–388

The C elegans genome consortium (1998) Genome sequence of the nematode *C elegans*: a platform for investigating biology. Science 282:2012–2018

The Heliconius genome consortium (2012) Butterfly genome reveals promiscuous exchange of mimicry adaptations among species. Nature 487:94–98

The honeybee genome consortium (2006) Insights into social insects from the genome of the honeybee *Apis mellifera*. Nature 443:931–949

The pea aphid genome consortium (2010) Genome sequence of the pea aphid *Acyrthosiphon pisum*. Plos Biol 8:e1000313. doi:10.1371/journal.pbio.1000313

The silkworm genome consortium (2008) The genome of a lepidopteran model insect, the silkworm *Bombyx mori*. Insect Biochem Mol Biol 38:1036–1045

Thorley JL, Wilkinson M (1999) Testing the phylogenetic stability of early tetrapods. J Theor Biol 200:343–344

Trewick SA (1998) Sympatric cryptic species in New Zealand Onychophora. Biol J Linn Soc 63:307–329

Vieira FG, Rozas J (2011) Comparative genomics of the odorant-binding and chemosensory protein gene families across the Arthropoda: origin and evolutionary history of the chemosensory system. Genome Biol Evol 3:476–490

Xia X, Xie Z (2001) DAMBE: software package for data analysis in molecular biology and evolution. J Hered 92:371–373

Zeng V, Villanueva KE, Ewen-Campen BS, Alwes F, Browne WE, Extavour CG (2011) De novo assembly and characterization of a maternal and developmental transcriptome for the emerging model crustacean *Parhyale hawaiensis*. BMC Genomics 12:581. doi: 10.1186/1471-2164-12-581

Zhang DX, Hewitt GM (1997) Insect mitochondrial control region: a review of its structure, evolution and usefulness in evolutionary studies. Biochem Syst Ecol 25:99–120

# Arthropod Embryology: Cleavage and Germ Band Development

**4**

Gerhard Scholtz and Carsten Wolff

## Contents

## 4.1 Developmental Diversity in Arthropods

The overwhelming diversity of arthropod morphology and lifestyles finds it correspondence in a comparatively impressive variety of developmental trajectories. These ontogenetic differences concern all embryonic stages, steps, and levels from gene expression, cleavage and gastrulation, germ band formation and growth, to segmentation and morphogenesis (Weygoldt 1960a, 1963; Anderson 1973; Scholtz 1997; Akam 2000; Hughes and Kaufman 2002a). Likewise, postembryonic development reveals all sorts of growth patterns, direct and indirect development and within the latter a great variety of larval types with a wide spectrum of lifestyles comparable to those of the adult forms (see Chap. 5). However, it has to be stressed that variation in development is not necessarily directly correlated or even causally linked to

G. Scholtz (✉) · C. Wolff
Humboldt-Universität zu Berlin, Institut für Biologie, Vergleichende Zoologie, Philippstr. 13, Berlin, 10115, Germany
e-mail: gerhard.scholtz@rz.hu-berlin.de

C. Wolff
e-mail: carsten.wolff@rz.hu-berlin.de

A. Minelli et al. (eds.), *Arthropod Biology and Evolution*,
DOI: 10.1007/978-3-642-36160-9_4, © Springer-Verlag Berlin Heidelberg 2013

adult diversity. Similar adult body organization and shapes can result from very different ontogenies, whereas similar ontogenies can result in highly diverse adults (Scholtz 2005).

This enormous variation in arthropod ontogenies has led to the formulation of numerous concepts in order to discriminate between different types of processes and ontogenetic structure and events. These relate to superficial versus total cleavage, various modes of gastrulation, short and long germ development, holometabolous versus hemimetabolous development, anamorphic versus epimorphic development, to name just a few. In addition, ontogenetic patterns have been considered in a phylogenetic context and the question has been addressed, how to use ontogenies for the reconstruction of arthropod phylogenetic relationships and the evolution of development itself.

Here, we restrict our comparative treatment to early arthropod development, namely the cleavage process and germ band formation and differentiation.

## 4.2 Cutting Continuous Processes into Slices: The Subdivision of Development

Development in metazoan animals is traditionally conceptualized as a series of characteristic stages or phases which subdivide the process of ontogeny: the early *cleavage* leads to a *blastula* which is the starting point for the subsequent *gastrulation*, etc. However, in many instances, in particular in arthropods, these stages are not so clearly separated, which hampers a strict definition and comparison of these stages between different animal ontogenies. Hence, for the purpose of this account of early development in arthropods, we have to use somewhat loose definitions of the corresponding embryogenetic processes. Here, we conceptualize cleavage in arthropods as the period of development between the zygote and blastoderm formation. During this period, the egg is subdivided into structural and functional compartments (blastomeres, energids). In several cases, however, the boundary

between the early cleavage and blastula/blastoderm stage is not so obvious. Moreover, in many arthropods gastrulation, here interpreted as 'separation of germ layers', is not just a single process but it comprises a number of various steps and processes which take place over a certain period of early development, sometimes including the phase of early cleavage divisions (Weygoldt 1960a, 1979; Fioroni 1970).

## 4.3 Cleavage Modes: Superficial, Total, and Mixed

The textbook example for arthropod cleavage is a *meroblastic cleavage* mode described as *superficial cleavage* or *intralecithal cleavage* (Fioroni 1987; Gilbert and Raunio 1997) (Fig. 4.1). This mode of cleavage is characterized by the absence of cytokinesis, which leads to the lack of cell membranes between the cleavage products, the *energids*. Hence, these energids lie embedded in the yolk and with the cleavage process they form a polynuclear cell (plasmodium). After a certain number of divisions, the energids migrate to the periphery of the egg and form an acellular blastoderm (*periblastula*), which is cellularized in a second step by membrane formation and surrounds the central yolk mass. Like most researchers, we use superficial cleavage and intralecithal cleavage synonymously. This stands in contrast to a recent article by Eriksson and Tait (2012) on onychophoran development. These authors make a distinction between intralecithal cleavage with the early energids lying in the yolk centre and superficial cleavage with a peripheral position of the energids from the onset. However, the latter situation is traditionally called *discoidal cleavage*, representing another kind of meroblastic cleavage (Fioroni 1987; Gilbert and Raunio 1997). Among arthropods, a discoidal cleavage mode is found, for instance, in scorpions (Anderson 1973).

The meroblastic (superficial) cleavage mode stands in contrast to a *holoblastic* or *total cleavage* with proper blastomeres which are separated by membranes from the onset (Fioroni 1987;

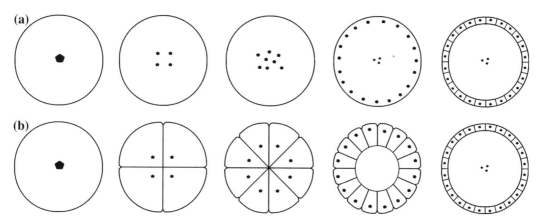

**Fig. 4.1** Schematic representation of cleavage modes. **a** A generalized meroblastic cleavage of the superficial mode with the undivided zygote, the four-energid stage, an advanced cleavage stage, an early blastoderm stage after the energids have migrated to the periphery but did not yet form cell membranes (sometimes some energids remain in the yolk), and a cellularized blastoderm surrounding a central yolk mass (the yolk mass may or may not contain some nuclei) (from *left* to *right*) (compare with Fig. 4.5). **b** A generalized holoblastic or total cleavage with the undivided zygote, the four-cell-stage, an advanced cleavage stage, and a blastula with liquid-filled hollow central space (from *left* to *right*). In the last picture a blastoderm surrounding a central yolk mass is shown (the yolk mass may or may not contain some nuclei). This stage, which is characteristic for superficial cleavage occurs in some arthropods despite an early total cleavage (compare with **a**)

Gilbert and Raunio 1997) (Fig. 4.1b). Eventually, the blastomeres form a central cavity, the liquid-filled blastocoel, and the whole stage is called a blastula. Total cleavage is found in animals as diverse as, for example, echinoderms, chordates, molluscs, annelids, ctenophores, and cnidarians and is considered as the ancestral cleavage type of metazoans (Siewing 1979).

Both cleavage modes, superficial cleavage and total cleavage, occur in arthropods (Anderson 1973; Scholtz 1997). However, these generalized terms do not appropriately describe the variations within these modes. For instance, the nuclei of a superficial cleavage can form a cellularized blastoderm at the 32–64 cell stages as in the spider *Achaearanea tepidariorum* (Kanayama et al. 2010) or after 13 rounds of divisions (8192 cells) as in *Drosophila melanogaster* (Campos-Ortega and Hartenstein 1985). Indeed, this difference raises the question whether the early cellularization in the spider species should be considered as part of cleavage or as early blastoderm formation. Likewise, there are differences concerning the structure of the central yolk mass. It might be either cellular or acellular, it can be a homogeneous mass or contain compartments (Fioroni 1970). In some cases of superficial cleavage, the membranes of the blastomeres, at least in some stages, do not reach the centre of the yolk and subdivide the egg only partly (Fig. 4.2e). This has been described for crustaceans (e.g. Weldon 1892; Vollmer 1912; von Baldass 1941; Scheidegger 1976).

In addition to these variations of superficial and total cleavage modes, very often aspects of both modes are found in combination. For this phenomenon, the term *mixed cleavage* was introduced by Dawydoff (1928) as *segmentation mixte* (see Fioroni 1970). However, mixed cleavage can mean quite different things. For instance, the term mixed cleavage is used if some stages show energids embedded in a yolk mass, whereas other stages are characterized by proper blastomeres separated by membranes. The shift between these modes can occur at any stage. There are ontogenies which begin with superficial cleavage and switch to total cleavage (e.g. Kühn 1912), or they begin with superficial cleavage, then switch to total and back to superficial cleavage (e.g. Jura 1965). Furthermore, a set of early total cleavages that nevertheless lead to a

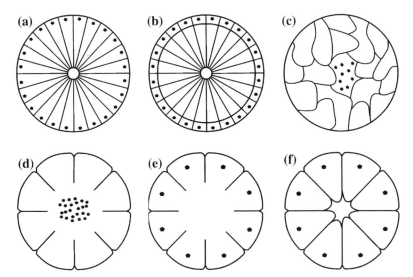

**Fig. 4.2** Various concepts of yolk pyramids. **a, b** Primary yolk pyramids as found in crayfish and other decapods. **a** Early stage after the energids have reached the egg's periphery. Each energid is forming a yolk pyramid probably by cell membranes growing towards the egg's centre. The apex of the yolk pyramids points to the centre of the egg. There is an undivided central yolk mass. **b** Advanced stage in which the blastoderm cells were separated from the yolk pyramids which do not contain nuclei. Shortly after this stage, the yolk pyramids disappear. **c** Yolk pyramids as observed in spider and other chelicerate embryos. The yolk is forming compartments during early cleavage. The latter takes place in the centre of the egg. It is unlikely that the boundaries of the yolk pyramids are formed by membranes. Rather they are plasmatic strands, along which the energids migrate to the periphery at a later stage. **d** The centipede yolk pyramids following the interpretation of the older studies. As in the chelicerate example, the pyramids occur during early cleavage and are thought to not be directly related to the cleavage process, which takes place in the egg's centre. It is not clear whether the pyramid boundaries are formed by cell membranes. According to Brena and Akam (2012), the yolk pyramids in the geophilomorph *Strigamia maritima* reach to the egg centre and they are the early blastomeres (as in **f**). Furthermore, Brena and Akam claim (2012) that the central cleavage products are cells and not just energids. **e** Superficial cleavage with membranes that cleave the yolk mass only partly. This has been observed in many crustaceans. Sometimes, it is considered as mixed cleavage. This situation is clearly different from what occurs in chilopods. **f** Total cleavage with large yolk containing blastomeres. This situation occurs in Progoneata and other arthropods and Tiegs (1940) called the blastomeres yolk pyramids

blastoderm stage with a central yolk mass are considered as mixed cleavage (Fioroni 1970) (Fig. 4.1). In addition, the incomplete penetration of yolk by cleavage membranes has been subsumed under the same term (Fioroni 1970) (Fig. 4.2e). Hence, the concept of mixed cleavage is somewhat vague and should be used with care in a comparative approach. Here, we restrict the term mixed cleavage to cases in which the early cleavage divisions are characterized by one or some total cleavages combined with one or some intralecithal cleavage divisions. We classify cases with incomplete cleavage furrows as mixed cleavages only when they occur in combination with total cleavage stages as is the case in many cladocerans (Kühn 1912). Furthermore, cases in which the total cleavage mode is given up during blastoderm formation (which we do not consider as part of cleavage) (examples are among others Progoneata, Amphipoda, Acari, and some Collembola) (Claypole 1898; Tiegs 1940, 1947; Dohle 1964; Scholtz and Wolff 2002; Laumann et al. 2010a) (Fig. 4.1) are not classified by us as mixed cleavage.

## 4.4 Dealing with a Yolk Mass: Yolk Pyramids

The term *yolk pyramids* has been applied to a number of different phenomena (Fig. 4.2). The original meaning deals with the pyramid-shaped yolk compartments that are formed after intralecithal cleavage and during blastoderm and gut differentiation in crayfish and other malacostracan crustaceans (Fig. 4.2a). These were first described by Rathke (1829), named by Lerebouillet (1862), elaborated by Reichenbach (1886), and classified by Fioroni (1970). In this case, membranes (this has in fact never been really tested) grow from each energid that has reached the egg margin towards the egg centre leaving a central compact undivided yolk mass. It is obvious that these yolk pyramids have nothing to do with cleavage divisions as such, although they have been interpreted as indication of a former total cleavage. In a second step, the blastoderm cells separate from the yolk pyramids (Fig. 4.2b), and with the formation of the germ disc, these yolk pyramids disappear (Reichenbach 1886). After gastrulation, 'secondary to quaternary yolk pyramids' are formed from the presumptive midgut cells (Reichenbach 1886; Fioroni 1970). These secondary to quaternary yolk pyramids form compartments subdividing the yolk situated in the crustacean embryonic gut.

A further instance of yolk pyramids has been described for Lithobiomorpha, Scolopendromorpha, and Geophilomorpha among the Chilopoda (Fig. 4.2d). Here, the intralecithal cleavage with its formation of energids in the egg centre is accompanied by putative membranes which reach from outer parts of the egg into the yolk for a certain distance. However, in contrast to superficial or mixed cleavage in crustacean eggs with partial cleavage furrows such as cladocerans or malacostracans (Vollmer 1912; Scheidegger 1976) (Fig. 4.2e), these partial yolk compartments do not appear to be related to the cleavage process. The figures in Heymons (1901), Hertzel (1985), and in Chipman et al. (2004) suggest that the number of

energids in the central yolk mass does not correspond to that of the yolk pyramids and that it increases more quickly than the number of yolk pyramids. Whether the boundaries between the yolk compartments are at all cell membranes needs a clarification; at least they contain actin filaments (Brena and Akam 2012). Furthermore, it has been documented that these compartment boundaries are used by energids for their migration towards the egg periphery during the formation of the blastoderm (Heymons 1901; Hertzel 1985; Chipman et al. 2004; Brena and Akam 2012). In contrast to older descriptions of centipede embryonic development, Brena and Akam (2012) suggested that the yolk pyramids in the geophilomorph *Strigamia maritima* reach to the egg centre and that they are in fact blastomeres indicating total cleavage as in pauropods and symphylans (see below). Moreover, these authors found evidence that the central cleavage products are cells rather than energids. These results are interesting because a central position of small blastomeres encircled by large yolky cells with apically situated nuclei is unique among arthropods.

Tiegs (1940, 1947) applied the term yolk pyramids again differently. He used it simply as a descriptive term for the pointed yolk-rich blastomeres of the total cleavages of pauropods and symphylans (Fig. 4.2f).

Another meaning of yolk pyramids is related to the yolk compartments which occur during the early development of spiders (e.g. Schimkewitsch and Schimkewitsch 1911; Anderson 1973; Wolff and Hilbrant 2011) (Fig. 4.2c). These yolk pyramids are apparently not separated by membranes but by plasmatic strands and are not related to early cleavage, but during blastoderm formation, they are associated with the energids that migrate along the plasmatic strands to the egg periphery (Seitz 1966).

These different meanings of yolk pyramids led to some confusion concerning the interpretation of embryo fossils of the early Cambrian. Chen et al. (2004) hypothesized that the structures found in these putative metazoan eggs are yolk pyramids which would suggest a much earlier indication of arthropod origins. In contrast to this, Donoghue et al. (2006) claimed that the same structures have to be interpreted as blastomeres of a total cleavage embryo, which according to their interpretation would preclude arthropod affinities. Neither suggestion is fully correct. Chen et al. (2004) used the derived chilopod example of yolk pyramids, but interpreted these as equivalent to the yolk pyramids of crustaceans. Irrespective of this confusion, the inference that yolk pyramids are indicative of arthropod affinities is problematic as such. There are cnidarians, in particular some anthozoan species, which possess yolky eggs with superficial cleavage and a blastoderm stage with pyramid-shaped blastoderm cells in combination with an undivided central yolk mass (Tardent 1978, Fig. 77), a pattern that is highly reminiscent to what we find in numerous arthropod embryos.

The conclusions of Donoghue et al. (2006) are also problematic, because total cleavage with proper blastomeres (sometimes even with low amounts of yolk) occurs in arthropods and might be even ancestral for the group (see below). Moreover, these authors argue with the instance of centipede type of yolk pyramids but the shape and, in particular, the pointed apex of the observed structures allow for the possibility that they are primary yolk pyramids as in crustaceans (Reichenbach 1886) or the yolk pyramids that correspond to blastomeres as in pauropods and symphylans (Tiegs 1940, 1947).

## 4.5 Determinate Versus Indeterminate Cleavage

Another perspective on classifying cleavages is dealing with the question of whether cleavage divisions reveal a stereotyped pattern of spindle orientations, blastomere sizes, blastomere positions, and blastomere fates or whether the blastomeres are irregularly arranged with no reproducible pattern of some sort. This is the discrimination between *determinate* and *indeterminate cleavages* (Fioroni 1987; Gilbert and Raunio 1997). The most obvious examples for

determinate cleavages are spiral cleavage, the cleavage of Tunicata, and that of a number of nematodes including *Caenorhabditis elegans* (Fioroni 1987; Gilbert and Raunio 1997). However, as with most categorizations in biology, the distinction between determinate and indeterminate cleavages is not straightforward, and numerous ambiguous cases occur in which reproducible patterns are combined with more irregular stages and processes. As with respect to the alternative superficial versus total cleavage, we find determinate and indeterminate cleavages and forms that show aspects of both among arthropods (see Figs. 4.4 and 4.5). At first sight, it appears that a determinate cleavage, i.e., a stereotyped arrangement of blastomeres with a traceable fate is restricted to total cleaving eggs. However, some indications suggest that even in superficial or mixed cleavage modes a determined position of energids occurs (Samter 1900; von Baldass 1941; Dohle 1970; Gerberding 1994; Wolff 2009) (see Figs. 4.4 and 4.5). Perhaps, the lack of suitable landmarks such as cell size and relative cell orientation, which help identifying individual cells in total cleaving eggs, is due to the concealment of the regular arrangement of cleavage products. Hence, it might just be a matter of perception if determinate cleavage is not recognized in superficial or mixed cleavage modes.

It has to be stressed that determinate cleavage as such cannot be used as a character for phylogenetic or evolutionary analyses (see e.g. Peterson and Eernisse 2001). Total or superficial cleavage modes are clearly characterized by structural features such as presence or absence of membranes. In contrast to this, determinate cleavage can only be homologized, when the specific patterns of cell divisions and cell fates are also taken into account.

Within arthropods superficial, total, and mixed cleavages, as well as those of determinate and indeterminate cleavage modes occur in a pattern that does not allow for an easy and entirely unambiguous possibility to reconstruct the ancestral mode for arthropods. There is apparently an opportunistic back and forth evolution between superficial and total as well as between determinate and indeterminate cleavage modes related to life style and reproduction. Yolk content is generally considered as a major aspect related to the cleavage mode. According to this view, a large amount of yolk prevents the membranes from penetrating the yolk mass and leads to meroblastic cleavage. However, there are examples in which heritage overrules the putative mechanistic effects of egg size and yolk mass (see Fioroni 1987, p. 173f.). Fioroni demonstrates that there are eggs of prosobranch gastropods that are much larger than those of some cephalopods. Nevertheless, the prosobranch eggs follow the mode of total spiral cleavage and those of cephalopods show meroblastic discoidal cleavage. Among arthropods, this is exemplified by the constant pattern of total cleavage among amphipod crustaceans despite a great amount of yolk and huge size differences (see Scholtz and Wolff 2002).

## 4.6 Early Arthropod Cleavage: Superficial or Total or a Mix of Both?

### 4.6.1 Chelicerata

Among Arachnida, there are several cases of total cleavage, most clearly within scorpions (see Anderson 1973; Moritz 1993), pseudoscorpions (Weygoldt 1964), and in Acari (Anderson 1973; Dearden et al. 2002; Laumann et al. 2010a, b). In contrast to older views, Laumann et al. (2010a, b) considered total cleavage as a general character of Acari. The yolky eggs of Xiphosura (e.g. Kingsley 1892; Ivanov 1933; Kimble et al. 2002) and most Arachnida (as far as known) undergo a superficial cleavage (e.g. Thelyphonidae: Schimkewitsch 1903; Amblypygi: Weygoldt 1975; Araneae: Schimkewitsch and Schimkewitsch 1911; Wolff and Hilbrant 2011; Mittmann and Wolff 2012; Opiliones: Moritz 1957). Data about the early development are lacking for groups such as Schizomida, Ricinulei, Solifugae, and Palpigradi. Nevertheless, irrespective of the detailed topology of

phylogenetic interrelationships within Euchelicerata (e.g. Weygoldt and Paulus 1979; Shultz 2007; Regier et al. 2010), the distribution of total versus superficial cleavage modes makes it likely that the cases of total cleavage within Euchelicerata are secondary. This is quite evident for scorpions, since total cleavage is related here to yolk reduction due to the placental nutrition of embryos and evolved most likely within the group showing a shift from meroblastic (in this case discoidal) cleavage (Anderson 1973; Moritz 1993). Given the phylogenetic position of Acari within arachnids, a transition from superficial to total cleavage seems likely as well. In summary, the last common ancestor of Xiphosura and Arachnida showed a superficial cleavage.

The situation in Pycnogonida is different. Total cleavage has been described in Pycnogonida without known exception (e.g. Morgan 1891; Dogiel 1913; Ungerer and Scholtz 2009). Sanchez's (1959) interpretation of callipallenid cleavage as being superficial contradicts the images in her own study (Sanchez 1959, Plate 4) and the results of Morgan (1891) which indicate a total cleavage mode. In any case, the large yolky eggs of nymphonids and callipallenids are apparently apomorphic within Pycnogonida (Ungerer and Scholtz 2009). Hence, all known cases show total cleavage and this permits the conclusion that total cleavage occurred in the pycnogonid stem species (e.g. Morgan 1891; Dogiel 1913; Ungerer and Scholtz 2009).

If pycnogonids are interpreted as being the sister group to the remaining chelicerates (Euchelicerata), the cleavage type of the chelicerate stem species remains ambiguous (Fig. 4.3).

### 4.6.2 Myriapoda

Among the myriapod groups, both total and superficial cleavages were described. An early total cleavage mode occurs in Pauropoda, Symphyla, and Diplopoda (Tiegs 1940, 1947; Dohle 1964), which suggest total cleavage in the

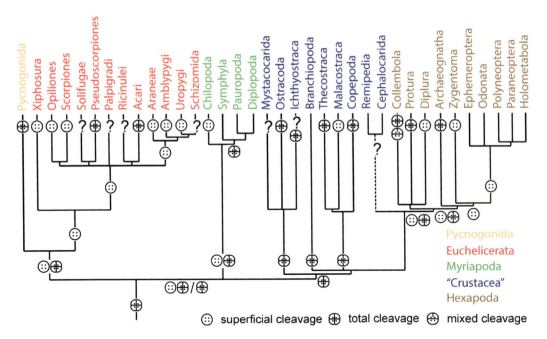

**Fig. 4.3** Cleavage modes of the various arthropod groups mapped on the consensus of arthropod phylogenetic relationships (adopted from Chap. 2). Taxa for which we found no data are flagged with a question mark. Ambiguous situations are indicated by the two alternative symbols. Irrespective of the alternative Paradoxopoda or Mandibulata, we gain total cleavage for the arthropod stem species (the lowest symbol). The putative Paradoxopoda stem species would be ambiguous, that of Mandibulata would show total cleavage (shown *left* and *right* of the slash under the *lowest horizontal line*)

stem species of Progoneata. In Chilopoda, intralecithal cleavage has been reported for representatives of Scutigeromorpha (Knoll 1974), Lithobiomorpha (Hertzel 1985), Scolopendromorpha (Heymons 1901), and Geophilomorpha (Sograff 1883; Chipman et al. 2004). A putative apomorphic condition of the Pleurostigmophora (the centipedes to the exclusion of Scutigeromorpha) is the early subdivision of yolk into so-called yolk pyramids as described above. The cleavage of energids takes place in a central undivided yolk mass (Sograff 1883; Heymons 1901; Hertzel 1985; Chipman et al. 2004). Scutigeromorpha does not show these characteristic early yolk compartments (see Knoll 1974). In contrast to older descriptions (e.g. Sograff 1883), Brena and Akam (2012) suggested that the yolk pyramids observed in *Strigamia maritima* might indicate that geophilomorph cleavage is total and that total cleavage occurs in all other Myriapoda. However, the conclusion that all myriapods show total cleavage is premature. In particular, the

study of Knoll (1974) suggests that scutigeromorphs undergo a classical superficial cleavage lacking yolk pyramids. Given that the previous studies on the early development of all major chilopod taxa reported superficial cleavage, and in the light of the current consensus on the phylogeny of Chilopoda with geophilomorphs nested within Pleurostigmophora (see Edgecombe and Giribet 2004), there is some indication for the conclusion that the ancestral cleavage mode of Chilopoda was superficial.

It is quite evident that monophyletic Progoneata and Chilopoda are sister groups (Regier et al. 2010; Shear and Edgecombe 2010). Hence, the reconstruction of the ancestral cleavage condition in Myriapoda is ambiguous (Fig. 4.3).

### 4.6.3 Pancrustacea/Tetraconata

Among Hexapoda, total cleavage or mixed cleavage has been reported from Collembola,

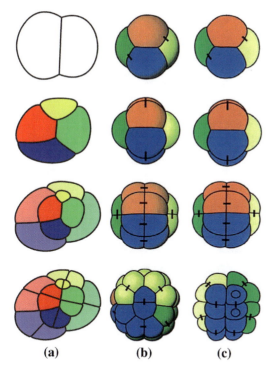

**Fig. 4.4** Examples for determinate cleavage in Crustacea. **a** Amphipoda, **b** Euphausiacea, **c** Decapoda. The colour code for the amphipod is arbitrary and does not imply homology between **a** and **b/c**. The pattern found in dendrobranchiate decapods and euphausiaceans is characterized by a specific arrangement of the blastomeres at the four-cell-stage with two cross-furrows and two interlocking cell bands. At the end of one of the cell bands, gastrulation is initiated by immigration of two large cells (*blue*).The amphipod pattern shows a smaller quadrant at the four-cell-stage (*yellow*) and macromeres and micromeres at the eight-cell-stage with a particular fate. [After Wolff and Scholtz 2002 (**a**); Alwes and Scholtz 2004 (**b**); Hertzler and Clark 1992 (**c**)]

Protura, and Machilidae (Claypole 1898; Jura 1965; Machida et al. 1990; Machida 2006). In contrast to this, Diplura, Lepismatidae, and Pterygota show intralecithal cleavage (e.g. Heider 1889; Anderson 1973; Campos-Ortega and Hartenstein 1985; Tojo and Machida 1998). Among Pterygota, only a few groups such as *Stylops*, aphids, or parasitoid wasps display total cleavage, certainly a secondary feature related to the derived reproduction mode in these groups (e.g. Pflugfelder 1962; Grbic 2003).

With respect to the paraphyly of the apterygote hexapods, in particular the paraphyletic arrangement of bristletails and zygentomans in relation to the Pterygota (Kristensen 1997; Giribet et al. 2004; Meusemann et al. 2010; von Reumont et al. 2012), the distribution allows for a reconstruction of the ancestral hexapod cleavage as total or at least as mixed cleavage (Fig. 4.3).

The following picture emerges for the various 'crustacean' groups. Nothing is known about the embryology of Remipedia, Cephalocarida, and Mystacocarida. Within Malacostraca, total cleavage has been described in Amphipoda (Bregazzi 1973; Gerberding et al. 2002; Scholtz and Wolff 2002), a number of Decapoda (e.g. Gorham 1895; Hertzler and Clark 1992; Biffis et al. 2009), Euphausiacea (Taube 1909; George and Strömberg 1985; Alwes and Scholtz 2004; Montuy-Gómez et al. 2012), Anaspidacea (Hickman 1937), and some parasitic Isopoda (Strömberg 1971). Mixed cleavage occurs in some decapods ( e.g. Weldon 1892; Scheidegger1976; Müller 1984) and Thermosbaenacea (Zilch 1974). In all other cases (Leptostraca, Stomatopoda, most Peracarida, most Decapoda), a clear superficial cleavage mode has been observed (Manton 1928, 1934; Zehnder 1934; Shiino 1942; Scholl 1963; Dohle 1970; Wolff 2009). Most phylogenetic analyses resolve the taxa with total cleavage as being deeply nested within the malacostracan tree (see Richter and Scholtz 2001) suggesting superficial cleavage as plesiomorphic condition. In Cirripedia, we find total cleavage throughout (Bigelow 1902; Delsman 1917; Anderson 1965, 1969; Turquier 1967; Scholtz et al. 2009a). The same is true for Copepoda (Schimkewitsch 1896, Amma 1911; Fuchs 1914; Kohler 1976). There is only limited knowledge about early development in Branchiopoda. This is, in particular, true for Notostraca and the paraphyletic conchostracans. The detailed study of Benesch (1969) on *Artemia* describes total cleavage. With some exceptions showing superficial cleavage (Samter 1900; Cannon 1921; Gerberding 1994), Cladocera exhibit the mixed cleavage mode with some early superficial stages switching to total cleaving stages (Kühn 1912; Vollmer 1912; von Baldass 1941; Kaudewitz 1950). With respect to the increasing agreement about branchiopod

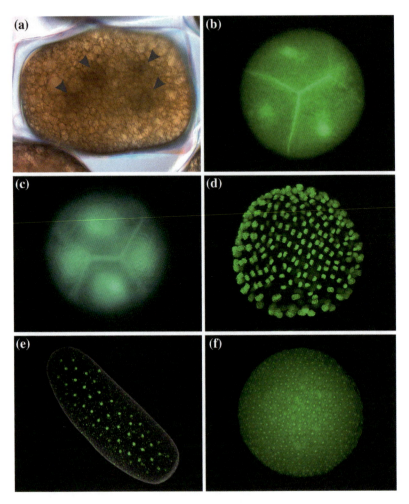

**Fig. 4.5** Cleavage and blastoderm stage. **a** An example of superficial cleavage. The second cleavage (four-energid-stage) of the egg of the isopod crustacean *Porcellio scaber* (unstained egg). Despite the fact that this is a superficial cleavage mode, the energids (arrowheads) occupy a stereotyped position. **b** Total cleavage. The four-cell-stage of the pycnogonid *Endeis spinosa* with the characteristic arrangement of blastomeres and the cross-furrow (compare with **c** and Fig. 4.4) (Sytox nuclear staining) (photograph by Petra Ungerer). **c** Total cleavage. The four-cell-stage of the euphausiacean crustacean *Meganyctiphanes norvegica* with the characteristic arrangement of blastomeres and the cross-furrow (compare with **b** and Fig. 4.4) (Sytox nuclear staining) (photograph by Frederike Alwes). **d** Blastoderm stage of the euphausiacean crustacean *Meganyctiphanes norvegica* (Sytox nuclear staining) (photograph by Frederike Alwes). **e** Superficial cleavage in *Drosophila melanogaster* (about 128 nuclei) (Sytox nuclear staining). **f** The blastoderm stage of the spider *Cupiennius salei* (compare with Fig. 4.1) (Sytox nuclear staining)

phylogeny (Braband et al. 2002; Richter et al. 2007; Olesen 2009), it seems reasonable to assume a total or at least a mixed cleavage for the branchiopod stem species. If a clade comprising Malacostraca, Cirripedia, Copepoda, and Branchiopoda is accepted, as has been suggested in recent analyses (Regier et al. 2010; see Jenner 2010 for discussion), then the last common ancestor of this group underwent a total cleavage during early development. The cleavage of Branchiura is not known, but their putative sister group, the Pentastomida (see Møller et al. 2008; Regier et al. 2010; Sanders and Lee 2010; see also Castellani et al. 2011 for a different view), undergoes total cleavage (Osche 1963). The same is true for Ostracoda (Müller-Calé 1913;

Weygoldt 1960b), which equips the putative ancestor of Branchiura, Pentastomida, and Ostracoda with a total cleaving embryo.

The analysis of the distribution of cleavage modes based on the current views of tetraconate phylogenetic relationships (e.g. Edgecombe 2010; Regier et al. 2010; von Reumont et al. 2012) leads clearly to the inference that the stem species of Tetraconata underwent total cleavage (Fig. 4.3).

## 4.7  Total Cleavage in the Arthropod Stem Species

If the Mandibulata hypothesis is correct and Myriapoda is the sister group to Tetraconata (Edgecombe 2010; see also Chap. 2), then we must assume that the last common ancestor of mandibulates showed total cleavage. As a logical consequence, if we accept a sister-group relationship between Mandibulata and Chelicerata, a total cleavage mode would have been present in the ancestral arthropod as well. Under the assumption of the Paradoxopoda hypothesis (Mallatt et al. 2004; Pisani et al. 2004), the sister taxa Myriapoda and Chelicerata would both be ambiguous with respect to their ancestral cleavage mode, which in turn would result in an open situation for the putative paradoxopodan stem species. Nevertheless, the reconstruction of the common ancestor of all arthropods would then still show total cleavage as the most likely pattern. The same would be the case under the assumption of a sister-group relationship between Pycnogonida and remaining Arthropoda (Cormogonida) (Zrzavý et al. 1998; Giribet et al. 2001).

In general, a very plausible assumption is that the ancestral cleavage mode was total (compare with Scholtz 1997, who came to a more ambiguous conclusion) and this inference is independent of the current controversies about the topology of the arthropod tree. However, this does not allow for any statement on whether a certain pattern was present in the early development of the ancestral arthropod (Fig. 4.3).

### 4.7.1  Determinate Cleavage or Irregular Blastomere Arrangement

Among Arthropoda, a stereotyped arrangement of blastomeres is mainly a feature of crustaceans (Figs. 4.4 and 4.5). In Chelicerata with total or mixed cleavage (e.g. Weygoldt 1964; Dearden et al. 2002; Laumann et al. 2010a, b), Myriapoda (Tiegs 1940, 1947; Dohle 1964), a number of crustaceans (e.g. Benesch 1969; Strömberg 1971; Scheidegger 1976), and Hexapoda (e.g. Jura et al. 1987; Claypole 1898), the first cleavage divisions result sometimes in a characteristic pattern of blastomere positions and size differences, but this cannot be observed in subsequent stages (whether there is no determination or whether this is a matter of perception is not clear). The early development of Pycnogonida is difficult to classify. Two different modes have been described. One is characterized by a more or less equal cleavage with no recognizable defined blastomere arrangement and division sequence (see Ungerer and Scholtz 2009). The other mode is unequal, forming micromeres and macromeres after the third or even after the first cleavage division (Morgan 1891; Dogiel 1913; Sanchez 1959). Nevertheless, no regularity has been found in later stages. However, this issue needs to be re-investigated. It is evident from all phylogenetic analyses that the unequal cleavage pattern, which is restricted to nymphonids and callipallenids, is representing the derived condition within pycnogonids, whereas the equal cleaving mode is likely ancestral (Ungerer and Scholtz 2009). In contrast to this, various degrees of stereotyped cleavage patterns have been described for cirripedes (Bigelow 1902; Delsman 1917; Anderson 1969), copepods (Fuchs 1914), cladocerans (Kühn 1912; von Baldass 1941), ostracods, and decapods (Gorham 1895; Hertzler and Clark 1992; Biffis et al. 2009), euphausiaceans (Taube 1909; Alwes and Scholtz 2004), amphipods (Gerberding et al. 2002; Wolff and Scholtz 2002) among Malacostraca (Fig. 4.4). This is perhaps also true for Anaspidacea (Hickman 1937), but needs to be re-investigated.

There are good reasons to assume that the cleavage patterns within some of the monophyletic taxa are homologous or even apomorphic. The latter is likely for Amphipoda (Scholtz and Wolff 2002) (Fig. 4.4). It is also quite straightforward to homologize the cleavage patterns within some arthropodan groups. For instance, there seems to be a cladoceran pattern shared by most subtaxa (von Baldass 1941). Also, the cleavage described in several copepods follows a stereotyped pattern (Fuchs 1914). Furthermore, there is a cirripede pattern (Bigelow1902; Anderson 1969) which nevertheless became considerably altered within a clade including Iblomorpha and Rhizocephala (either separately or in a common stem lineage) (see Scholtz et al. 2009a). Likewise, the patterns of early cleavages of dendrobranchiate decapods and euphausiaceans seem homologous, although it is ambiguous as to whether it is a synapomorphy of Decapoda and Euphausiacea or a plesiomorphic character within Malacostraca (Alwes and Scholtz 2004; Biffis et al. 2009; Scholtz et al. 2009b) (Fig. 4.4).

However, the attempts to homologize these various early developmental patterns at a broader scale or even with those of malacostracans were so far not convincing (Taube 1909; Kühn 1912; von Baldass 1941; Anderson 1973). This is partly due to the underlying assumption of many of these authors that crustacean development is related to spiral cleavage (Taube 1909; von Baldass 1941; Anderson 1973). Thus, an unbiased and detailed comparison of the data at hand is urgently required.

### 4.7.2 Determinate Cleavage in the Arthropod Stem Species?

The distribution of the various determinate cleavage patterns does not allow formulating an unambiguous answer to the question of whether the arthropod stem species showed a stereotyped arrangement and fate of blastomeres. However, with the data at hand it is more likely that this

was not the case if we consider determinate cleavage in the strict sense of annelid spiral cleavage or the early development of *Caenorhabditis elegans*. However, with a more relaxed approach, we arrive at a different perspective. There is a certain arrangement of the blastomeres at the 4-cell stage which is widespread among arthropods. It occurs in a number of crustaceans (Müller-Calé 1913; Fuchs 1914; Hertzler and Clark 1992; Alwes and Scholtz 2004), in myriapods (Tiegs 1940), hexapods (Claypole 1898), and pycnogonids (Dogiel 1913; Ungerer and Scholtz 2009) (Fig. 4.5). This pattern is characterized by a more or less equal blastomere size and by cross-furrows between the two non-sister cells of both poles (Figs. 4.4 and 4.5). It has been shown in some crustaceans that this blastomere arrangement leads to a characteristic pattern of two interlocking cell bands and that gastrulation is initiated at the end of one of these cell bands (Müller-Calé 1913; Fuchs 1914; Hertzler and Clark 1992; Alwes and Scholtz 2004) (Fig. 4.4).

### 4.8 The Arthropod Sister Group

There is a continuous debate about the identity of the arthropod sister group. Tardigrada, Onychophora or both taxa together have been proposed as candidates (reviewed in Zantke et al. 2008; Campbell et al. 2011). Hence, the cleavage modes described for these groups are of crucial interest for the question of the ancestral cleavage in arthropods. There are not many data on early cleavage in tardigrades. All described cases show total cleavage and a blastoderm with a cellular central yolk mass (Marcus 1929; Hejnol and Schnabel 2005; Gabriel et al. 2007). According to Hejnol and Schnabel (2005), the cleavage in *Thulinia stephaniae* is irregular and indeterminate. In contrast to this, Gabriel et al. (2007) state that the cleavage of *Hypsibius dujardini* has a stereotyped cleavage pattern. However, the proposed stereotypy is not based on a cell-by-cell basis but is rather related to equivalence groups. All we can deduce from this is that perhaps the ancestral cleavage pattern of

Tardigrada was total, but no more detail can be added. Concerning Onychophora, superficial cleavage resembling that in arthropods is found in a number of cases, in particular in ovovivip-arous forms (Sheldon 1887; Anderson 1966, 1973; Eriksson and Tait 2012). However, there is contradictory evidence as to whether the en-ergids are situated in the egg's centre (Anderson 1966) or at the margin of the egg (Eriksson and Tait 2012), the latter being rather a kind of discoidal cleavage as in scorpions (see above). Moreover, apart from the fact that the eggs are yolky, nothing is known about the early devel-opment of the eggs of oviparous onychophorans. Total cleavage occurs as well but is restricted to those species with vivipary, of which some evolved placenta-type nutrition of yolk-free eggs and embryos in their uterus (Manton 1949; Pflugfelder 1962; Anderson 1973). The general opinion is that superficial cleavage is plesio-morphic within Onychophora, because ovipa-rous or ovoviviparous reproduction is considered as plesiomorphic (Anderson 1973; Reid 1996; Nielsen 2001). There is no indication for deter-minate cleavage in onychophorans. Given that Onychophora is most probably the sister group of Arthropoda, the cleavage mode of the last common ancestor would be ambiguous, either superficial or total. However, in any case, a determinate cleavage is unlikely. If one accepts that Tardigrada is the sister group to a clade of Onychophora and Arthropoda, the ancestor of this assemblage would have had a total cleavage.

## 4.9 Once More, the Question of Spiral Cleavage

Despite the view that superficial cleavage might be the characteristic, typical (Paulus 2007) or even apomorphic (for Onychophora plus Arthropoda, Weygoldt 1986) developmental mode of Arthropoda, there have been frequent attempts to relate arthropod cleavage patterns to that of spiral cleavage as is found in Annelida or Mollusca (e.g. Anderson 1973; Nielsen 2001, but see the new edition, Nielsen 2012). In particular, the total cleavage that occurs in a number of crustacean groups such as Cirripedia, Malacostraca, or Branchiopoda has been inter-preted as closely related to spiral cleavage (e.g. Taube 1909; Anderson 1973). This interpreta-tion is based on the traditional view that panar-thropods are the sister group or at least close relatives of Annelida within the Articulata (see Scholtz 2002) and thus that arthropods are dee-ply nested within Spiralia. Nevertheless, despite the past general acceptance of the Articulata, there were critical voices which expressed seri-ous doubts on the spiralian affinities of crusta-cean cleavages (e.g. Dohle 1979; Siewing 1979; Zilch 1979; Scholtz 1997; Wolff and Scholtz 2002; Alwes and Scholtz 2004). This view is based on the lack of correspondence between the determinate cleavage patterns observed in arthropods, crustaceans in particular, and spiral cleavage. Even if these arthropod cleavage modes were derived from a plesiomorphic spiral pattern, they would be so greatly altered that no resemblance to spiral cleavage would have been left.

The currently dominating perspective of metazoan phylogeny puts panarthropods in close relationship to cycloneuralians with both groups together forming the Ecdysozoa (see Edge-combe et al. 2011, Chap. 2). This renders any interpretation of arthropod cleavage as following the spiral mode obsolete. None of the cyclone-uralian taxa show spiral cleavage (e.g. Malakhov and Spiridonov 1984; Kozloff 2007; Wennberg et al. 2008; Schulze and Schierenberg 2011) and hence, there is no indication for spiral cleavage in the ecdysozoan stem species. Finally, if the reconstruction is correct that the arthropod (or even panarthropod) stem species did not show a determinate cleavage, this automatically excludes the occurrence of spiral cleavage as well.

In general, it is not sufficient to show that arthropod and panarthropod cleavage may not be of the spiral mode. More important is to relate characteristics of arthropod cleavage to those of other metazoan groups, in this case cycloneura-lians. This requires a more detailed comparison of ontogenetic patterns in arthropods and those of priapulans, kinorhynchs, loriciferans, nematodes, and nematomorphs. The perspective is to arrive eventually at an (ancestral) ecdysozoan cleavage pattern.

## 4.10  The Germ Band

One of the characteristic features of arthropod development is the so-called germ band (see Krause 1939; Anderson 1973; Sander 1983; Scholtz 1997). A germ band is a concentrated elongate field of blastoderm cells lying at the surface of one side of the yolky egg representing the embryo proper (Fig. 4.6). The rest of the egg is formed by extra-embryonic tissue sometimes containing specializations such as dorsal organs (Anderson 1973; Meschenmoser 1996) or a serosa (Anderson 1973). In arthropods, a germ band is mostly formed by cell migration and aggregation at one side of the yolky egg (e.g. Scholtz and Wolff 2002). Initially, this aggregation is called germ disc, but with longitudinal growth, it turns into the germ band (Fig. 4.6). The germ band stretches along the longitudinal axis of the embryo and marks the future ventral side. Accordingly, ventral ectodermal aspects of segments such as segmental furrows, limbs, central nervous system, and ventral mesoderm structures are first formed (Fig. 4.6).

Germ bands comprise the ectoderm and the mesoderm germ layers. Both germ layers form metamerically arranged structures which contribute to the segments. There is a long-standing question of whether there is a certain hierarchy between ecto- and mesoderm with respect to segmentation, i.e., whether mesodermal segmentation is intrinsic and induces ectodermal segmental structures or *vice versa*. Molecular and cell ablation experiments in insects and crustaceans suggest that segmentation is an intrinsic property of the ectoderm, whereas metameric structures in the mesoderm are induced by the ectoderm (Bock 1942; Haget 1953; Frasch 1999; Hannibal et al. 2012). Whether this holds true for arthropods in general remains to be seen.

The germ band lies on the the surface of the yolk mass in most arthropods. In contrast to this, it is covered by an amnion throughout its development in dicondylian hexapods. The amnion is formed by folds of the lateral part of the germ band (Anderson 1973). With advanced development, the lateral sides grow towards the dorsal side and begin with the differentiation of lateral and dorsal structures and eventually leading to the dorsal closure. This way the yolk is internalized and the embryo achieves its 3-dimensional shape.

Since the occurrence of a germ band is widespread among all arthropod groups, it has been considered as apomorphy for arthropods (plus onychophorans) (Scholtz 1997) or as the arthropod 'phylotypic stage', i.e., the embryonic stage in which the basics of the arthropod body plan is set up and which is somehow buffered against evolutionary change (Sander 1983; Peel et al. 2005).

## 4.11  Short Germ Versus Long Germ Development

Based on his comparative studies in insect embryology, Krause (1939) developed the concept of *long germ* (Langkeim) versus *short germ* (Kurzkeim) development. Short germ development in its extreme expression is characterized by a germ band comprising only the material for the head lobes, and the rest is successively budded by a posterior growth zone. The segments are differentiated in a general anteroposterior sequence (Figs. 4.6, 4.7). Short germ development is found in beetles such as *Tribolium castaneum* and in grass hoppers such as *Schistocerca americana* (Patel et al. 1994). In contrast to this, in the extreme long germ, the material for the whole length of the embryo appears from the onset of germ band formation and segments are formed almost simultaneously along the whole germ band (see Scholtz 1997; Liu and Kaufman 2005). The honey bee *Apis mellifera* and the fruit fly *Drosophila melanogaster* are the best known examples for long

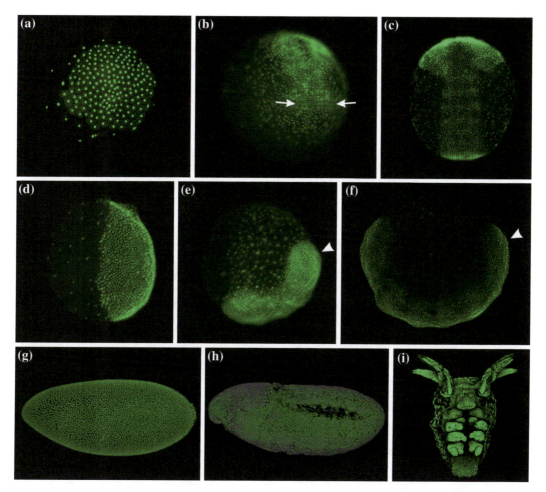

**Fig. 4.6** Germ bands in arthropods (Sytox staining). **a–c** Short to intermediate germ band development of the isopod crustacean *Porcellio scaber*. **a** Germ disc. **b** Early germ band with the anterior head lobes (*top*) and a posterior growth zone formed by a horizontal row of large teloblasts (*arrow*). **c** Advanced elongated germ band with segmental structures such as intersegmental furrows, limb buds, and ganglion anlagen (ventral view, anterior is on top). **d–f** Intermediate germ band development of the spider *Parasteatoda tepidariorum*. **d** Germ disc. **e** Early germ band with the anterior head lobes (*left*) and a posterior growth zone (arrowhead). **f** Advanced elongated germ band with segmental structures such as intersegmental furrows, limb buds, and ganglion anlagen (lateral view, anterior is *left*). The growth zone is marked by an arrowhead. **g, h** Long germ band development in the insect *Drosophila melanogaster*. **g** Blastoderm stage (lateral view, anterior on the *left*). **h** Elongated germ band stage with visible segmentation (lateral view, anterior on the *left*). **i** Metanauplius stage of the euphausiid crustacean *Meganyctiphanes norvegica*. There is no ventral germ band, but the whole larva develops three-dimensionally (ventral view, anterior is on top) (photograph by Frederike Alwes)

germ development (Fleig 1990; Patel et al. 1994) (Fig. 4.6). Between these extremes, all sorts of transitions are found, called intermediate or semi-long germ bands (Fig. 4.6). Although Krause described these phenomena in insects, the concept of long and short germs can be applied to other arthropods as well. For instance, the germ band of freshwater crayfish is a short to intermediate germ band (Scholtz 1992; Alwes and Scholtz 2006). The same holds true for isopods (Wolff 2009) (Fig. 4.6), whereas the development of *Gammarus pulex* and other amphipods is characterized by an intermediate to long germ (Scholtz 1990). The germ band of *Daphnia* and other water fleas is considered as representing the long germ type (Schwartz

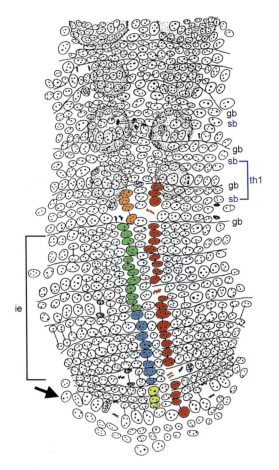

**Fig. 4.7** Semi-schematic *camera lucida* drawing of the nuclei of the ectoderm of the post-naupliar germ band of the mysid crustacean *Neomysis integer* (after Scholtz 1984). Anterior is on top showing the limb buds of the mandibles). The *arrow* marks the ectoteloblasts that are relatively large and form a transverse row. *Red* the complete progeny of one ectoteloblast formed by eleven asymmetric divisions of the ectoteloblast cell and subsequent divisions of the ectoteloblast daughter cells. The cells anterior to the cells marked in red are not descendants of the ectoteloblasts. Nevertheless, most of them show the same behaviour (transverse row formation via two waves of division with longitudinal orientation and differential cleavages). *ie* the region of intercalary germ band elongation by divisions in longitudinal direction of the ectoteloblast daughter cells. *Yellow* cells budded by an ectoteloblast; *blue* ectoteloblast descendants, which underwent their first division with a longitudinally oriented spindle; *green* ectoteloblast descendants which underwent their second division with a longitudinally oriented spindle; *orange* ectoteloblast descendants starting the phase of differential cleavage which are oriented in various directions and contribute only partly to germ band elongation. All ectoteloblast descendants show the same pattern in an anteroposterior gradient, i.e., in the next division cycle, the *yellow cells* would show the same pattern as the *blue cells*, and the *blue cells* would show the same pattern as the *green cells*, and these would enter the stage of the *orange cells*. This reveals that in fact most of the germ band elongation is caused by the intercalary divisions and not by the budding of the ectoteloblasts. *gb* genealogical boundaries marking the polyclones (= parasegments) formed by the ectoteloblast descendants (only some anterior boundaries are designated, showing that non-ectoteloblastic cells and teloblastic cells show a corresponding pattern). *sb* segmental boundaries as indicated by the forming intersegmental furrows. It is obvious that the segmental boundaries do not match the genealogical boundaries and that the descendants of two ectoderm rows contribute to one morphological segment. *th*1 the first thoracic segment. This is composed by non-ectoteloblastic cells (anterior part) and ectoteloblast descendants (posterior part)

1973). In Chelicerata, we find a short to inter-mediate germ development in xiphosurans and scorpions starting with a germ band comprising the head lobes and three segments (Kingsley 1892; Brauer 1895). In spiders, the whole area of the prosoma is formed by the germ disc and only the opisthosoma is generated by an elongation process (Wolff and Hilbrant 2011; Mittmann and Wolff 2012) (Fig. 4.6). A similar pattern occurs in mites (with a delay of the fourth walking limb segment, which develops subsequently together with the anlagen of two opisthosoma segments) (Barnett and Thomas 2012). Likewise, the germ band of some pycnogonid species comprises the cellular material of all prosomal segments from the onset, but the remaining segments are formed during post-embryonic growth (Brenneis et al. 2011). Hence, this has to be considered as intermediate to long germ development. The myriapods show a short to intermediate or an intermediate germ development. Centipedes and diplopods show a short to intermediate type of development (Dohle 1964; Hertzel 1985; Hughes and Kaufman 2002b; Chipman et al. 2004; Brena and Akam 2012; Janssen 2012), symphylans and pauropods differentiate a germ band which comprises the anlagen of all seg-ments that are formed during embryogenesis (Tiegs 1940, 1947). However, there is also a post-embryonic addition of segments and, as in mites, pycnogonids and in many crustacean nauplii, this mode of development has to be considered as an intermediate growth but strictly speaking as a long germ band, if a germ band is not involved in post-embryonic growth. More-over, growth and elongation of the embryo is not restricted to the occurrence of a proper germ band. In crustacean nauplii, only the anterior-most three segments are formed and hence, they show a longitudinal growth (although in this case a post-embryonic growth) despite the lack of a germ band (Fig. 4.6).

One would expect that long and short germ developments involve very different molecular mechanisms related to growth and segmentation. Indeed, it has been found that the expression and function of several genes related to segmentation such as a number of pair-rule genes differ between long and short germ insects (see Liu and Kaufman 2005). On the other hand, how-ever, surprising correspondences in gene expression and function occur and in some aspects, such as the expression of the pair-rule gene *even-skipped* differences among short germ insects are even greater than between some short germ and the long germ developments (Patel et al. 1994). This suggests that the morphologi-cal mode of germ band growth is not strictly correlated with a specific molecular mechanism. Based on morphological data, it has been sug-gested that there is no necessary fundamental difference between long germ and short germ band development. Rather a heterochronic shift between germ band growth and the onset of segmentation may cause the shift from a short (intermediate) germ to a long germ, i.e., a long germ is formed if there is a delayed beginning of the segmentation process with respect to germ band growth (Scholtz 1992). Even if it can be shown that there are some specific molecular mechanisms involved in the various germ band types, the hypothesis of a heterochronic shift between germ band growth and the segmenta-tion process could explain the evolutionary change from short- to long germ bands and its frequent independent occurrence. The above-mentioned inconsistencies between the mor-phological modes of germ band development and the molecular level underpin this view (see Liu and Kaufman 2005).

## 4.12  What is a Growth Zone?

It is an almost generally accepted view that germ band elongation or growth is caused by a pos-terior pre-anal proliferation or growth zone. Despite this view, it is not clear what the term growth zone actually implies (see Davis and Patel 2002; Liu and Kaufman 2005). Often, it has been defined as a region in the posterior of the germ band with a high cell division activity. However, investigations dealing with cell divi-sion markers could not resolve a subterminal region with high mitotic activity and exemplified in Onychophora, the existence of a subterminal proliferation zone has even been denied (Mayer et al. 2010). Moreover, biological elongation processes do not necessarily imply cell division; cell migration, cell rearrangements, and changes in cell shape have also been described as mechanism for growth in animal embryos (see Scholtz 2002, Fig. 5; Keller 2006) including arthropods (Liu and Kaufman 2005; Chipman 2008). Finally, many authors conceptualize the growth zone as a subterminal region that is adding segments by a combination of cell proliferation and pattern formation (e.g. Martin and Kimelman 2009). In the following, we discuss these problems from an analytical morphological perspective.

The clearest expression of a growth zone is apparently seen in germ bands with teloblasts (Fig. 4.7). Teloblasts are relatively large stem cells, which divide asymmetrically and only in anterior direction, i.e., the budded smaller daughter cells lie anterior to the teloblasts (Dohle et al. 2004). This way teloblasts produce chains of cells along the longitudinal axis of the germ band (Fig. 4.7). Teloblasts have been described in the ectoderm and the mesoderm layers. Among arthropods, the occurrence of ectoteloblasts and mesoteloblasts seems restricted to malacostracan crustaceans. Reports on teloblasts in other crustaceans are either based on the view that every growth of germ bands is called teloblastic, even in the absence of specialized teloblast cells, or the putative existence of teloblast cells requires confirmation (Dohle et al. 2004). Hence, it is most likely that teloblasts are an apomorphy of Malacostraca (Fischer et al. 2010). In the course of malacostracan evolution, number and arrangement of ectoteloblasts underwent some changes from a circular arrangement of 19 to a circular arrangement of about 40 or a linear arrangement of variable numbers up to an apomorphic loss of ectoteloblasts in Amphipoda (Dohle and Scholtz 1988; Scholtz and Dohle 1996; Dohle et al. 2004). In contrast to this, mesoteloblasts are more conservative. Despite the change from a plesiomorphic circular to a linear arrangement, their number is always eight. Since the post-

naupliar germ band of malacostracans is also characterized by a stereotyped cell division pattern and a grid-like arrangement of cells, malacostracan embryos offer a high resolution of the processes related to germ band growth. It is clear that at the cellular level two processes are involved in germ band elongation (Dohle and Scholtz 1988; Scholtz and Dohle 1996). One is the proliferation of cells from the posteriorly situated teloblast cells; the other is a kind of intercalary growth within each of the transverse cell rows. This intercalary elongation is achieved by two rounds of longitudinally oriented cell division cycles of the daughter cells of the ectodermal teloblasts (Bergh 1893; McMurrich 1895; Manton 1928; Dohle and Scholtz 1988; Scholtz and Dohle 1996; Fischer et al. 2010) (Fig. 4.7). Hence, the contribution of these intercalary cell divisions to longitudinal germ band growth is three times larger than that of the ectoteloblasts, if cell numbers are considered. Even during the subsequent differential cleavage, which leads to the formation of segmental structures, some elongation takes place. Similar processes have been described for the cells in annelid germ bands (Shankland 1999; Shimizu and Nakamoto 2001). Hence, there is no known germ band elongation involving only posterior growth. It is always a combination of posterior proliferation and intercalary divisions.

If these findings are extrapolated to short germ development in other arthropods, which do not differentiate teloblasts of some sort, then the posterior growth zone, even if present, would be difficult to identify with the application of markers of cell division because the intercalary division would be marked as well, resulting in a more or less homogeneous distribution. A way to recognize the existence of a posterior growth zone would be to analyze the size and distribution of clones produced by single cells in the posterior region of the early germ band. If these clones are found along a larger portion of the length of the germ band, then one can possibly speak of a growth zone. This approach requires single cell labelling, 4D microscopy, or similar techniques. If a scattered distribution of clones is found, then there is no regionalized growth zone.

It might well be the result of this kind of investigations that the concept of a subterminal growth zone has to be restricted to teloblastic growth. However, given the large amount of intercalary elongation, the teloblastic proliferation is just one (although a special one) of several areas of cell division activity involved in longitudinal germ band growth. Hence, this weakens the concept of a subterminal growth or proliferation zone even more. Nevertheless, the term 'growth zone' can perhaps be rescued if it is meant to characterize the embryonic region from which the cells originate (encompassing some growth) that subsequently are involved in elongation and segment formation.

## 4.13 Germ Band Formation and Differentiation

It is often stated that the posterior growth zone buds or produces segments, i.e., growth and patterning is initiated simultaneously by the growth zone. However, for several reasons this view is problematic. First, it is evident that longitudinal growth of the embryo is phylogenetically older than segmentation. It is likely that longitudinal growth evolved together with the adult bilaterian body organization (Jacobs et al. 2005), whereas an arthropod-like segmentation appeared only later in evolution. Second, it is by no means obvious to what degree information for segmental patterning, if at all, is implemented in the growth zone. This is not a trivial problem. Experiments affecting the growth zone by cell ablation (Price et al. 2010) or suppression of gene expression (Copf et al. 2004) necessarily have an impact on segmentation as well. If there are no cells that can form segments, segmental structures cannot be formed either. Third, as mentioned above, it is not at all clear whether in most arthropods there is a posterior growth zone.

Based on comparative morphological analyses in malacostracan crustaceans, it appears rather that the growth zone or longitudinal growth in general generates competent cell material, which eventually becomes segmented (Dohle 1972; Scholtz 1992). Indications for this

are found in the heterochronies between cell proliferation and molecular and morphological segmentation within the germ band of a species or between those of different species (Scholtz 1992; Scholtz and Dohle 1996). Hence, morphologically, two processes can be discriminated that are involved in germ band development. One is the formation and growth of the germ band. The other is germ band differentiation into metamerically arranged structures, which in combination lead to what is called segmentation. This postulated dissociation of the growth and segmentation processes finds support by recent studies in anostracan crustacean larvae (Williams et al. 2012). An experimental disruption of Notch signalling leads to suppression of segmentation, but longitudinal growth is largely unaffected (Williams et al. 2012). Furthermore, not all metamerically formed structures are situated in a strictly registered position. There are some examples for so-called segmental mismatch such as the polypody in notostracan crustaceans and the diplopody in diplopod myriapods (Linder 1952; Dohle 1964; Janssen et al. 2006; see also Chap. 9). In both cases, there is more than one leg pair and other ventral structures such ganglia and muscles associated with one tergite; up to ten in the notostracan post-genital region and two in most of the diplopod trunk segments. If posterior growth was inextricably linked to segmental patterning, segmental mismatch would be difficult to achieve.

The posterior growth zone produces cells towards anterior. In contrast, differentiation follows a general anterior-posterior direction, i.e., the anterior segmental structures differentiate before more posterior ones. However, it is not necessarily the anteriormost head segment the one in which differentiation starts; sometimes it is the mandibular, maxillary or first thoracic segment (Patel et al. 1989; Fleig 1990; Scholtz et al. 1994; Manzanares et al. 1996; Chipman et al. 2004; Janssen 2012). These differences suggest again that longitudinal growth and segmentation show a certain degree of independence. Otherwise, one would expect that differentiation begins in relation to the timing of the cells budded by the growth zone, i.e., the earliest cells that originated from the growth zone are the first to differentiate into segmental structures.

## 4.14  Segment Precursors

As a general mode, arthropod segments are formed one by one in a general anteroposterior sequence. This is congruently indicated by the expression of segment polarity genes, cell division patterns, and morphogenesis. An antero-posterior gradient is even found in *Drosophila* with its almost simultaneous segmentation of its long germ band (DiNardo et al. 1985). Segments, however, are not the first signs of a metameric subdivision of the germ band. Based on clonal analyses, morphogenesis, and gene expression data in *Drosophila*, the concept of the parasegment as the primary metameric unit has been developed (see Lawrence 1992). The parasegment does not match the segment but contributes to parts of two adjacent segments, the posterior compartment of the anterior segment and the anterior compartment of the posterior segment (Lawrence 1992). The row formation and the clonal situation in malacostracan germ bands resemble the *Drosophila* parasegment (Fig. 4.7) and investigations on the expression patterns of a variety of segmental genes in a spider indicate that the parasegment might be a general arthropod feature (Damen 2002). In contrast, ideas about merosegmentation, i.e., the subdivision of primary segments into secondary segments (Minelli 2001, 2005) suffer from a lack of developmental evidence and depend very much on what we define as a segment. For instance, if the expression of pair-rule genes with their characteristic two-segmented periodicity is considered as segmentation, then the subsequent subdivision into the final number of morphological segments can be seen as merosegmentation. Likewise, if the external cuticle between two moults does not correspond to the underlying segment anlagen at the cellular level, then the impression occurs that one segment is suddenly subdivided into two or

more secondary segments, although the segment anlagen were formed one by one. It is the question whether this shift between different levels (molecular-cellular and cellular-cuticular) is meaningful in terms of explanatory power and whether any of these subdivisions should be called segment. For instance, the mismatch between tergites and ventral structures in Diplopoda can be explained with an independent ventral and dorsal patterning system (Janssen et al. 2006) and does not require the assumption of underlying merosegmentation. Malacostracan crustacean development with teloblastic germ band growth and the stereotyped cell division pattern shows that the assumption of a general field subdivided by merosegmentation is unlikely. It would mean that a field extending over 15 segments (segments of 2nd maxilla to uropods) would be condensed within each ectotel-oblast cell. This is unlikely and ablation experiments with intra-germ layer compensation of ablated tissue contradict this view (Price et al. 2010).

## 4.15  Embryonic and Post-Embryonic Growth

Longitudinal growth and the addition of segments occur either during embryogenesis, in a post-embryonic phase, or as a mixture of both. All these modes are found in arthropods (see also Chap. 9). Species with epimorphic development hatch with the complete number of segments. This is the case in most hexapods (with the notable exception of the Protura), in direct developing crustaceans such as amphipods, isopods, freshwater crayfish, and cladocerans, in geophilomorph and scolopendromorph centipedes among myriapods, and in most arachnid groups. In contrast to this, anamorphic development is characterized by hatchlings, which lack some or most of the segments. Anamorphosis is found in all crustaceans with larval development including Euphausiacea and Dendrobranchiata among the malacostracans, in xiphosurans, in mites, and most myriapods. There seems to be no fundamental difference between embryonic and post-embryonic growth and segment formation (see Hughes and Kaufman 2002c), neither at the level of gene expression, nor concerning morphogenesis (e.g. compare Scholtz et al. 1994 and Manzanares et al. 1996). Furthermore, it has been shown in some cases that the anlagen of those segments that occur during post-embryonic anamorphic growth are already formed and differentiated to some degree in the embryo (e.g. Barnett and Thomas 2012). This renders the difference between embryonic and post-embryonic segmentation, at least partly, an issue of different resolution. On the one hand, we have the cellular and morphogenetic level, on the other hand, there is the observation of segment addition correlated to moulting. It appears that the major difference is the timing of hatching and the differentiation of structures which allow the free-living mode.

## 4.16   A Germ Band in the Ground Pattern of Arthropods?

Since a germ band is found in the ontogeny of representatives of every large arthropod subgroup, it has been considered as part of the arthropod ground pattern (Scholtz 1997). However, there are a number of cases, in particular among crustaceans such as Dendrobranchiata and Euphausiacea where after total cleavage of yolk-free eggs the larvae form directly and three-dimensionally. Hence, a germ band in a strict sense is not formed (Fig. 4.6). A corresponding phenomenon occurs in annelids in which the three-dimensional trochophore development has been switched to germ bands in yolky eggs, for example, in clitellates or among polychaetes (von Wistinghausen 1891; Seaver et al. 2005; Mayer et al. 2010). Based on these examples, one can conclude that the occurrence of a germ band is related to a certain amount of yolk. Hence, the occurrence or absence of a germ band is prone to convergence and in arthropods, there is a back and forth change between yolky and less yolky eggs. If we assume that the pattern shared by larval crustaceans and

pycnogonids might be ancestral within arthropods (see Waloszek and Maas 2005), then a germ band is not part of the arthropod ground pattern (see Scholtz 1997). In any case, the extreme short germ development as observed in *Schistocerca* and relatives seems to be derived within arthropods. The plesiomorphic condition appears to be the intermediate growth mode with the anlagen of a few anterior segments, irrespective of the existence of a true germ band (Scholtz 1997). Apart from the fact that the phylotypic stage is a problematic typological concept (see Scholtz 2005), the putative absence of a germ band in the arthropod stem species renders the concept of the germ band as phylotypic stage even more problematic.

**Acknowledgments**  The authors thank the editors for the invitation to contribute to this book. We are grateful to Frederike Alwes and Petra Ungerer for a number of photographs. The valuable comments of Carlo Brena and an anonymous reviewer are greatly appreciated.

## References

Akam M (2000) Arthropods: developmental diversity within a (super) phylum. Proc Natl Acad Sci USA 97:4438–4441

Alwes F, Scholtz G (2004) Cleavage and gastrulation of the euphausiacean *Meganyctiphanes norvegica* (Crustacea, Malacostraca). Zoomorphology 123:125–137

Alwes F, Scholtz G (2006) Stages and other aspects of the embryology of the parthenogenetic Marmorkrebs (Decapoda, Reptantia, Astacida). Dev Genes Evol 216:169–184

Amma K (1911) Über die Differenzierung der Keimbahnzellen bei den Copepoden. Arch Zellforsch 6:497–576

Anderson DT (1965) Embryonic and larval development and segment formation in Ibla quadrivalvis (Cuv.). Austr J Zool 13:1–15

Anderson DT (1966) The comparative early embryology of the Oligochaeta, Hirudinea and Onychophora. Proc Linn Soc New South Wales 91:10–43

Anderson DT (1969) On the embryology of the cirripede crustaceans *Tetraclita rosea* (Krauss), *Tetraclita purpurascens* (Wood), *Chthamalus antennatus* Darwin and *Chamaesipho columna* (Spengler) and some considerations of crustacean phylogenetic relationships. Phil Trans R Soc B 256:183–235

Anderson DT (1973) Embryology and Phylogeny in Annelids and Arthropods. Pergamon, Oxford

Barnett AA, Thomas RH (2012) The delineation of the fourth walking leg segment is temporally linked to posterior segmentation in the mite *Archegozetes longisetosus* (Acari: Oribatida, Trhypochthoniidae). Evol Dev 14:383–392

Benesch R (1969) Zur Ontogenie und Morphologie von *Artemia salina* L. Zool Jb Anat 86:307–458

Bergh RS (1893) Beiträge zur Embryologie der Crustaceen. I. Zur Bildungsgeschichte des Keimstreifens von *Mysis*. Zool Jb Anat 6:491–528

Biffis C, Alwes F, Scholtz G (2009) Cleavage and gastrulation of the dendrobranchiate shrimp *Penaeus monodon* (Crustacea, Malacostraca, Decapoda). Arthropod Struct Dev 38:527–540

Bigelow MA (1902) The early development of *Lepas*. A study of cell lineage and germ layers. Bull Mus Comp Zool Harvard 40:61–144

Bock E (1942) Wechselbeziehungen zwischen den Keimblättern bei der Organbildung von *Chrysopa perla* (L.) I. Die Entwicklung des Ektoderms in mesodermdefekten Keimteilen. W Roux's Arch EntwMech Org 141:159–279

Braband A, Richter S, Hiesel R, Scholtz G (2002) Phylogenetic relationships within the Phyllopoda (Crustacea, Branchiopoda) based on mitochondrial and nuclear markers. Mol Phylogenet Evol 25:229–244

Brauer A (1895) Beiträge zur Kenntnis der Entwicklungsgeschichte des Skorpions II. Z Wiss Zool 59:351–435

Bregazzi PK (1973) Embryological development in *Tryphosella kergueleni* (Miers) and *Cheirimedon femoratus* (Pfeffer) (Crustacea: Amphipoda). Br Antarct Surv Bull 32:63–74

Brena C, Akam M (2012) The embryonic development of the centipede *Strigamia maritima*. Dev Biol 363:290–307

Brenneis G, Arango CP, Scholtz G (2011) Morphogenesis of *Pseudopallene* sp. (Pycnogonida, Callipallenidae) I: Embryonic development. Dev Genes Evol 221:309–328

Campbell LI, Rota-Stabelli O, Edgecombe GD, Marchioro T, Longhorn SJ, Telford MJ, Philippe H, Rebecchi L, Peterson KJ, Pisani D (2011) MicroRNAs and phylogenomics resolve the relationships of Tardigrada and suggest that velvet worms are the sister group of Arthropoda. Proc Natl Acad Sci USA 108:15920–15924

Campos-Ortega JA, Hartenstein V (1985) The Embryonic Development of *Drosophila melanogaster*. Springer, Berlin

Cannon HG (1921) The early development of the summer egg of a cladoceran (*Simocephalus vetulus*). Quart J Micr Sci 65:627–642

Castellani C, Maas A, Waloszek D, Haug JT (2011) New pentastomids from the late Cambrian of Sweden–deeper insights of the ontogeny of fossil tongue worms. Palaeontographica A 293:95–145

Chen J, Braun A, Waloszek D, Peng Q, Maas A (2004) Lower Cambrian yolk-pyramid embryos from southern Shaanxi, China. Progr Nat Sci 14:167–172

Chipman AD (2008) Thoughts and speculations on the ancestral arthropod segmentation pathway. In: Minelli A, Fusco G (eds) Evolving Pathways. Cambridge University Press, Cambridge, pp 343–358

Chipman AD, Arthur W, Akam M (2004) Early development and segment formation in the centipede, *Strigamia maritima* (Geophilomorpha). Evol Dev 6:78–89

Claypole AM (1898) The embryology and oogenesis of *Anurida maritima*. J Morphol 14:219–300

Copf T, Schröder R, Averof M (2004) Ancestral role of *caudal* genes in axis elongation and segmentation. Proc Natl Acad Sci USA 101:17711–17715

Damen WGM (2002) Parasegmental organization of spider embryos implies that the parasegment is an evolutionary conserved entity in arthropod embryogenesis. Development 129:1239–1250

Davis GK, Patel NH (2002) Short, long, and beyond: molecular and embryological approaches to insect segmentation. Annu Rev Entomol 47:669–699

Dawydoff C (1928) Traité d'embryologie comparée des Invertébrés. Masson, Paris

Dearden PK, Donly C, Grbic M (2002) Expression of pair-rule gene homologues in a chelicerate: early patterning of the two-spotted spider mite *Tetranychus urticae*. Development 129:5461–5472

Delsman HC (1917) Die Embryonalentwicklung von *Balanus balanoides* Linn. Tijdschr Ned Dierk Ver 2nd series 15:419–520

DiNardo S, Kuner JM, Theis J, O'Farell PH (1985) Development of embryonic pattern in *D. melanogaster* as revealed by accumulation of the nuclear *engrailed* protein. Cell 43:59–69

Dogiel V (1913) Embryologische Studien an Pantopoden. Z wiss Zool 107:575–741

Dohle W (1964) Die Embryonalentwicklung von *Glomeris marginata* (Villers) im Vergleich zur Entwicklung anderer Diplopoden. Zool Jb Anat 81:241–310

Dohle W (1970) Die Bildung und Differenzierung des postnauplialen Keimstreifs von *Diastylis rathkei* (Crustacea, Cumacea) I. Die Bildung der Teloblasten und ihrer Derivate. Z Morph Ökol Tiere 67:307–392

Dohle W (1972) Über die Bildung und Differenzierung des postnauplialen Keimstreifs von *Leptochelia* spec. (Crustacea, Tanaidacea). Zool Jb Anat 89:503–566

Dohle W (1979) Vergleichende Entwicklungsgeschichte des Mesoderms bei Articulaten. Fortschr zool Syst Evolutionsf 1:120–140

Dohle W, Gerberding M, Hejnol A, Scholtz G (2004) Cell lineage, segment differentiation, and gene expression in crustaceans. In: Scholtz G (ed) Evolutionary Developmental Biology of Crustacea (Crustacean Issues 15). Balkema, Lisse, pp 95–133

Dohle W, Scholtz G (1988) Clonal analysis of the crustacean segment: the discordance between genealogical and segmental borders. Development 104 Suppl.:147–160

Donoghue PCJ, Bengtson S, Dong X-P, Gostling NJ, Huldtgren T, Cunningham JA, Yin C, Yue Z, Peng F,

Stampanoni M (2006) Synchrotron X-ray tomographic microscopy of fossil embryos. Nature 442:680–683

Edgecombe GD (2010) Arthropod phylogeny: an overview from the perspectives of morphology, molecular data and the fossil record. Arthropod Struct Dev 39:74–87

Edgecombe GD, Giribet G (2004) Adding mitochondrial sequence data (16S rRNA and cytochrome *c* oxidae subunit I) to the phylogeny of centipedes (Myriapoda: Chilopoda): an analysis of morphology and four molecular loci. J Zool Syst Evol Res 42:89–134

Edgecombe GD, Giribet G, Dunn CW, Hejnol A, Kristensen RM, Neves RC, Rouse GW, Worsaae K, Sørensen MV (2011) Higher-level metazoan relationships: recent progress and remaining questions. Org Divers Evol 11:151–172

Eriksson BJ, Tait NN (2012) Early development in the velvet worm *Euperipatoides kanangrensis* Reid 1996 (Onychophora: Peripatopsidae). Arthropod Struct Dev 41:483–493

Fioroni P (1970) Am Dotteraufschluß beteiligte Organe und Zelltypen bei höheren Krebsen; der Versuch einer einheitlichen Terminologie. Zool Jb Anat 87:481–522

Fioroni P (1987) Allgemeine und vergleichende Embryologie der Tiere. Springer, Berlin

Fischer A, Pabst T, Scholtz G (2010) Germ band differentiation in the stomatopod *Gonodactylaceus falcatus* and the origin of the stereotyped cell division pattern in Malacostraca (Crustacea). Arthropod Struct Dev 39:411–422

Fleig R (1990) *Engrailed* expression and body segmentation in the honeybee *Apis mellifera*. Roux's Arch Dev Biol 198:467–473

Frasch M (1999) Intersecting signaling and transcriptional pathways in *Drosophila* heart specification. Semin Cell Dev Biol 10:61–71

Fuchs K (1914) Die Keimblätterentwicklung von *Cyclops viridis* Jurine. Zool Jb Anat 38:104–156

Gabriel WN, McNuff R, Patel SK, Gregory TR, Jeck WR, Jones CD, Goldstein B (2007) The tardigrade *Hypsibius dujardini*, a new model for studying the evolution of development. Dev Biol 312:545–559

George RY, Strömberg J-O (1985) Development of eggs of Antarctic krill *Euphausia superba* in relation to pressure. Pol Biol 4:125–133

Gerberding M (1994) Superfizielle Furchung, Bildung des Keimstreifs und Differenzierung von Neuroblasten bei Leptodora kindti Focke 1844 (Cladocera, Crustacea). Diplom thesis, Freie Universität Berlin

Gerberding M, Browne WE, Patel NH (2002) Cell lineage analysis of the amphipod crustacean *Parhyale hawaiensis* reveals an early restriction of cell fates. Development 129:5789–5801

Gilbert SF, Raunio AM (eds) (1997) Embryology: constructing the organism. Sinauer Associates, Sunderland

Giribet G, Edgecombe GD, Carpenter JM, D'Haese CA, Wheeler WC (2004) Is Ellipura monophyletic? A combined analysis of basal hexapod relationships with emphasis on the origin of insects. Org Divers Evol 4:319–340

Giribet G, Edgecombe GD, Wheeler WC (2001) Arthropod phylogeny based on eight molecular loci and morphology. Nature 413:157–161

Gorham FP (1895) The cleavage of the egg of *Virbius zostericola*, Smith—a contribution to crustacean cytogeny. J Morph 11:741–746

Grbic M (2003) Polyembryony in parasitic wasps: evolution of a novel mode of development. Int J Dev Biol 47:633–642

Haget A (1953) Analyse expérimentale des facteurs de la morphogenèse embryonnaire chez le coléoptère *Leptinotarsa*. Bull Biol Fr Belg 87:123–217

Hannibal RL, Price AL, Patel NH (2012) The functional relationship between ectodermal and mesodermal segmentation in the crustacean, *Parhyale hawaiensis*. Dev Biol 361:427–438

Heider K (1889) Die Embryonalentwicklung von *Hydrophilus piceus* L. Fischer, Jena

Hejnol A, Schnabel R (2005) The eutardigrade *Thulinia stephaniae* has an indeterminate development and the potential to regulate early blastomere ablations. Development 132:1349–1361

Hertzel G (1985) Die Embryonalentwicklung von *Lithobius forficatus* (L.) im Vergleich zur Entwicklung anderer Chilopoden. Dissertation, Pädagogische Hochschule Erfurt-Mühlhausen

Hertzler PL, Clark WH (1992) Cleavage and gastrulation in the shrimp *Sicyonia ingentis*: invagination is accompanied by oriented cell division. Development 116:127–140

Heymons R (1901) Die Entwicklungsgeschichte der Scolopender. Zoologica 33:1–244

Hickman VV (1937) The embryology of the syncarid crustacean *Anaspides tasmaniae*. Pap Proc Roy Soc Tasmania 1936:1–35

Hughes CL, Kaufman TC (2002a) Hox genes and the evolution of the arthropod body plan. Evol Dev 4:459–499

Hughes CL, Kaufman TC (2002b) Exploring the myriapod body plan: expression patterns of the ten Hox genes in a centipede. Development 129:1225–1238

Hughes CL, Kaufman TC (2002c) Exploring myriapod segmentation: the expression patterns of *even-skipped*, *engrailed*, and *wingless* in a centipede. Dev Biol 247:47–61

Ivanov PP (1933) Die embryonale Entwicklung von *Limulus moluccanus*. Zool Jb Anat 56:163–348

Jacobs DK, Hughes NC, Fitz-Gibbon ST, Winchella CJ (2005) Terminal addition, the Cambrian radiation and the Phanerozoic evolution of bilaterian form. Evol Dev 7:498–514

Janssen R (2012) Segment polarity gene expression in a myriapod reveals conserved and diverged aspects of early head patterning in arthropods. Dev Genes Evol 222:299–309

Janssen R, Prpic N-M, Damen WGM (2006) A review of the correlation of tergites, sternites, and leg pairs in diplopods. Front Zool 2:3; doi:10.1186/1742-9994-3-2

Jenner RA (2010) Higher-level crustacean phylogeny: consensus and conflicting hypotheses. Arthropod Struct Dev 39:143–153

Jura Cz (1965) Embryonic development of *Tetrodontophora bielanensis* (Waga) (Collembola) from oviposition till germ band formation stage. Acta Biol Cracov Ser Zool 8:141–157

Jura Cz, Krzystofowizc A, Kisiel E (1987) Embryonic development of *Tetrodontophora bielanensis* (Collembola): Descriptive with scanning electron micrographs. In: Ando H, Jura CZ (eds) Recent Advances in Insect Embryology in Japan and Poland. Arthropod Embryolog Soc Japan, Tsukuba, pp 77–124

Kaudewitz F (1950) Zur Entwicklungsphysiologie von *Daphnia pulex*. Roux' Arch Entwickl Mech 144:410–447

Kanayama M, Akiyama-Oda Y, Oda H (2010) Early embryonic development in the spider *Achaearanea tepidariorum*: Microinjection verifies that cellularization is complete before the blastoderm stage. Arthropod Struct Dev 39:436–445

Keller R (2006) Mechanisms of elongation in embryogenesis. Development 133:2291–2302

Kimble M, Coursey Y, Ahmad N, Hinsch GW (2002) Behavior of the yolk nuclei during embryogenesis, and development of the midgut diverticulum in the horseshoe crab *Limulus polyphemus*. Invert Biol 12:365–377

Kingsley JS (1892) The embryology of Limulus. J Morph 7:35–66

Knoll HJ (1974) Untersuchungen zur Entwicklungsgeschichte von *Scutigera coleoptrata* L. (Chilopoda). Zool Jb Anat 92:47–132

Kohler H-J (1976) Embryologische Untersuchungen an Copepoden: die Entwicklung von *Lernaeocera branchialis* L. 1767 (Crustacea, Copepoda, Lernaeoida, Lernaeidae). Zool Jb Anat 95:448–504

Kozloff EN (2007) Stages of development, from first cleavage to hatching, of an *Echinoderes* (Phylum Kinorhyncha: Class Cyclorhagida). Cah Biol Mar 48:199–206

Krause G (1939) Die Eitypen der Insekten. Biol Zentralbl 59:495–536

Kristensen NP (1997) The groundplan and basal diversification of the hexapods. In: Fortey RA, Thomas RH (eds) Arthropod Relationships. Chapman & Hall, London, pp 281–293

Kühn A (1912) Die Sonderung der Keimesbezirke in der Entwicklung der Sommereier von *Polyphemus pediculus* de Geer. Zool Jb Anat 35:243–340

Laumann M, Bergmann P, Norton RA, Heethoff M (2010a) First cleavages, preblastula and blastula in the parthenogenetic mite *Archegozetes longisetosus* (Acari, Oribatida) indicate holoblastic rather than superficial cleavage. Arthropod Struct Dev 39:276–286

Laumann M, Norton RA, Heethoff M (2010b) Acarine embryology: inconsistencies, artificial results and misinterpretations. Soil Organ 82:217–235

Lawrence PA (1992) The Making of a Fly. Blackwell, Oxford

Lerebouillet A (1862) Recherches d'embryologie comparée sur le développement du brochet, de la perche et de l'écrevisse. Mém Acad Sci Inst Fr 17:1–356

Linder F (1952) Contributions to the morphology and taxonomy of the Branchiopoda Notostraca, with special reference to the north American species. Proc U S Nat Mus 102:1–69

Liu PZ, Kaufman TC (2005) Short and long germ segmentation: unanswered questions in the evolution of a developmental mode. Evol Dev 7:629–646

Machida R (2006) Evidence from embryology for reconstructing the relationships of hexapod basal clades. Arthropod Syst Phyl 64:95–104

Machida R, Nagashima T, Ando H (1990) The early embryonic development of the jumping bristletail *Pedetontus unimaculatus* Machida (Hexapoda: Microcoryphia, Machilidae). J Morph 206:181–195

Malakhov VV, Spiridonov SE (1984) The embryogenesis of *Gordius* sp. from Turkmenia, with special reference to the position of the Nematomorpha in the animal kingdom. Zool Zh 63:1285–1297 [Russian with English summary]

Mallatt JM, Garey JR, Shultz JW (2004) Ecdysozoan phylogeny and Bayesian inference: first use of nearly complete 28S and 18S rRNA gene sequences to classify the arthropods and their kin. Mol Phylogenet Evol 31:178–191

Manton SM (1928) On the embryology of a mysid crustacean *Hemimysis lamornae*. Phil Trans R Soc B 216:363–463

Manton SM (1934) On the embryology of *Nebalia bipes*. Phil Trans R Soc B 223:168–238

Manton SM (1949) Studies on the Onychophora VII. The early embryonic stages of *Peripatopsis* and some general considerations concerning the morphology and phylogeny of the Arthropoda. Phil Trans R Soc B 233:483–580

Manzanares M, Williams TA, Marco R, Garesse R (1996) Segmentation in the crustacean *Artemia*: *engrailed* expression studied with an antibody raised against the *Artemia* protein. Roux's Arch Dev Biol 205:424–431

Marcus E (1929) Zur Embryologie der Tardigraden. Zool Jb Anat 50:333–384

Martin BL, Kimelman D (2009) Wnt signaling and the evolution of embryonic posterior development. Curr Biol 19:R215–R219

Mayer G, Kato C, Quast B, Chisholm RH, Landman KA, Quinn LM (2010) Growth patterns in Onychophora (velvet worms): lack of a localised posterior Proliferation zone. BMC Evol Biol 10:339; doi:10.1186/1471-2148-10-339

McMurrich JP (1895) Embryology of the isopod Crustacea. J Morph 11:63–154

Meschenmoser M (1996) Dorsal- und Lateralorgane in der Embryonalentwicklung von Peracariden (Crustacea, Malacostraca). Cuvillier Verlag, Göttingen

Meusemann K, von Reumont BM, Simon S, Roeding F, Strauss S, Kück P, Ebersberger I, Walzl M, Pass G, Breuers S, Achter V, von Haeseler A, Burmester T, Hadrys H, Wägele JW, Misof B (2010) A phylogenomic approach to resolve the arthropod tree of life. Mol Biol Evol 27:2451–2464

Minelli A (2001) A three-phase model of arthropod segmentation. Dev Genes Evol 211:509–521

Minelli A (2005) A morphologist's perspective on terminal growth and segmentation. Evol Dev 7:568–573

Mittmann B, Wolff C (2012) Embryonic development and staging of the cobweb spider *Parasteatoda tepidariorum* C. L. Koch, 1841 (syn.: *Achaearanea tepidariorum*; Araneomorphae; Theridiidae). Dev Genes Evol 222:189–216

Møller OS, Olesen J, Avenant-Oldewage A, Thompson PF, Glenner H (2008) First maxillae suction discs in Branchiura (Crustacea): development and evolution in light of the first molecular phylogeny of Branchiura, Pentastomida, and other "Maxillopoda". Arthropod Struct Dev 37:333–346

Montuy-Gómez D, Gómez-Gutiérrez J, Rodríguez-Jaramillo C, Robinson CJ (2012) *Nyctiphanes simplex* (Crustacea: Euphausiacea) temporal association of embryogenesis and early larval development with female molt and ovarian cycles. J Plankt Res 34:531–547

Morgan TH (1891) A contribution to the embryology and phylogeny of the Pycnogonida. Stud Biol Lab J Hopkins Univ 5:1–76

Moritz M (1957) Zur Embryonalentwiclung der Phalangiiden (Opiliones, Palpatores) unter besonderer Berücksichtigung der äußeren Morphologie, der Bildung des Mitteldarmes und der Genitalanlage. Zool Jb Anat 76:331–370

Moritz M (1993) Unterstamm Arachnata. In: Gruner H-E (ed) Lehrbuch der speziellen Zoologie, Band I, 4.Teil Arthropoda. Gustav Fischer, Jena, pp. 64–442

Müller YMR (1984) Die Embryonalentwicklung von *Macrobrachium carcinus* (L.) (Malacostraca, Decapoda, Natantia). Zool Jb Anat 112:51–78

Müller-Calé K (1913) Die Entwicklung von *Cypris incongruens*. Zool Jb Anat 36:113–170

Nielsen C (2001) Animal Evolution: Interrelationships of the Living Phyla, 2nd edn. Oxford University Press, Oxford

Nielsen C (2012) Animal Evolution: Interrelationships of the Living Phyla, 3rd edn. Oxford University Press, Oxford

Olesen J (2009) Phylogeny of Branchiopoda (Crustacea): character evolution and contribution of uniquely preserved fossils. Arthr Syst Phyl 67:3–39

Osche G (1963) Die systematische Stellung und Phylogenie der Pentastomida–embryologische und vergleichend-anatomische Studien an *Reighardia sternae*. Z Morph Ökol Tiere 52:487–596

Patel NH, Condron BG, Zinn K (1994) Pair-rule expression patterns of *even-skipped* are found in both short- and long-germ beetles. Nature 367:429–434

Patel NH, Kornberg TB, Goodman CS (1989) Expression of *engrailed* during segmentation of grasshopper and crayfish. Development 107:201–212

Paulus H (2007) Arthropoda. In: Westheide W, Rieger R (eds) Spezielle Zoologie, Teil 1: Einzeller und wirbellose Tiere. Spektrum, München, pp 438–446

Peel AD, Chipman AD, Akam M (2005) Arthropod segmentation: beyond the *Drosophila paradigm*. Nature Rev Gen 6:905–916

Peterson KJ, Eernisse DJ (2001) Animal phylogeny and the ancestry of bilaterians: inferences from morphological and 18S rDNA sequences. Evol Dev 3:170–205

Pflugfelder O (1962) Lehrbuch der Entwicklungsgeschichte und Entwicklungsphysiologie der Tiere. VEB Gustav Fischer, Jena

Pisani D, Poling LL, Lyons-Weiler M, Hedges SB (2004) The colonization of land by animals: molecular phylogeny and divergence times among arthropods. BMC Biol 2:1; doi:10.1186/1741-7007-2-1

Price AL, Modrell MS, Hannibal RL, Patel NH (2010) Mesoderm and ectoderm lineages in the crustacean *Parhyale hawaiensis* display intra-germ layer compensation. Dev Biol 341:256–266

Rathke H (1829) Untersuchungen über Bildung und Entwicklung des Flusskrebses. Leopold Voss, Leipzig

Regier JC, Shultz JW, Zwick A, Hussey A, Ball B, Wetzer R, Martin JW, Cunningham CW (2010) Arthropod relationships revealed by phylogenomic analysis of nuclear protein-coding sequences. Nature 463:1079–1083

Reichenbach H (1886) Studien zur Entwicklungsgeschichte des Flusskrebses. Abhandl Senckenb Naturforsch Gesellsch 14:1–137

Reid AL (1996) Review of the Peripatopsidae (Onychophora) in Australia, with comments on peripatopsid relationships. Invert Tax 10:663–936

Richter S, Olesen J, Wheeler WC (2007) Phylogeny of Branchiopoda (Crustacea) based on a combined analysis of morphological data and six molecular loci. Cladistics 23:301–336

Richter S, Scholtz G (2001) Phylogenetic analysis of the Malacostraca (Crustacea). J Zool Syst Evol Res 39:113–136

Samter M (1900) Studien zur Entwicklungsgeschichte der *Leptodora hyalina* Lillj. Z wiss Zool 63:169–260

Sanchez S (1959) Le développement des Pycnogonides et leurs affinités avec les Arachnides. Arch Zool Exp Gén 98:1–102

Sander K (1983) The Evolution of patterning mechanisms: gleanings from insect embryogenesis and spermatogenesis. In: Goodwin BC, Holder N, Wylie CG (eds) Development and Evolution. Cambridge University Press, Cambridge, pp 137–158

Sanders KL, Lee MS (2010) Arthropod molecular divergence times and the Cambrian origin of pentastomids. Syst Biodiv 8:63–74

Scheidegger G (1976) Stadien der Embryonalentwicklung von *Eupagurus prideauxi* Leach (Crustacea, Decapoda Anomura), unter besonderer

Berücksichtigung der Darmentwicklung und der am Dotterabbau beteiligten Zelltypen. Zool Jb Anat 95:297–353

Schimkewitsch L, Schimkewitsch W (1911) Ein Beitrag zur Entwicklungsgeschichte der Tetrapneumones, Teil I. Bull Acad Imp Sci St Petersb 6th series 5:637–654

Schimkewitsch W (1896) Studien über parasitische Copepoden. Z wiss Zool 61:339–362

Schimkewitsch W (1903) Über die Entwicklung von Telyphonus caudatus (L.). Zool Anz 26:665–685

Scholl G (1963) Embryologische Untersuchungen an Tanaidaceen (Heterotanais oerstedi Kröyer). Zool Jb Anat 80:500–554

Scholtz G (1984) Untersuchungen zur Bildung und Differenzierung des postnauplialen Keimstreifs von Neomysis integer Leach (Crustacea, Malacostraca, Peracarida). Zool Jb Anat 112:295–349

Scholtz G (1990) The formation, differentiation and segmentation of the post-naupliar germ band of the amphipod Gammarus pulex L. (Crustacea, Malacostraca, Peracarida). Proc R Soc Lond B 239:163-211

Scholtz G (1992) Cell lineage studies in the crayfish Cherax destructor (Crustacea, Decapoda): germ band formation, segmentation, and early neurogenesis. Roux's Arch Dev Biol 202:36–48

Scholtz G (1997) Cleavage, germ band formation and head segmentation: the ground pattern of the Euarthropoda. In: Fortey RA, Thomas RH (eds) Arthropod Relationships. Chapman & Hall, London, pp 317–332

Scholtz G (2002) The Articulata hypothesis—or what is a segment? Org Divers Evol 2:197–215

Scholtz G (2005) Homology and ontogeny: pattern and process in comparative developmental biology. Theory Biosci 124:121–143

Scholtz G, Abzhanov A, Alwes F, Biffis C, Pint J (2009a) Development, genes, and decapod evolution. In: Martin JW, Crandall KA, Felder DL (eds) Decapod Crustacean Phylogenetics (Crustacean Issues 18). CRC Press & Taylor and Francis, Boca Raton, pp 31–46

Scholtz G, Dohle W (1996) Cell lineage and cell fate in crustacean embryos—a comparative approach. Int J Dev Biol 40:211–220

Scholtz G, Patel NH, Dohle W (1994) Serially homologous engrailed stripes are generated via different cell lineages in the germ band of amphipod crustaceans (Malacostraca, Peracarida). Int J Dev Biol 38:471–478

Scholtz G, Ponomarenko E, Wolff C (2009b) Cirripede cleavage patterns and the origin of the Rhizocephala (Crustacea: Thecostraca). Arthropod Syst Phyl 67:219–228

Scholtz G, Wolff C (2002) Cleavage pattern, gastrulation, and germ disc formation of the amphipod crustacean Orchestia cavimana. Contrib Zool 71:9–28

Schulze J, Schierenberg E (2011) Evolution of embryonic development in nematodes. EvoDevo 2:18. doi: 10.1186/2041-9139-2-18

Schwartz V (1973) Vergleichende Entwicklungsgeschichte der Tiere. Georg Thieme, Stuttgart

Seaver EC, Thamm K, Hill SD (2005) Growth patterns during segmentation in two polychaete annelids, Capitella sp. I and Hydroides elegans: comparisons at distinct life history stages. Evol Dev 7:312–326

Seitz K-A (1966) Normale Entwicklung des Arachniden-Embryos Cupiennius salei (Keyserling) und seine Regulationsfähigkeit nach Röntgenbestrahlungen. Zool Jb Anat 83:327–447

Shankland M (1999) Anteroposterior pattern formation in the leech embryo. In: Moody SA (ed) Cell lineage and cell fate determination. Academic Press, San Diego, pp 207–224

Shear WA, Edgecombe GD (2010) The geological record and phylogeny of Myriapoda. Arthropod Struct Dev 39:174–190

Sheldon L (1887) On the development of Peripatus novae-zealandiae. Q J Microsc Sci 28:205–237

Shiino SM (1942) Studies on the embryonic development of Squilla oratoria de Haan. Mem Coll Sci Kyoto Univ B 28:77–174

Shimizu T, Nakamoto A (2001) Segmentation in annelids: cellular and molecular basis for metameric body plan. Zool Sci 18:285–298

Shultz JW (2007) A phylogenetic analysis of the arachnid orders based on morphological characters. Zool J Linn Soc 150:221–265

Siewing R (1979) Homology of cleavage-types? Fortschr zool Syst Evolutionsf 1:7–18

Sograff N (1883) Materials toward the knowledge of the embryonic development of Geophilus ferrugineus and Geophilus proximus. Izvêstya Imperatorskago Obshchestva Lyubitelei Estestvoznaniya, Antropologii i Etnografii pri Imperatorskom Moskovskom Universitete 2:1–77 (in Russian)

Strömberg JO (1971) Contribution to the embryology of bopyrid isopods with special reference to Bopyrides, Hemiarthrus and Pseudione (Isopoda, Epicaridea). Sarsia 47:1–46

Tardent P (1978) Coelenterata, Cnidaria. In: Seidel F (ed) Morphogenese der Tiere, Erste Reihe: Deskriptive Morphogenese, Lieferung I: A-I. VEB Gustav Fischer, Jena

Taube E (1909) Beiträge zur Entwicklungsgeschichte der Euphausiiden. I. Die Furchung des Eies bis zur Gastrulation. Z wiss Zool 92:427–464

Tiegs OW (1940) The embryology and affinities of the Symphyla, based on a study of Hanseniella agilis. Q J Microsc Sci 82:1–225

Tiegs OW (1947) The development and affinities of the Pauropoda, based on a study of Pauropus silvaticus. Q J Microsc Sci 88:275–336

Tojo K, Machida R (1998) Early embryonic development of the mayfly Ephemera japonica McLachlan (Insecta: Ephemeroptera, Ephemeridae). J Morph 238:327–335

Turquier Y (1967) L'embryogénèse de Trypetesa nassarioides Turquier (Cirripède Acrothoracique). Ses rapports avec celle des autres Cirripèdes. Arch Zool Exp Gén 108:111–137

Ungerer P, Scholtz G (2009) Cleavage and gastrulation in *Pycnogonum litorale* (Arthropoda, Pycnogonida): morphological support for the Ecdysozoa? Zoomorphology 128:263–274

Vollmer C (1912) Zur Entwicklung der Cladoceren aus dem Dauerei. Z wiss Zool 102:646–700

von Baldass F (1941) Entwicklung von *Daphnia pulex*. Zool Jb Anat 67:1–60

von Reumont BM, Jenner RA, Wills MA, Dell'Ampio E, Pass G, Ebersberger I, Meyer B, Koenemann S, Iliffe TM, Stamatakis A, Niehuis O, Meusemann K, Misof B (2012) Pancrustacean phylogeny in the light of new phylogenomic data: Support for Remipedia as the possible sister group of Hexapoda. Mol Biol Evol 29:1031–1045

von Wistinghausen C (1891) Untersuchungen über die Entwicklung von *Nereis dumerilii*. Mitt Zool Stat Neapel 10:41–74

Waloszek D, Maas A (2005) The evolutionary history of crustacean segmentation: a fossil-based perspective. Evol Dev 7:515–527

Weldon WFR (1892) The formation of the germ-layers in *Crangon vulgaris*. Q J Microsc Sci 33:343–363

Wennberg SA, Janssen R, Budd GE (2008) Early embryonic development of the priapulid worm *Priapulus caudatus*. Evol Dev 10:326–338

Weygoldt P (1960a) Mehrphasige Gastrulation bei Arthropoden. Zool Anz 164:381–395

Weygoldt P (1960b) Embryonaluntersuchungen an Ostracoden. Die Entwicklung von *Cyprideis litoralis*. Zool Jb Anat 78:369–426

Weygoldt P (1963) Grundorganisation und Primitiventwicklung bei Articulaten. Zool Anz 171:363–376

Weygoldt P (1964) Vergleichend-embryologische Untersuchungen an Pseudoskorpionen (Chelonethi). Z Morph Ökol Tiere 54:1–106

Weygoldt P (1975) Untersuchungen zur Embryologie und Morphologie der Geißelspinne *Tarantula marginemaculata* C. L. Koch (Arachnida, Amblypygi, Tarantulidae). Zoomorphologie 82:137–199

Weygoldt P (1979) Gastrulation in arthropods? Fortschr zool Syst Evolutionsf 1:73–81

Weygoldt P (1986) Arthropod interrelationships–the phylogenetic-systematic approach. Z Zool Syst Evolut-forsch 24:19–35

Weygoldt P, Paulus HF (1979) Untersuchungen zur Morphologie, Taxonomie und Phylogenie der Chelicerata. II. Cladogramme und die Entfaltung der Chelicerata. Morphologische Untersuchungen. Z zool Syst Evolut-forsch 17:117–200

Williams T, Blachuta B, Hegna TA, Nagy LM (2012) Decoupling elongation and segmentation: Notch involvement in anostracan crustacean segmentation. Evol Dev 14:372–382

Wolff C (2009) The embryonic development of the malacostracan crustacean *Porcellio scaber* (Isopoda, Oniscidea). Dev Genes Evol 219:545–564

Wolff C, Hilbrant M (2011) The embryonic development of the Central American wandering spider *Cupiennius salei*. Front Zool 8:15. doi:10.1186/1742-9994-8-15

Wolff C, Scholtz G (2002) Cell lineage, axis formation, and the origin of germ layers in the amphipod crustacean *Orchestia cavimana*. Dev Biol 250:44–58

Zantke J, Wolff C, Scholtz G (2008) Three-dimensional reconstruction of the central nervous system of *Macrobiotus hufelandi* (Eutardigrada, Parachela): implications for the phylogenetic position of Tardigrada. Zoomorphology 127:21–36

Zehnder H (1934) Über die Embryonalentwicklung des Flusskrebses. Acta Zool 15:261–408

Zilch R (1974) Die Embryonalentwicklung von *Thermosbaena mirabilis* Monod (Crustacea, Malacostraca, Pancarida). Zool Jb Anat 93:462–576

Zilch R (1979) Cell lineage in arthropods? Fortschr zool Syst Evolutionsforsch 1:19–41

Zrzavý J, Hypša V, Vlášková M (1998) Arthropod phylogeny: taxonomic congruence, total evidence and conditional combination approaches to morphological and molecular data sets. In: Fortey RA, Thomas RH (eds) Arthropod Relationships. Chapman & Hall, London, pp 97–107

# Arthropod Post-embryonic Development

**Alessandro Minelli and Giuseppe Fusco**

## Contents

A. Minelli (✉) · G. Fusco
Department of Biology, University of Padova,
Via U. Bassi 58 B, I 35131, Padova, Italy
e-mail: alessandro.minelli@unipd.it

G. Fusco
e-mail: giuseppe.fusco@unipd.it

## 5.1 A Neglected Developmental Time and Its Periodization

The study of arthropod post-embryonic development is a chapter of biology that requires a new conceptual framework. Some of the reasons behind the inadequacy of the prevailing approach are common to the study of post-embryonic development of all, or most, animals groups; others are specific to the Arthropoda.

Despite tremendous progress in developmental biology, in comparative developmental genetics especially, over the last three decades, the study of post-embryonic development still remains somehow neglected compared to the two segments of ontogeny on which attention is preferentially focussed: one is embryonic development, an increasingly obvious target of research, as a consequence of the increasing appreciation of the early developmental stage at which many body features are essentially determined, including those that will be eventually expressed at a much later date, sometimes in the adult phase only; and the other is the adult phase, too often acritically selected as the vantage point from which to read the whole course of development, thus reducing the embryonic and post-embryonic development to a sequence of preparatory stages. This perspective is encapsulated by the widespread use of the term 'imago'—to conceptualize the real, legitimate or ultimate expression of the species' essence—for the last developmental stage of insects and, less frequently,

A. Minelli et al. (eds.), *Arthropod Biology and Evolution*,
DOI: 10.1007/978-3-642-36160-9_5, © Springer-Verlag Berlin Heidelberg 2013

for the last developmental stage of arachnids (cf. Canard and Stockmann 1993).

Up to now, a justified focus on the developmentally earliest expression patterns and control cascades of genes responsible for different aspects of body patterning has resulted in unjustified neglect for the expression patterns and control cascades of developmental genes along the post-embryonic segment of life, with few exceptions, as for instance the studies on the metamorphosis of model species among the holometabolous insects (Diptera: *Drosophila melanogaster*; Lepidoptera: *Manduca sexta, Bombyx mori* and *Helicoverpa armigera*) (Arbeitman et al. 2002; Li and White 2003; Riddiford et al. 2003; Dong et al. 2007; Tanaka and Truman 2007; Ando et al. 2011) and a few pioneering studies on the expression of *Sex combs reduced* in the post-embryonic development of hemimetabolous insects (Chesebro et al. 2009; Hrycaj et al. 2010).

Another major problem with the study of arthropod post-embryonic development is the acritical adoption of a standard periodization based on the succession of moults (Minelli et al. 2006). To be sure, moults subdivide arthropod development into a variable number of stages, whose number can be—nearly always—unambiguously determined. Stages seem therefore obvious units to be compared (in terms of number, body organization, etc.) between different arthropod species, but this is not always the case, as we will see in this chapter.

To contribute to a better systematization of studies in the post-embryonic developmental biology of the Arthropoda, we start by 'dissecting' arthropod post-embryonic development into a few concurrent processes, whose description will be based on the standard periodization provided by the succession of moults. Then, we will provide examples of the mutual independence of these concurrent processes. This will show that the 'cuticular view' imposed by the moult-based periodization of arthropod development is not always the best framework for analysing the interactions between different developmental processes and their evolution. We will end with a short account of the evolutionary patterns of arthropod post-embryonic development.

## 5.2 Dissecting Arthropod Post-embryonic Development

Arthropod post-embryonic development is customarily described with reference to the succession of stages, rather than with reference to absolute time, as frequently done in non-moulting animals.

A *stage* is any segment of the post-embryonic development of an arthropod between two moults, or following the last moult, in the life of an individual. More precisely, a stage is delimited by ecdysis. Although some authors (e.g. Snodgrass 1935) have suggested that a stage should be considered as the period in between two apolyses, this choice is quite unpractical and is applied very rarely. The last part of a stage, when apolysis has occurred and a new cuticle is visible underneath the old one, is called the *pharate phase* of that stage. Instead of stage, many authors prefer to use the terms *instar* (especially in insects) or *stadium* (in millipedes), although Snodgrass (1935) distinguishes between stage, as temporal segment of life, and instar, to denote an animal in a given developmental stage. A uniform nomenclature would be preferable in principle, but these differences in terminology are more or less deeply entrenched in tradition for each major arthropod subtaxon.

Quite different definitions of stage have been suggested, often of limited application. Very recently, Fritsch and Richter (2012), following von Lieven's (2005) proposal for nematodes, have advocated a concept of stage based on any subjectively relevant, recognizable difference in external and/or internal morphological structures, independent of moults (see also Anger 2001).

### 5.2.1 Development of Segmentation

Arthropod post-embryonic development involves two aspects of these animals' segmental organization: the production schedule of segments, here intended as sets of serial elements of different segmental structures (see Chap. 9), and the differentiation of these segments whose result is the patterning of the main body axis. Neither of these processes is necessarily completed at the beginning of post-embryonic life, and the different arthropod clades exhibit diverse modes of development in this respect. This is illustrative of very divergent, sometimes opposing, evolutionary trends in arthropod segmentation. In the following sections, we will mainly discuss segment production schedules, as the patterning of the main body axis is so diverse as to escape easy classification and cannot be summarized here. For these aspects of the development of segmentation, we refer to zoology treatises and taxon-specific literature, although the discussion in this chapter may help to critically revise the evidence therein.

### 5.2.1.1 Anamorphosis and Epimorphosis

In some holometabolous insects, the number of segments recognizable in the adult is lower than the number present at the beginning of post-embryonic development. To our knowledge, there is no technical term to indicate this mode of segmentation, where the number of body segments decreases during ontogeny. In fact, in most arthropods, the number of body segments recognizable at the beginning of the post-embryonic development is either lower or identical to the number eventually found in the adult. Development by *anamorphosis* (with segment number increasing moult after moult) is thus distinguished from development by *epimorphosis* (no post-embryonic increment in segment number) (Minelli and Fusco 2004; Fusco 2005). In anamorphic development, the new segments appear sequentially in an antero-posterior progression from a subterminal region, often referred to as the *proliferative zone* (also, *generative zone*; see Fusco 2005), but information about morphogenesis and gene expression associated with anamorphosis is scarce. Evidence of a conserved role of the segment polarity gene *engrailed* across the embryonic/post-embryonic boundary was found in the crustaceans *Artemia* (Manzanares et al. 1993)

and *Sacculina* (Gibert et al. 2000) and in the centipede *Lithobius* (Bortolin et al. 2011).

The diversity of developmental schedules found among anamorphic arthropods has suggested a further subdivision, first proposed by Enghoff et al. (1993) for the Diplopoda (Fig. 5.1). In *euanamorphosis*, segment number increases at each moult throughout the whole post-embryonic life to terminate only with the death of the animal, thus both segment number and the number of moults are not fixed at the species level, or within each sex. Intraspecific variation in the number of segments added at each moult contributes further to the total variation in segment numbers at later stages. In *teloanamorphosis*, segment number also increases throughout the animal's life, but both the number of moults and the schedule of segment addition at each moult are fixed, thus there is a final number of segments, which is constant for a given species and sex (often also at family level, or even above). Finally, in *hemianamorphosis*, the post-embryonic development includes a first anamorphic phase, comprising a first batch of moults, followed by an epimorphic phase where moults take place without further increase in the number of body segments. Thus, a final number of segments is obtained at the end of the anamorphic phase, the number of anamorphic moults and the per-moult schedule of segment addition being generally, but not always, strictly fixed for a given species and sex. All three kinds of anamorphosis are found among the Diplopoda. Hemianamorphic are Polyxenida, Glomerida, Sphaerotheriida, Glomeridesmida and most of the Spirobolida: the number of anamorphic stages is quite different in the different clades, for example, five in *Glomeris* and seven in *Polyxenus*. The Polydesmida, Chordeumatida and most of the Callipodida are teloanamorphic, while the Stemmiulida, Julida, Epinannolenidea, Cambalida (with one possible exception), and also probably the Polyzoniida and Platydesmida are euanamorphic.

Hemianamorphic development is also characteristic for three of the five major clades of Chilopoda, that is, Scutigeromorpha, Lithobiomorpha and Craterostigmomorpha; of Symphyla and

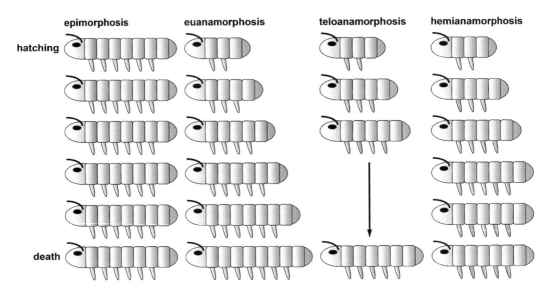

**Fig. 5.1** Schematic representation of segmentation post-embryonic schedules in arthropods. Along each ontogeny, a line is a stage. There is no intended alignment among the stages in the four schedules, but for the first and last stages. The penultimate and the last stage in teloanamorphosis are connected by an *arrow*

Pauropoda; of many Crustacea [Cephalocarida (Sanders 1963; Addis et al. 2007), the Upper Cambrian branchiopod *Rehbachiella* (Walossek 1993), Anostraca (Benesch 1969), Copepoda (Ferrari and Dahms 2007), Euphausiacea and Decapoda Dendrobranchiata (Scholtz 2000)]; of the Protura among the Hexapoda; of Pycnogonida, Trilobita and some extinct Chelicerata.

Epimorphosis is characteristic of scolopendromorph and geophilomorph centipedes, of hexapods to the exception of the Protura, of many crustaceans, and the Arachnida to the exception of the actinotrichid mites, which show (likely secondary) anamorphosis, adding up to 3 opisthosomal segments during post-larval development (Evans 1992; Alberti and Coons 1999).

The phylogenetic distribution of these different modes of segment addition during post-embryonic development suggests that hemianamorphosis may represent the ancestral condition of post-embryonic development in Euarthropoda (Hughes et al. 2006).

### 5.2.1.2  A Law of Anamorphosis?

With reference to polydesmidans, Fabre (1855) suggested a 'law of anamorphosis' according to

which all apodous rings of a given stage are fated to become pedigerous in the following stage. Halkka (1958) and Sahli (1969) found that this rule applies also to the Julida. The rule, however, has no universal application within the Diplopoda—indeed, it is limited to two major derived clades, the Polydesmida and the Julida + Spirobolida + Spirostreptida (Enghoff et al. 1993).

Interestingly, all three types of anamorphosis (hemi-, telo- and eu-) are represented both within the clades whose post-embryonic schedule of segmental increase follows Fabre's rule, and those that do not follow it. A further demonstration of the basic independence of the temporal deployment of different aspects of post-embryonic development is given by the frequent independence of the formation of new tergites versus the production of new sternites and new pairs of legs. This has been observed, for instance, in representatives of the Platydesmida, Glomeridesmida, Sphaerotheriida and Glomerida (Mauriès 1980; Enghoff et al. 1993). This is in accord with the evidence from developmental genetics for independent mechanisms of segmentation of the dorsal and the ventral aspects of the *Glomeris* embryo (Janssen et al. 2004).

The starting phase of the post-embryonic development of most millipedes is broadly equivalent, irrespective of the schedule of addition of new segments over the following moults: characteristic of stage I is the presence of three pairs of legs (a condition that in the past has invited speculation about the possible origin of hexapods from neotenic millipede-like ancestors; cf. Dohle 1988). However, species with a different number of leg pairs in stage I have been recorded in a few species: four pairs in *Polyzonium germanicum* (Polyzoniida) (Enghoff et al. 1993) and *Brachycybe nodulosa* (Platydesmida) (Murakami 1962), 27 pairs in *Pachyiulus flavipes* (Julida) (Dirsh 1937), 41 pairs in *Dolistenus savii* (Platydesmida) (Silvestri 1949).

No rule comparable to the 'law of anamorphosis' of millipedes is recognized in anamorphic crustaceans. Schedules of addition of segments and appendages may differ even between closest relatives, as between different species of *Derocheilocaris* (Mystacocarida) (summarized in Schram 1986).

### 5.2.1.3 Segment Production Versus Segment Articulation in Trilobites

Trilobites developed by hemianamorphosis over a series of juvenile and mature stages. New segments appeared progressively from a subterminal posterior region and become morphologically recognizable before becoming fully articulated units of the trunk. Traditional descriptions of ontogenetic phases of trilobites are based on the presence and number of dorsally fully articulated segments, the set of which is called the thorax. The set of more posterior, dorsally conjoined segments is called the pygidium. Three phases of post-embryonic development are thus distinguished: (1) *protaspid*, with no dorsal articulations and no thoracic segments, (2) *meraspid*, with a number of articulations and thoracic segments smaller than the final number, and (3) *holaspid*, with the final number of thoracic segments (N) and dorsal trunk articulations (N + 1). But anamorphosis in the proper sense, that is, the

emergence of new, distinct although still dorsally non-articulated segments in the pygidium, did not necessarily proceed in pace with the release of new fully articulated segments from the pygidium into the thorax. Thus, Hughes et al. (2006) distinguished three kinds of segmentation schedules in trilobites: *protarthrous* when the onset of the holaspid phase precedes the onset of the epimorphic phase, *synarthromeric* when both phases have synchronous onset, and *protomeric* when the onset of the epimorphic phase precedes the onset of the holaspid phase. In other words, in trilobites, the fixation of tagmosis was independent from segment production (Dai and Zhang 2011).

## 5.2.2 Development of Reproductive Maturity

It is common knowledge that the correlation between somatic development to obtain an *adult phenotypic condition* (or *adulthood*) and achievement of *reproductive maturity* is not necessarily strict. Indeed, the uncoupling of these two classes of developmental phenomena is usually a main evolutionary issue in terms of heterochronic change. In the case of arthropods, it is thus sensible to distinguish between *mature stages*, characterized by reproductive maturity, from *adult stages*, characterized by a morphologically invariant final condition. Adulthood can in general precede maturity, but the two typically co-occur in the clades where the adult/mature animal cannot undergo additional moults, as in all insects with the exception of the Ephemeroptera. Otherwise, maturity can precede adulthood in those derived taxa where the reproductive faculty is achieved at a stage that in their closest relatives corresponds to an immature larva or juvenile.

### 5.2.2.1 Onset of Maturity

Conceptual and terminological problems are not necessarily solved just by keeping the two issues distinct, achieving adult morphology versus onset of sexual maturity, because defining the latter is also not necessarily straightforward. Should we equate it with the first availability of mature gametes or, alternatively, restrict the concept to the time at which ripe gonads are present in an individual that is morphologically, physiologically and behaviourally ready to start reproductive activity? This conceptual conundrum, however, is perhaps less disturbing than the limited data thus far collected on the actual progress in gamete maturation and differentiation with respect to other developmental events, in particular, the moult to (first or only) adult stage. We will therefore only go through some of the more conspicuous patterns of the relationship between morphological (somatic) adulthood and onset of reproductive maturity (Fig. 5.2). The first aspect worth mentioning is that there is not necessarily a correlation between the stage at which the adult number of segments is obtained and the stage of the onset of maturity, as well as the number of stages through which it extends.

Reproduction, indeed, is limited to one stage in the teloanamorphic Diplopoda, Pauropoda and Crustacea (Copepoda, podocopan Ostracoda) and in the Protura, but also in many epimorphic Arthropoda (the majority of the Arachnida including Scorpiones, Uropygi, Palpigradi, Pseudoscorpiones, Solifugae, Ricinulei and most but not all representatives of Araneae, Acari and Opiliones) and the vast majority of the Pterygota.

Mature-to-mature moults occur (not necessarily in all species) in the Pantopoda, in the hemianamorphic and euanamorphic Diplopoda, but also in epimorphic clades: among the Arachnida in the Amblypygi, in the basal clades of the Araneae (Mygalomorphae, Filistatidae, perhaps also the Mesothelae), in at least some Grassatores among the Opiliones (Gnaspini et al. 2004; Gnaspini 2007), in some Prostigmata (Michener 1946; Immamura 1952) and possibly also in the Notostigmata (Coineau and Legendre 1975) among the Acari, in some at least of the epimorphic Chilopoda (Scolopendromorpha and Geophilomorpha), in the Symphyla, in many anomopodans and Malacostraca among the

crustaceans, as well as in the Collembola, Diplura, Archaeognatha, Zygentoma and (limited to the subimago-to-imago moult) in the Ephemeroptera among the Hexapoda.

The number of mature stages in hemianamorphic and euanamorphic millipedes is variable. In the hemianamorphic *Polyxenus lagurus*, there are at least five stages (Schömann 1956), among the euanamorphic Julida their number ranges from three (*Ophyiulus pilosus*, *Julus scandinavius*) to at least eight or nine (e.g. *Proteroiulus fuscus*) (more data in Enghoff et al. 1993).

In most representatives of the Polydesmida, sexual maturity is reached in both sexes at stage VIII (with 18 leg-bearing segments); in others, at stage VII (with 17 leg-bearing segments), a few species are sexually dimorphic, with males having 17 leg-bearing segments and females 18. The post-embryonic development of representatives of the polydesmidan genus *Devillea* (Xystodesmidae) is not known, but in this case, maturity is very likely reached through a longer series of immature stages, and adult-to-adult moults (otherwise unknown in Polydesmida) cannot be ruled out (Enghoff et al. 1993). However, a few individuals of species of Polydesmida undergoing a supplementary moult to stage IX have been recorded in laboratory cultures of *Polydesmus complanatus* (in both males and females) (Verhoeff 1916, 1928) and *P. angustus* (David and Geoffroy 2011). Intraspecific variability in the number of segments of adult males has been also recorded in nature for another polydesmidan, the pyrgodesmid *Muyudesmus obliteratus* (Adis et al. 2000).

Millipedes provide good examples of the fundamental uncoupling between the progression in the production of body segments and the achievement of sexual maturity. In the hemianamorphic species, the end of the anamorphic phase of post-embryonic development does not correspond with maturity, which is only attained a few moults later. The same happens in the hemianamorphic Chilopoda (Scutigeromorpha, Lithobiomorpha, Craterostigmomorpha). On the other hand, euanamorphic millipedes continue adding segments after attaining sexual maturity.

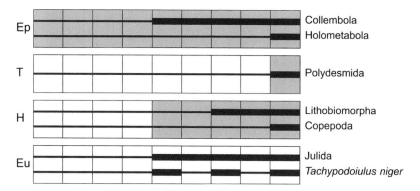

**Fig. 5.2** Schematic representation of the observed relationships between segmentation post-embryonic schedule and the onset of reproductive maturity in arthropods. The four rows of *boxes* represent the four segmentation modes, Epimorphosis (Ep), Teloanamorphosis (T), Hemianamorphosis (H) and Euanamorphosis (Eu). Each box is a stage. *Empty boxes* are anamorphic stages, *grey boxes* are epimorphic stages. *Thin purple lines* go over immature stages, *thick purple lines* go over mature stages. For each combination of segmentation and maturity schedule, a taxon is indicated by way of an example

Extending reproductive activity over two or more stages does not necessarily mean than the mature stages form a continuous series in the advanced part of the animal's post-embryonic development. In some arthropods, subsequent mature stages are actually separated by intercalary stages which are unable to reproduce and often show different degrees of reduction of their sexual appendages. This phenomenon has been recorded in certain Collembola (Cassagnau 1985), but is better known for female isopods and male millipedes, although under different names. In female isopods, there is a regular alternation between mating-ready stages with fully developed oostegites and intervening stages with reduced oostegites, ending with intercalary and parturial moults, respectively.

In millipedes, the presence of intercalary stages along the mature phase of a male's post-embryonic life has been described for a few species of Julidae (*Allajulus nitidus, Cylindroiulus caeruleocinctus, Ommatoiulus sabulosus, Tachypodoiulus niger*) and Blaniulidae (*Proteroiulus fuscus, Nopoiulus kochii, Blaniulus guttulatus, B. lorifer consoranensis*: in the latter subspecies, intercalary stages have been found in the female too). This phenomenon is known as *periodomorphosis*. The number of intercalary stages between two reproductive stages is usually one or two, but in the blaniulid *Proteroiulus*

*fuscus* Rantala (1974) described as many as six consecutive intercalaries. As many as four mature stages separated by single intercalary stages have been recorded in julids (Verhoeff 1916, 1923, 1933, 1934, 1939; Sahli 1985, 1989, 1990).

### 5.2.2.2  Changes Accompanying the Moult to the Only or First Reproductive Stage

In arthropods with only one (final) reproductive stage, the physiological onset of maturity may broadly coincide with the last moult, but sometimes precedes it or, on the contrary, will be reached only at a later time, following adequate feeding, as in female mosquitoes. Independence of maturity from adult feeding is often associated with a very short adult life, as in the case of Ephemeroptera and Plecoptera. Non-feeding adults are probably more widespread than generally perceived, examples being found in different groups such as wind scorpions or Solifugae (male *Othoes saharae*; Junqua 1966), gnathiid Isopoda (Monod 1926), aphids (the gynoparae of *Rhopalosiphum padi*; Walters et al. 1984) and also many Lepidoptera (Miller 1996).

The actual transition to maturity is accompanied by physiological processes that may translate into conspicuous changes, both internal and external, during the adult stage. This is the case of those insects, such as many ant queens, where flight muscles undergo histolysis after the mating flight, or the occasional changes in the colour of the cuticle only manifested several days after complete hardening of the latter. This phenomenon has been studied in adult males of the desert locust *Schistocerca gregaria* where, in crowded conditions, a redistribution of $\beta$-carotene mediated by a protein carrier (Goodwin and Srikukh 1949) determines the conspicuous colour changes characteristic of the gregarious morph, ca. 10 days after the last moult (De Loof et al. 2010).

### 5.2.3    Moulting

#### 5.2.3.1    Number of Moults: An Overview

The number of post-embryonic stages varies very conspicuously among the arthropods and often also among close relatives, but it is generally less than 15. But counting moults is not always easy, for two reasons at least. First, in a few arthropods with extremely abbreviated development, for example, some bathysciine leiodid beetles and termitoxeniine phorid flies (cf. Sect. 5.4.1) one or more 'virtual stages' may be recognized just because of the simultaneous shedding of multiple cuticles. Second, moulting may begin during the embryonic phase of development, with two or three embryonic cuticles being shed before hatching (Konopová and Zrzavý 2005), so the first moult undergone after hatching by many arthropods could correspond to the last embryonic moult of others. In many arthropod groups, the hatchling is more or less inactive and 'embryoid'. This stage is sometimes termed as a *pronymph* (cf. Sect. 5.3.1), which moults into the first active nymph or larva. Here, we will consider the series of stages following the pronymph, if present, and preceding the adult.

Within the Chelicerata, two different kinds of post-embryonic development can be distinguished. In what is arguably the more primitive condition, the animal goes through a series of moults whose number is variable even between closely related taxa, separating nymphal stages of increasing size, which progressively display differentiation of adult (and sexual) characters. The number of moults is 4–7 in Scorpiones, 8 or more in Amblypygi, mostly 5–14 in Araneae, ca. 10 in Solifugae, and 6 or more in Opiliones. Strong reduction in the number of moults accompanies miniaturization: for example, the male of *Glyptocranium cornigerum* (Araneidae) is only 1.5–2 mm long and develops through only two moults (Gertsch 1955). The flexibility of this moulting schedule is further demonstrated by the occasional occurrence of intraspecific variation in the number of post-embryonic moults, as recorded in the spider *Thomisus onustus* (Thomisidae) (Levy 1970).

In contrast, a very stereotypic series of stages is found in arachnids belonging to other clades. There are 4 nymphal stages (proto-, deuto-, trito- and tetranymph) in Thelyphonida and Schizomida, 3 in Pseudoscorpiones and a larval plus 3 nymphal stages (pro-, deuto- and tritonymph) in Ricinulei and Acari. This at least is the primitive scheme, although one or more stages can be suppressed in different clades of Acari (André and Van Impe 2012).

Comparative data on the number of post-embryonic stages in non-hexapod crustaceans and in hexapods are summarized in Tables 5.1 and 5.2, respectively.

Fixation of stage number varies from the condition of species where even siblings, hatching from eggs laid by the same mother, pass through differing numbers of post-embryonic moults (21–30 in *Cloeon simile*; Degrange 1959), to the condition of many large clades throughout which the number is fixed. More common is the intermediate condition, where a large clade generally exhibits a fixed number of stages, but with a small number of exceptions. A few examples follow.

In Coleoptera, there are 3 or 4 larval stages in most families, but the number can be unstable even within a genus and may be as low as 1 or more than 10. In particular, there are generally 3 larval stages in Adephaga, Staphylinidae, Hydrophilidae and Scarabaeidae, but in the largest adephagan family (Carabidae), a few species

**Table 5.1** Number and kinds of post-embryonic (pre-adult) stages in crustaceans

| | | |
|---|---|---|
| Cephalocarida | 18 (in *Hutchinsoniella*) | Metanauplius (13), juvenile (5) |
| Anostraca | 17+ (in *Artemia*) | Nauplius (in *Artemia* and *Branchinecta*, not in *Tanymastix* and *Siphonophanes*); metanauplius; juvenile |
| Notostraca | Up to 40 (in *Triops cancriformis*) | Metanauplius; juvenile |
| 'Conchostraca' | Ca. 12 | Nauplius + metanauplius (4–5); juvenile (7) |
| Cladocera | 2-6 | Metanauplius (only in *Leptodora kindtii*); juvenile |
| Podocopa | 8 | Nauplius (atypical); metanauplius; juvenile |
| Myodocopa | 5 | Metanauplius; juvenile |
| Mystacocarida | 7–10 | Metanauplius; juvenile |
| Copepoda | Typically 11 | Nauplius (6); copepodid (5) |
| Tantulocarida | 1 | Tantulus (equivalent to copepodid) |
| Branchiura | 9 | Copepodid |
| Ascothoracida | Several | Nauplius; ascothoracid stadium |
| Cirripedia | Various | Nauplius (6); cypris (1); juvenile |
| Leptostraca | | Juvenile |
| Stomatopoda | 6–13+ | Gonodactyloidea, Squilloidea: pseudozoea (3); alima (9); post-larva Lysiosquilloidea: antizoea |
| Anaspidacea | | Post-larva |
| Bathynellacea | 7–8 | Parazoea (3); bathynellid (4–5) |
| Mysidacea | 12–13 | 'nauplius' (1); 'post-nauplius' (1); juvenile (10–11) |
| Amphipoda | Various, for example, 7 (post-manca) juveniles in *Hyalella* spp., 5–6 in *Caprella grandimana* | Manca; juvenile |
| Cumacea | 4–8 | Manca (2), juvenile (2–6) |
| Tanaidacea | 6+ | Manca (2); neutrum (2); female: preparatory (2) – male: juvenile (several) |
| | | Adult females become secondary male in Neotanaidae |
| Isopoda[a] | Various, for example, 12 (post-manca) juveniles in *Porcellio scaber*, 5–7 in *Idotea viridis*, 9 in *Janaira gracilis* | Manca; juvenile |
| Euphausiacea | About 10–15 | Nauplius (2); metanauplius (1); calyptopis (3); furcilia (2–5); cyrtopia; juvenile |
| Amphionidacea | Probably 11 | Amphion |
| Decapoda | 3–16 | Nauplius (only in Dendrobranchiata); zoea or equivalent (0–15); megalopa or equivalent (0-many); juvenile |

[a] Diverging developmental schedules in some subgroups, especially in the parasitic Bopyridae, where the post-embryonic development runs through three unique larval stages, the epicaridium, microniscium and cryptoniscium, in the order

References: mainly after Gruner 1993; data on Peracarida partly after Howes 1939; Michel and Manning 1972; Soares Moreira and Setubal Pires 1977; Zimmer 2002; da Silva Castiglioni et al. 2007; Baeza-Rojano et al. 2011

have only 2 larval stages (some species of *Amara*, some species of *Harpalus*, *Thermophilum sex-maculatum*), while others have four (*Eurycoleus*) *macularis*) or five (*Brachinus* spp.). In the hyperspecialized troglobitic genus *Aphaenops*, there is probably only one larval stage, and a

**Table 5.2** Number of post-embryonic moults before the achievement of the adult condition in hexapods, extreme exceptional values in brackets

| Protura | 5–6 |
|---|---|
| Collembola | 4–11 |
| Diplura (*Campodea*) | 8–11 |
| Archaeognatha | 8–10 |
| Zygaentoma | 9–13 |
| Ephemeroptera | (12)15–30(52) |
| Odonata | 10–16 |
| Orthoptera | 4–12 |
| Phasmatodea | 4–12 |
| Grylloblattodea | 8 |
| Mantophasmatodea | 5 |
| Dermaptera | 4–9 |
| Isoptera | 5–11 |
| Blattodea | 3–14 |
| Mantodea | 4–9 |
| Zoraptera | 5 |
| Plecoptera | 13–34 |
| Embioptera | 4 |
| Thysanoptera | 4–5 |
| Hemiptera | (3)5(8) |
| Psocoptera | 6 |
| Phthiraptera | 3 |
| Coleoptera | 3–24 |
| Strepsiptera | 5 |
| Neuroptera | 3–5 |
| Megaloptera | 11–13 |
| Raphidioptera | 11–16 |
| Hymenoptera | 2–7 |
| Trichoptera | 6–8 |
| Lepidoptera | (3)5–6(31) |
| Mecoptera | 5 |
| Siphonaptera | 3–4 |
| Diptera | 4–5(11) |

References: mainly after Heming 2003; with additional data from Stehr 1987, 1991; Jarjees and Merritt 2002; Dunger 2003; Groll and Günther 2003; Esperk et al. 2007a; Hockman et al. 2009; Shimizu and Machida 2011

parallel reduction is found in some taxa, also hyperspecialized troglobionts, belonging to the Bathysciinae subfamily of the staphyliniform family Leiodidae. Other leiodids have 2 larval stages, as have the Histeridae (also belonging to the Staphyliniformia) and some staphylinine and paederine Staphylinidae, but other staphylinids, such as the Aleocharinae and Oxytelinae, have instead 4 and 5 larval stages, respectively. Numbers greater than three are the rule in other clades, for example, the fireflies (Lampyridae) with 4 or 5 larval stages, and the ladybirds (Coccinellidae) with 4 (but only three in *Hyperaspis lateralis*; Mckenzie 1932); while 8–10 larval stages are probably the rule in *Cantharis* and *Rhagonycha* (Cantharidae).

Within the Diptera, most of the paraphyletic 'Nematocera' have 4 larval stages, but black flies (Simuliidae) have 6–8; most of the Orthorrhapha have 5–8 (but up to 11 in Tabanidae), whereas the Cyclorrhapha only 3. Clades with variable numbers of larval stages are also known from the Hymenoptera, most members of the Vespidae, for example, have 5 larval stages, but in the Stenogastrinae, the number is reduced to 4.

### 5.2.3.2    Number of Moults: Male Versus Female

Differences between the sexes in the number of post-embryonic moults are not rare and commonly translate into size differences in the adult (Esperk et al. 2007b).

In insects, a higher number of moults in the female have been recorded for instance in representatives of the Blattodea, Coleoptera (several Dermestidae), Diptera (*Tabanus lineola*), Hymenoptera (*Arge* spp.) and Lepidoptera (Arctiidae: *Apantesis vittata*; Crambidae: *Diatraea* spp.; Lasiocampidae: *Streblote panda*; Lymantriidae; Cossidae: *Prionoxystus robiniae*; Tortricidae: *Choristoneura retiniana*, *Ctenopseustis obliquana*, *Platynota stultana*) (Esperk et al. 2007a). In contrast, in the earwig *Labidura riparia*, the number of moults to maturity is higher in the male (6 versus 5–6) (Esperk et al. 2007a). In a sample of 40 species of acridomorph Orthoptera studied by Dirsh (1967), 19 species did not show any sex-related differences in the number of stages, while in the remaining 21 species, the females had either one or two stages more than conspecific males. Similarly, in the Xiphosura, the normal number of moults to maturity is identical in one species (*Carcinoscorpius rotundicauda*), but slightly higher in the females in the other three species (Sekiguchi et al. 1988).

### 5.2.3.3 Number of Moults: Within-Sex Intraspecific Variation

In several arthropod groups, no intraspecific variation has been ever recorded in the number of larval/juvenile stages of the same sex. This is more common when this number is low (tentatively, up to 5) and often extends to whole families and orders (e.g. nearly always 5 in Heteroptera; Štys and Davidova-Vilimova 1989). No variation in the number of larval stages has been reported for *Drosophila* spp. (N = 3) and the Colorado beetle *Leptinotarsa decemlineata* (N = 4) (Esperk et al. 2007a), two insects where these developmental traits have been very extensively studied.

Intraspecific variation in the number of post-embryonic moults (*developmental polymorphism*; Schmidt and Lauer 1977) is common in some other groups, for example, Orthoptera (especially, some families of the Ensifera) (Calvert 1929; Uvarov 1966) and Odonata (Wigglesworth 1972; Corbet 1999). Extreme variation (from 8 to 31 larval stages) has been recorded in the cossid moth *Prionoxystus robiniae* (Solomon 1973). In some groups, starvation may release an undetermined number of additional moults (Beck 1971), and the same effect has been observed in larvae reared in groups (Quennedey et al. 1995), or on poor-quality diet (Titschack 1926), or under conditions inducing larval diapause (Chippendale and Yin 1973; Kfir 1991).

In insects, intraspecific variation in the number of post-embryonic stages has been shown to depend on a variety of external influences: a short summary is given in Table 5.3, mainly based on the detailed survey of Esperk et al. (2007a).

### 5.2.3.4 Duration of Stages

Data on the average duration of individual stages of post-embryonic development, and on its dependence on food availability and diversity of environmental conditions, are available for many species, especially for a substantial number of those of economic importance. We cannot deal with this topic in detail here, but we report on one observation from the perspective of evolutionary developmental biology. As a rule of thumb, the length of individual post-embryonic stages increases through ontogeny. However, this is only true of the active stages preceding maturity, whereas the length of the mature, not necessarily feeding stage(s) is not generally correlated with the length of the preceding stages, especially when there is just one mature/adult stage. The same is true for the resting (e.g. pupal) phase. One can speculate that the progressive length increase of juvenile stages is, by default, simply a 'mechanical' correlate of size increase (see Sect. 5.2.4), whereas the extreme diversity in length of the mature stage(s) requires group-by-group, if not species-by-species explanations in terms of reproductive strategies.

### 5.2.3.5 Homology of Stages

Terms used for the different post-embryonic stages of arthropods are confused and confusing, not only because traditions were often established without any regard for homology of stages across related clades. Hastily determined or poorly supported homology often provided the basis for the terminology. In this respect, the most fundamental question is, whether the individual stages can be legitimately regarded (1) as units of homology, that is, as independently evolving units that can be added, deleted or modified (Scholtz 2008), or (2) as a more or less integrated set of character state combinations, related to developmental processes that do not necessarily progress as a unit in development, but that can be temporally and functionally associated together in evolution (thus becoming a modular target of selection) and that can possibly become dissociated again. The latter view informs the arguments of the whole chapter (cf. Sect. 5.3).

Particularly well-researched is the question of the equivalence between the pre-adult stages in the post-embryonic development of hemimetabolous and holometabolous insects. An exhaustive

**Table 5.3** Factors determining an increase in the number of post-embryonic moults, and exemplary insect taxa in which the effect has been recorded, based on data compiled by Esperk et al. (2007a)

*Diet*
Heteroptera: Miridae: *Cyrtorhinus lividipennis*
Orthoptera: Acrididae: *Chorthippus brunneus, Melanoplus sanguinipes*
Lepidoptera: Arctiidae: *Hyphantria cunea*; Bombycidae: *Bombyx mori*; Lasiocampidae; Lymantriidae: *Orgyia antiqua*; Noctuidae; Pyralidae: *Galleria mellonella, Samea multiplicalis*; Tineidae: *Tineola bisselliella*; Tortricidae

*Temporary starvation (in some species, only the first stage is sensitive to this condition)*
Coleoptera: Dermestidae: *Trogoderma glabrum*
Lepidoptera: Bombycidae: *Bombyx mori*; Crambidae: *Chilo partellus*; Lymantriidae: *Lymantria dispar, Orgyia antiqua*; Noctuidae: *Spodoptera littoralis*; Notodontidae: *Syntypistis punctatella*; Sphingidae: *Manduca sexta*

*Lower population density*
Orthoptera: Acrididae: *Melanoplus differentialis, Nomadacris septemfasciata, Ornithacris turbida, Schistocerca gregaria*

*Higher population density*
Coleoptera: Tenebrionidae: *Zophobas atratus*
Hymenoptera: Argidae: *Arge* spp.
Lepidoptera: Lymantriidae: *Lymantria dispar*

*Low temperature*
Orthoptera: Gryllidae: *Pteronemobius nitidus* (under short daylength)
Odonata: Libellulidae: *Brachythemis contaminata*
Coleoptera: Cerambycidae: *Psacothea hilaris*
Lepidoptera: Cossidae: *Prionoxystus robiniae*; Crambidae: *Diatraea grandiosella*; Lycaenidae: *Lycaena* spp.;
Noctuidae: *Spodoptera frugiperda*; Pieridae: *Pieris brassicae*; Pyralidae: *Herpetogramma licarsisalis*;
Zygaenidae: *Harrisina brillians*; Tortricidae: *Adoxophyes orana*

*High temperature*
Odonata: Libellulidae: *Orthetrum sabina*
Orthoptera: Acrididae: *Chorthippus brunneus, Melanoplus ferrumrubrum*; Gryllidae: *Gryllus bimaculatus, Pteronemobius nitidus* (under long daylength)
Coleoptera: Cerambycidae: *Monochamus carolinensis*; Tenebrionidae: *Tenebrio molitor*
Lepidoptera: Notodontidae: *Syntypistis punctatella*

*Both low and high temperature*
Homoptera Aphididae: *Nasanovia ribisnigri*
Coleoptera: Dermestidae: *Attagenus sarnicus, Dermestes lardarius*; Silvanidae: *Oryzaephilus surinamensis*
Hymenoptera: Tenthredinidae: *Nematus oligospilus*
Lepidoptera: Crambidae: *Diatraea lineolata*; Noctuidae: *Agrotis ipsilon, Copitarsia decolora*; Tortricidae: *Acleris minuta, Platynota stultana*

*Shorter day-length*
Blattodea: Blaberidae: *Opisthoplatia orientalis* (female)
Orthoptera: Gryllidae: *Modicogryllus siamensis, Pteronemobius nitidus* (at low temperatures)
Coleoptera: Cerambycidae: *Psocothes hilaris*
Lepidoptera: Crambidae: *Diatraea grandiosella*; Geometridae: *Selenia tetralunaria*; Lycaenidae: *Lycaena phlaeas*; Noctuidae: *Sesamia nonagrioides, Xestia c-nigrum*; Nymphalidae: *Coenonympha pamphilus, Sasakia charonda*; Tortricidae: *Platynota idaeusalis*

*Longer day-length*
Blattodea: Blaberidae: *Opisthoplatia orientalis* (male)
Orthoptera: Gryllidae: *Allonemobius fasciatus, Pteronemobius nitidus* (at normal temperatures)
Lepidoptera: Arctiidae: *Pyrrarctia isabella*

*Low humidity*
Coleoptera: Dryophthoridae: *Sitophilus oryzae*; Dermestidae: *Dermestes lardarius, Trogoderma glabrum, T. variabile*; Tenebrionidae: *Tenebrio molitor*

*Extreme high humidity*
Coleoptera: Dermestidae: *Attagenus sarnicus*

survey of the many interpretations advanced thus far was provided by Heming (2003). Among those disparate views, a sensible evolutionary scenario is provided by Berlese's (1913) model, as revived and perfected by Truman and Riddiford (1999), according to which the active (larval) stages of holometabolans would collectively correspond to the 'embryoid' pronymph of hemimetabolans.

Ultimately, with regard to homologies between developmental stages, it is advisable to adopt a factorial approach, as suggested for homology of the morphological traits (Minelli 1998; Minelli and Fusco 2013). For example, stage A in arthropod X and stage B in arthropod Y can be homologous with respect to certain developmental or morphological features, but not necessarily to others. An interesting example is provided by spiders: araneoideans build sticky webs from the second stage onwards, whereas cribellates only show this ability from the third stage on (Szlep 1961; Yu and Coddington 1990): in such a case, homology of stages in terms of ordinal position in the sequence of stages does not match what we know about functional homology, in terms of silk gland activity.

## 5.2.4 Growth

Growth, like other aspects of arthropod post-embryonic development, is generally described with reference to the ontogenetic succession of stages. Indeed, the moult cycle has observable effects on the growth dynamics of several tissues and organs: in particular, size increase of external structures occurs mainly in stepwise manner, in accord with the occurrence of ecdyses (see also Chap. 6). Basically, while tissue growth and nutrient reserve accumulation occur throughout the intermoult period, only limited parts of the animal exoskeleton can stretch and extend between successive moults. In contrast, the sclerotized parts, which usually cover most of the body, can grow only during a short interval preceding the ecdysis itself. In consequence, size growth is to a large extent discontinuous, while growth in biomass is nearly

continuous, punctuated by stasis, when the animal stops feeding during the moult, or even loses weight, when the old cuticle is shed at the ecdysis (Nijhout 1994).

Not all body parts of all arthropods are provided with strongly sclerotized cuticle. Arthrodial membranes remain flexible and, to some extent, extensible, as for example, most of the cuticle of the so-called soft-bodied holometabolous insect larvae (Nijhout 1994) and of the adults of many parasitic copepods. Size increase during the intermoult period can be generated in different ways: (1) through simple stretching of the cuticle (*isovolumetric* cuticle expansion), or through a real growth of the cuticle, either (2) by *intussception*, that is, by diffuse incorporation of new protein and chitin molecules into the already deposited cuticle, or (3) by *appositional growth*, that is, addition of new cuticle layers underneath the older layers, while these become thinner as a consequence of stretching (Nijhout 1994). However, for such species, ontogenetic stages or body parts for which the exoskeleton is essentially inextensible after the post-ecdysial hardening, intermoult growth is in practice ignored (Hartnoll 1982).

As a consequence of the stage-based description of growth, time is often treated as an independent variable of the process, in the form of *stage duration* (or *length of the intermoult interval*), as if this were an additional property of each stage. Both size increase and intermoult interval duration at each stage are discrete elements of growth, often with very different responses to intrinsic and extrinsic changes (Hartnoll 1982); however, they can also combine in different ways, in particular in the regulation of growth (see Sect. 5.2.4.2).

### 5.2.4.1 Growth Modes

In arthropods, several *growth modes* (or *formats*) can be distinguished, which are combinations of the modalities of three *growth mode variables*, as proposed by Hartnoll (1982) for crustaceans: (1) the presence, or not, of a terminal moult (*determinate* versus *indeterminate growth*); (2) in determinate growth, the number of stages (fixed or variable); and (3) in determinate growth, the onset of reproductive maturity (at the last stage or before). Four basic growth modes are observed in arthropods: (1) indeterminate growth (e.g. euanamorphic millipedes and many crustaceans); (2) determinate growth, with a variable number of stages, and maturity occurring before the terminal stage (e.g. many hemianamorphic myriapods and many decapod crustaceans); (3) determinate growth with a variable number of stages, and maturity obtained at the terminal stage (e.g. some Odonata and Blattodea among the insects and Majidae (spider crabs) among the decapod crustaceans); (4) determinate growth with a fixed number of stages, and maturity attained at the terminal stage (e.g. teloanamorphic millipedes, most pterygote insects, most spiders and many crustaceans). It is not certain whether there are arthropods with determinate growth, constant stage number and maturity obtained before the terminal stage. Mature gametes indeed are often available, in the males especially, before the moult to adult, as in the males of the cellar spider *Pholcus phalangioides* (Michalik and Uhl 2005) as well as in the males and females of mayflies, stoneflies and many lepidopterans. But in a majority of these taxa, the number of post-embryonic stages shows some variability within the species.

While in some clades (e.g. holometabolous insects) the growth mode is the same for all species, in others, different modes can be observed among related taxa, as for instance in brachyuran Decapoda and Isopoda (Hartnoll 1982).

### 5.2.4.2 Absolute Growth

*Absolute growth* is the stage-by-stage progression in size of the body, or of a part of it, through the ontogenetic sequence of stages. The ratio between post-moult and pre-moult size is called *per-moult growth rate*, while the *average per-moult growth rate* over a given section of ontogeny (a set of successive stages) is calculated as the geometric mean of the post-moult/pre-moult size ratios among the given stages.

### Size Increment

Of several models formulated for describing discrete size increment, the so-called Dyar's rule (Dyar 1890) is considered a 'null model' of arthropod growth (Klingenberg and Zimmermann 1992). Dyar's rule assumes a constant rate of size increase between moults. Accordingly, by numbering developmental stages in their temporal sequence, the growth progression assumes the form of a geometric progression (the discrete counterpart of an exponential function): as a formula, $X_{i+1}=rX_i$, where $X_i$ is the value of a linear size variable at the $i$th stage, $X_{i+1}$ is the value of the same variable at the following stage, and $r$ is the growth rate. For a linear size variable (e.g. the distance between two landmarks), this growth rate is called *Dyar's coefficient*. Differential growth rates in distinct body parts of the same animal (*allometric growth*) produce ontogenetic changes in body shape (see Sect. 5.2.4.3).

Comparisons of Dyar's coefficients, or average per-moult growth rates, across the Arthropoda should be handled with caution because observed values depend on several factors, such as the segment of ontogeny considered, the size characters measured, and the environmental condition under which growth has occurred. As an indication of the range of magnitude of growth rates for specific body parts and specific moults in normal development, values vary from 1.00, denoting zero growth (something observable, for example, on occasional supernumerary moults), to more that 2.00 (more than doubling in size per moult). However, modal values for overall size growth are generally within the interval 1.15–1.50 for most taxa: around 1.50 for the larvae of holometabolous insects (Cole 1980), 1.25 for hemimetabolous insects (Cole 1980), 1.20 for decapod larvae (Rice 1968), spiders (Enders 1976) and trilobites (Fusco et al. 2012), and 1.15 for lithobiomorph centipedes (Albert 1982).

Among arthropods, several instances of deviation from Dyar's rule have been recorded (e.g. Albert 1982). Sometimes the departure from Dyar's rule is simply a tendency for growth ratios to decrease progressively with stage

(Hartnoll 1982). A metric able to quantify the fit of ontogeny to growth progression at a constant rate (*index of conformity to Dyar's rule*) has recently been devised by Fusco et al. (2012).

Several explanations for growth at a constant rate and for specific values of Dyar's coefficient have been suggested. Some of these appeal to 'external causes' based on interactions of the organism with its environment, such as competitive exclusion (Horn and May 1977; Maiorana 1978), food-finding strategy (Enders 1976), habitat stability (Cole 1980) and maximization of growth efficiency (Hutchinson and Tongrid 1984). In such instances, a constant growth rate (eventually with a definite value) would be promoted and actively maintained by natural selection. Other explanations appeal to 'internal causes' related to the organism's developmental system, anatomy and physiology, such as the mechanism of intermoult hypodermal growth (Bennet-Clark 1971; Freeman 1991) and cell proliferation (Przibram and Megušar 1912; Bodenheimer 1933). In such instances, a constant or specific growth rate would result from effective developmental constraints.

## Growth Compensation

An individual *target phenotype* is the phenotype that would be specified by the individual genetic make-up and environmental conditions in the absence of perturbing factors of any kind (Nijhout and Davidowitz 2003). We think it is useful to expand the concept of target phenotype, which refers to a combination of character states at a given developmental stage, into a more general concept of *target ontogenetic trajectory*—the series of target character states through all the developmental stages of an individual.

During post-embryonic development, body growth (or the growth of specific body parts) tends to depart from the target trajectory as a result of variation in the external factors that are known to influence growth rates (such as temperature, nutrition or parasitism; reviewed in Hartnoll 1982), or because of developmental instability (Nijhout and Davidowitz 2003). However, arthropod developmental systems generally incorporate compensating mechanisms that can buffer, to some extent, the effects of perturbing factors. There are two main ways in which a target trajectory can be followed, or a target final size eventually attained, via a compensation system: (1) by altering the number of stages, or (2) through a stage-by-stage feedback mechanism that continuously corrects the increments in size. Combinations of the two are also possible, and both could involve correlated variation in the duration of stages.

*Regulation of the number of moults* is possibly related to one or more 'size check-points' along the ontogeny. Typically, the tobacco hawkmoth *Manduca sexta* develops through five larval stages, but if a larva grows poorly and does not reach a threshold size by the fifth stage, it will moult to a sixth larval stage before pupating (Nijhout 1975). Females of the German cockroach *Blattella germanica* reach the adult stage after either five or six nymphal stages, and this developmental pattern is determined at the end of the third nymphal stage (Tanaka 1981). In the damselfly *Enallagma vernale*, the size of the last nymph does not depend on the number of nymphal stages, either 13 or 14 (Corbet 1999).

*Compensatory growth* (also termed *targeted growth* or *convergent growth*) is accomplished when individuals adjust their growth trajectories stage-by-stage, thus keeping their ontogenetic size trajectory close to the target trajectory, eventually reaching a target final body size. Exceedingly large specimens tend to grow less than average, while those whose size is smaller than expected, tend to grow more (Klingenberg 1996). Compensatory growth, eventually limited to a subset of ontogenetic stages, has been reported for many arthropod taxa, in particular among insects (e.g. Tanaka 1981; Klingenberg 1996) and crustaceans (e.g. Hartnoll and Dalley 1981; West and Costlow 1987; Freeman 1991; Twombly and Tisch 2000), but also in a trilobite species (Fusco et al. 2004). However, as some (but not all) statistical procedures applied for detecting compensatory growth in morphometric data can produce false positives, the supposed prevalence of targeted growth in arthropod development has been questioned (Lytle 2001).

In species with determinate growth, the level of compensation is reflected in the magnitude of size variation at the terminal reproductive stage, as in the insect imago. In some taxa, adults present limited size variation, obtained through either a variable or a fixed number of immature stages, in the latter case only regulating stage duration. In other taxa, for instance in many beetles whose larvae feed on wood (e.g. Cerambycidae and Lucanidae), the size of the adult shows great individual variation, reflecting the variable quality of food resources, thus displaying an ineffective regulation of stage duration.

Irrespective of the effectiveness of arthropod growth compensation and of its taxonomic distribution, with the exception of a few systems (e.g. *Drosophila*, *Manduca* and *Rhodnius*, among the insects), the underlying developmental mechanism are poorly understood (see Nijhout 2003; Stern 2003; Nijhout et al. 2010; Grewal 2012).

### 5.2.4.3    Relative Growth

*Relative growth* (or *allometric growth*, or *ontogenetic allometry*) is the change in proportional size of one body part with respect to another, or to the whole body, with growth. In other words, relative growth is about change of shape with growth. This is generally expressed by means of the scaling function known as the *allometric equation*; $X_1 = bX_2{}^k$, a power law where $X_1$ and $X_2$ are measures of two body parts at the same stage, while $b$ and $k$ are constant parameters. If $k = 1$ (and the measures of the two parts have the same geometric dimension, for example, two surfaces), the relation is said to be *isometric* and the two parts do not change in relative proportions during ontogeny. If $k \neq 1$, the relation is *allometric* and the proportions between the two parts change as these grow in size.

There is a relationship between allometry and Dyar's rule. The relationships between two body parts, $X_1$ and $X_2$, both growing according a geometric progression with rates $r_1$ and $r_2$, respectively, can be expressed as a power law with exponent $k = \log(r_1)/\log(r_2)$ (Huxley 1932). For a deeper discussion on the

dependence of allometric relations from growth kinetics of different body parts, see Nijhout (2011).

Shape change can occur gradually over a series of moults, or abruptly, even at a single moult. Different allometric relationships between the same body parts can characterize different sections of ontogeny of the same animal, thus subdividing post-embryonic development into a small number of *growth phases*, each comprising a given number of contiguous stages (Hartnoll 1982). Allometric relationships typically change at specific events during ontogeny, such as the onset of reproductive maturity (sometimes differentially in the two sexes, to produce sexual dimorphism, for example, in *Metacancer magister*; Hartnoll 1982), the transition from the anamorphic to the epimorphic developmental phase in hemianamorphic arthropods (e.g. *Lithobius mutabilis*; Albert 1982), or the passage from the larval to the post-larval phase (e.g. *Corystes cassivelaunus*; Hartnoll 1982).

Different patterns of relative growth are responsible for different forms of polymorphism (e.g. sexual dimorphism) and polyphenism (e.g. differences among horned beetle morphs; Tomkins and Moczek 2009). Allometric growth can also affect homologous body parts on either side of the body, producing different forms of bilateral asymmetry. This is for instance the case of *heterochely* of decapod crustaceans, where the left and the right claws of the first pair of legs differ in size and shape, sometimes conspicuously (Hartnoll 1982).

Beyond general and theoretical works on allometry (see Nijhout 2011 and references therein), the literature on arthropod ontogenetic allometry is vast and scattered. Comprehensive accounts for major groups include Hartnoll (1982) for crustaceans and Heming (2003) for insects.

### 5.2.4.4    Meristic Growth

*Meristic growth* is the variation in the number of serially homologous features with growth. Post-embryonic development is generally accompanied by changes in the numbers of 'body units' of different kinds, from simple structures consisting of one or a few cells (*organules*, sensu Lawrence 1966) to complex anatomical modules. These units can be segmental structures whose number changes through anamorphosis (see Sect. 5.2.1.1), but also limb pairs, articles of appendages, sensory bristles, epidermal glands or units of superficial structural or pigment patterns. Such features occurring in multiple copies may represent independent developmental units, or unitary targets of selection, or both (Minelli et al. 2010).

Certain body features, comprising only one cell or a fixed set of very few cells, tend to vary, during ontogeny, exclusively in number, rather than in size. This is, for instance, the case for the ommatidia of Tetraconata, which at each moult are added in rows at the margin of the growing compound eye (Harzsch et al. 2007), or of the sparse sensory setae found on most body sclerites. In contrast, other features, consisting of an indeterminate number of cells, can in principle change with growth in number and in size. Besides the obvious cases of body segments and appendage articles, a lesser known example is provided by the ocelli of lithobiomorph centipedes and eye-bearing millipedes. On each side of the head, there is a field of photoreceptive units whose number increases regularly at each moult, thus providing a quantitative meristic trait useful for identification of an individual's developmental stage (e.g. Andersson 1981; Enghoff et al. 1993). However, in these animals, also the size of at least some ocelli increases with stage (Minelli et al. 2010).

### 5.2.4.5 Degrowth and Anti-Growth

Occasionally, moults occurring under stressful condition are accompanied by reduction in body size, as observed in some crustaceans and insects. This phenomenon has been described as *degrowth*. For instance, the dermestid beetle *Trogoderma glabrum* normally develops through five or six larval stages, but prolonged

lack of water and food may induce the larva to go through extra moults, with a progressive decrease in body size (Beck 1971).

Degrowth should not to be confounded with *anti-growth* (Kluge 2004), which is the morphological regression of specific structures accompanying a moult. Examples are provided by the mouthparts of Ephemeroptera, which regress from the last nymphal stage to the subimago, and by several kinds of trunk appendages that experience a morphological (sometimes reversible) regression during ontogeny (see Sects. 5.2.5.1 and 5.2.5.3).

## 5.2.5 Larvae and Metamorphosis

A butterfly's life cycle is a textbook example of development marked by metamorphosis. Morphological and functional changes separating the mature caterpillar larva from the pupa and the pupa from the adult are indeed enormous, although the actual reorganization depending on histolysis of larval structures and building up of adult structures out of imaginal discs and histoblasts is less dramatic than in other insects such as *Drosophila* (Held 2002).

In a life cycle marked by metamorphosis, the post-embryonic stages preceding the event are generally called *larvae*. However, there is a huge disparity in the use of this term across the arthropods. The reasons are not merely the different taxonomic traditions of the main taxa, but they often reflect actual disparity in the mode by which key developmental transitions occur. This is the subject of the following sections.

### 5.2.5.1 Arthropod Larvae

What is a larva? Criteria for the use of the term in many different metazoan clades, and warnings about its abuse, have been extensively discussed (e.g. Hall and Wake 1999; Minelli 2009). Focussing on the Arthropoda, we will discuss here four different, partly but not necessarily overlapping criteria that would justify the use of the term with respect to a given post-embryonic stage.

## Larva as a Stage with Incomplete Segment Number

The presence of a number of segments lower than in the adult is commonly assumed (e.g. in crustaceans and Chilopoda but not in Diplopoda) as a criterion to distinguish a larval from a post-larval (juvenile to adult) stage. However, there are several reasons to regard this as inadequate. First, in hemianamorphic centipedes and millipedes, morphological differences—other than number of segments and leg pairs—between anamorphic and post-anamorphic stages are minor if any. Second, the anamorphic phase in the development of some hemianamorphic arthropods is inconspicuous, as in Protura and Craterostigmomorpha. In these arthropods, should we use larva for an early juvenile simply because it has one or two segments less than the adult?

Third, nothing but tradition explains why 'larva' is in use for the anamorphic stages of Scutigeromorpha and Lithobiomorpha, but not for Diplopoda.

Fourth, as a consequence of this use, 'post-larva' is used for the juvenile (epimorphic) stages with full complement of segments in the anamorphic Chilopoda, a term otherwise used only in certain lineages of crustaceans, where the use of larva for the earliest stages is arguably better justified than in the hemianamorphic centipedes.

## Larva as a Stage Lacking Some Pairs of Adult Appendages

In the absence of differences in the number of body segments, an early stage lacking one or more pairs of appendages eventually found in the adult is also called a larva. This is the case of the hexapod 'larva' of Ricinulei and anactinotrichid mites, or the manca larva of peracaridan Malacostraca.

From a comparative developmental perspective, however, this choice is disputable. In mites, the fourth pair of legs is in fact present in the embryo and temporarily disappears in the so-called larval stage, only to re-appear (in many mites) in the nymphal and adult stages. Temporarily disappearing 'Lazarus appendages' (Minelli 2003) are also found in some decapod

crustaceans (Balss et al. 1940–1961; Schram 1986). For example, in the mastigopus stage of *Sergestes*, the appendages of the pereion, previously full-developed in the mysis stage, loose their exopodites and the last two pairs of locomotory legs regress completely, only to reappear at later stages. In several pycnogonids, the three first pairs of appendages present in the earliest stages are completely lost in stage VII, but later the second and third pair of larval appendages grow again, with new form and function (Dogiel 1913).

We should also mention here the fact that the larvae of some holometabolous insects are polypodous, that is, possess abdominal appendages of which no trace remains in the adult (e.g. *Sialis* among the Megaloptera; the majority of lepidopteran caterpillars and of sawflies larvae).

Similar to the difficulty in establishing a larval/post-larval divide based on incomplete versus complete segment number, distinguishing larvae from post-larvae is often arbitrary, and also based on the lack versus presence of appendages typical of the adult. This problem is quite serious in the case of decapod crustaceans (Rabalais and Gore 1985). One reason for concern, with respect to the traditional scheme, is that the number of larval (zoeal) stages varies even among congeners, and the zoea even disappears in some species, as in *Pinnotheres* (0–5 zoeas, according to species) (Rabalais and Gore 1985). Secondly, 'typically larval' traits are often found in otherwise post-larval stages or vice versa. Gore (1985) introduced the concept of a *continuum of developmental types* to comprehensively describe the diversity of post-embryonic developmental schedules in decapods, with abbreviated development resulting, other than by reduction of larval stages, by variation in the post-larval stages, for example, by suppression of the megalopa. Remarkably, individual traits are affected by such reductions independently, for example, in *Upogebia savignyi*, there is only a very short larval stage (Gurney 1937), a kind of 'advanced zoea' whose appendages, except for the antenna, antennule and the first three pairs of pereopods, are like those of the adult. The reasons for classifying a developmental stage (or phase) as larval or post-larval are sometimes completely arbitrary. For example, in the shrimp *Macrobrachium* (Palaemonidae), Shokita (1977) distinguished a 'megalopal phase' from a second 'zoeal phase' despite the fact that the 'megalopa' exhibits some zoeal characters such as an incompletely formed telson, combined with many more post-larval characters such as the presence of uropods and functional pleopods.

## Larva as a Stage Preceding Major Metamorphic Change

In the absence of differences in the number of body segments and in the number of appendages, a larva is recognized whenever there are 'major differences' between the early and the adult stages, so that the morphological change associated with the moult to adult is described as a metamorphosis. This is more sensible when larval organs are discarded, reduced or remodelled during the metamorphosis. But remodelling associated with the presence and activation at metamorphosis of specialized sets of 'set-aside' cells (imaginal discs, histoblasts) is not universal in these cases of conspicuous to dramatic metamorphosis. One might thus question the validity of classifying under the same name of larva a fly maggot little of which will go unmodified into the adult, and the aquatic juvenile (sometimes admittedly called a *nymph*, or a *najad*, rather than a larva) of a damselfly (Odonata Zygoptera), irrespective of the fact that with the moult to adult the latter will eventually abandon its leaf-like trachaeobronchial appendages and turn the mask into a much more conventional labium. Also within the holometabolan clade, the role of imaginal discs in generating the adult epidermis is very diverse, and often much less conspicuous than in Lepidoptera or Diptera (Grimaldi and Engel 2005).

Adult organs of holometabolans often have a mixed origin, with various contributions from more or less modified, or respecified, surviving larval components, plus newly formed components deriving from the proliferation and differentiation of the set-aside cells forming the imaginal discs. For example, the adult central nervous system derives from both larval neuroblasts that differentiate into adult neurons and from respecified larval neurons (Truman et al. 1993; Truman and Reis 1995; Tissot and Stocker 2000; Knittel and Kent 2005; Williams and Truman 2005), while the peripheral nervous system of the adult is derived entirely from imaginal disc cells (Tissot and Stocker 2000).

## Larva as an Active Stage Preceding a Resting One

Larvae are also recognized when a morphological discontinuity between the pre-adult post-embryonic stages and the adult is accompanied by the presence of an intercalary resting stage, as in the Holometabola, where the resting stage is generally known as the *pupa*. But it is not that easy to generalize or, especially, to extend this description outside the Holometabola. Two difficulties arise even within the latter clade. First, a resting, non-feeding stage is not necessarily followed by the adult. In the blister beetles (Meloidae), a resting, non-feeding 'coarctate larva', or *hypnotheca*, is intercalated between two active larval phases. Second, the pupae of several holometabolan clades are far from being totally immobile and some of them (*decticous* pupae) possess well-articulated, movable mandibles.

More controversial is the categorization of the post-embryonic developmental stages of the Thysanoptera. The first two stages, both of them active, feeding and devoid of external wing pads, are generally known as larvae; these are followed by two (Terebrantia) or three (Tubulifera) resting stages, with increasingly longer wing pads, of which the first is called a pronymph and the last (or last two) nymph(s). As a rule, these resting stages take no food, except for *Caliothrips indicus* (Wilson 1975). Clearly, there are both similarities and differences between the developmental phases of thysanopterans and those of holometabolans, and this helps to explain the somehow confusing stage terminology adopted in the former taxon. Thysanopterans and holometabolans share the presence of resting stages between the earliest, actively moving and feeding post-embryonic stages and the final adult stage, but only in thysanopterans does the resting phase comprise multiple stages separated by moults. Second, the use of the terms larva and nymph in thysanopterans is in principle questionable if we require some degree of homology to underpin the choice of terms. It is not our aim in this chapter to suggest a new terminology, but to invite further thought and eventual future stabilization of a better terminology. Nymph, anyway, is currently used in a quite consistent way for the pre-adult stages (at least for those with visible wing pads; for example, Sehnal 1985) of several hemimetabolous insect taxa, for example, Orthoptera, but less frequently for those of other taxa, such as Heteroptera or Odonata, which are often called larvae instead. We do not see any morphological or developmental reason for this inconsistency. Eventually, if we accept usage of the term larva for the pre-metamorphic stages of holometabolans, it seems sensible to use a different term—nymph—for the juvenile stages of hemimetabolans, including all those of the thysanopterans, then perhaps distinguishing the active (currently, 'larval') ones from the resting ('pronymphal' and 'nymphal) ones could be achieved through the adoption of a prefix or an adjective. This choice would eventually leave the term pronymph for unambiguous usage as the name of the embryoid hatchlings of many arthropod groups as mentioned above (see Sect. 5.2.3.1). We must admit that even if this were done, we would still be some distance from a completely satisfactory categorization of thysanopteran stages. Morphologically, their resting stages, with their external wing pads, are not that different from advanced nymphs of typical hemimetabolans, but they are affected by massive histolytic and histogenetic processes that are much more profound than those found in most other hemimetabolans. Last but not least, female thysanopterans are often fertilized during their 'nymphal' phase (Zur Strassen and Göllner-Scheiding 2003), thus further blurring the boundaries between conventional developmental phases.

### 5.2.5.2 Larval Kinds Across the Arthropoda (and their Too Many Names)

The difficult problem of providing a consistent approach to the diversity of arthropod larvae is the compound result of their divergent evolutionary pathways and of the heterogeneity of criteria by which these post-embryonic stages have been characterized and named.

The number of terms currently in use for the different kinds of arthropod post-embryonic stages is enormous, in crustaceans especially. For the Decapoda only, Williamson (1969, 1982) listed some 140 names that have been proposed for their larvae and post-larvae. Names used for decapod larvae include *nauplius* (Penaeida only), *elaphocaris* (Sergestidae), *prezoea*, *protozoea*, *zoea*, *naupliosoma*, *phyllosoma* (Palinuridae and Scyllaridae); names used for post-larvae are *megalopa*, *acanthosoma* (Sergestidae), *decapodid*, *eryoneicus* (Polychelidae), *glaucothoe*, *grimothea*, *mastigopus* (Sergestidae), *mysis* (Dendrobranchiata; also in the Euphausiacea), *nisto* (Scyllaridae), *parva*, *pseudibacus* (Scyllaridae), *puerulus* (Palinuridae), and there are many others.

Characteristic of Tanaidacea and Isopoda is the *manca* stage lacking the last pair of pereopods, whereas specific to the parasitic bopyrid isopods are the stages called *cryptoniscium*, *epicaridium* and *microniscium*. The Stomatopoda have their *alima*, *antizoea*, *erichthus* and *pseudozoea*. Several clades of 'entomostracan' crustaceans (in addition to the penaeid decapods) have a *nauplius* (*orthonauplius*, *metanauplius*), followed in the copepods by post-larval juveniles called the *copepodids*. Cirripeds have nauplius and *cyprid* larvae, rhizocephalans also the male larva *trichogon* and the female larva *kentrogon*; facetotectans have *Y larvae*, whose adult is still unknown despite the recent success in inducing these larvae to metamorphosis (Glenner et al. 2008).

Despite their huge diversity, the number of terms available for the active stages of holometabolous insects is quite limited. For most of them, the single term larva is the technical equivalent of many vernacular names, such as maggot, caterpillar, grub. A few main kinds of insect larvae are simply distinguished by descriptive adjectives pointing to the number of leg pairs (apodous, oligopodous, polypodous), the conspicuousness of the cephalic capsule (eucephale, acephale), or the general shape (e.g. campodeiform, eruciform, scarabaeiform or melolonthoid, onisciform, vermiform). More specific terms, such as *triungulin* and *planidium*, are limited to characteristic stages of some hypermetabolous insects (cf. Sect. 5.4.2.1). Post-embryonic stages or phases are also distinguished in termites, for example, *male minor worker* or *presoldier*.

### 5.2.5.3 Non-systemic Metamorphosis

The development of male *gonopods*, the specialized appendages involved in sperm transfer in julid millipedes, is an extreme case of complex and highly species-specific structures differentiating in a very advanced phase of an arthropod's post-embryonic development (Drago et al. 2008, 2011). These appendages develop through a kind of metamorphosis with dramatic effects on the external morphology and internal anatomy of the trunk diplosegment (or ring) bearing the eighth and ninth pair of legs, but leave the sections of the trunk that both precede and follow it completely unaffected. We introduced the term *non-systemic metamorphosis* for this kind of significant post-embryonic transformation confined to a circumscribed body district (Drago et al. 2008).

Millipede gonopods replace, in the adult, one or two pairs of normal walking legs, whose place is first taken, following a moult, by rudimentary (squamiform) appendages and eventually, following one or more additional moults, by the fully formed gonopods, whose shape is often highly complex and totally unlike an articulated arthropod leg. The segmental position of the gonopods along the main body axis is virtually the same in all helminthomorph millipedes and is very likely marked at an early developmental stage, several moults (i.e. a few to many moults) before the gonopods are actually formed.

Identifying the genes specifically involved in this process, and revealing their spatial and temporal patterns of expression would be an interesting model for the comparative developmental biology of arthropods.

## 5.3 Combining Developmental Events

The usual periodization of arthropod post-embryonic development in terms of stages separated by moults is based on a 'cuticular view' (Minelli et al. 2006), that is, on a comparison of periodically frozen morphological features, with total disregard for the morphogenesis of internal organs, which is mostly continuous, or follows a periodization other than that suggested by moults, and major physiological events, such as the achievement of reproductive maturity.

There are two reasons why the moult-based stages should not be taken unquestioned as temporal units of development.

One reason is that the moults that an individual arthropod undergoes throughout its life are not equivalent. As it is most obvious in the case of holometabolous insects, some moults are little more than punctuations during growth, whereas other moults mark a true metamorphosis, consisting of more or less extensive histolytic and morphogenetic events. At a closer inspection, we realize that moults marking major structural changes are not limited to the larva to pupa, or the pupa to adult transitions. For example, in many Coleoptera, there are systematic differences between first and later larval stages (Lawrence 1991), for instance in general body shape and in the tergal pigmentation and armature (Sphaerosomatidae, some Bothrideridae), in the form of the tergum IX (Lymexylidae, Tenebrionidae), or in the number of antennal articles (Cupedidae, Dytiscidae, Helodidae). Another example is provided by the nymphs of locusts (Acrididae), whose wing pads change orientation with the third moult (Brusven 1987). Important physiological transitions during post-embryonic development are like to remain unnoticed unless gene expression studies are performed. Thus, during the third larval stage of *Drosophila*, a 'mid-3rd transition' (Andres and Cherbas 1992; Andres et al. 1993) occurs, which has no morphological or behavioural correlates, but is flagged by marked changes (increase or decrease by at least a factor 10) in the expression level of over 1,500 genes (Graveley et al. 2011).

The other reason is that throughout the postembryonic development of all arthropods different processes occur, which may or may not be associated, and whose association can differ both between clades and through ontogeny. Most of these processes are independent from the moult cycle, running in an essentially continuous way, or beginning and/or ending at times not correlated with any moult cycle event.

To be sure, association of events referable to two or more processes is mechanistically likely (e.g. the temporal coupling of moults and mitotic waves in the epidermis), or functionally advantageous (e.g. the association between feeding phases and intermoult periods), and the occasional convergence of many different events within a short time span, as in the pupal phase of holometabolans, fully justifies singling out certain stages as significant temporal modules in an arthropod's life. However, generalizing to all stages, as developmental segments comprised between two subsequent moults, would be unwarranted, as the examples in the following sections will illustrate. Accordingly, the evolution and evolvability of arthropod post-embryonic development do not rest on an evolutionary independence of developmental stages (Scholtz 2008), but rather on the association and dissociation of concurring processes throughout arthropod phylogeny (cf. Minelli et al. 2006).

### 5.3.1 The Embryonic/Post-embryonic Divide

Invariant features of the hatching event belong more to the eggshell than to the little arthropod that until then was encased within it. Minelli et al. (2006) remarked that there is no reliable phylum-wide correlation between hatching and (1) moulting (variable number of moults preceding hatching), (2) degree of morphological differentiation of the hatchling (from embryoid to adult-like), and (3) body segmentation (from anamorphic to epimorphic). Thus, hatching fails to be an unequivocal reference point for the comparison of ontogenetic schedules.

The animal emerging from a freshly opened eggshell can either be embryoid, with incompletely articulated appendages and thus incapable of motion, or have all the features of an active animal in its post-embryonic life, soon leaving what remains of the eggshell. An embryoid newborn generally undergoes a moult (in some groups two moults) following which it becomes an active, evidently post-embryonic animal. Embryoid hatchlings are often called *pronymphs* or *prenymphs* or *prelarvae*, according to the different traditions for the individual groups; even more specific terms have been coined for some groups (for a summary see Minelli et al. 2006).

Within the Chelicerata, still embryoid hatchlings with incompletely articulated appendages—variously known as larva, pronymph or (Solifugae only) *post-embryo*—occur is many clades of Chelicerata—Scorpiones, Thelyphonida, Amblypygi, Araneae (but not all of them), Solifugae, Ricinulei, Acari, cyphophthalm Opiliones. However, a corresponding stage does not exist in the Pseudoscorpiones and in the majority of Opiliones. More or less embryoid pronymphs are found also in millipedes (a *pupoid* with unarticulated anlagen of antennae and legs), pauropods (one (*Pauropus*) or two (*Gravieripus*) pupoid stages), symphylans (usually called a prelarva, with non-functional mouthparts), epimorphic centipedes (a pupoid and a *peripatoid* stage), several branchiopods (Fritsch and Richter 2012) (especially the 'embryonized larva' of anomopod cladocerans; Kotov and Boikova 1998; Boikova 2008), several decapods (a non-feeding and non-swimming *prezoea*) and also insects such as dragonflies, grasshoppers (a prelarva) and mayflies in which the prolarva lacks the external gills found in the next stage.

The presence or absence of pronymphs is often characteristic of higher taxa, but there are exceptions. For example, pronymphs occur in many spiders, but not in all; in many pycnogonids, but not in all. Even among the holometabolous insects, which as a rule hatch at a similar stage of development, there are a few exceptions, as among the endoparasitoid hymenopterans (Sehnal et al. 1996).

### 5.3.2 Hatching Versus Change of Internal Structure

Little is known of the changes in internal anatomy that accompany hatching, but these are likely to vary from minor, in many cases, to relatively conspicuous, in others, as in the vermiform pronymph of the grasshopper *Schistocerca gregaria*. This earliest post-embryonic stage uses a number of muscles in opening the chorion, digging up through the soil to reach the surface, and finally in shedding the last embryonic cuticle. Of these muscles, a majority will be preserved throughout the nymphal stages, although only few of them will be retained in the adult, with others degenerating following the moult to nymph I (Bernays 1972).

### 5.3.3 Mitosis Versus Moulting and Shape Change

Moulting is generally accompanied by a burst of mitotic activity in the epidermis, but this rule is not universal. In several mites, for example, tetranychids, no post-larval mitosis has been reported (Evans 1992).

Size and shape changes can occur between moults or after the last moult, most probably in the absence of mitosis, although, to the best of our knowledge, firm evidence about the presence/absence of mitoses is lacking. Examples are provided by the enormous, physogastric abdomen of several social insect queens, among the termites especially, about which there are ultrastructural and histochemical studies (Bordereau 1982).

The opposite is also true, with mitoses not correlated with ontogenetic growth through the moult cycle, but with physiological growth responses. The underlying epidermis of the abdominal cuticle of ticks and blood-feeding reduviid bugs undergoes cell division during feeding, accompanied by secretion of new cuticle, to sustain the very rapid expansion of the body wall (see Hackman 1975, 1982; Sehnal 1985) produced by the abundant liquid meal.

### 5.3.4 Moulting Versus Change of Internal Structure

'The use of ecdyses for staging insect development should not detract attention from the continuity of insect ontogeny' (Sehnal 1985, pp. 21). This is all too often overlooked, but it is certainly true for arthropods generally.

In the centipede *Lithobius*, production and differentiation of new neuromeres precede the external appearance of the corresponding segments, and essentially proceed continuously, irrespective of the moult cycle (Minelli et al. 2006). In some insects, neurogenesis continues well into the adult stage. This has been documented in *Agrotis ipsilon* (Lepidoptera: Noctuidae) (Dufour and Gadenne 2006), *Aleochara curtula* (Coleoptera: Staphylinidae) (Bieber and Fuldner 1979), *Acheta domesticus* (Orthoptera: Gryllidae) (Cayre et al. 1994) and in the cockroach *Diploptera punctata* (Gu et al. 1999), but there is no adult neurogenesis in the honeybee (Malun et al. 2003) or in grasshoppers (Orthoptera Acrididae) (Cayre et al. 1996).

### 5.4 Evolutionary Patterns in Post-embryonic Development

The evolution of post-embryonic developmental schedules deserves detailed analysis in the context of phylogeny (arguably available for many interesting nodes) and developmental genetics (about which current knowledge is still too fragmentary), but this is beyond the scope of this chapter. Instead, as a hint on to the evolvability of arthropod post-embryonic development, we will present two examples of opposite trends, towards reduction of the 'normal' sequence of post-embryonic stages and towards increasing complexity of this schedule.

### 5.4.1 Reduction of Post-embryonic Stages

The number of post-embryonic stages is high and variable in many plesiotypic, or 'basal' arthropod clades, and low and fixed in many apotypic, or 'derived' clades. However, it is not possible to trace a consistent trend towards reduction and fixation of the number of post-embryonic moults—right as we cannot describe the morphological evolution of arthropods as dominated by a consistent trend towards a reduction in number and number variation accompanied by an increase in individual differentiation of body segments ('Williston's law'; for a discussion, see Minelli 2003).

However, within individual clades, the evolution of very specialized life styles has been repeatedly accompanied by extreme reduction in the number of post-embryonic stages. For example, the termitophilous beetle *Corotoca* (Staphylinidae: Aleocharinae) is ovoviviparous and deposits larvae that are already so advanced in development that they are almost ready to pupate (Seevers 1957). Other insects retain oviparity, but lay enormous eggs which develop into larvae that undergo a reduced number of moults. In the most specialized members of the Bathysciinae (Coleoptera: Leiodidae) (Deleurance-Glaçon 1963a) and Trechinae (Coleoptera: Carabidae) (Deleurance-Glaçon 1963b), the larva does not feed, and in some species there is no larval moult. A parallel behaviour has been recorded in four Western Australian bolboceratine beetles (Geotrupidae), *Blackburnium reichei*, *Blackbolbus frontalis*, *Bolborhachium inclinatum* and *B. trituberculatum*, which lay one gigantic egg at a time. In *Bl. reichei*, the egg

is up to 56 % of the female weight. Larval feeding can be ruled out at least in some of these species (Houston 2010).

Extreme reduction of the larval stage is also observed in some Termitoxeniinae, a very specialized clade within the diverse dipteran family Phoridae (Disney 1994). The larvae of *Clitelloxenia hemicyclia*, for example, do not feed at all (Franssen 1933, 1936) during their extremely short existence, a few minutes only. In this species, the hatching larva has a double skin (Bridarolli 1937) but it is not clear whether these correspond to the I and II stages, with suppression of the III, or to the II and III, with the first stage either suppressed or passed when still in the egg, the second alternative being favoured by Ferrar (1987) and Disney (1994).

One larval stage only is also present in the tiny parasitoid wasp *Trichogramma australicum* (Jarjees and Merritt 2002).

### 5.4.2 Increasing Complexity

In evolutionary terms, there have been several distinct routes towards increasing complexity of arthropod life cycles, including (1) increasing differentiation between pre-adult and adult stages, as with the transitions to pterygote insects; (2) increasing differentiation within the series of pre-adult stages, as in many crustaceans; (3) specialization of a (the) pre-adult stage as a resting stage allowing body reorganization through extensive histolysis and histogenesis, as, most conspicuously, in the holometabolous insects; (4) evolution of a multigeneration life cycle (Minelli and Fusco 2010) with the regular alternation between amphigonic and parthenogenic generations (heterogony) (e.g. many aphids).

Widespread, and often very conspicuous, also is increasing complexity through polyphenism, that is, the development of alternative phenotypes in response to specific environmental physical, chemical or biological cues (West-Eberhard 2003). The most popular examples among arthropods are the solitary and gregarious phases of some locusts and the castes of social insects. In a few instances, the animal's reproductive mode also falls within the scope of polyphenism, as in several heteropezine midges (Nikolei 1961; Wyatt 1961, 1964) and the beetle *Micromalthus debilis* (Pollock and Normark 2002). In these insects, external cues induce a larva to complete metamorphosis into a conventional adult, or, alternatively, to lay, paedogenetically, mature eggs and/or larvae. Far from being simple expression of environmental sensitivity of development, the effects of polyphenism can eventually turn into a genetically controlled polymorphism. If this transition affects different developmental stages differentially, this can possibly contribute to the evolution of more complex life cycles (Minelli and Fusco 2010).

### 5.4.2.1 Hypermetamorphosis

A sequence of post-embryonic stages is conventionally regarded as 'normal' if the only phases recognizable along the sequence of stages correspond to one of the following divides: (1) early morphologically simple stages versus late morphologically complex stages (arthropods whose life cycle includes a metamorphosis); (2) early active (feeding) stages versus late resting (non-feeding) stages (generally, one) (e.g. holometabolous insects); (3) anamorphic stages versus epimorphic stages (hemianamorphic arthropods); (4) segment articulation forming stages versus post-segment articulation forming stages (trilobites). But several arthropods, among the holometabolous insects especially, have more complex schedules of post-embryonic development, whose sequences deserve the name of *hypermetamorphosis*.

Hypermetamorphosis has evolved not less than a dozen times, that is, in the Strepsiptera and in a number of representatives of Coleoptera (Micromalthidae, Meloidae and Rhipiphoridae, plus a few genera in other families), Neuroptera (Mantispidae), Hymenoptera (Chalcidoidea), Lepidoptera (Gracillariidae) and Diptera (Bombyliidae, Acroceridae, Nemestrinidae).

This phenomenon has been most intensively studied in meloid beetles. Here, a common type of hypermetamorphosis involves a campodeiform first larval stage, which is involved in active or passive dispersal; this larva, usually called a triungulin, is generally followed by a grub-like larva (four stages), followed in turn by a resting coarctate larva, which is not to be compared to a pupa: it does not metamorphose into an adult, but it gives rive to one more grub-like stage, which is the last (relatively) active stage preceding the pupa and, at last, the adult. Variations on this theme are, however, common. For example, in response to high temperature, some *Epicauta* develop from the last stage of the first grub phase directly into the pupa, whereas in response to adverse conditions, the larvae of some Lyttini may revert to a coarctate phase after reaching the second grub phase.

Similarly complex, but less easy to interpret, is the hypermetamorphic developmental schedule of the Strepsiptera. Here, in the male, we recognize a free-living primary larva (similar to the blister beetles' triungulin), an endoparasitic secondary larva (a phase punctuated by several moults), a tertiary larva, a prepupa, a pupa and finally the winged adult. Less clear is the development of the eventually sac-like female, where it is uncertain if a pupation really happens—it does not according to Kinzelbach (1971).

Interestingly, a strong morphological contrast between the first larval stage (usually, more active and even with an exploratory behaviour) and the following stages is found also in the bombyliid flies and in the Perilampidae and Eucharitidae among the tiny chalcidoid wasps. It is easy to suggest an interpretation in terms of parallel adaptive scenarios: mobility and exploratory behaviour of the first larva are of vital importance in host location, whereas the following stages can quietly feed on it without needing to move. However, one wonders whether this parallelism between the ontogenies of independently derived examples of hypermetamorphosis may rest upon shared patterns of evolvability. To shed light onto such aspects of this peculiar kind of post-embryonic development, transcriptomic data for the individual developmental stages would be very useful.

### 5.4.2.2 Morphological Complexity Versus Developmental Complexity

Minelli (1996, 2003) remarked that some arthropods with unusually complex life cycle are also characterized by the presence of more complex patterning along their main body axis or, more frequently, along the longitudinal axis of some of their appendages. For example, in multisegmented appendages like insect antennae, the most proximal and also the most distal segments are often of unique shape (some terminal articles building, for example, a club), but uniquely specialized articles occur very seldom at mid-length, to be followed by non-descript distal articles similar to those preceding the few specialized ones. A most conspicuous example of antenna with unique intermediate articles is found in the males of many meloine blister beetles—interestingly, members of a hypermetabolous insect clade. Whether developmental and morphological complexity evolved based on shared genetic or genomic conditions for evolvability is, at the moment, only matter for speculation.

## 5.5 Final Remarks

The huge diversity of both patterns and processes in arthropod post-embryonic development is still largely known only from descriptive, comparative evidence. Only a few scattered studies go beyond this level of investigation, and for a handful of model species. These results do not allow generalizations about the expression of developmental genes at nymphal, larval or pupal stages. In most instances, we do not know how much a morphological feature characteristic of a given post-embryonic stage, the adult included, depends on persistent (or renewed) expression of developmental genes of which we know the patterns of embryonic expression, or on more or less remote downstream effects of embryonically expressed genes whose transcription and translation products are long disappeared at the time a new feature is expressed in post-embryonic development. A comparative study of post-embryonic development at the cellular and genetic level would be a precious target for future research.

**Acknowledgments** We are grateful to Fred Schram for comments on a draft of this chapter.

## References

Addis A, Biagi F, Floris A, Puddu E, Carcupino M (2007) Larval development of *Lightiella magdalenina* (Crustacea Cephalocarida). Mar Biol 152:733–743

Adis J, Golovatch SI, Wilck L, Hansen B (2000) On the identities of *Muyudesmus obliteratus* Kraus, 1960 versus *Poratia digitata* (Porat, 1889), with first biological observations on parthenogenetic and bisexual populations (Diplopoda: Polydesmida: Pyrgodesmidae). Fragmenta Faunistica (Warszawa) Suppl 43:149–170

Albert AM (1982) Deviation from Dyar's rule in Lithobiidae. Zool Anz 208:192–207

Alberti G, Coons LB (1999) Acari: mites. In: Harrison FW (ed) Microscopic anatomy of invertebrates, Vol 8c. Wiley-Liss, New York, pp 515–1265

Andersson G (1981) Taxonomical studies on the post-embryonic development in Swedish Lithobiomorpha (Chilopoda). Ent Scand 16(Suppl):105–124

Ando T, Kojima T, Fujiwara H (2011) Dramatic changes in patterning gene expression during metamorphosis are associated with the formation of a feather-like antenna by the silk moth, *Bombyx mori*. Dev Biol 357:53–63

André HM, Van Impe G (2012) The missing stase in spider mites (Acari: Tetranychidae): when the adult is not the imago. Acarologia 52:3–16

Andres AJ, Cherbas P (1992) Tissue-specific ecdysone responses: regulation of the *Drosophila* genes *Eip28/29* and *Eip40* during larval development. Dev 116:865–876

Andres AJ, Fletcher JC, Karim FD, Thummel CS (1993) Molecular analysis of the initiation of insect metamorphosis: a comparative study of *Drosophila* ecdysteroid-regulated transcription. Dev Biol 160:388–404

Anger K (2001) The biology of the decapod crustacean larvae (Crustacean Issues 14). Balkema, Rotterdam

Arbeitman MN, Furlong EE, Imam F, Johnson E, Null BH, Baker BS, Krasnow MA, Scott MP (2002) Gene expression during the life cycle of *Drosophila melanogaster*. Science 297:2270–2275

Baeza-Rojano E, Guerra-García M, Pilar Cabezas M, Pacios I (2011) Life history of *Caprella grandimana* (Crustacea: Amphipoda) reared under laboratory conditions. Mar Biol Res 7:85–92

Balss HV, Buddenbrock W, Gruner H-E, Korschelt E (1940–1961) Decapoda. In: Bronn's Klassen und Ordnungen des Tierreichs, vol 5(1). Akademische Verlagsgesellschaft Geest and Portig, Leipzig

Beck SD (1971) Growth and retrogression in larvae of *Trogoderma glabrum* (Coleoptera Dermestidae). 1. Characteristics under feeding and starvation conditions. Ann Entomol Soc Am 64:149–155

Benesch R (1969) Zur Ontogenie und Morphologie von *Artemia salina* L. Zool Jahrb Anat 86:307–458

Bennet-Clark HC (1971) The cuticle as a template for growth in *Rhodnius prolixus*. J Insect Physiol 17:2421–2434

Berlese A (1913) Intorno alle metamorfosi degli insetti. Redia 9:121–136

Bernays EA (1972) The muscles of newly hatched *Schistocerca gregaria* larvae and their possible functions in hatching, digging and ecdysial movements (Insecta: Acrididae). J Zool 166:141–158

Bieber M, Fuldner D (1979) Brain growth during the adult stage of a holometabolous insect. Naturwissenschaften 66:426

Bodenheimer FS (1933) The progression factor in insect growth. Quart Rev Biol 8:92–95

Boikova OS (2008) Comparative investigation of the later embryogenesis of *Leptodora kindtii* (Focke, 1844) (Crustacea: Branchiopoda), with notes on types of embryonic development and larvae in Cladocera. J Nat Hist 42:2389–2416

Bordereau C (1982) Ultrastructure and formation of the physogastric termite queen cuticle. Tissue Cell 14:371–396

Bortolin F, Benna C, Fusco G (2011) Gene expression during post-embryonic segmentation in the centipede *Lithobius peregrinus* (Chilopoda, Lithobiomorpha). Dev Genes Evol 221:105–111

Bridarolli A (1937) Los termitoxenidos y los estadios de su periodo larval. Estudios, Buenos Aires 56:121–138

Brusven MA (1987) Superfamily Acridoidea. In: Stehr FW (ed) Immature insects, vol 1. Kendall/Hunt, Dubuque, IA, pp 162–166

Calvert PP (1929) Different rates of growth among animals with special reference to the Odonata. Proc Am Philos Soc 68:227–274

Canard A, Stockmann R (1993) Comparative post-embryonic development of arachnids. Mem Queensland Mus 33:61–468

Cassagnau P (1985) Le polymorphisme des femelles d'*Hydroisotoma schaefferi* (Krausbauer): un nouveau cas d'épitoquie chez les collemboles. Ann Soc Entom France NS 21:287–296

Cayre M, Strambi C, Charpin P, Augier R, Meyer MR, Edwards JS, Strambi A (1996) Neurogenesis in adult insect mushroom bodies. J Comp Neurol 371:300–310

Cayre M, Strambi C, Strambi A (1994) Neurogenesis in an adult insect brain and its hormonal control. Nature 368:57–59

Chesebro J, Hrycaj S, Mahfooz N, Popadic A (2009) Diverging functions of Scr between embryonic and post-embryonic development in a hemimetabolous insect, *Oncopeltus fasciatus*. Dev Biol 329:142–151

Chippendale GM, Yin C-M (1973) Endocrine activity retained in diapause insect larvae. Nature 246:511–513

Coineau Y, Legendre R (1975) Sur un mode de régénération appendiculaire inédit chez les arthropodes: la régénération des pattes marcheuses chez les opilioacariens (Acari: Notostigmata). CR Hebd Séances Acad Sci, Paris 280D, pp 41–43

Cole BJ (1980) Growth ratios in holometabolous and hemimetabolous insects. Ann Entom Soc Am 73:489–491

Corbet PS (1999) Dragonflies. Harley Books, Colchester

Dai T, Zhang X (2011) Ontogeny of the eodiscoid trilobite *Tsunyidiscus acutus* from the lower Cambrian of South China. Palaeontol 54:1279–1288

da Castiglioni DS, Garcia-Schroeder D, Barcelos DF, Bond-Buckup G (2007) Intermolt duration and post-embryonic growth of two sympatric species of *Hyalella* (Amphipoda, Dogielinotidae) in laboratory conditions. Nauplius 15:57–64

David J-F, Geoffroy J-J (2011) Additional moults into '*elongatus*' males in laboratory-reared *Polydesmus angustus* Latzel, 1884 (Diplopoda, Polydesmida, Polydesmidae): implications for taxonomy. ZooKeys 156:41–48

Degrange C (1959) Nombre de mues et organe de Palmén de *Cloeon simile* Etn. (Ephéméroptères), vol 249. C R Hebd Seances Acad Sci, Paris, pp 2118–2119

Deleurance-Glaçon S (1963a) Recherches sur les coléoptères troglobies de la sous-famille des Bathysciinae. Ann Sci Nat Zool 12:1–173

Deleurance-Glaçon S (1963b) Contribution à l'étude des coléoptères cavernicoles de la sous-famille des Trechines. Ann Spéléol 18:227–265

De Loof A, Huybrechts J, Geens M, Vandermissen T, Boerjan B, Schoofs L (2010) Sexual differentiation in adult insects: male-specific cuticular yellowing in *Schistocerca gregaria* as a model for evaluating some current (neuro)endocrine concepts. J Insect Physiol 56:919–925

Dirsh VM (1937) Postembryonic growth in the *Pachyiulus flavipes* C. L. Koch (Diplopoda). Zool Zh 16:324–335 (in Russian)

Dirsh VM (1967) The post-embryonic ontogeny of Acridomorpha (Orthoptera). Eos, Madrid 43:413–514

Disney RHL (1994) Scuttle flies: the Phoridae. Chapman and Hall, London

Dogiel V (1913) Embryologische Studien an Pantopoden. Ztsch wiss Zool 107:575–741

Dohle W (1988) Myriapoda and the ancestry of insects. Manchester Polytechnic, Manchester

Dong D-J, He H-J, Chai L-Q, Jiang X-J, Wang J-X, Zhao X-F (2007) Identification of genes differentially expressed during larval molting and metamorphosis of *Helicoverpa armigera*. BMC Dev Biol 7:73; doi: 10.1186/1471-213X-7-73

Drago L, Fusco G, Garollo E, Minelli A (2011) Structural aspects of leg-to-gonopod metamorphosis in male helminthomorph millipedes (Diplopoda). Front Zool 8: 19; doi:10.1186/1742-9994-8-19

Drago L, Fusco G, Minelli A (2008) Non-systemic metamorphosis in male millipede appendages: long delayed, reversible effect of an early localized positional marker? Front Zool 5: 5; doi: 10.1186/1742-9994-5-5

Dufour MC, Gadenne C (2006) Adult neurogenesis in a moth brain. J Comp Neurol 495:635–643

Dunger W (2003) Ordnung Diplura, Doppelschwänze. In: Dathe HH (ed) Lehrbuch der speziellen Zoologie. Band I: Wirbellose Tiere, 5. Teil: Insecta. Spektrum, Heidelberg, pp 87–96

Dyar HG (1890) The number of molts of lepidopterous larvae. Psyche 5:420–422

Enders F (1976) Size, food-finding, and Dyar's constant. Envir Entomol 5:1–10

Enghoff H, Dohle W, Blower JG (1993) Anamorphosis in millipedes (Diplopoda)—the present state of knowledge with some developmental and phylogenetic considerations. Zool J Linn Soc 109:103–234

Esperk T, Tammaru T, Nylin S (2007a) Intraspecific variability in number of larval instars in insects. J Econ Entomol 100:627–645

Esperk T, Tammaru T, Nylin S, Teder T (2007b) Achieving high sexual size dimorphism in insects: females add instars. Ecol Entomol 32:243–256

Evans GO (1992) Principles of acarology. CAB International, Wallingford

Fabre JL (1855) Recherches sur l'anatomie des organes reproducteurs et sur le développement des Myriapodes. Ann Sc nat Zool 4:257–316

Ferrar P (1987) A guide to the breeding habits and immature stages of Diptera Cyclorrhapha. (Entomonograph vol 8). Brill and Scandinavian Science Press, Leiden-Copenhagen

Ferrari FD, Dahms HU (2007) Postembryonic development of the Copepoda. Crust Issues 8:1–232

Franssen CJH (1933) Biologische Untersuchungen an Termitoxenia hemicyclia Schmitz, Termitoxenia punctiventris Schmitz und Odontoxenia brevirostris Schmitz. Biol Zbl 53:337–358

Franssen CJH (1936) Aanteekeningen over de ontwikkelingscyclus der Termitoxeniidae (Dipt.). Ent Meded Nederlandsch-Indië 2:62–65

Freeman JA (1991) Growth and morphogenesis in crustacean larvae. Mem Queensland Mus 31: 309–319

Fritsch M, Richter S (2012) Nervous system development in Spinicaudata and Cyclestherida (Crustacea, Branchiopoda): comparing two different modes of indirect development by using an event pairing approach. J Morphol 273:672–695

Fusco G (2005) Trunk segment numbers and sequential segmentation in myriapods. Evol Dev 7:608–617

Fusco G, Garland T Jr, Hunt G, Hughes NC (2012) Developmental trait evolution in trilobites. Evolution 66:314–329

Fusco G, Hughes NC, Webster M, Minelli A (2004) Exploring developmental modes in a fossil arthropod: growth and trunk segmentation of the trilobite Aulacopleura konincki. Am Nat 163:167–183

Gertsch WJ (1955) The north American bolas spiders of the genera Mastophora and Agatostichus. Bull Amer Mus Nat Hist 106:225–254

Gibert J-M, Mouchel-Vielh E, Quéinnec E, Deutsch JS (2000) Barnacle duplicate engrailed genes: divergent expression patterns and evidence for a vestigial abdomen. Evol Dev 2:1–9

Glenner H, Hoeg JT, Grygier MJ, Fujita Y (2008) Induced metamorphosis in crustacean y-larvae: towards a solution to a 100 years-old riddle. BMC Biology 6:21; doi:10.1186/1741-7007-6-21

Gnaspini P (2007) Development. In: Pinto-da-Rocha R, Machado G, Giribet G (eds) Harvestmen: the biology of Opiliones. Harvard University Press, Cambridge, pp 455–472

Gnaspini P, Da Silva MB, Pioker FC (2004) The occurrence of two adult instars among Grassatores (Arachnida: Opiliones): a new type of life-cycle in arachnids. Invert Repr Dev 45:29–39

Goodwin TW, Srikukh S (1949) The biochemistry of locusts I. The carotenoids of the integument of two locust species [Locusta migratoria migratorioides R and F, and Schistocerca gregaria (Forsk.)]. Biochem J 45:263–268

Gore RH (1985) Molting and growth in decapod larvae. In: Wenner AM (ed) Larval growth (Crustacean Issues 2). Balkema, Rotterdam, pp 1–65

Graveley BR, Brooks AN, Carlson JW, Duff MO, Landolin JM, Yang L, Artieri CG, Van Baren MJ, Boley N, Booth BW, Brown JB, Cherbas L, Davis CA, Dobin A, Li R, Lin W, Malone JH, Mattiuzzo NR, Miller D, Sturgill D, Tuch BB, Zaleski C, Zhang D, Blanchette M, Dudoit S, Eads B, Green RE, Hammonds A, Jiang L, Kapranov P, Langton L, Perrimon N, Sandler JE, Wan KH, Willingham A, Zhang Y, Zou Y, Andrews J, Bickel PJ, Brenner SE, Brent MR, Cherbas P, Gingeras TR, Hoskins RA, Kaufman TC, Oliver B, Celniker SE (2011) The developmental transcriptome of Drosophila melanogaster. Nature 471:473–479

Grewal SS (2012) Controlling animal growth and body size: does fruit fly physiology point the way? F1000 Biol Rep 4:12; doi: 10.3410/B4-12

Grimaldi D, Engel MS (2005) Evolution of the insects. Cambridge University Press, Cambridge

Groll EK, Günther KK (2003) Ordnung Saltatoria (Orthoptera), Heuschrecken, Springschrecken. In: Dathe HH (ed) Lehrbuch der speziellen Zoologie. Band I: Wirbellose Tiere, 5. Teil: Insecta. Spektrum, Heidelberg, 17:261–290

Gruner H-E (1993) Klasse Crustacea. In: Gruner H-E, Moritz M, Dunger W (eds) Lehrbuch der speziellen Zoologie. Band I: Wirbellose Tiere, 4. Teil: Arthropoda (ohne Insecta). Fischer, Jena, 1:448–1030

Gu SH, Tsia WH, Chiang AS, Chow YS (1999) Mitogenic effect of 20-hydroxyecdysone on neurogenesis in adult mushroom bodies of the cockroach Diploptera punctata. J Neurobiol 39:264–274

Gurney R (1937) Notes on some decapod Crustacea from the Red Sea I. The genus *Processa*. II. The larvae of *Upogebia savignyi* Strahl. Proc Zool Soc London 1937:85–101

Hackman RH (1975) Expanding abdominal cuticle in the bug *Rhodnius* and the tick *Boophilus*. J Insect Physiol 21:1613–1623

Hackman RH (1982) Structure and function in tick cuticle. Annu Rev Entomol 27:75–95

Halkka R (1958) Life history of *Schizophyllum sabulosum* (L.) (Diplopoda, Iulidae). Anns Zool Soc Zool Bot Fenn Vanamo 19(4):1–72

Hall BK, Wake MH (1999) The origin and evolution of larval forms. Academic Press, San Diego

Hartnoll RG, Dalley R (1981) The control of size variation withininstars of a crustacean. J Exp Mar Biol Ecol 53:235–239

Hartnoll RG (1982) Growth. In: Bliss DE (ed) The biology of Crustacea, vol 2. Academic Press, New York, pp 111–196

Harzsch S, Melzer RR, Müller CHG (2007) Mechanisms of eye development and evolution of the arthropod visual system: the lateral eyes of myriapoda are not modified insect ommatidia. Org Divers Evol 7: 20–32

Held LI Jr (2002) Imaginal discs: the genetic and cellular logic of pattern formation. Cambridge University Press, Cambridge

Heming BS (2003) Insect development and evolution. Comstock, Ithaca-London

Hockman D, Picker MD, Klass K-D, Pretorius L (2009) Postembryonic development of the unique antenna of Mantophasmatodea (Insecta). Arthropod Struct Dev 38:125–133

Horn HS, May RM (1977) Limits to similarity among coexisting competitors. Nature 270:660–661

Houston TF (2010) Egg gigantism in some Australian earth-borer beetles (Coleoptera: Geotrupidae: Bolboceratinae) and its apparent association with reduction or elimination of larval feeding. Austral J Entomol 50:164–173

Howes NH (1939) Observations on the biology and post-embryonic development of *Idotea viridis* (Slabber) (Isopoda, Valvifera) from New England Creek, South-east Essex. J mar biol Ass UK 23:279–310

Hrycaj S, Chesebro J, Popadic A (2010) Functional analysis of *Scr* during embryonic and post-embryonic development in the cockroach, *Periplaneta americana*. Dev Biol 341:324–334

Hughes NC, Minelli A, Fusco G (2006) The ontogeny of trilobite segmentation: a comparative approach. Paleobiology 32:602–627

Hutchinson GE, Tongrid N (1984) The possible adaptive significance of the Brooks-Dyar rule. J Theor Biol 106:437–439

Immamura T (1952) Notes on the moulting of the adult of the water mite, *Arrenurus uchidai* n. sp. Annotat Zool Jap 25:447–451

Janssen R, Prpic N-M, Damen WGM (2004) Gene expression suggests decoupled dorsal and ventral segmentation in the millipede *Glomeris marginata* (Myriapoda: Diplopoda). Dev Biol 268:89–104

Jarjees EA, Merritt DJ (2002) Development of *Trichogramma australicum* Girault (Hymenoptera: Trichogrammatidae) in *Helicoverpa* (Lepidoptera: Noctuidae) host eggs. Austral J Entomol 41:310–315

Junqua C (1966) Recherches biologiques et histophysiologiques sur un solifuge saharien *Othoes saharae* Panouse. Mém Mus Natn Hist Nat A 43:1–124

Kfir R (1991) Effect of diapause on development and reproduction of the stem borers *Busseola fusca* (Lepidoptera: Noctuidae) and *Chilo partellus* (Lepidoptera: Pyralidae). J Econ Entomol 84:1677–1680

Kinzelbach R (1971) Strepsiptera (Fächerflügler). Handbuch der Zoologie 4 (2, 2/24). de Gruyter, Berlin

Klingenberg CP (1996) Individual variation of ontogenies: a longitudinal study of growth and timing. Evol 50:2412–2428

Klingenberg CP, Zimmermann M (1992) Dyar's rule and multivariate allometric growth in nine species of waterstriders (Heteroptera: Gerridae). J Zool (Lond) 227:453–464

Kluge NJ (2004) Larval/pupal leg transformation and a new diagnosis for the taxon Metabola Burmeister, 1832 = Oligoneoptera Martynov, 1923. Russian Entomol J 13:189–229

Knittel LM, Kent KS (2005) Remodeling of an identified motoneuron during metamorphosis: hormonal influences on the growth of dendrites and axon terminals. J Neurobiol 63:106–125

Konopová B, Zrzavý J (2005) Ultrastructure, development, and homology of insect embryonic cuticles. J Morphol 264:339–362

Kotov AA, Boikova OS (1998) Comparative analysis of the late embryogenesis of *Sida crystallina* (O.F. Müller, 1776) and *Diaphanosoma brachyurum* (Lievin 1848) (Crustacea: Branchiopoda: Ctenopoda). Hydrobiologia 380:103–125

Lawrence JF (1991) Order Coleoptera. In: Stehr FW (ed) Immature insects, vol 2. Kendall/Hunt, Dubuque, pp 144–658

Lawrence PA (1966) Development and determination of hairs and bristles in the milkweed bug, *Oncopeltus fasciatus* (Lygaeidae, Hemiptera). J Cell Sci 1:475–498

Levy G (1970) The life cycle of *Thomisus onustus* (Thomisidae: Araneae) and outlines for the classification of the life histories of spiders. J Zool (Lond) 160:523–536

Li TR, White KP (2003) Tissue-specific gene expression and ecdysone-regulated genomic networks in *Drosophila*. Dev Cell 5:59–72

Lytle DA (2001) Convergent growth regulation in arthropods: biological fact or statistical artifact? Oecologia 128:56–61

Maiorana VC (1978) An explanation of ecological and developmental constants. Nat 273:375–377

Malun D, Moseleit AD, Grunewald B (2003) 20-Hydroxyecdysone inhibits the mitotic activity of neuronal precursors in the developing mushroom bodies of the honeybee, *Apis mellifera*. J Neurobiol 57:1–14

Manzanares M, Marco R, Garesse R (1993) Genomic organization and developmental pattern of expression of the *engrailed* gene from the brine shrimp *Artemia*. Dev 118:1209–1219

Mauriès J-P (1980) Diplopodes chilognathes de la Guadeloupe et ses dépendances. Bull Mus Natn Hist Nat (4) 2A:1059–1111

Mckenzie HL (1932) The biology and feeding habits of *Hyperaspis lateralis* Mulsant (Coleoptera – Coccinellidae). Univ California Publ Entomol 6:9–21

Michalik P, Uhl G (2005) The male genital system of the cellar spider *Pholcus phalangioides* (Fuesslin 1775) (Pholcidae, Araneae): Development of spermatozoa and seminal secretion. Front Zool 2:12; doi: 10.1186/1742-9994-2-12

Michel A, Manning RB (1972) The pelagic larvae of *Chorisquilla tuberculata* (Borradaile, 1907). (Stomatopoda). Crustaceana 22:113–126

Michener CD (1946) The taxonomy and bionomics of some Panamanian trombidiid mites (Acarina). Annals Entom Soc Am 39:349–380

Miller WE (1996) Population behavior and adult feeding capability in Lepidoptera. Envir Entomol 25:213–226

Minelli A (1996) Segments, body regions and the control of development through time. Mem California Acad Sci 20:55–61

Minelli A (1998) Molecules, developmental modules and phenotypes: a combinatorial approach to homology. Mol Phyl Evol 9:340–347

Minelli A (2003) The development of animal form: ontogeny, morphology, and evolution. Cambridge University Press, Cambridge

Minelli A (2009) Perspectives in animal phylogeny and evolution. Oxford University Press, Oxford

Minelli A, Brena C, Deflorian G, Maruzzo D, Fusco G (2006) From embryo to adult. Beyond the conventional periodization of arthropod development. Dev Genes Evol 216:373–383

Minelli A, Fusco G (2004) Evo-devo perspectives on segmentation: model organisms, and beyond. Trends Ecol Evol 19:423–429

Minelli A, Fusco G (2010) Developmental plasticity and the evolution of animal complex life cycles. In: Fusco G, Minelli A (eds) From polyphenism to complex metazoan life cycles. Phil Trans R Soc B 365:631–640

Minelli A, Fusco G (2013) Homology. In: Kampourakis K (ed) Philosophical issues in biology education. Springer, Heidelberg-Berlin

Minelli A, Maruzzo D, Fusco G (2010) Multi-scale relationships between numbers and size in the evolution of arthropod body features. Arthropod Struct Dev 39:468–477

Monod T (1926) Les Gnathiidae. Essai monographique (morphologie, biologie, systématique). Mém Soc Sci Nat Maroc 13:1–661

Murakami Y (1962) Postembryonic development of the common Myriapoda of Japan. XI–XII Life history of *Bazillozonium nodulosum* Verhoeff (Colobognatha,

Platydesmidae). 1–2. Zool Mag (Dobutsugaku Zasshi) 71:250–255, 291–294

Nijhout HF (1975) A threshold size for metamorphosis in the tobacco hornworm *Manduca sexta* (L.). Biol Bull 149:214–225

Nijhout HF (1994) Insects hormones. Princeton University Press, Princeton

Nijhout HF (2003) The control of body size in insects. Dev Biol 261:1–9

Nijhout HF (2011) Dependence of morphometric allometries on the growth kinetics of body parts. J Theor Biol 288:35–43

Nijhout HF, Davidowitz G (2003) Developmental perspectives on phenotypic variation, canalization, and fluctuating asymmetry. In: Polak M (ed) Developmental instability: causes and consequences. Oxford University Press, New York, pp 3–13

Nijhout HF, Roff DA, Davidowitz G (2010) Conflicting processes in the evolution of body size and development time. Phil Trans R Soc B 365:577–591

Nikolei E (1961) Vergleichende Untersuchungen zur Fortpflanzung der heterogenen Gallmücken unter experimentellen Bedingungen. Ztschr Morphol Ökol Tiere 50:281–329

Pollock DA, Normark BB (2002) The life cycle of *Micromalthus debilis* LeConte (1878) (Coleoptera: Archostemata: Micromalthidae): historical review and evolutionary perspective. J Zool Syst Evol Res 40:105–112

Przibram H, Megušar F (1912) Wachstumsmessungen an *Sphodromantis bioculata* Burm. I. Länge und Masse. Arch Entw Mech Org 34:680–741

Quennedey A, Aribi N, Everaerts C, Delbecque J-P (1995) Post-embryonic development of *Zophobas atratus* Fab. (Coleoptera: Tenebrionidae) under crowded or isolated conditions and effects of juvenile hormone analogues applications. J Insect Physiol 41:143–152

Rabalais NN, Gore RH (1985) Abbreviated development in decapods. In: Wenner AM (ed) Larval growth (Crustacean Issues 2). Balkema, Rotterdam, pp 67–126

Rantala M (1974) Sex ratio and periodomorphosis of *Proteroiulus fuscus* (Am Stein) (Diplopoda, Blaniulidae). Symp Zool Soc Lond 32:463–469

Rice AL (1968) Growth 'rules' and the larvae of decapod crustaceans. J Nat Hist 2:525–530

Riddiford LM, Hiruma K, Zhou X, Nelson CA (2003) Insights into the molecular basis of the hormonal control of molting and metamorphosis from *Manduca sexta* and *Drosophila melanogaster*. Insect Biochem Mol Biol 33:1327–1338

Sahli F (1969) Contribution à l'étude de développement post-embryonnaire des Diplopodes Iulides. Ann Univ Saraviensis math-naturwiss Fak 7:1–154

Sahli F (1985) Periodomorphose et mâles intercalaires des Diplopodes Julida: une nouvelle terminologie. Bull sci Bourgogne 38:23–31

Sahli F (1989) The structure of two populations of *Tachypodoiulus niger* (Leach) in Burgundy and some remarks on periodomorphosis. Rev Ecol Biol Sol 26:355–361

Sahli F (1990) On post-adult moults in Julida (Myriapoda: Diplopoda). Why do periodomorphosis and intercalaries occur in males? In: Minelli A (ed) Proceedings of the 7th International Congress of Myriapodology. Leiden, Brill, pp 135–156

Sanders HL (1963) The Cephalocarida. functional morphology. larval development, comparative external anatomy. Mem Conn Acad Arts Sci 15:1–80

Schmidt FH, Lauer WL (1977) Developmental polymorphism in *Choristoneura* spp. (Lepidoptera: Tortricidae). Ann Entom Soc Am 70:112–118

Scholtz G (2000) Evolution of the nauplius stage in malacostracan crustaceans. J Zool Syst Evol Res 38:175–187

Scholtz G (2008) On comparisons and causes in evolutionary developmental biology In: Minelli A, Fusco G (eds) Evolving pathways. Cambridge University Press, Cambridge, pp 144–159

Schömann K (1956) Zur Biologie von *Polyxenus lagurus* (L. 1758). Zool Jahrb Syst 84:195–256

Schram FR (1986) Crustacea. Oxford University Press, New York

Seevers CH (1957) A monograph on the termitophilous Staphylinidae (Coleoptera). Fiediana Zoology 40:1–334

Sehnal F (1985) Growth and life cycles. In: Kerkut GA, Gilbert LI (eds) Comprehensive insect physiology, biochemistry and pharmacology, vol 2. Pergamon Press, Oxford, pp 1–86

Sehnal F, Svácha P, Zrzavý J (1996) Evolution of insect metamorphosis. In: Gilbert LI, Tata JR, Atkinson NG (eds) Metamorphosis. Postembryonic reprogramming of gene expression in amphibian and insect cells. Academic Press, New York, pp 3–58

Sekiguchi K, Seshimo H, Sugita H (1988) Post-embryonic development of the horseshoe crab. Biol Bull 174:337–345

Shimizu S, Machida R (2011) Reproductive biology and postembryonic development in the basal earwig *Diplatys flavicollis* (Shiraki) (Insecta: Dermaptera: Diplatyidae). Arthropod Syst Phylog 69:83–97

Shokita S (1977) Abbreviated metamorphosis of landlocked fresh-water prawn, *Macrobrachium asperulum* (Von Martens) from Taiwan. Annot Zool Jap 50:110–122

Silvestri F (1949) Segmentazione del corpo dei Colobognati (Diplopodi). Boll Lab Entom Agr Portici 9:115–121

Snodgrass RE (1935) Principles of insect morphology. McGraw-Hill, New York

Soares Moreira P, Setubal Pires AM (1977) Aspects of the breeding biology of *Janaira gracilis* Moreira and Pires (Crustacea, Isopoda, Asellota). Bol Inst Oceanogr 26:181–199

Solomon JD (1973) Instars in the carpenterworm, *Prionoxystus robiniae*. Ann Entom Soc Am 66: 1258–1260

Stehr FW (ed) (1987, 1991) Immature insects. Vol 1 (1987), Vol 2 (1991). Kendall/Hunt, Dubuque

Stern D (2003) Body-size control: how an insect knows it has grown enough. Curr Biol 13:R267–R269

Štys P, Davidova-Vilimova J (1989) Unusual numbers of instars in Heteroptera: a review. Acta Entomol Bohemoslov 86:1–32

Szlep R (1961) Developmental changes in the web-spinning instinct of Uloboridae: construction of the primary type web. Behaviour 27:60–70

Tanaka A (1981) Regulation of body size during larval development in the German cockroach, *Blattella germanica*. J Insect Physiol 27:587–592

Tanaka K, Truman JW (2007) Molecular patterning mechanism underlying metamorphosis of the thoracic leg in *Manduca sexta*. Dev Biol 305: 539–550

Tissot M, Stocker RF (2000) Metamorphosis in *Drosophila* and other insects: the fate of neurons throughout the stages. Prog Neurobiol 62:89–111

Titschack E (1926) Untersuchungen über das Wachstum, den Nahrungsverbrauch und die Eierzeugung. II. *Tineola bisselliella* Hum. Gleichzeitig ein Beitrag zur Klärung der Insektenhäutung. Ztschr wiss Zool 128:509–569

Tomkins JL, Moczek AP (2009) Patterns of threshold evolution in polyphenic insects under different developmental models. Evolution 63:459–468

Truman JW, Reiss SE (1995) Neuromuscular metamorphosis in the moth *Manduca sexta* –hormonal regulation of synapse loss and remodeling. J Neurosci 15:4815–4826

Truman JW, Riddiford LM (1999) The origins of insect metamorphosis. Nature 401:447–452

Truman JW, Taylor BJ, Awad TA (1993) Formation of the adult nervous system. In: Bate M, Martinez Arias A (eds) The development of *Drosophila melanogaster*. Cold Spring Harbor Laboratory Press, New York, pp 1245–1275

Twombly S, Tisch N (2000) Body size regulation in copepod crustaceans. Oecologia 122:318–326

Uvarov BP (1966) Grasshoppers and locusts. a handbook of general acridology. Vol 1. Cambridge University Press, Cambridge

Verhoeff KW (1916) Abhängigkeit der Diplopoden und besonders der Juliden-Schaltmännchen von äußeren Einflüssen. Ztschr Wiss Zool 116:535–586

Verhoeff KW (1923) Periodomorphose. Zool Anz 56(233–238):241–254

Verhoeff KW (1928) Durch Zucht erhaltene Formen des *Polydesmus complanatus, illyricus* Verh. und ihre Bedeutung, sowie Beurteilung der Elongation. Ztschr Morphol Ökol Tiere 12:684–705

Verhoeff KW (1933) Wachstum und Lebensverlängerung bei Blaniuliden und über die Periodomorphose. Ztschr Morph Ökol Tiere 27:732–748

Verhoeff KW (1934) Über die Diplopoden der Allgäuer Alpen, deutsche Craspedosomen und Periodomorphose. Zool Anz 108:27–40

Verhoeff KW (1939) Wachstum und Lebensverlängerung bei Blaniuliden und über die Periodomorphose, II Teil. Ztschr Morph Ökol Tiere 36:21–40

von Lieven AM (2005) The embryonic moult in diplogastrids (Nematoda): homology of developmental stages and heterochrony as a prerequisite for morphological diversity. Zool Anz 244:79–91

Walossek D (1993) The Upper Cambrian *Rehbachiella kinnekullensis* and the phylogeny of Branchiopoda and Crustacea. Fossils Strata 32:1–202

Walters FFA, Dixon AFG, Eagles G (1984) Non-feeding by adult gynoparae of *Rhopalosiphum padi* and its bearing on the limiting resource in the production of sexual females in host alternating aphids. Entom Exp Appl 36:9–12

West TL, Costlow JD (1987) Size regulation in the crustacean *Balanus eburneus* (Cirripedia: Thoracica). Mar Biol 96:47–58

West-Eberhard MJ (2003) Developmental plasticity and evolution. Oxford University Press, New York

Wigglesworth VB (1972) The principles of insect physiology. Chapman and Hall, London

Williams DW, Truman JW (2005) Remodeling dendrites during insect metamorphosis. J Neurobiol 64:24–33

Williamson DI (1969) Names of larvae in the Decapoda and Euphausiacea. Crustaceana 16:210–213

Williamson DI (1982) Larval morphology and diversity. In: Abele LG (ed) Biology of the Crustacea, vol 2. Academic Press, New York, pp 43–110

Wilson TH (1975) A monograph of the subfamily Panchaetothripinae (Thysanoptera: Thripidae). Mem Am Entom Inst 23:1–354

Wyatt IJ (1961) Pupal paedogenesis in the Cecidomyidae (Diptera), I. Proc R Entom Soc A 36:133–143

Wyatt IJ (1964) Immature stages of Lestremiinae (Diptera: Cecidomyidae) infesting cultivated mushrooms. Trans R Entom Soc 116:15–27

Yu L, Coddington J (1990) Ontogenetic changes in the spinning fields of *Nuctenea cornuta* and *Neoscona theisi* (Araneae, Araneidae). J Arachnol 18:331–345

Zimmer M (2002) Postembryonic ontogenetic development in *Porcellio scaber* (Isopoda: Oniscidea): the significance of food. Invert Repr Dev 42:75–82

zur Strassen R, Göllner-Scheiding U (2003) Ordnung Thysanoptera (Physopoda), Fransenflügler, Thripse, Blasenfüße. In: Dathe HH (ed) Lehrbuch der speziellen Zoologie. Band I: Wirbellose Tiere, 5, 21. Teil: Insecta. Spektrum, Heidelberg, pp 331–342

# Arthropod Developmental Endocrinology

**6**

## H. Frederik Nijhout

## Contents

## 6.1    Introduction

Embryonic development in arthropods appears to be largely controlled by gene regulatory cascades and networks, and gene products that move by diffusion. By contrast, postembryonic growth and differentiation are controlled almost entirely by circulating hormones and secreted growth factors. Also, in contrast to the dozens and perhaps hundreds of genes that control early stages of embryonic specification and differentiation, only a very small handful of developmental hormones control an extraordinarily diverse array of postembryonic developmental processes ranging from growth, to moulting, metamorphosis, and the development of alternative phenotypes in response to environmental signals. Hormones such as ecdysone and JH can have many categorically different effects, depending on the species, stage of the life cycle and target tissue. Some hormones, such as ecdysone, appear to be used universally across the Arthropoda whereas others such as JH and androgenic hormone are taxon restricted (to the Insecta and decapod Crustacea, respectively).

The developmental endocrinology of arthropods is concerned with the control of growth and the control of form. Growth can be partitioned into two somewhat independent processes: growth that occurs at each moult, which is concerned primarily with growth of the exoskeleton, and growth during the intermoult, which is mostly concerned with the growth of internal tissues. These two forms of growth are interrelated in that the amount of growth that occurs at a moult depends on the mass accumulated during the intermoult period, and both growth processes are stimulated by the same

H. Frederik Nijhout (✉)
Department of Biology, Duke University,
125 Science Drive, Duke University, Durham,
NC 27708, USA
e-mail: hfn@duke.edu

A. Minelli et al. (eds.), *Arthropod Biology and Evolution*,
DOI: 10.1007/978-3-642-36160-9_6, © Springer-Verlag Berlin Heidelberg 2013

hormones, though under different control mechanisms. The regulation of growth is also intimately involved with regulation of size, both body size and the relative sizes of body parts. Size regulation is a matter of when to stop growing, and different species have found different solutions. When arthropods moult, they not only become larger but they can also change form, either gradually with small morphometric changes from moult to moult, or dramatically, with profound changes in morphology during metamorphosis. In addition, many species have the ability to moult into one of the two or more alternative morphologies, depending on environmental signals they receive, a kind of phenotypic plasticity called polyphenism. In this chapter, I will deal with the endocrine regulation of each of these developmental processes in turn.

## 6.2    Growth and Moulting

Insects and arachnids grow to a species-characteristic final adult size, but many crustaceans and myriapods continue to grow as adults. In species with *determinate growth* sexually mature adults do not moult, whereas in species with *indeterminate growth* sexually mature adults continue to moult and often grow larger at each moult. Lobsters and spider crabs can grow to gigantic size simply because they continue to moult and grow as adults. Among the insects, only the Thysanura and Collembola are known to moult as adults (Christiansen 1964; Watson 1964; Joosse and Veltkamp 1969). In *Thermobia domestica,* moulting cycles and reproductive cycles alternate (Watson 1964; Rohdendorf and Watson 1969; de la Paz et al. 1983; Bitsch et al. 1985), but there is little or no growth during the moult. All Arthropoda have a non-living external skeleton made primarily of chitin (a carbohydrate polymer: poly-N-acetylglucosamine) and proteins (Andersen 1979; Hopkins and Kramer 1992), generally called the cuticle. This cuticle is hardened or sclerotized by crosslinking and chemical transformation of its components and/or by mineralization (e.g. calcification, as in

diplopods and decapods),and cannot grow so that somatic growth requires occasional shedding of the cuticle and the manufacture of a new larger exoskeleton. In soft-bodied larvae of holometabolous insects, the soft cuticle can grow by intercalation of protein and chitin (Wolfgang and Riddiford 1981), and such larvae need not moult in order to grow.

### 6.2.1    The Moulting Process

The body wall of arthropods is made up of a single cell layer, the epidermis (called hypodermis or epithelium in some of the non-insectan literature), that secretes an extracellular cuticle. The cuticle is made up of several layers. From the epidermis outward, there is a thick endocuticle, a somewhat thinner exocuticle and a very thin epicuticle. Although only a few microns thick, the epicuticle has a complex chemical makeup including lipoproteins, glycoproteins and waxes and its primary function appears to be to limit water and chemical permeability of the cuticle. The exocuticle is made up of a mixture of chitin and protein, which can become crosslinked and sclerotized to different degrees depending on the location. Where the protein is only lightly cross-linked, the cuticle is resilient and flexible (for instance at the joints, or the body wall of soft-bodied arthropods). Highly cross-linked and sclerotized exocuticles are stiff and hard. The endocuticle is typically the thickest layer and is also composed of a mixture of chitin and protein, but there is generally little sclerotization, and the endocuticle is typically tough and flexible. When the cuticle is pigmented, the pigment typically is deposited in the exocuticle; the endocuticle is, with few exceptions, colourless. In the decapod Crustacea and Diplopoda, the cuticle becomes calcified by incorporation of calcium carbonate mostly in the form of crystalline and amorphous calcite inserted between the protein and chitin fibres (Roer and Dillaman 1984).

Because of sclerotization and calcification, the hard cuticle of arthropods cannot grow and must occasionally be shed in order to build a new larger cuticle. The shedding of an old cuticle and synthesis of a new larger cuticle is referred to as the moulting process. The first step in the moulting cycle is apolysis: the separation of the epidermis from the cuticle. In the space, the epidermis then secretes a gelatinous material, the moulting gel, that contains inactive enzymes that will later digest the old cuticle. The epidermis then undergoes a round of cell division, which enlarges its surface area and the epidermis becomes finely folded or corrugated within the old cuticle. Next, the epidermis secretes the new epicuticle, which is followed by activation of the enzymes in the moulting gel. The enzymes now begin to digest the old endocuticle (the epidermis is protected from digestion by the new epicuticle), and the epidermis begins to secrete the new exocuticle. The breakdown products of the old cuticle are resorbed and stored and will be used for the synthesis of the new cuticle. Only the endocuticle is digested, and it is the exocuticle that will be shed and lost. In Decapoda, the calcium salts from both the endocuticle and exocuticle are resorbed and stored in the hepatopancreas (Waterman 1960). Ecdysis, shedding of the old cuticle, begins by swallowing air or water to increase the internal volume, and peristaltic movements that break and shed the old exocuticle. The old exocuticle breaks along specific lines of weakness where the exocuticle is very thin, called ecdysial sutures. Immediately after ecdysis, the new expanded exocuticle undergoes sclerotization (and, in Decapoda, calcification) and hardens. The epidermis then begins to secrete the new endocuticle, and endocuticle deposition continues throughout the intermoult period.

## 6.2.2 Growth

Because of the need to moult, the growth of arthropods appears to be discontinuous, with a discrete increase in size of the exoskeleton at each ecdysis. Body mass, however, increases steadily during the intermoult period, in part by filling out the space occupied by air and water

during ecdysis, and in part by expanding folds in articulating membranes. The unsclerotized cuticle of soft-bodied insects (such as the larvae of Lepidoptera) can actually grow by intercalation of chitin and protein during the intermoult period (Wolfgang and Riddiford 1986).

Growth of arthropods thus has two components: episodic growth of the exoskeleton, and more or less continuous growth of biomass. There is a great deal of diversity in the amount of growth that occurs by each of these mechanisms. Some species grow by small increments and need many moults to reach adult size, whereas others grow by large increments and reach the same size in few moults (Cole 1980). Species also differ in the rate of growth between moults: some grow rapidly and moult frequently whereas others grow slowly and have a much longer interval between moults. There appears to be no systematic relationship between adult size and the number of moults required to reach adult size: some of the largest insects, the African Goliath beetles (Coleoptera: Scarabaeidae), undergo only two larval moults whereas flour beetles of the genus *Tribolium* (Coleoptera: Tenebrionidae) that are only a few millimetres long undergo 5–7 larval moults.

Somatic growth is controlled by hormones and secreted growth factors that regulate the onset, rate and duration of growth. The developmental regulation of growth and size can be partitioned into several semi-independent questions: (1) how is moulting controlled; (2) what controls the size increment at each moult; (3) what controls the growth rate between moults; (4) what controls the timing of a moult; (5) what controls the cessation of moulting when the final size is reached (or, alternatively, why does growth and moulting stop when a species-characteristic adult size is reached)? We treat each of these control mechanisms in the sections below.

## 6.2.3 Control of Moulting

In all arthropods, the moulting cycle is controlled by the steroid hormone, ecdysone. Ecdysone is secreted by the prothoracic glands in insects and by the X organ in Decapoda (see below). The ecdysteroid secreted by these glands is actually a relatively inactive prohormone (formerly referred to as α-ecdysone, now simply called ecdysone). This prohormone is activated in peripheral tissues by hydroxylation at the $C_{20}$ location to form the active 20-hydroxyecdysone (20E, formerly called β-ecdysone and ecdysterone). Many other variants of ecdysone, collectively called ecdysteroids, have been detected in arthropods, such as makisterone A in insects and Decapoda, ponasterone A in some decapods, bombykosterol in the silkworm *Bombyx mori* and many differently hydroxylated forms of ecdysone in many insects, including *Drosophila* (Dinan and Hormann 2005). To date, several hundred ecdysteroids have been isolated from animals, plants and fungi (an exhaustive database can be found at http://ecdybase.org/).

The biological functions of the multiplicity of arthropod ecdysteroids are not at all clear. Their biological activity is often assayed in tissue or cell cultures (Dinan et al. 1990; Smagghe and Degheele 1995; Harmatha and Dinan 1997) or by their ability to induce a moult (Krishnakumaran and Schneiderman 1969; Karlson and Koolman 1973; Hoffmann et al. 1974; Berghiche et al. 2007), but these tests provide no information about the natural roles of the different ecdysteroids. Most ecdysteroids can induce moulting, but ecdysteroids have many other functions, particularly in insect reproduction, and it is possible that different tissues have different sensitivities to different ecdysteroids during metamorphosis (see below).

In regard to the moulting cycle, it is clear, however, that in all Arthropoda, the entire process from apolysis through cell division and deposition of the new cuticle is controlled by a single more or less prolonged pulse of 20-hydroxyecdysone. The most direct evidence comes from tissue culture studies in which a piece of integument (epidermis plus cuticle) can be induced to undergo the entire sequence of moulting events when exposed to an ecdysteroid (Locke 1970; Mitsui and Riddiford 1976; Freeman and Costlow 1979). Thus, the proximate control of moulting is entirely attributable

to ecdysone, and it does not matter whether it is a larval moult or a metamorphic moult (insect larva to pupa, pupa to adult, larva to adult), or a stationary adult moult as happens in many Crustacea and in Thysanura (Watson 1964; Chang 1985). Embryonic moults in insects and Crustacea also appear to be controlled by ecdysone (Beydon et al. 1989; Charmantier and Charmantier-Daures 1998; Erezyilmaz et al. 2004).

Although ecdysone is the sole proximate cause for moulting, the secretion of ecdysone is regulated by higher centres, typically associated with the central nervous system, via tropic neurohormones with some evidences for direct nervous control as well (Yamanaka et al. 2006). The regulation of ecdysone release has been best studied in the decapod Crustacea and Hexapoda, each of which appears to have evolved a very different mechanism of neuroendocrine regulation.

### 6.2.3.1 Moulting: Crustacea

More than a century ago, Zeleny (1905) showed that ablating the eyestalks of the fiddler crab *Uca pugilator* could induce a premature moult. This finding was confirmed by many investigators over the years and led to the eventual discovery in the eyestalks of decapod crustaceans of the Y Organ, a cluster of neurosecretory cells that release their product into the blood via a neurohaemal organ called the Sinus Gland (Bruce and Chang 1984; Chang 1985; Hopkins 2012). The Y organ produces the moult-inhibiting hormone (MIH), a polypeptide that acts on the X organ, a compact organ anterior in the cephalothorax usually in the antennal or maxillary segments (Nakatsuji and Sonobe 2004; Nakatsuji et al. 2006, 2009). The X organ may be the homologue of the insect prothoracic gland and secretes ecdysone. As in insects, ecdysone is a relatively inactive prohormone that is hydroxylated in peripheral tissues into the active 20-hydroxyecdysone.

The control of ecdysone secretion and moulting in Decapoda appear to be negative, that is, via the inhibition of ecdysone secretion by MIH.

When inhibition by MIH disappears, ecdysone secretion begins and initiates apolysis and the moult. Studies have shown a reasonable though not perfect inverse correlation between MIH and ecdysone titres (Chang and Mykles 2011), which suggests that additional levels of control perhaps by variation in sensitivity of the X organ to MIH or another endocrine regulator.

In a variety of Crustacea, moulting is inhibited by stressful environmental factors including extreme temperature and crowding (Chang and Mykles 2011). In *Gecarcinus*, moulting can be induced by autotomy of legs. Although there has been no direct demonstration of positive control over ecdysone secretion in Crustacea, when a moult is induced by eyestalk ablation, the ecdysone titres decline after the moult has occurred, even though MIH is absent, which indicates that additional stimulatory or permissive regulators must exist.

### 6.2.3.2 Moulting: Insects

In insects, ecdysone is secreted by the prothoracic glands. As their name implies, in most insects, this gland is located in the first thoracic segment. In the Suborder Cyclorrhapha (the so-called "higher" Diptera, that include *Drosophila*), the prothoracic glands are fused with the *corpora allata* and *c. cardiaca* in a compact structure called the ring gland that encircles the anterior part of the dorsal vessel (Redfern 1983).

The earliest experiments that revealed a control over moulting were carried out by Kopec who showed (Kopec 1917, 1922) that placing a blood-tight ligature at mid-body of caterpillars of *Lymantria dispar* inhibited moulting in the posterior part, and moulting could be entirely inhibited by a blood-tight ligature at the neck (because insects have a distributed respiratory system, such ligations are perfectly survivable; they do prevent feeding, however, and that restricts long-term survival). Wigglesworth (1934, 1940) subsequently demonstrated by parabiosis and brain transplant experiments that the brain exerted its effect via the blood and not through its connection to the nervous system. The fact that two factors were actually required

for moulting was discovered by Williams (1947, 1948), who showed, that isolated abdomens of pupae of *Hyalophora cecropia*, could not be stimulated to moult by a brain implant alone, but also required simultaneous implantation of a fragment of the prothoracic gland. Subsequent experimental work revealed that the brain secretes a tropic hormone, initially called the "brain hormone", and today the prothoracicotropic hormone (PTTH) that stimulates the prothoracic glands to secrete ecdysone. PTTH is a neurosecretory hormone synthesized in neurosecretory cells of the brain and released by neurohaemal organs: the *c. cardiaca* (and in Lepidoptera by the *c. allata*; Nijhout 1975a).

PTTH is a peptide hormone whose only known molecular function is to stimulate ecdysone secretion by the prothoracic glands. The PTTH receptor is a receptor tyrosine kinase of the Torso family (Rewitz et al. 2009; Smith and Rybczynski 2012) that acts via the Ras-Raf-ERK pathway to stimulate ecdysteroidogenesis. In addition, there is some evidence that PTTH may also stimulate G-protein-coupled receptor (Rybczynski and Gilbert 2003), and calcium and cAMP signalling, and signalling via the insulin pathway (Gu et al. 2010, 2011), but the relative roles of these signalling systems in the regulation of ecdysteroid synthesis or release are not yet clear (Smith and Rybczynski 2012). It is possible that different portions of the ecdysteroid biosynthesis pathway are controlled independently, or that secretion of ecdysone is controlled independently from its synthesis. Alternatively, it is possible that several intracellular signalling pathways need to be activated that then act additively or synergistically in the control of ecdysone biosynthesis and release (Gu et al. 2011; Smith and Rybczynski 2012). Unlike the situation in decapod Crustacea, there is no evidence of negative control over the secretion of ecdysone in insects, with the exception that in the last larval instar PTTH and ecdysone secretion are inhibited by JH (Nijhout and Williams 1974; Rountree and Bollenbacher 1984, 1986; Gu et al. 1997; Nijhout et al. 2006).

## Role and Control of PTTH

Although the function of PTTH is restricted to the stimulation of ecdysone biosynthesis and secretion, the secretion of PTTH itself is regulated by many different physiological factors at different points in the life cycle and in different groups of insects. The immediate cause of PTTH secretion is cholinergic signalling in the brain (Lester and Gilbert 1987; Agui 1989; Smith and Rybczynski 2012). A possible involvement of serotonin and dopamine in the positive and negative regulation of PTTH secretion has been suggested (Agui 1989; Shirai et al. 1995).

The control over stimulated release of PTTH resides at higher levels and is quite diverse because the regulation of when to moult is often an adaptation for a particular life cycle strategy. In many insects, moults are stimulated when the larva reaches a particular size. In those cases, moult is tightly linked to some measure of body size and is independent of time since previous moult or the growth rate. The exact way in which this "size" is measured is known for a few insects. In bloodsucking reduviid bugs like *Rhodnius prolixus* and *Dipetalogaster maximus,* the moult is triggered by a single large blood meal that activates abdominal stretch receptors which send neural signals to the brain to initiate secretion of PTTH (Nijhout 1984; Chiang and Davey 1988). In the bug *Oncopeltus fasciatus*, a similar abdominal stretch is achieved gradually by progressive feeding. A moult can be triggered artificially by simply inflating the abdomen by an injection of saline or air (Nijhout 1979). Caterpillars of the tobacco hornworm, *Manduca sexta*, measure size by a completely different mechanism. In *Manduca*, as in all insects, the tracheal system does not grow during the intermoult period, but only increases periodically each time the larva moults. As the larva grows, the rate of oxygen supply by the tracheal system cannot keep up with the increasing demand of the growing body, and when a larva reaches a particular critical weight (4.8 times the initial mass of the instar, which is the time at which the size of the tracheal system is established), the maximum delivery capacity is reached and the

endocrine events that lead to a moult (see also Sect. 6.2.4 below) are triggered (Callier and Nijhout 2011). In the dung beetle, *Onthophagus taurus*, it is not body size but exhaustion of the food supply that triggers the metamorphic moult. *Onthophagus* larvae are buried with a small ball of dung, and when they have eaten that there is no chance of obtaining more. In the laboratory, simply removing a larva from its food triggers the moult two days later (Shafiei et al. 2001). Larvae of *Drosophila*, which live in an unstable habitat, likewise slightly accelerate the timing of pupariation when starved, relative to larvae of the same age that are allowed to continue feeding (Mirth et al. 2005). In these species, starvation stimulates a premature moult and this is presumably an adaption to a larval environment where food is limiting and the larva is unable to move to a new food source (Tobler and Nijhout 2010).

There is one circumstance in which there is a well-known negative control over PTTH secretion and moulting. This occurs in the last larval instar in Lepidoptera and Coleoptera (and probably most insects) where the JH inhibits PTTH secretion and moulting (Nijhout and Williams 1974; Quennedey et al. 1995; Connat et al. 1984; Rountree and Bollenbacher 1984, 1986; Gu et al. 1997; Emlen and Nijhout 1999). This also appears to be an adaptation to prevent the metamorphic moult from occurring whilst there is still some JH circulating, because a moult in the presence of a small amount of JH results in a partial metamorphosis and a monstrous non-viable mosaic of larval and pupal structures (Williams 1961). In some insects, the last larval instar can be prolonged by application of exogenous JH or a JH-analogue. Exogenous JH does not inhibit PTTH secretion or ecdysone secretion in earlier larval instars, when JH is normally high and maintains the larval *status quo* (see Sect. 6.3.2 below). Interestingly, JH does not have this inhibitory effect in the cyclorrhaphan Diptera, including *Drosophila*, nor does exogenous JH have any morphogenetic effects in larvae of this group (Riddiford and Ashburner 1991; Srivastava and Gilbert 1969).

A final mechanism of regulation of PTTH secretion is by means of biological clocks. In many insects, moults are initiated at particular times of day, often during the dark phase of the photoperiod. Presumably, this is an adaptation for the fact that after apolysis, arthropods are immobile and defenceless because muscle attachment to the cuticle is lost. In *M. sexta* and other Lepidoptera, PTTH secretion is "gated" by the photoperiod, which means that it only occurs during a relatively brief window of time that is controlled by a photoperiodic clock (Truman 1972; Truman and Riddiford 1974; Fujishita and Ishizaki 1982; Mizoguchi and Ishizaki 1984).

## Role of Ecdysone

The only function of PTTH is to stimulate ecdysone secretion, but ecdysone can have a broad diversity of effects that depend on the stage of the life cycle and target tissue. In addition to inducing moulting, ecdysone stimulates context-dependent gene transcription and controls cell division, tissue growth, the switch in commitment prior to metamorphosis and the development of several seasonal polyphenisms.

One of the earliest known effects of ecdysone is the induction of chromosomal puffs in the polytene salivary gland chromosomes of *Drosophila* (Ashburner 1972, 1973). Puffing in polytene chromosomes is associated with gene transcription, and the location of chromosomal puffs indicates which genes are being transcribed. Exposure to ecdysone elicits a characteristic sequence of early and late genes (Fig. 6.1) that reflect a complex combination of sensitivities to ecdysone, transcription rates, differential responses to rising and falling ecdysone concentrations and hierarchical gene activation (Huet et al. 1995). Although the genes in the ecdysone-activated hierarchy have been extensively described, their specific functions are still poorly understood.

The ecdysteroid receptor is a member of the nuclear hormone receptor superfamily that acts as transcriptional regulators upon binding to a ligand. The insect ecdysone receptor is a

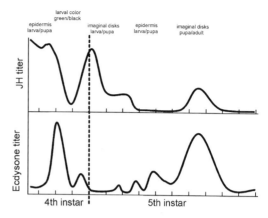

**Fig. 6.1** Juvenile hormone and ecdysone levels in late larval development of *Manduca sexta* (Lepidoptera: Sphingidae). Major elevations in ecdysone titres stimulate moults. The functions of the minor fluctuations are not understood, although some may be associated with the timing of juvenile hormone-sensitive periods. Juvenile hormone titres likewise fluctuate substantially. Juvenile hormone-sensitive periods are indicated by grey regions; labels indicate the relevant developmental switch with the fate of each tissue shown as the consequence high/low levels of juvenile hormone during the sensitive period. Different tissues and responses, such as epidermis, imaginal discs and pigmentation have distinct sensitive periods (after Nijhout 1994)

heterodimer of two gene products: one from the *EcR* gene and the other from the USP (*Ultraspiracle*) gene. USP is a member of the vertebrate retinoid X receptor (RXR) family of nuclear receptors that bind retinoic acid. Retinoids have been implicated in the regulation of regeneration in *Drosophila* and may be involved in the regulation of PTTH expression (Mansfield et al. 1998; Halme et al. 2010), though the functional relationships are not yet clear. There are at least three isoforms of the EcR protein each with a different function that is presumably due to differences in specificity of the genes they activate (Talbot et al. 1993; Truman et al. 1994; Thummel 1995; Schubiger et al. 2003). Expression of EcR is sometimes upregulated in response to ecdysteroids so that the amount of receptor increases with exposure to the hormone (Jindra et al. 1996; Hegstrom et al. 1998). However, the EcR expression pattern does not always closely follow the pattern of ecdysone titres in the blood (Jindra et al. 1996, 1997;

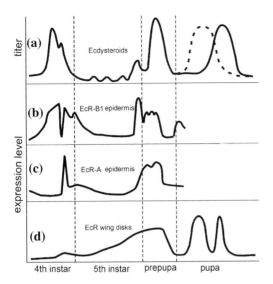

**Fig. 6.2** Ecdysone levels and expression levels of various isoforms of the ecdysone receptor in different tissues of *Manduca sexta* (Lepidoptera: Sphingidae). Different ecdysone isoforms in the epidermis are believed to target expression of different gene sets. Each isoform has a different temporal expression pattern, and epidermis and wing discs each have characteristic patterns of ecdysone receptor expression (after Jindra et al. 1997, 1996; Riddiford et al. 2000)

Hiruma and Riddiford 2010). In *Manduca,* the EcR isoforms and USP isoforms are expressed in complex tissue-specific temporal patterns (Fujiwara et al. 1995; Jindra et al. 1996, 1997). Different tissues can express different temporal patterns of ecdysteroid receptor (Fig. 6.2), and in this way, a complex profile of ecdysone pulses during the intermoult period can target some tissues whilst leaving others unaffected. As we will see below (Sect. 6.3.2), the USP component of the ecdysone receptor also appears to be a receptor for JH that may enable JH to modify the action of ecdysone by retargeting the receptor complex to different genes.

### 6.2.3.3 Growth at the Moult

A pulse of ecdysteroids induces the episodic growth and stepwise increase in surface area of the exoskeleton during the moulting cycle. The increase in area is due to enlargement of the epidermal cell layer by ecdysone-stimulated cell division (Kato and Riddiford 1987). If ecdysone induces a round of cell division in every epidermal cell one would expect a doubling of the surface area at each moult.

If surface area doubles at each moult then growth from moult to moult would be exponential with an exponent (for surface area) of 2; linear measures, such as the length of a leg segment or width of a carapace, would increase from moult to moult with an exponent of $\sqrt{2}$ (=1.41). In larval arthropods, there is often a regular exponential increase in size from moult to moult (Cole 1980; Freeman 1980; Anger 1985; Klingenberg and Zimmermann 1992; Cnaani and Hefetz 2001; Nijhout et al. 2006; Fusco et al. 2012). Indeed, the phenomenon is so common that it has acquired a name: Dyar's rule (Dyar 1890; Hutchinson et al. 1997). Many authors have noted that strict adherence to Dyar's rule occurs only in arthropods growing under ideal conditions. In nature, and in animals growing under suboptimal or variable conditions, growth ratios tend to deviate from the ideal (Enders 1976; Klingenberg and Zimmermann 1992; Hutchinson et al. 1997). In Crustacea, which have indeterminate growth and continue to moult as adults, there is a gradual decrease in the size increment at each successive moult (Hankin et al. 1989).

Size increments at moulting are seldom perfect doublings in surface area (or 1.41 times in linear measures). In many arthropods, shape changes as the animal becomes larger (Huxley 1932; Thompson 1942), which means that growth ratios of different body parts change with successive moults. This implies that not all cells in the epidermis respond to ecdysone the same way. If the increase in area is less than implied by cell doubling, then either not all cells divide, or some cells change shape (e.g. by becoming columnar). In *Drosophila* (and perhaps all cyclorrhaphan Diptera), epidermal cells do not divide but undergo endomitosis and become progressively more polyploid, and larger, with each moult (Edgar and Nijhout 2004). In *Manduca,* the epidermis in the last larval instar is a mixture of diploid and polyploid cells (Kato and Riddiford 1987; Kato et al. 1987; Wielgus et al. 1979), so ecdysone must stimulate both cell division and endomitosis.

Thus, growth at each moult can be due to a combination of cell division, endomitosis and change in cell shape (and possibly patterned cell death). This spatial diversity of responses is additional evidence that not all cells respond the same way to the pulse of ecdysone that stimulates a moult. The response to ecdysone is location- and stage-specific, suggesting that ecdysone acts as a general systemic stimulus, and that other factors control whether and how epidermal cells will respond locally to ecdysone.

If arthropod preadult stages are undernourished, then they increase little in size when they moult. Indeed, some insects can undergo regressive moults, becoming smaller in size after the moult, when they are starved of nutrition (Beck 1971, 1973). The way in which nutrition during the intermoult period controls the amount of ecdysone-induced epidermal cell growth during the moult is not understood.

### 6.2.3.4 Intermoult Growth

Growth of the integument is an ecdysone-triggered episodic event, but growth of internal organs is more or less continuous and nutrition dependent. Growth of internal organs thus requires cell growth and proliferation that is independent of the ecdysone-induced round of cell division of the epidermal cells.

The hormonal control of intermoult growth has been studied primarily in insects, where the focus has been on whole body growth and the regulation of final body size and on the growth of imaginal discs in the Holometabola. Body growth, obviously, depends on nutrition, but nutrients do not have a direct effect on cell and tissue growth. Rather, their effect is indirect, mediated by hormones. The primary hormones involved in regulation of growth are the ecdysteroids and insulin-like hormones. Much of the focus in the past decade has been on the role of insulin in insect growth. This work emerged from the discovery that mutations in the insulin receptor signalling pathway, or defects in the expression of insulin-like peptides in

*Drosophila*, greatly reduced adult body size, and that this reduction resembled that obtained by starvation in the larval stage (Bohni et al. 1999; Ikeya et al. 2002; Oldham et al. 2002).

The best demonstration that tissue growth is not simply a response to nutrient availability comes from tissue culture experiments with imaginal discs. When wing imaginal discs are taken from a feeding and growing larva of *Precis coenia* or *M. sexta*, and placed in a nutrient-rich culture medium, their growth ceases instantly. Such cultured discs can be induced to grow at a normal rate by adding 20-hydroxyecdysone plus insulin (either as bombyxin: the lepidopteran insulin; or as human insulin) to the culture medium (Nijhout and Grunert 2002; Nijhout et al. 2007). Studies with *Bombyx mori* have shown that nutrition regulates the secretion of bombyxin: starvation causes a decline in the level of circulating bombyxin and injection of glucose raises the level of bombyxin (Satake et al. 1997; Masumura et al. 2000). Insulin and ecdysone appear to act synergistically: either hormone alone stimulates little or no growth, but together they can support a normal rate of growth. The relative roles of the two hormones are not understood. Ecdysone is known to act as a mitogen in the epidermis (Kato and Riddiford 1987), and insulin is known to stimulate protein synthesis and cytoplasmic growth. A simple possibility is that insulin controls cell enlargement and ecdysone controls cell division. In *M. sexta*, there is a low and slightly fluctuating level of ecdysone in the blood throughout the intermoult period (Wolfgang and Riddiford 1986), well below the level required to induce a moult but exactly at the level that tissue culture studies have shown ecdysone is required to support normal growth of imaginal discs (Kawasaki 1988; Champlin and Truman 2000; Nijhout and Grunert 2002; Nijhout et al. 2007).

Insulin stimulates growth via PI3K signalling pathway leading to uptake of glucose and certain amino acids into cells (Escher and Rasmuson-Lestander 1999; Colombani et al. 2003; Reynolds et al. 2007). It appears that both glucose and amino acids stimulate cellular growth by regulating protein synthesis, and studies on a broad diversity of animals, from arthropods to vertebrates, have revealed that the protein kinase, target of rapamycin (TOR), is an important waypoint in the pathway that connects insulin and amino acid signalling to protein synthesis and cell growth (Oldham and Hafen 2003; Guertin et al. 2004). TOR can be stimulated by insulin signalling pathway via activation of TSC and Rheb GTPase (Avruch et al. 2006). But amino acids can also act in an insulin-independent effect on Rheb activation of TOR, and by controlling S6K activation (Kozma and Thomas 2002). Therefore, both insulin-dependent and insulin-independent pathways lead to enhanced protein synthesis and growth. In *Drosophila*, most internal tissues except the imaginal discs and histoblasts are polyploid and grow by cell enlargement and endomitosis. Growth of polyploid cells is controlled by insulin, and endomitoses appear to be a direct consequence of this growth and require no additional stimulation. The degree of cell enlargement and polyploidization during intermoult growth is tissue-specific (Britton and Edgar 1998; Johnston et al. 1999; Edgar and Nijhout 2004).

Insects express many different kinds of insulin-like proteins, most are neurosecretory hormones produced in the central nervous system, and some are produced in peripheral tissues such as the gut and imaginal discs. Some insulins circulate as endocrines whereas others act locally via paracrine and autocrine signalling. *Bombyx mori* has more than 30 different insulin-like proteins, called bombyxins (Kondo et al. 1996; Aslam et al. 2011), whose evolution has been elucidated by Kondo et al. (1996). The sphingid moth *Agrius cingulatus* has six insulin-like proteins (Iwami et al. 1996), and *Drosophila* has seven insulin-like proteins (Brogiolo et al. 2001; Ikeya et al. 2002; Rulifson et al. 2002). In spite of this overabundance and diversity of insulin-like ligands, there is only a single gene for the insulin receptor in these species, and presumably only a single form of the receptor. The functional significance of the diversity of insulin-like ligands is not entirely clear.

Thus, intermoult growth of internal tissues requires nutrition, which stimulates growth via insulin and TOR signalling. Imaginal discs also require low levels of ecdysone for normal growth, but whether ecdysone is also required for the growth of other tissues is not clear. The rate and amount of growth is modulated by nutrition, presumably via the modulation of insulin or ecdysone levels. As we will see in the next section, modulation of ecdysone secretion plays a significant role in regulating the proportional growth of body parts.

### 6.2.4 Body Size

Final body size and the relative sizes of body parts are the arguably the most distinguishing characteristics of a species, yet the developmental mechanisms by which a particular body size is achieved, and those that regulate the proportional sizes of body parts are still poorly understood. In Crustacea with indeterminate growth, the growth increment at each moult and moult frequency gradually decline (Pratten 1980; Hankin et al. 1989; Comeau and Savoie 2001). In arthropods with determinate growth, such as insects and most arachnids, adults do not grow and the size at which an immature larva moults to the pupa or to the adult reproductive form defines the final body size of the individual. Adult body size is determined by an interplay of both genetic and environmental factors. Nutrition affects growth rate and can have a profound effect on adult size, as can mutations that affect the insulin signalling pathway or the cell cycle.

The developmental regulation of adult size is basically the regulation of when to stop growing. In arthropods, growth stops when ecdysteroids are secreted that induce a moult. In larval moults, growth resumes again after the moult, of course, but at the metamorphic moults of insects and other arthropods with determinate growth, growth stops entirely. In insects, because of metabolic and tissue losses during metamorphosis, the mass of the adult is usually less than the mass of the larva when it stopped feeding and growing. Thus in an important sense, the mechanism that controls the secretion of ecdysone is also the mechanism that controls size, since growth stops when ecdysone secretion stimulates a moult. The triggers insects use to initiate ecdysone secretion were discussed above under the control of PTTH secretion (Sect. 6.2.3.2). These involve a diversity of size-sensing and nutrient-sensing mechanisms and illustrate that the mechanisms that control of body size have undergone a great deal of evolution and many may be adaptations to particular ecological conditions and modes of life.

What distinguishes an insect's larval moult (after which growth continues) from a metamorphic moult (which perpetually stops growth) is the juvenile hormone (JH) (see Sect. 6.3.2 below). JH is secreted throughout larval life, and whilst JH is being secreted and present in the blood, ecdysone induces a larval moult. At the end of larval life, JH secretion stops and the secretion of ecdysone in the absence of JH initiates metamorphosis. The mechanism that controls the cessation of JH secretion at the end of larval life remains one of the great puzzles in the developmental endocrinology of insects. It is associated with the attainment of a particular narrowly defined absolute size called the "threshold size" (Nijhout 1975b). Larvae of *Manduca* can be made to undergo 4–6 moults by manipulating diet and nutrition in early larval life and will continue larval moults as long as the size at the moult (measured as mass) is below the threshold size, independent of how many moults have preceded (Nijhout 1975b). Larvae whose size at a moult is below this threshold size will moult to another larval instar when next they moult, whereas those that are above that threshold are now in the last larval instar and will metamorphose when they next moult. A threshold size for metamorphosis has also been demonstrated in *Tribolium castaneum* (Preuss and Nijhout 2009). It is still not clear exactly what physiological process in the animal corresponds to what we can detect as threshold size in the laboratory.

At the beginning of the last larval instar in *Manduca*, the level of JH is still high, but the secretion of JH stops when the larva reaches the critical weight (Nijhout and Williams

1974). The critical weight (not to be confused with the threshold size) is a discrete multiple (4.8 times) of the initial weight of the instar and is sensed when the non-growing tracheal system is unable to maintain an adequate oxygen supply for the growing larva (Nijhout et al. 2006; Callier and Nijhout 2011). Whilst JH is still high, it inhibits PTTH and ecdysone secretion and the moult can be delayed indefinitely by exogenous JH (Nijhout and Williams 1974; Rountree and Bollenbacher 1986; Nijhout et al. 2006). Once JH has been cleared, the brain is disinhibited from secreting PTTH, and PTTH secretion will occur at the next photoperiodic gate (Truman 1972; Nijhout and Williams 1974; Truman and Riddiford 1974). The inactivation of JH secretion at the critical weight and the inhibition of PTTH and ecdysone secretion by JH are endocrine events that are unique to the last larval instar. In earlier larval instars, JH is continuously present and exogenous JH has no effect on the secretion of PTTH or ecdysone or the timing of the moult. These premetamorphosis endocrine events appear to be a physiological property seen only in larvae that have passed the threshold size for metamorphosis. Thus in *Manduca*, body size is regulated by the critical weight in the last larval instar, coupled with a complex cascade of endocrine events that involve the decay of JH and photoperiodic gating of PTTH and ecdysone secretion (Nijhout et al. 2006).

In *Drosophila*, body size is affected by both insulin signalling and ecdysone. Increasing insulin signalling in the prothoracic gland increases its size and increases expression of genes involved in ecdysone biosynthesis (Caldwell et al. 2005; Mirth et al. 2005), and this results in an accelerated metamorphic moult and a smaller adult body size, presumably because enhanced ecdysone synthesis caused a premature elevation of the ecdysone titre (Riddiford 2011). Ecdysone production by the prothoracic glands also requires an intact nutrient-sensing system that involves TOR. When TOR activity is reduced, the moult is delayed and this results in larger adults (Layalle et al. 2008). Thus, the rate and timing of ecdysone synthesis and secretion are sensitive to the direct effects of nutrition on the prothoracic glands via both the insulin and TOR pathways. The direct regulation of prothoracic gland size and activity provides an additional route by which nutrition can affect body size.

### 6.2.5 Relative Sizes of Appendages

Appendages need to develop in the correct proportion to body size. The regulation of proportional growth is complicated by the fact that in holometabolous insects, and other arthropods with complex metamorphoses, the appendages do not grow in simple proportion to the body as a whole throughout ontogeny. In holometabolous insects whose appendages develop from imaginal discs, the appendages undergo most of their growth during the prepupal and pupal stages, after feeding has ceased and the body has stopped growing.

Although body size is often thought of as being genetically determined, it is actually a highly plastic character that is affected by nutrition and temperature. In *Manduca*, as in many other insects, it is possible to obtain more than a two-fold variation in adult size by environmental manipulation alone. Yet, during the prepupal and pupal stages, the appendages develop in the correct proportion to the body, so there must exist a mechanism that regulates their growth to be in the proper proportion to a plastically variable body.

The developmental control of this mechanism has been studied in the context of wing-body scaling in *Manduca*. In this species, the proportional growth of the wing is regulated by variation in ecdysone secretion during the prepupal stage, after somatic growth has stopped but before pupation (Nijhout and Grunert 2010). Both the duration and level of ecdysone secretion are different in individuals of different sizes. Ecdysone controls cell division in the developing wing, and in small individuals, ecdysone secretion and cell division stop earlier than they do in larger individuals. Thus, the duration of wing

growth is shorter in smaller individuals. In addition, in small individuals, the wing also grows at a slower rate that is controlled by the concentration of ecdysone. Cell division and growth of the developing wing are stimulated by an optimal concentration of ecdysone and are inhibited at levels below and above this optimum. In small individuals, ecdysone titres rise to a higher level than they do in large individuals, and this elevated level of ecdysone has an inhibitory effect on the rate of cell division (Nijhout and Grunert 2010). Thus, ecdysone controls both the duration and rate of growth, and both are modulated to ensure wings grow in the correct proportion to a variable body size. The control over this modulated pattern of ecdysone secretion lies in the central nervous system and is exercised via the secretion of PTTH. The mechanism by which the brain becomes "aware" of body size during the prepupal stage is not understood.

## 6.3 Metamorphosis

Metamorphosis is one of the most profound developmental transformations in nature. During metamorphosis, a well-developed highly specialized form rapidly transforms into a very different-looking form specialized for a different mode of life. Metamorphosis occurs during postembryonic development, when the body plan is already complex and fully developed. Metamorphosis requires massive changes in gene expression that controls the breakdown of larva-specific tissues and the development and growth and morphogenesis of adult-specific traits. The transformations are often profound, such as that of a caterpillar into a butterfly, a maggot into a fly or a planktonic zoea larva into a benthic crab.

The metamorphosis of arthropods is functionally tied to the moulting cycle, since that is the only time when an exoskeleton with a new morphology can be made. Ecdysone is universally used to stimulate moulting, but the morphology of the new exoskeleton that is made depends on many other factors. In insects, JHs play an important (but not universal) role in controlling metamorphosis, and in decapod Crustacea, it is methyl farnesoate, a terpenoid hormone that is structurally similar to JH.

## 6.3.1  Crustacea and Insects

When an arthropod moults, it can moult to a larger version of the current form, as happens during larval–larval moulting insects (e.g. when a small caterpillar moults to a larger one). Alternatively, the moult can be progressive/transformative to a form with very different morphology. The larvae of crabs and lobsters moult to a successive series of distinctive form (in crabs called zoea, metazoea and megalopa, which are all small planktonic larval stages) (Gurney 1942). Some insects also have a series of distinctive larval stages, for example, those with hyper-metamorphosis go from an active long-legged triungulin to a poorly mobile grub-like form; a less extreme form is the progressive alteration in pigmentation and morphometrics in successive larval stages of many larval insects.

Little is known about how the progression through morphologically distinctive larval stages is controlled. In larvae of swallowtail butterflies (Papilionidae), small larvae in the first two instars have a black and white pattern that acts as bird-drop mimicry, and larger larvae are bright green. This progression of larval form and pigment pattern is controlled by JH (Futahashi and Fujiwara 2008). In larvae of silk moths (Saturniidae), by contrast, exogenous JH does not inhibit or alter this progression (Willis 1969, 1974).

In insects, the term metamorphosis is usually reserved for the moults that transform the non-reproductive larva to the final reproductive adult form. In insects, the moult to the adult is the last moult, and with the exception of Thysanura and Collembola, adult insects do not moult or grow. In decapod crustaceans, by contrast, there is a succession of morphologically distinct planktonic larval forms, and metamorphosis to the adult morphology occurs at a very small body size, after settlement of the planktonic larvae. The small adults are still sexually immature but continue to grow and moult, eventually reaching sexual maturity. Many species of lobster, crab and shrimp continue to grow and moult after reaching sexual maturity, some apparently indefinitely.

## 6.3.2  Insects and Juvenile Hormone

The hormonal control of metamorphosis was discovered by Wigglesworth in his studies (Wigglesworth 1936, 1940, 1948) on the control of moulting in *R. prolixus*. When he joined a 4th (penultimate)-instar larva and a 5th (final)-instar larva in parabiosis and induced them to moult, he found that the 4th-instar larva moulted to a normal 5th instar, but the 5th-instar larva that would have moulted to an adult instead moulted to a supernumerary 6th larval instar. Evidently, the 4th-instar larva inhibited metamorphosis of its parabiotic partner and caused it to moult to a never-before-seen 6th larval instar. Because the two larvae were joined only by their circulatory systems, the implication was that the inhibitor was a blood borne factor. The source of the active factor was eventually traced to a small pair of glands in the neck called the *c. allata* and identified as an epoxidized sesquiterpenoid ester (Röller et al. 1967) that came to be called the JH because it seemed to cause larvae to retain their immature or juvenile morphology.

The *c. allata* are attached to the *c. cardiaca* and the brain via conventional and neurosecretory neurons. Secretion of JH is controlled by neurosecretory hormones called allatostatins and allatotropins (Kataoka et al. 1989; Pratt et al. 1989; Kramer et al. 1991; Nijhout 1994; Audsley et al. 2000). At least five molecular forms of JH have been described that differ in the pattern of side-chain substitutions (Nijhout 1994). Most species make more than one of these forms, and many insect taxa have characteristic mixes of JHs. There is, however, no evidence that the different forms of JH have different functions.

The developmental effect of JH is to maintain the current developmental state, not to make insects "juvenile" as the name might imply. This interpretation comes from the finding that when JH is injected into a pupa, it causes the synthesis of pupal cuticle at the next moult and thus produces a second pupal stage, not an adult (Williams 1959; Zhou and Riddiford 2002). Accordingly, JH is often referred to as a *status quo* hormone (Willis 1981; Riddiford 1996;). In addition to its developmental effects during larval life and metamorphosis, JH is also deployed in adult insects as a reproductive hormone that, depending on the species, controls the ovarian maturation, yolk uptake and the synthesis of yolk proteins by the fat body (Nijhout 1994).

As a developmental hormone, JH controls tissue-level switches between alternative developmental pathways during metamorphosis. JH does not act in a concentration-dependent manner, as was once thought (Willams 1961; Schneiderman and Gilbert 1964), but acts during tissue-specific and stage-specific JH-sensitive periods (Nijhout and Wheeler 1982; Nijhout 1994). If JH is present during a particular sensitive period, then that tissue maintains its developmental state, and if JH is absent during that sensitive period, the pattern of gene expression changes and the tissue is set on a new developmental path.

JH has no overt developmental effect by itself, but typically requires ecdysone to express or reveal its effects. This is in part due to the fact that a moult is required to express the new fate of a tissue, and in part because in some cases, the JH-sensitive periods (Fig. 6.3) are associated with pulses of ecdysone secretion (Nijhout 1994, 2003). A decline in JH during the last larval stage causes a change in commitment of tissues to pupal development (Riddiford 1978, 1981; Kremen and Nijhout 1989, 1998). This change in commitment can be revealed by an injection or infusion of ecdysone, or by implanting a tissue into a larva undergoing a moult, which provokes a moult. Such an induced moult produces an individual with a mosaic mix of pupal and larval characters that reveals the state of commitment of different tissues (Kremen 1989;

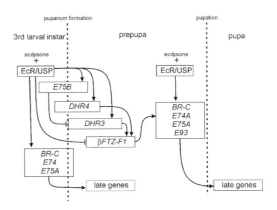

**Fig. 6.3** Cascade of ecdysone-activated genes in *Drosophila melanogaster* (Diptera; Drosophilidae). Two pulses of ecdysone successively induce puparium formation and pupation. Ecdysone binds to the ecdysone receptor (EcR/USP dimer), which stimulates a cascade of transcription factors, the early ecdysone-response genes. The early response genes repress their own function, thus terminating the response to ecdysone, and also stimulate a set of late-response genes, which are believed to be the effectors of the various physiological responses to ecdysone. The names of some of the transcription factors are derived from the bands or puffs in polytene salivary gland chromosomes (after Thummel 2001; King-Jones et al. 2005)

Koyama et al. 2008). Such studies have shown that in the Lepidoptera, the imaginal discs switch to pupal commitment very early in the last larval instar. Within a disc, there is a progressive spatial pattern of commitment, and different discs initiate and terminate this progression at different times (Kremen 1989; Kremen and Nijhout 1998). In *Manduca*, the commitment switch of the imaginal discs requires the absence of JH and appears to be stimulated by insulin signalling that is initiated when the larva begins to feed in the last larval instar (Koyama et al. 2008). The general epidermis of *Manduca* switches to pupal commitment much later in the last larval instar, and this switch is caused by a slight elevation of ecdysone that also causes the larva to stop feeding and enter the "wandering" stage in preparation for pupation (Riddiford 1978, 1981).

The succession from larva to pupa to adult in holometabolous insects was once thought to require progressively lower levels of JH (Willams 1961; Schneiderman and Gilbert 1964), but it now appears that this progression unfolds automatically in the absence of JH without requiring specific endocrine control of each successive stage (Nijhout 1994). This is in keeping with the *status quo* action of JH (Willis 1981; Riddiford 1996), which implies that the function of JH is to maintain the current state of differentiation, and not to direct the progressive differentiation of metamorphosis. A critical molecular event associated with the disappearance of JH is the expression of Broad Complex (BR–C) transcriptional regulators (Zhou et al. 1998; Zhou and Riddiford 2002). Broad expression is stimulated by ecdysone and only occurs when ecdysone acts in the absence of JH and then only in the epidermis of the pupa stage, not in that of the larva or adult (Zhou and Riddiford 2002). Broad expression appears to be the first step in the specification of pupal traits (Fig. 6.4).

JH has many other developmental functions besides its *status quo* effect in the control of metamorphosis. JH also regulates the development of alternative morphologies after a moult, that include the control of caste determination in social insects, horn development in scarab beetles, seasonal forms in butterflies and colour development in caterpillars (see Sect. 6.4 on polyphenisms below). After metamorphosis, JH plays a diversity of roles in the regulation of adult reproduction, migration and diapause (Rankin and Riddiford 1977; Chang 1993; Nijhout 1994; Denlinger 2002). It has been difficult to explain this diversity of developmental and physiological effects of a single hormone, a problem that is made even more challenging by the fact that JH does not appear to act as a classical hormone. JH does not seem to belong to the class of hormones that have either membrane receptors or nuclear receptors. Indeed, it has been difficult to find any receptor at all. JH binds with moderate affinity to a broad range of proteins, and this binding has generally been assumed to be non-functional. One of those proteins is USP, part of the ecdysteroid receptor (Jones and Sharp 1997). The binding affinity is rather low, with a $K_d$ in the 0.5–1 μM range which is well above the normal concentration of

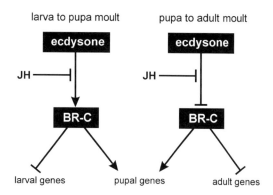

**Fig. 6.4** The role of juvenile hormone (JH) and Broad Complex (BR–C) in the regulation of gene switching during metamorphosis. During the larva to pupa moult, ecdysone (via the EcR/USP complex; see Fig. 6.1) induces expression of the BR–C transcriptional regulator, which controls activation of pupal genes and inhibition of larval gene expression. In the presence of JH, transcription of BR–C is inhibited and a larva to larva moult ensues. In the pupal stage, ecdysone inhibits expression of BR–C, which results in the inhibition of pupal genes and the activation (inhibition of an inhibition) of adult genes. JH in the pupal stage maintains BR–C and causes a moult to another pupal stage (after Zhou and Riddiford 2002)

JH. Nevertheless, it is possible that JH binding could alter the ecdysone receptor's affinity for its gene targets and change expression of the genes controlled by ecdysone. More recently, JH has been found to bind rather more strongly to the product of the methoprene-tolerant (Met) gene. Methoprene is a JH mimic, and mutations in Met make *Drosophila* insensitive to the deleterious effects of high doses of methoprene. The Met protein is a transcription factor that could be activated or retargeted by JH. JH binds to the Met protein with a $K_d$ in the 4–5 nM range (Miura et al. 2005; Charles et al. 2011); the range expected for classical hormone binding. JH also binds with moderate affinity to the retinoid receptor and to a variety of proteins in intracellular signalling pathways (Wheeler and Nijhout 2003). Even though the binding affinity to these various proteins is lower than expected for a classical hormone, it is possible that JH interaction with proteins in signalling pathways could modulate or retarget those pathways to produce JH-specific responses (Wheeler and Nijhout 2003).

### 6.3.3 Crustacea and Methyl Farnesoate

Ablation of eyestalks in crab larvae not only accelerates the next moult by removing the source of MIH, but also inhibits metamorphic progression (Costlow 1968; Hinsch 1972; Laufer et al. 1987; Laufer and Biggers 2001). In adult crabs and shrimp, removal of the eyestalk accelerates ovarian maturation (Laufer et al. 1987; Laufer and Biggers 2001). Thus, the endocrine centres in the eyestalk appear to have an inhibitory effect on metamorphosis and ovarian maturation. In insects, both metamorphosis and ovarian maturation are regulated by JH. Treatment with JH or with JH analogues like methoprene can delay early larval development in crabs and shrimp, but interestingly, accelerates metamorphosis in the later, premetamorphic, developmental stages of shrimp, and in barnacles (Laufer and Biggers 2001). Hinsch (1980) discovered that implantation of mandibular organs into young adult female crabs stimulated ovarian maturation. This finding led to the discovery that the mandibular organs secrete an analogue of JH, called methyl farnesoate (MF) (Laufer et al. 1987; Ding and Tobe 1991).

The secretory activity of the mandibular organ is regulated by inhibitory neurosecretory hormones produced in the eyestalk, the mandibular organ inhibitory hormones (MOIHs) which belong to the CHH family of neuropeptides (Borst et al. 1987, 2001, 2002; Laufer et al. 1987; Tsukimura and Borst 1992; Wainwright et al. 1996, 1999). MOIH has been shown to inhibit the activity of farnesoic acid O-methyltransferase, the last enzyme in the MF biosynthetic pathway (Borst et al. 2001, 2002).

MF has been detected in many species of decapod crustaceans, as well as in insects, where it is a biosynthetic precursor for JH-III but is itself without normal physiological function. Treatment of crustacean larvae with MF has revealed similar effects to those found with exogenous JH (except that JH treatment also induces developmental abnormalities that are not typically seen with MF). Exogenous MF appears to have different effects in early and later larval stages. There appears to be a critical period, before which MF retards developmental progression to a subsequent larval stage, and after which MF accelerates metamorphosis (Laufer and Biggers 2001). In addition, MF has been shown to have a moult-accelerating effect in crabs and shrimp (Homola and Chang 1997).

Like ecdysone secretion, the secretion of MF appears to be largely under negative (inhibitory) control in the decapod Crustacea. Because MF can have a diversity of developmental and physiological effects, many of which appear to be dependent on the developmental stage, it is likely that additional levels of control exist. MF secretion as well as the developmental effects of MF may be co-regulated by one or more of the many neuroendocrine factors discovered in Decapoda (Christie 2011) whose functions are as yet unknown.

## 6.4 Polyphenism

Developmental hormones of arthropods are intimately involved in the regulation of adaptive phenotypic plasticity. Phenotypic plasticity is a developmental response to an environmental variable that results in a change in morphology, physiology and/or behaviour. Arthropods are able to change form when they moult, which is the basis of metamorphosis and of progressive changes in shape during the growth phase. In addition, many arthropods can moult into one of the several alternative morphologies, depending on environmental cues received during the preceding stage. The development of alternative phenotypes in response to environmental signals is called polyphenism and is especially widespread among the arthropods. The alternative phenotypes are often adaptations to a contingent and variable environment or are forms with specialized functions in a social context as we see in the castes of social insects. In all cases investigated so far, the switch between alternative developmental pathways is controlled by hormones (Nijhout 1999, 2003).

## 6.4.1 Crustacea

Water fleas of the genus *Daphnia* have a well-known adaptive polyphenism. In the presence of potential predators, *Daphnia* develop an elongated helmet-shaped spine on the head and a more elongated caudal spine that serve as predator deterrents (Woltereck 1909; Grant and Bayly 1981; Dodson 1989). Exposure of neonate *Daphnia* to MF, the crustacean analogue of JH, induces development and enlargement of the helmet in a concentration-dependent manner (Oda et al. 2011). The length of the tail spine, by contrast, decreases after exposure to MF. Decrease of the tail spine is also seen under crowding conditions, and an increase in helmet size can be induced by a variety of environmental stressors, such as pesticides (Oda et al. 2011), so it is not entirely clear whether the effects of MF are physiological.

Males of the crayfish *Procambarus clarkii* have two adult morphotypes: a reproductive morph that has large spiny chelae and spines on some of its walking legs, and a non-reproductive morph with smaller chelae and no spines (Laufer et al. 2005). Males can switch back and forth between these two morphs in successive moults. Eyestalk ablation inhibits the transformation of the non-reproductive morph into the reproductive morph, and exogenous MF inhibits transformation of the non-reproductive morph into the reproductive morph. Assay of blood levels of MF revealed that reproductive morph development was associated with low levels of MF, whereas non-reproductive morph development was associated with high blood levels of MF (Laufer et al. 2005).

## 6.4.2 Social Insects

The social wasps, bees, ants and termites all have castes that are specialized for certain tasks. All social insects have at least a reproductive caste and a more or less non-reproductive worker caste, and the worker caste can be further subdivided into forms that are specialized for brood rearing and foraging or defence (Wilson 1971). The reproductive, worker and soldier castes differ in body size, body proportions, physiology and behaviour, yet this diversity is not due to a genetic polymorphism. Any larva can develop into any of the castes depending entirely on the nutrients and pheromones to which it is exposed.

In all cases that have been studied, the developmental switch that sets a larva onto the developmental pathway that leads to one or another caste is mediated by JH. In the ant *Pheidole bicarinata,* there is a brief JH-sensitive period in the last larval instar during which exogenous JH can induce a larva of whatever body size to develop into a soldier (Wheeler and Nijhout 1981, 1983; Wheeler 1991). It is believed that JH is elevated at that time in larvae receiving high-nutrient food, and this causes a developmental switch that raises the critical weight of the larva and also reprogrammes the development of the imaginal disc of the head to produce a large-bodied soldier with a disproportionally large head. Adult soldiers produce a soldier-inhibiting pheromone that raises the threshold of sensitivity of larvae to JH and prevents too many soldiers from being produced even under good nutrition (Wheeler and Nijhout 1984). In honeybees, *Apis mellifera,* there is a JH-sensitive period during the 4th and early 5th larval instars when exogenous JH can induce queen traits in any larva, and during which larvae that have been fed royal jelly have a naturally elevated JH titre that induces them to develop into queens (Wirtz and Beetsma 1972; Rachinsky and Hartfelder 1990). After larvae pass the JH-sensitive period, there is a massive change in the pattern of gene expression with very different patterns in larvae that will become queens from those that will become workers (Evans and Wheeler 1999, 2000). Thus, a different sequence of gene expression ensues depending on whether JH is above or below the threshold for queen induction during a brief JH-sensitive period. Insulin signalling is also different in presumptive worker and queen larvae (Wheeler et al. 2006; de Azevedo and Hartfelder 2008), indicating that insulin signalling mediates the nutritional cues that trigger the endocrine switch between the queen and worker castes. In the adult stage, worker honeybees have a progressive division of labour, called age polyethism, that progresses from cleaning, to nursing to comb building and to foraging, as they get older (Wilson 1971). This progression is controlled by a changing titre of JH in worker bees and can be altered and accelerated with exogenous JH (Robinson 1987). Insulin and nutrition can also alter the rate of this behavioural progression (Ament et al. 2008). JH has also been shown to control caste development in social wasps (Giray et al. 2005; Shorter and Tibbetts 2009) and bumblebees (Cnaani et al. 2000; Bortolotti et al. 2001). In termites, caste determination is also controlled by JH with different JH-sensitive periods for the development of soldier and reproductive characters (Luscher 1972; Nijhout and Wheeler 1982; Mao et al. 2005; Cornette et al. 2008).

### 6.4.3 Seasonal Polyphenisms

Insect species that have many generations per year often develop distinctive season-specific morphologies and behaviours. Sometimes, the alternative forms are so divergent that they have been initially described as different species. Aphids can have a seasonal alternation between winged and wingless forms and between sexual and parthenogenetic forms depending on nutrition, crowding and photoperiod (Hardie 2010). The switch between some of these forms can be induced with exogenous JH (Mittler et al. 1976; Hardie 1980; Corbitt and Hardie 1985). In addition, there are differences in the expression of neuroendocrine hormones that may also be required for the specification of alternative sets of traits (Hardie 2010). Species of aphids differ considerably in life cycle and the kinds of alternative morphologies they express, but overall it appears that the switch between various forms is controlled by the endocrine and neuroendocrine system.

Many species of butterflies have distinctive seasonal forms, and here, it is not JH but ecdysone that controls the switch. In *P. coenia* and

*Araschnia levana,* there is an ecdysone-sensitive period early in the pupal stage that determines the colour pattern of the adult butterfly. The presence or absence of ecdysone during that sensitive window determines the colour form of the adult. In nature, the timing of ecdysone secretion in the early pupal stage depends on the photoperiod and temperature experienced by the caterpillar (Koch and Bückmann 1987; Smith 1991; Rountree and Nijhout 1995).

## 6.5 Concluding Remarks

Hormones control almost every aspect of post-embryonic development in arthropods, from growth, moulting and metamorphosis, to the development of alternative adaptive phenotypes in response to environmental signals. Three classes of hormones appear to dominate: the insulins, the ecdysteroids and the juvenoids (JH and MF). Little is known about the mechanisms that control the secretion of insulins, but the secretion of ecdysteroids and juvenoids is regulated by the central nervous system via the secretion of tropic neurohormones. This in effect puts the CNS in control of growth and development. Control by higher integrative centres makes it possible to coordinate growth and development with variation in the environment and to use environmental signals to cue alternative developmental pathways.

Much of what is known about the developmental endocrinology of arthropods has been learned using only a small handful of species. There are large taxonomic groups, even among the insects, that have not been studied at all, and so, it is difficult to tell which aspects of developmental endocrinology are general, and which represent the adaptations and specializations of a particular taxon. The most generally applicable feature is that ecdysone is probably universally used to control the moulting cycle. Even though their function has been studied in only a few arthropods, it is also probably safe to assume that insulins play a general role in the regulation of growth, based on their universal role in animals outside the arthropods. That said, it still remains to be explicitly demonstrated that this assumption is correct. If *both* insulin and ecdysone are required for normal growth, then variation in either of these hormones could be used to control growth, and there could be much taxonomic diversity and specialization in which hormone plays the major role.

One of the main difficulties in developing a general and taxonomically broad-based understanding of arthropod endocrinology is that most species are small, and many are difficult to rear in the laboratory. Small size makes it difficult or impossible to get the volume of haemolymph (typically several microlitres) required for hormone assays. Difficulty in laboratory culture makes it hard to obtain developmentally synchronized animals, which are essential for elucidating the control of developmental processes and developmental transitions. One area where a comparative approach is highly feasible, though as yet little exploited, is the molecular biology of developmental endocrinology. Studies of hormone receptors, intracellular signalling pathways and gene expression patterns during development can take advantage of tools and technologies developed in model systems that are now widely applicable to non-model organisms (Zera et al. 2007).

There is much we do not yet know about the developmental endocrinology of arthropods. There is every reason to believe that in this great and ancient evolutionary radiation, there have evolved many special and unique developmental control mechanisms that remain to be discovered. This expectation is strengthened by the fact that much of developmental endocrinology is concerned with the regulation of postembryonic development, which has been little studied. Morphological and life-history evolution has occurred almost entirely by changes in postembryonic development, and this evolution must have been accompanied by the evolution of the processes that control postembryonic development. These processes, in turn, are mainly controlled by hormones.

**Acknowledgments** I am grateful to Diana Wheeler, and to the editors, for many perceptive and helpful comments on the manuscript. This work was supported by grants from the National Science Foundation.

# References

Agui N (1989) *In vitro* release of prothoracicotropic hormone (PTTH) from the cultured brain of *Mamestra brassicae* L.; effects of neurotransmitters on PTTH release. In: Mitsuhashi J (ed) Invertebrate cell system applications. CRC Press, Boca Raton, pp 111–119

Ament SA, Corona M, Pollock HS, Robinson GE (2008) Insulin signaling is involved in the regulation of worker division of labor in honey bee colonies. Proc Natl Acad Sci USA 105:4226–4231

Andersen S (1979) Biochemistry of insect cuticle. Annu Rev Entomol 24:29–59

Anger K (1985) Development and growth in larval and juvenile *Hyas coarctatus* (Decapoda, Majidae) reared in the laboratory. Mar Ecol Progr Ser 19:115–123

Ashburner M (1972) Patterns of puffing activity in the salivary gland chromosomes of Drosophila. Chromosoma 38:255–281

Ashburner M (1973) Sequential gene activation by ecdysone in polytene chromosomes of *Drosophila melanogaster*: I. Dependence upon ecdysone concentration. Dev Biol 35:47–61

Aslam AFM, Kiya T, Mita K, Iwami M (2011) Identification of novel bombyxin genes from the genome of the silkmoth *Bombyx mori* and analysis of their expression. Zool Sci 28:609–616

Audsley N, Weaver RJ, Edwards JP (2000) Juvenile hormone biosynthesis by corpora allata of larval tomato moth, *Lacanobia oleracea*, and regulation by *Manduca sexta* allatostatin and allatotropin. Insect Biochem Mol Biol 30:681–689

Avruch J, Hara K, Lin Y, Liu M, Long X, Ortiz-Vega S, Yonezawa K (2006) Insulin and amino-acid regulation of mTOR signaling and kinase activity through the Rheb GTPase. Oncogene 25:6361–6372

Beck SD (1971) Growth and retrogression in larvae of *Trogoderma glabrum* (Coleoptera: Dermestidae). 1. Characteristics under feeding and starvation conditions. Ann Entom Soc Am 64:149–155

Beck SD (1973) Growth and retrogression in larvae of *Trogoderma glabrum* (Coleoptera: Dermestidae). 4. Developmental characteristics and adaptive functions. Ann Entom Soc Am 66:895–900

Berghiche H, Smagghe G, Van de Velde S, Soltani N (2007) In vitro cultures of pupal integumental explants to bioassay insect growth regulators with ecdysteroid activity for ecdysteroid amounts and cuticle secretion. Afr J Agr Res 2:208–213

Beydon P, Permana A, Colardeau J, Morinière M, Lafont R (1989) Ecdysteroids from developing eggs of *Pieris brassicae*. Arch Insect Biochem Physiol 11:1–11

Bitsch C, Baehr J, Bitsch J (1985) Juvenile hormones in *Thermobia domestica* females: Identification and quantification during biological cycles and after precocene application. Cell Mol Life Sci 41:409–410

Bohni R, Riesgo-Escovar J, Oldham S, Brogiolo W, Stocker H, Andruss BF, Beckingham K, Hafen E (1999) Autonomous control of cell and organ size by CHICO, a *Drosophila* homolog of vertebrate IRS1-4. Cell 97:865–875

Borst D, Laufer H, Landau M, Chang E, Hertz W, Baker F, Schooley D (1987) Methyl farnesoate and its role in crustacean reproduction and development. Insect Biochem 17:1123–1127

Borst D, Ogan J, Tsukimura B, Claerhout T, Holford K (2001) Regulation of the crustacean mandibular organ. Am Zool 41:430–441

Borst D, Wainwright G, Rees H (2002) *In vivo* regulation of the mandibular organ in the edible crab, *Cancer pagurus*. Proc R Soc B 269:483–490

Bortolotti L, Duchateau MJ, Sbrenna G (2001) Effect of juvenile hormone on caste determination and colony processes in the bumblebee *Bombus terrestris*. Ent Exp Appl 101:143–158

Britton J, Edgar B (1998) Environmental control of the cell cycle in *Drosophila*: nutrition activates mitotic and endoreplicative cells by distinct mechanisms. Development 125:2149–2158

Brogiolo W, Stocker H, Ikeya T, Rintelen F, Fernandez R, Hafen E (2001) An evolutionarily conserved function of the *Drosophila* insulin receptor and insulin-like peptides in growth control. Curr Biol 11:213–221

Bruce M, Chang E (1984) Demonstration of a molt-inhibiting hormone from the sinus gland of the lobster, *Homarus americanus*. Comp Biochem Physiol A Physiol 79:421–424

Caldwell PE, Walkiewicz M, Stern M (2005) Ras activity in the *Drosophila* prothoracic gland regulates body size and developmental rate via ecdysone release. Curr Biol 15:1785–1795

Collier V, Nijhout H (2011) Control of body size by oxygen supply reveals size-dependent and size-independent mechanisms of molting and metamorphosis. Proc Natl Acad Sci USA 108:14664–14669

Champlin D, Truman J (2000) Ecdysteroid coordinates optic lobe neurogenesis via a nitric oxide signaling pathway. Development 127:3543–3551

Chang E (1985) Hormonal control of molting in decapod Crustacea. Am Zool 25:179–185

Chang E (1993) Comparative endocrinology of molting and reproduction: insects and crustaceans. Annu Rev Entomol 38:161–180

Chang E, Mykles D (2011) Regulation of crustacean molting: a review and our perspectives. Gen Comp Endocrinol 172:323–330

Charles J-P, Iwema T, Epa V, Takaki K, Rynes J, Jindra M (2011) Ligand-binding properties of a juvenile hormone receptor, Methoprene-tolerant. Proc Natl Acad Sci USA 108:21128–21133

Charmantier G, Charmantier-Daures M (1998) Endocrine and neuroendocrine regulations in embryos and larvae of crustaceans. Invert Repr Dev 33:273–287

Chiang RG, Davey KG (1988) A novel receptor capable of monitoring applied pressure in the abdomen of an insect. Science 241:1665–1667

Christiansen K (1964) Bionomics of Collembola. Annu Rev Entomol 9:147–178

Christie A (2011) Crustacean neuroendocrine systems and their signaling agents. Cell Tissue Res 345:41–67

Cnaani J, Hefetz A (2001) Are queen *Bombus terrestris* giant workers or are workers dwarf queens? Solving the 'chicken and egg' problem in a bumblebee species. Naturwissenschaften 88:85–87

Cnaani J, Robinson G, Hefetz A (2000) The critical period for caste determination in *Bombus terrestris*; and its juvenile hormone correlates. J Comp Physiol A: Neuroethol Sens Neur BehavPhysiol 186:1089–1094

Cole B (1980) Growth ratios in holometabolous and hemimetabolous insects. Ann Entom Soc Am 79:489–491

Colombani J, Raisin S, Pantalacci S, Radimerski T, Montagne J, Leopold P (2003) A nutrient sensor mechanism controls *Drosophila* growth. Cell 114:739–749

Comeau M, Savoie F (2001) Growth increment and molt frequency of the American lobster (*Homarus americanus*) in the southwestern Gulf of St. Lawrence. J Crust Biology 21:923–936

Connat JL, Delbecque JP, Delachambre J (1984) The onset of metamorphosis in *Tenebrio molitor* L.: effects of a juvenile hormone analogue and of 20-hydroxyecdysone. J Insect Physiol 30:413–419

Corbitt TS, Hardie J (1985) Juvenile hormone effects on polymorphism in the pea aphid, *Acyrthosiphon pisum*. Ent Exp App 38:131–135

Cornette R, Gotoh H, Koshikawa S, Miura T (2008) Juvenile hormone titers and caste differentiation in the damp-wood termite *Hodotermopsis sjostedti* (Isoptera, Termopsidae). J Insect Physiol 54:922–930

Costlow J (1968) Metamorphosis in crustaceans. In: Etkin W, Gilbert LI (eds) Metamorphosis, Appleton, New York, pp 3–41

de Azevedo SV, Hartfelder K (2008) The insulin signaling pathway in honey bee (*Apis mellifera*) caste development–differential expression of insulin-like peptides and insulin receptors in queen and worker larvae. J Insect Physiol 54:1064–1071

de la Paz A, Delbecque J, Bitsch J, Delachambre J (1983) Ecdysteroids in the haemolymph and the ovaries of the firebrat *Thermobia domestica* (Packard) (Insecta, Thysanura): correlations with integumental and ovarian cycles. J Insect Physiol 29:323–329

Denlinger DL (2002) Regulation of diapause. Annu Rev Entomol 47:93–122

Dinan L, Hormann R (2005) Ecdysteroid agonists and antagonists. In: Gilbert LI, Iatrou K, Gill SS (eds) Comprehensive molecular insect science, vol 3. Elsevier Pergamon, Oxford, pp 197–242

Dinan L, Spindler-Barth M, Spindler K-D (1990) Insect cell lines as tools for studying ecdysteroid action. Invert Repr Dev 18:43–53

Ding Q, Tobe S (1991) Production of farnesoic acid and methyl farnesoate by mandibular organs of the crayfish, *Procambarus clarkii*. Insect Biochem 21:285–291

Dodson S (1989) Predator-induced reaction norms. Bioscience 39:447–452

Dyar H (1890) The number of molts of lepidopterous larvae. Psyche 5:420–422

Edgar B, Nijhout H (2004) Growth and cell cycle control in *Drosophila*. In: Hall MN, Raff M, Thomas G (eds) Cell growth. Control of cell size. Cold Spring Harbor Press, Cold Spring Harbor, pp 22–83

Emlen D, Nijhout H (1999) Hormonal control of male horn length dimorphism in the dung beetle *Onthophagus taurus* (Coleoptera: Scarabaeidae). J Insect Physiol 45:45–53

Enders F (1976) Size, food-finding, and Dyar's constant. Envir Entomol 5:1–10

Erezyilmaz D, Riddiford L, Truman J (2004) Juvenile hormone acts at embryonic molts and induces the nymphal cuticle in the direct-developing cricket. Dev Genes Evol 214:313–323

Escher S, Rasmuson-Lestander Å (1999) The *Drosophila* glucose transporter gene: cDNA sequence, phylogenetic comparisons, analysis of functional sites and secondary structures. Hereditas 130:95–103

Evans J, Wheeler D (1999) Differential gene expression between developing queens and workers in the honey bee, *Apis mellifera*. Proc Natl Acad Sci USA 96:5575–5580

Evans J, Wheeler D (2000) Expression profiles during honeybee caste determination. Genome Biology 2:research0001.1 - research0001.6

Freeman J (1980) Molt increment, molt cycle duration, and tissue growth in *Palaemonetes pugio* Holthuis larvae. J Exp Mar Biol Ecol 143:47–71

Freeman J, Costlow J (1979) Hormonal control of apolysis in barnacle mantle tissue epidermis, in vitro. J Exp Zool 210:333–345

Fujishita M, Ishizaki H (1982) Temporal organization of endocrine events in relation to the circadian clock during larval-pupal development in *Samia cynthia ricini*. J Insect Physiol 28:77–84

Fujiwara H, Jindra M, Newitt R, Palli S, Hiruma K, Riddiford L (1995) Cloning of an ecdysone receptor homolog from *Manduca sexta* and the developmental profile of its mRNA in wings. Insect Biochem Mol Biol 25:845–856

Fusco G, Garland JT, Hunt G, Hughes NC (2012) Developmental trait evolution in trilobites. Evolution 66:314–329

Futahashi R, Fujiwara H (2008) Juvenile hormone regulates butterfly larval pattern switches. Science 319:1061

Giray T, Giovanetti M, West-Eberhard MJ (2005) Juvenile hormone, reproduction, and worker behavior in the neotropical social wasp *Polistes canadensis*. Proc Natl Acad Sci USA 102:3330–3335

Grant JWG, Bayly IAE (1981) Predator induction of crests in morphs of the *Daphnia carinata* king complex. Limnol Oceanogr 26:201–218

Gu S-H, Chow Y-S, Yin C-M (1997) Involvement of juvenile hormone in regulation of prothoracicotropic hormone transduction during the early last larval instar of *Bombyx mori*. Mol Cell Endocr 127:109–116

Gu S-H, Lin J-L, Lin P-L (2010) PTTH-stimulated ERK phosphorylation in prothoracic glands of the silkworm, *Bombyx mori*: Role of $Ca^{2+}$/calmodulin and receptor tyrosine kinase. J Insect Physiol 56:93–101

Gu S-H, Young S-C, Lin J-L, Lin P-L (2011) Involvement of PI3K/Akt signaling in PTTH-stimulated ecdysteroidogenesis by prothoracic glands of the silkworm, *Bombyx mori*. Insect Biochem Mol Biol 41:197–202

Guertin D, Kim D-H, Sabatini D (2004) Growth control through the mTOR network. In: Hall MN, Raff M, Thomas G (eds) Cell growth. Control of cell size. Cold Spring Harbor Laboratory Press, Cold Spring Harbor, pp 193–234

Gurney R (1942) The larvae of decapod Crustacea. Ray Society, London

Halme A, Cheng M, Hariharan IK (2010) Retinoids regulate a developmental checkpoint for tissue regeneration in *Drosophila*. Curr Biol 20:458–463

Hankin D, Diamond N, Mohr M, Ianelli J (1989) Growth and reproductive dynamics of adult female Dungeness crabs (*Cancer magister*) in northern California. J Mar Sci 46:94–108

Hardie J (1980) Juvenile hormone mimics the photoperiodic apterization of the alate gynopara of aphid, *Aphis fabae*. Nature 286:602–604

Hardie J (2010) Photoperiodism in insects: aphid polypenism. In: Nelson RJ, Denlinger DL, Somers DE (eds) Photoperiodism. The biological calendar. Oxford University Press, Oxford, pp 342–363

Harmatha J, Dinan L (1997) Biological activity of natural and synthetic ecdysteroids in the B11 bioassay. Arch Insect Biochem Physiol 35:219–225

Hegstrom C, Riddiford L, Truman J (1998) Steroid and neuronal regulation of ecdysone receptor expression during metamorphosis of muscle in the moth, *Manduca sexta*. The J Neuroscience 18:1786–1794

Hinsch G (1972) Some factors controlling reproduction in the spider crab, *Libinia emarginata*. Biol Bull 143:358–366

Hinsch G (1980) Effect of mandibular organ implants upon spider crab ovary. Trans Am Microsc Soc 99:317–322

Hiruma K, Riddiford L (2010) Developmental expression of mRNAs for epidermal and fat body proteins and hormonally regulated transcription factors in the tobacco hornworm, *Manduca sexta*. J Insect Physiol 56:1390–1395

Hoffmann J, Koolman J, Karlson P, Joly P (1974) Molting hormone titer and metabolic fate of injected ecdysone during the fifth larval instar and in adults of *Locusta migratoria* (Orthoptera). Gen Comp Endocrinol 22:90–97

Homola E, Chang E (1997) Methyl farnesoate: crustacean juvenile hormone in search of functions. Comp Biochem Physiol B: Biochem Mol Biol 117:347–356

Hopkins PM (2012) The eyes have it: a brief history of crustacean neuroendocrinology. Gen Comp Endocrinol 175:357–366

Hopkins T, Kramer K (1992) Insect cuticle sclerotization. Annu Rev Entomol 37:273–302

Huet F, Ruiz C, Richards G (1995) Sequential gene activation by ecdysone in *Drosophila melanogaster*: the hierarchical equivalence of early and early late genes. Development 121:1195–1204

Hutchinson J, McNamara J, Houston A, Vollrath F (1997) Dyar's Rule and the Investment Principle: optimal moulting strategies if feeding rate is size-independent and growth is discontinuous. Phil Trans R Soc B 352:11–138

Huxley J (1932) Problems of relative growth. Methuen, London

Ikeya T, Galic M, Belawat P, Nairz K, Hafen E (2002) Nutrient-dependent expression of insulin-like peptides from neuroendocrine cells in the CNS contributes to growth regulation in *Drosophila*. Curr Biol 12:1293–1300

Iwami M, Furuya I, Kataoka H (1996) Bombyxin-related peptides: cDNA structure and expression in the brain of the hornworm *Agrius convolvuli*. Insect Biochem Mol Biol 26:25–32

Jindra M, Huang J, Malone F, Asahina M, Riddiford L (1997) Identification and mRNA developmental profiles of two ultraspiracle isoforms in the epidermis and wings of *Manduca sexta*. Insect Mol Biol 6:41–53

Jindra M, Malone F, Hiruma K, Riddiford L (1996) Developmental profiles and ecdysteroid regulation of the mRNAs for two ecdysone receptor isoforms in the epidermis and wings of the tobacco hornworm, *Manduca sexta*. Dev Biol 180:258–272

Johnston L, Prober D, Edgar B, Eisenman R, Gallant P (1999) Drosophila myc regulates cellular growth during development. Cell 98:779–790

Jones G, Sharp P (1997) Ultraspiracle: an invertebrate nuclear receptor for juvenile hormones. Proc Natl Acad Sci USA 94:13499–13503

Joosse ENG, Veltkamp E (1969) Some aspects of growth, moulting and reproduction in five species of surface dwelling Collembola. Netherl J Zool 20:315–328

Karlson P, Koolman J (1973) On the metabolic fate of ecdysone and 3-dehydroecdysone in *Calliphora vicina*. Insect Biochem 3:409–417

Kataoka H, Toschi A, Li JP, Carney RL, Schooley DA, Kramer SJ (1989) Identification of an allatotropin from adult *Manduca sexta*. Science 243:1481–1483

Kato Y, Nair K, Dyer K, Riddiford L (1987) Changes in ploidy level of epidermal cells during last larval instar

of the tobacco hornworm, *Manduca sexta*. Development 99:137–143

Kato Y, Riddiford L (1987) The role of 20-hydroxyecdysone in stimulating epidermal mitoses during the larval-pupal transformation of the tobacco hornworm, *Manduca sexta*. Development 100:227–236

Kawasaki H (1988) Studies on the wing disc morphogenesis according to the *Bombyx* metamorphosis. Spec Bull Coll Agric Utsunomiya Univ 63:1–4

King-Jones K, Charles J-P, Lam G, Thummel C (2005) The ecdysone-induced DHR4 orphan nuclear receptor coordinates growth and maturation in *Drosophila*. Cell 121:773–784

Klingenberg CP, Zimmermann M (1992) Dyar's rule and multivariate allometric growth in nine species of waterstriders (Heteroptera: Gerridae). J Zool 227:453–464

Koch P, Bückmann D (1987) Hormonal control of seasonal morphs by the timing of ecdysteroid release in *Araschnia levana* L. (Nymphalidae: Lepidoptera). J Insect Physiol 33:823–829

Kondo H, Ino M, Suzuki A, Ishizaki H, Iwami M (1996) Multiple gene copies for bombyxin, an insulin-related peptide of the silkmoth *Bombyx mori*: structural signs for gene rearrangement and duplication responsible for generation of multiple molecular forms of bombyxin. J Mol Biol 259:926–937

Kopec S (1917) Experiments in metamorphosis of insects. Bull Int Acad Cracov B:57–60

Kopec S (1922) Studies on the necessity of the brain for the inception of insect metamorphosis. Biol Bull 42:323–342

Koyama T, Syropyatova M, Riddiford L (2008) Insulin/IGF signaling regulates the change in commitment in imaginal discs and primordia by overriding the effect of juvenile hormone. Dev Biol 324:258–265

Kozma SC, Thomas G (2002) Regulation of cell size in growth, development and human disease: PI3K, PKB and S6K. BioEssays 24:65–71

Kramer SJ, Toschi A, Miller CA, Kataoka H, Quistad GB, Li JP, Carney RL, Schooley DA (1991) Identification of an allatostatin from the tobacco hornworm *Manduca sexta*. Proc Natl Acad Sci USA 88:9458–9462

Kremen C (1989) Patterning during pupal commitment of the epidermis in the butterfly, *Precis coenia*: the role of intercellular communication. Dev Biol 133:336–347

Kremen C, Nijhout H (1989) Juvenile hormone controls the onset of pupal commitment in the imaginal disks and epidermis of *Precis coenia* (Lepidoptera: Nymphalidae). J Insect Physiol 35:603–612

Kremen C, Nijhout H (1998) Control of pupal commitment in the imaginal disks of *Precis coenia* (Lepidoptera: Nymphalidae). J Insect Physiol 44:287–296

Krishnakumaran A, Schneiderman HA (1969) Induction of molting in crustacea by an insect molting hormone. Gen Comp Endocrinol 12:515–518

Laufer H, Biggers W (2001) Unifying concepts learned from methyl farnesoate for invertebrate reproduction and post-embryonic development. Am Zool 41:442–457

Laufer H, Demir N, Pan X, Stuart J, Ahl J (2005) Methyl farnesoate controls adult male morphogenesis in the crayfish, *Procambarus clarkii*. J Insect Physiol 51:379–384

Laufer H, Landau M, Homola E, Borst D (1987) Methyl farnesoate: Its site of synthesis and regulation of secretion in a juvenile crustacean. Insect Biochem 17:1129–1131

Layalle S, Arquier N, Léopold P (2008) The TOR pathway couples nutrition and developmental timing in *Drosophila*. Dev Cell 15:568–577

Lester DS, Gilbert LI (1987) Characterization of acetyl-cholinesterase activity in the larval brain of *Manduca sexta*. Insect Biochem 17:99–109

Locke M (1970) The molt/intermolt cycle in the epidermis and other tissues of an insect *Calpodes ethlius* (Lepidoptera, Hesperiidae). Tissue Cell 2:197–223

Luscher M (1972) Environmental control of juvenile hormone (JH) secretion and caste differentiation in termites. Gen Comp Endocrinol 3(Suppl):509–514

Mansfield SG, Cammer S, Alexander SC, Muehleisen DP, Gray RS, Tropsha A, Bollenbacher WE (1998) Molecular cloning and characterization of an invertebrate cellular retinoic acid binding protein. Proc Natl Acad Sci USA 95:6825–6830

Mao L, Henderson G, Liu Y, Laine RA (2005) Formosan subterranean termite (Isoptera: Rhinotermitidae) soldiers regulate juvenile hormone levels and caste differentiation in workers. Ann Entom Soc Am 98:340–345

Masumura M, Satake S, Saegusa H, Mizoguchi A (2000) Glucose stimulates the release of bombyxin, an insulin-related peptide of the silkworm *Bombyx mori*. Gen Comp Endocrinol 118:393–399

Mirth C, Truman J, Riddiford L (2005) The role of the prothoracic gland in determining critical weight for metamorphosis in *Drosophila melanogaster*. Curr Biol 15:1796–1807

Mitsui T, Riddiford L (1976) Pupal cuticle formation by *Manduca sexta* epidermis in vitro: Patterns of ecdysone sensitivity. Dev Biol 54:172–186

Mittler TE, Nassar SG, Staal GB (1976) Wing development and parthenogenesis induced in progenies of kinoprene-treated gynoparae of *Aphis fabae* and *Myzus persicae*. J Insect Physiol 22:1717–1725

Miura K, Oda M, Makita S, Chinzei Y (2005) Characterization of the *Drosophila* Methoprene-tolerant gene product. FEBS J 272:1169–1178

Mizoguchi A, Ishizaki H (1984) Circadian clock controlling gut-purge rhythm of the saturniid *Samia cynthia ricini*: its characterization and entrainment mechanism. J Comp Physiol A: Neuroethol Sens Neur Behav Physiol 155:639–647

Nakatsuji T, Lee C-Y, Watson R (2009) Crustacean molt-inhibiting hormone: Structure, function, and cellular mode of action. Comp Biochem Physiol A: Mol Integr Physiol 152:139–148

Nakatsuji T, Sonobe H (2004) Regulation of ecdysteroid secretion from the Y-organ by molt-inhibiting hormone in the American crayfish, *Procambarus clarkii*. Gen Comp Endocrinol 135:358–364

Nakatsuji T, Sonobe H, Watson R (2006) Molt-inhibiting hormone-mediated regulation of ecdysteroid synthesis in Y-organs of the crayfish (*Procambarus clarkii*): involvement of cyclic GMP and cyclic nucleotide phosphodiesterase. Mol Cell Endocr 253:76–82

Nijhout H (1975a) Axonal pathways in the brain-retrocerebral neuroendocrine complex of *Manduca sexta* (L.) (Lepidoptera: Sphingidae). Int J Insect Morphol Embryol 4:529–538

Nijhout H (1975b) A threshold size for metamorphosis in the tobacco hornworm, *Manduca sexta* (L.). Biol Bull 149:214–225

Nijhout H (1979) Stretch-induced moulting in *Oncopeltus fasciatus*. J Insect Physiol 25:277–282

Nijhout H (1984) Abdominal stretch reception in *Dipetalogaster maximus* (Hemiptera: Reduviidae). J Insect Physiol 30:629–633

Nijhout H (1994) Insect hormones. Princeton University Press, Princeton

Nijhout H (1999) Control mechanisms of polyphenic development in insects. Bioscience 49:181–192

Nijhout H (2003) Development and evolution of adaptive polyphenisms. Evol Dev 5:9–18

Nijhout H, Davidowitz G, Roff D (2006) A quantitative analysis of the mechanism that controls body size in *Manduca sexta*. J Biol 5:16

Nijhout H, Grunert L (2002) Bombyxin is a growth factor for wing imaginal disks in Lepidoptera. Proc Natl Acad Sci USA 99:15446–15450

Nijhout H, Grunert L (2010) The cellular and physiological mechanism of wing-body scaling in *Manduca sexta*. Science 330:1693–1695

Nijhout H, Smith W, Schachar I, Subramanian S, Tobler A, Grunert L (2007) The control of growth and differentiation of the wing imaginal disks of *Manduca sexta*. Dev Biol 302:569–576

Nijhout H, Wheeler D (1982) Juvenile hormone and the physiological basis of insect polymorphisms. Quart Rev Biol 57:109–133

Nijhout H, Williams C (1974) Control of moulting and metamorphosis in the tobacco hornworm, *Manduca sexta* (L.): cessation of juvenile hormone secretion as a trigger for pupation. J Exp Biol 61:493–501

Oda S, Kato Y, Watanabe H, Tatarazako N, Iguchi T (2011) Morphological changes in *Daphnia galeata* induced by a crustacean terpenoid hormone and its analog. Envir Toxicol Chem 30:232–238

Oldham S, Hafen E (2003) Insulin/IGF and target of rapamycin signaling: a TOR de force in growth control. Trends Cell Biol 13:79–85

Oldham S, Stocker H, Laffargue M, Wittwer F, Wymann M, Hafen E (2002) The *Drosophila* insulin/IGF receptor controls growth and size by modulating PtdInsP(3) levels. Development 129:4103–4109

Pratt GE, Farnsworth DE, Siegel NR, Fok KF, Feyereisen R (1989) Identification of an allatostatin from adult *Diploptera punctata*. Biochem Biophys Res Comm 163:1243–1247

Pratten D (1980) Growth in the crayfish *Austropotamobius pallipes* (Crustacea: Astacidae). Freshwater Biol 10:401–402

Preuss K, Nijhout H (2009) The importance of threshold size for the initiation of metamorphosis in the insect *Tribolium castaneum*. Integr Comp Biol 49:E291–E291

Quennedey A, Aribi N, Everaerts C, Delbecque J-P (1995) Postembryonic development of *Zophobas atratus* Fab. (Coleoptera: Tenebrionidae) under crowded or isolated conditions and effects of juvenile hormone analogue applications. J Insect Physiol 41:143–152

Rachinsky A, Hartfelder K (1990) Corpora allata activity, a prime regulating element for caste-specific juvenile hormone titre in honey bee larvae (*Apis mellifera carnica*). J Insect Physiol 36:189–194

Rankin M, Riddiford L (1977) Hormonal control of migratory flight in *Oncopeltus fasciatus*: The effects of the corpus cardiacum, corpus allatum, and starvation on migration and reproduction. Gen Comp Endocrinol 33:309–321

Redfern C (1983) Ecdysteroid synthesis by the ring gland of *Drosophila melanogaster* during late-larval, prepupal and pupal development. J Insect Physiol 29:65–71

Rewitz K, Yamanaka N, Gilbert L, O'Connor M (2009) The insect neuropeptide PTTH activates receptor tyrosine kinase Torso to initiate metamorphosis. Science 326:1403–1405

Reynolds B, Laynes R, Ögmundsdóttir M, Boyd C, Goberdhan D (2007) Amino acid transporters and nutrient-sensing mechanisms: new targets for treating insulin-linked disorders? Biochem Soc Trans 35:1215–1217

Riddiford L (1978) Ecdysone-induced change in cellular commitment of the epidermis of the tobacco hornworm, *Manduca sexta*, at the initiation of metamorphosis. Gen Comp Endocrinol 34:438–446

Riddiford L (1981) Hormonal control of epidermal cell development. Am Zool 21:751–762

Riddiford L (1996) Juvenile hormone: The status of its "status quo" action. Arch Insect Biochem Physiol 32:271–286

Riddiford L (2011) When is weight critical? J Exp Biol 214:1613–1615

Riddiford L, Ashburner M (1991) Effects of juvenile hormone mimics on larval development and metamorphosis of *Drosophila melanogaster*. Gen Comp Endocrinol 82:172–183

Riddiford L, Cherbas P, Truman J (2000) Ecdysone receptors and their biological actions. In: Litwack G (ed) Vitamins and hormones Vol 60. Academic Press, San Diego, pp 1–73

Robinson GE (1987) Regulation of honey bee age polyethism by juvenile hormone. Behav Ecol Sociobiol 20:329–338

Roer R, Dillaman R (1984) The structure and calcification of the crustacean cuticle. Am Zool 24:893–909

Rohdendorf EB, Watson JAL (1969) The control of reproductive cycles in the female firebrat, *Lepismodes inquilinus*. J Insect Physiol 15:2085–2101

Röller H, Dahm KH, Sweely CC, Trost BM (1967) The structure of the juvenile hormone. Angewandte Chemie (Int Ed) 6:179–180

Rountree D, Bollenbacher W (1984) Juvenile hormone regulates ecdysone secretion through inhibition of PTTH release. Am Zool 24:A31–A31

Rountree D, Bollenbacher W (1986) The release of the prothoracicotropic hormone in the tobacco hornworm, *Manduca sexta*, is controlled intrinsically by juvenile hormone. J Exp Biol 120:41–58

Rountree D, Nijhout H (1995) Hormonal control of a seasonal polyphenism in *Precis coenia* (Lepidoptera: Nymphalidae). J Insect Physiol 41:987–992

Rulifson EJ, Kim SK, Nusse R (2002) Ablation of insulin-producing neurons in flies: growth and diabetic phenotypes. Science 296:1118–1120

Rybczynski R, Gilbert L (2003) Prothoracicotropic hormone stimulated extracellular signal-regulated kinase (ERK) activity: the changing roles of $Ca^{2+}$- and cAMP-dependent mechanisms in the insect prothoracic glands during metamorphosis. Mol Cell Endocr 205:159–168

Satake S, Masumura M, Ishizaki H, Nagata K, Kataoka H, Suzuki A, Mizoguchi * (1997) Bombyxin, an insulin-related peptide of insects, reduces the Major storage carbohydrates in the silkworm Bombyx mori. Comp Biochem Physiol B Biochem Mol Biol 118:349–357

Schneiderman HA, Gilbert Ll (1964) Control of growth and development in insects. Science 143:325–333

Schubiger M, Tomita S, Sung C, Robinow S, Truman J (2003) Isoform specific control of gene activity in vivo by the *Drosophila* ecdysone receptor. Mech Dev 120:909–918

Shafiei M, Moczek A, Nijhout H (2001) Food availability controls the onset of metamorphosis in the dung beetle *Onthophagus taurus* (Coleoptera: Scarabaeidae). Physiol Entom 26:173–180

Shirai Y, Shimazaki K, Iwasaki T, Matsubara F, Aizono Y (1995) The in vitro release of prothoracicotropic hormone (PTTH) from the brain-corpus cardiacum-corpus allatum complex of silkworm, *Bombyx mori*. Comp Biochem Physiol C: Pharmacol Toxicol Endocrinol 110:143–148

Shorter J, Tibbetts E (2009) The effect of juvenile hormone on temporal polyethism in the paper wasp *Polistes dominulus*. Insectes Soc 56:7–13

Smagghe G, Degheele D (1995) Biological-activity and receptor-binding of ecdysteroids and the ecdysteroid agonists RH-5849 and RH-5992 in imaginai wing discs of *Spodoptera exigua* (Lepidoptera, Noctuidae). Eur J Entomol 92:333–340

Smith K (1991) The effects of temperature and daylength on the rosa polyphenism in the buckeye butterfly, *Precis coenia* (Lepidoptera: Nymphalidae). J Research on the Lepidoptera 30:237–244

Smith W, Rybczynski R (2012) Protoracicotropic hormone. In: Gilbert LI (ed) Insect endocrinology. Academic Press, New York, pp 1–61

Srivastava US, Gilbert LI (1969) The influence of juvenile hormone on the metamorphosis of *Sarcophaga bullata*. J Insect Physiol 15:177–189

Talbot W, Swyryd E, Hogness D (1993) *Drosophila* tissues with different metamorphic responses to ecdysone express different ecdysone receptor isoforms. Cell 73:1323–1337

Thompson D (1942) On growth and form. Cambridge University Press, Cambridge

Thummel C (1995) From embryogenesis to metamorphosis: The regulation and function of *Drosophila* nuclear receptor superfamily members. Cell 83:871–877

Thummel C (2001) Molecular mechanisms of developmental timing in *C. elegans* and *Drosophila*. Dev Cell 1:453–465

Tobler A, Nijhout H (2010) A switch in the control of growth of the wing imaginal disks of *Manduca sexta*. PLoS ONE 5:e10723. doi:10.1371/journal.pone.0010723

Truman J (1972) Physiology of insect rhythms. 1, Circadian organization of the endocrine events underlying the molting cycle of larval tobacco hornworms. J Exp Biol 57:805–820

Truman J, Riddiford L (1974) Physiology of insect rhythms. 3. The temporal organization of the endocrine events underlying pupation of the tobacco hornworm. J Exp Biol 60:371–382

Truman J, Talbot W, Fahrbach S, Hogness D (1994) Ecdysone receptor expression in the CNS correlates with stage-specific responses to ecdysteroids during *Drosophila* and *Manduca* development. Development 120:219–234

Tsukimura B, Borst DW (1992) Regulation of methyl farnesoate in the hemolymph and mandibular organ of the lobster, *Homarus americanus*. Gen Comp Endocrinol 86:297–303

Wainwright G, Webster S, Rees H (1999) Involvement of adenosine cyclic-3-monophosphate in the signal transduction pathway of mandibular organ-inhibiting hormone of the edible crab, *Cancer pagurus*. Mol Cell Endocr 154:55–62

Wainwright G, Webster S, Wilkinson M, Chung J, Rees H (1996) Structure and significance of mandibular organ-inhibiting hormone in the crab, *Cancer pagurus*. J Biol Chem 271:12749–12754

Waterman T (1960) The physiology of Crustacea. Academic Press. London

Watson J (1964) Moulting and reproduction in the adult firebrat, *Thermobia domestica* (Packard) (Thysanura, Lepismatidae) - I. The moulting cycle and its control. J Insect Physiol 10:305–317

Wheeler D (1991) The developmental basis of worker caste polymorphism in ants. Am Nat 138:1218–1238

Wheeler D, Buck N, Evans J (2006) Expression of insulin pathway genes during the period of caste determination in the honey bee, *Apis mellifera*. Insect Mol Biol 15:597–602

Wheeler D, Nijhout H (1981) Soldier determination in ants: new role for juvenile hormone. Science 213:361–363

Wheeler D, Nijhout H (1983) Soldier determination in *Pheidole bicarinata*: effect of methoprene on caste and size within castes. J Insect Physiol 29:847–854

Wheeler D, Nijhout H (1984) Soldier determination in *Pheidole bicarinata*: inhibition by adult soldiers. J Insect Physiol 30:127–135

Wheeler D, Nijhout H (2003) A perspective for understanding the modes of juvenile hormone action as a lipid signaling system. BioEssays 25:994–1001

Wielgus JJ, Bollenbacher WE, Gilbert LI (1979) Correlations between epidermal DNA synthesis and haemolymph ecdysteroid titre during the last larval instar of the tobacco hornworm, *Manduca sexta*. J Insect Physiol 25:9–16

Wigglesworth V (1934) The physiology of ecdysis in *Rhodnius prolixus* (Hemiptera). II. Factors controlling moulting and 'metamorphosis'. Quart J Microsc Sci 77:191–222

Wigglesworth V (1936) The function of the corpus allatum in the growth and reproduction of *Rhodnius prolixus* (Hemiptera). Quart J Microsc Sci 79:91–121

Wigglesworth V (1940) The determination of characters at metamorphosis in *Rhodnius prolixus* (Hemiptera). J Exp Biol 17:201–223

Wigglesworth V (1948) The functions of the corpus allatum in *Rhodnius prolixus* (Hemiptera). J Exptl Biol 25:1–15

Williams C (1947) Physiology of insect diapause. II. Interaction between the pupal brain and prothoracic glands in the metamorphosis of the giant silkworm, *Platysamia cecropia*. Biol Bull 93:89–98

Williams C (1948) Physiology of insect diapause. III. The prothoracic glands in the Cecropia silkworm, with special reference to their significance in embryonic and postembryonic development. Biol Bull 94:60–65

Williams C (1959) The juvenile hormone. I. Endocrine activity of the corpora alata of the adult cecropia silkworm. Biol Bull 116:323–338

Williams C (1961) The juvenile hormone. II. Its role in the endocrine control of molting, pupation, and adult development in the Cecropia silkworm. Biol Bull 121:572–585

Willis J (1969) The programming of differentiation and its control by juvenile hormone in saturniids. J Embryol Exp Morphol 22:27–44

Willis J (1974) Morphogenetic action of insect hormones. Annu Rev Entomol 19:97–115

Willis J (1981) Juvenile hormone: the status of *status quo*. Am Zool 21:763–773

Wilson E (1971) The insect societies. Harvard University Press, Cambridge

Wirtz P, Beetsma J (1972) Induction of caste differentiation in the honeybee (*Apis mellifera*) by juvenile hormone. Ent Exp Appl 15:517–520

Wolfgang W, Riddiford L (1981) Cuticular morphogenesis during continuous growth of the final instar larva of a moth. Tissue Cell 13:757–772

Wolfgang W, Riddiford L (1986) Larval cuticular morphogenesis in the tobacco hornworm, *Manduca sexta*, and its hormonal regulation. Dev Biol 113:305–316

Woltereck R (1909) Weitere experimentelle Untersuchungen ueber Artveraenderung, speziell über das Wesen quantitativer Artunterschiede bei Daphnien. Verh dtsch zool Ges 19:110–172

Yamanaka N, Dusan Z, Kim Y-J, Adams ME, Hua Y-J, Suzuki Y, Suzuki M, Suzuki A, Satake H, Mizoguchi A, Asaoka K, Tanaka Y, Kataoka H (2006) Regulation of insect steroid hormone biosynthesis by innervating peptidergic neurons. Proc Natl Acad Sci USA 103:8622–8627

Zeleny C (1905) Compensatory regulation. J Exp Zool 2:1–102

Zera A, Harshman L, Williams T (2007) Evolutionary endocrinology: the developing synthesis between endocrinology and evolutionary genetics. Ann Rev Ecol Evol Syst 38:793–817

Zhou B, Hiruma K, Shinoda T, Riddiford L (1998) Juvenile hormone prevents ecdysteroid-induced expression of broad complex RNAs in the epidermis of the tobacco hornworm, *Manduca sexta*. Dev Biol 203:233–244

Zhou X, Riddiford L (2002) Broad specifies pupal development and mediates the *status quo* action of juvenile hormone on the pupal-adult transformation in *Drosophila* and *Manduca*. Development 129: 2259–2269

# Arthropod Regeneration

7

## Diego Maruzzo and Francesca Bortolin

## Contents

## 7.1 Introduction

Regeneration or restoration of a lost body part may occur at different levels: it can involve a part of the trunk, a specific organ or structure (such as a limb), tissues (such as the gut lining)

D. Maruzzo (✉) · F. Bortolin
Department of Biology, University of Padova, Via U. Bassi 58 B, I 35131, Padova, Italy
e-mail: maruzzo@bio.unipd.it

F. Bortolin
e-mail: francesca.bortolin@unipd.it

or cells (such as axons or muscle fibres) (Bely and Nyberg 2009).The vast majority of studies on arthropod regeneration are about limbs. Usually, arthropods do not regenerate parts of the trunk (but see Sect. 7.2), and specific accounts on regeneration of organs other than limbs (e.g. eyes; see Pastre 1960; Joly and Herbaut 1968), tissues and cells are restricted to few arthropod subgroups. This chapter will thus be mainly about arthropod limb regeneration, with a less detailed discussion on regeneration of arthropod trunk and tissues.

Regeneration is mostly intended as the restoration of a body part lost by some kind of external damage (e.g. amputations performed by researchers or following a predatory attack), but some physiological events may also involve regeneration. Indeed, many arthropods exhibit regular degeneration and regeneration of body parts during development; this is the case, for example, of limbs that are physiologically lost and formed again later in development (called 'Lazarus appendages' by Minelli 2003), of the male reproductive system of balanomorph cirripedes which normally degenerates at the end of a reproductive season to be regenerated before the following one (e.g. Klepal et al. 2008), and of the insect midgut epithelium which is more or less continuously lost and regenerated (see Sect. 7.4.2). This review will consider not only regeneration following external damage, subsequently called *reparative regeneration*, (e.g. Vorontsova and Liosner 1960; Seifert et al. 2012), but also the *physiological regeneration*

A. Minelli et al. (eds.), *Arthropod Biology and Evolution*,
DOI: 10.1007/978-3-642-36160-9_7, © Springer-Verlag Berlin Heidelberg 2013

(e.g. Morgan 1901; Seifert et al. 2012) that follows programmed degeneration or loss of a body part to produce a duplicate of it.

In all cases of reparative regeneration and in some cases of physiological regeneration, the process follows a breaking of the epidermis (with its cuticle) that separates the internal environment from the external one. A complete and functional regeneration will thus require (1) a proper wound healing and (2) one or more moults. There are many studies on arthropod wound healing (see, e.g. Adiyodi 1972; Truby 1983, 1985; Clare et al. 1990; Hopkins et al. 1999; Galko and Krasnow 2004; Theopold et al. 2004; Vafopoulou 2009; Belacortu and Paricio 2011; Repiso et al. 2011; and references therein), but we will not deal with this process here. Much has also been written about the relationship between moulting and regeneration in arthropods, and we will briefly treat this aspect.

There are many accounts on putative regeneration in fossil taxa, especially trilobites (e.g. McNamara and Tuura 2011; and references therein). While many of these descriptions are most likely based on regenerating specimens, the information that can be obtained from them is not comparable with evidence from living taxa that have been mostly studied experimentally. Thus, this review will consider only regeneration in extant species.

Finally, it must be acknowledged that arthropod regeneration is a topic that has received considerable attention, although not in the last decades, and many interesting reviews covering all arthropods or major subgroups are available; to our knowledge, the most informative are those of Korschelt (1907), Przibram (1909), Bodenstein (1953), Bliss (1960), Vorontsova and Liosner (1960), Needham (1965), Goss (1969), Bullière and Bullière (1985), Skinner (1985), Vernet and Charmantier-Daures (1994), and Maruzzo et al. (2005).

## 7.2  Trunk Regeneration

It is usually acknowledged that arthropods do not regenerate parts of the trunk (e.g. Minelli 2003; Bely and Nyberg 2009; Bely 2010). Some authors consider this not possible for intrinsic developmental reasons; for example, Minelli (2003) proposed that arthropods are not able to express the molecular markers needed to establish the anterior and posterior end of the trunk axis except in early embryonic development. It must be noted, however, that a few reports of limited arthropod trunk regeneration do exist.

Regeneration of the post-anal trunk part has been reported for the telson of horseshoe crabs (Clare et al. 1990) and malacostracan crustaceans (e.g. Needham 1952; Kraus and Weis 1988; Kahn et al. 1993; Mees et al. 1995) and for the 'caudal appendage' of the beetle *Scraptia fuscula* (Švácha 1995). Furthermore, many insects can regenerate small parts of the last segment, and some reports of complete regeneration of one or more posterior segment(s) (Fig. 7.1a–d) are available (see Abeloos 1932; Vorontsova and Liosner 1960; and references therein). Abnormal trunk regeneration, with something more similar to a trunk limb replacing an ablated trunk part, has been reported in pantopods (Fig. 7.1 e-g; Loeb 1905) and brachyurans (Kocian 1930). Most of these works (apparently, except those on pantopods) experienced high mortality, and this clearly suggests that both wound healing and intrinsic developmental reasons must be considered as explanations of the very limited, but not absolutely nil, arthropod trunk regenerative potential.

## 7.3  Limb Regeneration

Arthropod limb regeneration has been studied with many different purposes and focusing on a number of different aspects. Here, we will try to

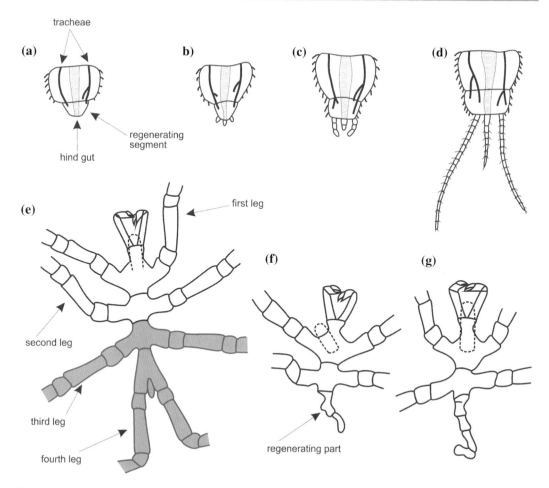

**Fig. 7.1** Examples of arthropod trunk regeneration. **a–d** Stages of regeneration of the last trunk segment in mayfly (redrawn from Abeloos 1932). **e–g** Trunk regeneration in pantopod. **e** Whole undamaged body with the subsequently ablated trunk part in *grey*. **f–g** Successive stages of regeneration (redrawn from Loeb 1905)

limit the discussion to the phylogenetic diversity of limb regenerative potential and to the relevant developmental and physiological aspects. The main topics we will not discuss are regeneration following experimental grafting (often performed with the purpose of studying limb pattern formation) and abnormal regeneration.

Many arthropods have specific mechanisms facilitating limb loss through limb breakage at specific points along the limb. Processes of limb loss involving such mechanisms are usually referred to as *autotomy*, and the specific breakage point is called the *autotomy plane*. However, there is conspicuous variation in the mechanisms facilitating limb loss. 'True autotomy', that is, limb loss by means of a specific nervous reflex activating a specific muscle, has been observed only in malacostracans and hexapods. Other forms of autotomy are *autospasy*, that is, limb loss by the action of an outside agent against resistance provided by the animal, and *autotilly*, when limb loss is obtained with the assistance of other limbs of the same animal (Bliss 1960; Maruzzo et al. 2005). For some taxa, there is no evidence of any mechanism facilitating limb loss (Maruzzo et al. 2005). More specific information about autotomy and its phylogenetic distribution can be found in Bliss (1960), Maruzzo et al. (2005), and Fleming et al. (2007)

## 7.3.1 Limb Reparative Regeneration

Many arthropod species, but not all, can regenerate a lost limb, and many factors (including developmental stage, limb type and amputation level) have been shown to influence the outcome of the regeneration process (Maruzzo et al. 2005). For descriptive reasons, we recognize here four levels of regenerative potential: absent (no regeneration), poor (regeneration fails to produce a structurally normal limb, or only a small portion of the limb is regenerated), good (regeneration produces a structurally normal limb under some specific conditions) and very good (regeneration produces a structurally normal limb under most conditions). A structurally normal regenerated limb is defined here as a limb that can perform most of the functions of an undamaged one. A review of the phylogenetic distribution of arthropod limb regenerative potential is provided in Table 7.1 for pre-oral and trunk limbs; the mouth parts are not considered here only because these have not been the object of specific studies and even general observations are very rare (Maruzzo et al. 2005).

### 7.3.1.1 Lack of Regenerative Potential

A complete absence of regeneration is found in some arthropod taxa, including some mites and geophilomorph centipedes (Table 7.1). Here, the remaining limb stump can assume different morphologies after moulting (some examples in Fig. 7.2).

Explaining why some species cannot regenerate at all is an interesting evolutionary problem, and both wound healing and developmental reasons should be considered. For example, mortality following leg amputation is very high in species of mites that do not regenerate (reaching more than 90 % and even 100 % in some species; Rockett and Woodring 1972), but there can be also a developmental reason for the different regenerative potential exhibited by different mite species. For example, Rockett and Woodring (1972) noted that the only species for which post-larval cell divisions have been

observed are those that can regenerate, while those that cannot regenerate apparently do not have any post-larval cell division.

### 7.3.1.2 Poor Regenerative Potential

Several arthropod taxa, for instance scorpions and springtails (Table 7.1), are unable to regenerate a structurally normal limb and regenerate only one or few segments of the limb or a small abnormal one (Fig. 7.3).

Factors such as developmental stage and amputation level usually have some influence. Scorpion can usually regenerate only the terminal segment (pre-tarsus) of the legs (Fig. 7.3a; Rosin 1964), and the only report of a specimen regenerating two segments (tarsus and pre-tarsus) is from a specimen that had these two segments removed during the first post-embryonic ('pre-nymphal') stage (Vachon 1957). Heteropterans have a poor regenerative potential, but amputation level has some influence: amputations performed in the proximal part of the leg produce at most one regenerated segment (the terminal one), but amputations distal to the tibia can produce up to three regenerated tarsal segments (Lüscher 1948; Shaw and Bryant 1974). Limb type can also be another factor explaining poor regenerative potential; the jumping legs of several tettigoniids, acridids and gryllids never regenerate except for the tarsus, while the first two pairs of legs have better regeneration ability (Bordage 1905).

### 7.3.1.3 Good Regenerative Potential

Many arthropod species are able to regenerate a structurally normal limb only under specific conditions (Table 7.1) the most critical factor usually being the level of the amputation. In the black widow spider *Latrodectus variolus*, for example, regeneration occurs always following amputations at or distal to the femur–patella joint; amputations at the femur mid-point lead to regeneration only in 70 % of the cases, while in 30 % of the cases and for amputations proximal to the femur mid-point, there is only wound

**Table 7.1** Regenerative potential of arthropod limbs

| Taxa | | Regeneration | | Notes |
|------|------|------|------|------|
| | | Pre-oral limbs | Trunk limbs | |
| Chelicerata | Acari | Absent to very good | | Regenerative potential differs according to subgroup |
| | Amblypygi | No information | Good | No regeneration following breakage proximal to the patella–tibia joint of the leg |
| | Araneae | No information | Poor to good | Regenerative potential differs according to subgroup |
| | Opiliones | Absent | | – |
| | Pycnogonida | Absent? | Very good? | Only old observations available; no specific studies, but some species lose their chelicerae in the last pre-adult instar and do not regenerate them |
| | Scorpiones | Poor? | Poor | Following leg amputation only the pre-tarsus (distal leg article) is usually regenerated; unclear if also the distal part of chelicerae regenerates |
| | Xiphosura | No information | Good? | Trunk limbs regenerate during larval stage only? |
| | Other taxa | No information | | – |
| Myriapoda | Diplopoda | Good | Very good? | Proximal antennal amputation leads to a regenerate with reduced segmentation; only old observations available for leg regeneration |
| | Geophilomorpha | Absent | | – |
| | Lithobiomorpha | Very good | | No regeneration following breakage proximal to the coxa–trochanter joint of the leg? |
| | Scolopendromorpha | Absent to good? | Absent | Antennal regeneration depending on subgroup; questionable indirect evidence for leg regeneration in some species |
| | Scutigeromorpha | No information | Very good | No regeneration following breakage proximal to the coxa–trochanter joint of the leg? |
| | Other taxa | No information | | – |
| Crustacea | Amphipoda | Very good | | – |
| | Branchiopoda | Absent? | Absent to very good? | Antennal regeneration absent in Cladocera but no information for other groups; trunk limb regenerative potential differs according to subgroup |
| | Cirripedia | – | Very good | No pre-oral limbs in juvenile and adults |
| | Copepoda | Absent? | | Some old indirect evidence suggestive of regeneration |
| | Decapoda | Very good | | – |
| | Isopoda | Very good | | – |
| | Phyllocarida | Very good? | No information | Only old observations available on antennal regeneration |
| | Other taxa | No information | | |

(continued)

**Table 7.1** (continued)

| Taxa | | Regeneration | | Notes |
|---|---|---|---|---|
| | | Pre-oral limbs | Trunk limbs | |
| Hexapoda | Archaeognatha | Very good? | | Only general observations available |
| | Blattodea | Very good | | – |
| | Collembola | Poor | No information | – |
| | Dermaptera | Very good? | No information | Only general observations available for the antennae |
| | Diplura | Poor? | Good? | Only general observations available |
| | Endopterygota | Very good? | Absent to good | Leg regenerative potential differs according to subgroup |
| | Ephemeroptera | No information | Good | Often only a leg of reduced size regenerated |
| | Hemiptera | Poor | Absent to poor | Leg regenerative potential differs according to subgroup |
| | Isoptera | No information | Very good | – |
| | Mantodea | No information | Very good | Unknown if the raptorial legs also regenerate |
| | Odonata | No information | Very good | – |
| | Orthoptera | Good | Absent to good | Abnormalities if antenna amputated proximally; leg regenerative potential differs according to subgroup but also within a given species according to leg type |
| | Phasmatodea | Poor | Very good | Proper antennal regeneration depends on amputation level |
| | Zygentoma | Very good? | Very good | Only general observations available for the antennae |
| | Other taxa | No information | | |

The regenerative potential for each taxon is rated as explained in the text. Pre-oral limbs are chelicerae in chelicerates, first and second antennae in crustaceans and antennae in myriapods and hexapods; trunk limbs are legs, thoracopods, pleopods, etc. In many instances, only old or fragmentary observations are available and the quality of the available observations for different taxa is very uneven; thus, we have not been able to completely avoid some arbitrary evaluation. More detail and specific references in Maruzzo et al. (2005); useful references not included in Maruzzo et al. (2005) are Nguyen Duy-Jacquemin (1972) and Petit (1974) for diplopod antennae, Jegla (1982) for xiphosuran legs, Buck and Edwards (1990) for zygentoman legs, Harvey et al. (2003) for cirripede cirri, Maruzzo et al. (2007), (2008) for isopod second antennae and Lüdke and Lakes-Harlan (2008) for orthopteran legs (one species with poor regenerative potential)

healing but no regeneration (Fig. 7.4; Randall 1981).

### 7.3.1.4 Very Good Regenerative Potential

Most studies on arthropod limb regeneration have been performed on species that are able to regenerate a structurally normal limb under most conditions, such as decapod crustaceans and cockroaches (Table 7.1).

A structurally normal limb, however, is not necessarily a perfect duplicate of an undamaged one. Cockroaches are a good example of this. These insects are famous for their high regenerative potential and have been extensively studied in this respect (reviewed in Bullière and Bullière 1985; Maruzzo et al. 2005). However, a

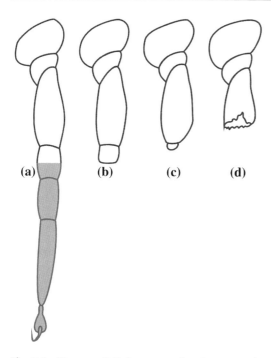

**Fig. 7.2** Absence of limb regeneration in some mite taxa. **a** Generalized mite leg with the subsequently ablated part in *grey*. **b–d** The remaining leg after one moult following the amputation in Gamasina (**b**), Astigmata (**c**) and Prostigmata (**d**), (redrawn from Rockett and Woodring 1972)

regenerated cockroach leg is usually not a perfect duplicate of an undamaged one. The tarsus is often composed of a different number of segments (see below), and there are also differences in internal anatomy. Kaars et al. (1984) observed that in the regenerated femur, one small muscle is missing and there are differences in the arrangement and branching pattern of nerves and tracheae (Fig. 7.5). These authors suggested that these differences are most likely due to different tissue-level interactions during regeneration, compared with normal development (Kaars et al. 1984). Among the anatomical differences noted by Kaars et al. (1984), the missing muscle may appear as an important one for the limb functional morphology; however, the very careful study of Alsop (1978; see the muscle named "5(fe)ung") showed that (1) the main function of this muscle is to flex the pretarsal claws (i.e. the leg can probably walk normally without it) and (2) there are other muscles originating from more distal points

along the leg that concur to perform the same function (i.e. without the muscle, the function is not entirely lost). Thus, while even species with high regenerative potential may not be able to regenerate a perfect duplicate of a missing limb, the functional morphology of the regenerated limb is usually not compromised. Since comparable fine anatomical studies are not available for other arthropod species, it is currently not possible to determine whether a perfect duplicate can be regenerated at all.

As noted above, limb reparative regeneration can also produce a structurally normal limb with an abnormal number of segments. These cases involve limb parts which do not exhibit autonomous movements; thus, a different segment number does not deeply affect limb function. Examples are the multisegmented tarsus of insect legs and the multisegmented tibia and tarsus of the foreleg of whip spiders (Amblypygi). In several insects, a leg with a reduced number of tarsal segments is not uncommonly regenerated (Maruzzo et al. 2005; and references therein). This has been more carefully studied in cockroaches (e.g. Bullière and Bullière 1985; Tanaka et al. 1992; Tanaka and Ross 1996), but a possible developmental explanation of this occurrence has been provided by a study on *Drosophila* leg imaginal disc regeneration which provided evidence that this could be the result of molecular interactions occurring during regeneration, but not during normal development, in the establishment of the limb proximo-distal axis (Bosch et al. 2010).

Tibia and tarsus of the foreleg of whip spiders, which do not increase in segment number after the nymphal stage during normal development, always regenerate with a higher number of segments. Igelmund (1987) noted that regenerated forelegs of older (and thus larger) specimens have more segments than those of younger (and thus smaller) ones. Analysing data for 52 whip spider species, Minelli et al. (2010) found a positive correlation between the species' adult body size and the number of tibial and tarsal segments of undamaged foreleg. These observations suggest that this seemingly puzzling case of hyper-segmentation during

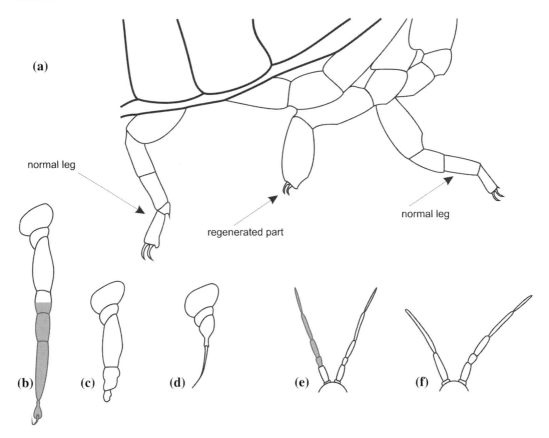

**Fig. 7.3** Examples of poor regenerative potential. **a** Regeneration in scorpion leg (drawn from photograph in Rosin 1964). **b-d** Poor regenerative potential in some mite taxa. **b** Generalized mite leg with the subsequently ablated part in *grey*. **c–d** The regenerated leg after one moult following the amputation in Uropodida (**c**) and Cryptostigmata (**d**) (redrawn from Rockett and Woodring 1972). **e–f** Poor regenerative potential in the antennae of collembolans. **e** Anterior part of the head with two undamaged antennae with the subsequently ablated part in *grey*. **f** Subsequent regeneration (redrawn from Ernsting and Fokkema 1983)

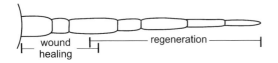

**Fig. 7.4** Regenerative potential in the black widow spider leg. The picture shows the levels along the leg where amputations are followed by regeneration or only by wound healing (redrawn from Randall 1981)

regeneration can be easily explained as a result of size-related tibial and tarsal segmentation; that is, the specimen producing tibial and tarsal segmentation during regeneration is larger than the juvenile where tibial and tarsal segmentation was originally established during normal development.

## 7.3.2 Physiological Limb Regeneration

There are several examples of arthropod species where a limb or a limb part is physiologically lost or degenerated and subsequently regenerated. Minelli (2003) named 'Lazarus appendages' those limbs that are lost at a given developmental stage but are subsequently regrown after one or more developmental stage(s) lacking those limbs. Examples are from mite legs, copepod first antennae, various limbs of decapod crustaceans, second and third pair of pantopod larval limbs and millipede gonopods (Minelli 2003).

**(a)**

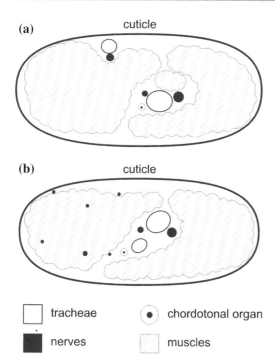

**(b)**

| □ | tracheae | ◉ | chordotonal organ |
|---|---|---|---|
| ■ | nerves | ▨ | muscles |

**Fig. 7.5** Anatomical differences between normal and regenerated cockroach legs. Cross section at about one-third of the femur from its proximal end of a leg produced during normal development (**a**) and during regeneration following amputation at the trochanter–femur joint (**b**). Note that the anatomy shown in (**b**) is consistently reproduced following amputations at the trochanter–femur joint (modified from Kaars et al. 1984)

A different example of physiological limb regeneration is provided by the antennal flagellum of some malacostracan crustaceans and insects. In crayfish, lobster and cockroach, for example, the flagellum of the antennae from a given developmental stage onwards regularly loses some of its more distal segments (usually called annuli), while new ones are produced in the proximal part (Campbell and Priestley 1970; Schafer 1973; Sandeman and Sandeman 1996; Steuellet et al. 2000).

### 7.3.3 Development of the Regenerating Limb

#### 7.3.3.1 Morphological Aspects

In most cases of limb reparative regeneration, the process takes place within the limb stub

(Fig. 7.6a). The only exceptions are found in decapod crustaceans and, possibly, in horseshoe crabs (see Jegla 1982) where the regenerating limb grows outside the limb stub, within a thin and transparent cuticular sac (or cuticular sheath); this outgrowing structure is usually referred to as a papilla (Fig. 7.6b). Different decapod species exhibit different morphology and development of the papilla (Bliss 1960; Goss 1969) and regeneration following amputations in the distal part of a limb are accomplished within the limb stub, as in all other arthropods (Bliss 1960). As the regenerate grows, either within the limb stub or within an 'external' cuticular sac, it may become folded (e.g. Bliss 1960; Needham 1965; Goss 1969; Bullière and Bullière 1985). In decapod crustaceans, growth of the regenerate before its appearance after a moult is not a continuous process, but happens in phases separated by periods of stasis (Bliss 1960; Adiyodi 1972; Skinner 1985). Tissue and segment differentiation are usually completed within the first growing period, the so-called basal growth (e.g. Adiyodi 1972).

In most arthropods, limb regeneration follows an all-or-nothing principle, that is, after the first moult following damage or amputation, the regenerate is either not present at all or structurally in its final form (although it may be not functional and of reduced size, it may require several additional moults to reach the size of an undamaged limb). The former case is usually found when damage or amputation happens too close to the next moult, and thus, only wound healing happens, while the full regeneration process will occur after that moult (e.g. Truby 1985). The 'critical time' in the intermoult period, after which an amputation does not lead immediately to a regenerate at the following moult, is species specific; for example, in the house centipede *Scutigera coleoptrata*, it is at about three-quarters of the intermoult period, while in the isopod *Asellus aquaticus* is at about half and in both the shrimp *Palaemon serratus* and the cockroach *Blattella germanica* is at about two-thirds of the period (Goss 1969). There are exceptions, however. The kissing bug

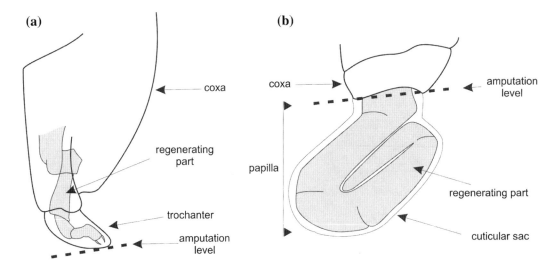

**Fig. 7.6** Different types of limb regeneration. **a** Regeneration within the limb stub in the cockroach leg (redrawn from Goss 1969). **b** Regeneration within an 'outgrowing' cuticular sac in the fiddler crab leg (redrawn from Hopkins et al. 1999). In both cases, folding of the regenerating tissue due to growth is common

*Rhodnius prolixus*, which has poor limb regenerative potential, does not follow the all-or-nothing principle. This species has a critical point after which leg amputation leads only to wound healing after the next moult, but amputation before the critical point leads to different results (in terms of both length and segmentation of the regenerate) according to how many days before the moult the amputation was performed (Lüscher 1948).

When the regenerated limb first appears, it is usually composed of its final number of segments. There are very few cases where the regenerate first appears as a limb bud and, after a further moult, as a segmented limb, or where there is an increase in segment number with new moults (Maruzzo et al. 2005). A limb part that regularly increases in segment number in subsequent moults during both normal development and regeneration is the antennal flagellum of several malacostracan crustaceans and insects; in these cases, new segments are produced in the proximal part of the flagellum, in the so-called growth zone (Maruzzo and Minelli 2011). In some species (i.e. shrimps and cockroaches), during normal development, the flagellum becomes very long and, from a given developmental stage onwards, regularly loses some of its more distal segments; these cases should be considered as physiological limb regeneration (see Sect. 7.3.2), but this does not happen in all species and throughout post-embryonic development. In the few cases where the segmentation during regeneration has been studied, it was found that the mechanism producing new segments during normal development is simply accelerated during regeneration (Schafer 1973; Maruzzo et al. 2007, 2008; but, apparently, no acceleration following distal amputation).

### 7.3.3.2 Histological Aspects

Histological aspects of arthropod limb regeneration have been studied in some depth only in regenerating legs of malacostracans and insects. After amputation of a limb, wound healing (which involves extensive cell migration and differentiation) is the first histological process that takes place, but the very beginning of the regeneration process is the formation of the blastema, the mass of cells responsible for producing the regenerate.

Separating cellular events involved in wound healing from those involved in blastema formation is not so obvious (e.g. Truby 1985), and in our opinion, this has caused much confusion and conflicting interpretations of the histology of regenerating arthropod limbs (see e.g. Needham 1965; Adiyodi 1972; Truby 1983, 1985; Lumb et al. 1991; Read and Govind 1998; Hopkins et al. 1999; and references therein). Truby (1983, 1985) demonstrated that in the cockroach leg, the blastema is mainly formed by dedifferentiation of the epidermis and subsequent proliferation at and close to the wound site. Instead, other authors (e.g. Adiyodi 1972; Read and Govind 1998) gave more attention to the role of inner mesenchymal cells in the development of the regenerate. Regardless of the different cell types present and their role, there is no evidence of any dedifferentiation to a pluripotent cell stage. Most likely, in the formation and development of the blastema, cells undergo limited dedifferentiation and produce only cells of the same type. While there has never been a question that the epidermis and epidermal-derived structures of the regenerate are produced by blastema epidermis, contrasting opinions have been put forward for the origin of muscles. We agree with Goss (1969; and see also Needham 1965) that evidence for an epidermal origin of muscle cells is not convincing, but it is currently unclear whether they derive from former muscles that have undergone histolysis or from immigrant blastocytes of unknown origin (e.g. Needham 1965; Lumb et al. 1991; Read and Govind 1998). Nerves are regenerated from nervous tissues from the stump (e.g. Lumb et al. 1991; Cooper 1998; and references therein for decapods).

There is both direct and indirect evidence that a remarkable proportion of the limb stub tissue close to the amputation dedifferentiates and/or degenerates and becomes involved in the regeneration process. The best observations on this are from regenerating cockroach legs where, for example, Bodenstein (1955) provided evidence for muscle degeneration in the limb stub proximal to the amputation and Truby (1985) showed that for amputations between trochanter and femur, the formation of the blastema involves epidermal cells in the whole trochanter and the distal part of the coxa (this is the segment proximal to the trochanter; thus, even the joint between coxa and trochanter is lost and formed again during regeneration). Indirect evidence for similar processes in other arthropods is widespread. For example, in scorpion legs, the segment proximal to the amputation site changes setal pattern after regeneration (Rosin 1964). Interestingly, even in decapods, where the regenerating limb develops within the outgrowing cuticular sac (Sect. 7.3.1.1), uninjured muscles proximal to the amputation level atrophy after amputation (Moffett 1987; Schmiege et al. 1992).

All data discussed so far for the histological aspects of arthropod limb regeneration are from malacostracan and insect legs. Clearly, there may be differences in other taxa and/or for other limb types. The antennal flagellum of malacostracans and insects discussed above is an example; from morphological observations, it was shown that amputations in the flagellum of the isopod *Asellus aquaticus* induced only a rearrangement in the amputation site, possibly corresponding to wound healing at the histological level, and the regeneration is accomplished by acceleration of the normal postembryonic development (activity of the growth zone) of this limb part (Maruzzo et al. 2008).

### 7.3.3.3  Developmental Genetic Aspects

There are currently very few data concerning the developmental genetics of arthropod limb regeneration. Indeed, only three models have been studied in this respect: the leg imaginal disc of *Drosophila*, the legs of *Gryllus bimaculatus* nymphs and the legs of *Tribolium castaneum* larvae. Strictly speaking, limb imaginal discs are not limbs, they are just a tissue. However, it has been demonstrated that, at the molecular level, the legs of *Drosophila* are already essentially patterned in the imaginal discs, thus they have been extensively used in studies on arthropod limb development (Nagy and Williams 2001; Williams 2004; Angelini and Kaufman 2005b; Giorgianni and Patel 2005; Prpic and Damen 2008). In this section, we will discuss some data on imaginal discs that we consider relevant for limb regeneration (see Sect. 7.4.1 for imaginal disc regeneration as a form of tissue regeneration).

Several growth-controlling proteins and signalling pathways have been shown to regulate tissue growth not only during normal development, but also during regeneration. During normal development of insect legs, at least four kinds of morphogens, Hedgehog (Hh), the TGF-$\beta$ family member Decapentaplegic (Dpp), the Wnt family member Wingless (Wg) and ligand(s) for epidermal growth factor receptor (EGFR), are involved in regulating cell fate (e.g. Kojima 2004). These cell–cell signalling proteins are linked to a pathway involved in cell proliferation, the Fat–Warts signalling network (reviewed in Bando et al. 2011).

During cricket leg regeneration, *hh*, *wg* and *dpp* are expressed in the blastema of a leg amputated at a distal tibia level in a pattern comparable with that of normal development observed in both cricket embryo leg buds and *Drosophila* leg imaginal discs (Mito et al. 2002; Nakamura et al. 2007). During normal development, these genes function in setting up the presumptive proximo-distal axis in the leg imaginal disc of *Drosophila*, and their expression in the cricket embryo leg buds suggests a similar role in hemimetabolous insects (Niwa et al. 2000; although the only functional analysis available on *wg* in an hemimetabolous insect [*Oncopeltus fasciatus*; Angelini and Kaufman 2005a] showed that the silencing of this gene did not lead to complete loss of limb, as shown in *Drosophila* [Angelini and Kaufman 2005b] and *Tribolium* [Ober and Jockusch 2006]). Thus, it seems likely that the mechanism initiating proximo-distal patterning in the regenerating leg is similar to the corresponding process in normal development. In addition, Mito et al. (2002) suggested that during cricket leg regeneration, the region where *wg*-expressing cells abut *dpp*-expressing cells is established as the distal tip of the regenerate, as in the *Drosophila* leg imaginal disc during normal development. A functional analysis in the cricket revealed that the EGFR pathway, induced by the peak of activity of *wg* and *dpp* at the distal tip of the regeneration

blastema, is then required for the regeneration of the distal leg structures, the tarsus and the claws, as in normal development (Nakamura et al. 2008).

The $\beta$-catenin Armadillo is one of the key effectors of the canonical Wnt pathway involved in several processes during normal development. No leg regeneration occurs in the cricket when *armadillo* is knocked down by RNAi, and this result indicates that the canonical Wnt signalling pathway is involved in the regeneration process (Nakamura et al. 2007). The role of Wnt signalling in regenerating limbs was also shown in the *Drosophila* leg imaginal discs (Smith-Bolton et al. 2009; Schubiger et al. 2010) and in the legs of *T. castaneum* larvae (Shah et al. 2011). In the leg imaginal disc of *Drosophila*, however, the mechanism of proximo-distal patterning during regeneration (following amputation of the presumptive distal leg part) shows some differences compared with normal development (Bosch et al. 2010). In this case, normal development and regeneration are served by a set of identical genes, but the way their proximo-distal patterns are eventually obtained is different; this is possibly due to specific requirement to develop distal fates in the presence of pre-existing proximal ones during regeneration (Bosch et al. 2010).

The Salvador–Warts–Hippo (SWH) signalling pathway has recently emerged as a key modulator of tissue growth during normal development by regulating key proteins involved in cell proliferation (Tapon et al. 2002). The core SWH pathway proteins are regulated by inputs from several upstream regulatory branches, including the protocadherins Fat and Dachsous (Ft–Ds). All these proteins have a role within interconnected signalling pathways, the principal function of which is the control of growth and polarity of developing tissues (Reddy and Irvine 2008). Recently, the Ds–Ft and Wts–Hpo signalling pathways were demonstrated to be involved in cricket leg regeneration, regulating size and shape (Bando et al. 2009). The important role of Wts–Hpo in the specification of organ size during development involves an ability to control proliferation in response to tissue damage, as also shown during *Drosophila* imaginal disc regeneration (Grusche et al. 2011).

### 7.3.4 Nervous Control of Limb Regeneration

In many animals, the regeneration process is dependent on intact nerve supply (e.g. Seifert et al. 2012; and references therein). The role of nerves in arthropod limb regeneration used to be a highly controversial topic with many conflicting results (see Bodenstein 1953, 1955; Needham 1965; Nüesch 1968; Goss 1969; Bullière and Bullière 1985); however, it has not been investigated in recent decades. In general, a denervated arthropod limb is not prevented from regenerating, but usually it does not regenerate normally (Goss 1969).

### 7.3.5 Moulting, Hormones and Limb Regeneration

#### 7.3.5.1 Moulting and Regeneration
As noted above, arthropod limbs cannot regenerate without undergoing one or more moults; this means that arthropods that do not moult throughout their whole adult life (see Chap. 5) inevitably lose their regenerative capabilities after the last moult. However, this does not mean that the regenerative potential is completely lost, but, rather, that the production of a functional regenerate is not possible anymore. Bodenstein (1955) demonstrated that an adult cockroach leg, when amputated, is still able to regenerate if a moult can be induced, and a small regenerate within the leg stub has been observed in adult embiopterans and cockroaches (Vorontsova and Liosner 1960; and references therein).

Regeneration of lost limbs can affect timing and number of moults depending on the species, on the time of injury within the intermoult period and on the number of damaged limbs (Bullière and Bullière 1985; Skinner 1985). Limb regeneration delays moulting in several

insects (Maleville and De Reggi 1981; Bullière and Bullière 1985), but leg amputation immediately following a moult shortens the intermoult period in adult silverfish (Buck and Edwards, 1990; note that silverfish regularly moults throughout their life). In cockroaches, limb amputation during the first larval stages leads to an increase in the number of moults, which in *Blattella germanica* seems approximately proportional to the severity of the damage (Bullière and Bullière 1985). Amputations of several legs induce precocious moults in the house centipede *Scutigera* (Cameron 1926) and in various decapod crustacean species (reviewed in Skinner 1985), although not in all (e.g. spider crabs, Skinner and Graham 1972). In the silverfish *Thermobia domestica* and in some crabs, the loss of increasing numbers of limbs accelerates the next moult incrementally (Hopkins 1982; Buck and Edwards 1990), while in the crab *Gecarcinus lateralis*, the autotomy of five legs triggers a precocious moult, but the loss of additional limbs does not shorten the cycle further (Skinner and Graham 1972). Goss (1969) proposed that the influence of limb regeneration on moulting has been largely shaped by natural selection: the moulting cycle is accelerated when the most important thing is to regenerate as soon as possible (as is the case when many limbs are missing), and it is delayed when it is not critical to have a regenerate immediately, and longer time may allow proper regeneration.

#### 7.3.5.2 Ecdysteroids and Other Signals Involved in Regeneration
Current knowledge about hormones and regeneration in arthropods is from decapods and insects, whereas information about other taxa is missing. Moulting and limb regeneration are tightly coupled processes, both of which are regulated by ecdysteroid hormone synthesized and secreted by the Y-organ in decapods and by prothoracic gland in insects (see Chap. 6). Usually, only limbs amputated before a critical point in the intermoult period are able to regenerate at the subsequent moult (Bullière and Bullière 1985; Skinner 1985; see also Sect.

7.3.3.1). In crickets, quantitative hormonal analyses performed in parallel with regeneration have demonstrated that the first ecdysteroid peak during the 10th stage coincides with the critical point. Furthermore, after an amputation at the beginning of the cycle, the first and the second hormone peaks (that triggers the moult) are delayed and their amplitude is reduced (Maleville and De Reggi 1981). Similar perturbations in ecdysteroid level have been also observed in *Blattella germanica* (Roberts et al. 1983) and *Rhodnius prolixus* (Knobloch and Steel 1988), although with species-specific differences.

In insects, it has been shown that the presence of high levels of ecdysteroids inhibits the initiation of regeneration (Bullière and Bullière 1985). The hypothesis is that the low ecdysteroid levels observed following limb lost facilitate the initial phases of regeneration delaying the process of cellular activation in the epidermis normally associated with the beginning of the moult (Knobloch and Steel 1988). The longer intermoult period observed in regenerating insects seems to be related to the delay in the appearance of ecdysteroid peaks, but the mechanism by which regeneration influences hormone titre is unknown (Maleville and De Reggi 1981; Bullière and Bullière 1985; Knobloch and Steel 1988).

In decapod crustaceans, significant research has been performed by removing regenerating limbs still enclosed in the cuticular sac (limb buds). Primary limb buds are those produced after amputation of a limb; secondary limb buds are those produced after amputation of a primary limb bud. In decapods, circulating ecdysteroid drops in response to primary limb bud autotomy and remains low during the subsequent basal growth; hormone titres begin to increase during the subsequent growth of the secondary limb buds (McCarthy and Skinner 1977). Primary limb buds removed after the critical point are not regenerated at the following moult, and haemolymph ecdysteroid levels remain elevated (Holland and Skinner 1976).

Ecdysteroids affect their target tissues by interacting with nuclear receptors. The ecdysone receptor *EcR* and retinoid *X* receptor *RXR* have been indentified and sequenced in several insects and crustaceans, but studies on gene expression during regeneration have been performed only in the crab *Uca pugilator* (Hopkins 2001; Durica et al. 2002, 2006; Wu et al. 2004). In limb buds, levels of *EcR* and *RXR* transcripts remain very low during early basal growth, correlating with lower ecdysteroid titres in the haemolymph. The transition to proecdysial growth (the intense tissue growth restricted to the pre-moult period and characterized by an intense muscle protein synthesis and water uptake) requires a transitory pulse of ecdysteroids, and levels of *EcR* and *RXR* rise. During the end of proecdysial growth, there is a significant increase in both transcripts, and subsequently, a series of large pulses of circulating ecdysteroids are necessary for a successful ecdysis (review in Hopkins 2001).

In decapods, primary and secondary limb buds respond differently to the same concentration of ecdysteroid, which suggests that limb regeneration is not regulated only by ecdysteroid levels. It has been suggested that it probably involves interactions between ecdysteroid and peptide factors on tissues that may have different sensitivities to these molecules; the identity of these peptides, however, remains unknown (Yu et al. 2002). Autotomy of one or more limb buds before the critical point delays proecdysis until secondary limb buds differentiate and grow to the approximate size of the primary buds (Holland and Skinner 1976). On the basis of these observations, it has been proposed (Skinner 1985) that limb buds produce two factors that control proecdysial events: limb autotomy factor–anecdysis (LAFan) and limb autotomy factor–proecdysis (or LAFpro). The LAFan (produced by primary limb buds when at least five legs are autotomized) stimulates anecdysial animals to enter proecdysis, thus initiating a precocious moult. The LAFpro (produced by secondary limb buds when at least one primary limb bud is autotomized) inhibits the proecdysial processes. To date, there is no evidence on the identity of the stimulatory factor LAFan, whereas a putative LAFpro has been isolated from extracts of secondary buds in *Gecarcinus lateralis*. It seems to be a small peptide, related to the moult-inhibiting hormone, that inhibits primary limb bud growth by lowering haemolymph ecdysteroid level (Yu et al. 2002), but a more detailed characterization is necessary to understand its role in regeneration.

## 7.4 Tissue Regeneration

Usually, arthropods show good tissue regenerative potential and exhibit both reparative and physiological tissue regeneration. Moreover, regeneration can be experimentally induced in many tissues that, under natural conditions, would never have the need to regenerate, such as imaginal discs. We cannot provide a comprehensive review of arthropod tissue regeneration here; however, we will discuss some examples focusing on the most studied models.

### 7.4.1 Reparative and Experimentally Induced Tissue Regeneration

Although only scattered observations are available, it seems that many external, single-tissue structures may break off and regenerate in several arthropod species. This is found, for example, in the spines of decapod crustacean larvae (Freeman 1983) and in wing buds of several hemimetabolous insects (e.g. Vorontsova and Liosner 1960). Even species with poor limb regenerative potential are able to regenerate these structures (e.g. spines and denticles in scorpions, Rosin 1964; sensory setae in cladocerans, Agar 1930). Clearly, the regeneration of these structures requires at least one moult.

While many works are available on nerve regeneration following experimental damage (see Pearse and Govind 2002; Krüger et al. 2011; Stern et al. 2012; and references therein), the most studied arthropod model for experimentally damaged inner tissues is *Drosophila* imaginal discs. The regeneration of experimentally damaged imaginal discs has long been known for different holometabolous insects (e.g. Vorontsova and Liosner 1960), although it does

not happen in all (Njihout and Grunert 1988); however, studies have been limited until recently, when *Drosophila* imaginal discs also became a model for regenerative studies (e.g. Bergantiños et al. 2010b; Belacortu and Paricio 2011; Repiso et al. 2011). *Drosophila* imaginal discs regenerate missing cells by proliferation of cells close to the wound edge (Smith-Bolton et al. 2009; Bosch et al. 2010; Bergantiños et al. 2010a), and many molecules involved or necessary for regeneration are now known (e.g. McClure and Schubiger 2008; Bergantiños et al. 2010b; Belacortu and Paricio 2011; Repiso et al. 2011). Unfortunately, there are no comparable observations for any other holometabolous insect, and thus, nothing can be said about the phylogenetic diversity of this process.

### 7.4.2 Physiological Regeneration

Arthropods provide a large number of cases of physiological loss or degeneration and subsequent regeneration of different tissues. However, most of these cases are taxon specific, such as the decapod claw muscles that partially degenerate and regenerate with the moulting cycle (e.g. El Haj 1999; Mykles 1999; and references therein). To some extent, even the extensive tissue degeneration and regrowth observed during metamorphosis in, for example, holometabolous insects and cirripede crustaceans could be described as physiological regeneration. The main difference between these events and conventional physiological regeneration lies in the fact that during metamorphosis, the regrown organs and tissues are not duplicates of the initial ones, but are more or less morphologically different. Clearly, physiological regeneration and metamorphosis may not be completely unrelated processes after all, especially considering that during, for example, limb regeneration, in most cases, a perfect duplicate is never obtained (see Sect. 7.3).

A model that received considerable attention in recent time is the insect midgut. The insect midgut is usually composed of at least two cell types, digestive cells and regenerative cells (stem cells); additionally, endocrine cells and goblet cells have also been described in some taxa (e.g. Rost-Roszkowska 2008; and references therein). During the insect's life, digestive cells (as well as endocrine and goblet cells if present) regularly degenerate and new ones are produced from regenerating cells. Midgut morphology, degeneration and regeneration have been described for a number of taxa, and regenerative midgut cells have been known and described since long (as far as 1897 by Rengel according to Nardi et al. 2010). However, these cells have recently received much greater attention after they were carefully described in lepidopterans and *Drosophila*, and their name was changed to that of midgut stem cells (Corley and Lavine 2006; Micchelli and Perrimon 2006; Ohlstein and Spradling 2006; Jiang and Edgar 2012).

There is phylogenetic variation in several aspects of the morphology and development of regenerating cells. For example, regenerating cells can occur as single cells among digestive cells (e.g. Rost 2006a) or in groups; if in group, they can be found among digestive cells (e.g. Rost-Roszkowska et al. 2010a, c) or partially among digestive cells and partially protruding into the haemocoel (e.g. Rost-Roszkowska 2008; Nardi et al. 2010; Nardi and Bee 2012), but there are even examples where regenerating cells can be found both singly or in groups in the same species. For example, after metamorphosis in the beetle *Epilachna*, regenerating cells are initially isolated; they then proliferate to form a group all cells of which, afterwards, differentiate to digestive cells except a single cell that can start the cycle again (Rost-Roszkowska et al. 2010b).

There are even exceptions to the presence or activity of these regenerating cells. In some springtail species (but not in all), regenerating cells are missing and it has been suggested that the regeneration is accomplished by divisions of digestive cells (Rost-Roszkowska et al. 2007; Rost-Roszkowska and Undrul 2008). In *Drosophila* larvae, regenerating cells proliferate but do not differentiate into digestive cells (Jiang and Edgar 2009); however, from metamorphosis

onwards, they both proliferate and differentiate (Micchelli and Perrimon 2006; Ohlstein and Spradling 2006; Jiang and Edgar 2009; Strand and Micchelli 2011).

The degenerative and regenerative events of the midgut may be cyclical, usually linked to the moult cycle, or more or less continuous. There is no apparent phylogenetic trend as even closely related species may behave differently; for example, among the Zygentoma, the midgut is regenerated cyclically in *Lepisma saccarina*, but continuously in *Thermobia domestica* (Rost 2006b).

The molecular control of midgut regeneration has been investigated only in some holometabolous insects (reviewed in Hakim et al. 2010; and see also Strand and Micchelli 2011; Jiang and Edgar 2012), but since different studies focused on different aspects in different taxa, it is hard to compare the evidence. It is also hard to make comparisons with other arthropods where less is known. Regenerating cells have been described in several crustaceans (although with some specific exceptions; see Rost-Roszkowska et al. 2012; and references therein), in myriapods (de Godoy and Fontanetti 2010; Chajec et al. 2012), in a few chelicerates (e.g. Becker and Peters 1985; Šobotník et al. 2008; Filimonova 2009) and, among the close arthropod relatives, in some tardigrades but not in all (Greven 1976; Rost-Roszkowska et al. 2011; and references therein).

## 7.5 Conclusions

Many questions regarding arthropod regeneration are still unanswered. While only a few arthropod species are able to regenerate parts of the trunk, and to a very limited extent, most arthropods are able to regenerate organs and tissues to some degree. Limb regenerative potential fluctuates across the Arthropoda. Why some taxa regenerate well and others do not is clearly an important evolutionary question; more comparative data are needed, and a simple explanation may be unlikely.

Many factors, not only phylogeny, influence the outcome of a regeneration process. Developmental stage, limb type and amputation level are very often critical for limb regeneration. However, our knowledge of the developmental events connected to limb regeneration is based on relatively few studies, and this is not only true for developmental genetics, but also for histological studies. Limb regeneration seems to involve only limited cell dedifferentiation, and most likely, slightly dedifferentiated cells produce only cells of the same type. In other cases, however, specific regenerative (or stem) cells are present, as in the case of insect midgut.

Moulting influences regeneration, and in many instances, regeneration influences moulting. However, the relationship between moulting and regeneration is not yet clear, and while moulting is necessary for proper limb regeneration, at least for some insects there is no evidence of any loss of regenerative potential after the final moult.

**Acknowledgments** In our effort to keep the reference list within a reasonable length, we often preferred to cite reviews or research articles that include references to previous useful works, instead of citing all original works directly. We apologize with all the authors whose original, valuable work was not directly cited here. Frank D. Ferrari, Giuseppe Fusco, Alessandro Minelli and H. Frederik Nijhout provided useful comments on earlier drafts of this chapter.

# References

Abeloos M (1932) La régénération et les problèmes de la morphogenèse. Gauthier-Villars, Paris

Adiyodi RG (1972) Wound healing and regeneration in the crab *Paratelphusa hydrodromus*. Int Rev Cytol 32:257–289

Agar WE (1930) A statistical study of regeneration in two species of Crustacea. J Exp Biol 7:349–369

Alsop DW (1978) Comparative analysis of the intrinsic leg musculature of the American cockroach, *Periplaneta americana* (L.). J Morph 158:199–242

Angelini DR, Kaufman TC (2005a) Functional analyses in the milkweed bug *Oncopeltus fasciatus* (Hemiptera) support a role for Wnt signaling in body segmentation but not appendage development. Dev Biol 283:409–423

Angelini DR, Kaufman TC (2005b) Insect appendages and comparative ontogenetics. Dev Biol 286:57–77

Bando T, Mito T, Maeda Y, Nakamura T, Ito F, Watanabe T, Ohuchi H, Noji S (2009) Regulation of leg size and shape by the Dachsous/Fat signalling pathway during regeneration. Development 136:2235–2245

Bando T, Mito T, Nakamura T, Ohuchi H, Noji S (2011) Regulation of leg size and shape: involvement of the Dachsous-Fat signalling pathway. Dev Dyn 240:1028–1041

Becker A, Peters W (1985) Fine structure of the midgut gland of *Phalangium opilio* (Chelicerata, Phalangida). Zoomorphol 105:317–325

Belacortu Y, Paricio N (2011) *Drosophila* as a model of wound healing and tissue regeneration in vertebrates. Dev Dyn 240:2379–2404

Bely AE (2010) Evolutionary loss of animal regeneration: pattern and process. Integr Comp Biol 50:515–527

Bely AE, Nyberg KG (2009) Evolution of animal regeneration: Re-emerging of a field. Trends Ecol Evol 25:161–170

Bergantiños C, Corominas M, Serras F (2010a) Cell death-induced regeneration in wing imaginal discs requires JNK signalling. Development 137:1169–1179

Bergantiños C, Vilana X, Corominas M, Serras F (2010b) Imaginal discs: renaissance of a model for regenerative biology. BioEssays 32:207–217

Bliss DE (1960) Autotomy and regeneration. In: Waterman TH (ed) The physiology of Crustacea, vol 1. Academy Press, New York, pp 561–589

Bodenstein D (1953) Regeneration. In: Roeder KD (ed) Insect physiology. Wiley, New York, pp 866–878

Bodenstein D (1955) Contributions to the problem of regeneration in insects. J Exp Zool 129:209–224

Bordage E (1905) Recherches anatomiques et biologiques sur l'anatomie et régénération chez diverses arthropodes. Bull scient Fr Belg 39:307–454

Bosch M, Bishop S-A, Baguñà J, Couso J-P (2010) Leg regeneration in *Drosophila* abridges the normal developmental program. Int J Dev Biol 54:1241–1250

Buck C, Edwards JS (1990) The effect of appendage and scale loss on instar duration in adult firebrats, *Thermobia domestica* (Thysanura). J Exp Biol 151:341–347

Bullière D, Bullière F (1985) Regeneration. In: Kerkut GA, Gilbert LI (eds) Comprehensive insect physiology, biochemistry and pharmacology, vol 2. Pergamon Press, Oxford, pp 371–424

Cameron JA (1926) Regeneration in *Scutigera forceps*. J Exp Zool 46:169–179

Campbell FL, Priestley JD (1970) Flagellar annuli of *Blattella germanica* (Dictyoptera: Blattellidae)—changes in their number and dimensions during postembryonic development. Ann Entomol Soc Am 63:81–88

Chajec Ł, Rost-Roszkowska MM, Vilimova J, Sosinka A (2012) Ultrastructure and regeneration of midgut epithelial cells in *Lithobius forficatus* (Chilopoda, Lithobiidae). Invertebr Biol 131:119–132

Clare AS, Lumb G, Clare PA, Costolow JD Jr (1990) A morphological study of wound response and telson regeneration in postlarval *Limulus polyphemus* (L.). Invertebr Reprod Dev 17:77–87

Cooper RL (1998) Development of sensory processes during limb regeneration in adult crayfish. J Exp Biol 201:1745–1752

Corley LS, Lavine MD (2006) A review of insect stem cell types. Semin Cell Dev Biol 17:510–517

de Godoy JAP, Fontanetti CS (2010) Diplopods as bioindicators of soils: analysis of midgut of individuals maintained in substract [sic] containing sewage sludge. Water Air Soil Pollut 210:389–398

Durica DS, Wu X, Anilkumar G, Hopkins PM, Chung AC (2002) Characterization of crab EcR and RXR homologs and expression during limb regeneration and oocyte maturation. Mol Cell Endocrinol 189:59–76

Durica DS, Kupfer D, Najar F, Lai H, Tang Y, Griffin K, Hopkins PM, Roe B (2006) EST library sequencing of genes expressed during early limb regeneration in the fiddler crab and transcriptional responses to ecdysteroid exposure in limb bud explants. Integr Comp Biol 46:948–964

El Haj AJ (1999) Regulation of muscle growth and sarcomeric protein gene expression over the intermolt cycle. Am Zool 39:570–579

Ernsting G, Fokkema DS (1983) Antennal damage and regeneration in springtails (Collembola) in relation to predation. Neth J Zool 33:476–484

Filimonova SA (2009) The ultrastructure investigation of the midgut in the quill mite *Syringophilopsis fringilla* (Acari, Trombidiformes: Syringophilidae). Arthropod Struct Dev 38:303–313

Fleming PA, Muller D, Bateman PW (2007) Leave it all behind: a taxonomic perspective of autotomy in invertebrates. Biol Rev 82:481–510

Freeman JA (1983) Spine regeneration in the larvae of the crab, *Rhithropanopeus harrisii*. J Exp Zool 225:443–448

Galko MJ, Krasnow MA (2004) Cellular and genetic analysis of wound healing in *Drosophila* larvae. PLoS Biol 2:e239. doi:10.1371/journal.pbio.0020239

Giorgianni M, Patel NH (2005) Conquering land, air and water: the evolution and development of arthropod appendages. In: Briggs DEG (ed) Evolving form and function: fossils and development, Peabody Museum of natural history. Yale University, New Haven, pp 159–180

Goss RJ (1969) Principles of regeneration. Academic Press, New York and London

Greven H (1976) Some ultrastructural observations on the midgut epithelium of *Isohypsibius augusti* (Murray, 1907) (Eutardigrada). Cell Tiss Res 166:339–351

Grusche FA, Degoutin JL, Richardson HE, Harvey KF (2011) The Salvador/Warts/Hippo pathway controls regenerative tissue growth in *Drosophila melanogaster*. Dev Biol 350:255–266

Hakim RS, Baldwin K, Smagghe G (2010) Regulation of midgut growth, development, and metamorphosis. Annu Rev Entomol 55:593–608

Harvey R, Burrows MT, Speirs R (2003) Cirral regeneration following non-lethal predation in two intertidal barnacle species. J Mar Biol Ass UK 83:1229–1231

Holland CA, Skinner DM (1976) Interactions between molting and regeneration in the land crab. Biol Bull 150:222–240

Hopkins PM (1982) Growth and regeneration patterns in the fiddler crab, *Uca pugilator*. Biol Bull 163:301–319

Hopkins PM (2001) Limb regeneration in the fiddler crab, *Uca pugilator*: hormonal and growth factor control. Am Zool 41:389–398

Hopkins PM, Chung AC-K, Durica DS (1999) Limb regeneration in the fiddler crab, *Uca pugilator*: histological, physiological and molecular consideration. Am Zool 39:513–526

Igelmund P (1987) Morphology, sense organs, and regeneration of the forelegs (whips) of the whip spider *Heterophrynus elaphus* (Arachnida, Amblypygi). J Morph 193:75–89

Jegla TC (1982) A review of the molting physiology of the trilobite larva of *Limulus*. In: Bonaventura J, Bonaventura C, Tesh S (eds) Physiology and biology of horseshoe crabs: Studies on normal and environmentally stressed animals. Alan R Liss, New York, pp 83–101

Jiang H, Edgar BA (2009) EGFR signaling regulates the proliferation of *Drosophila* adult midgut progenitors. Development 136:483–493

Jiang H, Edgar BA (2012) Intestinal stem cell function in *Drosophila* and mice. Curr Opin Gen Dev 22:354–360

Joly R, Herbaut C (1968) Sur la régénération oculaire chez *Lithobius forficatus* L. (Myriapode Chilopode). Arch Zool Exp Gén 109:591–612

Kaars C, Greenblatt S, Fourtner CR (1984) Patterned regeneration of internal femoral structures in the cockroach, *Periplaneta americana* L. J Exp Biol 230:141–144

Kahn AT, Weis JS, Saharig CE, Polo AE (1993) Effect of tributylin on mortality and telson regeneration of grass shrimp, *Palaemonetes pugio*. Bull Environ Contam Toxicol 50:152–157

Klepal W, Gruber D, Pflugfelder B (2008) Natural cyclic degeneration by a sequence of programmed cell death modes in *Semibalanus balanoides* (Linnaeus, 1767) (Crustacea, Cirripedia, Thoracica). Zoomorphol 127:49–58

Knobloch CA, Steel CGH (1988) Interactions between limb regeneration and ecdysteroid titres in last larval instar *Rhodnius prolixus* (Hemiptera). J Insect Physiol 34:507–514

Kocian V (1930) Un cas d'hétéromorphose chez *Argulus foliaceus* L. Arch Zool Exp Gen 70:23–27

Kojima T (2004) The mechanism of *Drosophila* leg development along the proximodistal axis. Dev Growth Differ 46:115–129

Korschelt E (1907) Regeneration und Transplantation. Fischer, Jena

Kraus ML, Weis JS (1988) Differences in the effects of mercury on telson regeneration in two populations of the grass shrimp *Palaemonetes pugio*. Bull Environ Contam Toxicol 17:115–120

Krüger S, Haller B, Lakes-Harlan R (2011) Regeneration in the auditory system of nymphal and adult bush crickets Tettigonia viridissima. 36:235–246

Loeb J (1905) Studies in general physiology. The University of Chicago Press, Chicago

Lüdke J, Lakes-Harlan R (2008) Regeneration of the tibia and somatotopy of regenerated hair sensilla in *Schistocerca gregaria* (Forskål). Arthropod Struct Dev 37:210–220

Lumb G, Clare AS, Costlow JD (1991) Cheliped regeneration in the megalopa of the mud crab, *Rhithropanopeus harrisii* (Gould). Invertebr Reprod Dev 20:87–96

Lüscher M (1948) The regeneration of legs in *Rhodnius prolixus* (Hemiptera). J Exp Biol 25:334–343

Maleville A, De Reggi M (1981) Influence of leg regeneration on ecdysteroid titres in *Acheta* larvae. J Insect Physiol 27:35–40

Maruzzo D, Minelli A (2011) Post-embryonic development of amphipod crustacean pleopods and the patterning of arthropod limbs. Zool Anz 250:32–45

Maruzzo D, Bonato L, Brena C, Fusco G, Minelli A (2005) Appendage loss and regeneration in arthropods: a comparative view. In: Koenemann S, Jenner R (eds) Crustacea and arthropod relationships, Crustacean issues 16. CRC Press, Boca Raton, pp 215–245

Maruzzo D, Minelli A, Ronco M, Fusco G (2007) Growth and regeneration of the second antennae of *Asellus aquaticus* (Isopoda) in the context of arthropod antennal segmentation. J Crust Biol 27:184–196

Maruzzo D, Egredzija M, Minelli A, Fusco G (2008) Segmental pattern formation following amputations in the flagellum of the second antennae of *Asellus aquaticus* (Crustacea, Isopoda). Ital J Zool 75:225–231

McCarthy JF, Skinner DM (1977) Interruption of proecdysis by autotomy of partially regenerated limbs in the land crab, *Gecarcinus lateralis*. Dev Biol 61:299–310

McClure KD, Schubiger G (2008) A screen for genes that function in leg disc regeneration in *Drosophila melanogaster*. Mech Dev 125:67–80

McNamara KJ, Tuura ME (2011) Evidence for segment polarity during regeneration in the Devonian asteropygine trilobite *Greenops widderensis*. J Paleontol 85:106–110

Mees J, Fockedy N, Dewicke A, Janssen CR, Sorbe J-C (1995) Aberrant individuals of *Neomysis integer* and other Mysidacea: intersexuality and variable telson morphology. Neth J Aquat Ecol 29:161–166

Micchelli CA, Perrimon N (2006) Evidence that stem cells reside in the adult *Drosophila* midgut epithelium. Nature 439:475–479

Minelli A (2003) The development of animal form. Cambridge University Press, Cambridge

Minelli A, Maruzzo D, Fusco G (2010) Multi-scale relationships between number and size in the evolution of arthropod body features. Arthropod Struct Dev 39:468–477

Mito T, Inoue Y, Kimura S, Miyawaki K, Niwa N, Shinmyo Y, Ohuchi H, Noji S (2002) Involvement of *hedgehog*, *wingless*, and *dpp* in the initiation of proximodistal axis formation during the regeneration of insect legs, a verification of the modified boundary model. Mech Dev 114:27–35

Moffett S (1987) Muscles proximal to the fracture plane atrophy after limb autotomy in decapod crustaceans. J Exp Zool 244:485–490

Morgan TH (1901) Regeneration. Macmillan, New York

Mykles DL (1999) Proteolytic processes underlying molt-induced claw muscle atrophy in decapod crustaceans. Am Zool 39:541–551

Nagy LM, Williams TA (2001) Comparative limb development as a tool for understanding the evolutionary diversification of limbs in arthropods: challenging the modularity paradigm. In: Wagner GP (ed) The character concept in evolutionary biology. Academic Press, San Diego, pp 455–488

Nakamura T, Mito T, Tanaka Y, Bando T, Ohuchi H, Noji S (2007) Involvement of the canonical Wnt/Wingless signaling in determination of the proximodistal positional values within the leg segment of the cricket *Gryllus bimaculatus*. Dev Growth Differ 49:79–88

Nakamura T, Mito T, Bando T, Ohuchi H, Noji S (2008) Dissecting insect leg regeneration through RNA interference. Cell Mol Life Sci 65:64–72

Nardi JB, Bee CM (2012) Regenerative cells and the architecture of beetle midgut epithelia. J Morph 273:1010–1020

Nardi JB, Bee CM, Miller LA (2010) Stem cells of the beetle midgut epithelium. J Insect Physiol 56:296–303

Needham AE (1952) Regeneration and wound healing. Methuen, London

Needham AE (1965) Regeneration in the Arthropoda and its endocrine control. In: Kiortsis V, Trampusch HAL (eds) Regeneration in animals and related problems. North-Holland, Amsterdam, pp 283–323

Nguyen Duy-Jacquemin M (1972) Régénération chez les larves et les adultes de *Polyxenus lagurus* (Diplopode, Pénicillate). C R Hebdo Séances Acad Sci Paris D 274:1323–1326

Niwa N, Inoue Y, Nozawa A, Saito M, Misumi Y, Ohuchi H, Yoshioka H, Noji S (2000) Correlation of diversity of leg morphology in *Gryllus bimaculatus* (cricket) with divergence in *dpp* expression pattern during leg development. Development 127: 4373–4381

Njihout HF, Grunert LW (1988) Color pattern regulation after surgery on the wing disks of *Precis coenia* (Lepidoptera: Nymphalidae). Development 102: 377–385

Nüesch H (1968) Role of nervous system in insect morphogenesis and regeneration. Annu Rev Entomol 13:27–44

Ober KA, Jockusch EL (2006) The roles of *wingless* and *decapentaplegic* in axis and appendage development in the red flour beetle, *Tribolium castaneum*. Dev Biol 294:391–405

Ohlstein B, Spradling A (2006) The adult *Drosophila* posterior midgut is maintained by pluripotent stem cells. Nature 439:470–474

Pastre S (1960) Sur la régénération de l'œil de l'Isopode *Idotea baltica* (Aud.). C R Acad Sci Fr 250: 3738–3739

Pearse J, Govind CK (2002) Remodeling of the proximal segment of crayfish motor nerves following transaction. J Comp Neurol 450:61–72

Petit G (1974) Sur les modalités de la croissance et la régénération des antennes de larves de *Polydesmus angustus* Latzel. Symp Zool Soc London 32:301–315

Prpic N-M, Damen WGM (2008) Arthropod appendages: a prime example for the evolution of morphological diversity and innovation. In: Minelli A, Fusco G (eds) Evolving pathways. Cambridge University Press, Cambridge, pp 381–398

Przibram H (1909) Regeneration. Deuticke, Leipzig

Randall JB (1981) Regeneration and autotomy exhibited by the black widow spider, *Latrodectus variolus* Walckenaer. Wilhelm Roux' Arch Dev Biol 190: 230–232

Read AT, Govind CK (1998) Cell types in regenerating claws of the snapping shrimp, *Alpheus heterochelis*. Can J Zool 76:1080–1090

Reddy BV, Irvine KD (2008) The Fat and Warts signaling pathways: new insights into their regulation, mechanism and conservation. Development 135:2827–2838

Repiso A, Bergantiños C, Corominas M, Serras F (2011) Tissue repair and regeneration in *Drosophila* imaginal discs. Dev Growth Differ 53:177–185

Roberts B, Wentworth SL, Kotzman M (1983) The levels of ecdysteroids in uninjured and leg-autotomized nymphs of *Blattella germanica* (L.). J Insect Physiol 29:679–685

Rockett CL, Woodring JP (1972) Comparative studies of acarine limb regeneration, apolysis, and ecdysis. J Insect Physiol 18:2319–2336

Rosin R (1964) On regeneration in scorpions. Isr J Zool 13:177–183

Rost MM (2006a) Ultrastructural changes in the midgut epithelium in *Podura aquatica* L. (Insecta, Collembola, Arthropleona) during regeneration. Arthropod Struct Dev 35:69–76

Rost MM (2006b) Comparative studies on regeneration of the midgut epithelium in *Lepisma saccarina* and *Thermobia domestica*. Ann Entomol Soc Am 99:910–916

Rost-Roszkowska MM (2008) Ultrastructural changes in the midgut epithelium of *Acheta domesticus* (Orthoptera: Gryllidae) during degeneration and regeneration. Ann Entomol Soc Am 101:151–158

Rost-Roszkowska MM, Undrul A (2008) Fine structure and differentiation of the midgut epithelium of *Allacma fusca* (Insecta, Collembola, Symphypleona). Zool Stud 47:200–206

Rost-Roszkowska MM, Poprawa I, Swiatek P (2007) Ultrastructural changes in the midgut epithelium of the first larva of *Allacma fusca* (Insecta, Collembola, Symphypleona). Invertebr Biol 126:366–372

Rost-Roszkowska MM, Jansta P, Vilimova J (2010a) Fine structure of the midgut epithelium in two Archaeognatha, *Lepismachilis notata* and *Machilis hrabei* (Insecta), in relation to its degeneration and regeneration. Protoplasma 247:91–101

Rost-Roszkowska MM, Poprawa I, Klag J, Migula P, Mesjasz-Przybyłowicz J, Przybyłowicz W (2010b) Differentiation and regenerative cells in the midgut epithelium of *Epilachna cf. nylanderi* (Mulsant 1850) (Insecta, Coleoptera, Coccinellidae). Folia Biol 58:209–216

Rost-Roszkowska MM, Vilimova J, Chajec Ł (2010c) Fine structure of the midgut epithelium of *Atelura formicaria* (Hexapoda: Zygentoma: Ateluridae), with special reference to its regeneration and degeneration. Zool Stud 49:10–18

Rost-Roszkowska MM, Poprawa I, Wójtowicz M, Kaczmarek Ł (2011) Ultrastructural changes of the midgut epithelium in *Isohypsibius granulifer granulifer* Thulin, 1928 (Tardigrada: Eutardigrada) during oogenesis. Protoplasma 248:405–414

Rost-Roszkowska MM, Vilimova J, Sosinka A, Skudlik J, Franzetti E (2012) The role of autophagy in the midgut epithelium of *Eubranchipus grubii* (Crustacea, Branchiopoda, Anostraca). Arthropod Struct Dev 41:271–279

Sandeman RE, Sandeman DC (1996) Pre- and postembryonic development, growth and turnover of olfactory receptor neurons in crayfish antennules. J Exp Biol 199:2409–2418

Schafer R (1973) Postembryonic development in the antenna of the cockroach, *Leucophaea maderae*: Growth, regeneration and the development of the adult pattern of sense organs. J Exp Zool 183: 353–364

Schmiege DL, Ridgway RL, Moffett SB (1992) Ultrastructure of autotomy-induced atrophy of muscles in the crab *Carcinus maenas*. Can J Zool 70:841–851

Schubiger M, Sustar A, Schubiger G (2010) Regeneration and transdetermination: the role of *wingless* and its regulation. Dev Biol 347:315–324

Seifert AW, Monaghan JR, Smith MD, Pasch B, Stier AC, Michonneau F, Maden M (2012) The influence of fundamental traits on mechanisms controlling appendage regeneration. Biol Rev 87:330–345

Shah MV, Namigai EKO, Suzuki Y (2011) The role of canonical Wnt signaling in leg regeneration and metamorphosis in the red flour beetle *Tribolium castaneum*. Mech Dev 128:342–358

Shaw VK, Bryant PJ (1974) Regeneration of appendages in the large milkweed bag, *Oncopeltus fasciatus*. J Insect Physiol 20:1849–1857

Skinner DM (1985) Molting and regeneration. In: Bliss DE (ed) The biology of Crustacea, vol 9. Academic Press, New York, pp 43–146

Skinner DM, Graham DE (1972) Loss of limbs as a stimulus to ecdysis in Brachyura (true crabs). Biol Bull 143:222–233

Smith-Bolton R, Worley M, Kanda H, Hariharan I (2009) Regenerative growth in *Drosophila* imaginal discs is regulated by Wingless and Myc. Dev Cell 16:797–809

Šobotník J, Alberti G, Weyda F, Hubert J (2008) Ultrastructure of the digestive tract in *Acarus siro* (Acari: Acaridida). J Morph 269:54–71

Stern M, Scheiblich H, Eickhoff R, Didwischus N, Bicker G (2012) Regeneration of olfactory afferent axons in the locust brain. J Comp Neurol 520:679–693

Steuellet P, Cate HS, Derby CD (2000) A spatiotemporal wave of turnover and functional maturation of olfactory receptor neurons in the spiny lobster *Panulirus argus*. J Neurosci 20:3282–3294

Strand M, Micchelli CA (2011) Quiescent gastric stem cells maintain the adult *Drosophila* stomach. Proc Natl Acad Sci USA 108:17696–17701

Švácha P (1995) The larva of *Scraptia fuscula* (P.W.J. Müller) (Coleoptera: Scraptiidae): autotomy and regeneration of the caudal appendage. In: Pkaluk J, Ślipiński SA (eds) Biology, phylogeny, and classification of Coleoptera. Muzeum i Instytut Zoologii PAN, Warszawa, pp 473–489

Tanaka A, Ross MH (1996) Regeneration of tarsomeres in the fused tarsi trait of the German cockroach. Jpn J Ent 64:429–441

Tanaka A, Akahane H, Ban Y (1992) The problem of the number of tarsomeres in the regenerated cockroach leg. J Exp Zool 262:61–70

Tapon N, Harvey KF, Bell DW, Wahrer DCR, Schiripo TA, Haber DA, Hariharan IK (2002) *Salvador* promotes both cell cycle exit and apoptosis in *Drosophila* and is mutated in human cancer cell lines. Cell 110:467–478

Theopold U, Schmidt O, Söderhäll K, Dushay MS (2004) Coagulation in arthropods: defence, wound closure and healing. Trends Immunol 25:289–294

Truby PR (1983) Blastema formation and cell division during cockroach limb regeneration. J Embryol Exp Morph 75:151–164

Truby PR (1985) Separation of wound healing from regeneration in the cockroach leg. J Embryol Exp Morph 85:177–190

Vachon M (1957) La régénération appendiculaire chez le scorpions (Arachnides). C R Hebdo Séances Acad Sci Paris 244:2556–2559

Vafopoulou X (2009) Mechanisms of wound repair in crayfish. Invertebr Surviv J 6:125–137

Vernet G, Charmantier-Daures M (1994) Mue, autotomie et régénération. In: Grassé PP (ed) Traité de Zoologie, Tome 7, Fascicule 1. Masson, Paris, pp 153–194

Vorontsova MA, Liosner LD (1960) Asexual propagation and regeneration. Pergamon Press, London

Williams TA (2004) The evolution and development of crustacean limbs: an analysis of limb homologies. In: Scholtz G (ed) Evolutionary developmental biology of Crustacea, Crustacean issues 15. AA Balkema, Lisse, pp 169–193

Wu X, Hopkins PM, Palli SR, Durica DS (2004) Crustacean retinoid-X-receptor isoforms: distinctive DNA binding and receptor–receptor interaction with a cognate ecdysteroid receptor. Mol Cell Endocrinol 218:21–38

Yu XL, Chang ES, Mykles DL (2002) Characterization of limb autotomy factor proecdysis (LAFpro), isolated from limb regenerates, that suspends molting in the land crab *Gecarcinus lateralis*. Biol Bull 202:204–212

# The Arthropod Cuticle

8

## Bernard Moussian

## Contents

B. Moussian (✉)
Interfaculty Institute for Cell Biology,
Animal Genetics, Eberhard-Karls University
of Tübingen, Auf der Morgenstelle 28,
72076 Tübingen, Germany
e-mail: bernard.moussian@medgen.gu.se

## 8.1 Introduction

What accounts for the beauty and singularity of arthropods is the cuticle that enables them to compete in their small world. What we see is the surface but what does it look like inside? In the past two centuries, starting with the discovery of chitin as a major component of the arthropod cuticle by Odier (1823), a vast number of publications contributed to the understanding of cuticle architecture and composition (reviewed in Locke 2001; Moussian 2010). The arthropod cuticle is a multifunctional coat that defines and stabilises the shape of the body, appendages and internal organs including the hindgut, the foregut and, in insects, the tracheae, preventing dehydration and infection, and protecting against predators of the same scale. As an exoskeleton, additionally, it allows locomotion and flight. Witnessing the ecological success and relevance of arthropods, the cuticle is a highly versatile device facilitating formation of many different body shapes that reflect habitat adaptation, and indeed, arthropods populate a broad range of ecological habitats ranging from oceans to deserts.

In a given species, environmental constraints may also dictate stage- and tissue-specific differences in the physical properties of the cuticle. Usually, for instance, in caterpillars and other insect larvae, the body cuticle is soft and elastic serving as a hydrostatic jacket withstanding the internal pressure of the haemolymph, thereby allowing locomotion (Fig. 8.1). In the same

A. Minelli et al. (eds.), *Arthropod Biology and Evolution*,
DOI: 10.1007/978-3-642-36160-9_8, © Springer-Verlag Berlin Heidelberg 2013

**Fig. 8.1** Cuticle architecture. *Upper image* The typical arthropod cuticle is a layered extracellular structure produced by a monolayer of epithelial cells at their apical side. The polarity of these cells is illustrated by the presence of adherens junctions (*AJ*) at apicolateral positions of the lateral membrane and the septate junctions (*SJ*) underneath. The outermost layer is the envelope (*env*), a relatively new term for this structure. In the literature, it has been described as the lipid-bearing outer epicuticle or the cement layer. The epicuticle (*epi*), formerly called inner epicuticle, is an ultrastructurally distinct layer beneath the envelope, and also contains lipids and proteins, but is devoid of chitin. The innermost procuticle (*pro*) is a chitin–protein matrix attached to the surface of the epithelial cell. *Lower image* Different types of cuticles are present within one animal, for example, in the *Panorpa vulgaris* (mecoptera) first instar larva with abdominal soft cuticle and a coloured and hard thoracic and head cuticle

animal, the head skeleton consists of hard cuticle required for mastication and probably to shield the brain. Hard body cuticle, by contrast, is the prevalent cuticle type of mainly adult animals, especially covering their dorsal side that is usually more exposed to the environment than the ventral side. Sclerites of hard cuticle are joined by soft cuticle rendering the exoskeleton pliable.

Along with its relative advantages, the cuticle makes an arthropod's life also more complicated: to accommodate possible habitat changes during the life cycle of an organism and to allow growth from one developmental stage to the next, the cuticle has to be detached from the epithelial surface, shed and replaced by a new one (see Chap. 6). This implies stage-specific composition reflecting the required physical properties.

Commonly, cuticles are composed of lipids and waxes, glycosylated and unglycosylated proteins, the polysaccharide chitin and catecholamines. Additionally, especially in crustaceans, minerals such as calcite may be incorporated. Species-, stage- and tissue-specific differences mainly rely on lipid and wax composition, different albeit related proteins, the amounts of chitin and the degree of covalent cross-links by, for example, catecholamines. Analogous to the vertebrate skin, lipids and waxes are implicated in preventing water loss and are mainly coating the surface of the animal. Whereas vertebrates employ sphingolipids such as ceramides (Madison 2003; Harding 2004; Jensen and Proksch 2009), insects apply neutral lipids (n-alkanes and n-alkenes) and wax esters as water repellents. The genomic sequences of many arthropods, mainly insects, have led to the discovery and bioinformatics characterisation of several classes of putative structural cuticle proteins, many of which harbour chitin-binding domains. These classes have been excellently described recently by Willis (2010). Concerning chitin, the second most abundant polysaccharide on earth, molecular biology of arthropod cuticle chitin synthesis has been inspired by advances in research on fungal chitin (Merzendorfer 2006). However, since chitin is highly organised in arthropods, while it seems not to be particularly organised in fungal cell walls, insights from this side are rather limited. Ordered packing of cuticle components involves covalent and non-covalent interactions between them. Major covalent linkages are mediated by catecholamines that eventually also cause cuticle tanning.

Looking at the literature published in the last century or so cuticle organisation is, in spite of the variety of components, basically pretty well conserved between distantly related arthropod species. This observation in turn implies that the molecular mechanisms of cuticle formation are largely conserved as well, permitting the possibility of using model arthropods to answer this fundamental biological problem. In the last few years, molecular and genetic approaches in the insects *Drosophila melanogaster* and *Tribolium castaneum* have indeed boosted our understanding about cuticle differentiation. Classic histology paired with recent molecular data together draw an exciting scheme of cuticle differentiation that is summarised in this chapter.

## 8.2 Architecture and Composition of the Cuticle

The common denominator of virtually all cuticles is, with very few exceptions, their stereotypic organisation in three ultrastructurally distinct horizontal layers (Fig. 8.2). There are numerous terms for the different cuticle layers, and in this chapter, the newest unifying nomenclature proposed by Locke (2001) is used.

### 8.2.1 The Surface Envelope

The outermost layer composed of neutral lipids, wax esters and proteins is the envelope, which is a composite structure with a thickness of around 25 nm consisting of several alternating electron-dense and electron-lucid sheets. Lipids and waxes predominantly localise to the body surface. Some lipids seem to be free molecules and are easily washed out by organic solvents such as hexane. This has allowed the identification of the molecules in various insects by gas chromatography and mass spectrometry (Nelson et al. 2001, 2002, 2003, 2004; Patel et al. 2001; Nelson and Charlet 2003; Everaerts et al. 2010). The majority of molecules at the insect surface are neutral lipids like long-chain alkanes and alkenes, long-chain alcohol and fatty acid esters. For instance, the most abundant neutral lipids in *D. melanogaster* imagines are 7-tricosene (male) and 7,11-heptacosadiene (female). The obvious role of lipids and waxes is to protect the animal against dehydration and soaking (Gibbs 1998, 2011). In addition, they are reported to act as pheromones in various insects (Tillman et al. 1999; Howard and Blomquist 2005). In an exciting work using matrix-assisted laser desorption/ionisation (MALDI) imaging, that combines mass spectrometric identification of molecules with their localisation in the tissue, lipids (e.g. heptacosane and nonacosane) were identified on the surface of insect wings (Vrkoslav et al. 2010).

In 1933, Wigglesworth named the major component at the surface of *Rhodnius prolixus* cuticulin, which he proposes to be composed of lipids and sclerotin, a protein–quinone complex (Wigglesworth 1933, 1990). In his earlier work, Locke termed the outermost layer cuticulin (Locke 1966). Sclerotin can be visualised by silver precipitation after harsh chloroform extraction of masking surface lipids (Wigglesworth 1985). This argentaffin staining method reveals that in addition to a surface localisation, sclerotin is also present between the lamellae of the procuticle and in pore canals, nano-scaled tubes that run through the entire cuticle from the apical surface of the cell

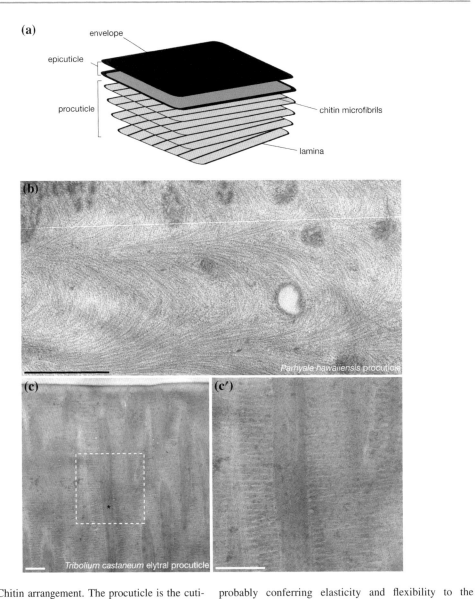

**Fig. 8.2** Chitin arrangement. The procuticle is the cuti-cle layer harbouring chitin that associates with proteins, which are necessary both as structural and as functional co-factors. In electron micrographs, chitin microfibrils that consist of around 18 chitin fibres appear as grey parabolic fibres although chitin itself is not contrasted by lead or uranium, suggesting that electron density of these fibres is due to associated proteins (Neville 1975; Neville et al. 1976). In the procuticle, chitin fibres are bundled as microfibrils, which in turn are arranged in parallel to each other forming horizontal chitin sheets, the so-called laminae (**a**). Often, laminae are stacked helicoidally, probably conferring elasticity and flexibility to the procuticle. Oblique sections of such procuticles give the impression that chitin microfibrils are oriented as para-bolic arches (**b**, compare to **a**). This architecture of the procuticle was first described by Yves Bouligand in 1965 in crustaceans. In 1969, Neville and Luke ascribed chitin organisation in the insect procuticle to Bouligand's model (Neville and Luke 1969a, b). Helicoidal arrangement of chitin laminae is not a paradigm. In some hard cuticle, chitin–protein complexes are organised as bricks probably making the cuticle stiff (**c, c′** shows a magnification of the framed region in **c**). Scale bars are 500 nm

to the surface of the cuticle (Wigglesworth 1990). Hence, lipids seem not only to form a sheet-like barrier at the surface of the animal but also to impregnate the entire cuticle either to prevent water loss or to contribute to cuticle architecture (Wigglesworth 1975). The wall of pore canals also

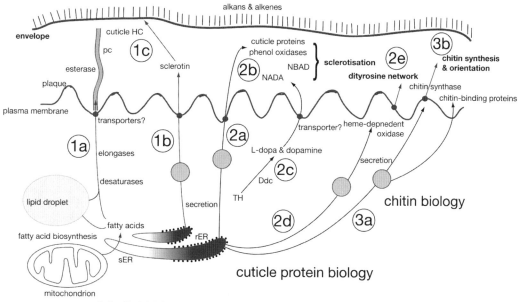

**Fig. 8.3** Cuticle production pathways. Cellular and molecular mechanisms of cuticle production can be subsumed in three pathways. Cuticle lipid biology (*1*). Lipids are either provided by lipid droplets or synthesised in mitochondria and the smooth endoplasmic reticulum (*sER*) by elongases and desaturases. Lipid deposition and organisation involve transfer by as yet unidentified transporters into extracellular pore canals (*pc*) that transport lipids to their destination (*1a*). For lipid organisation, lipid-binding proteins (sclerotin) are secreted through the canonical secretory pathway (*1b*). Sclerotin, through a yet unidentified reaction, forms a complex with polyphenols and lipids to form a waterproof barrier (*1c*). Free long-chain alkanes and alkenes are also present at the surface of the animal. Cuticle protein biology (*2*). Proteins are delivered to the

extracellular space via the canonical secretory pathway (*2a*). Here, during sclerotisation and melanisation, they react with catecholamines (*NADA* and *NBAD*, *2b*) that are transported to the extracellular space by yet unknown transporters. Synthesis of catecholamines starts in the cytoplasm, where dopamine and L-Dopa are formed by the tyrosine hydroxylase (*TH*) and Ddc (*2c*). Protein cross-linking comprises as yet unidentified membrane-bound or extracellular peroxidases (*2d*) that catalyse dityrosine formation between proteins (*2e*). Chitin biology (*3*). The canonical secretory pathway localises the chitin synthase to plaques at the tip of membrane corrugations and proteins assisting chitin synthesis and organisation to the membrane or the extracellular space (*3a*). Chitin organisation occurs in the extracellular space (*3b*). Secretory vesicles are depicted as grey circles

displays esterase activity that probably contributes to wax synthesis (Locke 1961). Taken together, production of lipids and waxes is conceivably initiated in the cytoplasm, followed by deposition into the pore canals by an unknown mechanism. Some of these lipids and waxes interact with proteins like sclerotin, others persist as free molecules. Both are subsequently modified and travel through the pore canals to their final site. One should be aware that these conclusions are based on fixed material; hence, for a dynamic view on cuticle, lipid biochemistry, molecular and genetic data are important to confirm or reject the working model presented in Fig. 8.3.

The molecular and biochemical pathways of cuticular hydrocarbon synthesis and transport have been studied in some insects, and the enzymes responsible have been identified in a few model species. For instance, biosynthesis of bombykol from *Bombyx mori*, a twice desaturated C16 alcohol which was the first lipid pheromone isolated by Butenandt (Butenandt et al. 1961a, b) branches out from the canonical fatty acid biosynthesis pathway (Matsumoto 2010). In brief, palmitic acid (C16) is desaturated at C10 and C12 by the specific acyl-CoA desaturase Bmpgdesat1, and the carboxyl group is reduced to an alcohol by the reductase

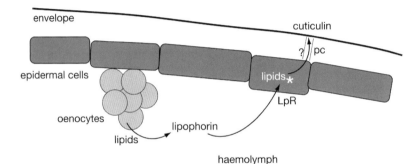

**Fig. 8.4** Lipids and cuticle formation. Many cuticle lipid precursors are produced in sub-epidermal clusters of oenocytes. Lipids are bound to lipophorin and transported through the haemolymph to the basal side of cuticle producing epithelial cells. Here, they are internalised via the lipophorin receptor (*LpR*). In these cells, they may be modified (*lipids**) and transported to the cuticle by a yet unidentified mechanism involving pore canals (*pc*) to form the protein–lipid complex cuticulin

pgFAR. This is a rather simple pathway that also seems to be present in other lepidopterans. Other fatty acid derivatives may require more complex modifications such as chain shortening. Nevertheless, it is possible, in principle, that the biosynthesis of many, if not all, cuticular hydrocarbons may follow this scheme (Fig. 8.3).

Lipid production occurs predominantly in sub-epidermal oenocytes, which are considered to be hepatocyte-like cells involved in lipid homoeostasis (Gutierrez et al. 2007). In *D. melanogaster*, these cells are organised as paired clusters of five cells each in every abdominal segment (Wigglesworth 1970; Gutierrez et al. 2007). Their activity as secretory organs correlates with cuticle moulting (Wigglesworth 1970). In the frame of their systemic function as lipid relays, they supply epidermal cells with cuticle lipid precursors. How are lipids that are produced and stored in oenocytes delivered to epidermal cells? In a simple yet unproven scenario depicted in Fig. 8.4, cuticle lipid precursors are produced and stored in oenocytes (and the fat body), as lipid droplets and crystal-like inclusions, and released into the haemolymph as lipophorin complexes, that are taken up by epidermal cells at their basal side by lipophorin receptors (LpR), through an endocytosis-independent pathway, as described for *D. melanogaster* nurse cells (Parra-Peralbo and Culi 2011). The lipid precursors are modified and processed accordingly within the epidermis and transported to the differentiating cuticle by unknown transporters. Pore canals that connect the epidermal apical plasma membrane with the cuticle surface are involved in further lipid modification and delivery, mainly to the cuticle surface. Hence, oenocytes systemically participate at cuticle differentiation.

## 8.2.2 The Epicuticle

Underneath the envelope, there is the epicuticle, which is mainly composed of largely unidentified proteins and lipids, probably covalently cross-linked. The interaction between the components of this layer does not result in a conspicuous texture, as indicated by electron micrographs of the epicuticle, which rather displays an amorphous ultrastructure. Deposition of the epicuticle in *D. melanogaster* depends on the activity of the steroid hormone ecdysone, since ecdysone-deficient larvae do not produce an epicuticle (Gangishetti et al. 2012). Obviously, expression of epicuticle producing enzymes and structural proteins is triggered by the ecdysone-signalling pathway. Candidate proteins are those that do not contain a chitin-binding domain directing them to the procuticle (see below). For instance, there are glycine-rich proteins with GGYGG or GGxGG repeats, called "cuticle protein glycine-rich" (CPG) in lepidopterans (Futahashi et al. 2008a), alanine-rich cuticle proteins (CPLCA) confined to dipterans, and apidermins in hymenopterans (Kucharski et al. 2007), that do not have obvious chitin-binding domains. Remarkably, many of these proteins are specific to single arthropod orders. Despite this specificity, they share several common features. First, they are relatively small proteins of around 10 kDa. Second, despite some conserved sequences, they are characterised by low structural complexity, that is, they probably do not adopt a complex tertiary structure (Andersen 2011). To verify whether these proteins are indeed components of the epicuticle, immuno-detection on thin sections should be performed, along with genetic and RNA interference (RNAi) analyses to elucidate their role.

### 8.2.3 The Procuticle

The procuticle is the innermost cuticle layer and harbours the N-acetylglucosamine (GlcNAc) polymer chitin in association with proteins. Usually, chitin orientation in the procuticle is not random as in fungi, but crystalline (Neville 1965a; Neville et al. 1976). The core molecule of chitin crystals is a bundle of on average 20 chitin fibres arranged antiparallel to each other (Vincent and Wegst 2004), an arrangement that is named α-chitin. These nanofibers, which have a diameter of around 30 Å, associate with proteins to form microfibrils with a diameter of around 100 nm. These microfibrils are arranged in parallel, to form two-dimensional sheets called laminae, which are stacked, with each lamina twisted by a small angle with respect to the lamina below. This helicoid pattern (Fig. 8.2) was described for the first time by Bouligand in 1965 through extensive ultra-structural analyses of crustacean cuticles (Bouligand 1965). Subsequently, Luke and Neville found that chitin in insect cuticles adopts the Bouligand arrangement as well (Neville and Luke 1969a, b). An interesting tissue with a specialised procuticle is the eye lens of insects. It consists of twisted chitin laminae that are arranged as a spherical extracellular matrix (Yoon et al. 1997). The lens cuticle serves as a protective structure, especially for digging insects, but may also be a light collector. Alternatively, in some cases, laminae may also be arranged like plywood (Neville and Luke 1969a, b; Neville et al. 1976; Cheng et al. 2009). For instance, the elytral cuticle of the red flour beetle *T. castaneum* is characterised by tightly packed protein–chitin brick-like units that do not display a helicoidal organisation (Fig. 8.2). In cockroaches and water bugs, Neville has found that chitin orientation changes from lamellate to non-lamellate following a circadian rhythm of light and dark (Neville 1965b). In crustaceans, nanofibrils do not run straight but meander, creating a honeycomb-like structure when viewed from above (Raabe et al. 2005). This structure is thought to prevent crack progression. Whether this pattern is present in other arthropods, including insects, remains to be investigated. Occasionally, pore canals, which are probably useful for cuticle repair, interrupt the crystalline organisation of the chitin–protein matrix.

What are the cellular and molecular requirements of chitin assembly? Ordered chitin synthesis at the apical plasma membrane of epithelia certainly has an important impact on chitin organisation. Indeed, the chitin synthase complex, visible as electron-dense plaques in electron micrographs, resides at the crest of repetitive plasma membrane corrugations. In *D. melanogaster*, these corrugations, called apical undulae, are longitudinal structures that are stabilised by microtubules (Moussian et al. 2006). In other arthropods, these structures have not been described, and rather microvilli-like units are regarded as the sites of chitin synthesis (Locke 2001, 2003). Elimination or reduction of chitin synthase activity are lethal in *D. melanogaster* and *T. castaneum* and cause cuticle disorganisation and collapse (Arakane et al. 2004, 2005b, 2008; Moussian et al. 2005a, b; Tonning et al. 2006). Cuticular proteins coagulate and are unable to ensure the formation of a uniform cuticle. Hence, the presence of chitin is essential for uniform thickness of the cuticle.

Arthropods possess more than one chitin synthase, but these enzymes do not have redundant functions (Merzendorfer 2006, 2011). The epidermal and tracheal chitin synthase 1 or A (CS-1 or CHS-A) is required for cuticle production, whereas the midgut chitin synthase 2 or B (CS-2 or CHS-B) contributes to the formation of the midgut peritrophic matrix that protects the midgut epithelial cells from pathogens and digestive enzymes (Lehane and Billingsley 1996). It should also be pointed out that partner proteins of chitin synthase itself, that is, other constituents of the plaques have not been identified to date.

Fusion of vesicles carrying cuticle proteins occurs in the valleys between microvilli-like membrane corrugations. Thus, chitin synthesis and secretion of chitin-binding proteins are spatially separated processes, and probably, some extracellular self-assembly mechanisms are needed for a stereotypic association of chitin with its partners. Recent genetic data underline that along with structural proteins, secreted enzymes and membrane-inserted factors are required for chitin organisation. Four chitin organising proteins, Knickkopf (Knk), Retroactive (Rtv), Serpentine (Serp) and Vermiform (Verm) have been identified and to some extent characterised in the last few years in *D. melanogaster* and *T. castaneum* (Moussian et al. 2005a, b, 2006a, b; Luschnig et al. 2006; Tonning et al. 2006; Arakane et al. 2009; Chaudhari et al. 2011). Collectively, respective mutations in the genes coding for these factors provoke an unordered mass of chitin in the procuticle. Serp and Verm contain a chitin-binding domain and have significant similarity with chitin deacetylases from bacteria, suggesting that deacetylation of chitin to chitosan is a central trimming reaction modifying chitin. However, biochemical proof for this function of Serp and Verm is lacking. Assuming that they may truly deacetylate chitin, their enzymatic activity argues that deacetylated chitin is essential for chitin organisation. Indeed, partially deacetylated chitin has been proposed to raise accessibility of chitin to proteins (Neville 1975). Serp and Verm have another domain that may be crucial for their function in organising the procuticle. N-terminal to the chitin deacetylase signature is a low-density lipoprotein (LDL) domain that presumably enables these enzymes to bind to lipids, including cholesterol. Does this point to a connection between Wigglesworth's cuticulin and chitin organisation? In summary, it is obvious that the (pro)cuticle has a certain capacity for self-assembly, that is, it is not a simple structure deposited by the epithelial cell as a finished and ready to function product.

Knk and Rtv are membrane-associated factors of unknown biochemical function. Knk is inserted into the apical plasma membrane via a C-terminal Glycosylphosphatidylinositol (GPI) anchor, while Rtv has a C-terminal transmembrane domain (Moussian et al. 2005a, b, 2006a, b). Knk has three conserved domains: at the N-terminus, just after the signal peptide, there is a tandem of DM13 motifs, followed by a middle DOMON domain (Aravind 2001; Iyer et al. 2007). These domains have been proposed to transport electrons to a yet unidentified substrate. Since chitin organisation is severely disrupted in *knk* mutant *D. melanogaster* larvae, it is possible that chitin may be the substrate of Knk. However, a requirement for chitin as an electron receiver has not been reported. Experimental evidence for these hypotheses is still missing. In *T. castaneum*, besides its implication in chitin organisation, Knk has a second function which is to protect chitin from degradation by chitinases (Chaudhari et al. 2011). This finding as illustrated in Fig. 8.5 indicates that organisation of chitin may be the result of balanced chitin production, degradation, that is, shortening and modification by deacetylases, while structural proteins eventually stabilise and conserve an optimal status. Rtv is a small protein (151 aa), which is conserved in arthropods (Moussian et al. 2005b; Havemann et al. 2008). It belongs to the Ly6-type protein family and is characterised by three loops exposing highly

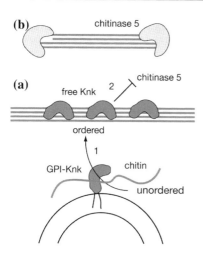

**Fig. 8.5** Knickkopf function. Knickkopf (Knk) has two functions during cuticle formation (**a**). Knk is a GPI-anchored protein that assists in chitin organisation (*1*). Ordered chitin is protected from degradation through chitinase-5 by free Knk (*2*). During moulting, Knk is removed from the cuticle allowing chitinases to degrade chitin (**b**)

conserved aromatic amino acids that are hypothesised to bind to partners. Other Ly6-type proteins appear to be important for sorting events during secretion of proteins of the lateral plasma membrane (Hijazi et al. 2009; Nilton et al. 2010). Based on these findings, one may speculate that Rtv is needed for trafficking of chitin organising factors to the apical plasma membrane. Indeed, this has recently been demonstrated to be the case in *T. castaneum* (Chaudhari et al. 2013).

The crystalline configuration of chitin suggests a non-random association of chitin with proteins at each level of organisation. In the past, an arsenal of peptide sequences of cuticle proteins was identified biochemically and through intensive efforts of localised genome sequencing (Chihara et al. 1982; Snyder et al. 1982; Silvert et al. 1984; Doctor et al. 1985; Fristrom et al. 1986; Wolfgang et al. 1986; Andersen et al. 1995). Today, using sequence information retrieved by classical biochemical work, sequenced insect genomes are being consulted to identify the whole complement of cuticle proteins. Among these, over 100 chitin-binding proteins are classified in two groups: cuticle

proteins with Riddiford and Rebers motif (cuticle protein R&R, CPR) and Tweedle proteins (Tang et al. 2010; Willis 2010). CPRs constitute the group with most members. In addition to an N-terminal signal peptide, which directs their deposition to the extracellular space via the canonical secretory pathway, they contain at least one R&R domain that has been shown to bind chitin in vitro (Rebers and Willis 2001; Togawa et al. 2004; Tang et al. 2010). *D. melanogaster* has 102, *Anopheles gambiae* 156 and *Aedes aegypti* 240 CPR coding genes that are organised in distinct clusters. In general, CPR proteins are small, and besides their R&R domain, their sequences are very diverse. This indicates that they play a structural rather than an enzymatic role in chitin organisation. One may also argue that structural diversity of CPR proteins ensures non-perfect, but flexible chitin organisation. The large number of CPRs may also suggest tissue- and stage-dependent expression of different clusters. The Tweedle group of cuticle proteins is less diverse and comprises 27 members in *D. melanogaster*, and only 12 and 9 in *An. gambiae* and *Ae. aegypti*, respectively. They share a domain of unknown function (DUF243) preceded by a signal peptide underscoring that they are extracellular proteins. In contrast to CPR proteins, and although they may bind to chitin in vitro (Tang et al. 2010), Tweedle proteins lack a known chitin-binding domain. Some insight into their function comes from *D. melanogaster* genetic approaches. Dominant mutations in some *Tweedle* genes in *D. melanogaster* cause a tubby phenotype, indicating that these proteins are involved in cuticle structuring that has an impact on body shape (Guan et al. 2006). Tweedle proteins do not participate in basic cuticle organisation as they are insect specific and seem to be absent in crustaceans such as *Daphnia pulex*. *Tweedle* genes are expressed in a tissue and stage-specific manner, suggesting that they are key components in different cuticle types. In a recent bioinformatics approach, Cornman defined two motifs, GYR and YLP with exposed tyrosine residues that are present in members of CPRs and Tweedle proteins (Cornman 2010). These motifs are also present in other cuticle proteins such as CPLCGs and CPF/CPFLs. Corman proposes that these motifs are involved in protein–protein interaction. Extensive interaction between cuticle proteins, via their GYR and YLP motifs, and association with chitin is indeed an attractive model for a chitin-cuticular network where all components are linked together. In Sect. 8.4, we will encounter types of cross-linking of cuticular components that classically has been viewed as a stabilising element of cuticle structure.

### 8.2.4 Cuticle Irregularities

The naked cuticle is an extracellular matrix with uniform thickness. Cuticular protrusions like scutes, denticles and bristles disrupt cuticle uniformity, affecting the texture or thickness of distinct cuticle layers. Scutes in centipedes, for instance, are ridges in the epicuticle coinciding with the outline of hypodermal cells, suggesting that cells dictate epicuticle irregularity (Fusco et al. 2000). In *D. melanogaster*, chitin in larval ventral denticles and adult sensory bristles is not textured as in the naked cuticle, but is unorganised (Uv and Moussian 2010; Nagaraj and Adler 2012). Most cuticle irregularities probably arise from different cell identities that were determined during early pattern formation and from cell-immanent planar polarity. Cell- and tissue-position-specific programs may instruct efficiency and duration of chitin synthesis and deposition of specific proteins and lipids in cuticle protrusions. During bristle formation in the thorax of *D. melanogaster*, for instance, correct localisation of the bristle-specific membrane-bound zona pellucida (ZP) protein Dusky-like (Dyl) depends on the small GTPase Rab11 that functions in all epidermal cells (Nagaraj and Adler 2012). In this work, it was also proposed that Dyl is the effector of Rab11 in chitin synthesis and cuticle deposition. Since Dyl is not expressed in non-bristle cells, these findings imply that another Rab11 effector acts in these cells to mediate Rab11's role in cuticle formation.

## 8.3 Cuticle Producing Epithelial Cells

What are the cellular programs that bring about the 3D architecture fundamental for cuticle function as a protective and supporting extracellular tunic? Evidently, molecular pathways of cuticle differentiation are deployed in the cytoplasm and organelles of polarised epithelial cells, producing cuticle components or precursors that travel through the active apical plasma membrane to the extracellular space where they are eventually modified and assembled.

### 8.3.1 Properties of Cuticle Producing Epithelial Cells

The cuticle is an extracellular matrix produced by most ectodermal tissues, comprising the epidermis, the fore- and hindgut epithelia, and respiratory organs (i.e. tracheae and book lungs) at their apical side. At stage transitions, to accommodate growth, these cells either divide or expand their apical areas in order to enlarge the surface of the respective tissue. Concomitantly, they shed their cuticle (apolysis) and produce a new one underneath that replaces the old one (ecdysis). The regulation of these processes is reviewed in Chap. 6. Besides the persistent ectodermal tissues, the extraembryonic serosal cells also produce a cuticle, called the serosal cuticle during embryogenesis that transiently prevents embryo desiccation and allows survival during periods of drought (Rezende et al. 2008).

Cuticle producing epithelial cells display the standard polar organisation in apical, lateral and basal domains (Fig. 8.1). At their basal side, they are covered by the basal lamina, an extracellular matrix consisting of a network of collagens and laminins, which are contributed by haemocytes. Studies using the fruit fly *D. melanogaster* have revealed that the integrity of the basal lamina is a prerequisite for an intact cuticle. Interference with basal lamina assembly through mutations in *sparc*, for instance, coding for a collagen-binding protein that is provided by haemocytes and stabilises collagen IV in the basal lamina, causes fragmentation of the cuticle (Martinek et al. 2008). A functional basal lamina may indirectly support cuticle production by ensuring cell vitality. A functional basal lamina could also directly influence cuticle production by mediating interaction between the basal plasma membrane of cuticle producing cells and free-floating haemocytes which deliver cuticle proteins and enzymes (Sass et al. 1993, 1994).

The lateral membrane of cuticle producing epithelial cells is decorated by three kinds of junctions. The apicalmost adherens junctions contact neighbouring cells and interact with the actin cytoskeleton, stabilising the tissue. The main players of these structures are the membrane-inserted E-cadherin and the cytoplasmic β-catenin. The extracellular domains of clustered E-cadherin molecules in neighbouring cells bind to each other, while their intracellular parts of the protein, via associated factors, serve as anchors to span a panel of stabilising actin cytoskeleton. Mutations in the *D. melanogaster* genes coding for E-cadherin *shotgun* (*shg*) and β-catenin *armadillo* (*arm*) do not affect cuticle differentiation. Ruptures in the cuticle are not due to defective cuticle formation but to the failure to renew cell contact after delamination of neuroblasts from the epidermal primordium (Tepass et al. 1996). However, a possible requirement may be masked by maternally provided proteins. Hence, genetic analyses of the function of adherens junctions during cuticle differentiation is difficult and have not been conducted.

Basal to these junctions, there are the septate junctions that seal the paracellular space, thus preventing free diffusion of water and solutes between the two separated milieus. The assembly of septate junctions seems to be modular. At least two complexes—the Gliotactin–Discs large (Dlg) complex and the core complex including Coracle, Neurexin IV and Nervana 2—come together for full establishment of septate junctions (Schulte et al. 2006; Laprise et al. 2009; Oshima and Fehon 2011). At the septate junction domains, the plasma membrane meanders, enlarging the contact zone and enforcing the belt

like character of the epithelium. It is, however, not known whether the septate junction proteins themselves are responsible for membrane curvature. Work in *D. melanogaster* demonstrates that apical secretion may be compromised in cells with disrupted septate junctions. For instance, Knk, an apical membrane-inserted protein fails to be delivered quantitatively to the apical plasma membrane of tracheal cells in embryos mutant for *Fas2* and *sinuous* that code for septate junction components. Secretion control at the septate junctions and their function as paracellular diffusion barriers and cell shape determinants seems to be independent from each other (Laprise et al. 2010).

The septate junctions are occasionally interrupted by gap junctions that are required for cell–cell communication. In invertebrates, innexins are the major constituents of gap junctions (Bauer et al. 2005). The role of innexins and gap junctions in epidermal differentiation during cuticle production has not been investigated.

Epithelial cell polarity is, as expected, a prerequisite for correct deposition of the cuticle at the apical side of the cell. Cells that fail to polarise do not produce cuticle at random sides, but mostly undergo apoptosis. This is best exemplified by the phenotype of *crumbs* (*crb*) mutant larvae in *D. melanogaster*, which is characterised by patches of cuticle produced by surviving epidermal cells (Tepass et al. 1990). Genetic analyses have revealed that cell polarity is not a stable state but needs maintenance during cuticle differentiation. Abrogation of ER-born vesicle formation through mutations in *sec23* and *sec24* encoding respective COPII components, for instance, results in cuboidal cells that gradually lose their polarity (Norum et al. 2010). Thus, the canonical secretory pathway is necessary for sustaining polarity.

## 8.3.2    The Plasma Membrane of Cuticle Producing Cells

During differentiation, cuticle material is deposited into the extracellular space across the apical plasma membrane. However, the apical plasma membrane does not serve simply as an interface, but seems to actively contribute to cuticle assembly. In the *D. melanogaster* embryo, the envelope, which is the first layer to be formed, is produced as fragments at the tips of irregular protrusions of the apical plasma membrane (Moussian et al. 2006a). Deposition of the envelope is effectuated by the canonical secretory pathway (Norum et al. 2010). Envelope fragments eventually fuse together to give rise to a continuous layer consisting of parallel electron-dense and electron-lucid sheets. The parallel course of the envelope sheets suggests an invariant structural coupling of the components.

Classically, later during pro- and epicuticle production, regular microvilli-like protrusions of the plasma membrane are formed during deposition of the pro- and epicuticle (Fig. 8.6). These structures are somewhat different from midgut cell microvilli. A midgut cell microvillus in arthropods is a cylindrical membraneous structure which is stabilised by a core of actin and associated proteins that are homologous to actin-binding proteins in vertebrates. Indeed, the major microvillar actin-binding and actin-organising proteins such as Espin, Fascin, Villin, Myosin 1A and calmodulin are present in *D. melanogaster* (Bartles 2000; Tilney et al. 2004; Hegan et al. 2007). Epidermal microvilli are stunted and carry electron-dense plaques at their tips that harbour the chitin synthesis and probably organisation factors. Microvilli formation has been studied especially at the site of bristle formation. The *D. melanogaster* Espin Forked, for instance, determines the number of microvilli during bristle formation in the pupa by regulating the thickness of actin bundles (Tilney et al. 1998, 2000, 2004). In *forked* mutant animals, actin bundle number at the plasma membrane is increased and more microvilli are formed, although actin bundle diameter is reduced. Interestingly, mutations in *forked* and the other microvillus factors are not lethal indicating either functional redundancy between them or involvement of yet unidentified factors.

At the cytoplasmic side the microvillar plaques are nourished by chitin monomers GlcNAc

that are synthesised from glucose through the Leloir pathway comprising six cytoplasmic enzymes (Moussian 2008). One may speculate that extensive chitin synthesis may necessitate concentration of GlcNAc production at the membrane protrusions. An isoform of the last enzyme of the Leloir pathway in *D. melanogaster*, UDP-N-acetylglucosamine pyrophosphorylase, has a 37 amino acid N-terminal domain that could mediate its localisation to the apical plasma membrane (Tonning et al. 2006). Mutations in the gene coding for this enzyme are lethal and cause a chitin-less and collapsed cuticle.

Secretion of cuticle proteins occurs at the depression between microvilli. The physical separation of chitin synthesis and protein secretion implies firstly that cuticle assembly takes place in the extracellular space. Secondly, coordination of chitin synthesis and protein secretion is not simply effectuated by direct physical contact between the effectors, but that communication is required at some other as yet unknown hub.

At sites of denticles of the *D. melanogaster* embryo, the apical plasma membrane forms large protrusions at the posterior half of the cell (Fig. 8.6). The formation of these protrusions obeys cues from planar polarity defined by the asymmetric distribution of Strabismus, Dishevelled, Diego, Prickle and Frizzled at cell junctions (Goodrich and Strutt 2011). The denticle forming protrusion is stabilised by an actin core that involves several factors important for microvilli formation, as well. The apical plasma membrane of denticle forming protrusions is decorated by several membrane-inserted zona pellucida (ZP) proteins that specify distinct domains of the future denticle (Chanut-Delalande et al. 2006; Fernandes et al. 2010). Zye (Zye) and Trynity (Tyn) mark the basis of a denticle, and Miniature (M) separates the apical part of the denticle occupied by Dusky-like (Dyl).

At the end of larval cuticle production during embryogenesis, the apical plasma membrane smoothens. This process is a critical step in differentiation. Mutations in *wollknäuel* (*wol*)

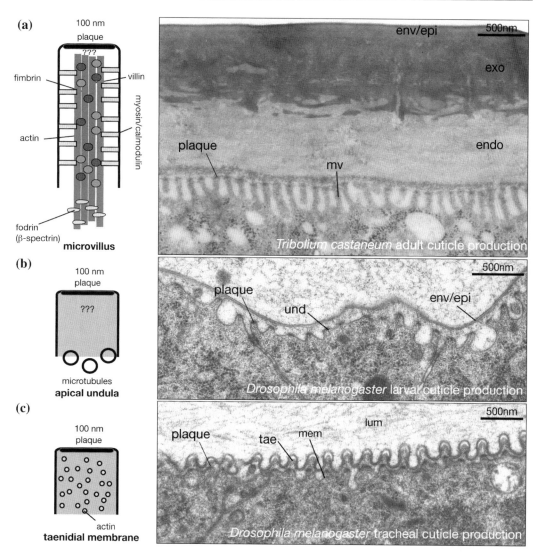

**Fig. 8.6** The plasma membrane of cuticle producing epithelial cells. Classical microvilli (*mv*) are cylindrical extensions of the apical plasma membrane that have an inner skeleton consisting of actin bundles (**a**). Organisation of actin involves several actin-binding proteins. At the tip of microvilli-like structures that are protruded during cuticle deposition in various arthropods, an electron-dense plaque harbours the chitin synthesis apparatus. Apical undulae have been demonstrated to form during cuticle deposition in the *D. melanogaster* embryo (**b**). These longitudinal protrusions are traced by microtubule filaments at their basis. Actin has not been observed within the undulae, and their supporting inner cytoskeleton is unknown, but does not seem to contain high amounts of Fodrin also known as Spectrin, which accumulates rather at the lateral membrane (Thomas and Kiehart 1994; Das et al. 2008). The taenidia (*tae*) of insect tracheae follow the spiral extrusion of the apical plasma membrane of tracheal cells (**c**). The plasma membrane (*mem*) is stabilised by actin cables that run in parallel to the taenidia. They prevent collapse of the tracheal tube lumen (*lum*). Scale bar 500 nm

that codes for the *D. melanogaster* ALG5 (UDP-glucose:dolichyl-phosphate glucosyltransferase) cause uncontrolled protrusions of the apical plasma membrane at the end of cuticle differentiation (Shaik et al. 2011). This phenotype is associated with persistent apical localisation of Crb, that is, a determinant of the apical plasma membrane identity (Assemat et al. 2008). Thus, removal of Crb from its apical position may be a prerequisite for plasma membrane smoothening.

### 8.3.3 Secretion of Cuticle Material

The cuticle is an extracellular matrix and is naturally the product of secretion and deposition of its components. Some components are secreted directly via the canonical secretory pathway from the ER tubules, via the Golgi apparatus and secretory vesicles to the plasma membrane, where they are released to the extracellular space. Most if not all proteins such as Verm, Serp and cuticle proteins follow this route. Some other components are produced at or transferred across the apical plasma membrane through membrane-localised enzymes, such as the chitin synthase, that links GlcNAc monomers together and extrudes the polymer chitin probably through a pore formed by the enzyme itself. The apical plasma membrane also harbours those factors required for chitin organisation such as Knk and Rtv (Moussian et al. 2005a, b, 2006b). Like *bona fide* secreted proteins, the membrane-inserted factors are also positioned within the membrane by the secretory pathway. Hence, many different components and enzymes travel through the secretory pathway at the same time with the same destination, the apical plasma membrane. The topology of the apical plasma membrane argues that at some point, these factors have to be sorted. Where does this take place? In *D. melanogaster*, we are beginning to understand this process. The *Maclura pomifera* agglutinin (MPA) recognises some epitopes at the envelope and some within the procuticle of the *D. melanogaster* larva (Moussian et al. 2007). In larvae mutant for the plasma membrane t-SNARE Syntaxin 1A (Syx1A), some secretory vesicles that erroneously accumulate beneath the apical plasma membrane are MPA positive, but others are not. This indicates that some factors are separated at the exit of the Golgi apparatus.

One enzyme that may define a class of secretory vesicles is the chitin synthase. In fungi, chitin synthases localise to specialised intracellular vesicles of 40–70 nm diameter, the chitosomes that deliver the chitin synthesis complex to the site where this is required during cell division (e.g. Bartnicki-Garcia 2006). Several specific factors are associated with chitosomes

and are required for ordered positioning of the chitin synthesis complex. Fungal chitin synthases are, however, not active within chitosomes, suggesting that activation has to be triggered. It seems that localisation to the plasma membrane is mandatory for chitin synthase activity. In yeast, the CaaX protease Ste24 facilitates the localisation of Chs3, the major chitin synthase in yeast, to the plasma membrane through trimming of the Chs3 partner Chs4, which is absent in arthropods (Meissner et al. 2010).

Do Arthropods have chitosomes? In the moth *Manduca sexta*, it was found that a chitin synthase–specific antibody recognises an epitope at the apical surface but also within the midgut cell (Zimoch and Merzendorfer 2002). Obviously, as membrane-inserted enzymes, chitin synthases travel through the secretory pathway to reach the plasma membrane. Therefore, sorting at some level is imperative. The central question is whether plasma membrane plaques of chitin synthesis are preformed within vesicles, or whether they travel to the plasma membrane where they are assembled. In the *D. melanogaster* embryo, the plasma membrane t-SNARE Syx1A is dispensable for delivery of chitin synthases and chitin organising factors, such as Knk, to the plasma membrane (Moussian et al. 2007). This finding suggests that another t-SNARE may mediate localisation of plaques to the plasma membrane, in turn arguing (and confirming the MPA data) that different Golgi-born vesicles deliver distinct factors to decorate the apical plasma membrane and to produce the cuticle. In other words, the Golgi apparatus is the main compartment where sorting of cuticle structural and production components takes place.

## 8.4 Cross-Linking of Cuticle Components

### 8.4.1 A Dityrosine Transcellular Barrier

Soft body cuticle of caterpillars and larvae has to withstand the internal hydrostatic pressure in order to serve as an exoskeleton. Indeed, in *D. melanogaster*, mutations in the chitin synthase gene or in *knk* and *rtv* that are needed for chitin organisation result in loss of body shape and inability to move, suggesting that chitin is an essential element of the soft exoskeleton (Moussian et al. 2005a, b, 2006a, b). In view of the elaborate interaction between chitin and chitin-binding proteins, one may assume that chitin on its own is not the barrier component opposing water pressure at the cuticle. Indeed, it seems that a network of proteins at the basal side of the procuticle covalently bind to each other via tyrosine residues, probably constructing a network adjacent to the plasma membrane that confers elasticity and stiffness to resist the internal hydrostatic pressure (Shaik et al. 2012). The establishment of this dityrosine network depends on a haem-dependent enzyme, which is yet unknown. Likewise, the sequence of the linked proteins is unknown, as well. Candidate cuticle proteins may be those low complex proteins with GYR- and YLP-like motifs that are characterised by invariant tyrosine (Y) residues (Cornman 2010). The membrane-inserted dual oxidase Duox could be considered as a good candidate for being the haem-dependent enzyme involved in dityrosine formation. It has an intracellular flavoprotein domain that accepts electrons from NADPH, which are transferred across the membrane to the extracellular peroxidase domain that catalyses $H_2O_2$ production (Donko et al. 2005). Tyrosines are oxidised and spontaneously react with each other to link neighbouring proteins. It is difficult to assume that this last reaction of dityrosine formation is ordered and specific. Rather, within the range of $H_2O_2$ production, tyrosines from all proteins present are potential targets. Indeed, Duox stimulates the production of a dityrosine-based barrier that protects the midgut epithelium in mosquitoes against pathogen entry (Kumar et al. 2010). Consistently, reduction of Duox activity by RNAi in the *D. melanogaster* wing results in pale wings, suggesting that melanisation and probably sclerotisation are impaired in this tissue (Anh et al. 2011). Arguing against an

involvement of Duox in barrier formation, however, reduced Duox activity does not give rise to a respective haem-deficient phenotype. In fungi, a cytochrome 56 protein has been shown to drive extracellular dityrosine formation required to render the cell wall impermeable against loss of proteins. Hence, one cannot exclude that functional redundant enzymes may catalyse tyrosine oxidation in arthropods.

## 8.4.2   Resilin

In 1960, Weis-Fogh published his discovery that the long-range elastic behaviour of parts of the thoracic cuticle in the locust and dragonfly depends on the presence of a glycine-rich rubber-like protein he named resilin (Weis-Fogh 1960). Other cuticles with extreme extensibility like the cuticle covering the allosctum of ticks, such as *Ixodes ricinus*, also contain large amounts of resilin. Resilin visualisation is comparably simple as resilin fluoresces upon excitation with UV light by conventional fluorescence or confocal laser scanning microscopy (Michels and Gorb 2012). This characteristic of resilin is due to the presence of di- and trityrosines (Andersen 1964, 1966; Malencik and Anderson 2003). In the following decades, the in vivo physical properties of resilin were investigated in detail mainly by Weis-Fogh himself, Andersen, Edwards, Bennet-Clark, Neville and others (Bennet-Clark 2007). Finally, in the genomic era of insect biology, the full sequence of resilin was uncovered in different insect species (Andersen 2010b; Lyons et al. 2011). Presence of resilin has been reported in crustaceans, as well (Burrows 2009). It was Andersen who first identified the sequence of the resilin monomer, proresilin, encoded by the gene *CG15920* in the *D. melanogaster* genome, by using the amino acid sequence of three tryptic peptides he had obtained from the elastic cuticle of the wing hinges and prealar arms from the desert locust *Schistocerca gregaria* (Ardell and Andersen 2001). *D. melanogaster* proresilin has an N-terminal signal peptide that directs it to the

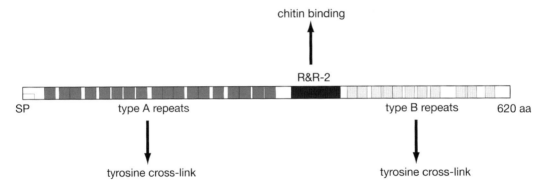

**Fig. 8.7** Resilin. Proresilin has an N-terminal signal peptide (*SP*) that allows the protein to be secreted to the extracellular space through the canonical secretory pathway. It harbours two types of repeats, the type A and the type B repeat with the consensus sequences GGRPSDSY-GAPGGGN and GYSGGRPGGQDLG, respectively, that flank an R&R chitin-binding motif. Tyrosine residues in the repeat sequences may covalently link to tyrosines of neighbouring proteins. Resilin is a polymer of proresilins that are linked to each other via dityrosine bridges. Upon illumination with ultraviolet light (maximum 315 nm), dityrosines emit blue light (409 nm maximum)

apical extracellular space (Fig. 8.7). In addition, the protein is composed of 18 repeats of a 15 residue motif (type A repeat) and 11 repeats of a 13 residue motif (type B repeat), which flank a type 2 Riddiford and Rebers chitin-binding domain (R&R-2) of 62 amino acids. Prolines and glycines occupy distinct positions within both repeats, probably forming β-turns separated by irregular loops, overall yielding a β-spiral conformation, which accounts for protein elasticity. Proresilin's type A repeats from *D. melanogaster* and *An. gambiae* in synthetic polypeptides have elastic and resilient properties comparable to those of recombinant full-length proresilin, underscoring the significance of sequences with low complexity for protein elasticity and resilience (Lyons et al. 2009). Proresilin sequence information allows us to model resilin function within the cuticle: proresilin monomers are polymerised to resilin via di- and trityrosine bridges that were discovered some decades ago, tyrosine residues being present, especially in the N-terminal type A repeats, and associate with chitin via their chitin-binding domain of the R&R-2 type, which in *D. melanogaster* is indeed able to bind to chitin in vitro (Qin et al. 2009). *D. melanogaster* possesses one gene coding for two alternatively spliced proresilin isoforms (620 and 575 amino

acids). The implication of two isoforms is unclear, as it is not known whether they are expressed in different tissues or stages. The proresilin isoform from the shorter mRNA lacks the R&R-2 domain, suggesting that it may be present in cuticle types with low, or no chitin (Andersen 2010b). Association and non-association with chitin may confer specific elastic properties to the respective cuticle. Low complexity of the proresilin sequence suggests that other low complex cuticle proteins may have similar physical properties. Indeed, another protein encoded by the CG9036 locus in the *D. melanogaster* genome may constitute a proresilin paralog (Ardell and Andersen 2001).

Resilin as a biomaterial has been extensively investigated since the identification of its sequence. Similarly, resilin's function in whole organisms and tissues is well understood. By contrast, resilin cell biology including regulation of its cell-specific expression and extracellular polymerisation is less well analysed. To advance in these problems, a genetic approach in *D. melanogaster* would be helpful. During *D. melanogaster* embryogenesis, *proresilin* is expressed in segmental clusters of epidermal cells and in stretch receptors (Wong et al. 2012). Later, as expected, *proresilin* expression is detectable in cells at the base of the wing. Mutations in *D.*

*melanogaster proresilin* should reveal the importance of resilin function in these cells.

## 8.4.3   Transglutaminase

Another type of protein covalent cross-links in the cuticle, the $\varepsilon$-($\gamma$-glutamyl) lysine bonds, is catalysed by extracellular transglutaminases. Generally, expression and activity of transglutaminases are known to be induced upon injury in both vertebrates and invertebrates. In *D. melanogaster*, transglutaminase is also robustly expressed during late developmental stages, that is, in L3 larvae, in pupae and the adult flies (Shibata et al. 2010). Abrogation of transglutaminase translation by RNAi causes deformation and tanning failure of the adult abdominal cuticle and wrinkling of the wing. In contrast to the dityrosine network, the transglutaminase cross-links do not constitute a water barrier. Several cuticle proteins have been identified as targets of transglutaminase activity. Fondue, for example, an unknown extracellular protein is deposited into the cuticle during clot formation after wounding (Lindgren et al. 2008). Moreover, extractability of the cuticle proteins Cpr47Ef, Cpr64Ac, Cpr76Bd and Cpr97Eb depends on transglutaminase activity, suggesting that these proteins are normally fixed within the cuticle by $\varepsilon$-($\gamma$-glutamyl) lysine bonds (Shibata et al. 2010). Interestingly, the wing phenotype of transglutaminase depleted animals can at least partially be attributed to the function of Cpr97Eb, as knock-down of this factor causes a similar phenotype. Taken together, transglutaminase-based cross-linking of cuticle components seems to be needed for cuticle pigmentation and stiffness.

## 8.4.4   Sclerotisation and Melanisation

Since Pryor's work in the 1940s, hardening (sclerotisation) of the cuticle is known to depend on cross-linking of proteins with phenolic substances (Pryor 1940; Pryor et al. 1946, 1947). Recent advances in the field are excellently summarised by Andersen (2010a, 2012). The basic feature of the sclerotised cuticle is the covalent linkage of cuticle components by the acyldopamines N-acetyldopamine (NADA) and N-$\beta$-alanyldopamine (NBAD). Production of phenolic substances starts in the cytoplasm where tyrosine is converted to Dopa by tyrosine dehydroxylase (Pale in *D. melanogaster*), which is subsequently used to produce dopamine by Dopa-decarboxylase (Ddc). Dopamine is the substrate of the dopamine N-acetyl-transferase and NBAD synthase (Ebony in *D. melanogaster*) that catalyse the formation of NADA and NBAD, respectively. The catecholamines NADA and NBAD are transported to the extracellular space through as yet unknown transporters. In the differentiating extracellular matrix, they are oxidised to their ortho-quinones that may react with free amino groups of proteins and possibly also deacetylated chitin. The incorporation of these ortho-quinones results in brown cuticle, whereas usage of the oxidation intermediate dehydro-NADA as preferred by insects only lightly colours the cuticle. The extracellular sclerotisation reactions are catalysed by extracellular laccases and tyrosinases (Suderman et al. 2006). Multi-copper-containing Laccase 2 in the beetles *T. castaneum* and *Monochamus alternatus* has been reported to affect cuticle integrity, and this can be attributed to sclerotisation defects underlining the importance of these modifications (Arakane et al. 2005a; Niu et al. 2008). Only few cuticle proteins have been experimentally shown to be cross-linked via catecholamines. Pioneering work has been performed in *Manduca sexta*: lysyl groups of the cuticle protein MsCP36 contribute to protein oligomerisation (Suderman et al. 2010). Interestingly, dityrosine bounds were also found to be involved in MsCP36 cross-linking.

Animals suffering Laccase 2 reduction are also pale compared to siblings with normal laccase 2 activity. This nicely underlines that sclerotisation (hardening) and melanisation (pigmentation, tanning) share extracellular enzymes. However, there are also melanisation-specific enzymes (Sugumaran 2009). The extracellular Yellow protein represents a prominent class of melanisation enzymes. Mutations in the *D. melanogaster yellow* gene are not lethal, but provoke a yellow body colour. The two members of Yellow protein family Yellow-f1 and Yellow-f2 have been shown to be dopachrome-conversion enzymes (Han et al. 2002). Dopachrome, the intramolecular cyclisation product of dopa, is converted to 5,6-dihydroxyindole-2-carboxylic acid (DHICA) that is subsequently used for polymerisation of melanin. An interesting mechanism of melanisation control has been analysed in the *D. melanogaster* wing (Riedel et al. 2011). Melanin is usually produced at the distal region of the wing procuticle. Confinement of melanisation reactions to this region necessitates timed removal of Yellow from the cuticle by endocytosis regulated by Rab5 and Megalin. Failure to clear the procuticle from Yellow results in extension of melanisation to proximal regions of the procuticle.

Taken together, defects in the melanisation pathway, if they do not impair sclerotisation, are not lethal but cause body colour changes. Indeed, the differential activity of melanising enzymes can be used by nature to create colour patterns. The antagonistic functioning of Yellow and Ebony, for example, generates the striped pattern in the abdomen of adult fruit flies (Wittkopp et al. 2002). Not only in *D. melanogaster* but also in other insects, Yellow and Ebony are involved in pigmentation patterns (Futahashi et al. 2008b; Arakane et al. 2010).

### 8.4.5 Calcite Deposition in Crustacean Cuticle

The cuticle of crustaceans is stiffened through internal deposition of calcite. The molecular mechanisms of calcification are being intensively studied in terrestrial isopods by Ziegler and his group. Storage of calcite in the extracellular space and its resorption are highly regulated during moulting in *Porcellio saber* (Ziegler et al. 2004). A central enzyme of this process is the smooth endoplasmic reticulum $Ca^{2+}$-ATPase (SERCA), the expression of which

is up-regulated at late pre-moult and intermoult stages (Ziegler et al. 2002; Hagedorn et al. 2003). Before the synthesis of the new cuticle, $Ca^{2+}$ and $HCO_3^-$ ions are reabsorbed from the posterior half of the cuticle, that moults first, and are transported through the haemolymph and across the epithelium to the apical extracellular space of anterior sternites, where storage calcite is formed. Concomitantly, protons produced during $CaCO_3$ formation are pumped into the haemolymph through the V-type $H^+$-ATPase (VHA) that localises to the basolateral plasma membrane of these cells at this stage. Abundance of the enzyme is regulated at the transcriptional level. To mobilise $Ca^{2+}$ and $HCO_3^-$ for cuticle remineralisation, the extracellular space is acidified by the VHA that now localises to the apical plasma membrane. Hence, induction of *vha* and *serca* transcription and sorting of VHA are the major mechanisms for calcite storage and recycling and cuticle calcification in these animals.

## 8.5 Tracheal Cuticle

The epidermal cuticle is an even structure and produces protrusions only at distinct sites where trichomes are formed. By contrast, the tracheal cuticle in insects and myriapods, that consists of an envelope, a thin epicuticle and a conspicuous procuticle, is uneven, following regular protrusions of the apical plasma membrane of tracheal epithelial cells (Fig. 8.6) (Lewis 1981; Uv and Moussian 2010). These protrusions, the taenidia, are supported by actin cables that, as spirals, run perpendicularly to the length of the tracheal tube. The formin dDAAM, that organises the polymerisation of actin bundles, has been shown to be required for this pattern in *Drosophila* (Matusek et al. 2006). In *ddaam* mutant embryos, the taenidia are disordered but present, suggesting that dDAAM is not directing taenidia formation per se but their orientation. Likewise, mutations in *polished rice* (*pri*), a polycistronic gene coding for several short peptides, dramatically impair actin organisation in epidermal and tracheal cells (Kondo et al. 2007). In consequence, epidermal denticles and taenidia are not formed. Thus, the Pri peptides can be considered as master regulators of actin-based membrane protrusions in cuticle producing epithelial cells.

These genes required for epidermal cuticle formation are evidently also acting during tracheal cuticle formation. What then makes the difference between these two types of cuticles? Possibly, chitin organisation dictates the shape of the cuticle. In the epidermal procuticle, chitin preferably adopts the Bouligand pattern, whereas in the taenidial procuticle chitin is rather amorphous. This difference may be due to cuticle-specific chitin-binding proteins that pack chitin in distinct cuticle-typical ways. It is also imaginable that chitin fibre length has an influence on cuticle shape. Consistently, the length of chitin fibres varies in the epidermis and the trachea. In the epidermal cuticle, chitin fibres are rather long, whereas in the taenidia, they are comparably short and therefore indiscernible. Length differences in turn may reflect tissue-specific differences in chitin synthase processivity, which implies tissue-specific co-factors of chitin synthesis.

Tracheael chitin synthesis is subdivided into two phases. First, a luminal chitin rod is produced that is needed for tube diameter adjustment (Devine et al. 2005; Tonning et al. 2005; Moussian et al. 2006a, b). Abrogation of chitin synthesis or organisation results in irregular and cystic tracheal lumen. The cellular mechanisms, that is, the localisation of the chitin synthesis machinery during this process has not been investigated. Second, ordered activity of chitin synthase complexes at the crests of spiral corrugations of the apical plasma membrane produces the tracheal cuticle chitin.

Less is known about the thin cuticle lining book lungs in arachnids and the epithelial cells that produce these organs. Like insect and myriapod tracheae, book lungs are ectodermal and have a lumen interspersed with lamellae that are partially stabilised by cellular structures (pillars) and cuticular trabeculae (Kamenz et al. 2005; Scholtz and Kamenz 2006). A thorough molecular and histological description of book lung development in different arachnids would teach us about the importance of the cuticle for the stability and function of internal structures.

## 8.6 Control of Cuticle Differentiation

Cuticle differentiation during embryogenesis is the last process to occur before hatching. Construction of the layers presupposes controlled expression of the components. Expression analyses of 6003 genes in *D. melanogaster* (44 % of all genes) by in situ hybridisation and microarray time-course data revealed that epidermal differentiation at the end of embryogenesis, that is, cuticle formation does not require maternal input, but is largely accomplished by zygotic factors (Tomancak et al. 2007). Several transcription factors cooperate in this mission as shown in Fig. 8.8. The evolutionary conserved CP2-type transcription factor Grainyhead (Grh), for instance, transcribes a subset of cuticle genes including *knk* in *D. melanogaster* (Gangishetti et al. 2012). The expression of another subset of cuticle genes including *serp* and *verm* is regulated by ecdysone-induced transcription factors such as βFtzF1 (Gangishetti et al. 2012). These two transcription factors cooperate in regulating the expression of a third subset of genes like *dsc73*. The relatively mild mutant phenotype of βFtzF1 mutant larvae compared to the rather strong mutant phenotype of ecdysone-deficient larvae indicates that other ecdysone-induced transcription regulators act together with βFtzF1 to drive cuticle differentiation. Indeed, βFtzF1 and DHR3 have overlapping functions during late *D. melanogaster* embryonic development (Ruaud et al. 2010). Expression of *serp* and *verm* also depends on Ribbon (Rib), a BTB/POZ domain nuclear factor with a broad range of functions during embryogenesis (Luschnig et al. 2006). Interestingly, Rib also controls the expression of *crb*, which codes the membrane-inserted determinant of apical plasma membrane, thereby indirectly influencing cuticle formation (Kerman et al. 2008). Tramtrack (Ttk), a zinc-finger transcription factor, is essential for correct cuticle architecture, as demonstrated in a

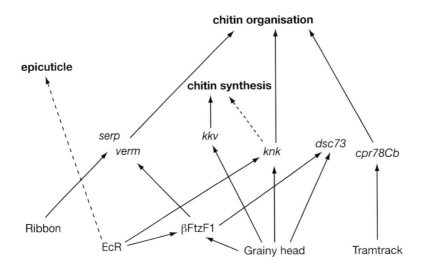

**Fig. 8.8** Transcriptional regulation of cuticle differentiation during *D. melanogaster* embryogenesis. Five transcription factors—Ribbon, the Ecdysone receptor EcR, βFtzF1, Grainyhead and Tramtrack—have been reported to date to drive expression of cuticle genes in the embryo of *D. melanogaster*. EcR, Ribbon and Tramtrack are regulators of major morphogenetic and differentiation events, whereas Grainyhead is rather specific to epidermal differentiation. The relatively mild mutant phenotype of *βFtzF1* mutant larvae suggests that it may act redundantly to other ecdysone-induced transcription factors (*see text*)

microarray experiment by its regulatory activity on several cuticle genes, including *cpr78Cb*, which codes for an R&R protein (Araujo et al. 2007; Rotstein et al. 2012). It is clear from this short summary that a master regulator of cuticle differentiation does not seem to exist.

Post-embryonic cuticle formation during moulting is more complicated. Generally, ecdysone is a central regulator of cuticle renewal (Charles 2010). In fruit flies, the ecdysone inducible transcription factor DHR38 is essential for adult cuticle formation, by regulating the transcription of cuticle genes, such as *Acp65A*, which is also regulated by the Broad complex transcription factor isoform BR–C Z1, and *Ddc*, which is essential for cuticle sclerotisation and melanisation (Cui et al. 2009; Kozlova et al. 2009). Interestingly, DHR38 also regulates the formation of glycogen (Ruaud et al. 2011), which may be the source of the chitin monomer GlcNAc. It remains to be shown whether chitin synthesis relies on DHR38 activity. In imaginal discs of *B. mori*, βFtzF1 and the ecdysone-induced Broad complex transcription factor isoform BR–C Z4 together regulate the expression of *BMWCP5*, which codes for a cuticle protein (Wang et al. 2009). Another Broad complex transcription factor isoform, BR–C Z2, has been proposed to regulate the expression of another cuticle gene, *BmorCPG11* in *B. mori* that is probably not under the control of *βFtzF1* (Ali et al. 2012). In summary, ecdysone-signalling induces the expression of cuticle genes in a locus-specific manner. Moreover, impact of ecdysone on cuticle differentiation is stage specific. For example, mutations in *dhr38* do not affect larval cuticle formation, but impair adult cuticle integrity. Moreover, mutations in the Broad complex genes in *D. melanogaster* are not larval lethal, and therefore, the respective transcription factors are not essential for cuticle formation in the larvae. Overall, regulation of cuticle differentiation is obviously a stage-specific phenomenon.

# References

Ali MS, Wang HB, Iwanaga M, Kawasaki H (2012) Expression of cuticular protein genes, BmorCPG11 and BMWCP5 is differently regulated at the pre-pupal stage in wing discs of *Bombyx mori*. Comp Biochem Physiol B 162:44–50

Andersen SO (1964) The cross-links in resilin identified as dityrosine and trityrosine. Biochim Biophys Acta 93:213–215

Andersen SO (1966) Covalent cross-links in a structural protein, resilin. Acta Physiol Scand Suppl 263:1–81

Andersen SO (2010a) Insect cuticular sclerotization: a review. Insect Biochem Mol Biol 40:166–178

Andersen SO (2010b) Studies on resilin-like gene products in insects. Insect Biochem Mol Biol 40:541–551

Andersen SO (2011) Are structural proteins in insect cuticles dominated by intrinsically disordered regions? Insect Biochem Mol Biol 41:620–627

Andersen SO (2012) Cuticular sclerotization and tanning. In: Gilbert LI (ed) Insect molecular biology and biochemistry. Elsevier, Amsterdam, pp 167–192

Andersen SO, Hojrup P, Roepstorff P (1995) Insect cuticular proteins. Insect Biochem Mol Biol 25:153–176

Anh NT, Nishitani M, Harada S, Yamaguchi M, Kamei K (2011) Essential role of Duox in stabilization of Drosophila wing. J Biol Chem 286:33244–33251

Arakane Y, Hogenkamp DG, Zhu YC, Kramer KJ, Specht CA, Beeman RW, Kanost MR, Muthukrishnan S (2004) Characterization of two chitin synthase genes of the red flour beetle, Tribolium castaneum, and alternate exon usage in one of the genes during development. Insect Biochem Mol Biol 34:291–304

Arakane Y, Muthukrishnan S, Beeman RW, Kanost MR, Kramer KJ (2005a) Laccase 2 is the phenoloxidase gene required for beetle cuticle tanning. Proc Natl Acad Sci USA 102:11337–11342

Arakane Y, Muthukrishnan S, Kramer KJ, Specht CA, Tomoyasu Y, Lorenzen MD, Kanost M, Beeman RW (2005b) The Tribolium chitin synthase genes TcCHS1 and TcCHS2 are specialized for synthesis of epidermal cuticle and midgut peritrophic matrix. Insect Mol Biol 14:453–463

Arakane Y, Specht CA, Kramer KJ, Muthukrishnan S, Beeman RW (2008) Chitin synthases are required for survival, fecundity and egg hatch in the red flour beetle, Tribolium castaneum. Insect Biochem Mol Biol 38:959–962

Arakane Y, Dixit R, Begum K, Park Y, Specht CA, Merzendorfer H, Kramer KJ, Muthukrishnan S, Beeman RW (2009) Analysis of functions of the chitin deacetylase gene family in Tribolium castaneum. Insect Biochem Mol Biol 39:355–365

Arakane Y, Dittmer NT, Tomoyasu Y, Kramer KJ, Muthukrishnan S, Beeman RW, Kanost MR (2010) Identification, mRNA expression and functional analysis of several yellow family genes in Tribolium castaneum. Insect Biochem Mol Biol 40:259–266

Araujo SJ, Cela C, Llimargas M (2007) Tramtrack regulates different morphogenetic events during Drosophila tracheal development. Development 134:3665–3676

Aravind L (2001) DOMON: an ancient extracellular domain in dopamine beta-monooxygenase and other proteins. Trends Biochem Sci 26:524–526

Ardell DH, Andersen SO (2001) Tentative identification of a resilin gene in Drosophila melanogaster. Insect Biochem Mol Biol 31:965–970

Assemat E, Bazellieres E, Pallesi-Pocachard E, Le Bivic A, Massey-Harroche D (2008) Polarity complex proteins. Biochim Biophys Acta 1778:614–630

Bartles JR (2000) Parallel actin bundles and their multiple actin-bundling proteins. Curr Opin Cell Biol 12:72–78

Bartnicki-Garcia S (2006) Chitosomes: Past, present and future. FEMS Yeast Res 6:957–965

Bauer R, Loer B, Ostrowski K, Martini J, Weimbs A, Lechner H, Hoch M (2005) Intercellular communication: the Drosophila innexin multiprotein family of gap junction proteins. Chem Biol 12:515–526

Bennet-Clark H (2007) The first description of resilin. J Exp Biol 210:3879–3881

Bouligand Y (1965) Sur une architecture torsadée répandue dans de nombreuses cuticules d'Arthropodes. C R Acad Sci (Paris) D 261:4864–4867

Burrows M (2009) A single muscle moves a crustacean limb joint rhythmically by acting against a spring containing resilin. BMC Biol 7:27

Butenandt A, Beckmann R, Hecker E (1961a) Über den Sexuallockstoff des Seidenspinners. I. Der biologische Test und die Isolierung des reinen Sexual-Lockstoffes Bombykol. Hoppe Seylers Z Physiol Chem 324:71–83

Butenandt A, Beckmann R, Stamm D (1961b) Über den Sexuallockstoff des Seidenspinners. II. Konstitution und Konfiguration des Bombykols. Hoppe Seylers Z Physiol Chem 324:84–87

Chanut-Delalande H, Fernandes I, Roch F, Payre F, Plaza S (2006) Shavenbaby couples patterning to epidermal cell shape control. PLoS Biol 4(9):e290. doi:10.1371/journal.pbio.0040290

Charles JP (2010) The regulation of expression of insect cuticle protein genes. Insect Biochem Mol Biol 40:205–213

Chaudhari SS, Arakane Y, Specht CA, Moussian B, Boyle DL, Park Y, Kramer KJ, Beeman RW, Muthukrishnan S (2011) Knickkopf protein protects and organizes chitin in the newly synthesized insect exoskeleton. Proc Natl Acad Sci USA 108:17028–17033

Chaudhari SS, Arakane Y, Specht CA, Moussian B, Kramer KJ, Muthukrishnan S, Beeman RW (2013) Retroactive Maintains Cuticle Integrity by Promoting the Trafficking of Knickkopf into the Procuticle of Tribolium castaneum. PLoS Genet 9(1):e1003268

Cheng L, Wang L, Karlsson AM (2009) Mechanics-based analysis of selected features of the exoskeletal microstructure of Popillia japonica. J Mater Res 24:3253–3267

Chihara CJ, Silvert DJ, Fristrom JW (1982) The cuticle proteins of Drosophila melanogaster: stage specificity. Dev Biol 89:379–388

Cornman RS (2010) The distribution of GYR- and YLP-like motifs in Drosophila suggests a general role in cuticle assembly and other protein–protein interactions. PLoS ONE 5(9):e12536. doi:10.1371/journal.pone.0012536

Cui HY, Lestradet M, Bruey-Sedano N, Charles JP, Riddiford LM (2009) Elucidation of the regulation of an adult cuticle gene Acp65A by the transcription factor broad. Insect Mol Biol 18:421–429

Das A, Base C, Manna D, Cho W, Dubreuil RR (2008) Unexpected complexity in the mechanisms that target assembly of the spectrin cytoskeleton. J Biol Chem 283:12643–12653

Devine WP, Lubarsky B, Shaw K, Luschnig S, Messina L, Krasnow MA (2005) Requirement for chitin biosynthesis in epithelial tube morphogenesis. Proc Natl Acad Sci USA 102:17014–17019

Doctor J, Fristrom D, Fristrom JW (1985) The pupal cuticle of Drosophila: biphasic synthesis of pupal cuticle proteins in vivo and in vitro in response to 20-hydroxyecdysone. J Cell Biol 101:189–200

Donko A, Peterfi Z, Sum A, Leto T, Geiszt M (2005) Dual oxidases. Phil Trans R Soc B 360: 2301–2308

Everaerts C, Farine JP, Cobb M, Ferveur JF (2010) Drosophila cuticular hydrocarbons revisited: mating status alters cuticular profiles. PLoS ONE 5(3):e9607. doi:10.1371/journal.pone.0009607

Fernandes I, Chanut-Delalande H, Ferrer P, Latapie Y, Waltzer L, Affolter M, Payre F, Plaza S (2010) Zona pellucida domain proteins remodel the apical compartment for localized cell shape changes. Dev Cell 18:64–76

Fristrom D, Doctor J, Fristrom JW (1986) Procuticle proteins and chitin-like material in the inner epicuticle of the Drosophila pupal cuticle. Tissue Cell 18:531–543

Fusco G, Brena C, Minelli A (2000) Cellular processes in the growth of lithobiomorph centipedes (Chilopoda: Lithobiomorpha). A cuticular view. Zool Anz 239:91–102

Futahashi R, Okamoto S, Kawasaki H, Zhong YS, Iwanaga M, Mita K, Fujiwara H (2008a) Genome-wide identification of cuticular protein genes in the silkworm, Bombyx mori. Insect Biochem Mol Biol 38:1138–1146

Futahashi R, Sato J, Meng Y, Okamoto S, Daimon T, Yamamoto K, Suetsugu Y, Narukawa J, Takahashi H, Banno Y, Katsuma S, Shimada T, Mita K, Fujiwara H (2008b) yellow and ebony are the responsible genes for the larval color mutants of the silkworm Bombyx mori. Genetics 180:1995–2005

Gangishetti U, Veerkamp J, Bezdan D, Schwarz H, Lohmann I, Moussian B (2012) Grainyhead and ecdysone cooperate during differentiation of the Drosophila skin. Insect Mol Biol 21:283–295

Gibbs A (1998) Water-proofing properties of cuticular lipids. Amer Zool 38:471–482

Gibbs AG (2011) Thermodynamics of cuticular transpiration. J Insect Physiol 57:1066–1069

Goodrich LV, Strutt D (2011) Principles of planar polarity in animal development. Development 138:1877–1892

Guan X, Middlebrooks BW, Alexander S, Wasserman SA (2006) Mutation of TweedleD, a member of an unconventional cuticle protein family, alters body shape in Drosophila. Proc Natl Acad Sci USA 103:16794–16799

Gutierrez E, Wiggins D, Fielding B, Gould AP (2007) Specialized hepatocyte-like cells regulate Drosophila lipid metabolism. Nature 445:275–280

Hagedorn M, Weihrauch D, Towle DW, Ziegler A (2003) Molecular characterisation of the smooth endoplasmic reticulum Ca($^{2+}$)-ATPase of Porcellio scaber and its expression in sternal epithelia during the moult cycle. J Exp Biol 206:2167–2175

Han Q, Fang J, Ding H, Johnson JK, Christensen BM, Li J (2002) Identification of Drosophila melanogaster yellow-f and yellow-f2 proteins as dopachrome-conversion enzymes. Biochem J 368:333–340

Harding CR (2004) The stratum corneum: structure and function in health and disease. Dermatol Ther 17(Suppl 1):6–15

Havemann J, Muller U, Berger J, Schwarz H, Gerberding M, Moussian B (2008) Cuticle differentiation in the embryo of the amphipod crustacean Parhyale hawaiensis. Cell Tissue Res 332:359–370

Hegan PS, Mermall V, Tilney LG, Mooseker MS (2007) Roles for Drosophila melanogaster myosin IB in maintenance of enterocyte brush-border structure and resistance to the bacterial pathogen Pseudomonas entomophila. Mol Biol Cell 18:4625–4636

Hijazi A, Masson W, Auge B, Waltzer L, Haenlin M, Roch F (2009) Boudin is required for septate junction organisation in Drosophila and codes for a diffusible protein of the Ly6 superfamily. Development 136:2199–2209

Howard RW, Blomquist GJ (2005) Ecological, behavioral, and biochemical aspects of insect hydrocarbons. Annu Rev Entomol 50:371–393

Iyer LM, Anantharaman V, Aravind L (2007) The DOMON domains are involved in heme and sugar recognition. Bioinformatics 23:2660–2664

Jensen JM, Proksch E (2009) The skin's barrier. G Ital Dermatol Venereol 144:689–700

Kamenz C, Dunlop JA, Scholtz G (2005) Characters in the book lungs of Scorpiones (Chelicerata, Arachnida) revealed by scanning electron microscopy. Zoomorphology 124:101–109

Kerman BE, Cheshire AM, Myat MM, Andrew DJ (2008) Ribbon modulates apical membrane during tube elongation through Crumbs and Moesin. Dev Biol 320:278–288

Kondo T, Hashimoto Y, Kato K, Inagaki S, Hayashi S, Kageyama Y (2007) Small peptide regulators of actin-based cell morphogenesis encoded by a polycistronic mRNA. Nat Cell Biol 9:660–665

Kozlova T, Lam G, Thummel CS (2009) Drosophila DHR38 nuclear receptor is required for adult cuticle integrity at eclosion. Dev Dyn 238:701–707

Kucharski R, Maleszka J, Maleszka R (2007) Novel cuticular proteins revealed by the honey bee genome. Insect Biochem Mol Biol 37:128–134

Kumar S, Molina-Cruz A, Gupta L, Rodrigues J, Barillas-Mury C (2010) A peroxidase/dual oxidase system modulates midgut epithelial immunity in *Anopheles gambiae*. Science 327:1644–1648

Laprise P, Lau KM, Harris KP, Silva-Gagliardi NF, Paul SM, Beronja S, Beitel GJ, McGlade CJ, Tepass U (2009) Yurt, Coracle, Neurexin IV and the Na(+), K(+)-ATPase form a novel group of epithelial polarity proteins. Nature 459:1141–1145

Laprise P, Paul SM, Boulanger J, Robbins RM, Beitel GJ, Tepass U (2010) Epithelial polarity proteins regulate *Drosophila* tracheal tube size in parallel to the luminal matrix pathway. Curr Biol 20:55–61

Lehane MJ, Billingsley PF (1996) Biology of the insect midgut. Chapman & Hall, London

Lewis JGE (1981) The biology of centipedes. Cambridge University Press, Cambridge

Lindgren M, Riazi R, Lesch C, Wilhelmsson C, Theopold U, Dushay MS (2008) Fondue and transglutaminase in the *Drosophila* larval clot. J Insect Physiol 54:586–592

Locke M (1961) Pore canals and related structures in insect cuticle. J Biophys Biochem Cytol 10:589–618

Locke M (1966) The structure and formation of the cuticulin layer in the epicuticle of an insect, *Calpodes ethlius* (Lepidoptera, Hesperiidae). J Morphol 118:461–494

Locke M (2001) The Wigglesworth Lecture: Insects for studying fundamental problems in biology. J Insect Physiol 47:495–507

Locke M (2003) Surface membranes, Golgi complexes, and vacuolar systems. Annu Rev Entomol 48:1–27

Luschnig S, Batz T, Armbruster K, Krasnow MA (2006) Serpentine and vermiform encode matrix proteins with chitin binding and deacetylation domains that limit tracheal tube length in *Drosophila*. Curr Biol 16:186–194

Lyons RE, Nairn KM, Huson MG, Kim M, Dumsday G, Elvin CM (2009) Comparisons of recombinant resilin-like proteins: repetitive domains are sufficient to confer resilin-like properties. Biomacromolecules 10:3009–3014

Lyons RE, Wong DC, Kim M, Lekieffre N, Huson MG, Vuocolo T, Merritt DJ, Nairn KM, Dudek DM, Colgrave ML, Elvin CM (2011) Molecular and functional characterisation of resilin across three insect orders. Insect Biochem Mol Biol 41:881–890

Madison KC (2003) Barrier function of the skin: La raison d'etre of the epidermis. J Invest Dermatol 121:231–241

Malencik DA, Anderson SR (2003) Dityrosine as a product of oxidative stress and fluorescent probe. Amino Acids 25:233–247

Martinek N, Shahab J, Saathoff M, Ringuette M (2008) Haemocyte-derived SPARC is required for collagen-IV-dependent stability of basal laminae in *Drosophila* embryos. J Cell Sci 121:1671–1680

Matsumoto S (2010) Molecular mechanisms underlying sex pheromone production in moths. Biosci Biotechnol Biochem 74:223–231

Matusek T, Djiane A, Jankovics F, Brunner D, Mlodzik M, Mihaly J (2006) The *Drosophila* formin DAAM

regulates the tracheal cuticle pattern through organizing the actin cytoskeleton. Development 133:957–966

Meissner D, Odman-Naresh J, Vogelpohl I, Merzendorfer H (2010) A novel role of the yeast CaaX protease Ste24 in chitin synthesis. Mol Biol Cell 21:2425–2433

Merzendorfer H (2006) Insect chitin synthases: a review. J Comp Physiol B 176:1–15

Merzendorfer H (2011) The cellular basis of chitin synthesis in fungi and insects: common principles and differences. Eur J Cell Biol 90:759–769

Michels J, Gorb SN (2012) Detailed three-dimensional visualization of resilin in the exoskeleton of arthropods using confocal laser scanning microscopy. J Microsc 245:1–16

Moussian B (2008) The role of GlcNAc in formation and function of extracellular matrices. Comp Biochem Physiol B 149:215–226

Moussian B (2010) Recent advances in understanding mechanisms of insect cuticle differentiation. Insect Biochem Mol Biol 40:363–375

Moussian B, Schwarz H, Bartoszewski S, Nusslein-Volhard C (2005a) Involvement of chitin in exoskeleton morphogenesis in *Drosophila melanogaster*. J Morphol 264:117–130

Moussian B, Soding J, Schwarz H, Nusslein-Volhard C (2005b) Retroactive, a membrane-anchored extracellular protein related to vertebrate snake neurotoxin-like proteins, is required for cuticle organization in the larva of *Drosophila melanogaster*. Dev Dyn 233:1056–1063

Moussian B, Seifarth C, Muller U, Berger J, Schwarz H (2006a) Cuticle differentiation during Drosophila embryogenesis. Arthropod Struct Dev 35:137–152

Moussian B, Tang E, Tonning A, Helms S, Schwarz H, Nusslein-Volhard C, Uv AE (2006b) *Drosophila* Knickkopf and Retroactive are needed for epithelial tube growth and cuticle differentiation through their specific requirement for chitin filament organization. Development 133:163–171

Moussian B, Veerkamp J, Muller U, Schwarz H (2007) Assembly of the *Drosophila* larval exoskeleton requires controlled secretion and shaping of the apical plasma membrane. Matrix Biol 26:337–347

Nagaraj R, Adler PN (2012) Dusky-like functions as a Rab11 effector for the deposition of cuticle during *Drosophila* bristle development. Development 139:906–916

Nelson DR, Charlet LD (2003) Cuticular hydrocarbons of the sunflower beetle, *Zygogramma exclamationis*. Comp Biochem Physiol B 135:273–284

Nelson DR, Tissot M, Nelson LJ, Fatland CL, Gordon DM (2001) Novel wax esters and hydrocarbons in the cuticular surface lipids of the red harvester ant, *Pogonomyrmex barbatus*. Comp Biochem Physiol B 128:575–595

Nelson DR, Olson DL, Fatland CL (2002) Cuticular hydrocarbons of the flea beetles, *Aphthona lacertosa* and *Aphthona nigriscutis*, biocontrol agents for leafy spurge (*Euphorbia esula*). Comp Biochem Physiol B 133:337–350

Nelson DR, Adams TS, Fatland CL (2003) Hydrocarbons in the surface wax of eggs and adults of the Colorado potato beetle, *Leptinotarsa decemlineata*. Comp Biochem Physiol B: Biochem Mol Biol 134:447–466

Nelson DR, Hines H, Stay B (2004) Methyl-branched hydrocarbons, major components of the waxy material coating the embryos of the viviparous cockroach *Diploptera punctata*. Comp Biochem Physiol B 138:265–276

Neville AC (1965a) Chitin lamellogenesis in locust cuticle. Q J Microsc Sci 106:269–286

Neville AC (1965b) Circadian organization of chitin in some insect skeletons. Q J Microsc Sci 106:315–325

Neville AC (1975) Biology of the arthropod cuticle. Springer, Berlin

Neville AC, Luke BM (1969a) A two-system model for chitin-protein complexes in insect cuticles. Tissue Cell 1:689–707

Neville AC, Luke BM (1969b) Molecular architecture of adult locust cuticle at the electron microscope level. Tissue Cell 1:355–366

Neville AC, Parry DA, Woodhead-Galloway J (1976) The chitin crystallite in arthropod cuticle. J Cell Sci 21:73–82

Nilton A, Oshima K, Zare F, Byri S, Nannmark U, Nyberg KG, Fehon RG, Uv AE (2010) Crooked, coiled and crimpled are three Ly6-like proteins required for proper localization of septate junction components. Development 137:2427–2437

Niu BL, Shen WF, Liu Y, Weng HB, He LH, Mu JJ, Wu ZL, Jiang P, Tao YZ, Meng ZQ (2008) Cloning and RNAi-mediated functional characterization of MaLac2 of the pine sawyer, *Monochamus alternatus*. Insect Mol Biol 17:303–312

Norum M, Tang E, Chavoshi T, Schwarz H, Linke D, Uv A, Moussian B (2010) Trafficking through COPII stabilises cell polarity and drives secretion during *Drosophila* epidermal differentiation. PLoS ONE 5(5):e10802

Odier A (1823) Mémoires sur la composition chimique des parties cornées des insectes. Mém Soc Hist Nat Paris 1:29–42

Oshima K, Fehon RG (2011) Analysis of protein dynamics within the septate junction reveals a highly stable core protein complex that does not include the basolateral polarity protein Discs large. J Cell Sci 124:2861–2871

Parra-Peralbo E, Culi J (2011) *Drosophila* lipophorin receptors mediate the uptake of neutral lipids in oocytes and imaginal disc cells by an endocytosis-independent mechanism. PLoS Genet 7(2):e1001297

Patel S, Nelson DR, Gibbs AG (2001) Chemical and physical analyses of wax ester properties. J Insect Sci 1:4

Pryor MG (1940) On the hardening of cuticle of insects. Proc R Soc B 128:393–407

Pryor MG, Russell PB, Todd AR (1946) Protocatechuic acid, the substance responsible for the hardening of the cockroach ootheca. Biochem J 40:627–628

Pryor MG, Russell PB, Todd AR (1947) Phenolic substances concerned in hardening the insect cuticle. Nature 159:399

Qin G, Lapidot S, Numata K, Hu X, Meirovitch S, Dekel M, Podoler I, Shoseyov O, Kaplan DL (2009) Expression, cross-linking, and characterization of recombinant chitin binding resilin. Biomacromolecules 10:3227–3234

Raabe D, Romano P, Sachs C, Al-Sawalmih A, Brokmeier H-G, Yi S-B, Servos G, Hartwig HG (2005) Discovery of a honeycomb structure in the twisted plywood patterns of fibrous biological nanocomposite tissue. J Cryst Growth 283:1–7

Rebers JE, Willis JH (2001) A conserved domain in arthropod cuticular proteins binds chitin. Insect Biochem Mol Biol 31:1083–1093

cRezende GL, Martins AJ, Gentile C, Farnesi LC, Pelajo-Machado M, Peixoto AA, Valle D (2008) Embryonic desiccation resistance in *Aedes aegypti*: presumptive role of the chitinized serosal cuticle. BMC Dev Biol 8:82

Riedel F, Vorkel D, Eaton S (2011) Megalin-dependent yellow endocytosis restricts melanization in the *Drosophila* cuticle. Development 138:149–158

Rotstein B, Molnar D, Adryan B, Llimargas M (2012) Tramtrack is genetically upstream of genes controlling tracheal tube size in *Drosophila*. PLoS ONE 6(12):e28985

Ruaud AF, Lam G, Thummel CS (2010) The *Drosophila* nuclear receptors DHR3 and beta FTZ-F1 control overlapping developmental responses in late embryos. Development 137:123–131

Ruaud AF, Lam G, Thummel CS (2011) The *Drosophila* NR4A nuclear receptor DHR38 regulates carbohydrate metabolism and gycogen storage. Mol Endocrinol 25:83–91

Sass M, Kiss A, Locke M (1993) Classes of integument peptides. Insect Biochem Mol Biol 23:845–857

Sass M, Kiss A, Locke M (1994) Integument and hemocyte peptides. J Insect Physiol 40:407–421

Scholtz G, Kamenz C (2006) The book lungs of Scorpiones and Tetrapulmonata (Chelicerata, Arachnida): evidence for homology and a single terrestrialisation event of a common arachnid ancestor. Zoology (Jena) 109:2–13

Schulte J, Charish K, Que J, Ravn S, MacKinnon C, Auld VJ (2006) Gliotactin and Discs large form a protein complex at the tricellular junction of polarized epithelial cells in *Drosophila*. J Cell Sci 119:4391–4401

Shaik KS, Pabst M, Schwarz H, Altmann F, Moussian B (2011) The Alg5 ortholog Wollknauel is essential for correct epidermal differentiation during *Drosophila* late embryogenesis. Glycobiology 21:743–756

Shaik KS, Meyer F, Vazquez AV, Flotenmeyer M, Cerdan ME, Moussian B (2012) delta-Aminolevulinate synthase is required for apical transcellular barrier formation in the skin of the *Drosophila* larva. Eur J Cell Biol 91:204–215

Shibata T, Ariki S, Shinzawa N, Miyaji R, Suyama H, Sako M, Inomata N, Koshiba T, Kanuka H, Kawabata S (2010) Protein crosslinking by transglutaminase controls cuticle morphogenesis in *Drosophila*. PLoS ONE 5(10):e13477. doi:10.1371/journal.pone.0013477

Silvert DJ, Doctor J, Quesada L, Fristrom JW (1984) Pupal and larval cuticle proteins of *Drosophila melanogaster*. Biochemistry 23:5767–5774

Snyder M, Hunkapiller M, Yuen D, Silvert D, Fristrom J, Davidson N (1982) Cuticle protein genes of *Drosophila*: Structure, organization and evolution of four clustered genes. Cell 29:1027–1040

Suderman RJ, Dittmer NT, Kanost MR, Kramer KJ (2006) Model reactions for insect cuticle sclerotization: cross-linking of recombinant cuticular proteins upon their laccase–catalyzed oxidative conjugation with catechols. Insect Biochem Mol Biol 36:353–365

Suderman RJ, Dittmer NT, Kramer KJ, Kanost MR (2010) Model reactions for insect cuticle sclerotization: participation of amino groups in the cross-linking of *Manduca sexta* cuticle protein MsCP36. Insect Biochem Mol Biol 40:252–258

Sugumaran M (2009) Complexities of cuticular pigmentation in insects. Pigment Cell Melanoma Res 22:523–525

Tang L, Liang J, Zhan Z, Xiang Z, He N (2010) Identification of the chitin-binding proteins from the larval proteins of silkworm, *Bombyx mori*. Insect Biochem Mol Biol 40:228–234

Tepass U, Theres C, Knust E (1990) Crumbs encodes an EGF-like protein expressed on apical membranes of *Drosophila* epithelial cells and required for organization of epithelia. Cell 61:787–799

Tepass U, Gruszynski-DeFeo E, Haag TA, Omatyar L, Torok T, Hartenstein V (1996) Shotgun encodes Drosophila E-cadherin and is preferentially required during cell rearrangement in the neurectoderm and other morphogenetically active epithelia. Genes Dev 10:672–685

Thomas GH, Kiehart DP (1994) Beta heavy-spectrin has a restricted tissue and subcellular distribution during *Drosophila* embryogenesis. Development 120:2039–2050

Tillman JA, Seybold SJ, Jurenka RA, Blomquist GJ (1999) Insect pheromones–an overview of biosynthesis and endocrine regulation. Insect Biochem Mol Biol 29:481–514

Tilney LG, Connelly PS, Vranich KA, Shaw MK, Guild GM (1998) Why are two different cross-linkers necessary for actin bundle formation in vivo and what does each cross-link contribute? J Cell Biol 143:121–133

Tilney LG, Connelly PS, Vranich KA, Shaw MK, Guild GM (2000) Regulation of actin filament cross-linking and bundle shape in *Drosophila* bristles. J Cell Biol 148:87–100

Tilney LG, Connelly PS, Guild GM (2004) Microvilli appear to represent the first step in actin bundle formation in *Drosophila* bristles. J Cell Sci 117:3531–3538

Togawa T, Nakato H, Izumi S (2004) Analysis of the chitin recognition mechanism of cuticle proteins from the soft cuticle of the silkworm, *Bombyx mori*. Insect Biochem Mol Biol 34:1059–1067

Tomancak P, Berman BP, Beaton A, Weiszmann R, Kwan E, Hartenstein V, Celniker SE, Rubin GM (2007) Global analysis of patterns of gene expression during *Drosophila* embryogenesis. Genome Biol 8:R145

Tonning A, Hemphala J, Tang E, Nannmark U, Samakovlis C, Uv A (2005) A transient luminal chitinous matrix is required to model epithelial tube diameter in the *Drosophila* trachea. Dev Cell 9:423–430

Tonning A, Helms S, Schwarz H, Uv AE, Moussian B (2006) Hormonal regulation of mummy is needed for apical extracellular matrix formation and epithelial morphogenesis in *Drosophila*. Development 133:331–341

Uv A, Moussian B (2010) The apical plasma membrane of *Drosophila* embryonic epithelia. Eur J Cell Biol 89:208–211

Vincent JF, Wegst UG (2004) Design and mechanical properties of insect cuticle. Arthropod Struct Dev 33:187–199

Vrkoslav V, Muck A, Cvacka J, Svatos A (2010) MALDI imaging of neutral cuticular lipids in insects and plants. J Am Soc Mass Spectrom 21:220–231

Wang HB, Nita M, Iwanaga M, Kawasaki H (2009) betaFTZ-F1 and Broad-Complex positively regulate the transcription of the wing cuticle protein gene, BMWCP5, in wing discs of *Bombyx mori*. Insect Biochem Mol Biol 39:624–633

Weis-Fogh T (1960) A rubber-like protein in insect cuticle. J Exp Biol 37:889–907

Wigglesworth VB (1933) The physiology of the cuticle and of ecdysis in *Rhodnius prolixus* (Triatomidae, Hemiptera); with special reference to the oenocytes and function of the dermal glands. Quart J Microscop Soc 76:270–318

Wigglesworth VB (1970) Structural lipids in the insect cuticle and the function of the oenocytes. Tissue Cell 2:155–179

Wigglesworth VB (1975) Distribution of lipid in the lamellate endocuticle of *Rhodnius prolixus* (Hemiptera). J Cell Sci 19:439–457

Wigglesworth VB (1985) Sclerotin and lipid in the waterproofing of the insect cuticle. Tissue Cell 17:227–248

Wigglesworth VB (1990) The distribution, function and nature of cuticulin in the insect cuticle. J Insect Physiol 36:307–313

Willis JH (2010) Structural cuticular proteins from arthropods: annotation, nomenclature, and sequence characteristics in the genomics era. Insect Biochem Mol Biol 40:189–204

Wittkopp PJ, True JR, Carroll SB (2002) Reciprocal functions of the *Drosophila* yellow and ebony proteins in the development and evolution of pigment patterns. Development 129:1849–1858

Wolfgang WJ, Fristrom D, Fristrom JW (1986) The pupal cuticle of *Drosophila*: differential ultrastructural immunolocalization of cuticle proteins. J Cell Biol 102:306–311

Wong DC, Pearson RD, Elvin CM, Merritt DJ (2012) Expression of the rubber-like protein, resilin, in developing and functional insect cuticle determined using a *Drosophila* anti-Rec 1 resilin antibody. Dev Dyn 241:333–339

Yoon CS, Hirosawa K, Suzuki E (1997) Corneal lens secretion in newly emerged *Drosophila melanogaster* examined by electron microscope autoradiography. J Electron Microsc (Tokyo) 46:243–246

Ziegler A, Weihrauch D, Towle DW, Hagedorn M (2002) Expression of $Ca^{2+}$-ATPase and $Na^+/Ca^{2+}$-exchanger is upregulated during epithelial $Ca^{2+}$ transport in hypodermal cells of the isopod *Porcellio scaber*. Cell Calcium 32:131–141

Ziegler A, Weihrauch D, Hagedorn M, Towle DW, Bleher R (2004) Expression and polarity reversal of V-type $H^+$-ATPase during the mineralization-demineralization cycle in *Porcellio scaber* sternal epithelial cells. J Exp Biol 207:1749–1756

Zimoch L, Merzendorfer H (2002) Immunolocalization of chitin synthase in the tobacco hornworm. Cell Tissue Res 308:287–297

# Arthropod Segmentation and Tagmosis — 9

Giuseppe Fusco and Alessandro Minelli

## Contents

## 9.1 Basic Features of Arthropod Body Architecture

According to a well-consolidated tradition, the body of arthropods is described in terms of segments and tagmata. Even the oldest names for these animals, Aristotle's ἔντομα (*entoma*, internally (sub)divided) and Linnaeus' Latin equivalent *Insecta*, now restricted to one of the major arthropod subgroups, already referred to the modular organization of the body. In the idealistic perspective of the past, this trait, more than the presence of articulated appendages to which the current name of arthropods refers, was considered the defining attribute for the body plan of these animals.

Accordingly, the arthropod body is traditionally interpreted as comprising an antero-posterior array of potentially articulated modules, the *segments*, with blocks of contiguous segments, morphologically or functional integrated, forming a smaller number of body regions, or *tagmata* (singular, *tagma*). Equivalent, at least in terms of descriptive morphology, is the alternative view of the arthropod body as divided into a few regions along its main axis, each region being further divided into a number of segments. Departures from this scheme are reckoned to be evolutionarily derived.

G. Fusco (✉) · A. Minelli
Department of Biology, University of Padova, Via U. Bassi 58 B, I 35131, Padova, Italy
e-mail: giuseppe.fusco@unipd.it

A. Minelli
e-mail: alessandro.minelli@unipd.it

A. Minelli et al. (eds.), *Arthropod Biology and Evolution*,
DOI: 10.1007/978-3-642-36160-9_9, © Springer-Verlag Berlin Heidelberg 2013

This basic architecture has been often credited with being at the heart of the evolutionary success (however measured) of the arthropods. The modular organization of their body is reputedly highly evolvable through a mechanism of 'multiplication and change' of modules that allows acquisition of new functions without losing the original ones. This generic evolutionary mechanism, that also applies at other levels of biological organization (e.g., at the level of genes and genomes), would open doors to an apparently unbounded increase in morphological complexity and functional specialization. This view of arthropod morphology and morphological evolution is well captured by the metaphor of 'the arthropods as Swiss Army knives,' where each species is equipped with a unique set of specialized tools, corresponding to the series of its functional body modules (e.g., Ruppert et al. 2004, or, for an educational animation, see http://shapeoflife.org/video/animation/arthropod-animation-swiss-army-knife). This view can be roughly classified among the 'internalist' explanations for arthropod evolution and diversification, that is, among those based on intrinsic qualities of arthropod biology, including body organization, rather than those based on the interaction with other components of the ecosystems of which they are part.

This commonplace scenario of arthropod body architecture and morphological evolution has evident limitations. First, the apparently obvious evolutionary potential of a modular organization of the body rests on patterns of morphological integration and developmental modularity (Klingenberg 2008) that allow the expression of such a potential. Thus, beyond considerations on the 'technologically clever' body organization of arthropods, modern discussion around this theme tends to be less superficial and more based on the properties of the developmental systems in generating phenotypic variation or on the character of the genotype-phenotype map in this group (see Chap. 18). Second, most of the discussion on arthropod morphological evolvability and evolution rests on an implicit but unnecessary

assumption: that the wide spectrum of complex and diverse arthropod forms evolved from the body structure of a homonomously segmented legless worm-like ancestor (e.g., Snodgrass 1935; Raff and Kaufman 1983). This is in part the result of an outdated phylogenetic view of arthropods and annelids as close relatives within a superphylum Articulata, but has also its roots in a more basic unwarranted prejudice, namely that evolution always, or preferentially, moves from the simple to the complex, for example, according to the so-called Williston's rule (critically discussed, for example, in Minelli 2003). Irrespective of the phylogenetic scenario, the hypothesis of a protostome-grade arthropod ancestor that was segmented but completely unpatterned along its antero-posterior (A-P) body axis conflicts with evidence from comparative morphology and developmental genetics alike, as A-P trunk patterning controlled by *Hox* gene expression is older than arthropod origins (Minelli and Fusco 2005).

Aside from specific phylogenetic hypotheses and post hoc evolutionary interpretations, this chapter will illustrate different 'dimensions' of the 'space' of arthropod segmentation and tagmosis, beyond simplifying idealizations and with special focus on the points where textbook simplifications (e.g., insects have eleven abdominal segments) conflict with factual diversity (many holometabolous insects have ten or less abdominal segments) and where too superficial notions of segment and tagma conflict with observed morphology, as in the instances of segmental mismatch (see Sect. 9.3.3).

Uncritical use of a descriptive framework easily obscures homologies and phylogenetic relationships and can be an obstacle to understanding the evolution of phenotypes and developmental processes. As a general rule, a descriptive model for a given form cannot be used to address questions on the generative processes producing the same form (Fusco 2008). On the contrary, a less idealized view of arthropod morphology can help framing more insightful questions about arthropod development and evolution.

## 9.2 Subject Circumscription and Operational Definitions

Arthropod segmentation and tagmosis are currently discussed in a wide range of contexts, from developmental biology to phylogeny and taxonomy, that tangibly affect even the most fundamental concepts of segment and tagma. Also, arthropod species richness and diversity have prompted the development of different independent traditions of study targeted on individual subtaxa, conspicuously reflected in the disparity of morphological nomenclature related to segmentation and tagmosis in the different groups. Some preliminary remarks are thus in order.

### 9.2.1 Subject Circumscription

The terms 'segmentation' and 'tagmosis' are used to describe morphological features (a form of body symmetry and a form of body organization, respectively) as well as the developmental processes that generate them. These different concepts are obviously related, but should not be confused (Fusco 2008).

As developmental processes, segmentation and tagmosis are not restricted to embryogenesis, as these aspects of body organization can also change through post-embryonic life. In pycnogonids, in most actinotrichid mites, in myriapods to the exclusion of geophilomorph and scolopendromorph centipedes, in most crustaceans and in proturans, the final segmental composition of the body is attained during post-embryonic life through a series of moults. In some cases, as in many millipedes and perhaps in remipedes among the crustaceans, new trunk segments are added throughout life. In many holometabolous insects, there is a reduction in the number of abdominal segments from the larva to the adult. Similarly, body regionalization can vary during ontogeny, especially when this includes metamorphosis. This chapter is restricted to patterns of segmentation and tagmosis in adult forms, while some developmental aspects are discussed in Chaps. 4 and 5.

Assuming the homology of the most anterior six body segments in all extant arthropods, as suggested by *Hox* genes expression domains (Manuel et al. 2006; Eriksson et al. 2010), the anterior border of the following 'post-cephalic' region coincides with that of the trunk, thorax or pereion in mandibulates, and with the region posterior to the second or third leg-bearing segment in pycnogonids and chelicerates, respectively. This chapter is mainly focused on the post-cephalic section of the body, while head segmentation, including segmentation of the cephalic region of pycnogonid and chelicerate prosoma, is discussed in Chap. 10 in the context of head organization.

Finally, highly derived morphologies, often associated with parasitic or sessile life styles, are considered here only marginally. The different forms of deviation from the organization of their free-living (or non-sessile) ancestors are so diverse and complexly varied that they simply cannot be summarized here. For these groups, we refer to zoology treatises and taxon-specific literature, although the topics discussed in this chapter can provide a critical background to revisit the evidence therein. A list of morphologically highly derived arthropods includes pentastomids, branchiurans, rhizocephalans, caprellid amphipods, parasitic copepods, endoparasitic ascothoracidans, parasitic isopods, parthenogenetic female tantulocarids, female scale insects and many mites.

### 9.2.2 Segment: Definition and Use of the Term

In many bilaterians, including the arthropods, the body has the appearance of a series of more or less differentiated, repeated anatomical units (Minelli and Fusco 2004). In arthropods, these units are called *segments* or, less frequently, *metameres* or *somites*. However, although for many descriptive purposes it is useful to consider the arthropod body as series of modules, under more accurate scrutiny this is possible only up to a point, and that point is not the same for all arthropod taxa. Not all putatively segmental structures (especially those of internal anatomy) are in register, as they can have different period, phase or domain of extent along the main body axis, or even more complex arrangements.

A more realistic and accurate depiction of arthropod body organization is obtained by dissociating the serial homology of individual periodic structures, or segmentation, from the concept of the segment as a body module (e.g., Budd 2001; Minelli and Fusco 2004; Fusco 2005, 2008). A segmental pattern can be defined as the occurrence of serially homologous structures along a specified body axis or direction (Fusco 2005). Accordingly, we will indicate a structure (e.g., the ventral nerve cord) with a periodic organization as a *segmental structure* and each repetitive element (e.g., a neuromere) as a *segmental* (or *serial*) *element*.

That said, we will take a pragmatic approach here (Minelli et al. 2010). When not otherwise specified, segments are intended as *reference body landmarks*, coinciding with the body units traditionally recognized by descriptive morphology within each taxon. These are generally defined on the basis of a small number of 'leading' segmental structures with concordant period (e.g., tergites and sternites) or on the basis of the assumed primitive segmental arrangement of the body. This is exactly how to read a sentence such as 'in most polydesmidan millipedes, there is one ozopore on each side of trunk segments V, VII, IX, X, XII, XIII and XV–XVIII, while genital openings are on trunk segment III.' The main trunk sclerites provide a positional grid for locating segmental structures with a different periodicity, as well as for non-segmental structures. In this framework, segmental boundaries are conventionally established by the actual or putative position of the arthrodial membranes between the main dorsal and/or ventral sclerites. Thus, periodic structures developing within such borders are said to be *segmental* (e.g., the nerves originating from the ventral ganglia of a given trunk segment that innervate structures of the same segment), while those developing across them are said *intersegmental* (e.g., the nerves originating from the ventral ganglia of a given trunk segment that innervate structures in other segments). However, it must be remarked that while anatomical elements (e.g., sclerites) or gene expression domains can have objective boundaries, segments, that is, the sections into which the main body axis would ideally be subdivided, can have only conventionally stipulated boundaries, since, geometrically, the number of alternative repeating motifs for one and the same periodic pattern is actually infinite (Fusco 2008).

To further complicate the picture, beyond these basic questions, there are different traditions in the way the segmental composition of the trunk is described and interpreted in different taxonomic groups. For instance, in centipedes, the segmental composition of the trunk is generally reported in terms of the number of leg-bearing segments (LBS), excluding one anterior segment that bears a pair of maxillipedes, and a posterior ano-genital region of uncertain segmental composition (possibly, 2–3 segments plus the telson; Minelli and Koch 2011). In groups with marked segmental mismatch (see Sect. 9.3.3), the term segment is often carefully avoided. In their review of millipede post-embryonic segmentation, Enghoff et al. (1993) decided to 'avoid the term "segment" as far as possible,' preferring more precise, although less comparable, terms for specific serial structures, such as those of tergite (T), pleurotergite (PT), ring (R) (depending on the taxon-specific organization of the main trunk sclerites) and leg pair (LP). Analogously, for a proper description of segmentation in notostracans, Linder (1952) explicitly replaced the term 'segment' with 'ring' for the units of the dorsal segmental series and 'leg pair' for the ventral ones.

Finally, a remark about the terminal trunk piece, the *telson*: the telson is generally defined as the most posterior piece of the body that, however, does not count as a 'true' segment. The non-segmental nature of the telson (and eventually also of the *acron*, the most anterior piece of the head) would be supported by the previously mentioned presumed close relationship between arthropods and annelids. In that framework, acron and telson in arthropods would be homologous to *prostomium* and *pygidium* in annelids, respectively, which, in contrast to 'true' segments, would not have a teloblastic origin. Irrespective of the question whether annelids and arthropods evolved segmentation independently or not, certainly a teloblastic mode of segmentation does not apply to the majority of arthropods (see Chap. 4), thus the special status of the telson, beyond being the posteriormost piece of the trunk, is doubtful at least. Actually, the telson can be post-anal (e.g., xiphosurans, scorpions, palpigrads, thelyphonids, schizomids) or peri-anal (most arthropods), conspicuous (e.g., many chelicerates, many malacostracans), inconspicuous (e.g., myriapods) or absent (e.g., insects). It can also take different names in different taxa: 'anal somite' in crustaceans (Brusca and Brusca 2002) and 'pygidium' in pauropods (Scheller 2011). The latter is an unfortunate choice, given that pygidium is also used for the terminal multi-segmented body region of trilobites (see Hughes et al. 2006) and for the last visible tergite of the beetle abdomen (especially in the cases where this is not covered by the elytra).

## 9.2.3 Tagma: Definition and Use of the Term

The body of many bilaterian taxa is traditionally described as divided into a small number of regions, or *tagmata*, along the main axis. This applies to segmented and non-segmented bilaterians alike; however, in the case of segmented animals, as in the arthropods, tagmata are defined as extending over a certain range of body segments, so that their boundaries are generally coincident with the boundary between two contiguous segments. In other words, a tagma corresponds to a specific set of contiguous segments, and, for a given scheme of tagmosis, a segment exclusively belongs to one tagma or another. This mode of body regionalization is clearly based on the concept of segments as fundamental body modules (see Sect. 9.2.2), and tagmosis represents a form of higher-level modularity along the main body axis.

There is little agreement on how tagmata should be defined and their boundaries characterized. Tagmata should in some way capture the articulation of the main body axis into main morpho-functional units. However, more often than not, units of descriptive morphology (e.g., the head in a crayfish) do not exactly overlap with those of functional biology (e.g., a crayfish's cephalothorax), not to say with observed or inferred developmental units (e.g., a crayfish's 'naupliar head' region in the embryonic germ-band) (Minelli and Fusco 1995).

Instead of representing actual body structures, tagmata, similar to segments, are more often used as units of description. Thus, sacrificing the appreciation of diversity to the benefit of ease of comparison, it is common practice to recognize a 'typical' tagmatic organization for each of the main arthropod taxa in traditional classifications, to eventually specify with further detail the nature and composition of each tagma, or to single out the exceptional body organization of clades with derived tagmosis. For instance, the insect thorax is defined as a three-segment leg-bearing tagma following the head and preceding the abdomen. However, in apocritan hymenopterans, this region incorporates four segmental units in front of the typical 'wasp waist,' despite the fact that it bears no more than the usual three pairs of legs. Rather than saying that the thorax of these insects comprises four segments, the standard description of this body region is that the first segment of the abdomen (here called the propodeum) has been added to the thoracic segments, to form the so-called *mesosoma*, an apocritan taxon-specific tagma.

Taxon-specific tagmata, often downgraded to *pseudotagmata* to save the accepted primacy of the 'true tagmata', provide ad hoc compromise solutions, adopted by arthropod comparative anatomists to account for a disparity of body architectures that cannot be shoehorned into the too rigid categories of classical tagmata. This is the case for the *gnathosoma* and *idiosoma* in mites and ricinuleans, the *proterosoma* and *hysterosoma* in actinotrichid mites, and the *prosoma* and *urosoma* in copepods. Parallel solutions are offered by the identification of taxon-specific *subtagmata*, like scorpions' *mes-osoma* and *metasoma* (parts of the opisthosoma), many holometabolans' *preabdomen* and *postabdomen* (parts of the abdomen), or amphipods' *pleosoma* and *urosoma* (parts of the pleon). However, it must remarked that these terms, used for descriptive purposes in groups with different traditions of study, provide a way of avoiding problems in tracing homologies, rather than a way to address them (Minelli 2003).

Admittedly, the concept of tagma is to a large extent arbitrary, and this is clearly reflected in the uncertainty of the morphological nomenclature in certain animal groups, where some zoologists recognize a smaller, others a larger number of body regions. For example, in millipedes, most authors recognize only one post-cephalic tagma, the trunk, but some (e.g., Verhoeff 1926) recognize a 'thorax' that comprises the first few segments with no more than one pair of legs each, and an 'abdomen' for the remaining set of 'diplosegments' with two pair of legs each, but eventually disregarding the presence of one or more apodous units at the posterior end. Another example is the notostracan 'trunk' of some authors, that others would divide into a 'thorax' of 11 segments with one pair of appendages each, followed by an 'abdomen' with a number of segments that varies both between and within species and is affected by marked dorsoventral segmental mismatch (Linder 1952). In copepods, two overlapping tagmatic subdivisions of the trunk are acknowledged in the specialist literature: one is based on the presence of limbs, which separates a limb-bearing thorax (including a posterior genital segment) from an abdomen without appendages to the exclusion of the terminal caudal furca; the other is based on the position of a conspicuous dorsal hinge joint, posterior to thoracic segment V (podoplean position, most taxa) or posterior to thoracic segment VI (gymnoplean position, Platycopioida and Calanoida), which separates the *prosoma* from the *urosoma* (Huys and Boxshall 1991). Other inconsistencies emerge when different authors locate differently the border between the same two tagmata in the same taxon, as in the case of

anostracans (abdomen of 6 vs. 8 segments; see McLaughlin 1980; Westheide and Rieger 2007), mystacocarids (abdomen of 3 vs. 5 segments; see Brenneis and Richter 2010) and cephalocarids (abdomen of 10 vs. 11 segments; see Olesen et al. 2011).

Pragmatically, as in the case of segments, we will follow the common practice of using the 'typical' tagmatic organization for each main arthropod taxon of the traditional classifications as a descriptive reference basis for discussion, without any implication for homology (but see Sect. 9.4.3). The body is divided into *prosoma* and *opisthosoma* in most chelicerates; *head* and *trunk* in myriapods and some crustaceans; *head*, *pereion* and *pleon* in malacostracans; *head*, *thorax* and *abdomen* in the remaining crustaceans and in the hexapods. In the latter two assemblages, the term *trunk* is used to indicate the 'supertagma' formed by all post-cephalic segments (thorax/pereion plus abdomen/pleon).

## 9.3 The Space of Variation of Arthropod Segmentation

With some caution about what to count as a segment and how to compare segments between distantly related arthropods, and recalling that segments in the same (ordinal) post-cephalic position in two different individuals of two different species (or in some cases even of the same species) are not necessarily homologous, we can attempt a survey of arthropod diversity regarding this aspect of their morphology. When not otherwise specified, the telson, if present, is excluded from the count.

### 9.3.1 Interspecific Variation in the Number of Post-cephalic Segments

The number of adult post-cephalic segments is highly diverse within the Arthropoda (Table 9.1), ranging from 6 in acrothoracican cirripeds (Westheide and Rieger 2007) to more than 380 in the siphonophorid millipede *Illacme plenipes* (Hoffman 1982). The number of post-cephalic segments in the common ancestor of crown-group arthropods is difficult to estimate, but on the basis of the very rough observation that the most basal groups of all the main arthropod assemblages have a number of post-cephalic segments in the range of 11–18, and disregarding the very derived morphology of pycnogonids, irrespectively of their phylogenetic relationship with the other arthropods (see Chap. 2), this range can also be tentatively considered to represent a primitive state for the whole group.

The number of segments is generally stable or shows little variability within high-rank clades, with the notable exceptions of (1) millipedes, which, varying between 14 and 380 trunk segments (counting their ventral segmental units) almost span the full range of post-cephalic segment numbers for the entire Arthropoda (Enghoff et al. 1993), (2) centipedes, where the number of segments in the leg-bearing trunk spans the odd values between 15 and 191 (Minelli and Koch 2011), with only a few aberrant specimens with an even number of leg pairs recorded from a single population of the geophilomorph *Haplophilus subterraneus* (Leśniewska et al. 2009), (3) remipedes, with 16–43 trunk segments (Koenemann et al. 2006) and (4) branchiopods, with 10–44 trunk segments (in notostracans, in this case counting their dorsal segmental units) (Westheide and Rieger 2007). Subclades of these four clades are also illustrative of a general association between interspecific variation in segment numbers among closely related species and intraspecific variation in the same character (see below). In geophilomorph centipedes and helminthomorph millipedes, this variation is also associated with significantly higher numbers of segments in comparison with relatively more basal clades (Fusco 2005).

## 9.3.2 Intraspecific Variation in the Number of Segments

Intraspecific sex differences in segment numbers are scattered across the arthropods. In some species of polydesmidan millipedes, females have two segments more than males, while the situation is reversed in glomerid and sphaerotheriid millipedes, where males have two segments more than females (Enghoff et al. 1993). Among branchiopods, females have one segment more than males in Laevicaudata (12 vs. 10) and Cyclestherida (16 vs. 15) (Westheide and Rieger 2007). The most posterior abdominal segments (*post-abdomen*) of several holometabolous insect taxa are variably reduced in number in the two sexes in relation to the structures of the genitalia (CSIRO 1991). For instance, males in Coleoptera generally have a higher number of abdominal segments than females (10 vs. 9).

In species with intra-sex variation in segment number (see below), sex differences are apparent as a phase shift between the distributions of segment numbers of the two sexes. In the Adesmata, one of the two major clades of geophilomorph centipedes, females generally have two segments more than the conspecific males, but in a few species, there is no sexual dimorphism and in other species, the phase shift is a multiple of two (recalling that segment number variation in centipedes is discrete, with a module of two), up to 16 segments (Minelli and Bortoletto 1988; Berto et al. 1997). An exception is the oryid geophilomorph *Orphnaeus heteropodus*, where, according to an old record, females possess about twice as many trunk segments as males (∼120 vs. ∼60; Lawrence 1963). Conversely, in notostracans, it is usual for males to have a few more trunk segments than the females of the same species, but there are exceptions (Linder 1952).

Within the same sex, intraspecific variation in segment number can have different origins that should not be confused. In species with epimorphic, hemianamorphic and teloanamorphic development (see Chap. 5), where adults are credited with a 'developmentally targeted' segmental composition of the trunk, it is possible to take apart static adult variation from ontogenetic variation. However, in euanamorphic species, the 'open' ontogenetic accumulation of new segments, which often does not follow an

**Table 9.1** Adult tagmosis and segment numbers in extant Arthropoda (taxa are not necessarily monophyletic)

| Taxon | Tagmata with number of segments | | Notes | References |
|---|---|---|---|---|
| | Prosoma | Opisthosoma | | |
| Pycnogonida | 8 | n.s. | 1, 2 | a |
| Xiphosura | 7 | 10 | 3 | b |
| Opiliones | 7 | 9 | 4 | b |
| Scorpiones | 7 | 12 | 4, 5 | a |
| Solifugae | 7 | 11 | 4 | b |
| Pseudoscorpiones | 7 | 12 | 4 | b |
| Palpigradi | 7 | 11 | 4 | b |
| Ricinulei | 7 | 10 | 4, 6 | a |
| Araneae | 7 | 12 | 4, 7 | b |
| Amblypygi | 7 | 12 | 4 | b |
| Thelyphonida | 7 | 12 | 4, 8 | b |
| Schizomida | 7 | 12 | 4, 8 | b |
| | Gnathosoma | Idiosoma | | |
| Acari Anactinotrichida | 3 | 17 | 9, 10, 11 | c |
| Acari Actinotrichida | 3 | 14 | 9, 10, 12 | c |
| | Head | Trunk | | |
| Chilopoda Scutigeromorpha | 6 | 18 | 13, 14, 15 | d |
| Chilopoda Lithobiomorpha | 6 | 18 | 13, 14 | d |
| Chilopoda Craterostigmomorpha | 6 | 18 | 13, 14, 15 | d |
| Chilopoda Scolopendromorpha | 6 | 24–46 | 13, 14, 16 | d |
| Chilopoda Geophilomorpha | 6 | 30–194 | 13, 14, 15, 17 | d |
| Symphyla | 6 | 14 | 13, 15 | e |
| Pauropoda | 6 | 12 | 15, 18 | f |
| Diplopoda Polyxenida | 6 | 14–18 | 15, 18, 19 | g |
| Diplopoda Glomeridesmida | 6 | 36-37 | 15, 18, 20 | h |
| Diplopoda Sphaerotheriida | 6 | ♀ 22, ♂ 24 | 15, 18, 20 | h |
| Diplopoda Glomerida | 6 | ♀ 18, ♂ 20 | 15, 18, 20 | h |
| Diplopoda Platydesmida | 6 | 66–216 | 15, 18, 21, 22 | i, j |
| Diplopoda Siphonophorida | 6 | 68–380 | 15, 18, 19, 22 | i |
| Diplopoda Polyzoniida | 6 | 36–166 | 15, 18, 19, 22 | j, k |
| Diplopoda Stemmiulida | 6 | 66–104 | 15, 18, 21, 22 | h |
| Diplopoda Callipodida | 6 | 70–122 | 15, 18, 21, 23 | h |
| Diplopoda Chordeumatida | 6 | 46–58 | 15, 18, 21 | h |
| Diplopoda Polydesmida | 6 | 30–52 | 15, 18, 24 | h |
| Diplopoda Julida | 6 | 44–198 | 15, 18, 22, 24 | h |
| Diplopoda Spirostreptida | 6 | 32–182 | 15, 18, 23, 25 | h |
| Diplopoda Spirobolida | 6 | 57–151 | 15, 18, 23, 25 | h |
| Ostracoda | 6 | unc. | 26, 27 | l |
| Remipedia | 6 | 16–43 | 22, 26, 28 | m |
| Tantulocarida | 6 | ♀ 6, ♂ 7 | 26, 29 | n |
| Branchiopoda Notostraca | 6 | 25–44 | 26, 30 | a |
| Branchiopoda Laevicaudata | 6 | ♀ 12, ♂ 10 | 26 | o |
| Branchiopoda Spinicaudata | 6 | 16–32 | 26 | o |
| Branchiopoda Cyclestherida | 6 | ♀ 16, ♂ 15 | 26 | o |

(continued)

**Table 9.1** (continued)

| Taxon | Tagmata with number of segments | | | Notes | References |
|---|---|---|---|---|---|
| | Head | Thorax | Abdomen | | |
| Branchiopoda Cladocera | 6 | 4–6 | n.s./unc. | 26, 31 | a |
| Branchiopoda Anostraca | 6 | 11–19 | 8 | 26, 32, 33 | p |
| Mystacocarida | 6 | 5 | 5 | 26, 34 | q |
| Ichthyostraca Branchiura | 6 | 4 | n.s. | 26, 35 | a |
| Ichthyostraca Pentastomida | n.s. | n.s. | n.s. | 36 | a |
| Thecostraca Facetotecta | – | – | – | 37 | a |
| Thecostraca Ascothoracida | 6 | 6 | 4–5 | 26, 38 | r |
| Thecostraca Acrothoracica | 6 | 6 | – | 26, 39 | a |
| Thecostraca Thoracica | 6 | 6 | n.s./unc. | 26, 40 | a |
| Thecostraca Rhizocephala | n.s. | n.s. | n.s. | 41 | a |
| Copepoda | 6 | 7 | 3 | 26, 42, 43 | s |
| Cephalocarida | 6 | 8 | 11 | 26, 44 | p |
| | Head | Pereion | Pleon | | |
| Malacostraca Phyllocarida | 6 | 8 | 7 | 26 | p |
| Malacostraca Eumalacostraca | 6 | 8 | 6 | 26, 45 | p |
| | Head | Thorax | Abdomen | | |
| Collembola | 6 | 3 | 6 | 46, 47 | a |
| Protura | 6 | 3 | 11 | 46, 48 | a |
| Diplura | 6 | 3 | 10 | 46, 49 | a |
| Insecta | 6 | 3 | 9–11 | 46, 49, 50, 51 | t |

Notes. These numbers should be taken with some caution, as (1) authors often disagree on interpreting body segmental composition, (2) segmental mismatch can severely invalidate the meaning of the count, (3) variation at lower taxonomic level can easily be overlooked and (4) some taxa are little known in this respect. The substantial list of notes accompanying the table is a consequence of these facts. Variation referable to cases interpreted as 'loss of articulation' between two or more contiguous segments (or sclerites), very common among several taxa across the Arthropoda, is not accounted for here. Segment counts do not include the telson, where recognizable
n.s. non-segmented
unc. uncertain
1. Prosoma includes an ocular segment followed by 7 segments, respectively bearing chelifores, pedipalps, ovigers and 4 leg pairs
2. Some species with 5 or 6 leg-bearing segments
3. Prosoma includes an ocular segment followed by 6 segments, respectively bearing chelicerae and 5 leg pairs
4. Prosoma includes an ocular segment followed by 6 segments, respectively bearing chelicerae, pedipalps and 4 leg pairs
5. Traditional interpretation of segmentation of the opisthosoma, comprising a mesosoma of 7 segments and a metasoma of 5 segments, but other interpretations have been suggested (see Shultz 2007)
6. Traditional interpretation of segmentation of the opisthosoma, comprising a mesosoma of 7 segments and metasoma of 3 segments, but other interpretations have been suggested (see Shultz 2007)
7. Putative primitive opisthosoma segment number, but in general segmentation is lost or segment number is reduced (6–8 tergites and sternites in the opisthosoma in Mesothelae)
8. Opisthosoma comprising a mesosoma of 9 segments and a metasoma of 3 segments
9. Gnathosoma includes an ocular segment followed by two segments, respectively bearing chelicerae and pedipalps
10. Putative primitive hysterosoma segment number, but the actual number of segments is problematic for most taxa due to extensive simplification or loss of segmental structures
11. A different scheme of tagmosis is also in use: prosoma (7 segments)/opisthosoma (13 segments)
12. Different schemes of tagmosis are also in use: prosoma (7 segments)/opisthosoma (10 segments) and proterosoma (5 segments)/hysterosoma (12 segments)
13. Head includes 6 segments, generally termed ocular, antennal, intercalary, mandibular, first maxillary, second maxillary

**Table 9.1** (continued)

14. Number of trunk segments is roughly calculated as #LBS + 3 (one anterior trunk segment bearing a pair of poisonous maxillipedes (forcipular segment) and two terminal apodous segments of the ano-genital region (telson excluded), but the segmental composition of post-pedal trunk is uncertain)

15. Dorsoventral mismatch: the number of trunk segments is based on the number of ventral segmental units

16. 21, 23, 39 or 43 LBS

17. 27–191 LBS, odd values only

18. Head includes 6 segments, generally termed ocular, antennal, intercalary, mandibular, first maxillary (with gnathochilarium), second maxillary (without appendages)

19. Number of trunk segments roughly calculated as 2(#T)-4 (the apodous collum and 3 tergites corresponding to one LP each)

20. Number of trunk segments roughly calculated as #LP + 1 (the apodous collum)

21. Number of trunk segments roughly calculated as 2(#PT)-4 (the apodous collum and 3 pleurotergites corresponding to one LP each)

22. Euanamorphosis

23. Euanamorphosis in some taxa

24. Number of trunk segments roughly calculated as 2(#R)-4 (the apodous collum and 3 rings corresponding to one LP each)

25. Number of trunk segments roughly calculated as 2(#R)-5 (the apodous collum and 4 rings corresponding to one LP each)

26. Head includes 6 segments, generally termed ocular, first antennal, second antennal, mandibular, first maxillary, second maxillary

27. Trunk generally non-segmented or with faint traces of segmentation; up to 11 segments recognizable in Podocopa, up to 7 in Myodocopa

28. Trunk includes one anterior segment bearing a pair of maxillipedes (fused to the head)

29. Parthenogenetic females are ectoparasites with a non-segmented trunk

30. Dorsoventral mismatch: the number of trunk segments is based on the number of dorsal segmental units

31. Abdomen with 3 segments in *Leptodora* (Haplopoda)

32. Most species have 11 thoracic segments, while species in the family Polyartemiidae have 17 or 19

33. For some authors (e.g., Westheide & Rieger 2007), the border between thorax and abdomen is two segments more posterior

34. For some authors (e.g., Huys 1991), the border between thorax and abdomen is two segments more posterior

35. Ectoparasites with partial loss of segmentation

36. Parasites with loss of segmentation

37. Adults unknown

38. Loss of segmentation in endoparasitic species

39. Abdomen absent

40. Sessile adults with partial loss of segmentation

41. Parasites with loss of segmentation as adults

42. Primitive condition, observable in most free-living species; trunk includes one anterior segment bearing a pair of maxillipedes (fused to the head to form a cephalosome)

43. An alternative tagmatic subdivision of the trunk sees a division between metasoma (from thoracic segment II to V or VI, depending on the subtaxa) and urosoma

44. For some authors (e.g., Olesen et al. 2011), the border between thorax and abdomen is one segment more posterior

45. Some species of Lophogastrida with 7 pleonites (telson excluded)

46. Head includes 6 segments, generally termed ocular, antennal, intercalary, mandibular, maxillary, labial

47. Telson present only in the embryo

48. A twelfth abdominal segment is interpreted as a telson

49. A telson is present only in the embryo, or in vestigial form in some taxa

50. In many taxa, the first 4–6 abdominal segments (depending on the taxon) are identified as *preabdomen*, which is followed by a telescopically retractile *post-abdomen*; abdominal segments from VIII or IX (occasionally more anteriorly) to the end of the abdomen form an ano-genital subtagma called *terminalia*

51. Fusions, reductions or loss of abdominal segments are not infrequent among the more anterior and the more posterior segments of the tagma

References: (a) Westheide and Rieger (2007); (b) Shultz (2007); (c) van der Hammen (1989); (d) Minelli and Koch (2011); (e) Szucsich and Scheller (2011); (f) Scheller (2011); (g) Nguyen Duy–Jacquemin et al. (2011); (h) Enghoff et al. (1993); (i) Hoffman (1982); (j) Brolemann (1935); (k) Mauriès (1964); (l) Horne et al. (2005); (m) Koenemann et al. (2006); (n) Huys et al. (1993); (o) Martin (1992); (p) McLaughlin (1980); (q) Brenneis and Richter (2010); (r) Schram (1986); (s) Huys and Boxshall (1991); (t) CSIRO (1991) invariant scheme of addition, does not allow us to completely disentangle variation in adult segment number from variation in post-embryonic segmentation schedule. Although ranges of variation are often reported in taxonomic literature, these are not directly comparable to those in species with targeted segmentation and will not be discussed further.

Within-sex intraspecific variation in the number of trunk segments has evolved independently several times within arthropods. In chilopods, (1) at least three times within the major geophilomorph clade Placodesmata (93–101 LBS in *Mecistocephalus microporus*, Bonato et al. 2003; 57–59 LBS in *M. diversisternus* and 63–65 LBS in *M. japonicus*, Uliana et al. 2007), (2) at the base of its sister clade Adesmata, recorded in almost all of the 1,100 or so species, with a range of variation between 2 LBS (e.g., *Hyphydrophilus adisi*, ♂ 39–41, ♀ 41–43, Pereira et al. 1994) to more than 80 LBS (*Himantarium gabrielis*, ♂ 87–171, ♀ 95–179 LBS, Simaiakis 2009) and (3) in the two scolopendromorph sister species *Scolopendropsis bahiensis* (21–23 LBS, Schileyko 2006) and *S. duplicata* (39 or 43 LBS, Chagas et al. 2008). Among non-euanamorphic Diplopoda, within-sex intraspecific variation is recorded in (1) one species of hemianamorphic Glomeridesmida (19–20 T), (2) several teloanamorphic species of Callipodida (range up to 3 PT), (3) several species of hemianamorphic Spirostreptida (range up to 18 R) and (4) several hemianamorphic species of Spirobolida (range up to 10 R) (Enghoff et al. 1993).

Within-sex intraspecific variation in the number of trunk segments is recorded also in notostracan crustaceans, both as the number of trunk rings (e.g., *Triops longicaudatus*, ♂ 38–44, ♀ 34–43, Linder 1952), as well as the number of ventral segmental units of the leg-bearing trunk.

## 9.3.3 Forms of Segmental Mismatch

Under the label of 'segmental mismatch,' one could list any case of discordance between different segmental series within the same animal.

In general, however, the use of this term is restricted to those cases where the mismatch is between 'comparable' segmental structures, for example, dorsal vs. ventral serial sclerites. Other cases of mismatch, which for instance involve discordance between segmental structures of the internal anatomy and serial structures of the exoskeleton, are discussed in Sects. 9.3.4 and 9.5.

A dorsoventral mismatch between sclerites in the opisthosoma is very common among the Opiliones, where the two series variably differ in number and alignment in different taxa (van der Hammen 1985; Fig. 9.1a). Most myriapods exhibit some form of dorsoventral mismatch along the trunk, the series of tergites failing to match the two generally concordant series of sternites and leg pairs (reviewed in Fusco 2005). In symphylans, there are more tergites than sternites (15, 17, 21 or 24 vs. 14, depending on the species, Fig. 9.1b), the same in craterostigmomorph (21 vs. 15) and geophilomorph centipedes (in the latter, the number of tergites is twice that of leg pairs). On the contrary, in scutigeromorph centipedes, the number of leg-bearing trunk tergites is smaller than the number of sternites (8 vs. 15), as in the trunk of tetramerocerate pauropods (6, 9, or 10 vs. 12, depending on species, Fig. 9.1c). In millipedes, for most of the trunk, there are two leg pairs for each tergal plate, but the first post-cephalic tergite apparently does not correspond to any ventral structure and the following three (in the Spirobolida, four) tergal plates correspond to one leg pair each. Moreover, in most millipedes with 'free sternites,' that is, sternites that are not fused to pleural and dorsal sclerites to form exoskeletal rings, the correlation between sternite and tergite numbers does not follow any consistent rule, so that the number of dorsal sclerites is not predictive of the number of ventral segmental units, and vice versa (Enghoff et al. 1993).

In notostracans, dorsal and ventral structures of the posterior part of the trunk (also called abdomen, from the twelfth trunk segment on) present marked differences in periodicity, length of the series and post-embryonic segmentation schedule (Linder 1952; Minelli and Fusco 2004). The series of leg pairs and the series of skeletal rings are quite independent of each other. The series of legs is shorter than the series of rings, and there is no strict correlation between the cardinality of the two series. The series of legs can end posteriorly at any place within the longitudinal extent of a given body ring, and a given pair of legs in the posterior part of the leg series can be found under a different ring in different individuals of the same population. Longitudinal dorsal and lateral muscles have the same periodicity as the rings, whereas ventral longitudinal muscles have the same periodicity as leg pairs.

In the posterior body segments (genital region) of many hexapod orders, it is not uncommon that during embryonic or post-embryonic development, some body segments apparently disappear, at least from the dorsal (or ventral) aspect (Minelli 2003). Remarkably, in the entomobryomorph Collembola, it is the dorsal sclerite of the first thoracic segment (pronotum) which disappears completely during embryonic development, so that the first dorsal sclerite behind the head is the mesonotum (Matsuda 1979).

Traditional descriptions tend to interpret these discordances with the occurrence of fusion, splitting or loss of segmental elements (especially, sclerites) putatively belonging to primitive segmental units (modules), each characterized by a full allocation of segmental elements: one tergite, one sternite, one pair of appendages, etc. Not only is this not always possible, for example, in the more radical cases of mismatch in millipedes described above, but in general, the search for a primitive (in evolution) or an early (in development) correspondence between elements of discordant segmental series can be of limited meaning because of the ways different segmental series in the same animal develop and potentially evolve (Fusco 2008). For instance, expression studies of segmentation genes in the pill millipede *Glomeris* showed that dorsal and ventral serial structures are independently established in the embryo, and that the A-P boundaries of the prospective dorsal

sclerites do not correlate with the A-P boundaries of the anlagen of either ventral or dorsal structures (Janssen et al. 2004). Expression patterns of segmentation genes in the prospective ventral and dorsal tissues are different as well (Janssen et al. 2008). Regrettably, nothing is known about the expression of segmentation genes during millipede anamorphosis (see Chap. 5), when the majority of segmental elements is formed.

A concept of segmentation independent from the concept of segment also helps to account for other forms of mismatch, in particular those that do not involve any difference in the number of segmental elements between discordant segmental series. For instance, in many tanaid and isopod crustaceans, the series of legs displays a different period with respect to both dorsal and ventral trunk sclerites. Leg insertion is relatively anterior with respect to the corresponding trunk sclerites in the more anterior elements of the series, but shifts progressively more posteriorly with respect to the trunk sclerites toward the posterior part of the trunk (Fig. 9.1d). More extreme shifts between segmental elements belonging to different series are possible, as in the case of sphecid wasps of the genus *Ammophila*, where the long peduncle of the usual wasp waist is formed entirely by the second segment of the abdomen, despite the fact that it seems to comprise two segmental units, rather than one. This is because the dorsal and ventral sclerites of this segment are longitudinally displaced in such a way that the anterior part of the peduncle is formed by the sternite only, closed along the dorsal midline, while the posterior part is formed by the tergite only, closed along the ventral midline (Fig. 9.1e). Another example of marked reciprocal sliding of segmental elements is provided by the thorax of dragonflies, where the two wing-bearing segments (meso- and metathorax) are strongly associated to form a complex called the *pterothorax*. Tergites of these two segments are quite small and do not extend over the whole dorsal surface of the pterothorax. Instead, the anterior half of this complex is occupied dorsally by the pleurites of the mesothorax, which make contact along the midline. In this way, the dorsal

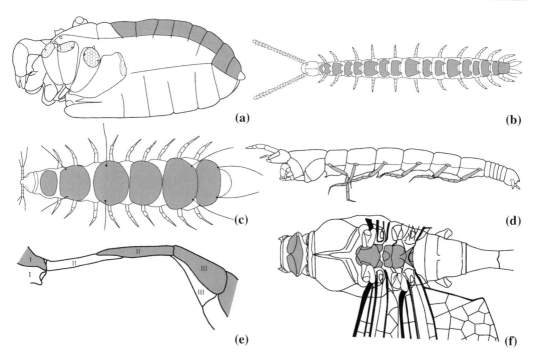

**Fig. 9.1** Forms of segmental mismatch. **a** In harvestmen (here *Paroligolophus agrestis*, *side view*, legs cut), the series of dorsal (*light gray*) and ventral sclerites of the opisthosoma differ in number and alignment (redrawn from van der Hammen 1985); **b** in symphylans (here *Symphylella* sp., *dorsal view*), there are more trunk tergites (17 in this species, *light gray*) than ventral segmental units (12, of which 11 leg bearing in this species; redrawn from Scheller and Adis 2002; **c** in tetramerocerate pauropods (here *Pauropus amicus*, *dorsal view*), the number of trunk tergites (6 in this species, *light gray*) is smaller than the number of ventral segmental units (12, of which 9 leg bearing in this species; redrawn from Harrison 1914); **d** in many tanaid crustaceans (here *Paranarthrura tenuimanus*, female, *side view*), the series of legs (*light gray*) displays a different period with respect to both dorsal and ventral

sclerites, as apparent by the variable position of leg insertions with respect to the corresponding trunk sclerites along the pereion (redrawn from Larsen 2005); **e** in some sphecid wasps (here *Ammophila sabulosa*, side view of the waist, tergites in *light gray*, abdominal segments with Roman numbers), the long peduncle is entirely formed by the second segment of the abdomen: the anterior part of the peduncle is formed by the sternite, closed along the dorsal midline, while the posterior part is formed by the tergite closed along the ventral midline (redrawn from Minelli 2003); **f** in dragonflies (here *Acanthaeschna parvistigma*, *dorsal view* of the anterior trunk, thoracic tergites in *light gray*), the tergite of the first thoracic segment is widely separated from the corresponding mesothoracic sclerites by the conspicuous pleurites of the mesothorax which make contact along the midline (redrawn from Watson and O'Farrell 1991)

sclerite of the first thoracic segment (pronotum) is widely separated from the corresponding mesothoracic sclerite (Minelli 2003) (Fig. 9.1f).

### 9.3.4 Periodic Patterns

Body patterns in segmented animals can be independent of any segmental arrangement of the structures over which they extend. However, it is quite common to observe that the segmental framework of a given structure provides the

basis for a concordant segmental pattern. For instance, different kinds of 'minor order' structures (e.g., microsculptures) or colour patterns often show a periodic pattern along the trunk which is in register with the segmental series of sclerites or legs on which they develop.

In arthropods, beyond these widespread forms of body pattern, there are also less common cases of structures and processes that show forms of periodic pattern with a less strict connection to segmental organization.

In the leg-bearing trunk of lithobiomorph and scolopendromorph centipedes, segments with long and with short tergites alternate until segment VII, which has a long tergite but is followed by segment VIII also with a long tergite, after which alternation resumes until the end of the leg-bearing trunk. This irregularity in an otherwise periodic pattern (known as the VII–VIII discontinuity) is mimicked by the tracheal system, whose openings (spiracles) are generally present on segments with long tergites only, to the exclusion of segments I and VII. However, there are exceptions. In Henicopini lithobiomorphs, there is a pair of spiracles also on LBS I; and in a few other lithobiomorph species, some spiracle pairs are lacking (up to four, out of the normal six pairs) (Hilken et al. 2011). In some or all species of eight scolopendromorph genera, indicative of at least four independent origins, there is also a pair of spiracles on LBS VII, and in the scolopendromorph *Plutonium zwierleini* spiracles are present from the second to the penultimate leg-bearing trunk segment (II–XX) (Bonato et al. 2011). Among the Hexapoda, in japygid diplurans, the three-segment thorax bears four pairs of spiracles (the extra pair is in the metathorax).

Most scolopendromorphs are more or less uniform in colour along the trunk (segmental organization ignored). However, (1) some species (e.g., *Scolopendra heros*) display different colours in different portions of the body (regional pattern), (2) other species (e.g., *S.cingulata*) have colour bands in each segment, in a specific position on a given sclerite (segmental pattern) and (3) *S. hardwickei* shows alternation between red-coloured segments (those with a long tergite) and melanic segments (those with a short tergite) which respects the VII–VIII anomaly and is concordant with the segmental distribution of spiracles (non-segmental periodic pattern; Fig. 9.2a).

Similar periodic patterns are observed in polydesmidan millipedes, whose trunk counts 19 rings in the adults of most species. Although in one species (*Prosopodesmus panporus*) the openings of the repugnatory glands, the ozopores, are present on all rings from V to the last one (a 'diplosegmental' pattern, as in all Helminthomorpha to the exclusion of Chordeumatida), in most species the ozopores are present only on rings V, VII, IX, X, XII, XIII, XV–XIX. This pattern with 11 pairs of ozopores is possibly the primitive condition for the Polydesmida, from which more derived conditions, with either loss or gain of ozopore pairs, have evolved (Hoffman 2005). Interestingly, the rings without ozopores are those that represent the most posterior ring in each batch of rings (when this is not composed of one ring only) added at each moult during anamorphosis. In some species, the ozopore pattern is correlated to cuticle colour pattern in rings V–XIX. In *Polydesmus collaris,* the lateral tergal expansions (paranota) of body rings which do not carry ozopores are markedly yellow in contrast to the dark background (Fig. 9.2b), while in *Prepodesmus ornatus* contrasting light spots are found instead on the paranota of the body rings with ozopores (Enghoff 2011).

Cases of 'trans-segmental colour patterning,' again possibly related to anamorphic development, have been recorded in other millipedes. For instance, in the 'var. *sexfasciatus*' of the spirobolid *Centrobolus vastus,* there are six transverse black bands on a bright red background, each transverse band covering two successive body rings, formed in connection with anamorphic moults and extending over the first two of the newly formed leg-bearing body rings (Fig. 9.2c) (Enghoff 2011). As in the case of the two polydesmidans, a periodicity in *space* (the main body axis) is related to a periodicity in developmental *time* (the periodic addition of segments during anamorphosis). Such an accretive-growth-related colour pattern is analogous to those observable in the shells of many mollusks, where colour patterns depend on local dynamics in the mantle along the growing margin of the shell itself (Meinhardt 1994). However, there are also cases of trans-segmental colour patterning in millipedes where this explanation probably does not apply (Enghoff 2011).

**Fig. 9.2** Periodic patterns with non-segmental periodicity. **a** In *Scolopendra hardwickei,* red-coloured segments (those with a long tergite) alternate with melanic segments (those with a short tergite) while still respecting the VII-VIII anomaly mentioned in the text (courtesy Damien Teo); **b** in *Polydesmus collaris,* the paranota of body rings IV, VI, VIII, XI and XIV (rings which do not carry ozopores) are yellow in contrast to the brown background (courtesy Dragiša Savić); **c** in the 'var. *sexfasciatus*' of *Centrobolus vastus,* there are six transverse black bands on a bright red background, formed in connection with posterior segment addition throughout the anamorphic phase of development (courtesy Guido Coza)

### 9.3.5  Homology of Segments

There is no single answer to the question of whether it is possible to homologize trunk segments with the same ordinal post-cephalic position in series that exhibit different numbers of elements. When a segmental framework is used as a grid for evaluating positional homology, conflicts with special homology can easily emerge.

If serial homology between segments can to some extent apply both within the same individual and between individuals of distantly related species (although a more accurate statement would be to consider the elements of specific serial structures, rather than those of the 'series of segments'), the precise identity of a given segment, qualified as the *i*th of a segmental series, in general does not go beyond the limits of a restricted clade, and is lost completely in the case of species with intraspecific variation in segment numbers. For instance, special homology of the terminal genital segments of centipedes is not in question, despite the fact that these can be separated from the last cephalic segment by a number of trunk segments varying between 16 and 192. On the other hand, treating the long segmental series of leg-bearing segments as a variable partition of a homologous domain cannot apply either, since in

lithobiomorph and scolopendromorph centipedes, the VII–VIII discontinuity has the same
segmental position irrespective of the length of
the series of leg-bearing trunk segments (15–43).
Restricting homology assessment to the segments of a limited part of the trunk, as a tagma,
cannot provide a general solution to the problem
also because this excludes the possibility of an
evolutionary change through variation in the
expression domain of the determinants of the
key features that characterize the morphological
and/or functional specificity of the given body
region.

Homology needs to be evaluated case by
case, clarifying the kind of relationship to be
assessed (e.g., positional vs. special homology)
and possibly adopting a combinatorial concept
of homology, where the relationship is not of the
all-or-nothing kind, but one that can account for
more complex developmental and evolutionary
correlations between the structures under scrutiny (Minelli 1998; Minelli and Fusco in press).

## 9.4   The Space of Variation of Arthropod Tagmosis

Recalling the multifaceted (or vague) nature of the
concept of tagma, we will develop our discussion
starting from a classical view (Table 9.1).

### 9.4.1   Delimitation of Tagmata

Although used to encapsulate the main features
that characterize body architecture, the tagmata
of a given animal often have boundaries whose
positions are somehow disputable, depending on
the definition of tagma adopted and on the
interpretation of the body structures present in
proximity to the putative boundaries themselves,
in the context of the animal life history, morphology, development and evolution.

Conflicting delimitations of tagmata are often
associated with the distinction between head and
trunk (or the most anterior trunk tagma, for
example, thorax, or pereion, if more post-cephalic
tagmata are recognized). This is the case for the

numerous arthropods where one or more pairs of
appendages, belonging to segments posterior to
those that bear the conventional mouthparts, are
morphologically and functionally transformed
into feeding tools (maxillipedes). Such are the
poisonous fangs of centipedes, the toothed raptorial appendages of mantids, mantispid neur-
opterans and mantid shrimps used to capture
prey, or the simpler maxilla-like appendages of
many crustaceans, used to select, clean, tear apart
or otherwise manipulate the food. Thus, from the
point of view of the specialization of the
appendages, one could say that the head of these
arthropods comprises one or a few additional
segments compared with their relatives that lack
maxillipedes.

Segments with maxillipedes may retain their
original articulations, as for instance, the forcipular segment of centipedes and the prothorax
of mantids. In many crustaceans, however, the
integration into the cephalic tagma of one or
more segments with maxillipedes is much more
pronounced, and a 'pseudotagma' called *cephalothorax* is often recognized. For instance, in
isopods, there is a single-dorsal sclerite corresponding to the cephalic shield plus the tergite of
the maxillipede-bearing segment. In some
groups (e.g., Mystacocarida), the term *thoraco-
abdomen* is used to indicate the portion of the
trunk posterior to the segment with the pair of
maxillipedes. On the other hand, the incorporation of post-cephalic segments in the head does
not necessarily require a functional integration
of their appendages with the mouthparts. For
instance, in the leptostracans, the first thoracic
segment is fused to the cephalon, but its
appendages do not differ from the other thoracic
appendages. This is also the case for copepods,
where the anterior region that includes the first
thoracic segment is called *cephalosoma*, while
the term *cephalothorax* is used when additional
leg-bearing thoracic segments are incorporated
in the most anterior region of the body (Huys
and Boxshall 1991). In the euphausiaceans, there
is a *cephalothorax* comprising the head and all
eight segments of the pereion (as in crabs, lobsters and shrimps) but none of these segments
bears food-processing maxillipedes.

Conflicting delimitations of tagmata can occur also at the boundary between two post-cephalic regions, thorax vs. abdomen, or pereion vs. pleon. Traditional descriptions assign a fixed number of segments to non-terminal tagmata (prosoma, head or head + thorax/pereion) at the level of the four classes of traditional classification, but for the crustaceans, where segmental composition of the middle tagma, thorax or pereion, can vary between different subgroups (e.g., 7 in mystacocarids, 8 in malacostracans, 9 in cephalocarids). Within each of these taxa, segments of one tagma which are lost to become incorporated into the adjacent tagma are qualified as 'migrants' that can contribute to functional units, or pseudotagmata, without invalidating the original 'true' regionalization. This is, for instance, the situation of the aforementioned mesosoma of wasps. Another is provided by the last pereion segment of gnathiid isopods that is limb-less and functionally associated to the adjoining pleon.

## 9.4.2 Dorsoventral Mismatch Between Tagmata

It is quite common that dorsal and ventral aspects of the same animal suggest a different tagmosis, that is, that domains of functional and/or morphological integration of dorsal structures do not correspond to those one would recognize on the ventral side.

Integration of dorsal segmental units often takes the form of a lack of articulation between contiguous segments. Depending on taxon-specific tradition, such a 'compound sclerite' is variably named *shield*, *plate* or *carapace*, although in crustaceans—but not in chelicerates—the latter term tends to be restricted to the cases where a marginal fold is present, which covers the exoskeleton of flanking, either anterior, pleural or posterior, structures (for the debate about the crustacean carapace, see for example, Newman and Knight 1984; Dahl 1991; Olesen in press). However, the borders of these dorsal structures, that can also extend to the pleural area, do not necessarily match with the borders of integrated sets of ventral structures, such as the appendages. This is a very common feature of arthropod body organization and a few examples will suffice to illustrate the point.

*Limulus* has only two dorsal articulations, one in the middle of the second segment of the opisthosoma and one in front of the telson, while the border between leg-like and plate-like appendages marks the boundary between prosoma and opisthosoma. In Solifugae, Palpigradi and Schizomida, the posterior border of the most anterior dorsal sclerite of the prosoma, the *propeltidium*, corresponds to a location within the homonomous series of legs. In several mites, there is only one dorsal sclerite for the whole body. In pill millipedes (Glomerida), the anterior border of the terminal dorsal sclerite, the *anal shield*, corresponds to a ventral position among the last leg pairs. In the females of some gnathiid isopods (Fig. 9.3a), a single dorsal sclerite extends over the leg-bearing pereion segments V–VII, while the leg-bearing pereion segments III–IV are fully articulated.

Beyond non-coincident dorsal and ventral boundaries, another kind of mismatch between dorsal and ventral body organization is found when a homonomous segmental series on one side is associated with more or less heteronomous segmental structures on the other side. For instance, in gammarid amphipods, the relatively homonomous series of pereion tergites contrasts with the serially diversified series of appendages (pereiopods, Fig. 9.3b). Another example, but with inverted roles for the ventral and dorsal sides, is provided by the leg-bearing trunk of lithobiomorph centipedes, where the alternation of long and short tergites, accompanied by alternating presence/absence of lateral spiracles (see Sect. 9.3.4), has no equivalent in the homonomous series of sternites and leg pairs.

## 9.4.3 Homology of Tagmata

Terms such as 'head,' 'thorax' and 'abdomen' are used for the body regions of the most diverse animals, from vertebrates to polychaetes to arthropods. Although tracing homologies

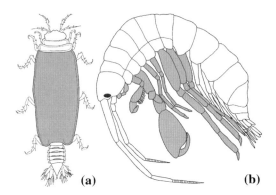

**Fig. 9.3** Forms of tagmatic mismatch. **a** In the females of some gnathiid isopods (here *Gnathia dentata*, redrawn from Kaestner 1970), a single-dorsal sclerite (*light gray*) extends over a subset of the pereion segments; **b** in many gammarid amphipods, the homonomous segmental series of pereion tergites is associated with a heteronomous series of appendages (pereiopods in *light gray*; redrawn from McLaughlin 1980)

between body regions across the whole of Bilateria seems to be a hopeless enterprise, homology of tagmata at the level of lower rank taxa is generally more or less tacitly assumed. This is the case of arthropods in textbooks, where tagmata are implicitly considered homologous at least within each of the classes of the traditional classification. However, as it is often the case with homology arguments, things are not as simple as description-aimed idealizations would suggest. The crustacean trunk will illustrate the point.

In most crustaceans, three main body regions are recognized: a head and two post-cephalic tagmata. However, in malacostracans, both post-cephalic regions are typically provided with appendages, while in most non-malacostracan crustaceans, appendages are present in the most anterior post-cephalic region only. The latter condition has been traditionally regarded as equivalent to that of the Hexapoda, which have a three-segment leg-bearing thorax and a nearly appendage-less abdomen. Accordingly, the two terms used for the two post-cephalic tagmata of the hexapods, 'thorax' and 'abdomen,' are also used for most non-malacostracan crustaceans, like copepods, cephalocarids, mystacocarids and others, but not for the malacostracans, where the two post-cephalic regions are termed instead

'pereion' and 'pleon.' According to some authors (e.g., Lauterbach 1975; Walossek and Müller 1997; Schram and Koenemann 2004), the whole trunk of the malacostracans, pereion and pleon together, would be homologous to the thorax of the other crustaceans, and the primitive abdomen would have disappeared or nearly so.

Beyond morphology (presence of appendages), arguments in favor of this hypothesis have been based on a different kind of evidence. Comparing post-cephalic *Hox* genes expression patterns in malacostracans and non-malacostracans, it appears that in *Artemia franciscana* (a non-malacostracan, branchiopod crustacean), the genes *Antennapedia*, *Ultrabithorax* and *abdominal-A* are expressed in largely overlapping domains in the thorax, which is followed by two genital segments expressing *Abdominal-B* (Averof and Akam 1995), while in malacostracans like *Procambarus clarkii* and *Porcellio scaber* (and in insects), the less extensively overlapping expression domains of the same genes specify distinct segment types over most of the trunk (Abzhanov and Kaufman 2000a, b). Thus, the thorax of *Artemia*, whose body organization is regarded as an example of a more primitive condition, would correspond to the entirety of the pregenital segments of the malacostracan pereion and pleon, or the insect thorax and abdomen. In this scenario, the leg-bearing insect thorax would not correspond to the leg-bearing anostracan thorax either. Schram and Koeneman (2004) arrive at defining a crustacean abdomen as a posterior region which (1) lacks limbs, (2) is posterior to the expression domain of *Abdominal-B* and (3) does not express *Hox* genes. However, other authors (e.g., Angelini and Kaufman 2005) have cast doubts on the reliability of patterns of *Hox* genes expression as providing a 'molecular definition of tagmata.' There are many instances where domains of *Hox* gene expression cross tagmatic boundaries, often at later stages of development, when they modify the fate of individual segments within a tagma (Angelini and Kaufman 2005). Also, gene expression studies across the Arthropoda have shown that variation in individual *Hox* gene domains is not the only way trunk tagmosis can evolve, as conspicuous transformations of body organization can also result through changes in the downstream target genes, for instance via modifications of the regulatory regions of the latter (Carroll et al. 2005).

Deutsch and Mouchel-Vielh (2003) suggested that the proper definition of an abdomen depends not only on the absence of limbs, but also on the absence of local ventral nerve cord ganglia, in consideration of the fact that the abdominal segments of non-malacostracan crustaceans with a limbless abdomen, for example, copepods and anostracans, are innervated from ganglionic masses localized in the immediately anterior region. Brenneis and Richter (2010) cast doubt on this criterion as an infallible guide to homologize body regions across crustacean taxa, as, for instance, in mystacocarids, the posterior end of the series of appendages (post-cephalic segment V) does not coincide with the posterior end of the series of ventral nerve cord ganglia (post-cephalic segment VIII).

It is generally assumed that two body regions in two different animals can be homologous irrespective of the number of segments of which they are composed, if this is not the result of an evident 'migration' from flanking regions. But what should be expected for the segmental patterns within each tagma? Are they based on a 'grid' of *absolute* segmental positions within the tagma, or on a grid of *relative* segmental positions within the tagma? To undermine even the possibility of a general answer to this question, conflicting evidence emerges from the leg-bearing trunk of two sister taxa of centipedes, the geophilomorphs and the scolopendromorphs. In the geophilomorph *Clinopodes flavidus*, comparing the segmental trend of metric traits on tergites and sternites along the trunk between specimens with different numbers of trunk segments, it appears that the segmental pattern is largely independent of the number of segments on which it develops (Berto et al. 1997; Fusco and Minelli 2000). Features of the segmental pattern appear not to be based the absolute ordinal position of a given trunk segment but rather to depend on its relative position within the trunk. In contrast, as

mentioned above, in scolopendromorphs, the VII-VIII discontinuity always has the same absolute position within the leg-bearing trunk, despite variation in the total number of leg-bearing segments (21, 23, 39 or 43).

An argument in favor of an actual identity of tagmata, irrespective of the number of segments on which they develop, is the higher evolvability of the segments at the extremities of the series forming a given tagma. Indeed, one of the 'general laws of arthropod metamerism' formulated by Lankester (1902) states that the most anterior and most posterior segments of a tagma are particularly liable to regressive evolution. This rule appears as a special case of a more general principle formulated by Bateson (1894), according to which taxonomic variation among the elements of a serial structure, not necessarily the segments of an arthropod tagma, is very often concentrated at one extremity of the series, or both. This is often true of the most anterior abdominal segments of insects, of the most anterior opisthosomal segments of arachnids and, in general, of the most posterior segments of the arthropod trunk. It is not uncommon for the regressive transformations of terminal segmental elements (e.g., sclerites) to occur differentially on the dorsal and ventral sides.

As in the case of segments (see Sect. 9.3.5), the question of homology of tagmata finds a more suitable setting for investigation and discussion by adopting a concept of homology that is not of the all-or-nothing kind, but instead proceeds through a factorial disentangling of the diverse pathways of evolutionary change involved in different body structures of the same animal (Minelli 1998; Minelli and Fusco in press).

## 9.5 Limits to Segmental and Tagmatic Organization

Segmentation and tagmosis can account for arthropod body architecture only up to a point. The archetypical model of a trunk formed by a series of body modules that despite specialization preserve a common structure is observable only in a few trunk portions of a few arthropod subtaxa. There are indeed many different ways in which this scheme is subverted.

Limits to segmentation and tagmosis are both very common, so much as to paradoxically risk being overlooked, and very diverse, so that a complete list or a detailed classification is beyond the scope of this chapter. Instead, we propose to sort them here on the basis of two coarse-grain qualities: *pervasivity*, which applies to both segmentation and tagmosis, and *extension of segmental domain*. These two qualities are not completely independent of each other, and although defined in a way to reduce overlap with segmental and tagmatic mismatches, they are to some extent related to them. These qualities apply independently to specific body domains or body structures, rather than to the whole body, although similar qualities of different body domains can contribute to the general appearance (more or less clearly segmented, more or less clearly regionalized) of the body.

### 9.5.1 Pervasivity of Segmentation

Segmental pervasivity is how much of the anatomy of a given domain of the body, not necessarily coincident with a tagma, presents segmental organization, irrespective of whether the different segmental structures are in register or not.

At one extreme of a continuum of morphological expression, there is the leg-bearing trunk of geophilomorph centipedes, with the segmental series of tergites, sternites, pleurites, legs, tracheae, ostia, ganglia, body wall and locomotory muscles (Fig. 9.4a). On the other extreme, there is the almost unpatterned opisthosoma of pycnogonids, or the trunk of many parasitic arthropods (Fig. 9.4b). Obviously, this scale captures only a very superficial aspect of the question, as each potential segmental series can be present or absent, well developed or reduced, independent of other series. In the males of the hymenopteran families Orussidae and Stephanidae, the number of abdominal spiracles

is reduced to one functional pair, located in the most anterior segment (Vilhelmsen 2003), while in the same region, the central nervous system is formed by nine distinct ganglia. In contrast, the abdominal tracheal system is well developed in most Muscoidea dipterans, while all ventral ganglia (both thoracic and abdominal) are fused into a single mass (synganglion; Yeates et al. 2002).

Concentration of the central nervous system into a number of gangliar masses smaller than the number of the credited segmental units is very common, although the segmental organization is in general maintained at the level of lateral nerves, and it could be hardly otherwise if the structures that are innervated are segmental. For instance, Yeates et al. (2002) identify six different patterns of thoracic (T) and abdominal (A) ganglion fusion in dipterans, most of which have evolved several times independently: (1) fusion of T1 + T2 (some Nemestrinidae and Asiloidea), (2) fusion of T3 + A1 (some Nemestrinidae and Asiloidea), (3) fusion of T1 + T2 and T3 + A1 (some Xylophagomorpha, Stratiomyomorpha, Tabanomorpha and Cyclorrhapha), (4) anterior incorporation of neuromeres in one-terminal abdominal ganglion (some Eremoneura), (5) fusion of anterior abdominal neuromeres (some Cyclorrhapha) and (6) complete fusion of thoracic and abdominal neuromeres into a synganglion (evolved at least four times in the Eremoneura).

In the idealized concept of an arthropod ancestor with a 'perfectly' segmented body, the dominant evolutionary trend cannot be other than one of decreasing pervasivity with time. However, this is not necessarily so. For instance, certain derived aspects of segmentation have evolved independently, such as the paired gut diverticula in pycnogonids (where these diverticula project into the appendages), pseudoscorpions, acari, amblypygi and schizomids. Also, 'secondary segmentation,' in the form of body wall annulations independent of 'true' segmentation, has evolved in a few taxa, as for instance in pentastomids and in demodicid and eriophyid mites. Finally, the possibility of the evolutionary regaining of pleon segmentation in caprogammarid

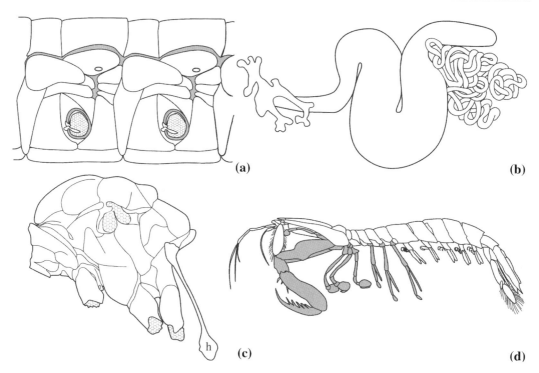

**Fig. 9.4** Segmental and tagmatic pervasivity. **a** High degree of segmental pervasivity in the leg-bearing trunk of a geophilomorph centipede (here, *Haplophilus subterraneus*, *side view*, legs cut; redrawn from Manton 1965); **b** low segmental pervasivity in the periodically unpatterned body of a parasitic copepod (here, *Lernaeocera branchialis*, female, redrawn from Kabata 1979); **c** high degree of tagmatic pervasivity in the morphologically highly integrated thorax of a tipulid dipteran (*side view*, legs and wing cut; h, halter; redrawn from Colless and McAlpine 1991); **d** *low degree* of tagmatic pervasivity in a stomatopod pereion with multifunctional appendages (pereiopods in *light gray*; redrawn from Westheide and Rieger 2007)

amphipods, after the reduction in size and segmentation at the basis of the more inclusive crown-group clade Caprellidea, has recently found some empirical support (Ito et al. 2011).

### 9.5.2 Segmental Domain

Segmental domain is the extent of a given segmental series within one or more tagmata. Segmental series of different body features can have different A-P domains of extent along the post-cephalic region. This is a very common and widespread situation. The most obvious case is the series of trunk appendages, that often extend over a limited portion of the post-cephalic region, either anterior (e.g., many crustaceans and all chelicerates and hexapods), or intermediate, with anterior and posterior appendage-less segments (e.g., symphylans and pauropods). But virtually all serial structures can extend over a limited A-P segmental domain, and examples can be found in all main arthropod taxa. To cite only a few examples involving the respiratory system: scorpion book lungs are only on opisthosoma segments III to VI, spiracles of geophilomorph centipedes are only present from the second leg-bearing segment to the penultimate one, there are no more than five pairs of pleurobranchs (side gills) in the pereion of decapod crustaceans and syrphid dipterans have only two pairs of spiracles, both of which are on the thorax and none on the abdomen.

### 9.5.3 Pervasivity of Tagmosis

Tagmatic pervasivity is the level of integration of the segmental elements of a given tagma. Structural and/or functional homogeneity of the segmental structures (e.g., appendages) and mechanical interlocking of structures within the tagma plus marked differentiation from flanking regions are all in accord to give a highly pervasive meaning to tagma identity. This is for instance the case of the thorax of many pterygote insects, especially in relation to flight performance (Fig. 9.4c). Otherwise, regional mismatch of different segmental series and/or low degrees of functional integration between segmental elements can almost nullify the significance of a given tagma, beyond its traditional designation. The latter is the case, for instance, of the stomatopod pereion, with the first pair of appendages used for cleaning, the second pair with raptorial function, followed by three pairs of chelate appendages for food manipulation and finally by three pairs of walking legs (Fig. 9.4d). Also the functional heterogeneity of segmental structures can somehow downgrade the significance of a tagma. This is true, for instance, of the thorax of dragonflies, mantids and beetles, where the prothorax is morphologically and functionally very divergent from the other two segments, and of the abdomen of many holometabolous insects, where the most anterior segments form a more or less globular subregion (*preabdomen*), while the most posterior tubular ano-genital subtagma (the *terminalia* complex) is telescopically retracted within the more anterior section of the abdomen at rest, to be evaginated during insemination or oviposition.

## 9.6 Final Remarks

As any zoologist knows, the actual body organization of arthropods is quite different from the archetypical model of a series of segmental modules organized in specialized groups that form the tagmata. The obviousness of this observation, however, can make it easy to overlook, to the point that idealized conceptions can subliminally creep into current investigations.

Segments, as modular partitions of the main body axis, are not inescapable products of the segmental organization of many different morphological structures, and tagmata are not unavoidable products of regional specializations of segmental structures. Segments and tagmata are to some extent epiphenomenal, emerging when particular, although not infrequent, conditions for segmentation and serial specialization are met. For instance, in cases of periodic concordance between several segmental structures (e.g., the series of leg pairs, tergites, sternites, spiracles, neuromers), the body can quite realistically be described as *comprised* of a certain number of segments, that is, modular partitions of the main body axis (Minelli and Fusco 2004). Otherwise, when many different segmental structures show discordant arrangement, segment delimitation and count should be acknowledged as at least arbitrary, if not questionably meaningful.

Segments and tagmata certainly have a fundamental role in descriptions and comparisons, but their value as developmental units or units of evolutionary change should not be uncritically assumed. With some attention, it is possible to exploit the undisputable descriptive value of segments and tagmata without falling in the trap of envisaging constraints to arthropod phenotypic evolution based on the limits posed by the evolution of these modules. In such a case, a morphological description of body units would be mistaken for a sort of 'internal description' (e.g., Lawrence 1992) of the animal capable of constraining evolutionary changes of body organization. Even conceding the existence of such units of internal description (see Fusco 2008 for an argument against), these do not necessarily match with the units we can detect by inspecting external morphology, internal anatomy or gene expression.

**Acknowledgments** Gerd Alberti, Geoff Boxshall, Diego Maruzzo, Marco Uliana and Lars Vilhelmsen kindly provided insightful comments on an early version of this chapter. Matteo Simonetti prepared the figure drawings.

# References

Abzhanov A, Kaufman TC (2000a) Embryonic expression patterns of the Hox genes of the crayfish *Procambarus clarkii* (Crustacea, Decapoda). Evol Dev 2:271–283

Abzhanov A, Kaufman TC (2000b) Crustacean (malacostracan) Hox genes and the evolution of the arthropod trunk. Development 127:2239–2249

Angelini DR, Kaufman TC (2005) Comparative developmental genetics and the evolution of arthropod body plans. Annu Rev Genet 39:95–119

Averof M, Akam M (1995) Hox genes and the diversification of insect and crustacean body plans. Nature 376:420–423

Bateson W (1894) Materials for the study of variation treated with especial regard to discontinuity in the origin of species. Macmillan, London

Berto D, Fusco G, Minelli A (1997) Segmental units and shape control in Chilopoda. Ent Scand Suppl 51:61–70

Bonato L, Foddai D, Minelli A (2003) Evolutionary trends and patterns in centipede segment number based on a cladistic analysis of Mecistocephalidae (Chilopoda: Geophilomorpha). Syst Ent 28:539–579

Bonato L, Edgecombe GD, Zapparoli M (2011) Chilopoda—taxonomic overview. In: Minelli A (ed) Treatise on zoology—anatomy, taxonomy, biology. The Myriapoda, vol 1. Brill, Leiden, pp 363–443

Brenneis G, Richter S (2010) Architecture of the nervous system in Mystacocarida (Arthropoda, Crustacea)—an immunohistochemical study and 3D reconstruction. J Morphol 271:169–189

Brolemann HW (1935) Faune de France 29. Myriapodes Diplopodes (Chilognathes I). Lechevalier, Paris

Brusca RC, Brusca GJ (2002) Invertebrates, 2nd edn. Sinauer, Sunderland

Budd GE (2001) Why are arthropods segmented? Evol Dev 3:332–342

Carroll SB, Grenier JK, Weatherbee SD (2005) From DNA to diversity: molecular genetics and the evolution of animal design, 2nd edn. Blackwell, Oxford

Chagas A Jr, Edgecombe GD, Minelli A (2008) Variability in trunk segmentation in the centipede order Scolopendromorpha: a remarkable new species of *Scolopendropsis* Brandt (Chilopoda: Scolopendridae) from Brazil. Zootaxa 1888:36–46

Colless DH, McAlpine DK (1991) Diptera. In: CSIRO (ed) The insects of Australia, 2nd edn. Melbourne University Press, Melbourne, pp 717–786

CSIRO (1991) The insects of Australia, 2nd edn. Melbourne University Press, Melbourne

Dahl E (1991) Crustacea Phyllopoda and Malacostraca: a reappraisal of cephalic and thoracic shield and fold systems and their evolutionary significance. Phil Trans R Soc B 334:1–26

Deutsch JS, Mouchel-Vielh E (2003) Hox genes and the crustacean body plan. BioEssays 25:878–887

Enghoff H (2011) Trans-segmental serial colour patterns in millipedes and their developmental interpretation (Diplopoda). Int J Myr 6:1–27

Enghoff H, Dohle W, Blower JG (1993) Anamorphosis in millipedes (Diplopoda)—the present state of knowledge and phylogenetic considerations. Zool J Linn Soc 109:103–234

Eriksson BJ, Tait NN, Budd GE, Janssen R, Akam M (2010) Head patterning and Hox gene expression in an onychophoran and its implications for the arthropod head problem. Dev Genes Evol 220:117–122

Fusco G (2005) Trunk segment numbers and sequential segmentation in myriapods. Evol Dev 7:608–617

Fusco G (2008) Morphological nomenclature, between patterns and processes: segments and segmentation as a paradigmatic case. In: Minelli A, Bonato L, Fusco G (eds) Updating the linnaean heritage: names as tools for thinking about animals and plants. Zootaxa 1950:96–102

Fusco G, Minelli A (2000) Measuring morphological complexity of segmented animals: centipedes as model systems. J Evol Biol 13:38–46

Harrison L (1914) On some Pauropoda from New South Wales. Proc Linn Soc NS Wales 39:615–634

Hilken G, Müller CHG, Sombke A, Wirkner CS, Rosenberg J (2011) Chilopoda—tracheal system. In Minelli A (ed) Treatise on zoology—anatomy, taxonomy, biology. The Myriapoda, vol 1. Brill, Leiden, pp 137–155

Hoffman RL (1982) Diplopoda. In: Parker SP (ed) Synopsis and classification of living organisms, vol 2. Mc Graw-Hill, New York, pp 689–724

Hoffman RL (2005) Monograph of the Gomphodesmidae, a family of African polydesmoid millipeds. Naturhistorisches Museum Wien, Wien

Horne DJ, Schön I, Smith RJ, Martens K (2005) What are Ostracoda? A cladistic analysis of the extant superfamilies of the subclasses Myodocopa and Podocopa (Crustacea Ostracoda). In: Koenemann S et al (eds) Crustacea and arthropod relationships (Crustacean Issues 16). Taylor and Francis, Boca Raton, pp 249–273

Hughes NC, Minelli A, Fusco G (2006) The ontogeny of trilobite segmentation: a comparative approach. Paleobiology 32:602–627

Huys R (1991) Tantulocarida (Crustacea: Maxillopoda): a new taxon from the temporary meiobenthos. Mar Ecol 12:1–34

Huys R, Boxshall GA (1991) Copepod evolution. Ray Society, London

Huys R, Boxshall GA, Lincoln R (1993) The tantulocaridan life cycle: the circle closed? J Crust Biol 13:432–442

Ito A, Aoki MN, Yahata K, Wada H (2011) Complicated evolution of the caprellid (Crustacea: Malacostraca: Peracarida: Amphipoda) body plan, reacquisition or multiple losses of the thoracic limbs and pleons. Dev Genes Evol 221:133–140

Janssen R, Prpic N-M, Damen WGM (2004) Gene expression suggests decoupled dorsal and ventral segmentation in the millipede *Glomeris marginata* (Myriapoda: Diplopoda). Dev Biol 268:89–104

Janssen R, Budd GE, Damen WGM, Prpic N-M (2008) Evidence for *Wg*-independent tergite border formation in the millipede *Glomeris marginata*. Dev Genes Evol 218:361–370

Kabata Z (1979) Parasitic Copepoda of British fishes. Ray Society, London

Kaestner A (1970) Invertebrate zoology: the Crustacea. Wiley, New York

Klingenberg CP (2008) Morphological integration and developmental modularity. Annu Rev Ecol Evol Syst 39:115–132

Koenemann S, Schram FR, Iliffe TM (2006) Trunk segmentation patterns in Remipedia. Crustaceana 79:607–631

Lankester ER (1902) Arthropoda. Encyclopaedia Britannica, 10th edn. Vol 25 pp 689–701

Larsen K (2005) Deep-sea Tanaidacea (Peracarida) from the Gulf of Mexico (Crustacean Monographs 5). Brill, Leiden

Lauterbach K-E (1975) Über die Herkunft der Malacostraca (Crustacea). Zool Anz 194:165–179

Lawrence RF (1963) New Myriapoda from Southern Africa. Ann Natal Mus 15:297–318

Lawrence PA (1992) The making of a fly. Blackwell, Oxford

Leśniewska M, Bonato L, Minelli A, Fusco G (2009) Trunk anomalies in the centipede *Stigmatogaster subterranea* provide insight into late-embryonic segmentation. Arthropod Struct Dev 38:417–426

Linder F (1952) Contributions to the morphology and taxonomy of the Branchiopoda Notostraca, with special reference to the North American species. Proc U S Nat Mus 102:1–69

Manton SM (1965) The evolution of arthropodan locomotory mechanisms. Part 8. Functional requirements and body design in Chilopoda, together with a comparative account of their skeleto-muscular system and Appendix on a comparison between burrowing forces of annelids and chilopods and it bearing upon the evolution of the arthropodan haemocoel. J Linn Soc (Zool) 46:251–483

Manuel M, Jager M, Murienne J, Clabaut C, Le Guyader H (2006) Hox genes in sea spiders (Pycnogonida) and the homology of arthropod head segments. Dev Genes Evol 6:481–491

Martin JW (1992) Branchiopoda. In: Harrison FW, Humes AG (eds) Microscopic anatomy of invertebrates, Crustacea, vol 9. Wiley-Liss, New York, pp 25–224

Matsuda R (1979) Morphologie du thorax et des appendices thoraciques des insectes. In: Grassé PP (ed) Traité de zoologie, vol 8(2). Masson, Paris, pp 1–289

Mauriès J-P (1964) Sur quelques diplopodes de la Peninsule Iberique. Bull Soc Hist nat Toulouse 99:157–170

McLaughlin PA (1980) Comparative morphology of recent Crustacea. Freeman, San Francisco

Meinhardt H (1994) The algorithmic beauty of sea shells. Springer, Heidelberg

Minelli A (1998) Molecules, developmental modules and phenotypes: a combinatorial approach to homology. Mol Phyl Evol 9:340–347

Minelli A (2003) The development of animal form: ontogeny, morphology and evolution. Cambridge University Press, Cambridge

Minelli A, Bortoletto S (1988) Myriapod metamerism and arthropod segmentation. Biol J Linn Soc 33:323–343

Minelli A, Fusco G (1995) Body segmentation and segment differentiation: the scope for heterochronic change. In: McNamara KJ (ed) Evolutionary change and heterochrony. Wiley, Chichester, pp 49–63

Minelli A, Fusco G (2004) Evo-devo perspectives on segmentation: model organisms, and beyond. Trends Ecol Evol 19:423–429

Minelli A, Fusco G (2005) Conserved versus innovative features in animal body organization. J Exptl Zool (Mol Dev Evol) 304B:520–525

Minelli A, Fusco G (in press) Homology. In: Kampourakis K (ed) The philosophy of biology—a companion for educators. Springer, Berlin

Minelli A, Koch M (2011) Chilopoda—general morphology. In: Minelli A (ed) Treatise on zoology—anatomy, taxonomy, biology. The Myriapoda, vol 1. Brill, Leiden, pp 43–66

Minelli A, Maruzzo D, Fusco G (2010) Multi-scale relationships between numbers and size in the evolution of arthropod body features. Arthropod Struct Dev 39:468–477

Newman WA, Knight MD (1984) The carapace and crustacean evolution—a rebuttal. J Crust Biol 4:682–687

Nguyen Duy–Jacquemin M, Uys C, Geoffroy J-J (2011) Two remarkable new species of Penicillata (Diplopoda, Polyxenida) from table Mountain National Park (Cape Town, South Africa). ZooKeys 156:85–103

Olesen J (2013) The crustacean carapace—morphology, function, development, and phylogenetic history. In: Watling L, Thiel M (eds) Functional morphology and diversity (Natural history of the Crustacea). Oxford University Press, Oxford, pp 103–139

Olesen J, Haug JT, Maas A, Waloszek D (2011) External morphology of *Lightiella monniotae* (Crustacea, Cephalocarida) in the light of Cambrian 'Orsten' crustaceans. Arthropod Struct Dev 40:449–478

Pereira LA, Minelli A, Barbieri F (1994) New and little known geophilomorph centipedes from Amazonian inundation forests near Manaus, Brasil (Chilopoda: Geophilomorpha). Amazoniana 13:163–204

Raff RA, Kaufman TC (1983) Embryos, genes, and evolution. MacMillan, New York

Ruppert EE, Fox RS, Barnes RD (2004) Invertebrate zoology, 7th edn. Thomson Brooks/Cole, Toronto

Scheller U (2011) Pauropoda. In: Minelli A (ed) Treatise on zoology—anatomy, taxonomy, biology. The Myriapoda, vol 1. Brill, Leiden, pp 467–508

Scheller U, Adis J (2002) Symphyla. In: Adis J (ed) Amazonian Arachnida and Myriapoda. Pensoft, Sofia, pp 547–554

Schileyko AA (2006) Redescription of *Scolopendropsis bahiensis* (Brandt, 1841), the relations between *Scolopendropsis* and *Rhoda*, and notes on some characters used in scolopendromorph taxonomy (Chilopoda: Scolopendromorpha). Arthropoda Selecta 15:9–17

Schram FR (1986) Crustacea. Oxford University Press, New York

Schram FR, Koenemann S (2004) Developmental genetics and arthropod evolution: on body regions of Crustacea. In: Scholtz G (ed) Evolutionary developmental biology of Crustacea (Crustacean Issues 15). Balkema, Lisse, pp 75–92

Shultz JW (2007) A phylogenetic analysis of the arachnid orders based on morphological characters. Zool J Linn Soc 150:221–265

Simaiakis SM (2009) Relationship between intraspecific variation in segment number and geographic distribution of *Himantarium gabrielis* (Linné, 1767) (Chilopoda: Geophilomorpha) in Southern Europe. Soil Org 81:359–371

Snodgrass RE (1935) Principles of insect morphology. McGraw-Hill, New York

Szucsich N, Scheller U (2011) Symphyla. In: Minelli A (ed) Treatise on zoology—anatomy, taxonomy, biology. The Myriapoda, vol 1. Brill, Leiden, pp 445–466

Uliana M, Bonato L, Minelli A (2007) The Mecistocephalidae of the Japanese and Taiwanese islands (Chilopoda: Geophilomorpha). Zootaxa 1396:1–84

van der Hammen L (1985) Comparative studies in Chelicerata III. Opilionida. Zool Verh Leiden 220:1–60

van der Hammen L (1989) An introduction to comparative arachnology. SPB Publishing, The Hague

Verhoeff KW (1926) Gliederfüßler: Arthropoda, II. Abteilung: Myriapoda. 2. Buch: Diplopoda. 3. Lieferung, in Bronn's Klassen und Ordnungen des Tierreichs, 5 (2). Akademische Verlagsgesellschaft, Leipzig, pp 289–480

Vilhelmsen L (2003) Phylogeny and classification of the Orussidae (Insecta: Hymenoptera), a basal parasitic wasp taxon. Zool J Linn Soc 139:337–418

Walossek D, Müller KJ (1997) Cambrian, 'Orsten'-type arthropods and the phylogeny of Crustacea. In: Fortey RA, Thomas RH (eds) Arthropod relationships. Chapman and Hall, London, pp 139–153

Watson, JAL, O'Farrell AF (1991) Odonata. In: CSIRO (ed) The insects of Australia, 2 edn. Melbourne University Press, Melbourne, pp 294–310

Westheide W, Rieger R (2007) Spezielle Zoologie. Teil 1: Einzeller und Wirbellose Tiere. 2. Auflage. Elsevier—Spektrum Akademischer, München

Yeates DK, David J, Merritt DJ, Baker CH (2002) The adult ventral nerve cord as a phylogenetic character in brachyceran Diptera. Org Divers Evol 2:89–96

# The Arthropod Head

## Stefan Richter, Martin Stein, Thomas Frase and Nikolaus U. Szucsich

## Contents

S. Richter (✉) · T. Frase
Allgemeine und Spezielle Zoologie, Universität
Rostock, Universitätsplatz 2, 18055, Rostock,
Germany
e-mail: stefan.richter@uni-rostock.de

T. Frase
e-mail: thomas.frase@uni-rostock.de

M. Stein
Danish Museum of Natural History, University of
Copenhagen, Universitetsparken 15, 2100,
Copenhagen, Denmark
e-mail: martin.stein@snm.ku.dk

N. U. Szucsich
Department of Integrative Zoology, University of
Vienna, Althanstrasse 14, 1090, Vienna, Austria
e-mail: nikola.szucsich@univie.ac.at

The anterior region of arthropods is profoundly influenced by effects of condensation and integration that has taken place in various character complexes. Prominent examples are the cerebralization of the central nervous system, the integration of anterior appendages to encompass sensory function and food uptake, the integration of anterior segments covered by a continuous dorsal shield, and a condensation of the endoskeleton which has resulted in the partial obscuring of the segmental organization. The borders between these different complexes, however, do not necessarily correspond. The exact composition and origin of the 'arthropod head' is an enduring problem in arthropod evolution. The discussion is heavily theory-laden, and any new account needs to consider a huge number of older theories and models (see Scholtz and Edgecombe 2005; 2006 for the most recent and detailed reviews). Although our understanding of and ideas about arthropod relationships have changed significantly over the last decade, the historical burden remains.

## 10.1    What is a Head?

In a recent debate, it has been suggested that morphological descriptions and terminology should be free of homology assumptions (Vogt 2008; Vogt et al. 2010). This approach is particularly challenging when it comes to a topic

A. Minelli et al. (eds.), *Arthropod Biology and Evolution*,
DOI: 10.1007/978-3-642-36160-9_10, © Springer-Verlag Berlin Heidelberg 2013

like the 'arthropod head problem' where almost every statement implies something about homology and evolutionary transformation polarity. The idea that homology statements should be avoided is not intended to deny the existence of homology; however, it simply seeks to separate the various steps in (evolutionary) morphology. A purely descriptive first step is properly followed by an evolutionary approach which encompasses the conceptualization of evolutionary characters (Wirkner and Richter 2010). In this context, it is important to point out that not only should the terminology used for description be free of homology assumptions, but also the underlying concepts themselves should be based on 'pure' description. This does not necessarily imply that the situation in adults should be considered in isolation because developmental data are at least as important, and gene expression data also play an obvious role. Even more crucial, gene expression data are primarily descriptive (although they certainly have a functional role) and become evolutionarily interpretative only in the framework of evolutionary developmental biology, as part of the new extended evolutionary synthesis (Pigliucci and Müller 2010).

Following on from Johann Wolfgang von Goethe's description of an insect head,[1] let us start with a concept of the arthropod head: the head (or cephalon) is always an anterior structure which should include primarily sensorial appendages and a brain which processes the sensory input. We also consider it important that the head be somehow separated from the trunk (otherwise, any imposed boundary would be based on non-descriptive concepts) and note that it may include appendages for feeding.

In segmented organisms like arthropods, the most conspicuous boundaries are constituted by segment boundaries (for a discussion of the segment problem, see Scholtz (2002) and Chap. 9). In the anterior part of the body of arthropods, traces of segmental organization are often restricted to some internal anatomical systems and the appendages. The dorsal surface usually fails to reflect segmental organization and is formed by a continuous sclerotization which covers a number of anterior segments. The exact number of segments involved differs. If this dorsal shield spans the regions/segments of proto-, deuto-, trito-cerebrum and the three following segments, it is usually referred to as the head (for stem lineage arthropods see below). In myriapods and hexapods, the term head capsule is usually used and refers to the entire cuticular envelope of the head. The posterior part of the head is often referred to as the gnathocephalon and comprises the mandibular and two maxillary segments (i.e., the segments of the maxillula and maxilla). We find it difficult to draw a clear distinction between the concept of the head capsule and that of the dorsal shield. The pattern, in any case, is obscured by the high level of disparity in the number of segments subsumed under a common dorsal shield/head capsule in crustaceans. This phenomenon is well known, and carcinologists differentiate between the cephalon and cephalothorax to describe the different conditions (Gruner and Scholtz 2004). Nevertheless, following the same concept as that applicable in insects and myriapods, a head would be present, for example, in Cephalocarida, Branchiopoda and Mystacocarida, and within Malacostraca at least in Bathynellacea. In other taxa, developmental data show the anterior tagma, or cephalothorax, to additionally incorporate one or more thoracic segments (Casanova 1991). In some cases—the Amphipoda, for example—a purely descriptive concept of a head, however, might well be applicable to a cephalothorax including the first thoracic segment with its maxilliped, which is clearly separated from the remaining thorax (Gruner 1993). In the Cephalocarida, a particularly interesting case, a dorsal head shield covers the segments of the

[1] Man betrachte die vollendeten Insecten! ... Das Haupt ist seinem Platze nach immer vorn, ist der Versammlungsort der abgesonderten Sinne und enthält die regierenden Sinneswerkzeuge, in einem oder mehreren Nervenknoten, die wir Gehirn zu nennen pflegen, verbunden. J.W. von Goethe—Erster Entwurf einer allgemeinen Einleitung in die vergleichende Anatomie ausgehend von der Osteologie. WA II, Bd 8, S. 13.

antennules, antennae, mandibles, maxillules and maxillae and clearly defines the border between the head and the trunk, although the maxillae and the first trunk limbs closely resemble each other (Fig. 10.1a). In the central nervous system, a well-demarcated brain can be distinguished from a subesophageal ganglion and the latter from the first thoracic ganglion (Stegner and Richter 2011; Fig. 10.1b). The sensorial appendages mentioned above as factors for inclusion in a concept of 'head' are the insect and myriapod antenna, and the crustacean antennules and antenna (Straus-feld 2012). Additional appendages are present (mandible, maxillule, maxilla/labium), though these are not primarily sense organs but rather are feeding organs.

Applying the concept of a head to chelicer-ates is most challenging. A prosomal shield covers eight regions/segments, including all the segments of the locomotory limbs (but note that in Xiphosura, the opisthosomal legs are used for

locomotion as well), and no head capsule as such is present. To complicate matters further, the appendages of the prosoma, such as the chelicerae, pedipalps and walking legs, are not primarily sensorial (although this concept itself is weak because almost all arthropod append-ages possess some kind of sensilla or sensory organs). Therefore, as most arachnologists would agree, the anterior tagma is not being considered as a head (but see below).

## 10.2   Endoskeleton

In addition to the exoskeleton, a number of endoskeletal structures in all arthropods both reinforce the head and serve as attachment sites for the cephalic musculature. Some of these structures are ingrowths from the cuticular invagination, whereas others are made up of connective tissue.

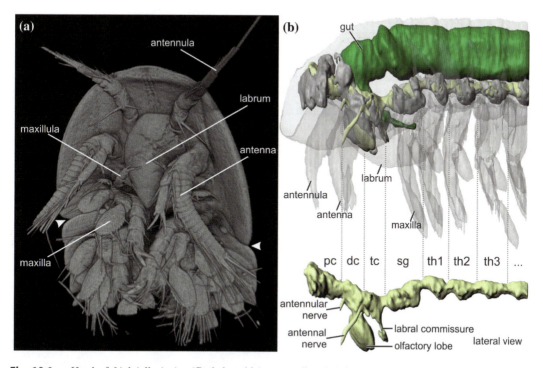

**Fig. 10.1**  **a** Head of *Lightiella incisa* (Cephalocarida). Note the distinct border between dorsal shield and thorax (*white arrow heads*), CLSM image. **b** Brain and ventral nerve cord of *Hutchinsoniella macracantha*. The brain clearly includes the tritocerebrum (tc); a subesophageal ganglion (sg) is present. Three-dimensional reconstruc-tions of somata (*gray*) and neuropil (*yellow*) in relation to body contours (semitransparent) and the gut (*green*), based on semi-thin sections; modified after Stegner and Richter (2011)

In most chelicerates, the main endoskeletal structures are made up of connective tissue, and segmental organization is retained to a high degree. A horizontal tendinous plate known as the endosternum is stabilized by more or less segmentally arranged dorsal, lateral and ventral suspensors (Firstman 1973; Shultz 1999, 2000, 2007).

The cephalic endoskeleton of myriapods is usually referred to as the *swinging tentorium* (Manton 1964; Koch 2003; Edgecombe 2004, 2010). As it is not fused to the head capsule, the tentorium has some degree of freedom of movement against it. Mandibular abduction is guided by the movements of the tentorium. In most Myriapoda, the tentorium is formed by a pair of internal cuticular processes (or tentorial arms) which are continuous with a number of exoskeletal bars integrated into membranous regions of the hypopharynx and the ventral head surface (Koch 2003; Szucsich et al. 2011). Most of the cuticular endoskeletal processes are associated with components made up of connective tissue. These tendinous structures either form bridges which link the cuticular processes or make up a horizontal framework suspended by muscles and tendons from the dorsal and lateral head capsule (Fig. 10.2).

In crustaceans, the cephalic endoskeleton is formed entirely of connective tissue. The segmental arrangement of components, which is easiest to follow in the trunk, is often still reflected in the cephalic endoskeleton (Fanenbruck 2003).

Although it displays some variation, the cephalic endoskeleton among groups of Hexapoda Ectognatha is always made up of a common set of components. This insect tentorium consists mainly of two pairs of cuticular invaginations. The anterior tentorial arms invaginate at the subgenal or epistomal ridges of the head exoskeleton and usually converge gradually in a caudal direction before merging to form the tentorial bridge. The tentorial bridge is formed by the fusion of the posterior tentorial arms, which invaginate at the ventral ends of the postoccipital ridge. In some groups, the central part of the resulting tentorium is enlarged, forming a plate-like structure known as the

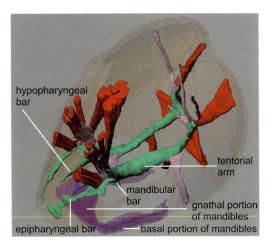

**Fig. 10.2** Cephalic endoskeleton of *Scutigerella immaculata* (Symphyla) in dorsolateral view (3D reconstruction). The cephalic endoskeleton encompasses cuticular components (*green*), components made up of connective tissue (*pink*) and muscular components (*red*). A pair of cuticular tentorial arms has a continuous connection to three strongly sclerotized bars of the exoskeleton, all of them lying at the anterior end of the tentorium. All three exoskeletal bars (epipharyngeal, hypopharyngeal and mandibular bar) are surrounded by weakly sclerotized regions. Thus, the tentorium can be moved against the head capsule (*transparent brown*) and is usually referred to as "swinging tentorium." Mandibular abduction is guided by movements of the tentorium

corpotentorium. In many groups of insects, additional cuticular components are present. A pair of dorsal arms often extends from the anterior arms to the dorsal head capsule. Reconstructing the hexapod ground pattern remains problematic, since the basally branching lineages display great disparity with regard to the cephalic endoskeleton. While all structures in Diplura and Collembola are made entirely of connective tissue, Protura exhibit cuticularized components, though these can hardly be homologized with structures of the ectognathan tentorium (Denis and Bitsch 1973; Koch 2000; Bitsch and Bitsch 2002).

Scenarios which address the evolution of the endoskeleton in the head and the trunk of arthropods usually proceed on the assumption that structures in postoral segments are homonomous. Among extant arthropods, the prosomal endoskeleton of chelicerates is usually judged to most closely reflect the plesiomorphic state for arthropods (Shultz 2007). Comparative investigations into extant and fossil groups indicate that a structure which Shultz (2001) termed the box-truss axial muscle system, which encompasses both muscular components and components made up of connective tissue, might be plesiomorphic for arthropods (Cisne 1974; Boudreaux 1979; Shultz 2001, 2007; Fanenbruck 2003). This structure is made up of a pair of longitudinal connectives situated dorsally of the nervous system. At the border of two segments, these longitudinal connectives are linked to each other by a transverse connective and connected to the dorsal and ventral cuticle by a pair of dorsal and ventral suspensors, respectively. The longitudinal connectives are additionally attached to the dorsal exoskeleton of the antecedent segment by a pair of anterior oblique suspensors and to the dorsal exoskeleton of the successive segment by a pair of posterior oblique suspensors. Objections to the scenario of a plesiomorphic box-truss axial muscle system mainly pertain to its underlying assumption that postoral structures are strictly homonomous, as mentioned above. In many arthropod groups in which the segmental organization of the endoskeleton is still clear, the anteriormost region of the postoral endoskeleton is far from being strictly homonomous, usually featuring an additional unpaired median component which is not present in more posterior segments (e.g., Hessler 1964; Fanenbruck 2003; Domínguez Camacho 2011). Most authors deem polarization to be unambiguous in only a few phylogenetic characters of the cephalic endoskeleton. An example is the presence of a 'swinging tentorium' in myriapods, frequently mentioned as one of the few synapomorphies supporting the monophyly of Myriapoda (Manton 1964; Koch 2003; Edgecombe 2004, 2010).

## 10.3   Brain

Arthropods possess a brain, known as the syncerebrum, which is generally interpreted to be the result of cephalization, that is, the structural and functional transformation of postoral neuromeres (usually considered to be ganglia) which are more or less fused to the pre-oral ancestral brain (Richter et al. 2010). A process of condensation can also be observed during development (e.g., Fritsch and Richter 2010), but the exact number of components involved is under dispute. A syncerebrum consisting of a protocerebrum, deutocerebrum and tritocerebrum can be identified in most adult mandibulate arthropods, though the borders between the three components are not always unambiguously recognizable. It could even be argued that a purely descriptive approach should omit the three terms. Generally, the three components are defined by the input of eyes, that is, compound eyes, median eyes, frontal eyes (protocerebrum), the input of the antennules (in crustaceans, but in the following, we include myriapod and insect antenna under the term antennule) (deutocerebrum) and the input of the (crustacean second) antenna (tritocerebrum).

Whether or not the tritocerebrum actually belongs to the arthropod brain is a matter of debate. Whereas Scholtz and Edgecombe (2006, p. 399) (see also Harzsch 2004) emphasize that 'its status as a brain neuromere' is evident, Mayer et al. (2010) suggest that the tritocerebrum (as a component of the syncerebrum) evolved in arthropod subgroups. Both views, obviously, are based on assumptions regarding the evolution of the syncerebrum and the arthropod ground pattern rather than simply describing an existing condition. Kirsch and Richter (2007), on the other hand, took a purely descriptive approach when considering the brain of the raptorial water flea *Leptodora kindtii* and concluded that it consists of a proto- and deutocerebrum only, because the tritocerebrum (or more precisely the ganglia which correspond to the tritocerebrum in taxa with a tripartite brain) is so far posterior in the head. Making the concept of the head independent from the concept of the syncerebrum affords a higher degree of freedom in discussions of possible evolutionary scenarios which involve both coupled and independent evolutionary events of cephalization and cerebralization. Whether or not the tritocerebrum is part of the brain varies among arthropods, and descriptions should not be concerned with whether or not the composition of the brain is plesiomorphic or derived.

The separation of the protocerebrum into an archicerebrum (comprising the mushroom bodies and the optic lobes) and a prosocerebrum (comprising the central complex)—one belonging to the acron, the other to a pre-antennal segment—appears to be hypothetical (Siewing 1969); on the basis of segmental gene expression data at least, there is no indication of an additional segment between the eyes and the antennules/chelicerae (reviewed by Scholtz and Edgecombe 2006). However, recent *Six3+* and *Otx* gene expression data support the notion that the protocerebrum is made up of two different portions, though the authors explicitly avoid calling them the archi- and prosocerebrum (Steinmetz et al. 2010). Even if the presence of a bipartite protocerebrum were to find support as an evolutionary concept (e.g., Strausfeld 2012), there is no unambiguous evidence that the two portions of the protocerebrum can be separated in adult arthropods (see also the discussion by Scholtz and Edgecombe 2006). We suggest that only the term protocerebrum has a place in a purely descriptive approach (or even better, as outlined above, more functionally defined subunits such as central complex, optic lobes, etc., should be used).

In chelicerates, no syncerebrum is present following the standard definitions. The prosoma instead contains a 'prosomal ganglion' which is usually separated into a supra- and subesophageal ganglion, though this distinction is predominantly a conceptual one because the entire ganglion is surrounded by somata which are not separated into somata clusters (Klussmann-Fricke and Wirkner, in progress). As a whole, the prosomal ganglion fits in with the concept of a brain as the most anterior condensation of neurites (see Richter et al. 2010).

To conclude, while it is plausible to assume the presence of a syncerebrum in the last common ancestor of all arthropods, the exact composition of the ground pattern in terms of the number of neuromeres it included remains, in our view, uncertain.

## 10.4 Gene Expression Data

Gene expression data provide important support for theories regarding the composition of arthropod heads (e.g., Scholtz 1997; Telford and Thomas 1998). That the arthropod head is composed of several units, generally considered to be segments, is beyond doubt. This is clearly recognizable in development and supported by gene expression data. Segment polarity genes such as *engrailed* and *wingless* in particular reveal the presence of six units in the head of most mandibulates (e.g., Scholtz 1997). The three anterior units are the protocerebral region (leaving it open whether this region is composed of two units or not) plus the segments of the antennules and antennae in crustaceans, or the antennal and intercalary segments in myriapods and insects. Schaeper et al. (2010) and Janssen et al. (2011) have recently detected *collier* expression in the intercalary segment of insects and myriapods, which the latter authors interpret as potential support for the traditional Atelocerata concept. The mandible, maxillule and maxilla (or labium) segments are distinguished by the expression of segment polarity genes (Scholtz 1997). In addition, certain segments are identifiable by the expression of Hox genes, a good indicator of homology of segments and their appendages. The proposed homology of the segments of the chelicera and antennule is based on the expression of Hox genes (Telford and Thomas 1998) and on evidence from axogenesis (Mittmann and Scholtz 2003). Moreover, the exact match in the anterior expression boundaries of the Hox genes *labial*, *proboscipedia* and *Deformed* supports the notion that the chelifore segment in Pycnogonida and the chelicera segment in the remaining euchelicerates are homologous (Manuel et al. 2006), making the hypothesis that the chelifore is innervated by the protocerebrum (Maxmen et al. 2005) improbable (see also Brenneis et al. 2008 for additional contradictory evidence based on axogenesis).

A comparison of all the Hox genes expressed in the head of mandibulates with the expression pattern in a spider led Averof (1998) to conclude that the spider prosoma indeed corresponds to the mandibulate head. On the basis of these data, the presence of a clear boundary between the prosoma and the opisthosoma coupled with the presence of a brain (i.e., the 'prosomal ganglion') might convince us to consider a head to be present in chelicerates too, despite the fact that this head is also used for walking and includes one more segment than the mandibulate head. This short summary shows that the gap between describing Hox gene expression and making assumptions about evolution is smaller than it may at first appear. Gene expression data certainly provide helpful arguments when it comes to establishing the homology of segments, but it must be borne in mind that the expression range of Hox genes might be subject to change, as in the case of crustacean maxillipeds (Averof and Patel 1997; Abzhanov and Kaufman 1999, 2000) to name but one example. Shared gene expression is not proof of segment homology. On the contrary, because Hox gene expression is responsible for certain aspects of, say, limb morphology, as in the case of the maxillipeds vs. non-specialized thoracopods (Liubicich et al. 2009; Pavlopoulos et al. 2009), Hox gene expression and the morphology of the limb cannot be used independently as support for homology hypotheses. We should also be aware that the developmental pathways of dorsal and ventral character systems may be decoupled (e.g., Janssen et al. 2006), an effect which might be especially pronounced in the anteriormost part of the body. This may account for at least some of the mismatches between the boundaries of different character systems.

## 10.5   Origin of the Arthropod Head

If a head is present in most arthropods and its components can reasonably be deemed to be homologous across all major taxa, the obvious question is how it evolved. Evolutionary scenarios need a starting point. The discussion of the origin of the arthropod head was once heavily influenced by the Articulata concept and

the idea of an annelid-like ancestor developing from a trochophoran larva with epi- and hyposphera. The Ecdysozoa concept (Aguinaldo et al. 1997; Giribet 2003) initially appeared to make the discussion obsolete, but aspects such as the presence of potentially homologous mushroom bodies in arthropods, annelids and other lophotrochozoans (Heuer and Loesel 2009; Heuer et al. 2010) and the similarities in the development of the anteriormost brain region in all bilaterian animals (Steinmetz et al. 2010) show that the debate is by no means at an end (see also Strausfeld 2012). Whatever the case, we take a less inclusive approach and start at the evolutionary level (i.e., the ground pattern) of Panarthropoda: the arthropods, onychophorans and tardigrades. Onychophorans in particular are a good starting point for understanding the evolution of the arthropod head (bearing in mind that they too display heterobathmy, a mixture of plesiomorphic and apomorphic characters). Although we have no doubt that the Tardigrada belong to the Panarthropoda as well (Dunn et al. 2008; Campbell et al. 2011), the current lack of consensus regarding the tardigrade head and the composition of the tardigrade brain lead us to exclude them from this discussion (Dewel and Dewel 1996; Zantke et al. 2008; Persson et al. 2012).

Although there is no distinct border on the surface of the body in Onychophora which would support a division into head and trunk, the differentiated appendages of the anterior body may indicate just this. The anterior body bears a pair of antennae, a pair of eyes, a mouth with a pair of jaws, and a pair of slime papillae. A head, then, is apparently present. On the basis of the expression of the anterior Hox genes *labial*, *proboscipedia*, *Hox3* and *Deformed*, the onychophoran jaws can be aligned with the chelicerae and the antennules, and the slime papillae with the pedipalps and the crustacean antennae (Eriksson et al. 2010). The slime papilla segment, therefore, corresponds to the intercalary segment in myriapods and insects (Eriksson et al. 2010; Mayer et al. 2010). This supports previous suggestions based on neuroanatomical data (Eriksson et al. 2003). Although the eyes might not correspond to the compound eyes but to the arthropod median eyes (Mayer 2006; but see Ma et al. 2012a supporting the idea that some fossil lobopodians possessed precursors of compound eyes), they belong to the corresponding region of the arthropod protocerebral region. However, it should be noted that Strausfeld (2012) suggests that the compound eyes are the structures which correspond to the segment associated with the onychophoran jaw, implying that the slime papilla corresponds in position to the chelicerae/antennules, which we do not hold to be very likely. Whatever the case, the onychophoran antenna is innervated by the anteriormost portion of the onychophoran brain (Eriksson et al. 2003). It has been suggested that the onychophoran brain is tripartite, as in arthropods, and that it features what Strausfeld et al. (2006) consider to be a protocerebrum, deutocerebrum and tritocerebrum. However, Mayer et al. (2010) performed backfills of cephalic segmental nerves in adult onychophorans and found that the somata of the neurons innervating the jaws and the slime papillae lie adjacent to the base of their nerves. While the neuron of the nerve innervating the jaws is situated in the posteriormost part of the brain (i.e., the deutocerebrum), the neurons innervating the slime papillae lie clearly separate from the brain in the ventral nerve cord. Following the definition of the brain in Richter et al. (2010), then, the onychophoran brain is clearly bipartite. Interestingly, while the onychophoran head consists of three units (protocerebral region and two segments), the brain encompasses two neuromeres only. However, we do not see any conceptual need for a strict correlation of these systems, that is, transformation of appendages and brain composition. Mayer and Harzsch (2008) considered the absence of ganglia in the ventral nerve cord of onychophorans to be the plesiomorphic condition, which could imply that the onychophoran brain is formed not by fused ganglia but by non-ganglionized neuromeres. In evolutionary terms, this assumption might imply that the cephalization of segmental units

preceded the formation of ganglia in the lineage leading to the arthropods. The syncerebrum could well represent a fusion of neuromeres but not of ganglia, potentially explaining why no clearly separated ganglia (corresponding to the proto-, deuto-, tritocerebrum) can be identified in the arthropod brain (see Richter et al. 2010 for more details, and Strausfeld (2012) for a different scenario). In an alternative scenario, the absence of distinct ganglia is interpreted as a secondary feature that is coupled with the probably secondary loss of the clear segmental organization of the body surface (something which in the main can now only be deduced from the distribution of the appendages).

Taking the onychophoran head as a starting point, the mandibulate head has three additional more posterior segments which are fused with the anterior part of the head. The question of when and how often the tritocerebrum became part of the brain remains open. The condition in the raptorial water flea *Leptodora* (Kirsch and Richter 2007) is certainly a secondary one. In Mystacocarida, for example, the tritocerebrum is only slightly separated from the proto–deutocerebral complex (Brenneis and Richter 2010). The presence of a brain featuring an incorporated tritocerebrum in myriapods seems to provide some support for the hypothesis that the mandibulate tripartite brain evolved only once (Sombke et al. 2012).

One remarkable transformation is that involving onychophoran jaws and mandibulate antennules. Taking into account the presence of the chelicerate chelicerae on the corresponding segment, a jaw-like structure might indeed represent the original condition. This would imply a major transformation in the stem lineage of Mandibulata from some kind of feeding structure to a 'secondary antenna' sensu Scholtz and Edgecombe (2005). On the basis of fossil lobopodians and arthropods, however, it seems more likely that the feeding structures evolved independently in onychophorans and chelicerates, with a non-specialized appendage as starting point (see Ou et al. 2012 and below).

## 10.6   The Fate of the Onychophoran Antenna

Another fascinating but problematic potential transformation is that from the onychophoran antenna (or any lobopodian antenniform appendage, see Ou et al. 2012) into the arthropod labrum. The homology of the labrum throughout arthropods appears to be strongly supported, particularly by the fact that its development is strikingly similar in chelicerates and mandibulates (Kimm and Prpic 2006). In many cases, the labrum anlage appears as a pair of structures at the front of the embryo, which later move backwards and fuse into a single organ (e.g., Ungerer and Wolff 2005; Mittmann and Wolff 2012). A comparable structure, however, is absent in Pycnogonida (Brenneis et al. 2011). There is some debate concerning the term 'labrum.' According to Maas et al. (2003) a 'fleshy labrum' evolved only in a taxon called Labrophora, which includes the extinct Phosphatocopina and a taxon which the authors call Eucrustacea, including all recent crustaceans and probably also all the hexapods (on the basis of molecular data; Regier et al. 2010; von Reumont et al. 2012). Non-Labrophora (particularly chelicerates and trilobites), then, are assumed to possess a structure called a hypostome, a sclerotized plate. Because of the detailed correspondences—as mentioned above—in the development of the 'upper lip' in chelicerates and crustaceans, a hypostome would also have to be present in crustaceans and the labrophoran labrum would have to be interpreted as a structure which evolved as part of the hypostome (Waloszek et al. 2007).

The segmental affinities of the labrum (or hypostome/labrum) have been debated intensively (see Scholtz and Edgecombe 2006 for a detailed discussion of labrum homology and segmental affinities). Recently, Posnien et al. (2009) showed that the labrum is formed by an appendage regulatory gene network and concluded as a result that the labrum is an appendage-like structure. Steinmetz et al. (2010) found *Six3* expression anterior to *Otx* expression

in the anteriormost region of the developing brain in both arthropods (the area where the antenna originates) and onychophorans (the area which innervates the antenna). Interestingly, this comes close to the test suggested by Scholtz and Edgecombe (2006) for obtaining direct support for the homology of the onychophoran antenna and the arthropod labrum. On the basis of these findings and the alignment of the onychophoran jaw segment with the mandibulate antennule segment using *lab*, *pb*, *Hox3* and *Dfd* expression, Eriksson et al. (2010) suggested that the onychophoran antenna is indeed homologous to the labrum. This view is supported by Strausfeld (2012) who hypothesized a complex scenario for the evolutionary transformation from the location of the frontal appendage into the more posterior position of the labrum.

Although we might not be able to solve the labrum problem, we do have some evidence to support the alternative hypothesis for the fate of the (onychophoran) primary antenna discussed by Scholtz and Edgecombe (2006), according to which the frontal filaments on the anterior part of the head in Remipedia and cirripedian nauplius larvae represent remnants of the primary antenna. In branchiopods, Fritsch et al. (2013) distinguish between the filamentous external 'frontal filament' and an internal region beneath the frontal filaments which they term the 'frontal filament organ' (also known as organ of Belonci). Although the two structures undoubtedly form one functional unit, we support this distinction, which reflects the history of discovery of the two structures (see Fritsch et al. 2013). A pair of frontal filaments is present in Notostraca, and other Phyllopoda.

In addition frontal filaments are also be present in certain copepods (Elofsson 1971) and certain ostracodes (Andersson 1977). They are apparently absent in the chelicerates, but *Cambropycnogon* (probably a representative of the stem lineage of Pycnogonida) possesses structures very similar to those of Notostraca (see Waloszek and Dunlop 2002). Frase and Richter (2013) show that nerves of the frontal filament organs (also known as cavity receptor organ,

Elofsson and Lake 1971) in Anostraca appear at the same time as the anlagen of the protocerebrum in the embryonic stages, when no evidence of functionality exists so early on (i.e., serotonergic immunoreactivity starts later). These neurite bundles are still present in the larval stages, but as the protocerebrum, the compound eyes and their nerves grow, and they cover the frontal filament organs and cause them to lose their prominence. In adults, the external part of the frontal filament organs are recognizable only as small cavities (Møller et al. 2004 for *Eubranchipus*). The correspondences between the nerves of the frontal filament organs and those of the onychophoran antenna are remarkable. Both originate in the anterolateral region of the protocerebrum and appear at the same time as the protocerebrum early on in development (Eriksson and Budd 2000; Mayer et al. 2010). If our suggestion of homology of the onychophoran antenna and the crustacean frontal filaments (Fig. 10.3) is correct, the labrum problem would remain unsolved but the need for a complicated scenario of transformation of the primary antenna into the labrum in the ancestral lineage

of arthropods would be obsolete (see Frase and Richter, 2013).

## 10.7 A Fossil Perspective on the Evolution of the Arthropod Head

Our view of the evolution of the arthropod head has been dominated by neontological data, but the rich fossil record of (pan)arthropods cannot be left unconsidered. Over the last two decades, fossils have played an increasingly central role in hypotheses concerning the evolution of the arthropod head (e.g., Chen et al. 1995; Budd 2002; Scholtz and Edgecombe 2005; Waloszek et al. 2005). Data retrieved from the fossil record are mostly limited to external morphology though internal structures have been reported in rare instances. Relatively common are segmental mid-gut diverticula (Butterfield 2002; Vannier and Chen 2002), which have been used to infer head segment numbers (Zhang et al. 2007; Stein and Selden 2012). Rarely, and sometimes controversially, other internal anatomical features

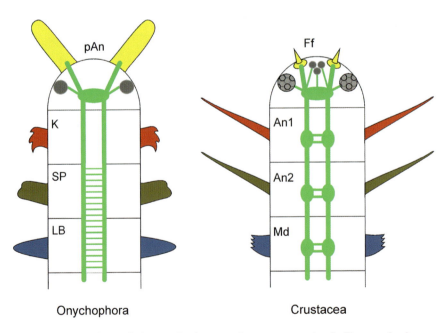

**Fig. 10.3** Schematic comparison of the onychophoran and a crustacean head. The onychophoran antenna is suggested as being homologous to the crustacean frontal filaments

are preserved which could be informative about the evolution of the arthropod head. These include putative elements of the endoskeletal system (Cisne 1975; Whittington 1993; Stein 2010), the musculature (Eriksson et al. 2012) and the nervous system (Bergström et al. 2008; Ma et al. 2012b). Just as the interpretation of morphological structures, in particular internal, is problematic in fossils so is the phylogenetic position of the taxa in question. In fact, the two problems are often linked, as our interpretation of morphological features can be influenced by expectations derived from the assumed phylogenetic position of the taxon studied (see the debate about the presence of lobopodous limbs in *Opabinia regalis*; Budd 1996; Zhang and Briggs 2007; Budd and Daley 2011). This phenomenon arises even when every attempt is made—as is desirable—to describe morphological structures independently of phylogenetic position.

An important aspect of any discussion involving both fossil and recent arthropods is the distinction between crown group Arthropoda (the last common ancestor of Chelicerata and Mandibulata and all its descendants) and stem group arthropods, that is, all representatives of the pan-Arthropoda (see Lauterbach 1989, sensu Meier and Richter 1992) which do not belong to the crown group Arthropoda (see Edgecombe 2010). The exact composition of the stem group of arthropods depends on the position of the Onychophora and Tardigrada, which to date remains unresolved. There are a number of fossil taxa, collectively referred to as lobopodians, which have a tubular body and unjointed tubular appendages. These taxa include possible stem group representatives of Panarthropoda, Onychophora, Tardigrada and part of the stem group of Arthropoda. Lobopodia is sometimes considered a paraphyletic assemblage which also includes the crown group of Onychophora (e.g., Liu et al. 2011) or those of both Onychophora and Tardigrada (e.g., Ma et al. 2009) but not the crown group of Arthropoda. The more crownward representatives of the arthropod stem, the Arthropoda *sensu stricto* of Waloszek et al. (2005), have pivot-jointed appendages and sclerotized segmental tergites. There is

consensus on the placement of some prominent fossil taxa, such as Trilobita or the more inclusive Artiopoda (Trilobita and closely related, non-biomineralizing forms; Stein and Selden 2012) in the crown group Arthropoda, and some taxa, such as *Fuxianhuia protensa* and similar forms from the Early Cambrian of Chengjiang, in the arthropod stem group (e.g., Budd 2002; Waloszek et al. 2005; Edgecombe 2010). However, there are still taxa which are subject of debate with regard to their phylogenetic position, one being the 'great appendage arthropods,' or Megacheira, which are considered to be either stem group Arthropoda (e.g., Budd 2002, 2008; Legg et al. 2012) or stem group Chelicerata (e.g., Chen et al. 2004; Haug et al. 2012a) (Fig. 10.4). The first cephalic appendage in the megacheirans is a large, ostensibly raptorial appendage termed the great appendage or multi-chela (Haug et al. 2012b). Another controversial taxon is *Canadaspis*, which is considered to belong to either the stem group Arthropoda (e.g., Budd 2002; Waloszek et al. 2007) or Mandibulata (e.g., Briggs et al. 2008). This is an important problem since the advocates of a stem group position afford these taxa a pivotal role in hypotheses regarding the early evolution of the arthropod head (Budd 2002, 2008). Regardless of the phylogenetic position of these taxa, the fossils do permit some inferences to be made about cephalization in the stem species of Arthropoda.

In all unambiguous fossil members of the arthropod crown group, and in the megacheirans, a single dorsal shield is present which covers a number of segments which are fused into one unit (comparable to the condition in Cephalocarida; see Fig. 10.1a). The number of appendage-bearing segments incorporated into this unit in the arthropod ground pattern and the constancy of this number among the fossil taxa is still a matter of debate. In the Megacheira, three appendage-bearing segments have been suggested for a number of species and four for others (see e.g., Edgecombe et al. 2011). A key taxon for the presence of only three is *Leanchoilia superlata*, but a recent revision revealed a small, specialized appendage posterior to the

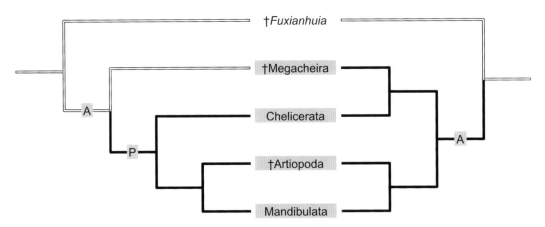

**Fig. 10.4** Two alternative phylogenetic placements of the Megacheira and consequences for the occurrence of a head shield incorporating more than two appendage-bearing segments. *Left* Megacheira as stem group arthropods; the head shield, an autapomorphy (*A*) of crown group Arthropoda + Megacheira, is retained as a plesiomorphy (*P*) in the ground pattern of Arthropoda. *Right* Megacheira as stem group chelicerates; the head shield is an autapomorphy (*A*) of crown group Arthropoda. *Solid lines*: crown group Arthropoda; *halftone fill*: head shield incorporating more than two appendage-bearing segments

great appendage, increasing the segment count to four (Haug et al. 2012b). This is the number found in Trilobita as well as the early representatives of Cambrian Crustacea *sensu* Stein et al. (2005) (presumably stem lineage representatives of Tetraconata). The number of segments in the head of Artiopoda has been claimed to be highly unstable (e.g., Zhang et al. 2007), but a number other than four can only be substantiated for Naraoiidae, at least some of which have five (Zhang et al. 2007), and Xandarellida, which have five to seven (Ramsköld et al. 1997). Stein and Selden (2012), for instance, found that only four segments are present in *Emeraldella brocki*, one of the key taxa cited for deviating segment counts. Megacheirans and artiopodans display a lesser degree of cephalic limb specialization than the extant mandibulatan subtaxa, although recent restudies show evidence of a gradual differentiation across the head–trunk boundary in some representatives, with the appendages of the head and anterior trunk more adapted to feeding than the mid- to posterior trunk appendages (Stein and Selden 2012). Regardless of the degree of appendage differentiation in the head, the head would, with its cohesive shield, still act as a single unit distinct from the free tergite-bearing segments of the trunk and thus be separate from the trunk (see

above). It is also true that this head bears important sensory structures in the form of the eyes as well as sensory appendages (antennae in artiopodans, long flagella on the great appendages of some megacheirans) and appendages suited to nutrition (albeit often coupled with a locomotory function).

To which segments the appendages observed in the anterior and cephalic region of some fossil arthropods belong is another point of contention, in particular with regards to the great appendages of megacheirans and the so-called frontal appendages of some lobopodians and of taxa such as *Kerygmachela kierkegaardi* (Fig. 10.5) and possibly the anomalocaridids. The latter taxa are situated on either side of the lobopod-arthropod transition (a character-based distinction within the arthropod stem lineage). Because of its ostensible position as the most anterior appendage flanking the mouth, the frontal appendage of *Kerygmachela* is considered to be protocerebral and homologized with the onychophoran antenna (e.g., Budd 2002). Going even further, the frontal appendage of the putative anomalocaridid *Parapeytoia yunnanensis* has been homologized with the megacheiran great appendage, with the latter consequently also interpreted as being protocerebral (e.g., Budd 2002; Daley et al. 2009). The

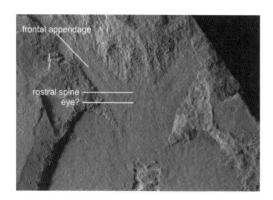

**Fig. 10.5** *Kerygmachela kierkegaardi* with prominent frontal appendages, rostral spines and bulbous structures that could represent the eyes

anomalocaridid affinities of *Parapeytoia*, however, are questionable, and the taxon could represent a *bona fide* megacheiran (Stein 2010). The alternative interpretation of the megacheiran great appendage as deutocerebral has gained support and wider acceptance recently (Stein 2010; Haug et al. 2012a). An argument for the deutocerebral interpretation comes from the position of the great appendage; it inserts laterally to the hypostome/labrum (e.g., Haug et al. 2012b) and not directly anteriorly. If, however, we accept that the megacheiran great appendage is homologous with the frontal appendages of *Kerygmachela* and anomalocaridids (Stein 2010), the latter would need to be reinterpreted as deutocerebral appendages as well. In this regard, it is interesting that the onychophoran antenna and the crustacean frontal filaments are sensorial organs (see above). Among the species considered to be important in the lobopod-arthropod transition (Liu et al. 2006, 2007; Dzik 2011), only *Megadictyon* cf. *haikouensis*, *Jianshanopodia decora* and *Siberion lenaicus* have prominent frontal appendages comparable to those in *Kerygmachela*. The anterior appendages of most other lobopodians are commonly interpreted as sensory (e.g., Ramsköld and Chen 1998), while the frontal appendages of *Kerygmachela* and anomalocaridids are usually regarded as feeding appendages. On the other hand, structures comparable to the onychophoran antenna might be present in

*Kerygmachela* in the form of the 'rostral spines' dorsal to the mouth (Budd 1998). Like the onychophoran antenna, these spines are associated with structures which Budd (1998) interprets as eyes, and they are annulated or 'segmentally divided' (Budd 1998). They are in a similar position to the onychophoran antennae, while the frontal appendages flanking the mouth would be positionally homologous to the deutocerebral onychophoran jaws. Little information is available about the anterior regions of *Megadictyon*, *Jianshanopodia*, and *Siberion*. A new, alternative interpretation, therefore, could be that the frontal appendages of some lobopodians and anomalocaridids indeed represent the homologous appendage to the megacheiran great appendage but are deutocerebral rather than protocerebral.

Recently, a specimen of the stem group arthropod *F. protensa* has been reported in which the nervous system is said to be preserved (Ma et al. 2012b), and it has been suggested that the syncerebrum already present in the ground pattern of (crown group) Arthropoda was tripartite, that is, that the tritocerebral part was fused to the proto- and deutocerebrum. There has been a contentious debate whether the first, antenna-like appendage of *Fuxianhuia*, was a sensory protocerebral 'primary' antenna (e.g., Scholtz and Edgecombe 2005) or a deutocerebral appendage that was largely unspecialized and served both in nutrition and as a sensory organ (e.g., Waloszek et al. 2005; Bergström et al. 2008). The new material seems to lend further support for the appendage being deutocerebral. The primary function (if any) remains unclear, but it is possible that a sensory 'secondary' antenna was already present in the arthropod ground pattern and not only in the mandibulate lineage. We hold the interpretation of this single specimen to be somewhat problematic, however, and would hesitate for the moment to reach such general conclusions on the basis of this specimen alone.

In summary, there is fossil evidence that the last common ancestor of Chelicerata and Mandibulata (i.e., crown group Arthropoda) had a head comprising the ocular region and at least three, but more likely four appendage-bearing segments. The anterior appendage inserts laterally to the hypostome/labrum and probably represents the deutocerebral appendage, but a smaller appendage-like structure might have been present anteriorly of this appendage. The postantennular appendages display little differentiation other than a gradual shift anteriorly toward limbs more adapted to feeding.

**Acknowledgments** We thank the editors for helpful comments improving the manuscript and Lucy Cathrow for a careful copy editing. The studies on the evolution of the arthropod brain have been supported by the DFG (RI 837/9–1, 2; 10–1, 2). MS is supported by the Carlsberg Foundation.

# References

Abzhanov A, Kaufman TC (1999) Novel regulation of the homeotic gene *Scr* associated with a crustacean leg-to-maxilliped appendage transformation. Development 126:1121–1128

Abzhanov A, Kaufman TC (2000) Embryonic expression patterns of the Hox genes of the crayfish *Procambarus clarkii* (Crustacea, Decapoda). Evol Dev 2:271–283

Aguinaldo AMA, Turbeville JM, Linford LS, Rivera MC, Garey JR, Raff RA, Lake JA (1997) Evidence for a clade of nematodes, arthropods and other moulting animals. Nature 387:489–493

Andersson A (1977) The organ of Bellonci in ostracodes: an ultrastructural study of the rod-shaped, or frontal, organ. Acta Zool (Stockh) 58:197–204

Averof M (1998) Origin of the spider's head. Nature 395:436–437

Averof M, Patel NH (1997) Crustacean appendage evolution associated with changes in Hox gene expression. Nature 388:682–686

Bergström J, Hou X, Zhang X, Clausen S (2008) A new view of the Cambrian arthropod *Fuxianhuia*. GFF 130:189–201

Bitsch C, Bitsch J (2002) The endoskeletal structures in arthropods: cytology, morphology and evolution. Arthropod Struct Dev 30:159–177

Boudreaux HB (1979) Significance of intersegmental tendon system in arthropod phylogeny and monophyletic classification of Arthropoda. In: Gupta AP (ed) Arthropod phylogeny. Van Nostrand Reinhold, New York, pp 551–586

Brenneis G, Arango CP, Scholtz G (2011) Morphogenesis of *Pseudopallene* sp. (Pycnogonida, Callipallenidae) I: embryonic development. Dev Genes Evol 221:309–328

Brenneis G, Richter S (2010) Architecture of the nervous system in Mystacocarida (Arthropoda, Crustacea)—an immunohistochemical study and 3D reconstruction. J Morphol 271:169–189

Brenneis G, Ungerer P, Scholtz G (2008) The chelifores of sea spiders (Arthropoda, Pycnogonida) are the appendages of the deutocerebral segment. Evol Dev 10:717–724

Briggs DEG, Lieberman BS, Hendricks JR, Halgedahl SL, Jarrard RD (2008) Middle Cambrian arthropods from Utah. J Paleontol 82:238–254

Budd GE (1996) The morphology of *Opabinia regalis* and the reconstruction of the arthropod stem-group. Lethaia 29:1–14

Budd GE (1998) The morphology and phylogenetic significance of *Kerygmachela kierkegaardi* Budd (Buen Formation, Lower Cambrian, N Greenland). Trans R Soc Edinb Earth Sci 89:249–290

Budd GE (2002) A palaeontological solution of the arthropod head problem. Nature 417:271–275

Budd GE (2008) Head structure in upper stem-group euarthropods. Palaeontology 51:561–573

Budd GE, Daley AC (2011) The lobes and lobopods of *Opabinia regalis* from the middle Cambrian Burgess Shale. Lethaia 45:83–95

Butterfield NJ (2002) *Leanchoilia* guts and the interpretation of three-dimensional structures in Burgess Shale-type fossils. Paleobiology 28:155–171

Campbell LI, Rota-Stabelli O, Edgecombe GD, Marchioro T, Longhorn SJ, Telford MJ, Philippe H, Rebecchi L, Peterson KJ, Pisani D (2011) MicroRNAs and phylogenomics resolve the relationships of Tardigrada and suggest that velvet worms are the sister group of Arthropoda. Proc Natl Acad Sci USA 108:15920–15924

Casanova B (1991) Origine protocéphalique antennaire de la carapace chez les Leptostracés, Mysidacés et Eucarides (Crustacés). Cr hebd Acad Sci 312(III):461–468

Chen J, Edgecombe GD, Ramsköld L, Zhou G (1995) Head segmentation in Early Cambrian *Fuxianhuia*: Implications for arthropod evolution. Science 268:1339–1343

Chen J, Waloszek D, Maas A (2004) A new 'great-appendage' arthropod from the Lower Cambrian of China and homology of chelicerate chelicerae and raptorial antero-ventral appendages. Lethaia 37:3–20

Cisne JL (1974) Trilobites and the origin of arthropods. Science 186:13–18

Cisne JL (1975) Anatomy of *Triarthrus* and the relationships of the Trilobita. Fossils Strata 4:45–63

Daley AC, Budd GE, Caron J, Edgecombe GD, Collins D (2009) The Burgess Shale anomalocaridid *Hurdia* and its significance for early euarthropod evolution. Science 323:1597–1600

Denis JR, Bitsch J (1973) Structure céphalique dans les ordres des insectes. In: Grassé PP (ed) Traité de zoologie: Anatomie, systématiques, biologie, tome VIII Insectes: tête, aile, vol. Masson, Paris, pp 101–593

Dewel RA, Dewel WC (1996) The brain of *Echiniscus viridissimus* Peterfi, 1956 (Heterotardigrada): A key to understanding the phylogenetic position of tardigrades and the evolution of the arthropod head. Zool J Linn Soc 116:35–49

Domínguez Camacho M (2011) Cephalic musculature in five genera of Symphyla (Myriapoda). Arthropod Struct Dev 40:159–185

Dunn CW, Hejnol A, Matus DQ, Pang K, Browne WE, Smith SA, Seaver E, Rouse GW, Obst M, Edgecombe GD, Sorensen MV, Haddock SHD, Schmidt-Rhaesa A, Okusu A, Kristensen RM, Wheeler WC, Martindale MQ, Giribet G (2008) Broad phylogenomic sampling improves resolution of the animal tree of life. Nature 452:745–749

Dzik J (2011) The xenusian-to-anomalocaridid transition within the lobopodians. Boll Soc Paleontol Ital 50:65–74

Edgecombe GD (2004) Morphological data, extant Myriapoda, and the myriapod stem-group. Contrib Zool 73(3):207–252

Edgecombe GD (2010) Arthropod phylogeny: an overview from the perspectives of morphology, molecular data and the fossil record. Arthropod Struct Dev 39:74–87

Edgecombe GD, García-Bellido DC, Paterson JR (2011) A new leanchoiliid megacheiran arthropod from the Lower Cambrian Emu Bay Shale, South Australia. Acta Palaeontol Polon 56:385–400

Elofsson R (1971) The ultrastructure of a chemoreceptor organ in the head of copepod crustaceans. Acta Zool 52:299–315

Elofsson R, Lake PS (1971) On the cavity receptor organ (X-organ or organ of Bellonci) of *Artemia salina* (Crustacea: Anostraca). Ztschr Zellforsch mikr Anat 326:319–326

Eriksson BJ, Budd GE (2000) Onychophoran cephalic nerves and their bearing on our understanding of head segmentation and stem-group evolution of Arthropoda. Arthropod Struct Dev 29:197–209

Eriksson BJ, Tait NN, Budd GE (2003) Head development in the onychophoran *Euperipatoides kanangrensis* with particular reference to the central nervous system. J Morphol 255:1–23

Eriksson BJ, Tait NN, Budd GE, Janssen R, Akam M (2010) Head patterning and Hox gene expression in an onychophoran and its implications for the arthropod head problem. Dev Genes Evol 220:117–122

Eriksson ME, Terfelt F, Elofsson R, Marone F (2012) Internal soft-tissue anatomy of Cambrian 'Orsten' arthropods as revealed by synchrotron x-ray tomographic microscopy. PLOSone 7(8):e42582. doi:10.1371/journal.pone.0042582

Fanenbruck M (2003) Die Anatomie des Kopfes und des cephalen Skelett-Muskelsystems der Crustacea, Myriapoda und Hexapoda: Ein Beitrag zum phylogenetischen System der Mandibulata und zur Kenntnis der Herkunft der Remipedia und Tracheata. Doctoral Thesis, Fakultät für Biologie, Ruhr-Universität Bochum, Bochum

Firstman B (1973) The relationship of the chelicerate arterial system to the evolution of the endosternite. J Arachnol 1:1–54

Frase T, Richter S (2013) The fate of the onychophoran antenna. Dev Genes Evol. doi:10.1007/s00427-013-0435-x

Fritsch M, Kaji T, Olesen J, Richter S (2013) The development of the nervous system in Laevicaudata (Crustacea, Branchiopoda): Insights into the evolution and homologies of branchiopod limbs and 'frontal organs'. Zoomorphology. doi:10.1007/s00435-012-0173-0

Fritsch M, Richter S (2010) The formation of the nervous system during larval development in *Triops cancriformis* (Bosc) (Crustacea, Branchiopoda): An immunohistochemical survey. J Morphol 271:1457–1481

Giribet G (2003) Molecules, development and fossils in the study of metazoan evolution; articulata versus Ecdysozoa revisited. Zoology 106:303–326

Gruner HE (1993) Arthropoda (ohne Insecta). In: Gruner HE (ed) Lehrbuch der speziellen Zoologie. Gustav Fischer Verlag, Jena. I(4):1–1279

Gruner HE, Scholtz G (2004) Segmentation, tagmata, and appendages. In: Forest J, von Vaupel Klein JC, Schram FR (eds) Treatise on Zoology—anatomy, taxonomy, biology. The Crustacea revised and updated from the Traité de Zoologie, vol 1. Brill, Leiden, pp13–57

Harzsch S (2004) Phylogenetic comparison of serotonin-immunoreactive neurons in representatives of the Chilopoda, Diplopoda and Chelicerata: implications for arthropod relationships. J Morphol 259:198–213

Haug JT, Briggs DE, Haug C (2012a) Morphology and function in the Cambrian Burgess Shale megacheiran arthropod *Leanchoilia superlata* and the application of a descriptive matrix. BMC Evol Biol 12:162. doi:10.1186/1471-2148-12-162

Haug JT, Waloszek D, Maas A, Liu Y, Haug C (2012b) Functional morphology, ontogeny and evolution of mantis shrimp-like predators in the Cambrian. Palaeontology 55:369–399

Hessler RR (1964) The Cephalocarida: comparative skeletomusculature. Mem Connect Acad Arts Sci 16:1–97

Heuer CM, Loesel R (2009) Three-dimensional reconstruction of mushroom body neuropils in the polychaete species *Nereis diversicolor* and *Harmothoe areolata* (Phyllodocida, Annelida). Zoomorphology 128:219–226

Heuer CM, Müller CHG, Todt C, Loesel R (2010) Comparative neuroanatomy suggests repeated reduction of neuroarchitectural complexity in Annelida. Front Zool 7:13. doi:10.1186/1742-9994-7-13

Janssen R, Damen WGM, Budd GE. (2011) Expression of *collier* in the premandibular segment of myriapods: support for the traditional Atelocerata concept or a case of convergence? BMC Evol Biol 11:50. doi:10.1186/1471-2148-11-50

Janssen R, Prpic N-M, Damen WGM (2006) A review of the correlation of tergites, sternites, and leg pairs in diplopods. Front Zool 3:2

Kimm MA, Prpic NM (2006) Formation of the arthropod labrum by fusion of paired and rotated limb-bud-like primordia. Zoomorphology 125:147–155

Kirsch R, Richter S (2007) The nervous system of *Leptodora kindtii* (Branchiopoda, Cladocera) surveyed with Confocal Scanning Microscopy (CLSM), including general remarks on the branchiopod neuromorphological ground pattern. Arthropod Struct Dev 36:143–156

Koch M (2000) The cuticular cephalic endoskeleton of primarily wingless hexapods: Ancestral state and evolutionary changes. Pedobiologia 44:374–385

Koch M (2003) Monophyly of the Myriapoda? Reliability of current arguments. Afr Invertebr 44:137–153

Lauterbach KE (1989) Das Pan-Monophylum—Ein Hilfsmittel für die Praxis der phylogenetischen Systematik. Zool Anz 223:139–156

Legg DA, Sutton MD, EdgecombeGD, Caron J-B (2012) Cambrian bivalved arthropod reveals origin of arthrodization. Proc R Soc B doi:10.1098/rspb.2012.1958

Liu J, Shu D, Han J, Zhang Z, Zhang X (2006) A large xenusiid lobopod with complex appendages from the Lower Cambrian Chengjiang Lagerstätte. Acta Pal Pol 51:215–222

Liu J, Shu D, Han J, Zhang Z, Zhang X (2007) Morphoanatomy of the lobopod *Magadictyon* cf. *haikouensis* from the Early Cambrian Chengjiang Lagerstätte, South China. Acta Zool 88:279–288

Liu J, Steiner M, Dunlop JA, Keupp H, Shu D, Ou Q, Han J, Zhang Z, Zhang X (2011) An armoured Cambrian lobopodian from China with arthropod-like appendages. Nature 470:526–530

Liubicich DM, Serano JM, Pavlopoulos A, Kontarakis Z, Protas ME, Kwan E, Chatterjee S, Tran KD, Averof M, Patel NH (2009) Knockdown of *Parhyale* Ultrabithorax recapitulates evolutionary changes in crustacean appendage morphology. Proc Natl Acd Sci USA 106:13892–13896

Ma X, Hou X, Aldridge RJ, Siveter DJ, Siveter DJ, Gabbott SE, Purnell MA, Parker AR, Edgecombe GD (2012a) Morphology of Cambrian lobopodian eyes from the Chengjiang Lagerstätte and their evolutionary significance. Arthropod Struct Dev 41:495–504

Ma X, Hou X, Bergström J (2009) Morphology of *Luolishania longicruris* (Lower Cambrian, Chengjiang Lagerstätte, SW China) and the phylogenetic relationships within lobopodians. Arthropod Struct Dev 38:271–291

Ma X, Hou X, Edgecombe GD, Strausfeld NJ (2012b) Complex brain and optic lobes in an early Cambrian arthropod. Nature 490:258–262

Maas A, Waloszek D, Müller KJ (2003) Morphology, ontogeny and phylogeny of the Phosphatocopina (Crustacea) from the Upper Cambrian 'Orsten' of Sweden. Fossils Strata 49:1–238

Manton SM (1964) Mandibular mechanisms and the evolution of arthropods. Phil Trans R Soc B 247:1–183

Manuel M, Jager M, Murienne J, Clabaut C, Le Guyade H (2006) Hox genes in sea spiders (Pycnogonida) and the homology of arthropod head segments. Dev Genes Evol 216:481–491

Maxmen A, Browne WE, Martindale MQ, Giribet G (2005) Neuroanatomy of sea spiders implies an appendicular origin of the protocerebral segment. Nature 437:1144–1148

Mayer G (2006) Structure and development of onychophoran eyes: what is the ancestral visual organ in arthropods? Arthr Struct Dev 35:231–245

Mayer G, Harzsch S (2008) Distribution of 5-HT-like immunoreactivity in the trunk of *Metaperipatus blainvillei* (Onychophora, Peripatopsidae): Implications for nervous system evolution in Arthropoda. J Comp Neurol 507:1196–1208

Mayer G, Whitington PM, Sunnucks P, Pflüger H-J (2010) A revision of brain composition in Onychophora (velvet worms) suggests that the tritocerebrum evolved in arthropods. BMC Evol Biol 10:255. doi: 10.1186/1471-2148-10-255

Meier R, Richter S (1992) Suggestions for a more precise usage of proper names of taxa. Ambiguities related to the stem lineage concept. Ztschr Zool Syst Evolforsch 30:81–88

Mittmann B, Scholtz G (2003) Development of the nervous system in the "head" of *Limulus polyphemus* (Chelicerata: Xiphosura): Morphological evidence for a correspondence between the segments of the chelicerae and of the (first) antennae of Mandibulata. Dev Genes Evol 213:9–17

Mittmann B, Wolff C (2012) Embryonic development and staging of the cobweb spider *Parasteatoda tepidariorum* C. L. Koch, 1841 (syn.: *Achaearanea tepidariorum*; Araneomorphae; Theridiidae). Dev Genes Evol 222:189–216

Møller OS, Olesen J, Høeg JT (2004) On the larval development of *Eubranchipus grubii* (Crustacea, Branchiopoda, Anostraca), with notes on the basal phylogeny of the Branchiopoda. Zoomorphology 123:107–123

Ou Q, Shu D, Mayer G (2012) Cambrian lobopodians and extant onychophorans provide new insights into early cephalization in Panarthropoda. Nat Commun 3:1261. doi:10.1038/ncomms2272

Pavlopoulos A, Kontarakis Z, Liubicich DM, Serano JM, Akam M, Patel NH, Averof M (2009) Probing the evolution of appendage specialization by Hox gene misexpression in an emerging model crustacean. Proc Natl Acad Sci USA 106:13897–13902

Persson DK, Halberg KA, Jørgensen A, Møbjerg N, Kristensen RM (2012) Neuroanatomy of *Halobiotus crispae* (Eutardigrada: Hypsibiidae): Tardigrade brain structure supports the clade Panarthropoda. J Morphol 273:1227–1245

Pigliucci M, Müller GB (2010) Elements of an extended evolutionary synthesis. In: Pigliucci M, Müller GB (eds) Evolution: the extended synthesis. MIT Press, Cambridge, pp 3–18

Posnien NF, Bashasab F, Bucher G (2009) The insect upper lip (labrum) is a nonsegmental appendage-like structure. Evol Dev 11:479–487

Ramsköld L, Chen J (1998) Cambrian lobopodians: morphology and phylogeny. In: Edgecombe GD (ed) Arthropod fossils and phylogeny. Columbia University Press, New York, pp 107–150

Ramsköld L, Chen J, Edgecombe GD, Zhou G (1997) *Cindarella* and the arachnate clade Xandarellida (Arthropoda, Early Cambrian) from China. Trans R Soc Edinb Earth Sci 88:19–38

Regier JC, Shultz JW, Zwick A, Hussey A, Ball B, Wetzer R, Martin JW, Cunningham CW (2010) Arthropod relationships revealed by phylogenomic analysis of nuclear protein-coding sequences. Nature 463:1079–1083

Richter S, Loesel R, Purschke G, Schmidt-Rhaesa A, Scholtz G, Stach T, Vogt L, Wanninger A, Brenneis G, Döring C, Faller S, Fritsch M, Grobe P, Heuer CM, Kaul S, Møller OS, Müller CHG, Rieger V, Rothe BH, Stegner MEJ, Harzsch S (2010) Invertebrate neurophylogeny—suggested terms and definitions for a neuroanatomical glossary. Front Zool 7:29. doi:10.1186/1742-9994-7-29

Schaeper ND, Pechmann M, Damen WG, Prpic NM, Wimmer EA (2010) Evolutionary plasticity of collier function in head development of diverse arthropods. Dev Biol 344:363–376

Scholtz G (1997) Cleavage, germ band formation and head segmentation: the ground pattern of the Euarthropoda. In: Fortey RA, Thomas RH (eds) Arthropod relationships. Chapman & Hall, London, pp 317–332

Scholtz G (2002) The Articulata hypothesis—or what is a segment? Org Divers Evol 2:197–215

Scholtz G, Edgecombe GD (2005) Heads, Hox and the phylogenetic position of trilobites. In: Koenemann S, Jenner R (eds) Crustacea and arthropod relationships (Crustacean Issues 16). CRC Press, Boca Raton, pp 139–165

Scholtz G, Edgecombe GD (2006) The evolution of arthropod heads: reconciling morphological, developmental and palaeontological evidence. Dev Genes Evol 216:395–415

Shultz JW (1999) Muscular anatomy of a whipspider, *Phrynus longipes* (Pocock) (Arachnida: Amblypygi), and its evolutionary significance. Zool J Linn Soc 126:81–116

Shultz JW (2000) Skeletomuscular anatomy of the harvestman *Leiobunum aldrichi* (Weed, 1893) (Arachnida: Opiliones) and its evolutionary significance. Zool J Linn Soc 128:401–438

Shultz JW (2001) Gross muscular anatomy of *Limulus polyphemus* (Xiphosura, Chelicerata) and its bearing on evolution in the Arachnida. J Arachnol 29:283–303

Shultz JW (2007) Morphology of the prosomal endoskeleton of Scorpiones (Arachnida) and a new hypothesis for the evolution of cuticular cephalic endoskeletons in arthropods. Arthropod Struct Dev 36:77–102

Siewing R (1969) Lehrbuch der vergleichenden Entwicklungsgeschichte der Tiere. Parey, Hamburg

Sombke A, Lipke E, Kenning M, Müller C, Hansson BS, Harzsch S (2012) Comparative analysis of deutocerebral neuropils in Chilopoda (Myriapoda): Implications for the evolution of the arthropod olfactory system and support for the Mandibulata concept. BMC Neurosci 13:1. doi:10.1186/1471-2202-13-1

Stegner MEJ, Richter S (2011) Morphology of the brain in Hutchinsoniella macracantha (Cephalocarida, Crustacea). Arthr Struct Dev 40:221–243

Stein M (2010) A new arthropod from the Early Cambrian of North Greenland with a 'great appendage' like antennula. Zool J Linn Soc 158:477–500

Stein M, Selden PA (2012) A restudy of the Burgess Shale (Cambrian) arthropod *Emeraldella brocki* and reassessment of its affinities. J Syst Palaeontol 10:361–383

Stein M, Waloszek D, Maas A (2005) *Oelandocaris oelandica* and the stem lineage of Crustacea. In: Koenemann S, Jenner RA (eds) Crustacea and arthropod relationships (Crustacean Issues 16). CRC/Taylor and Francis, Boca Raton, pp 55–72

Steinmetz PR, Urbach R, Posnien N, Eriksson J, Kostyuchenko RP, Brena C, Guy K, Akam M, Bucher G, Arendt D (2010) Six3 demarcates the anteriormost developing brain region in bilaterian animals. EvoDevo 1:14. doi:10.1186/2041-9139-1-14

Strausfeld NJ (2012) Arthropod brains: evolution, functional elegance and historical significance. Belknap Press, Cambridge

Strausfeld NJ, Strausfeld MC, Stowe S, Rowell D, Loesel R (2006) The organization and evolutionary implications of neuropils and their neurons in the brain of the onychophorans *Euperipatoides rowelli*. Arthropod Struct Dev 135:169–196

Szucsich NU, Pennerstorfer M, Wirkner CS (2011) The mouthparts of *Scutigerella immaculata*: correspondences and variation among serially homologous head appendages. Arthropod Struct Dev 40:105–121

Telford MJ, Thomas RH (1998) Expression of homeobox genes shows chelicerate arthropods retain their deutocerebral segment. Proc Natl Acad Sci USA 95:10671–10675

Ungerer P, Wolff C (2005) External morphology of limb development in the amphipod *Orchestia cavimana* (Crustacea, Malacostraca, Peracarida). Zoomorphology 124:89–99

Vannier J, Chen J (2002) Digestive system and feeding mode in Cambrian naraoiid arthropods. Lethaia 35:107–120

Vogt L (2008) Learning from Linnaeus: towards developing the foundations for a general structure concept for morphology. Zootaxa 1950:123–152

Vogt L, Bartolomaeus T, Giribet G (2010) The linguistic problem of morphology: structure versus homology and the standardization of morphological data. Cladistics 26:301–325

von Reumont BM, Jenner RA, Wills MA, Dell'Ampio E, Pass G, Ebersberger I, Meyer B, Koenemann S, Iliffe TM, Stamatakis A, Niehuis O, Meusemann K, Misof B (2012) Pancrustacean phylogeny in the light of new phylogenomic data: support for Remipedia as the possible sister group of Hexapoda. Mol Biol Evol 29(3):1031–1045

Waloszek D, Chen J, Maas A, Wang X (2005) Early Cambrian arthropods—new insights into arthropod head and structural evolution. Arthropod Struct Dev 34:189–205

Waloszek D, Dunlop J (2002) A larval sea spider (Arthropoda: Pycnogonida) from the Upper Cambrian "Orsten" of Sweden, and the phylogenetic position of pycnogonids. Palaeontology 45:421–446

Waloszek D, Maas A, Chen J, Stein M (2007) Evolution of cephalic feeding structures and the phylogeny of Arthropoda. Palaeogeogr Palaeocl 254:273–287

Whittington HB (1993) Anatomy of the Ordovician trilobite Placoparia. Phil Trans R Soc B 339:109–118

Wirkner CS, Richter S (2010) Evolutionary morphology of the circulatory system in Peracarida (Malacostraca; Crustacea). Cladistics 26:143–167

Zantke J, Wolff C, Scholtz G (2008) Three-dimensional reconstruction of the central nervous system of *Macrobiotus hufelandi* (Eutardigrada, Parachela): Implications for the phylogenetic position of Tardigrada. Zoomorphology 127:21–36

Zhang X, Briggs DEG (2007) The nature and significance of the appendages of *Opabinia* from the Middle Cambrian Burgess Shale. Lethaia 40:161–173

Zhang X, Shu D, Erwin DH (2007) Cambrian naraoiids (Arthropoda): morphology, ontogeny, systematics, and evolutionary relationships. J Paleontol 81(68):1–52

# Arthropod Limbs and their Development

**11**

Geoffrey Boxshall

## Contents

## 11.1 Introduction

Arthropods are characterized by bodies that are segmented and by the possession of paired ventral limbs carried on all, most or some of these body segments. These paired limbs are primitively segmented—and the name of the taxon Arthropoda refers to the jointed limbs of its members. While the origin of arthropods is not the focus of this chapter, it is relevant to note that the recent discovery of a Cambrian lobopodian, *Diania cactiformis*, possessing robust and probably sclerotized appendages with what Liu et al. (2011) interpret as articulating elements, led them to speculate whether arthropodization (sclerotization of limbs) preceded arthrodization (sclerotization of the body). In such a scenario, the acquisition of jointed limbs assumes centre stage as the key driver of arthropod evolution.

Historically, the intellectually intriguing task of reconstructing the evolutionary history of the arthropods has revolved around advances in understanding of structural diversity along two morphological axes: the tagmatization or functional division of the body along the antero-posterior (A-P) axis, and the segmentation and specialization of the jointed limbs along their proximo-distal (P-D) axis. Evolutionary trends along these two axes were separated by Boxshall (2004) in order to facilitate a morphological comparison of limbs between taxa exhibiting different tagmosis. However, limb specialization

G. Boxshall (✉)
The Natural History Museum, Cromwell Road, London, SW7 5BD, UK
e-mail: G.Boxshall@nhm.ac.uk

A. Minelli et al. (eds.), *Arthropod Biology and Evolution*,
DOI: 10.1007/978-3-642-36160-9_11, © Springer-Verlag Berlin Heidelberg 2013

reflects a developmental process that commences with specification of segmental identity along the A-P axis and is intimately bound up with the major developmental pathways that regulate tagmatization (Averof and Patel 1997; Mahfooz et al. 2007).

The spectacular diversity of limb morphology has long been regarded as a key component of the amazing adaptive radiation of the Arthropoda and our knowledge of the developmental patterning mechanisms that generate this diversity is expanding rapidly (see Pechmann et al. 2010; Angelini et al. 2011). The task of integrating data from developmental genetics and morphology is guided by our understanding of phylogenetic relationships and the iterative process of estimating phylogenies has been reinvigorated by the flood of molecular data from next generation sequencing. The availability of sequence data on a massive scale is not only transforming the phylogenomics of arthropods (Regier et al. 2010), but has also facilitated the application of some of the powerful new tools of developmental genetics. In particular, "knock-down" methods using RNA interference (RNAi) have allowed us to test the roles of specific genes more directly. No longer is it necessary to set up cultures and endlessly screen progeny for mutants of particular genes: now, we can directly interfere with the expression of a specific gene and observe the consequences (e.g. Liubicich et al. 2009; Mito et al. 2011). In addition, the discovery of important new fossil arthropods has continued (Siveter et al. 2007a, b; Zhang et al. 2007; Briggs et al. 2012) and the application of novel techniques for extracting fragmentary microfossils (Harvey and Butterfield 2008; Harvey et al. 2012) has widened our understanding of the morphological diversity of early Palaeozoic arthropods.

The primary goal of this chapter is to integrate the wealth of new data emerging from morphological and embryological studies, from novel fossils, and from developmental genetics, in order to address questions of interest to the communities of scientists involved in the study of arthropod morphology and phylogenetics. Answers to these questions will help us to begin to formulate a new understanding of the spectacular diversity in limb diversity structure that has been the key to their success.

## 11.2 The Distinction Between Segments and Annuli

Arthropod limbs are subdivided along the P-D axis into smaller units, either segments or annuli. The anatomical distinction between segments and annuli in arthropod limbs was emphasized by Boxshall (2004): true segments are characterized by the presence of intrinsic muscles that originate, insert or attach within the segment whereas annuli lack intrinsic muscle origins, intermediate attachments or insertions. Intrinsic muscles or their tendonous extensions may, however, pass through annuli to an insertion site in a more distally located segment (Fig. 11.1). Each articulation is typically

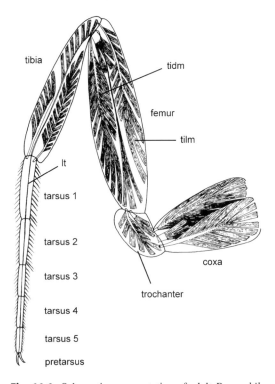

**Fig. 11.1** Schematic representation of adult *Drosophila* leg, showing intrinsic muscles and tendons, including tibia levator muscle (*tilm*), tibia depressor muscle (*tidm*) and the long tendon (*lt*) passing through the tarsal annuli (based on data from Soler et al. 2004)

provided with a hoop of arthrodial membrane which allows telescoping of the proximal rim of the more distal segment within the distal part of the more proximal segment. The appropriate terminology for the subdivisions of the main P-D axis of an arthropod limb is dependent upon their anatomy: subdivisions may be referred to variously as segments or articles, annuli or annulations, and the neutral term podomeres is often used when anatomical information about musculature is lacking, as in the case of the majority of fossils. Both segments and annuli can sometimes be incompletely expressed, particularly during larval development.

Maruzzo et al. (2009) examined segmental mismatch in the naupliar antennal exopodite of the branchiopod crustacean *Artemia*. The exopodite carries a series of natatory setae along its posterior-ventral side with each, apart from the apical seta, located on a transverse cuticular fold. Along the anterior side of the ramus is a P-D series of incomplete ringlets or sclerites separated by joint-like cuticular folds. The two series are not in register and there were, on average, more ringlets than setae. This phenomenon was also noted in the naupliar exopodites of the antenna and mandible of representatives of a few other taxa, including some fossil branchiopods, some phosphatocopines, and an extant thecostracan. However,

Maruzzo et al. (2009) showed that three exopodal muscles extend the length of the ramus and make intermediate attachments on both sides—in the ringlets (the anterior muscle) and in the setal-bearing cuticular folds (the two posterior muscles). Using the presence of intrinsic musculature as a rigid criterion, these naupliar rami could be regarded as multi-segmented, although the segments are incompletely expressed due to a decoupling of development in the two sides of the ramus analogous to the dorso-ventral decoupling in the development of diplopod body segments (see Damen et al. 2009).

Expressed segmentation can change significantly during development. In dendrobranchiate decapods, for example, the antenna of the naupliar and protozoeal phases initially has a multi-segmented exopodite (Fig. 11.2a). The exopodite gradually loses external segmentation until it has transformed into the characteristic, unsegmented antennal scale at the megalopa stage (Fig. 11.2b). The transition from segmented naupliar ramus to unsegmented antennal scale is unique to the caridoid malacostracans and is accompanied by a change in form of the endopodite, from a two-segmented ramus (Fig. 11.2a) to an annulate flagellum (Fig. 11.2c).

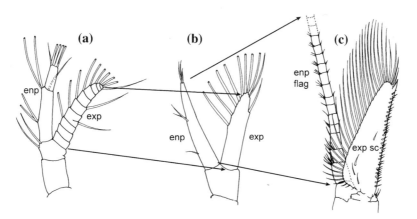

**Fig. 11.2** The antenna of *Pleoticus muelleri* (Decapoda). **a** Protozoea III stage. **b** Mysis I stage. **c** Megalopa stage. Showing transition of exopodite (*exp*) from multisegmented ramus to unsegmented antennal scale (*exp sc*), and of endopodite (*enp*) from two-segmented ramus to endopodal flagellum (*enp flag*) (redrawn from de Calazans 1992)

## 11.2.1 How are Segments Formed?

The early establishment of the P-D axis by the leg gap genes is a general feature of limb patterning during development in all arthropods (Angelini and Kaufman 2005; Williams 2008; Pechmann et al. 2010). However, the *Drosophila* leg is a useful comparative model since it comprises true segments proximally and tarsal annuli distally (Fig. 11.1). Leg formation in *Drosophila* depends upon the subdivision of the P-D axis into broad domains by leg gap genes: the early limb bud is subdivided into a distal domain expressing *Distal-less* (*Dll*) and a proximal domain expressing *extradenticle* (*exd*) and its co-factor *homothorax* (*hth*) (see Kojima 2004, for review). This proximal domain maintains expression of *hth* and *exd* and corresponds to the coxa and trochanter of the leg. Further differentiation along the P-D axis is mediated by the morphogens *Decapentaplegic* (*Dpp*) and *Wingless* (*wg*) which cooperate to induce the expression of *dachshund* (*dac*) in the intermediate region of the limb, between proximal and distal domains (Lecuit and Cohen 1997; Abu-Shaar and Mann 1998). These leg gap genes *hth*, *dac* and *Dll* control the formation of the proximal, middle and distal domains along the P-D axis, respectively (Fig. 11.3).

Downstream of the leg gap genes, the Notch signalling pathway plays a central role in segmentation along the P-D axis of the leg (de Celis et al. 1998; Bishop et al. 1999). The process of formation of true segments along the P-D axis of the limb of *Drosophila* takes place within the three leg gap gene domains, and the genes *Serrate* (*Ser*), *Delta* (*Dl*) and *fringe* are essential for joint formation (Rauskolb 2001; Mito et al. 2011). *Fringe* modulates Notch-ligand interactions (Panin et al. 1997). These induce expression of a set of transcriptional regulators that mediate joint morphogenesis and leg segment growth: *lines* and *bowl* act as a binary switch to generate a stable Notch signalling interface between *Dl*-expressing cells and adjacent distal cells (Greenberg and Hatini 2009).

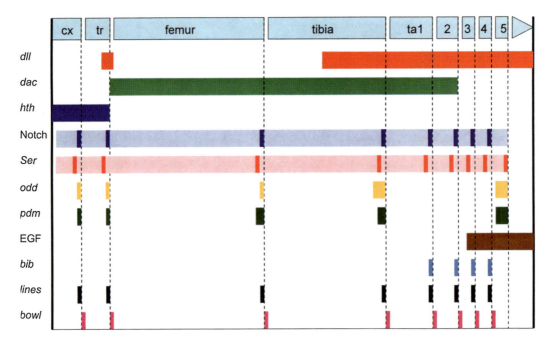

**Fig. 11.3** Schematic showing expression domains of genes along P-D axis of *Drosophila* leg, compiled from various sources. Proximal end on *left* commencing with coxa (*cx*), trochanter (*tr*), femur and tibia, and with first to fifth tarsal annuli (*ta1–5*) on *right*. Names of genes given on vertical axis, and coloured shading shows expression domains

A number of other genes are also known to be involved downstream of Notch signalling, including *nubbin* (*pdm*) (Rauskolb and Irvine 1999), *odd*-related genes (Hao et al. 2003), *h/ Enhancer of split*-related genes (Bishop et al. 1999) and *Activator Protein-*2 (Kerber et al. 2001). Although all originally discovered in *Drosophila*, a mandibulate, orthologues of these genes have also been found in chelicerates and Prpic and Damen (2009) concluded that, despite minor differences, the mechanisms regulating leg segmentation are likely to have been conserved from the common ancestor of the Arthropoda.

## 11.2.2 How are Annuli Formed?

Intercalary annulation of the endopodite is widespread in arthropods, most commonly in the tarsal region (Boxshall 2004). In the chelicerates, extreme tarsal annulation is found in the antenniform first walking legs of amblypygids, which may comprise as many as 28 tibial and 54 tarsal annuli (Weygoldt 1996), and tri-annulate femurs and bi-annulate trochanters are also known. Some pycnogonids also have distally annulate pedipalps and first walking legs. Stenopodoidean and caridean crustaceans such as processed shrimps can have a multi-annulate carpus on the fourth pereopod which has a normal chela at its tip. The trunk limbs of scutigeromorph centipedes exhibit extensive annulation of the tarsal region, interpreted by Manton (1977) as an adaptation for rapid running. Most insects exhibit some annulation in the tarsal region, with the number of tarsal annuli varying from one to five as in *Drosophila* (Fig. 11.1). Bitsch (2001) considered the pentameric tarsus a possible apomorphy for the dicondylian hexapods, with secondary reductions responsible for the variation, as found in the Zygentoma, for example. In arthropod locomotory limbs, annulations are typically intercalary, although there are examples of terminal annulation, such as the flagellate swimming exopodites of the Mysidacea or Anaspidacea (Fig. 11.8c). Most examples of terminal annulation in

arthropods involve sensory appendages, such as antennules and antennae (Fig. 11.2c).

In the *Drosophila* leg, there are five tarsal annuli and the patterning mechanism resulting in subdivision of the tarsus differs from that governing basic segmentation (Fig. 11.3). In the distal half of the leg is a zone of decreasing *dac* expression and increasing *Dll* expression extending from middle to tip of the leg. The genes *dpp* and *wg* together establish a secondary organizing centre towards the distal tip. Ligands from this centre activate the epidermal growth factor receptor pathway which controls the expression of the genes responsible for tarsal subdivision (Campbell 2002; Galindo et al. 2002). These tarsal genes, *bric-a-brac*, *apterous* and *BarH*1, act in combination with *dac* and *Dll*, to fine-pattern tarsal subdivision (see Greenberg and Hatini 2009). According to Greenberg and Hatini (2009), *lines* modulates the opposing expression landscapes of *dac* and the tarsal genes. Sharp boundaries in Dpp signalling trigger an episode of apoptosis that takes place during morphogenesis of tarsal joints in *Drosophila* (Manjón et al. 2007). Tarsal genes appear to be specific to the insects but little comparative research has been undertaken to either confirm their presence or determine the role of any orthologues in other arthropod taxa.

## 11.2.3 Is There a Difference in Timing of Appearance of Segments and Annuli During Development?

The distinction between limb segments and annuli is based on musculature. In the segmented antennules of copepods and ostracods, development follows a distal-to-proximal pattern with the articulations separating more distal segments typically appearing earlier than those separating the more proximal ones (Boxshall and Huys 1998; Smith and Tsukagoshi 2005). The adult antennules of copepods can possess up to 27 segments, and these are derived by a sequence of subdivisions of the three original segments present in the nauplius (Boxshall and

Huys 1998). The metamorphic moult from the sixth naupliar stage to the first copepodid stage was marked by the subdivision of the apical segment of the nauplius to form the distal eight segments of the adult antennule. No further subdivisions occur in this distal section throughout the subsequent moults. During the copepodid phase, the two proximal antennulary segments of the nauplius undergo a sequence of subdivisions to form segments 1 to 20 of the adult. Antennules with fewer expressed segments are envisaged as being generated by early cessation of the process of subdivision (Boxshall and Huys 1998; Schutze et al. 2000).

In limbs that possess a mix of segments and annuli, the segments tend to appear before the annuli. Unfortunately, *Drosophila* is not a good model here since both segments and annuli are everted simultaneously from the imaginal disc. In more basal hexapods such as symphypleone collembolans, Nayrolles (1991) showed that four true segments are initially expressed on the antennule; subsequently the distal segment undergoes annulation to generate the terminal flagellum (see Boxshall 2004: Fig. 1f–g). Minelli et al. (2000) showed that eosegments appear before merosegments in chilopod development and Boxshall (2004) considered this as analogous to the appearance of segments before annuli in other arthropods. In the decapod malacostracans, *Panulirus* and *Cherax,* the primary antennulary flagellum develops by the production of new annuli in a meristematic zone at the base of the flagellum (Sandeman and Sandeman 1996; Steullet et al. 2000). Subdivision takes place in annuli distal to the basal meristematic annulus and the process seems generally similar to that described for the endopodal flagellum of the antenna (i.e. the second antenna) of the isopod *Asellus*, which consists of a single segment divided into annuli devoid of intrinsic musculature (Wege 1911). The antennal flagellum comprises a proximal meristematic region, a central region composed of quartets (sets of 4 annuli each having a specific arrangement of setae), and an apical complex consisting of the apical annulus plus the four preceding annuli with specific setal patterns. The number of quartets in the central region is variable in *Asellus* since this isopod never ceases moulting and adds annuli throughout life (Maruzzo et al. 2007). The proximal meristematic annulus divides into a copy of itself (the meristem) and a distal annulus which is effectively an incomplete quartet,and divides following a set pattern each time, to produce the complete quartet. Maruzzo and Minelli (2011) found proximal growth zones on each of the elongate rami of the pleopods in amphipods. In these zones, new arthrodial membrane, separating newly differentiated annuli, and new setae were added during post-embryonic moults.

Proximal annulation is expressed transiently during the naupliar phase of some copepods but is lost by the first copepodid stage (Dahms 1992). Protozoeal larvae of some penaeid decapods similarly exhibit transient annulations in the proximal part of the antennule (Boxshall 2004: Fig. 2a-c), which are lost by the end of the zoeal phase. The proximal annulated part of the antennule of the fossil crustaceans *Rehbachiella* and *Bredocaris* may be interpreted as additional evidence of their larval status, but may also indicate that a proximal annulated zone is plesiomorphic for the Crustacea.

## 11.2.4 Are Segments Fundamentally Different from Annuli?

The patterning mechanisms generating segments and annuli are similar: knock-down of *Notch* in the cricket *Gryllus* resulted in a marked reduction in leg length and loss of joints along the P-D axis (Mito et al. 2011). The loss of joints is referred to as "fusion" by Mito et al. (2011) but is derived by failure of the joint to form and create a subdivision, rather than by fusion of subdivisions. Mito et al. (2011) found that the femur and tibia failed to separate and the tarsal annuli failed to subdivide normally, so Notch clearly plays a role in the formation of both segments and annuli. In contrast, *Dl* mutants of *Drosophila* showed shortened legs but only tarsal segments 2–4 of the wild type were not separated (Bishop et al. 1999). Similarly, knock-down of certain other genes is known to affect

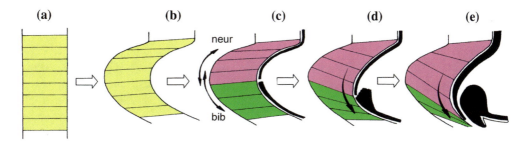

**Fig. 11.4** Schematic showing morphogenesis of ball-and-socket tarsal joint in *Drosophila* leg. **a** Undifferentiated epithelial cells (*yellow*) of leg prior to formation of joint. **b** Initiation of joint formation by invagination of epithelium. **c** Epithelial cells differentiate into those with low Notch activity (*pink*) that will produce the socket and those with high Notch activity (*green*) that will produce the ball. **d** Migration of socket-producing epithelial cells begins. **e** Cell migration continues and lip-like projection develops on ball (adapted from Tajiri et al. 2011)

tarsal subdivision but not basic leg segmentation, so the patterning mechanisms for leg segments and leg annuli, while similar, exhibit important differences in detail (Fig. 11.3).

During normal development differentiation of a typical "ball-and-socket" joint between tarsal annuli in *Drosophila* is dependent upon levels of Notch activity. High Notch signalling levels promote ball production whereas low levels are required for socket production (Fig. 11.4). Cells that produce the ball express *big brain* (*bib*) whereas socket cells express *neur* and tend to produce the more uniform, thinner cuticle of the socket. Elongation of the ball lip and the socket coincides with the migration of the cells that form them. Notch activity is also required for this cell motility, but it is probably under the control of an independent Notch-mediated pathway (Tajiri et al. 2011). Interestingly, disruption of the Notch signalling pathway during pupal development in *Drosophila* suppressed production of the normal ball-and-socket joint in the legs and resulted in the formation of a more uniform type of joint like that found in more basal hexapods such as Ephemeroptera according to Tajiri et al. (2011).

Morphologically, the key difference between segments and annuli is the presence of intrinsic musculature in segments. A huge body of literature is available describing limb musculature patterns in a wide range of arthropods (see Manton 1977 and references therein), and the precise sites of muscle origins and insertions have been considered as phylogenetically informative (e.g. Boxshall 1997). However, the key challenge is to integrate knowledge of the anatomy with what is known about the genetic mechanisms regulating myogenesis in arthropods. Unfortunately, most studies on myogenesis in *Drosophila* have focused on larval and flight muscles, so relatively little is known about the mechanisms governing adult leg myogenesis in the *Drosophila* leg model. Soler et al. (2004) summarized the stages of myogenesis: commencing in the leg imaginal discs of the third instar: myoblasts expressing *twist* (*twi*) and located in the vicinity of tendon precursors start to express the muscle founder cell marker *dumbfounded* (*duf*). Subsequently, epithelial tendon precursors invaginate within the developing leg segments, giving rise to the tendons. Tendon associated *duf*-expressing muscle founder cells become distributed along these developing tendons and fuse with surrounding myoblasts forming syncytial myotubes. Finally, these myotubes grow towards their epithelial insertion sites (the apodemes) and complete the link between internally located tendons and the leg epithelium. However, the process is understood only in outline.

Leg muscle patterning involves genes such as *ladybird early*, which is expressed in a subset of the *twi*-expressing myoblasts located dorsally and ventrally in the femur and giving rise to the tibia levator and depressor muscles (Fig. 11.1), respectively (Maqbool et al. 2006). Only fragmentary data on mechanisms responsible for P-D patterning of leg musculature are available for other arthropods. Recent work on muscle precursors in the developing limbs of isopod and decapod crustaceans showed that intrinsic limb muscles originate from single precursor cells which subsequently form multi-nucleate precursors, and this suggests fundamental similarities with the insects (Kreissl et al. 2008; Harzsch and Kreissl 2010).

These outlines of the sequence of events involved in myogenesis shed little light on how significant spatial aspects (i.e. the precise location of muscle origins and insertions) of myogenesis are determined. However, Park et al. (1998) showed that muscle founder cells arise from progenitor cells which are singled out by a lateral inhibition process mediated by the Notch–Delta signalling pathway. Given the central role of the Notch pathway in segmentation along the P-D axis of the arthropod leg, it seems probable that spatial regulation of muscle attachments is also linked to the existing framework of domains along the P-D axis of the limb.

## 11.3 Arthropod Limb Types

In a review, Boxshall (2004) concluded that there are two basic limb types in crown-group arthropods: a single-axis first cephalic appendage (the antennules/chelicerae) and biramous post-antennulary limbs. In the terminology of Scholtz and Edgecombe (2005), the first cephalic appendages of euarthropods represent the "secondary antennae", with innervation derived from the deutocerebrum, as distinct from the "primary antennae" associated with the protocerebrum and found in onychophorans (see Chap. 10). Scholtz and Edgecombe (2006) discuss possible fates for the missing "primary antennae" in euarthropods, but these are not of concern here. I am considering the first cephalic limb of arthropods, which is derived from the deutocerebral segment and is known as the antennule or first antenna in crustaceans, the antenna in insects, myriapods and trilobites, chelicera in crown-group chelicerates and the "great appendage" in megacheirans.

### 11.3.1 The First Cephalic Limb

Interpretations of antennules as possessing vestiges of an "exopodite" still crop up occasionally in crustacean taxa such as the podocope ostracods (Karanovic 2005; Marmonier et al. 2005) but lack credible supporting evidence according to Boxshall et al. (2010) who also concluded that the Remipedia, with an antennule comprising a single primary axis composed of segments (defined by the possession of intrinsic musculature), plus a proximally located ventral flagellum, and the Malacostraca, with a short segmented primary axis bearing two, occasionally three, distally located flagella, provide no evidence that contradicts the inference that the antennules of the Mandibulata are primitively single-axis limbs. This single axis may be either segmented, flagellate or a mix of segments and annuli but is essentially modular in construction, and this modularity confers important functional attributes, permitting, for example, the enhancement of a sensory array by the addition of modules or by the specialization of individual modules independent of others.

The first prosomal appendages of crown-group chelicerates are the paired chelicerae. The comparison of expression patterns of Hox genes in chelicerates and mandibulates has demonstrated that chelicerae are positional homologues of the antennules (Damen et al. 1998; Telford and Thomas 1998; Abzhanov et al. 1999), and the immunohistochemical analysis of neuroanatomy and neurogenesis has confirmed the deutocerebral derivation of the chelifores of pycnogonids (Brenneis et al. 2008).

The morphological gulf between an elongate sensory antennule and a short feeding chelicera seems profound, but recent analyses of cheliceromorph fossils have hypothesized how such transitions might have occurred (Fig. 11.5b–e). These analyses involve the Megacheira, the so-called short great appendage fossils, which are possible stem-group chelicerates (e.g. Chen et al. 2004; Cotton and Braddy 2004). The antennules of the megacheiran *Leanchoilia* (Fig. 11.5b) were considered as effectively triflagellate by

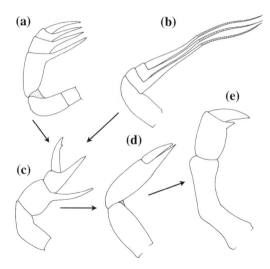

**(a)**

**(b)**

**(e)**

**(d)**

**(c)**

**Fig. 11.5** Schematic showing possible transition between raptorial great appendage and chelicera. **a** Great appendage of *Yohoia*. **b** Great appendage of *Leanchoilia*. **c** Great appendage of *Haikoucaris*. **d** Chelicera of *Limulus*. **e** Chelicera of pycnogonid (**a**, **c**, redrawn from Haug et al. (2012), **b**, redrawn from Edgecombe et al. (2011), **d,e**, drawn from photographs in Haug et al. (2012)

Boxshall (2004), but each flagellum is borne on a rigid spinous projection of the antennulomere (Bruton and Whittington 1983). In the evolutionary scenario constructed by Haug et al. (2012), the megacheirans, *Parapeytoia*, *Fortiforceps*, *Yohoia* (Fig. 11.5a) *Leanchoilia* and *Haikoucaris* (Fig. 11.5c), are all considered to be derivatives of the stem lineage of the Chelicerata, and a transition from triflagellate great appendage to chelicera is hypothesized as involving reduction and loss of the flagella, reduction and loss of segments, shortening of the spinous projections and the development of a special "elbow joint" between the two-segmented peduncle and the distal segments (Fig. 11.5a–e). However, this scenario needs further testing firstly because it was not supported by the phylogenetic analysis of Edgecombe et al. (2011), which recovered a monophyletic Megacheira as the sister-group of a poorly resolved group comprising chelicerates, aglaspids and other fossil cheliceromorph taxa such as *Cheloniellon* and *Sidneyia*. Secondly, the Silurian synziphosurine *Dibasterium durgae* has long flexible antenniform chelicerae (Briggs

et al. 2012) providing an elegant link between typical sensory antennule and feeding chelicerae.

Although bi-, tri- or multi-flagellate limbs are known in malacostracan crustaceans and in basal megacheirans, a truly biramous first limb (with two-segmented axes) is unknown in the Arthropoda. The only possible exception might be the Pauropoda which have two-branched antennules, but each branch is unsegmented and provided with musculature that inserts only around its proximal rim (Boxshall 2004: Fig. 2g) and so does not comprise a segmented axis.

The first cephalic limb of euarthropods has a single P-D axis and thus differs from post-antennulary limbs which are primitively biramous. How fundamental is this distinction, given that well-known homeotic mutations, such as the Antennapedia mutant of *Drosophila*, indicate that antennules and post-antennulary limbs can be viewed as serial homologues? Indeed, numerous homeotic mutations are now known that can transform maxillary palps, labial palps and genitalia into antennae or thoracic legs in a variety of insects, not just *Drosophila* (Angelini et al. 2011).

Less is known about patterning mechanisms in the developing arthropod antennule than in legs, but it is clear that early development is regulated by the activity of field-specific selector genes. The *Drosophila* antenna comprises only three segments and a terminal flagellate section, the arista. On the basis of gene expression domains, Postlethwait and Schneiderman (1971) concluded that the first antennal segment was "homologous" with the coxa of the leg; the second segment with the trochanter; and the third with the femur, tibia and first tarsal segment, and the arista with the second to fifth tarsal segments plus the tarsal claw. However, the homology is at the level of the shared early leg gap gene patterning mechanism, common to all arthropod limbs, and does not support an inference of homology between the segments themselves.

The basic patterning mechanism of the antenna is very similar to that of the leg, but differs in the extensive co-expression of the proximal and distal leg gap genes, *hth* and *Dll*, respectively, and in the absence of a functional intermediate domain specified by *dac* (Dong et al. 2001) (the *dac* expression domain lies completely within the *Dll* domain in the insect antenna). Downstream of the leg gap genes, the Notch pathway involving *Dl* has been reported for the antenna of the cricket *Gryllus* (Mito et al. 2011). Fine-scale mechanisms are also somewhat similar: the gene *lines*, for example, plays analogous roles in the subdivision of the flagellate arista of the antenna and of the annulate tarsal region on the leg, but again there are also some significant differences (Greenberg and Hatini 2009). In particular, in the antenna, *Dll* and *hth* cooperate in a secondary role, to impose identity on the antenna by activating antenna-specific genes in a cascade leading to *distal antenna*, a selector gene for antennal fate (Emerald et al. 2003).

There is no evidence from gene expression data to suggest that antennules (or chelicerae) are primitively anything other than single-axis limbs. So, for example, in the early embryo of the extant xiphosuran *Limulus,* the developing limb buds of the chelicerae do not develop a second lateral point of *Dll* expression even though transient laterally located expression points are shown by the developing buds of all the post-antennulary limbs on the prosoma, including pedipalps and walking legs (Mittmann and Scholtz 2001). There is a difference of interpretation concerning the homology of these transient *Dll*-expressing points (see Boxshall 2004) but the evidence relevant here is that the chelicerae lack such a point.

The shared common features between antenna and leg development in *Drosophila* indicate that, despite some significant differences, the antennules and post-antennulary limbs of all arthropods can be viewed as serial homologues, but specification of the anterior-most limb as the antennule ensures that it develops as a single axis rather than biramous limb.

## 11.3.2 The Post-antennulary Limbs

Boxshall (2004) concluded that the basic post-antennulary limb of crown-group arthropods comprises an undivided protopodite (also called the basipod), an endopodite of cylindrical segments and a more flattened exopodite probably of two segments.

### 11.3.2.1 Protopodite

The protopodite is the proximal part of the biramous limb and carries the rami. It is easy to recognize in biramous limbs, as found in crustaceans, trilobites and many other fossils such as the marrellomorph *Xylokorys*, but when limbs are uniramous, it can be difficult to identify the boundary between the protopodite and the endopodite (see Boxshall 2004).

The protopodite of all post-antennulary limbs of trilobites and most fossil and recent cheliceromorphans is entire and undivided, although a small, mobile proximal endite is present in xiphosurans, eurypterids and the Cambrian *Sidneyia* (a relative of *Aglaspis* according to the scheme in Edgecombe et al. 2011). In trilobites, the entire medial margin of the undivided protopodite was convex and provided with spines, forming a gnathobase. Similar undivided gnathobases are also retained on the pedipalps and walking limbs of *Limulus*, the pedipalps of spiders, in the first and second walking legs of scorpions and some harvestmen. The retention of protopodal endites (often referred to as gnathendites) in these taxa was a plesiomorphic character state in the analysis of Shultz (1990). The protopodite of chelicerates and cheliceromorphs in general appears to be short but very broad. However, the discovery of the cheliceromorph *Dibasterium* has revealed a biramous prosomal limb type in which the endopodite is carried on a recognizable protopodite but the well-developed and multi-segmented exopodite appears to originate separately on the adjacent ventral surface of the prosome (Briggs et al. 2012). The limbs of *Offacolus* were reinterpreted as similar to those of *Dibasterium* by Briggs et al. (2012).

In crustaceans, as representatives of basal mandibulates, the protopodites are more elongate in the P-D axis and retain gnathobases or endites in many members of the limb series: for

**Fig. 11.6** Trunk limbs of leptostracan and archaeostracan Malacostraca. **a** Trunk limb of *Nebalia* showing absence of endites in protopodal part (coxa and basis), musculature in exopodite (*exp*) and endopodite (*enp*) but none in the foliaceous epipodite (*epi*). **b** Trunk limb of *Cinerocaris* showing enditic margin of protopodite (*prp*) and array of foliaceous outer lobes (**a**, from Boxshall and Jaume (2009); **b**, redrawn from Briggs et al. (2004))

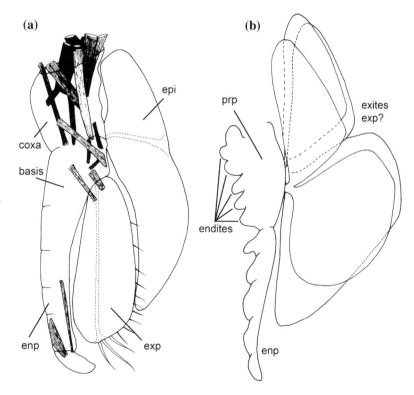

example, in mandibles, in the post-mandibular limbs of crustaceans such as cephalocaridans and branchiopods and in the maxillule of hexapods, chilopods and symphylans. Endites may also be transient features: the enditic process on the coxa of the antenna of planktotrophic crustacean nauplii is secondarily lost after the naupliar phase of development. In crustaceans, the medial surface of the enlarged protopodite typically carries a linear series of endites. The number of endites on the protopodite of postmandibular limbs varies: in Cambrian crustaceans such as *Rehbachiella* and *Dala,* it can be up to eight or nine (Walossek 1993; Walossek and Müller 1998). Only five or six endites are retained on the protopodal part of the trunk limbs of *Lepidocaris* and five endites has often been regarded as typical for extant branchiopods. However, Pabst and Scholtz (2009) regarded only three of the inner lobes as protopodal in origin, reinterpreting the two distal lobes as endopodal. The enditic margin only forms weak lobes in cephalocaridans rather than well-defined endites. The archaeostracan

*Cinerocaris* retains a series of endites on the undivided protopodite of the pereopods (Fig. 11.6b) and has endite-like expansions of the medial margin of the proximal endopodal segments (Briggs et al. 2004). Retention of an endite series along the protopodite was regarded by Walossek (1999) as characteristic of his Entomostraca, but is plesiomorphic for malacostracans also. The endites are lost in modern leptostracans (Fig. 11.6a).

Outside the crown-group crustaceans, only one endite per segment is typical. The lacinia and galea of the maxilla (first post-mandibular limb) in the basal hexapod *Thermobia* have been interpreted as representing the endites of two protopodal segments (Chaudonneret 1950). However, in *Tribolium*, two enditic lobes are present transiently in the early embryo but fuse before hatching to form the single endite present in the larva. This larval endite is presumed to give rise to the lacinia and galea of the adult (Jockusch et al. 2004), however, fusion and subsequent separation of endites derived from different protopodal segments seems unlikely.

Chaudonneret (1950) studied *Thermobia* but no relevant detailed genetic studies have yet been carried out on this species. Most hexapods and myriapods lack functional endites on their limbs.

The proximal endite on the protopodite of crustaceans has been regarded as of particular significance by Waloszek and co-authors (e.g. Maas et al. 2003; Waloszek 2003; Waloszek et al. 2007). As summarized by Waloszek et al. (2007, p. 284), the "proximal endite" is a "novelty of the ground pattern of the Crustacea" and is a "separately moveable" setose lobe "nested within the ample joint membrane medially below the basipod of the post-antennular limb". The proximal endite is clearly visible on the post-antennulary limb series in *Martinssonia* (Müller and Walossek 1986) and in the phosphatocopines (Maas et al. 2003) but is presented only in the mandible in *Oelandocaris* (Stein et al. 2005). The significance of this proximal endite in the phylogenetic debate is that it "is considered as a phylogenetic precursor of another limb portion developed in the Crustacea"—the coxa (Waloszek et al. 2007).

An alternative hypothesis, as summarized by Boxshall (2004), is that the protopodite (Waloszek's basipod) subdivided by the formation of a transverse articulation to form the proximal coxa and distal basis. This must have occurred in the stem lineage of the mandibulates at least in the antenna (first post-antennulary limb) and mandible (second), as well as in the maxillule (third) of crown-group Crustacea according to Boxshall (1997). In the maxilla and post-cephalic trunk limbs, the proximal endite is simply the proximal-most of the series of endites expressed along the medial margin of the protopodite. It may be capable of performing motions independent of the main promotor-remotor swing of the whole limb at the body-coxa articulation, but such multi-functionality is the hallmark of the crustacean limb.

Do gene expression data shed any light on the debate over the origin of the separate coxa and basis? Endites can express *Dll*. In the developing uniramous limbs of chelicerates and insects, a proximal zone of expression of *Dll* is found. It is localized in the gnathendite on the undivided

protopodite (the coxa) of the developing pedipalps of the mygalomorph spider *Acanthoscurria*, although not in the rudimentary gnathendites of the walking legs (Pechmann and Prpic 2009). Similar expression in the gnathendite of the pedipalps has also been observed in more derived spiders (Schoppmeier and Damen 2001; Prpic and Damen 2004). In insects, *Dll* is expressed on the maxilla of *Tribolium* in a distinct proximal domain that corresponds with the developing endite (Beermann et al. 2001) and in *Acheta*, in two domains corresponding with lacinia and galea (Angelini and Kaufman 2004). Interestingly, RNAi depletion of *Dll* did not affect the formation of the endites (the galea and lacinia) on the maxilla of another beetle, *Onthophagus*, although the palp became unsegmented (Simonnet and Moczek 2011).

In the phyllopodial limbs of anostracans, Williams (2008) demonstrated early *Dll* expression in the proximal regions of the limb in the series of endites carried on the medial margin, around the margins of both endopodite and exopodite, and in the pre-epipodite. Transient expression only was noted for the epipodite (which lacks setae in the adult anostracan). Williams (2008) noted that proximal *Dll* expression was found initially in general epithelial cells but subsequently became localized to setal-forming cells, irrespective of whether the setae were sensory or had a passive mechanical role as in the majority of enditic setae.

In the notostracan *Triops*, there is medially reiterated expression of *dac* in the very early limb bud that resolves to the endites. Each of the five endites carried along the medial margin of the *Triops* trunk limb (Fig. 11.7c) expresses *dac* in a zone along its lower (ventral) margin (Sewell et al. 2008). Localized *dac* expression was also noted in each endite on the trunk limbs of an anostracan branchiopod (Sewell et al. 2008). Four zones of *dac* expression were observed along the margin of the maxilla of the myriapod *Glomeris* (Prpic and Tautz 2003) and *dac* was also expressed in both endites present on the maxilla of the hexapod *Tribolium* (Prpic et al. 2001). Interestingly, the gnathendite of the

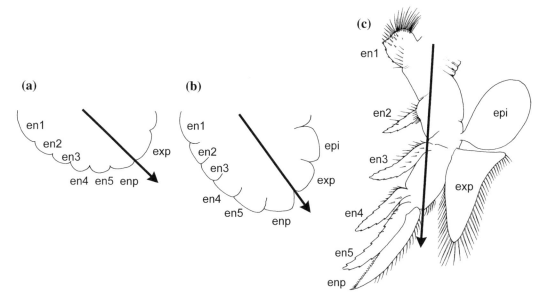

**Fig. 11.7** Development of trunk limb of Notostraca. **a** Schematic showing limb forming as transverse ridge with developing endites and rami as defined as lobes. **b** Later stage of limb development with epipodite lobe now present. **c** Adult limb of *Lepidurus* showing endites, rami and epipodite. Endite 4 (*en 4*) and endite 5 (*en 5*) plus endopodite (*enp*) of Sewell et al. (2008) were all interpreted as representing a tripartite endopodite in the scheme of Pabst and Scholtz (2009) (**a**, **b**, adapted from Sewell et al. (2008); **c**, redrawn from Sars (1896))

pedipalps of the model chelicerate *Cupiennius* lacks *dac* expression. Sewell et al. (2008) considered this as evidence consistent with the inference that the proximal endites of mandibulates and cheliceromorphs are non-homologous (Boxshall 2004).

These expression data can help us understand the derivation of the coxa and basis from an undivided ancestral protopodite in two respects. Firstly, the P-D subdivision of limbs is regulated by a patterning mechanism involving the leg gap genes, the Notch signalling pathway, and a downstream cascade of other genes, which is common to all arthropods. All subdivisions of the P-D axis appear to be regulated by this mechanism, and there is no evidence to suggest that the coxa-basis articulation in the mandibulate protopodite is different. I infer that the coxa and basis differentiate by a process of subdivision, as for every other segment and annulus along the limb, and that there is no special role for the proximal endite. Secondly, the proximal endite is one of a P-D series of protopodal

endites, all of which share a common expression pattern for the few genes (e.g. *dac* and *exd*) that have thus far been investigated. No unique expression pattern has yet been noted for the proximal endite: again there is nothing to suggest that the patterning mechanism responsible for the formation of this endite is different from that of the more distal endites in the series.

The enlarged proximal endite of the second post-antennary limb in the Mandibulata is modified as a gnathobase. Indeed, possession of the second post-antennary limb modified as a mandible has been used to characterize the Mandibulata, comprising the Crustacea, Hexapoda and Myriapoda (Snodgrass 1938). The limb carried on the homologous body segment in chelicerates is a walking leg (Damen et al. 1998; Telford and Thomas 1998)—the first walking leg in arachnids and the second in Xiphosura and Eurypterida. In trilobites and other fossils with homonomous post-antennary limbs, this limb exhibits no unique morphological specializations—resembling all other members of the series.

The mandibles of hexapods, myriapods and adult malacostracan and branchiopod crustaceans are gnathobasic and protopodal in origin (Popadić et al. 1996; Prpic et al. 2001), lacking a palp. The protopodal origin of the mandible in hexapods is confirmed by lack of *Dll* expression, and in myriapods by transient *Dll* expression (Popadić et al. 1996). The gnathobasic origin of the mandible in branchiopods, cephalocaridans, remipedes and malacostracans is not in question because they all possess a distal palp earlier in development. Most other crustacean taxa either retain a mandibular palp as an adult or lose the palp after the naupliar phase. Loss of the mandibular palp is shared with the Hexapoda and Myriapoda.

The mandibular gnathobase in Crustacea is formed from the proximal segment only (i.e. the coxa) of the two protopodal segments. The ostracod mandible with a basal endite as well as the coxal gnathobase is an exception (Boxshall 2004: Fig. 9b). When present, the palp comprises the distal protopodal segment (the basis) plus the rami. The mandible is homologous to all members of the Mandibulata, so the mandibular gnathobase is formed by the coxa only in hexapods and myriapods as well. In chelicerates and trilobites, the second post-antennulary limb has an undivided protopodite. The coxal gnathobase of the mandibulate mandible is not homologous with the gnathobase of the second post-antennulary limb of cheliceromorphs which is derived from the medial margin of the entire protopodite (as pointed out by many authors, see Boxshall 2004).

#### 11.3.2.2 Endopodite (= Telopodite)

There has been considerable confusion and debate concerning the number of endopodal segments in the phenotypic ground plan of each major arthropodan taxon, and numerous schemes have been proposed to establish homologous landmarks along the P-D axis of the various limbs. Manton (1966) referred to the "welter of assumptions" underpinning such schemes, and the key problem is that within every major arthropodan class, there is marked variation in number of endopodal segments expressed in the phenotype, so uncertainty remains despite the considerable attention devoted to this topic.

Numbers of apparent segments can be larger than a hypothesized ground plan due to subdivision of segments. In the diplopods, for example, the trunk legs were described by Manton (1954, 1958) as having a seven-segmented endopodite consisting of trochanter, pre-femur, femur, post-femur, tibia, tarsus and claw (pretarsus); however, the coxa and trochanter of Manton represent two annuli of a subdivided segment, and the femur and post-femur of Manton similarly represent a subdivided segment. Similarly, in some mysid malacostracans, for example, pereopodal endopodites have been described as having a total of six segments, with a pre-ischium located between the basis and ischium (e.g. Hansen 1925). This is also a secondary increase.

Oligomerization—the reduction in number of expressed limb segments in the phenotype—also seems to have been a common evolutionary trend in limb segmentation within taxa. The loss of segments typically results from failure of expression of articulations during development rather than from actual fusion (Boxshall and Huys 1998). These are different processes although both result in a compound segment originating from two or more ancestral segments. Articulations between true segments may fail to be expressed and in such cases, the plane of the ancestral articulation may be marked externally by a suture line in the integument, and/or internally by a muscle insertion or by the retention of a transverse tendonous section within a muscle (Boxshall 1985), or may be lost entirely.

The endopodite of branchiopod trunk limbs has often been interpreted as secondarily unsegmented, but new data on the development of *Limnadopsis* led Pabst and Scholtz (2009) to suggest that the endopodite is fundamentally three-segmented. They consider there to be good evidence supporting the view that a tri-partitite endopodite (either three-lobed or three-segmented) is the general pattern for the

Branchiopoda, as proposed earlier by Hansen (1925). By analogy with such an interpretation, the trunk limb of *Triops* would also have a tripartite endopodite (Fig. 11.7c).

It would be convenient if the wealth of emerging data on gene expression patterns were to provide any marker that could be used to unequivocally identify specific joints along the P-D axis to serve as landmarks for comparison between taxa. However, this seems unlikely since the comparative data that are available show homologous patterning domains do not necessarily mark homologous morphological domains (Abzhanov and Kaufman 2000; Sewell et al. 2008). There are, however, markers for very specific cellular functions which may be localized in particular limb parts, such as the epipodites.

In the biramous post-cephalic trunk limbs of barnacles (Crustacea: Thecostraca), the rami are transformed into cirri that form the food capture apparatus of the sessile adult. These cirri extend hydraulically and but flex using their intrinsic musculature (Cannon 1947). The intrinsic muscles form an intermediate attachment in each segment of the cirrus, indicating that these subdivisions are true segments rather than annuli. Similarly, both rami of the antenna of conchostracan crustaceans comprise multiple podomeres and appear flagellate, but both comprise segments defined by the presence of intrinsic muscles (Boxshall 2004: Fig. 8g). Such examples of secondary increases in true segmentation are relatively rare. Where both rami of a limb are similar and secondarily multi-segmented, as in the antennae of conchostracan branchiopods and the thoracopodal cirri of barnacles, the P-D patterning mechanism is presumably the same for both rami. In the pleopods of amphipod crustaceans, for example, both rami continue to add articulations in a proximal growth zone, at each post-embryonic moult (Maruzzo and Minelli 2011). In such cases, it can be inferred that the secondary segmentation would be controlled by a single, specialized patterning mechanism common to both rami.

### 11.3.2.3 Exopodite

Exopodites on post-antennulary limbs are a feature of the arthropod ground plan (Walossek 1999). The exopodite is the outer ramus and has a distal origin on the protopodite, lateral to the endopodite. It is typically provided with muscles originating in the protopodite and inserting within the ramus itself and when the exopodite is often two-segmented, the intrinsic musculature can move the segments relative to one another. Boxshall and Jaume (2009) looked at the diversity of exopodites, noting the prevalence of subdivided exopodites in branchiopods, branchiurans and cephalocaridans, but considered that the basic euarthropodan exopodite was two-segmented. However, multi-segmented exopodites are found in crustacean naupliar limbs (antennae and mandibles), in trunk limbs of copepods, thecostracans and remipedes, in certain phosphatocopines, and in *Agnostus*.

Foliaceous exopodites are present on the trunk limbs of branchiopods (Fig. 11.8b) and of most Palaeozoic fossil arthropods for which the limbs are known, including mandibulates such as *Bredocaris*, *Cinerocaris*, *Dala*, *Rehbachiella* and *Tanazios*; trilobites such as *Burgessia*, *Eoredlichia*, *Misszhouia* and *Olenoides*; and fossils of uncertain affinity such *Sapeiron* (see references in Boxshall 2004 and Boxshall and Jaume 2009). Indeed, arthropods with a series of uniramous post-antennulary limbs lacking exopodites, such as the Silurian pycnogonid *Haliestes* (Siveter et al. 2004), are the exception in the early to mid-Palaeozoic. The rare case of the fossil arthropod *Sarotrocercus* which apparently retains the exopodite only (see Boxshall 2004) may be better interpreted as lacking information on the endopodite (Haug et al. 2011). Within the extant Crustacea, each post-antennulary limb from the antenna to the uropod is biramous somewhere in crustacean morphospace. The exopodite is often lost from particular adult limbs, although larvae may retain an exopodite even if the adults secondarily lack one. In the Eumalacostraca, the distal segment of the pereopodal exopodite has been regarded as

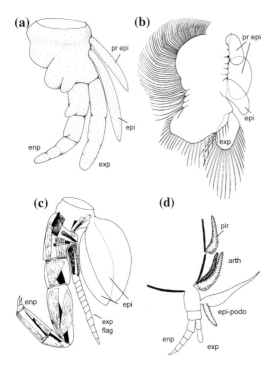

**Fig. 11.8** Diversity of biramous trunk limbs bearing epipodites. **a** *Tanazios* showing blade-like epipodite (*epi*) and pre-epipodite (*pr epi*) and cylindrical endopodite (*enp*) and exopodite (*exp*). **b** *Polyartemia* showing epipodite (*epi*), two pre-epipodites (*pr epi*) and foliaceous exopodite. **c** *Anaspides* showing double epipodite (*epi*) plus flagellate exopodite (*exp flag*). **d** schematic showing arthrobranch (*arth*), and pleurobranch (*plr*) gills, and epipodite-podobranch complex of dendrobranchiate decapod (**a**, simplified drawing from reconstruction in Siveter et al. (2007b), with enditic membranes omitted; **b**, redrawn from Sars (1896); **c**, adapted from Boxshall (2004); **d**, adapted from Boxshall and Jaume (2009))

primitively annulated (Fig. 11.8c) (Boxshall and Jaume 2009); however, the presence of muscles extending the length of the flagellate exopodite in larval decapods (Harzsch and Kreissl 2010) highlights the importance of obtaining better data for basal taxa such as the syncarids, the development of which is poorly documented. The exopodite is not expressed in extant hexapods and myriapods, so inferences on the form of the exopodite in the Mandibulata necessarily depend on evidence from the Crustacea and related fossils. The Silurian *Tanazios* has been interpreted as a probable stem-lineage crustacean (Siveter et al. 2007b) and as a labrophoran (Boxshall 2007), and it has slender type of

segmented exopodite on its trunk limbs (Fig. 11.8a).

The Silurian marrellomorph *Xylokorys* is of particular interest: it has a well-developed, single-axis antennule, followed by the first to fourth post-antennulary limbs each of which has a well-developed multi-segmented exopodite. The first and second post-antennulary limbs have exopodites comprising a basal part of two or three podomeres and a distal section of four or five podomeres carrying a conspicuous setal fan (Siveter et al. 2007a). In the third post-antennulary limb, the endopodite is reduced and the exopodite is very large with a distal section of up to 7 podomeres, each bearing a setal tuft (Fig. 11.9a).

This distinctive type of exopodite closely resembles that found in the Silurian cheliceromorph *Offacolus*, the second to fifth post-antennulary limbs of which each have a six-segmented exopodite (Fig. 11.9b) terminating in a setal fan (Sutton et al. 2002). *Dibasterium* also has a robust multi-segmented exopodite on the same prosomal limbs and Briggs et al. (2012) concluded that the exopodite of both *Offacolus* and *Dibasterium* inserts on the body surface separate from the endopodite-bearing protopodite.

The enigmatic Cambrian arthropod *Ercaia* has a very similar first post-antennulary limb, with an exopodite comprising a segmented cylindrical proximal part plus a flattened distal part bearing a conspicuous setal array (Chen et al. 2001). The presence of a well-developed, articulated cylindrical exopodite in these taxa suggests that this may represent a second basic exopodite type in Palaeozoic arthropods, in addition to the foliaceous type of exopodite.

Foliaceous exopodites are retained on the more posterior trunk limbs in these taxa and in other cheliceromorphs such as *Sanctacaris* and *Limulus* (Boxshall 2004: Fig. 4c), and other marrellomorphs such as *Marrella* and *Mimetaster*. Interestingly, *Xylokorys*, *Offacolus* and *Dibasterium* have the endopodites of the first few pairs of post-antennulary limbs terminating in a subchela. There appears to be a similar structure of the anteriormost pairs of limbs between these

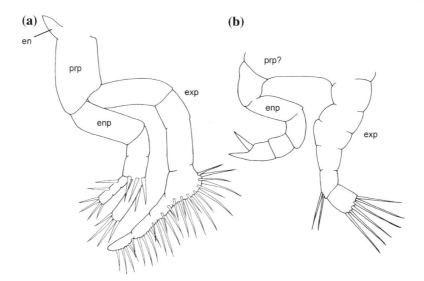

**Fig. 11.9** Anterior post-antennulary limbs from Silurian arthropods with well-developed exopodites. **a** Third post-antennulary limb of the marrellomorphan *Xylokorys*, showing well-developed cylindrical exopodite (*exp*) with distal part bearing setal array, and segmented endopodite with subchelate apex. **b** Post-antennulary limb of cheliceromorphan *Offacolous*, showing two multi-segmented rami, with setal tuft on apex of exopodite (*exp*) (**a**, drawn from reconstructions in Siveter et al. (2007a); **b**, redrawn from Sutton et al. (2002)). The form of the protopodite is uncertain (see Briggs et al. (2012))

two taxa, and, interestingly, both exhibit the biphasic arrangement of post-antennulary limbs into anterior and posterior homonomous blocks (Boxshall 2004).

*Dll* expression can be used to distinguish exopodites from lateral outgrowths, such as epipodites, which result from the establishment of new lateral axes. These usually do not express *Dll*. For example, the exopodal nature of the scaphognathite on the decapod maxilla (in the freshwater crayfish, Astacida) was confirmed by Scholtz et al. (2008) on the basis of such evidence. The earliest expression of *Dll* is in the tip of a crustacean limb bud irrespective of the form of the adult limb, that is, whether it is biramous or uniramous, stenopodial or phyllopodial (Olesen et al. 2001; Williams 2004; Wolff and Scholtz 2008). The endopodite and exopodite are formed by a secondary subdivision of the primary growth zone at the tip of the developing P-D limb axis (Wolff and Scholtz 2008). The subdivision of the primary limb axis is reflected by the transformation of the initially undivided *Dll* expression at the tip of the limb bud into two separate *Dll* domains representing the tips of the rami (Williams 2004; Wolff and Scholtz 2008). The mechanism producing this subdivision is unknown but likely scenarios are the suppression of *Dll* expression in the area between exopodal and endopodal domains, or apoptosis (Wolff and Scholtz 2008).

The loss of the exopodite from the thoracopods of the haplopodan branchiopod *Leptodora* resulted from suppression of the bifurcation of the early limb bud (Olesen et al. 2001). Wolff and Scholtz (2008) showed that uniramous pereopods of the amphipod *Orchestia* are formed by the suppression of the split into exopodite and endopodite of the primary growth zone of the main limb axis. Comparing the clonal composition of the embryonic pereopods and pleopods, Wolff and Scholtz (2008) demonstrated that a population of cells with the identical genealogical background to that which forms the exopodite in the biramous pleopods contributes to the outer part of the endopodite of the uniramous pereopods along most of the P-D axis but not to the tip. Boxshall and Jaume (2009) interpreted the failure of expression of the exopodite in development as resulting in the cells that would have comprised the exopodite anlage being conscripted to contribute to the endopodite.

### 11.3.2.4 Epipodites and Pre-epipodites

The crustacean epipodite is a lateral outgrowth from the coxal part of the limb protopodite. Epipodites are found on the post-maxillary trunk limbs in branchiopods (Figs. 11.7c, 11.8b) and on the thoracopods (maxillipeds and pereopods) in the Malacostraca (Figs. 11.6a, 11.8c,d). Epipodites are characterized by the lack of musculature (Boxshall and Jaume 2009). Transient rudiments of epipodites were also reported during the development of the anterior pleopods of the Leptostraca by Pabst and Scholtz (2009). Epipodites are rarely found on cephalic limbs within the extant crustaceans: exceptions include the presence of a setose lobate epipodite on the maxillule of copepods (Huys and Boxshall 1991) and the well-developed setose epipodite on the maxilla of the myodocopan ostracods. Myodocopans are the only crustaceans that possess an epipodite on the maxilla (Boxshall and Jaume 2009: Fig. 16).

In addition to the epipodite, a more proximally located pre-epipodite is also present in most anostracan Branchiopoda (Fig. 11.8b) and within the Malacostraca—in Leptostraca (Fig. 11.6a) and the Silurian archaeostracan *Cinerocaris* (Fig. 11.6b). Two pre-epipodites are present in chirocephalid Anostraca (Fig. 11.8b), and in other anostracans, the pre-epipodite shows clear evidence of a double origin (Williams 2007). Adult *Anaspides* has two very similar epipodites originating immediately adjacent to each other on the pereopodal coxa (Fig. 11.8c). Although neither shows any evidence of a double origin, one could represent the pre-epipodite. However, the presence of a single coxal epipodite in the Carboniferous *Palaeocaris* and in the bathynellaceans suggests the possibility that the presence of two lobes in *Anaspides* is a secondarily derived state within the Syncarida (Boxshall and Jaume 2009).

The epipodite typically appears very early in development as an unarmed, rounded lobate bud, and in the Branchiopoda (Fig. 11.7a–c), where post-maxillary limbs initially appear as transverse ridges, the epipodite bud appears just prior to the limbs commencing their swing down to the vertical, adult orientation (Møller et al. 2004). This pattern is common to anostracan and notostracan branchiopods. In leptostracan malacostracans, the epipodite on the pereopods appears somewhat later in development of the limbs, as the swing to vertical is taking place (Pabst and Scholtz 2009).

Ungerer and Wolff (2005) showed that the coxal plate and epipodite of amphipod pereopods arise from a common anlage in early development and considered it possible that the coxal plate of amphipods might be homologous with the pre-epipodite. Boxshall and Jaume (2009) questioned the widely assumed homology of the peracaridan oostegite with the pre-epipodite. Oostegites and pre-epipodites have different sites of origin on the protopodite, differ structurally, functionally and in orientation. More importantly, Boxshall and Jaume (2009) highlighted that oostegites are secondary sexual structures, often undergoing cyclical change in concert with the hormonally controlled, reproductive cycle of the female and hypothesized that their underlying genetic control mechanisms would also differ. Oostegites may well be a novel structure, apomorphic to the Peracarida.

The epipodite is characterized by distinctive gene expression patterns: strongly expressing *nubbin* (*pdm*), *apterous* (*ap*) (Averof and Cohen 1997), *trachealess* (Mitchell and Crews 2002) and *ventral veinless* (Franch-Marro et al. 2006), but only weakly expressing *Dll* in a transient manner (Williams 1998; Williams et al. 2002). Richter (2002) regarded the specific expression pattern of *pdm* and *ap* in the distal epipodite of *Artemia* and in the epipodite of *Pacifastacus* as a strong argument for homology of these two structures. Irrespective of shared ancestry, the expression of numerous genes by the epipodites of malacostracans and branchiopods probably reflects common functionality as osmoregulatory-gaseous exchange organ. Currently, there is little evidence available to suggest whether two pre-epipodites of chirocephalids or the double pre-epipodite of other anostracans are homologous with the pre-epipodite of anaspidacean malacostracans.

Boxshall (2004) concluded that epipodites on limb protopodites appeared relatively late in the Palaeozoic and were not present in the crustacean ground plan. The discovery of new fossils has challenged this conclusion: Zhang et al. (2007) reported "epipodites" on the trunk limbs of the Cambrian *Yicaris*, which they classified as a crown-group crustacean, and Siveter et al. (2007b) described the Silurian *Tanazios* which they interpreted as a stem-lineage crustacean. All post-mandibular limbs of *Tanazios* are biramous with two slender, blade-like, tapering exites on the outer margin of the protopodite (Fig. 11.8a), which were identified as epipodites by Siveter et al. (2007b). Boxshall (2007) considered that *Tanazios* should probably be classified as a member of the Labrophora but noted that the presence of two epipodites could be interpreted as evidence that such a state was basic to the Eucrustacea ground plan.

In *Yicaris,* three exites are present along the lateral margin of the protopodal part of the post-mandibular limbs. They were homologized with the epipodite plus pre-epipodite of anostracan Branchiopoda, and a ground plan of three epipodites per limb was suggested for the Eucrustacea (Zhang et al. 2007) or the Eubranchiopoda (Maas et al. 2009). Boxshall (2007) considered that the pattern of development in *Yicaris* (Fig. 11.10a–c) was significantly different from that of branchiopodan epipodites and regarded the evidence supporting the inference that these structures were homologues of the crustacean epipodite plus two pre-epipodites as weak. Boxshall and Jaume (2009) subsequently pointed to differences in form and in the timing of the appearance of the epipodite and pre-epipodite anlagen in anostracan embryos (Møller et al. 2004) and of the exites in *Yicaris* and inferred that the structures in the latter represent an independently derived exite series. Maas et al. (2009) reconsidered the evidence from the Cambrian fossils and concluded that the three exites were present in the ground

pattern of their Entomostraca and that these were retained in *Yicaris* and in the Branchiopoda.

The timing of appearance of these structures during development is very different (cf. Figs. 11.7, 11.10). In Branchiopoda, the epipodite (and pre-epipodite) appears very early when the limb primordium comprises a simple transverse ridge of tissue subdivided by slight indentations on the free margin (Fig. 11.7a). As this limb develops, the lobes (presumptive endites, rami, epipodite and pre-epipodite) become better defined (Fig. 11.7b), so by the time the developing limb swings from a transverse to a dorsoventral orientation, the epipodite is already clearly differentiated. In contrast, in *Yicaris* (Fig. 11.10a–c), the bilobate limb bud has a dorsoventral orientation (Fig. 11.10a) before any rudiment of any outer lobe appears (Fig. 11.10b). Assuming the posterior to anterior limb series serves as a surrogate for the development process in *Yicaris*, the three exites appear sequentially, together with the setation elements of the rami and the endites. The development of these exites on the outer margin of the protopodite of *Yicaris* has much in common with the sequential appearance of setation elements and raises doubts as to their homology with the epipodite and two pre-epipodites of the Branchiopoda.

## 11.4  Heteronomy of Post-antennulary Limbs

A corollary of the hypothesis that the arthropodan ground plan included only two limb types (a multi-segmented single-axis antennule and a biramous post-antennulary limb) is that the post-antennulary limbs formed an essentially homonomous series with little or no differentiation along the A-P axis except in relative size. This describes the trilobite condition: *Phacops,* for example, has paired antennules followed by a homonomous series of post-antennulary limbs (Bruton and Haas 1999). Other Cambrian arthropods, such as the xandarellid *Cindarella,* similarly show a homonomous series of post-antennulary limbs

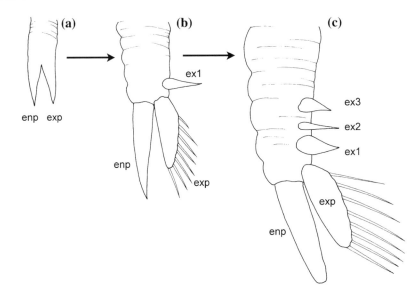

**Fig. 11.10** Simplified schematic showing development of exites on trunk limbs of *Yicaris*. **a** Early biramous limb bud in dorsoventral orientation showing endopodite (*enp*) and exopodite (*exp*). **b** More anterior limb with distalmost exite present on protopodite of limb. **c** More anterior limb with three exites present on protopodite (redrawn from data in Zhang et al. (2007) and Maas et al. (2009))

without significant A-P differentiation (Ramsköld et al. 1997).

Most arthropods exhibit some degree of cephalization in which one or more pairs of post-antennulary limbs are modified as specialized feeding appendages. A few Palaeozoic arthropods, such as *Marrella* and *Ercaia,* have just one pair of post-antennulary limbs differentiated from the posterior members of the series (Whittington 1971; Chen et al. 2001). Other fossil arthropods have the anterior two, three or more pairs of post-antennulary limbs differentiated. Mandibulates exhibit heteronomy of the post-antennulary limbs, as exemplified by the naupliar and the post-naupliar limbs in crustaceans. The naupliar limb series comprises the uniramous antennules plus the biramous antennae and mandibles: they differ markedly from the post-naupliar limbs (maxillules, maxillae and trunk limbs) which form a basically homonomous series. This progressive cephalization is the dominant processes underlying the trend towards increasing diversity of arthropodan limb types in the Palaeozoic (see Boxshall 2004). However, Boxshall (2004) also recognized that a basic biphasic arrangement of the

post-antennulary limbs into two homonomous series (anterior and posterior) is clearly expressed in early cheliceromorphs. In Palaeozoic forms, it is little modified by specialization within either block. The differences between this biphasic model and the cephalization model presumably reflect differences between the *Hox* genes control mechanisms of the cheliceromorphs and the mandibulates.

Specialization in limb structure along the A-P axis commences with the first post-antennulary limb. The subsequent process of cephalization is progressive, involving the differentiation of increasing numbers of limb pairs in different lineages. Cephalization in all extant arthropod lineages is under the control of *Hox* genes which specify the identity of segments along the A-P axis of arthropods and, thus, play a major role in determining limb morphology. The basic set of *Hox* genes common to extant members of the four major groups of Arthropoda comprises the following genes (orthologues of the *Drosophila* genes): *labial (lab)*, *proboscipedia (pb)*, *Hox*3, *Deformed (Dfd)*, *Sex combs reduced (Scr)*, *Antennapedia (Antp)*, *Ultrabithorax (Ubx)*, *abdominal A (abdA)* and *Abdominal B (AbdB)*

(Averof et al. 2010). The different *Hox* genes are expressed in different regions along the A-P axis of the body and comparative analysis of these expression patterns suggests that changes in regulation of *Hox* gene expression are correlated with segmental specialization and tagmosis in all arthropods (Akam et al. 1988; Damen et al. 1998; Telford and Thomas 1998; Abzhanov et al. 1999; Hughes and Kaufman 2002). The differences between limbs along the A-P axis reflect the functioning of *Hox* genes acting as selectors. For example, in *Drosophila,* specification of the antenna corresponds to the absence of *Hox* gene input; and thoracic leg identities reflect the action of a single gene: *Scr* for the first legs, *Antp* for the second legs and *Ubx* for the third (Struhl 1982).

Differences in fine-tuning of expression patterns can also occur within, as well as between, major arthropod taxa. Changes in expression of *pb* in the hemipteran *Oncopeltus fasciatus,* together with changes in function of the genes *Dfd*, *Dll* and *cap "n" collar* (*cnc*), correlated strongly with the evolutionary transformation of the haustellate or sucking type of labium from the more plesiomorphic limb found in orthopterans (Rogers et al. 2002). This transformation is slightly different in *Drosophila* where specification of the development of the labial imaginal disc yielding the adult proboscis involves the joint action of both *pb* and *Scr* (Percival-Smith et al. 1997; Joulia et al. 2006).

Changes in the function of *Hox* genes are correlated with changes in segmental organization or tagmosis (Averof and Patel 1997) and have probably played a key role in generating the diversity of arthropod limbs (Liubicich et al. 2009). *Ubx* provides one of the clearest examples. Shifts in the anterior boundary of *Ubx* expression are correlated with functional shifts in morphology within crustaceans (Averof and Patel 1997; Scholtz et al. 2008; Averof et al. 2010). The correlation between the anterior expression boundary of *Ubx* and the position and number of pairs of maxillipeds in crustaceans is striking, and knock-down methods have now been used to reduce *Ubx* expression in the model amphipod *Parhyale* resulting in transformation

of walking legs to a maxilliped-like identity (Liubicich et al. 2009).

In insects, the anterior boundary of *Ubx* expression lies in the third thoracic segment and expression extends back through most of the abdomen. *Ubx* expression is instrumental in specifying the boundary between thorax and abdomen, regulating segmental identities and repressing leg development on abdominal segments by repressing *Dll* (Angelini et al. 2005). In addition to this role in A-P axis patterning, *Ubx* also regulates other aspects of development of the third leg, such as the size of the enlarged jumping legs of orthopterans (Mahfooz et al. 2007). In myriapods *UbdA* (combined *Ultrabithorax* and *abdominal A*), expression starts in the second trunk segment and correlates with the morphological differences between the first and second trunk limbs (Grenier et al. 1997). The anterior boundary of *Ubx* expression starts in the second opisthosomal segment in chelicerates (Popadić and Nagy 2001) irrespective of the differences in the morphology of the anterior opisthosoma between spiders, scorpions and the xiphosuran *Limulus*. However, later in development, the anterior boundary of expression of *UbdA* in *Limulus* moves forward one segment to the first opisthosomal segment bearing the chilaria. In chelicerates, therefore, changes in morphology of the first opisthosomal segment are either not associated with changes in *UbdA* expression or correlate only with later changes in *UbdA* expression.

The basic *Hox* gene set is shared by all arthropods and was present in the common lobopodian/arthropodan ancestor, yet the ancestor of the arthropods is hypothesized as possessing a homonomous series of postantennulary trunk limbs. The original role of some of these *Hox* genes seems obscure with respect to limb differentiation, in an ancestral form with an undifferentiated, homonomous limb series behind the antennules. Given the primitive lack of differentiation along the limb series and the different pathways towards tagmosis exhibited across the Arthropoda, it seems likely that this will be reflected in a diversity of roles for *Hox* genes across arthropod lineages.

## 11.5 Conclusions

Modern arthropod phenotypes display an amazing diversity of limb types and their limbs are carried on segmented bodies that are patterned along their A-P axis by a basic set of *Hox* genes common to all four major groups of extant Arthropoda. These *Hox* genes play a pivotal role in specifying limb identity, regulating the cascade of genes that are responsible for patterning the limb itself. The early establishment of the P-D axis by the leg gap genes is also a general feature of limb patterning in the development of all arthropods, as is the Notch signalling pathway which is pivotal in the process of subdivision along the P-D axis. The mechanisms responsible for regulating subdivision of the P-D axis into segments or into annuli seem to diverge downstream of the Notch signalling pathway. Levels of Notch activity are central to the fine-scale regulation of joint production and the Notch signalling pathway is also involved in the specification of muscle founder cells. The domains established by the leg gap genes and the operation of the Notch signalling pathway within those domains appears to provide the basic P-D location information for all downstream processes that take place within the limb, including joint formation, muscle and tendon formation and attachment, endite formation and setal patterning.

Less is known about the patterning of the arthropod exopodite, but it seems likely that the same basic mechanism will regulate the P-D subdivision of both rami. New data emerging from fossil marrellomorphs and cheliceromorphs demonstrate that exopodal form was much more variable in early Palaeozoic arthropods than hitherto realized. The arthropodan exopodite exhibits significant morphological variation from multi-segmented to flagellate, and from cylindrical to foliaceous and it would benefit from more focused study.

Comparative data from different arthropod taxa show that homologous patterning domains do not necessarily mark homologous morphological domains. At present, it seems unlikely that gene expression patterns will provide us with reference points allowing the identification of homologies between the component segments of chelicerate, insect, myriapod and crustacean walking limbs. However, a possible exception might be limb components with very specific functional attributes that are reflected in cellular physiology. The epipodites of the branchiopodan trunk limb and malacostracan pereopod, for example, express several genes that are not expressed elsewhere; presumably, these are linked to specific cellular functions related to osmoregulation and gaseous exchange roles of the epipodite epithelial cells.

There remains a major gap in our knowledge—the gap between the new paradigm emerging from developmental genetics and the morphological study of phenotypes. In time, this gap will be filled by cell fate studies and clonal composition analysis and should transform our ability to understand the development of arthropod limbs through the entire timeline from specification to adult phenotype.

## References

Abu-Shaar M, Mann RS (1998) Generation of multiple antagonistic domains along the proximodistal axis during *Drosophila* leg development. Development 125:3821–3830

Abzhanov A, Kaufman TC (2000) Homologs of *Drosophila* appendage genes in the patterning of arthropod limbs. Dev Biol 227:673–689

Abzhanov A, Popadić A, Kaufman TC (1999) Chelicerate *Hox* genes and the homology of arthropod segments. Evol Dev 1:77–89

Akam M, Dawson I, Tear G (1988) Homeotic genes and the control of segment diversity. Development 104:123–133

Angelini DR, Kaufman TC (2004) Functional analyses in the hemipteran *Oncopeltus fasciatus* reveal conserved and derived aspects of appendage patterning in insects. Dev Biol 271:306–321

Angelini DR, Kaufman TC (2005) Insect appendages and comparative ontogenetics. Dev Biol 271:306–321

Angelini DR, Liu PZ, Hughes CL, Kaufman TC (2005) Hox gene function and interaction in the milkweed bug *Oncopeltus fasciatus* (Hemiptera). Dev Biol 287:440–455

Angelini DR, Smith FW, Aspiras AC, Kikuchi M, Jockusch EL (2011) Patterning of the adult mandibulate mouthparts in the red flour beetle, *Tribolium castaneum*. Genetics. doi:10.1534/genetics.111.134296

Averof M, Cohen SM (1997) Evolutionary origin of insect wings from ancestral gills. Nature 385:627–630

Averof M, Patel NH (1997) Crustacean appendage evolution associated with changes in Hox gene expression. Nature 388:682–686

Averof M, Pavlopoulos A, Kontarakis Z (2010) Evolution of new appendage types by gradual changes in Hox gene expression—the case of crustacean maxillipeds. Palaeodiversity 3(Suppl):141–145

Beermann A, Jay DG, Beeman RW, Hulskamp M, Tautz D, Jurgens G (2001) The *Short antennae* gene of *Tribolium* is required for limb development and encodes the orthologue of the *Drosophila* distal-less protein. Development 128:287–297

Bishop SA, Klein T, Martinez Arias A, Couso JP (1999) Composite signalling from *Serrate* and *Delta* establishes leg segments in *Drosophila* through *Notch*. Development 126:2993–3003

Bitsch J (2001) The hexapod appendage: basic structure, development and origin. In: Deuve T (ed) Origin of the Hexapoda, vol. 37, pp 175–193 (Annls Soc ent Fr)

Boxshall GA (1985) The comparative anatomy of two copepods, a predatory calanoid and a particle feeding mormonilloid. Phil Trans R Soc Lond B 311:303–377

Boxshall GA (1997) Comparative limb morphology in major crustacean groups: the coxa-basis joint in postmandibular limbs In: Fortey RA, Thomas R (eds) Arthropod phylogeny. Chapman and Hall, London, pp 155–167

Boxshall GA (2004) The evolution of arthropod limbs. Biol Rev 79:253–300

Boxshall GA (2007) Crustacean classification: on-going controversies and unresolved problems. Zootaxa 1668:313–325

Boxshall GA, Huys R (1998) The ontogeny and phylogeny of copepod antennules. Phil Trans R Soc Lond B 353:765–786

Boxshall GA, Jaume D (2009) Exopodites, epipodites and gills in crustaceans. Arthropod Syst Phylog 67:229–254

Boxshall GA, Danielopol DL, Horne DJ, Smith RJ, Tabacaru I (2010) A critique of biramous interpretations of the crustacean antennule. Crustaceana 83:153–167

Brenneis G, Ungerer P, Scholtz G (2008) The chelifores of sea spiders (Arthropoda, Pycnogonida) are the appendages of the deutocerebral segment. Evol Dev 10:717–724

Briggs DEG, Sutton MD, Siveter DJ, Siveter DJ (2004) A new phyllocarid (Crustacea: Malacostraca) from the silurian fossil-lagerstätte of herefordshire, UK. Proc R Soc Lond B 271:131–138

Briggs DEG, Siveter DJ, Siveter DJ, Sutton MD, Garwood RJ, Legg D (2012) Silurian horseshoe crab illuminates the evolution of arthropod limbs. Proc Natl Acad Sci USA. doi:10.1073/pnas.1205875109

Bruton DL, Haas W (1999) The anatomy and functional morphology of *Phacops* (Trilobita) from the hunsrück slate (Devonian). Palaeontographica A 253:29–75

Bruton DL, Whittington HB (1983) *Emeraldella* and *Leanchoilia*, two arthropods from the burgess shale, Middle Cambrian, British Columbia. Phil Trans R Soc B 300:553–582

Campbell G (2002) Distalization of the *Drosophila* leg by graded EGF-receptor activity. Nature 418:781–785

Cannon HG (1947) On the anatomy of the pedunculate barnacle *Lithotrya*. Phil Trans R Soc B 233:89–136

Chaudonneret J (1950) La morphologie céphalique de *Thermobia domestica* (Packard) (Insecte Aptérygote Thysanoure). Annls Sci nat 12:145–300

Chen J-Y, Vannier J, Huang D-Y (2001) The origin of crustaceans: new evidence from the early cambrian of China. Proc R Soc B 268:2181–2187

Chen J, Waloszek D, Maas A (2004) A new "great appendage" arthropod from the lower cambrian of China and the phylogeny of Chelicerata. Lethaia 37:3–20

Cotton TJ, Braddy SJ (2004) The phylogeny of arachnomorph arthropods and the origin of the Chelicerata. Trans R Soc Edinburgh: Earth Sci 94:169–193

Dahms H-U (1992) Metamorphosis between naupliar and copepodid phases in the Harpacticoida. Phil Trans R Soc B 335:221–236

Damen WGM, Hausdorf M, Seyfarth EA, Tautz D (1998) A conserved mode of head segmentation in arthropods revealed by the expression patterns of Hox genes in a spider. Proc Natl Acad Sci USA 95:10665–10670

Damen WGM, Prpic N-M, Janssen R (2009) Embryonic development and the understanding of the adult body plan in myriapods. Soil Org 81:337–346

de Calazans DK (1992) Taxonomy, distribution and abundance of protozoea, mysis and megalopa stages of penaeidean decapods from the southern Brazilian coast. Unpublished PhD thesis, University of London

de Celis JF, Tyler DM, de Celis J, Bray SJ (1998) Notch signalling mediates segmentation of the *Drosophila* leg. Development 125:4617–4626

Dong PD, Chu J, Panganiban G (2001) Proximodistal domain specification and interactions in developing *Drosophila* appendages. Development 128:2365–2372

Edgecombe GD, García-Bellido DC, Paterson JR (2011) A new leanchoiliid megacheiran arthropod from the lower Cambrian Emu Bay Shale, South Australia. Acta Palaeont Polon 56:385–400

Emerald BS, Curtiss J, Mlodzik M, Cohen SM (2003) *Distal antenna* and *distal antenna related* encode nuclear proteins containing pipsqueak motifs involved in antenna development in *Drosophila*. Development 127:1171–1180

Franch-Marro X, Martin N, Averof M, Casanova J (2006) Association of tracheal placodes with leg primordia in *Drosophila* and implications for the origin of insect tracheal systems. Development 133:785–790

Galindo MI, Bishop SA, Greig S, Couso JP (2002) Leg patterning driven by proximal-distal interaction and EGFR signalling. Science 297:256–259

Greenberg L, Hatini V (2009) Essential roles for *lines* in mediating leg and antennal proximodistal patterning and generating a stable Notch signalling interface at segmental borders. Dev Biol 330:93–104

Grenier JK, Garber TL, Warren R, Whittington PM, Carroll S (1997) Evolution of the entire arthropod hox gene set predated the origin and radiation of the onychophoran/arthropod clade. Curr Biol 7:547–553

Hansen HJ (1925) On the comparative morphology of the appendages in the Arthropoda. A. Crustacea. Gyldendalske, Copenhagen

Hao L, Green RB, Dunaevsky O, Lengyel JA, Rauskolb C (2003) The odd-skipped family of zinc finger genes promotes *Drosophila* leg segmentation. Dev Biol 263:282–295

Harvey THP, Butterfield NJ (2008) Sophisticated particle-feeding in a large early cambrian crustacean. Nature 452:868–871

Harvey THP, Vélez MI, Butterfield NJ (2012) Exceptionally preserved crustaceans from Western Canada reveal a cryptic cambrian radiation. Proc Natl Acad Sci USA 109:1589–1594

Harzsch S, Kreissl S (2010) Myogenesis in the thoracic limbs of the American lobster. Arthropod Struct Dev 39:423–435

Haug JT, Maas A, Haug C, Waloszek D (2011) *Sarotrocercus oblitus*—small arthropod with great impact on the understanding of arthropod evolution? Bull Geosci 86:725–736

Haug JT, Waloszek D, Maas A, Liu Y, Haug C (2012) Functional morphology, ontogeny and evolution of mantis shrimp-like predators in the cambrian. Palaeontology 55:369–399

Hughes CL, Kaufman TC (2002) Exploring the myriapod body plan: expression patterns of the ten Hox genes in a centipede. Development 129:1225–1238

Huys R, Boxshall GA (1991) Copepod evolution. The Ray Society, London

Jockusch EL, Willams T, Nagy LM (2004) The evolution of patterning of serially homologous appendages in insects. Dev Genes Evol 214:324–338

Joulia L, Deutsch J, Bourbon H-M, Cribbs DL (2006) The specification of a highly derived arthropod appendage, the *Drosophila* labial palps, requires joint action of selectors and signalling pathway. Dev Genes Evol 216:431–442

Karanovic I (2005) Comparative morphology of the Candoninae antennula, with remarks on the ancestral state in ostracods and a proposed new terminology. Spixiana 28:141–160

Kerber B, Monge I, Mueller M, Mitchell PJ, Cohen SM (2001) The *AP-2* transcription factor is required for joint formation and cell survival in *Drosophila* leg development. Development 128:1231–1238

Kojima T (2004) The mechanism of *Drosophila* leg development along the proximodistal axis. Dev Growth Differ 46:115–129

Kreissl S, Uber A, Harzsch S (2008) Muscle precursor cells in the developing limbs of two isopods (Crustacea, Peracarida): an immunohistochemical study using a novel monoclonal antibody against myosin heavy chain. Dev Genes Evol 218:253–265

Lecuit T, Cohen SM (1997) Proximal-distal axis formation in the *Drosophila* leg. Nature 388:139–145

Liu J, Steiner M, Dunlop JA, Keupp H, Shu D, Ou Q, Han J, Zhang Z, Zhang X (2011) An armoured Cambrian lobopodian from China with arthropod-like appendages. Nature 470:526–530

Liubicich DM, Serano JM, Pavlopoulos A, Kontarakis Z, Protas ME, Kwan E, Chatterjee S, Tran KD, Averof M, Patel NH (2009) Knockdown of *Parhyale Ultrabithorax* recapitulates evolutionary changes in crustacean appendage morphology. Proc Natl Acad Sci USA 106:13892–13896

Maas A, Waloszek D, Müller KJ (2003) Morphology, ontogeny and phylogeny of the Phosphatocopina (Crustacea) from the upper cambrian 'Orsten' of Sweden. Fossils Strata 49:1–238

Maas A, Haug C, Haug JT, Olesen J, Zhang X, Waloszek D (2009) Early crustacean evolution and the appearance of epipodites and gills. Arthropod Syst Phylog 67:255–273

Mahfooz N, Truchyn N, Mihajlovic M, Hrycaj S, Popadic A (2007) Ubx regulates differential enlargement and diversification of insect hind legs. PLoS ONE 2(9):e866. doi: 10.1371.journal.pone.0000866

Manjón C, Sánchez-Herrero E, Suzanne M (2007) Sharp boundaries of Dpp signalling trigger local cell death required for *Drosophila* leg morphogenesis. Nature Cell Biol 9:57–63

Manton SM (1954) The evolution of arthropodan locomotory mechanisms—part 4. The structure, habits and evolution of the Diplopoda. J Linn Soc 42:299–368

Manton SM (1958) The evolution of arthropodan locomotory mechanisms—part 6. habits and evolution of the Lysiopetaloidea (Diplopoda), some principles of leg design in Diplopoda and Chilopoda, and limb structure in Diplopoda. J Linn Soc 43:487–556

Manton SM (1966) The evolution of arthropodan locomotory mechanisms. Part 9. Functional requirements and body design in Symphyla and Pauropoda and the relationships between Myriapoda and pterygote Insects. J Linn Soc 46:103–141

Manton SM (1977) The Arthropoda: habits, functional morphology, and evolution. Clarendon Press, London

Maqbool T, Soler C, Jagla T, Daczewska M, Lodha N, Palliyil S, VijayRaghavan K, Jagla K (2006) Shaping leg muscles in *Drosophila*: role of *ladybird*, a conserved regulator of appendicular myogenesis. PLoS ONE 1(1):e122. doi:10.1371/journal.pone.0000122

Marmonier P, Boulal M, Idbennacer B (2005) *Maroccocandona*, a new genus of Candoninae (Crustacea, Ostracoda) from southern Morocco; morphological characteristics and ecological requirements. Ann Limnol Int J Limnol 41:57–71

Maruzzo D, Minelli A (2011) Post-embryonic development of amphipod crustacean pleopods and the
patterning of arthropod limbs. Zool Anz 250:32–45

Maruzzo D, Minelli A, Ronco M, Fusco G (2007)
Growth and regeneration of the second antennae of
*Asellus aquaticus* (Isopoda) in the context of arthro-
pod antennal segmentation. J Crust Biol 27:184–196

Maruzzo D, Minelli A, Fusco G (2009) Segmental
mismatch in crustacean appendages: the naupliar
antennal exopod of *Artemia* (Crustacea, Branchiopoda, Anostraca). Arthropod Struct Dev 38:163–172

Minelli A, Foddai D, Pereira LA, Lewis JGE (2000) The
evolution of segmentation of centipede trunk and
appendages. J Zool Syst Evol Res 38:103–117

Mitchell B, Crews ST (2002) Expression of the *Artemia
trachealess* gene in the salt gland and epipod. Evol
Dev 4:344–353

Mito T, Shinmyo Y, Kurita K, Nakamura T, Ohuchi H,
Noji S (2011) Ancestral functions of Delta/Notch
signaling in the formation of body and leg segments
in the cricket *Gryllus bimaculatus*. Development
138:3823–3833

Mittmann B, Scholtz G (2001) *Distal-less* expression in
embryos of *Limulus polyphemus* (Chelicerata, Xiphosura) and *Lepisma saccharina* (Insecta, Zygentoma)
suggests a role in development of mechanoreceptors,
chemoreceptors, and the CNS. Dev Genes Evol
211:232–243

Møller OS, Olesen J, Høeg JT (2004) On the development of *Eubranchipus grubii* (Crustacea, Branchiopoda, Anostraca), with notes on the basal phylogeny
of the Branchiopoda. Zoomorphology 123:107–123

Müller KJ, Walossek D (1986) *Martinssonia elongata*
gen. et sp. n., a crustacean-like euarthropod from the
upper cambrian 'Orsten' of Sweden. Zool Scripta
15:73–92

Nayrolles P (1991) La chetotaxie antennaire des Collemboles Symphypléones. Trav Lab Ecobiol Arthr
Edaph Toulouse 6(3):1–94

Olesen J, Richter S, Scholtz G (2001) The evolutionary
transformation of phyllopodous to stenopodous limbs
in the Branchiopoda (Crustacea)—is there a common
mechanism for early limb development in arthropods? Int J Dev Biol 45:869–876

Pabst T, Scholtz G (2009) The development of phyllopodous limbs in Leptostraca and Branchiopoda. J Crust
Biol 29:1–12

Panin VM, Papayannopoulos V, Wilson R, Irvine KD
(1997) Fringe modulates Notch-ligand interactions.
Nature 387:908–912

Park M, Yaich LE, Bodmer R (1998) Mesodermal cell
fate decisions in *Drosophila* are under the control of
the lineage genes numb, Notch, and sanpodo. Mech
Dev 75:117–126

Pechmann M, Prpic NM (2009) Appendage patterning in
the South American bird spider *Acanthoscurria
geniculata* (Aranae: Mygalomorphae). Dev Genes
Evol 219:189–198

Pechmann M, Khadjeh S, Sprenger F, Prpic NM (2010)
Patterning mechanisms and morphological diversity

of spider appendages and their importance for spider
evolution. Arthropod Struct Dev 39:453–467

Percival-Smith A, Weber J, Gilfoyle E, Wilson P (1997)
Genetic characterization of the role of the two HOX
proteins Proboscipedia and sex combs reduced, in
determination of adult antennal, tarsal, maxillary palp
and proboscis identities in *Drosophila melanogaster*.
Development 124:5049–5062

Popadić A, Nagy LM (2001) Conservation and variation
in *Ubx* expression among chelicerates. Evol Dev
3:391–396

Popadić A, Rusch D, Peterson M, Rogers BT, Kaufman
TC (1996) Origin of the arthropod mandible. Nature
380:395

Postlethwait JH, Schneiderman HA (1971) Pattern formation and determination in the antenna of the
homeotic mutant *Antennapedia* of *Drosophila melanogaster*. Dev Biol 25:606–640

Prpic N-M, Damen WG (2004) Expression patterns of leg
genes in the mouthparts of the spider *Cupiennius salei*
(Chelicerata: Arachnida). Dev Genes Evol 214:
296–302

Prpic N-M, Damen WG (2009) *Notch*-mediated segmentation of the appendages is a molecular phylotypic
trait of the arthropods. Dev Biol 326:262–271

Prpic N-M, Tautz D (2003) The expression of the
proximodistal axis patterning genes Distal-less and
dachshund in the appendages of Glomeris marginata
(Myriapoda: Diplopoda) suggests a special role of
these genes in patterning the head appendages. Dev
Biol 260:97–112

Prpic N-M, Wigand B, Damen WGM, Klinger M (2001)
Expression of *dachsund* in wild-type and *Distal-less*
mutant *Tribolium* corroborates serial homologies in
insect appendages. Dev Genes Evol 211:467–477

Ramsköld L, Chen J, Edgecombe GD, Zhou G (1997)
*Cindarella* and the arachnate clade Xandarellida
(Arthropoda, early cambrian) from China. Trans R
Soc Edinburgh: Earth Sci 88:19–38

Rauskolb C (2001) The establishment of segmentation in
the *Drosophila* leg. Development 128:4511–4521

Rauskolb C, Irvine KD (1999) Notch mediated segmentation and growth control of the *Drosophila* leg. Dev
Biol 210:339–350

Regier JC, Shultz JW, Zwick A, Hussey A, Ball B,
Wetzer R, Martin JW, Cunningham CW (2010)
Arthropod relationships revealed by phylogenomic
analysis of nuclear protein-coding sequences. Nature
463:1079–1083

Richter S (2002) The Tetraconata concept: hexapod-
crustacean relationships and the phylogeny of Crustacea. Org Divers Evol 2:217–237

Rogers BT, Peterson MD, Kaufman TC (2002) The
development and evolution of insect mouthparts as
revealed by the expression patterns of gnathocephalic
genes. Evol Dev 4:1–15

Sandeman RE, Sandeman DC (1996) Pre- and postembryonic development, growth and turnover of olfactory receptor neurons in crayfish antennules. J Exp
Biol 199:2409–2418

Sars GO (1896) Phyllocarida og Phyllopoda. Fauna Norvegiae 1. Mallingske Bogtrykkeri, Christiania

Scholtz G, Edgecombe GD (2005) Heads, Hox and the phylogenetic position of trilobites. In: Koenemann S, Jenner R (eds) Crustacea and arthropod relationships. CRC, Boca Raton, pp 139–165

Scholtz G, Edgecombe GD (2006) The evolution of arthropod heads: reconciling morphological, developmental and palaeontological evidence. Dev Genes Evol 216:395–415

Scholtz G, Abzhanov A, Alwes F, Biffis C, Pint J (2008) Development, genes, and decapod evolution. In: Martin JW, Crandall KA, Felder DL (eds) Decapod crustacean phylogenetics. CRC, Boca Raton, pp 31–46

Schoppmeier M, Damen WG (2001) Double-stranded RNA interference in the spider *Cupiennius salei*: the role of *Distal-less* is evolutionarily conserved in arthropod appendage formation. Dev Genes Evol 211:76–82

Schutze MLM, da Rocha CEF, Boxshall GA (2000) Antennular development during the copepodid phase in the family Cyclopidae (Copepoda, Cyclopoida). Zoosystema 22:749–806

Sewell W, Williams T, Coole J, Terry M, Ho R, Nagy L (2008) Evidence for a novel role for dachshund in patterning the proximal arthropod leg. Dev Genes Evol 218:295–305

Shultz JW (1990) Morphology of locomotor appendages in Arachnida: evolutionary trends and phylogenetic implications. Zool J Linn Soc 97:1–56

Simonnet F, Moczek AP (2011) Conservation and diversification of gene function during mouthpart development in *Onthophagus* beetles. Evol Dev 13:280–289

Siveter DJ, Sutton MD, Briggs DEG, Siveter DJ (2004) A Silurian sea spider. Nature 431:978–980

Siveter DJ, Fortey RA, Sutton MD, Briggs DEG, Siveter DJ (2007a) A Silurian 'marrellomorph' arthropod. Proc R Soc B 274:2223–2229

Siveter DJ, Sutton MD, Briggs DEG, Siveter DJ (2007b) A new probable stem lineage crustacean with three-dimensionally preserved soft parts from the Herefordshire (Silurian) Lagerstätte, UK. Proc R Soc B 274:2099–2107

Smith RJ, Tsukagoshi A (2005) The chaetotaxy, ontogeny and musculature of the antennule of podocopan ostracods (Crustacea). J Zool Lond 265:157–177

Snodgrass RE (1938) Evolution of the Annelida, Onychophora and Arthropoda. Smithson Misc Colln 97:1–159

Soler C, Daczewska M, De Ponte JP, Dastugue B, Jagla K (2004) Coordinated development of muscles and tendons of the *Drosophila* leg. Development 131:6041–6051

Stein M, Waloszek D, Maas A (2005) *Oelandocaris oelandica* and the stem lineage of Crustacea. In: Koenemann S, Jenner RA (eds) Crustacean issues 16: crustacea and arthropod relationships. Taylor & Francis, Boca Raton, pp 55–71

Steullet P, Cate HS, Michel WC, Derby CD (2000) Functional units of a compound nose: aesthetasc sensilla house similar populations of olfactory receptor neurons on the crustacean antennule. J Comp Neurol 418:270–280

Struhl G (1982) Genes controlling segmental specification in the *Drosophila* thorax. Proc Natl Acad Sci USA 79:7380–7384

Sutton MD, Briggs DEG, Siveter DJ, Siveter DJ, Orr PJ (2002) The arthropod *Offacolus kingi* (Chelicerata) from the Silurian of Herefordshire, England: computer based morphological reconstructions and phylogenetic affinities. Proc R Soc B 269:1195–1203

Tajiri R, Misaki K, Yonemura S, Hayashi S (2011) Joint morphology in the insect leg: evolutionary history inferred from *Notch* loss-of-function phenotypes in *Drosophila*. Development 138:4621–4626

Telford MJ, Thomas RH (1998) Expression of homeobox genes shows chelicerate arthropods retain their deutocerebral segment. Proc Natl Acad Sci USA 95:10671–10675

Ungerer P, Wolff C (2005) External morphology of limb development in the amphipod *Orchestia cavimana* (Crustacea, Malacostraca, Peracarida). Zoomorphology 124:89–99

Walossek D (1993) The upper cambrian *Rehbachiella* and the phylogeny of Branchiopoda and Crustacea. Fossils Strata 32:1–202

Walossek D (1999) On the Cambrian diversity of Crustacea. In: Schram FR, von Vaupel Klein JC (eds) Crustaceans and the biodiversity crisis. Brill, Leiden, pp. 3–27

Walossek D, Müller KJ (1998) Early arthropod phylogeny in light of the Cambrian "Orsten" fossils. In: Edgecombe GE (ed) Arthropod fossils and phylogeny. Columbia University Press, New York, pp 185–231

Waloszek D (2003) The 'Orsten' window—a three-dimensionally preserved upper cambrian meiofauna and its contribution to our understanding of the evolution of Arthropoda. Palaeont Res 7:71–88

Waloszek D, Maas A, Chen J, Stein M (2007) Evolution of cephalic feeding structures and the phylogeny of Arthropoda. Palaeogeo Palaeoclim Palaeoecol 254:273–287

Wege W (1911) Morphologische und experimentelle Studien an *Asellus aquaticus*. Zool Jb Allg Zool Physiol Tiere 30:217–320

Weygoldt P (1996) Evolutionary morphology of whip spiders: towards a phylogenetic system (Chelicerata: Arachnida: Amblypygi). J Zool Syst Evol Res 34:185–202

Whittington HB (1971) Redescription of *Marrella splendens* (Trilobitoidea) from the Burgess Shale, Middle Cambrian, British Columbia. Bull Geol Surv Can 209:1–24

Williams TA (1998) *Distal-less* expression in crustaceans and the patterning of branched limbs. Dev Genes Evol 207:427–434

Williams TA (2004) The evolution and development of crustacean limbs: an analysis of limb homologies. In: Scholtz G (ed) Evolutionary developmental biology of Crustacea. Balkema, Lisse, pp 169–193

Williams TA (2007) Limb morphogenesis in the branchiopod crustacean, *Thamnocephalus platyurus*, and the evolution of proximal limb lobes within Anostraca. J Zool Syst Evol Res 45:191–201

Williams TA (2008) Early *Distal-less* expression in a developing crustacean limb bud becomes restricted to setal-forming cells. Evol Dev 10:114–120

Williams TA, Nulsen C, Nagy LM (2002) A complex role for Distal-less in crustacean appendage development. Dev Biol 241:302–312

Wolff C, Scholtz G (2008) The clonal composition of biramous and uniramous arthropod limbs. Proc R Soc B 275:1023–1028

Zhang X-i, Siveter DJ, Waloszek D, Maas A (2007) An epipodite-bearing crown-group crustacean from the Lower Cambrian. Nature 449:595–598

# Insect Wings: The Evolutionary Development of Nature's First Flyers

**12**

Michael S. Engel, Steven R. Davis and Jakub Prokop

## Contents

M. S. Engel (✉) · S. R. Davis
Division of Entomology, Natural History Museum, and Department of Ecology and Evolutionary Biology, University of Kansas, 1501 Crestline Drive—Suite 140, Lawrence, Kansas 66045, USA
e-mail: msengel@ku.edu

S. R. Davis
e-mail: steved@ku.edu

J. Prokop
Department of Zoology, Faculty of Science, Charles University in Prague, Viničná 7, 12844, Praha 2, Czech Republic
e-mail: jprokop@natur.cuni.cz

## 12.1 Introduction

Powered flight is one of the more spectacular evolutionary novelties to have come about during the 4-billion-year history of life on Earth. Flight bestows upon the flyer another dimension in which to experience life. Suddenly, new avenues are available for dispersal, escape and avoidance, locating a suitable mate, and reaching once unobtainable resources. Moreover, wings can be so much more than merely a means to fly. Properly adapted the wings themselves may play a role in courtship, camouflage and mimicry, thermoregulation, and protection and defence. Despite the profound significance of flight, it is a challenging feat to achieve and control. Powered flight has evolved independently at least four times, three of which occur among the Amniota, while the last is far flung across the branches of the animal tree of life. It is this last lineage that was also the first to evolve this singularly successful means of locomotion, rivalling in numbers of species all other forms of life combined. Insects took to the skies perhaps as long as 400 million years ago, and some 170, 250, and 350 million years before pterosaurs, birds, and bats, respectively (Engel and Grimaldi 2004). The pterygote insects (Insecta: Pterygota), Nature's first flyers, have dominated the Earth's skies since the dawn of terrestrial animal life, and their origins are so remotely removed from our world today that it is their evolution that remains one of the more

A. Minelli et al. (eds.), *Arthropod Biology and Evolution*,
DOI: 10.1007/978-3-642-36160-9_12, © Springer-Verlag Berlin Heidelberg 2013

abominable mysteries in insect evolutionary biology.

Unlike the flight of vertebrates in which the homology of the wing with forelimbs is easily recognized and supported, the wings of insects are not merely a wholesale co-option of one or more legs. Indeed, the full complement of hexapodan legs is present and often unmodified, at least for purposes of flight, in all pterygotes just as it is in the primitively wingless insects. Thus, the question of wing origins in insects is more confounded than that of vertebrates. In addition to discovering for what purposes were wings or wing-like structures first employed or how they operated, we must also reveal from what morphological elements they were composed. The former questions regarding functional ancestry are seemingly simplistic to answer, but they are wholly dependent on first knowing from what wings were derived, yet this nature of critical reliance has evaded many in their quest for wing ancestry. Indeed, replies to these questions, both brilliant and ill-conceived, have abounded for more than a century, and the answers remain elusive. No satisfactory answer to the mechanical, behavioural, and physiological origins of insect flight will ever be produced until a conclusive answer is discovered regarding the morphological homology of the wing. It is in this context that the rise of evolutionary developmental biology offers one of the greatest opportunities to elucidate the homology of insect wings and, in turn, will permit a well-founded account of flight origins.

Wings arose once in insects, that is, the Pterygota are monophyletic and supported as such by abundant morphological and molecular evidence (Grimaldi and Engel 2005). This reality certainly simplifies the investigation of their origins, focusing our attention at a specific node representing the common ancestor of pterygotes as well as the associated transitional branch between that ancestor and its shared predecessor with the silverfish (Zygentoma). While wings have evolved a single time among insects, they have repeatedly been lost or become vestigial. Indeed, wings have been reduced or lost an innumerable number of times, even among close relatives within a single genus. In some lineages, the genetic architecture for developing wings has been turned off and on, resulting in a seemingly cyclical 'reevolution' of wings across the clade (e.g. among Phasmatodea). In all such instances, however, wings reappear wholesale with the same morphology, same arrangement of veins and crossveins (including the same arrangement of venational synapomorphies for the clade! and even in Phasmatodea when they are not well organized the same homologies can be recognized), and associated thoracic modifications belying the fact that these have not independently reevolved but instead have remained 'dormant' until such time as the entire genetic machinery has been reinitiated.

It is beyond the purposes of this review to provide a detailed account of pterygote comparative morphology and flight biomechanics and physiology in insects. For suitable reviews of these subjects, we refer the reader to Dudley (2000); Alexander (2002); Vigoreaux (2005); and Grimaldi and Engel (2005). Herein, we provide brief overviews of current developmental and palaeontological evidence for insect wing origins and diversity attempting to emphasize where present research has brought us and in what directions this field of inquiry might proceed to maximal benefit.

### 12.1.1 A General Word of Caution and Plea for Phylogeny

Given that wings have a single evolutionary origin among insects and that this event took place early in the hexapodan tree of life, wings, and their entire genetic architecture, are abundantly ancient. It is therefore all the more critical that any study be cognizant of phylogeny. The greatest insights will come from investigations as close to the base of the pterygote tree as is permissible with today's tools. Indeed, the most could be gleaned from suitably basal clades of the earliest extant winged insects, namely the mayflies (Ephemeroptera), and the dragonflies and damselflies (Odonata) (Grimaldi and Engel 2005). Of even greater interest are those stem-group Ephemeroptera from the Palaeozoic although they are unavailable for vital genetic and developmental work. Unfortunately, stem-group Pterygota remain unknown. Much fanfare and bravado have come from the study of stoneflies (Plecoptera), and yet these are by no means 'primitive' flyers and certainly not even close to being 'primitive' insects. Plecoptera may be relatively basal among extant lineages of Neoptera, but most evidence indicates that they are nested at the base of a subordinate clade within a monophyletic Polyneoptera or orthopterid group of orders (e.g. Haas and Kukalová-Peck 2001; Kjer et al. 2006; Ishiwata et al. 2011; Yoshizawa 2011; Trautwein et al. 2012). Even a cursory examination of hexapod phylogeny reveals the evolutionary distance between the extant Plecoptera and the stem of the pterygote node (Grimaldi and Engel 2005; Trautwein et al. 2012). All the more concerning it is then that virtually all of the model systems from which our knowledge of the developmental and genetic architecture of insect wings is derived are among the Neoptera and even derived species among highly derived families in highly derived orders of the holometabolan insects, themselves an apomorphic lineage of neopterans. Certainly, the homology of genetic systems for insect wing development is greatly conserved, but this architecture (genetic and morphological) still stems from a common ancestor nearly 400 million years removed from the extant species which we study. Deep insight into the developmental mechanisms of wings can be obtained, but we must temper our findings against phylogeny and, where possible, rely most heavily on truly comparative data, particularly those that come from independent comparisons with basal pterygote lineages. Naturally, model systems are used given the ease of working with them, and developing suitable models among mayflies, dragonflies, and the like is exceedingly difficult. Nonetheless, we must recognize that in terms of genetic systems, we are working with 'quaint' tools on less than ideal target organisms and should proceed cautiously and conservatively in our interpretations. In reviewing the developmental evidence for the origin of wings, it is

important to remain neutral to any previous hypotheses which may bias interpretation of these data. As is often the case in discussions of the origins of developmental features, such as wings, if such data are not scrutinized under an unbiased approach, it is easy to reach conclusions that unfurl beyond context and overlook direct evolutionary implications. As many authors have stated and reiterated, and since we are working in the bounds of comparative evolutionary biology, it is crucial not to dismiss key concepts of homology and phylogeny, particularly when synthesizing diverse forms of data as morphology and developmental genetics over long periods of evolutionary time.

## 12.2 Development of Insect Wings

Of the contending hypotheses put forward regarding wing origins, the most influential in driving relevant research have been renditions of the paranotal (Crampton 1916), gill/exite, and 'epipodite' theories. The former hypothesis, in the strict sense, regards wings as a novel feature derived from extensions of the thoracic tergites. The latter hypothesis, which also appears to be given greater support from developmental studies, suggests that winged insects have evolved from a common ancestor that possessed dorsal limb precursors of wings, likely in the form of some exite from the coxopodite (i.e. an epipodite, such as a crustacean gill). Any comparison of the wing with a coxal endite can be excluded; however, given its podite of derivation, the coxa is a distinctive part of the hexapod telopodite. The styli found in Zygentoma and Archaeognatha can also be disregarded as precursors to wings, as the thoracic styli originate from the coxae and the abdominal styli are hypothesized telopodites. Furthermore, while such structures are present in extant (derived) taxa of these orders, it is unknown whether they were plesiomorphic for hexapods or derived features within those lineages.

From an anatomical viewpoint, wings are essentially appendages. They develop as outgrowths from the body and articulate with the body at their bases. Unlike traditional append-ages such as legs, however, (yet similar to gills, exites, and other outgrowths from the pleural wall, itself originating from a basal appendage podite), muscles attach only to the base (axillary sclerites) and do not extend into the wing cavity. Genetically, wings are also definable as appendages, developing as a result of the expression of a basic set of gene networks involved in general appendage formation, while various other genes and gene networks are co-opted/induced to form the specific features that differentiate the wing from other appendages. Therefore, while it is clear that the wing is some rendition of an appendage, the nature of its development and genetic patterning has com-plicated attempts in formulating hypotheses of homology. However, in collating the evolution-ary developmental data gathered thus far, it appears that a combination of the paranotal and exite theories may be most plausible for explaining the origin and evolution of the hexapod wing. In order to elucidate such a complex topic, it is necessary to examine how wings develop, determine what genetic mecha-nisms are responsible for wing formation, and make comparisons with other insect appendages.

## 12.2.1 Embryology and Tissue Development

It has been known since Malpighi (1687) that the wings of adult pterygotes can be observed in the larval stages; however, it was not really until Weismann's work (1864) on muscid fly meta-morphosis that greater attention was devoted to studying wing development. In the late nineteenth century, it already had been discovered by many that the early wing primordium is already present in embryonic stages (Pratt 1900; Tower 1903). The cells of embryonic wing discs, in fact, are found occupying a very similar space to those of the meso- and metathoracic leg discs, as well as found among the epidermal cells that will form the main trunk of the longitudinal tracheae (Madhavan and Schneiderman 1977; Cohen et al. 1991; Williams and Carroll 1993;

Held 2002). This close association between the early wing and leg primordia, though, appears to be a derived feature only of some groups that possess true imaginal discs (Jockusch and Ober 2004). By the end of the embryonic stages, however, the leg and wing rudiments are well differentiated. The early developing wing is most frequently characterized as dorso-lateral in origin during tissue differentiation, in which the epi-dermal cells of the disc begin to thicken (Tower 1903; Powell 1904; Murray and Tiegs 1935). It must be noted, however, that this position is purely a description of relative location on the body trunk and not a definitive statement of tissue origin. This position is also always just lateral to or slightly dorso-lateral to the longitu-dinal thoracic tracheal trunk. Following differ-entiation and during proliferation of the disc cells, the slight dorsal migration of the disc appears to be an artefact of the reorganization of the adult trunk, including expansion of the pleural region. Although most recent works highlight the prom-inent invaginated form of *Drosophila* wing discs, at least five distinct forms of wing discs (or fields/bodies of proliferating wing tissue) have been described from the Holometabola (Tower 1903). These types range from completely evaginated (such as the form of beetle horn tissue prolifera-tion) to fully invaginated and stalked, including various intermediate forms of partially invagi-nated discs. In Coleoptera alone, several types of wing growth can be observed, from the invagi-nated to the fully evaginated (Fig. 12.1d–h) types (Powell 1904, 1905; Quennedey and Quenn-edey 1990). While the term imaginal disc is sometimes specifically applied to such invagi-nated pockets of ectoderm in the Holometabola from which certain imaginal structures are formed (mostly in regard to the observations made on *Drosophila*), this type of disc is apomorphic and appears to have evolved independently in several holometabolous lineages (e.g. Švácha 1992). In the broad sense, imaginal tissues (in this case, wings) that explicitly form as evaginations in holometabolous insects, therefore, should also be termed imaginal discs, though this terminology is avoided due to transparent ambiguities in delin-eation of wing growth types (Švácha 1992;

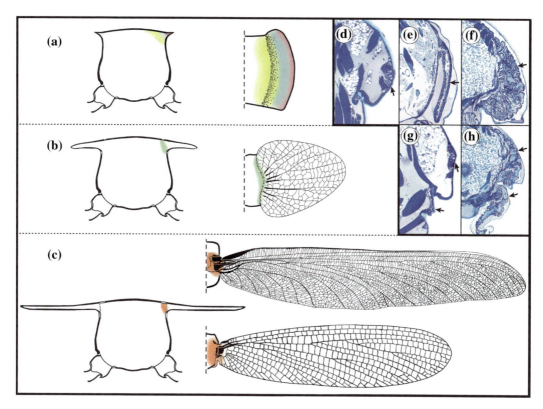

**Fig. 12.1** Hypothesized pathway illustrating the origin of hexapod wings based on current knowledge of palaeontological, neontological, and development data. **a–c** *Left* side, cross-sectional views of mesothorax, illustrating wing development. *Right* side, dorsal views of mesothoracic wing development corresponding to cross-sectional views on *left* side. Colours highlight gene expression as follows: *yellow* = *ap*, expressed in tergum and paranotal margin; *blue* = *vg*, expressed along paranotal margin; *red* = *wg*, expressed along paranotal margin; together, *ap*, *vg*, and *wg* allow for paranotal extension and development of paranotal lobe/primitive wing; *green* = *hth* and *exd* (as well as possibly many other genes), expressed along paranotal margin and base of paranotal lobe to produce primitive joint/articulation; *orange* = *hth* and *exd*, as well as induction of other elements, giving rise to more derived wing articulations. **d–h** Growth of wing tissue in *Tribolium castaneum* (Herbst) (Coleoptera: Tenebrionidae), showing an evaginated type of wing growth. **d** Mesothoracic wing bud of last larval instar. **e** Mesothoracic wing of mid-prepupa ('pharate pupa'). **f** Metathoracic wing of late prepupa. **g** Last larval instar, showing metathoracic wing bud and differentiating leg tissue. **h** Late prepupa, showing metathoracic wing and leg. *Arrows* indicate developing wing, as well as developing leg (**g, h**). Photomicrographs are of semi-thin sections (6 μm) embedded in LR White and stained with toluidine blue

Jockusch and Ober 2004). Thus, although it is unknown how many groups actually possess imaginal discs outside of Diptera (Pratt 1900; Waddington 1941; Wehman 1969; Bryant 1975; Fristrom and Rickoll 1982; Cohen 1993), Lepidoptera (Mercer 1900; Tannreuther 1910; Dixey 1931; Nardi et al. 1985; Niitsu 2003; Niitsu and Kobayashi 2008; Niitsu and Lobbia 2010), and some Coleoptera (Tower 1903), imaginal discs only include the invaginated *Drosophila* type, composed of columnar epithelial cells on one side and a peripodial membrane on the other (Milner et al. 1984). All other late-developing tissues that evaginate in the Holometabola, in addition to all hemimetabolous pterygotes, should be categorized otherwise, as Anlagen, which has been used in the past, or in reference to wing Anlagen, possibly as wing buds or wing fields. A functional explanation for the development of diverse forms of wing discs is unclear, though it has been hypothesized that the form of disc growth most likely is related to the life histories of the larval stages (Tower 1903; Švácha 1992; Truman and Riddiford 1999). Since it is clear, then, that invaginated imaginal discs evolved far after the origin of wings, it can be hypothesized that the close developmental association between leg and wing primordia is a derived feature. Also, as wing discs (as well as other appendage discs) have evolved independently numerous times throughout Holometabola, it can be expected that differences exist in their development and gene expression patterns.

Comparisons of wing disc development have been made to that of gill development in Ephemeroptera, despite the derived phylogenetic position of this group in the Insecta. Arguments here lay mainly with comparisons of similar tergal positions with larval gills (and even gill covers) on the thorax and abdomen of this group and wings, as well as some similarities in muscle arrangements (Wigglesworth 1972; Matsuda 1981). While it might be desired to make such associations, any recognition of absolute homology between wings and ephemeropteran gills stops at these observed similarities and the appendage patterning genes co-opted to form such appendicular outgrowths. Also, because of such associations between similarity in position and misinterpreted morphological features of fossil taxa, an observed ancestral presence of wings on all trunk segments has been dubiously postulated and propagated in the developmental literature (Carroll et al. 1995). It should also be noted that the thoracic wing-like structures of primitive aquatic immatures are in fact the developing wing buds, and those on the abdomen are the gills. Although gills may utilize similar appendage patterning genes and pathways as wings, with notable exceptions (Niwa et al. 2010), since they are features only of immature aquatic pterygotes, they are independent features from wings. It is unwise, therefore, to hypothesize that wings first evolved in the aquatic immature stages of pterygotes, such as paleodictyopteran nymphs (Carroll et al. 1995), some of which may not have been aquatic at all. It may still be possible to hypothesize that wings and gills share some degree of serial and/or developmental homology (Jockusch et al. 2004),

due to their sharing of similar developmental programmes; however, there is no evidence for gills and wings evolving together or during similar time periods, and it is more probable that gills arose independently, particularly considering that the earliest ephemeropteran immatures appear to lack gills (e.g. immature Protereismatidae: Grimaldi and Engel 2005).

## 12.2.2 Genes and Genetic Pathways

While traditional embryological studies have been able to determine that wing primordia first form in the embryo and that these ectodermal cells of the early wing disc are associated with the leg primordia, it was uncertain whether any of these cells are actually derived from the early leg disc. Together with developmental genetic techniques, it has become evident that in *Drosophila*, as revealed by early *vestigial* (*vg*) expression, the wing discs originate as a part of the leg discs and subsequently separate to migrate dorsally (Cohen et al. 1991, 1993; Williams and Carroll 1993). These data were used as further evidence that wings may be homologous to extensions from the coxal base (coxopodite/basicoxa), such as an epipodite or gill. Further studies outside of Diptera (in Hymenoptera) have shown, however, that this association between leg and wing primordia may yet be another derived feature in Diptera (Jockusch and Ober 2004), perhaps associated with the evolution of imaginal discs. Thus, as shared leg and wing primordia appear to not be the plesiomorphic state for Holometabola, they likely are not plesiomorphic for Pterygota. Although similar studies have yet to determine whether these primordia are also separate outside of Holometabola, it is intriguing that at least one holometabolous order (Coleoptera) shows such a pattern. Further support for this hypothesis of derived leg + wing primordia stems from observations in *Tribolium*, indicating that while there is anterior to posterior migration of early *snail* (*sna*) expressing wing primordia, dorsal migration of the wing primordia does not occur outside of Diptera (or perhaps outside of some other groups sharing wing disc development from a common ancestor) (Jockusch and Ober 2004). The hypothesis of the wing evolving specifically from a crustacean epipodite was first given evidential support by Averof and Cohen (1997), in which they demonstrated that *nubbin* (*pdm/nub*) and *apterous* (*ap*) appear to have similar expression patterns in insect wings as in the epipodite of *Artemia*. As Jockusch and Nagy (1997) explained in detail, such observations do not provide the evidence for such a precise conclusion (which explicitly excludes other hypotheses). It certainly demonstrates that appendages appear to require similar modes of development, but does not elucidate any differences in the targets that may be present downstream which differentiate various types of appendages.

It has been found that *Scr* is expressed in the first thoracic segment (T1) not only in derived pterygote groups, but also in basal insects (Rogers et al. 1997; Angelini and Kaufman 2005). It is uncertain whether it is expressed in non-insect apterygotes (i.e. Entognatha); however, because it appears to be present in basal Hexapoda (which lack wings), it is quite possible that *Scr* may have been exapted for the repression of prothoracic wings (Hughes and Kaufman 2002), possibly in several different ways and to varying extents, particularly as large 'winglets' (or 'paranotal lobes' as many were not articulating, or the evidence for their articulation is lacking) have been found throughout the extinct Odonatoptera, Palaeodictyopterida, and many other groups (e.g. Wootton 1972; Carpenter 1992; Grimaldi and Engel 2005). This hypothesis is further supported in Coleoptera (*Tribolium*: Tomoyasu et al. 2005 and *Onthophagus*: Wasik et al. 2010), whereby RNAi *scr⁻* mutants essentially develop mesothoracic wings on the prothorax. Orthoptera (*Gryllus*: Zhang et al. 2005) also show a similar expression pattern of *Scr*, probably indicating that *Scr* functions to repress wing formation in T1 in Orthoptera as well. In Hemiptera (*Oncopeltus*: Chesebro et al. 2009), while RNAi *Scr⁻* mutants also develop some aspect of ectopic mesothoracic wings on the prothorax, there

appears to be no indication of an articulation. A fascinating apomorphic derivation of this T1 pathway appears to have evolved in a different lineage of hemipterans (Membracidae). Here, *Scr* continues to be expressed in the ectoderm despite the formation of a dorsally derived appendage (Prud'homme et al. 2011). While this appendage likely is not homologous with meso- and metathoracic wings (Yoshizawa 2012), it is clear that it has co-opted portions of the wing/ appendage patterning pathways as has similarly occurred in many other insect groups. Because of such observations in similar expression patterns, it certainly is possible that similar genes have acquired different functions, mainly through changes in the regulation of downstream targets. Such data may also indicate that *Scr*, or likely downstream targets of *Scr*, has changed since the origin of basal hexapods and, particularly so, since the origin of pterygotes. These changes appear to differentially affect wing development, such as eliminating points of articulation, eliminating or reducing the laminate (paranotal) extension, or various degrees of both. As many studies are beginning to conclude, it may not be so much that differences in expression domains give rise to morphological novelties; rather, it is the differences in regulation and deployment of these genes that produce change (Averof 1997; Grenier et al. 1997). Such differentiation is difficult to detect with gene expression data for several reasons, such as topological conservation in expression (Bolker and Raff 1996). It also is equally likely that other undiscovered genes may play large roles in such seemingly conserved pathways. Since developmental studies have progressed largely in the light of candidate-gene approaches, including comparing expression patterns and functions of similar genes, it is quite possible that unstudied genetic architectures or features may have significant effects in producing the different outcomes we see in similarly expressed genes. Such genes might represent cascades of targets, downstream of conserved networks such as Hox genes, and could be influential in morphogenesis (Hughes and Kaufman 2002).

It is possible that the potential to develop embryonic wing primordia may be in every thoracic and abdominal segment; however, as defined by *snail* expression, since no definitive wings (or wing precursors) have been found on the abdomen of hexapods, it is incorrect to state that insects lost abdominal wings. This statement is at least consistent with the fossil record, because ancestrally, as said above, hexapods never had definitive wings on the abdomen. It is now evident that *Bithorax* complex (BX-C) genes have evolved the ability to regulate imaginal disc and imaginal tissue formation in segments, likely through various suites of target genes (Hughes and Kaufman 2002). Interestingly, while it appears that *Ubx* and *abd-A* have evolved the ability to repress abdominal wing (and leg) primordial development (i.e. to designate abdominal identity) (Simcox et al. 1991; Carroll et al. 1995), it has been demonstrated that $Ubx^-$ and $abd\text{-}A^-$ mutants of *Oncopeltus* (Hemiptera, hemimetabolous), although forming abdominal legs and dorsal pigmentation suggestive of early wing-pad development, have not been shown to form any definitive abdominal wing buds in the nymphs (Angelini et al. 2005). It is unknown in this case, however, whether embryonic wing primordia form. While RNAi studies have yet to be done in *Gryllus* (Orthoptera), the expression patterns of *Ubx* and *abd-A* are quite different from those in *Drosophila*, particularly with regard to *abd-A* during early and middle embryonic stages (Zhang et al. 2005). Interestingly, similar results are seen in *Tribolium* (Coleoptera) $Ubx^-/abd\text{-}A^-$ RNAi mutants as with those of *Drosophila* (Tomoyasu et al. 2005). In wild-type *Tribolium*, patches of cells expressing *sna* are observed not only in the thoracic segments, but also in nearly every abdominal segment (Jockusch and Ober 2004). Furthermore, as in *Tribolium*, it is fascinating that *Ubx* and *abd-A* knockouts of *Tenebrio molitor* have survived to the adult stage and demonstrate a homeotic transformation giving rise to the presence of wings (fore- or hindwing identity could not be confirmed) on all abdominal segments (Takahiro Ohde and Teruyuki Niimi, pers. comm.), though lacking signs of abdominal leg development. It should be noted, though, that *Ubx/abd-A* parental RNAi

induces abdominal leg formation in the larva of *Tribolium*. These fascinating results demonstrate that the genetic network and potential to form fully developed wings, though of questionable function, can be deployed in most (if not all) abdominal segments. Such findings, perhaps, should not be considered too extraordinary given the serially homologous ground plan of insect segmentation. Indeed, in addition to the results of Tomoyasu et al. (2005), this extant ability to produce wing-like structures on the abdomen is quite interesting, but is far beyond providing conclusive statements for early wing evolution and origins. As already mentioned, while the abdominal segments appear to also have a capacity for various types of dorsal appendage development and short tergal extensions or lobes (in addition to ventral appendage development in immatures and ancestral ventral leg development), wings have thus far not been found to have occurred naturally on these segments. It must be emphasized that, given current understanding of the functions of *Ubx* and *abd-A*, while they appear to remain broadly expressed in the abdomen throughout Hexapoda, several forms of appendages have evolved (mainly in immatures) on the abdomen in different hexapod orders; (Fig. 12.2). Such diversity in development suggests mechanisms of developmental drift, changes in downstream targets, and/or changes in expression patterns (e.g. Warren et al. 1994), modes of development which could utilize various components of an underlying appendage (though not necessarily and specifically a wing) formation programme. In addition, while such expression patterns could indicate possible serial homology of dorsal appendages in the thorax and abdomen in hexapods, early appendage patterning markers, such as *dpp*, *sna*, *vg*, and *wg*, do not necessarily dictate downstream processes such as wing formation. Therefore, such data may support observations contrary to Kukalová-Peck (1978) (e.g. Boxshall 2004; Grimaldi and Engel 2005) that wings may not have been a ground plan of the pterygote abdomen and, similar to the case of *Scr*, may represent a derived feature in these advanced holometabolous groups. As indicated by expression patterns of BX-C genes in some crustaceans, while superficially similar expression patterns may exist in distantly related taxa, such as is seen in various Hox genes, such expression patterns may likely serve divergent functions (Abzhanov and Kaufman 2000a, b). More data outside of Holometabola are direly needed to create an improved comparative framework.

While gills and wings may share similar patterning genes (given that they are both, at least in part, appendicular) and require similar pathways to define their antero-posterior (A/P), dorso-ventral (D/V), and proximo-distal (P/D) axes, this observation does not necessarily define them to be serially homologous (Jockusch et al. 2004). In other words, while similar expression patterns may be present in taxa that share a common ancestor, the morphological features that develop in those groups are not homologous unless they evolved through modification of the same structures present in the common ancestor (Hall 1994). Thus, similarity in patterns of gene expression may reflect conservation of gene function from a distant common ancestor, but it does not equate to homology of the derived structures in which the expression is seen (Bolker and Raff 1996). Furthermore, as there are gills of immatures that arise from ventral, pleural, and tergal regions in pterygotes, it is quite likely that such appendicular structures are independent, as hypothesized for many of the epipods and polyramous structure of crustacean limbs (Boxshall 2004). It is understandable that the general gestalt of such nymphal gills resemble wings; however, not only are their articulations completely different (Dürken 1907, 1923), early patterning genes, such as *apterous* (*ap*), show different expression patterns (Niwa et al. 2010). If we were to continue to be motivated by similarities of gene recruitment and co-option in forming our hypotheses of homology, then beetle horns, aside from their different form and location on the body, could also be hypothesized as derivatives of epipodites, styli, or gills and in some regards appear to show more similar expression patterns to wings than do styli or epipodites.

Appendage patterning genes, such as *dac*, *hth*, and *Dll*, show similar expression patterns in beetle horns to truly segmented appendages (Moczek and Rose 2009). Major signalling proteins for P/D patterning, such as *decapentaplegic* (*dpp*), which is required for leg outgrowth in most (but apparently not all; Jockusch and Ober 2004) hexapods, are also involved in horn formation, demonstrating recruitment of similar genes and pathways for apparently novel features (Wasik and Moczek 2011). In other words, aside from minor differences in gene expression and downstream targets, the main difference between ventral, lateral, and dorsal appendages is the site at which gene co-option/recruitment occurs.

### 12.2.3 Homologous Versus Novel: 'Epipodite' Versus Amalgamation

Although many structures that are said to be non-homologous to structures in ancestors may appear to be new, their formation and evolution typically originated from preexisting developmental architectures (Bowsher and Nijhout 2007; Prud'homme et al. 2007). For arthropods, this statement is now based on a wealth of developmental data on segmentation and appendage patterning. As already mentioned though, while expression of genetic pathways may be conserved (and the genes within them homologous), the deployment and functioning of these genes may be different and they may be expressed in non-homologous structures (Bolker and Raff 1996). Such differences in gene function serve only to further distort definitions of homology (Hall 2007), as may be the case in insect wing development.

Unlike the case for abdominal wings, the presence of definitive prothoracic wing-like structures has been documented (Crampton 1916; Ross 1964; Kukalová-Peck 1978; Grimaldi and Engel 2005), although evidence for articulations is lacking. As it has been demonstrated that nearly a full developmental programme for wing formation is present in the

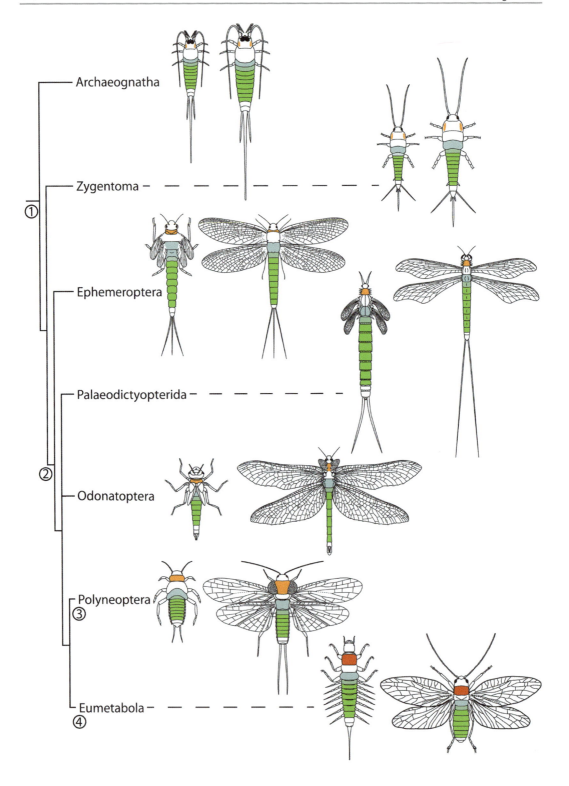

◀ **Fig. 12.2** Phylogenetic hypothesis of insect relationships and wing development. Abbreviated phylogeny of Hexapoda, focusing on basal orders to illustrate the distribution of various morphological features (paranotal lobes, gills, styli, and wings) in representative immatures and adults. Representative taxa are as follows: Archaeognatha (Meinertellidae); Zygentoma (Lepismatidae); Ephemeroptera (Protereismatidae); Palaeodictyopterida (Megasecoptera); Odonatoptera (Eugeropteridae); Polyneoptera (Lemmatophoridae); Eumetabola (Sialidae). Colours represent gene expression as follows: *orange/ red = Scr*, expressed in the prothorax; *blue = Ubx*, expressed primarily in the metathorax and first abdominal segment, but also in A2–8; *green = abd-A*, generally expressed in abdominal segments 2–8 (and partially in A1). Significance of numbers at nodes is as follows: **1** Origin of Hexapoda, loss of abdominal appendages through *Ubx/abd-A* regulation (although styli develop to various degrees on thorax and abdomen in Archaeognatha and Zygentoma), and expression of *Scr* along lateral margins of prothorax. **2** Appearance of wings on meso- and metathorax (Pterygota), broader expression of *Scr* in prothorax, and repression of wing formation on prothorax by *Scr* (though paranotal lobes begin to appear, indicating diverging functions of *Scr*; although *Ubx/abd-A* remains expressed in similar patterns throughout adult insects, gills, legs, and other such appendages develop on the abdomen in immatures of several orders. **3** Paranotal lobes remain in some lineages of Polyneoptera; however, they are lost in many other lineages, indicating diverging functions of *Scr* or induction of other genes/pathways. **4** While various abdominal gills (appendages) are present throughout pterygotes in the immature stages, more diversity appears in Eumetabola, including segmented gills and other abdominal appendages; diverging functions of *Scr*, and likely induction of other genes/pathways, also appear in Eumetabola, as modifications of the prothorax develop in various orders (e.g. wing-like appendages in Hemiptera)

prothorax of holometabolous and hemimetabolous insects, only it has been repressed at least by *Scr*, it appears more likely, at least given the results of developmental research thus far, that wings were a ground plan for the hexapods only in the thorax. Such a hypothesis is supported by expression patterns of the early limb induction module including the transcription factors *ap*, *wg*, and *vg* in taxa that are more representative of early pterygotes and hexapods (Niwa et al. 2010). These data also indicate that, contrary to popular citation, a wing is more likely an amalgamation of tergal and pleural outgrowths which develop according to the redeployment of limb patterning genes and portions of their pathways, as opposed to a modification of such structures as gills, epipodites, styli, or other limbs that share similar developmental modules. Contrasting expression patterns of these genes, while *ap* is expressed in a large dorsal area of the pterygote wing, it does not appear in ephemeropteran gills nor in archaeognathan styli. On the other hand, *wg* and *vg* were expressed in the gills and styli of the aforementioned taxa, as well as at their bases (as in wings), indicating regions of articulation (Buratovich and Wilder 2001; Niwa et al. 2010). From such investigations in early hexapod lineages, it is apparent that the interaction of at least *ap*, *wg*, and *vg*, as revealed through their expression at the tergal–pleural (coxopodite) boundary, functions as a module for paranotal extension (Ng et al. 1996; Niwa et al. 2010). Recent studies have also confirmed through RNAi that *vg* is involved in paranotal extension, in which adult *vg* knockdowns of *Tenebrio molitor* show paranotal extensions on all abdominal segments (Takahiro Ohde and Teruyuki Niimi, pers. comm.). Other major regulators of this outgrowth of the tergal margin may include *dpp* and *hedgehog* (*hh*) signalling, which are also important in later patterning of the veins (Celis 2003; de Celis and Diaz-Benjumea 2003), and possibly *scalloped* (*sd*). Subsequently, through the incorporation of existing appendage patterning genes (such as *dachshund* [*dac*], *exd*, *Distal-less* [*Dll*]), an articulating appendage (wing) is formed while also providing more refined patterning along the D/V, A/P, and P/D axes. As in legs, the complex of *homothorax* (*hth*) and *exd*, among their many functions, appears to play a role in defining the region of articulation (González-Crespo and Morata 1996; Azpiazu and Morata 2000; Casares and Mann 2000; Morata 2001) and patterning of the axillary sclerites, similar to the coxopodite of legs (Jockusch and Nagy 1997). It may be interesting, then, to hypothesize that the complex region forming the articulation of the wing may, in fact, involve a highly derived coxopodite (i.e.

the basal part of the appendage). Similar to how the ancestral basicoxa fragmented to form the areas of the insect pleuron, it may be possible that through recruitment of such coxopodite patterning genes, the primitive sclerotization surrounding the wing base in basal pterygotes also fragmented/fused to form the wing pteralia. While *hth* and *exd* are expressed at the margin of the base of the wing and extend onto the wing blade, *pdm* expression also extends into the area of the wing hinge, approximately covering its entire dorsal surface (Jockusch and Nagy 1997), as does the Iroquois complex (Iro-C) (Cavodeassi et al. 2002), *teashirt* (*tsh*) transcription factor, and *wg* signalling pathway (Klein and Martinez Arias 1998; Peterson et al. 1999; Klein 2001) and therefore also seem to function in axillary patterning (Ng et al. 1995), in addition to a number of other known and undescribed genes (Butler et al. 2003; Cho and Irvine 2004). This expression of *pdm* and the resulting mutant phenotype (which begins to resemble a paranotal lobe due to the near obliteration of the axillary sclerites), as well as its expression at leg joints, seems to provide further evidence that it may have been influential in providing the paranotal extension the needed articulation to produce a functional wing. Interestingly, extreme *pdm* mutants also lack nearly all venation (Ng et al. 1995).

As it has been observed in morphological studies, the wing not only is a paranotal extension, but it also appears to incorporate elements of the pleuron (coxopodite/basicoxa) (Grimaldi and Engel 2005; Hasenfuss 2008). While this observation has not been acknowledged with developmental data outside of Holometabola, it appears to be supported at least by *Drosophila* in the structure of the wing disc (Bryant 1975; Cohen 1993; Klein 2001). Since the dorsal elements of the pleuron are part of the wing disc, it is evident that some part of the coxopodite was integrated into the paratergal extension to give rise to the wing articulation. This observation provides support for the hypothesis put forward by Niwa et al. (2010), in which they postulated that the pterygote wing was formed through the intersection of two developmental modules, one

which produced a dorsal limb-like process and the other which created a planar extension of tergite at the tergal–pleural boundary (Šulc 1927, Kukalová-Peck 1983). Kukalová-Peck's (1983) assertion that wings 'may represent a fusion between the epicoxal segment' and zygentoman paranotal lobes are perhaps not far from accurate in essence, although the above demonstrates that ad hoc exites are superfluous. Interestingly enough, as the wing disc also includes tissue for the adult tergum, this association likely indicates an inseparable developmental module (as indicated by the early, uniform expression of *ap*; Jockusch and Nagy 1997; Klein and Martinez Arias 1998). If we accept the hypothesis that the wing is largely a paranotal extension, separating the two early developing tissues (that of the notum from that of the upper pleuron) would therefore abolish formation of the wing. While this is interesting from *Drosophila*, one wonders what differences might be found if such a study were to be undertaken in a basal, living mayfly where the sclerotized pleural surface is dramatically dissimilar.

## 12.2.4 Developmental Implications for Wing Origins?

We must keep in mind that while much excellent work has been accomplished in developmental genetics, much of the progress has been made in *Drosophila*, a highly apomorphic taxon in Diptera. Since much developmental evidence for wing origins stems from literature on *Drosophila*, while it certainly demonstrates many interesting developmental features, it is still difficult to separate gene expression features that may suggest deep homology and be indicative of ancestral wing origins from highly derived features in a lineage that is far removed from basal hexapods and certainly from ancestral Arthropoda. Straightforward conclusions drawn from such developmental data, then, must certainly be analysed in great detail and in a much broader comparative framework. Also, as it is becoming more evident that, while expression patterns of complexes such as HOM-C are thought to be largely conserved in diverse lineages such as crustaceans and hexapods (Hughes and Kaufman 2002), the functions of these genes may have shifted, likely through changes in downstream targets/pathways or upstream regulators (Prud'homme et al. 2007) and developmental drift (Angelini and Kaufman 2005), to give rise to the morphological diversity we see today. If such is the case, then even greater stress is placed on sampling diverse lineages, as well as examining expression patterns and performing functional assays. In compiling all palaeontological, neontological, and developmental evidence thus far (Fig. 12.2), and following the results of Niwa et al. (2010) and building upon their hypothesis, it appears that there is evidence for a developmental ground plan in Hexapoda that produced paranotal extensions of the thorax (Fig. 12.1a, b). Subsequently, through the integration of appendage patterning modules (e.g. those present in gills, exites, legs, and c.), a functional articulation (hinge) developed integrating the dorsal elements of the pleuron, providing a functional wing and providing a basis for which further refinements of the pterygote wing could be made, such as in wing shape, venation, structure of the articulation (axillary sclerites), and c (Fig. 12.1c).

Important avenues for understanding the intricacies of morphological change (and wing origins and evolution) will include emerging tools of genomics and methods for examining specific developing tissue subsets, such as in transcriptomics and proteomics (e.g. Alonso and Santarén 2005). As many of the more general patterning pathways are being elucidated, as Angelini and Kaufman (2005) note, it will be of great significance to focus closer on understanding the genetics behind the plethora of subtle morphological changes that occur through signalling of downstream targets of major appendage patterning pathways for example (e.g. Butler et al. 2003). More attention should also be devoted to understanding the genetic control of tergal outgrowths, as well as the formation of the axillary sclerites. It may be of great utility if future research also includes

examination of protein structure, which may provide insight into regulatory and functional changes that have occurred in such developmental genes.

## 12.3 Palaeontology of Insect Wings

We have purposely belaboured the point concerning phylogeny, and it is therefore of great interest to consider those taxa that are as close to the common ancestor of Pterygota as is possible. Naturally, any species living today is separated from this ancestral taxon by nearly 400 million years, and this creates several challenges. Palaeontological evidence is unique in its ability to bridge this gap, at least partially, and the investigation of phylogenetically relevant taxa from Palaeozoic deposits is of considerable interest in regard to the origins of flight. As critical as Palaeozoic insect fossils are, it must be admitted from the start that no fossil species of a stem-group pterygote with or without protowings has yet been recovered. Indeed, the pre-Late Carboniferous record of insects is amazingly sparse, and it is from the Devonian or earlier in which wings originated, meaning that the hunt continues for abundant, mid-Palaeozoic outcrops of completely preserved hexapods.

Insect wings are the most common source of data in palaeoentomology owing to their solidity and resistance to subsequent transportation and taphonomic processes. Generally, insect wings should be considered in conjunction with the remainder of the body for a reconstruction of the entire animal and comprehensive taxonomic and morphological treatment. However, in some cases, particularly for Palaeozoic taxa, isolated wings preserved as compressions or impressions provide the only evidence for past species richness [refer to Carpenter (1992) for the most recent comprehensive catalogue], and here, there is often a bias for well-sclerotized forewings modified for protection in certain clades, these being particularly durable for preservation. The use of these data in insect systematics varies dramatically by taxonomic group due to various adaptations, functional modifications, and polymorphisms and therefore must be reasonably evaluated by taxonomists and in a cladistic framework. Well-preserved insect fossils with complete body appendages and tiny morphological structures are known from amber inclusions, but such resin-entombed specimens are unfortunately unavailable prior to the Cretaceous except for a few fragmentary remains reported in Late Triassic amber from Italy (Schmidt et al. 2012). Admittedly, the fossil record of insects pales in comparison with the enormous numbers of Recent species, but the available evidence does highlight the dramatic number of lineages present in different epochs, serves as a reasonable proxy for diversity during these time periods, and gives a unique perspective on taxa with unique characters or character combinations, together reconstructing a profound understanding of insect evolution during its early phases. Grimaldi and Engel (2005) provide a cladistically framed overview of insect evolution and diversity, reflecting the available palaeontological evidence up to that date.

## 12.3.1 First Appearance of Winged Insects

The earliest hexapod fossils are known from the Early Devonian of Rhynie, Scotland, and preserved in chert formed in a silica-rich, volcanic spring and of Pragian age (ca. 407 million years old). Two definitive hexapods are known from the Rhynie chert, the first and most widely known being the collembolan *Rhyniella praecursor*, a species for which there is generally good knowledge of its overall morphology (Whalley and Jarzembowski 1981). The second, representing a true insect, is known only from the fragmentary remains of a head capsule and was dubbed *Rhyniognatha hirsti* (Tillyard 1928; Engel and Grimaldi 2004). The mandibles of *R. hirsti* were dicondylic, a synapomorphic trait placing them as more derived than the most basal order of wingless insects. Furthermore, the mandibles were of the typical metapterygotan organization, an apomorphic suite of traits found only among pterygote insects and in the Metapterygota (all winged

insects exclusive of Ephemeroptera) in particular (Engel and Grimaldi 2004). This cladistic placement indicated not only that *R. hirsti* was assuredly an insect but that the species belonged to the winged insects and was from a lineage that diverged subsequent to the divergence of the mayflies, implying that wing origins and diversification took place sometime prior to the Pragian. This revelation pushed back the presumed origins of wings by nearly 80 million years and also the origin of insects as a whole, highlighting that insects perhaps stemmed from the Silurian and were among the earliest forms of terrestrial animal life (Engel and Grimaldi 2004). Unfortunately, *R. hirsti* was fragmentary and no wings were preserved with the fossil, leaving open numerous questions regarding the putative wings of the species. Remarkably, the age and phylogenetic placement of *R. hirsti* are roughly in accord with estimates of divergence based on molecular data alone, which suggested an origin of pterygote insects anywhere from the latest Ordovician to the Silurian, and a later origin of the more derived neopteran insects, perhaps as early as the Early to mid-Devonian (Gaunt and Miles 2002; Rehm et al. 2011). Fossil evidence of a metapterygotan insect from the Early Devonian implies the acquisition of wings at least in the earliest Devonian (Lochkovian) or latest Silurian corresponds with the formation of the first trophic relationships between terrestrial arthropods and vascular plants, the latter having invaded land slightly earlier (Edwards et al. 1995).

Subsequent to *R. hirsti*, there are only a couple of definitive insect remains from the Devonian, the first being a relatively complete compression from Famennian strata near Strud, Belgium (Garrouste et al. 2012). Like *R. hirsti*, *Strudiella devonica* possessed metapterygotan mandibles and emphasized that the origination and diversification of pterygotes, at least into the most basal lineages, had already occurred. Again, similar to *R. hirsti*, *S. devonica* also lacked wings, either because it was a nymph and did not yet possess them because it was secondarily apterous or perhaps as a result of preservation. The sole specimen is too poorly preserved to permit analysis of fine details of the thorax to determine whether minute sclerites representing a point of articulation might have existed, and again critical questions regarding the form of wings in the earliest fossils assignable to Pterygota were left unresolved. The only other Devonian evidence for insects is the mid-Devonian (Givetian) bristletail fragments (Archaeognatha) from Gilboa, New York (Shear et al. 1984). Putative Eifelian remains of a remarkably modern-looking bristletail from Gaspé Bay in Quebec, Canada (Labandeira et al. 1988) have been revealed to be a modern contaminant (Jeram et al. 1990). No other records of hexapods are known from the Devonian, a scant record at best and the greatest hindrance to understanding early insect evolution.

The first wings preserved in the fossil record are much younger than any estimate of the age of Pterygota as well as than the fragmentary remains of pterygotes from the Devonian. Indeed, the earliest wings are known from the transition period between the Early and Late Carboniferous, approximately 318 million years ago and nearly 90 million years younger than the very incomplete remains of *R. hirsti*. This considerable gap is partly the result of a scarcity of Early Carboniferous freshwater deposits worldwide. These earliest Late Carboniferous insects have been attributed to the orthopterid lineage and thereby clearly derived from the Neoptera (Prokop et al. 2005). The Namurian is the earliest stage of the Late Carboniferous with a sudden occurrence of diverse winged insects comprising stem groups of the major lineages, highlighting that the extensive diversification of Pterygota had already taken place; those clades became well established and radiated themselves (Hennig 1981; Kukalová-Peck 1991; Grimaldi and Engel 2005; Prokop and Nel 2007). In terms of the fossil record, the Late Carboniferous reveals a world in which winged insects and flight were already ancient and this lineage had radiated into all of the higher (superordinal) clades which would persist to the present day, as well as a few which would not last beyond the end-Permian Event at the close of the Palaeozoic (ca. 251 million years ago). From the Palaeozoic, the fossil record currently provides two perspectives—one that is too scant to permit

much clarity (Devonian–Early Carboniferous) or one that is too late in regard to the window of time in which wings and flight originated (Late Carboniferous-Permian), a maddening situation for entomology and evolutionary biology.

## 12.3.2 Wing Flexion and Palaeoptery Versus Neoptery as Crucial Innovations

As mentioned previously, today's phylogenetic evidence universally supports a single origin for insect wings (e.g. Kukalová-Peck 1978, 1983, 1991; Boudreaux 1979; Hennig 1981; Kristensen 1991; Grimaldi and Engel 2005; Trautwein et al. 2012). At about the same time, Lameere (1922); Crampton (1924), and Martynov (1925) independently noted two fundamentally different means of wing flexion, this giving rise to the classificatory division between palaeopterous (those incapable of flexing the wing back over the abdomen) and neopterous (those capable of such flexion) insects. Those lineages with the palaeopterous condition were classified as the formal group Palaeoptera, the remainder in the Neoptera, and thus was born the debate over relationships between the basal orders of winged insects and whether or not the Palaeoptera are monophyletic and, if not, then which of its constituent groups were basal and which were more closely allied to the neopteran insects. In addition, the arrangement and form of the basal sclerites forming the wing base differ between the lineages in question. Indeed, the Odonatoptera differ notably from other pterygotes, something which led Matsuda (1970, 1981) and La Greca (1980) to reconsider pterygote monophyly. Despite the differences between odonates and other winged insects, the basal sclerites can be successfully homologized with those of Ephemeroptera and Neoptera (Ninomiya and Yoshizawa 2009). Furthermore, the thoracic musculature of the primitively wingless *Lepisma* (Zygentoma) and Pterygota was first established by Matsuda (1970) and again by Hasenfuss (2002). Hasenfuss (2002) provided a detailed comparative morphological study of the mesothorax of *Lepisma*, demonstrating the details of homology and transformation with the pterygote ground plan. The corresponding sclerites and muscles of three subcoxal leg elements present in lepismatids are recognizable in the pterygotan pterothorax (Hasenfuss 2008).

The fundamental debate has been over Palaeoptera monophyly. Palaeoptera was largely deconstructed for a long while. Börner (1904) arranged the basal winged orders with Ephemeroptera diverging from Metapterygota (all other winged insects), a position supported by morphology and molecular data sets (e.g. Staniczek 2000; Beutel and Gorb 2006; Cameron et al. 2006; Zhang et al. 2008). Schwanwitsch (1943) reversed this, with Odonata diverging first, establishing the Chiastomyaria hypothesis based on muscle arrangement (=Opisthoptera of Lemche 1940) and supported by initial phylogenomic data sets, albeit with understandingly limited taxon sampling (Simon et al. 2009). Palaeoptera monophyly was argued for by Kukalová-Peck (e.g. Kukalová-Peck and Brauckmann 1990; Kukalová-Peck 2009) in an expanded and revised system of Pterygota (although see Béthoux et al. 2008 for a discussion of some of her methods of character analysis). Palaeoptera has also been supported by limited molecular analyses (Kjer et al. 2006; Regier et al. 2010; Ishiwata et al. 2011) and by some morphological character systems (e.g. Blanke et al. 2012), although the most honest description of the available evidence is that there is ambiguity over relationships (e.g. Hovmöller et al. 2002; Ogden and Whiting 2003; Kjer et al. 2006; Whitfield and Kjer 2008). Kukalová-Peck (1997) proposed a strong convex brace 'cup-aa1' or a contact between AA and CuP (vein abbreviations used in text outlined in Table 12.1) as a putative synapomorphy of Ephemeroptera and Odonatoptera and subsequently proposed division of these groups into the 'Hydropalaeoptera' (=Ephemeroptera + Odonatoptera) and 'Rostropalaeoptera' (=Palaeodictyopterida) (Wootton and Kukalová-Peck 2000). Bechly (1996) proposed that the Ax0 in Odonatoptera was homologous to the subcostal brace ScA in Ephemeroptera, implying that it

**Table 12.1** Abbreviations for major wing veins discussed in text

| AA | Anal anterior |
|---|---|
| AP | Anal posterior |
| CP | Costa posterior |
| Cu | Cubitus |
| CuA | Cubitus anterior |
| CuP | Cubitus posterior |
| IN | Intercalary |
| M | Media |
| MA | Media anterior |
| MP | Media posterior |
| R | Radius |
| RA | Radius anterior |
| RP | Radius posterior |
| ScA | Subcosta anterior |
| ScP | Subcosta posterior |

was another potential synapomorphy supporting the same arrangement of orders. Haas and Kukalová-Peck (2001) purportedly identified 65 differences between Palaeoptera and Neoptera based on wing characters traceable in extant species, although the homology of some of these is tenuous. Later, Kukalová-Peck (2009), when describing the first Carboniferous protodonate immature, reviewed what she interpreted as synapomorphies of Ephemeroptera and Odonatoptera based on wing articulation and venation. Assuming Palaeoptera monophyly, it has been argued that the neopterous condition is plesiomorphic and that the palaeopterous condition is derived (Hasenfuss 2008; Kukalová-Peck 2009) and that the wing bases of Ephemeroptera and Odonata are secondarily stiffened (Willkommen 2009). In addition, it has been argued that wing development of Palaeozoic Palaeoptera proceeds gradually through numerous moults of nymphal instars to several subimaginal instars bearing articulated wings in comparison with Recent members (Kukalová-Peck 1978); although as noted by Béthoux et al. (2008), the evidence for moulting subimagos in the fossil record is tenuous. During the course of development, the wings of young nymphs of these fossil taxa apparently arch backward (Fig. 12.3a, f) and gradually become straightened in each subsequent instar until the wings are fully out-stretched, this putatively suggesting that palae-optery in the adult was secondarily derived (Kukalová-Peck 1978; Hubbard and Kukalová-Peck 1980). Relationships between Ephemer-opterida, Odonatoptera, Palaeodictyopterida, and Neoptera remain controversial and consti-tute a debate of lasting significance. Metaptery-gota (=Odonata + Neoptera) and Palaeoptera (Odonata + Ephemeroptera) are the most widely recovered suite of relationships, although many data are in conflict. Future work may well sup-port a monophyletic Palaeoptera, and it must remain a viable alternative solution to the arrangement of the basal winged lineages.

As mentioned, Palaeoptera are known for their inability to flex their wings back over the abdomen. Insects exhibiting the palaeopterous morphological condition were remarkably diverse and abundant in Late Palaeozoic eco-systems, indicating to some that they were the first of the flying insects (i.e. that neoptery is derived relative to palaeoptery), and these insects were decimated by the Permian/Triassic mass extinction (Labandeira and Sepkoski 1993). There is a single exception to the rule of permanently outstretched wings among Palae-optera, namely the extinct order Diaphano-pterodea whose species were capable of wing flexion in a roof-like position (Fig. 12.3h) owing to a unique arrangement of eight rows of movable sclerites at the wing base (Kukalová-Peck and Brauckmann 1990; Kukalová-Peck et al. 2009). Based on other characters, the Diaphanopterodea clearly belong within a monophyletic Palaeodictyopterida and their wing flexion is not only independent from that observed in Neoptera but also not indicative of the basal condition for the Palaeodictyopterida as they are not a primitive grade of this super-ordinal complex (Kukalová-Peck 1978).

Of course, critical to the aforementioned discussions of relationships are the Palaeo-dictyopterida, a diverse lineage of Palaeozoic palaeopterous insects. Obviously, no molecular study has included representatives of this lineage (or for that matter, any of the extensive stem-group representatives of the Ephemeroptera and

Odonatoptera known from the same time per-iod), and it remains unclear what influence the inclusion of palaeodictyopterids might have on modern phylogenetic interpretations. The only means of ascertaining the influence of Palaeo-dictyopterida on cladistic studies of basal lin-eages is the combination of molecular data with an extensive morphological data set coded for a suitably rich number of extinct species. Such a study would require direct observation from the fossils as there appears to be some misinterpre-tation of these Palaeozoic taxa in the literature (e.g. Béthoux and Briggs 2008; Béthoux et al. 2008; Kukalová-Peck and Beutel 2012; Shcher-bakov 2011), and data mining from such papers may conflate problems. The establishment of a robust phylogeny for basal pterygotes including all of the fossil taxa is one of the ripest chal-lenges for future research.

Another challenge resides in the assumption that the basal condition observed for extant members of a lineage holds true for stem groups. For example, coding the earwigs as having trimerous tarsi, the same as stick insects and webspinners, fails to consider palaeontological evidence that the trimerous condition is not homologous between these orders. Stem-group earwigs share abundant synapomorphies with crown-group Dermaptera but have fully pen-tamerous tarsi, the presumed plesiomorphic condition for Neoptera, if not all Insecta (Grim-aldi and Engel 2005). Similar evidence exists from stem-group stick insects that they inde-pendently arrived at the trimerous condition, and thus, any analysis treating these orders as pos-sessing the same character state in their ground plans is based on faulty data. Such is also a challenge for the basal winged lineages. For example, there is a widespread assumption that all basal and extinct groups of winged insects are aquatic in their immature stages based on crown-group Ephemeroptera and Odonata. The puta-tively plesiomorphic appearance of stoneflies among the Neoptera has led some to postulate that the ground plan condition for this clade is similarly aquatic. Yet, there remains no con-vincing evidence that this is the case. Indeed, stem-group Odonata lack a clear indication of the

**Fig. 12.3** A Palaeozoic bestiary of early winged insects. **a** Nymph of *Idoptilus onisciformis* Wootton (Palaeodictyoptera). **b** Prothoracic articulated winglets of *Lithomantis carbonarius* Woodward (Palaeodictyoptera: Lithomanteidae). **c** Prothoracic articulated winglets of *Stenodictya pygmaea* Meunier (Palaeodictyoptera: Dictyoneuridae). **d** *Arctotypus sylvaensis* Martynov (Protodonata: Meganeuridae). **e** Nygmata in wing membrane of *Lithomantis bohemica* Novák (Palaeodictyoptera: Lithomanteidae). **f** Nymph of *Protereisma americana* Demoulin (Ephemeroptera: Protereismatidae). **g** Wing venation of *Protereisma permianum* Sellards (Ephemeroptera: Protereismatidae). **h** Habitus of *Permuralia maculata* Kukalová-Peck and Sinitshenkova (Diaphanopterodea: Parelmoidae). **i** Wing venation of the stem-group mayfly relative *Lithoneura lameeri* Carpenter (Ephemeropterida: Syntonopterodea). **j** Habitus of *Permohymen schucherti* Tillyard (Megasecoptera: Permohymenidae). **k** Habitus of *Kemperala hagenensis* Brauckmann (Neoptera: Paoliidae). **l** Wing venation of *Diathemidia monstruosa* Sinitshenkova (Dicliptera: Diathemidae). **m** Paranotal extensions of the prothorax of *Lemmatophora typa* Sellards (Lemmatophoridae). Images a, b © The Natural History Museum, London; images f, g, i, j, m © Museum of Comparative Zoology, Harvard University.

life history for their immatures, and quite interestingly, nymphs of stem-group Ephemeroptera of the family Protereismatidae (Fig. 12.3f) do not possess abdominal gills (or gills of any kind!), suggesting that they were not aquatic (Grimaldi and Engel 2005). The same can be said for putative stem-group stoneflies among families such as Lemmatophoridae (Fig. 12.3m). Immatures of Palaeodictyopterida are widely known (Fig. 12.3a), but again no obvious gill structures are present, and the morphology of many is convergent with immature beetles living in moist detritus layers on tropical forest floors. There is no overwhelming evidence that any of these lineages have aquatic nymphs in their ground plans when fossils are considered. This does not rule out the possibility that they could have been aquatic but merely emphasizes that the underlying assumption that they must have been is entirely ad hoc. We desperately require a modern, revised understanding of the immature stages of Palaeozoic insect clades.

Obviously, the resolution of these relationships and early life histories has significant consequences for the interpretation of wing and flight origins and the ground plan reconstruction of basal wing structure and articulation. The developmental studies outlined above seem to be converging on a consistent picture of wing formation and homology but cannot resolve which form of articulation and flexion (or lack thereof) is basal. Such a polarization of the alternative differences in articulation requires the integration of such comparative developmental evidence with a robust phylogeny for early Pterygota. Thus, any changes in reconstruction for the base of

Pterygota will have profound influences on how we interpret the stages in wing origins as well as the associated scenarios proposed for the production of early powered flight. The phylogeny infused with palaeontological evidence will also permit a more precise timing for flight origins, the life history of those stem-group taxa involved, and those abiotic factors of the ancient ecosystem (one profoundly different from the world in which we live!) that influenced the evolutionary development of wings.

### 12.3.3 Principal Lineages of Palaeozoic Pterygota

The attribution of particular fossils to higher-rank taxa has been challenging, particularly given the apparent presence of convergent characters in wing venation across unrelated clades. For example, the Syntonopteridae (Fig. 12.3i) were first attributed to the Palaeodictyoptera and later considered as Ephemeroptera based on the presence of Y-shaped intercalary veins (Edmunds and Traver 1954; Edmunds 1972; Wootton 1981; Kukalová-Peck 1985; Carpenter 1992; Willmann 1999; Prokop et al. 2010). However, the presence or absence of intercalary veins cannot be considered a unique autapomorphy of the so-called Hydropalaeoptera since this also occurs in Palaeodictyoptera such as the families Calvertiellidae and Namuroningxiidae (Béthoux et al. 2007; Prokop and Ren 2007). In addition to many convergences, the wings across these lineages are clearly plesiomorphic. The wing venation of palaeopterous insects has a prominent alternation between convex and concave longitudinal veins including prominently developed convex MA and stem of M always present. Further putative plesiomophies in venation are veins ScP and RA terminating at the wing apex and RA and RP beginning as separate stems (Lameere 1922; Kukalová-Peck 1991). Thus, there are significant challenges to properly placing particular fossils, exacerbated by the abundance of isolated wings which must be interpreted in the absence of body characters. Here, we outline the principal lineages as they are presently understood in the hope that this characterization will fuel future cladistic treatments of both molecular and morphological (including palaeontological!) data.

The mayflies (Ephemeroptera) are considered a basal lineage of winged insects with wings bearing a full set of deeply corrugated main veins, intercalary veins universally present, and a prominent arched subcostal brace ScA at the wing base as synapomorphies. Several stem groups are allied to the Ephemeroptera to form the superorder Ephemeropterida. The fore- and hindwings of Palaeozoic species were nearly homonomous in comparison with most Mesozoic, Tertiary, and modern taxa whose hindwings are much smaller or completely reduced (as with the exception of the Jurassic families Mesephemeridae and Mickoleitiidae). Nymphs of Permian taxa attributable to the Protereismatidae exhibit wings freely articulated with the thorax and a venation pattern similar to adults (Fig. 12.3f, g) (Kukalová 1968; Hubbard and Kukalová-Peck 1980). Syntonopterodea are the oldest and most plesiomorphic members of the Ephemeropterida and are known from the Late Carboniferous to the Middle Permian. Syntonopterodea share with other mayflies the presence of a distinct anterior curve or 'zigzag' of AA1 + 2, constituting a potential synapomorphy for the clade (Kukalová-Peck 1985, 1997; Willmann 1999). The systematic position of this group is critical for the resolution of phylogenetic relationships between major pterygote lineages (e.g. Edmunds and Travers 1954, Kukalová-Peck 1985; Willmann 1999; Grimaldi and Engel 2005; Prokop et al. 2010). The most prominent diagnostic features are the

constriction of the area between AA1 + 2 and AA3 + 4 in the hindwing, the presence of a concave longitudinal vein IN (intercalary) between them, and a constriction of the area between AA3 + 4 and the first branch of the concave AP at the same point (Prokop et al. 2010). Unfortunately, most syntonopterids are based on isolated wings, but *Lithoneura lameeri* (Fig. 12.3i), known from a siderite nodule from Mazon Creek, Illinois, is an exceptionally preserved fossil with exquisite details also of body structures in addition to wing venation (Carpenter 1938, 1987; Kukalová-Peck 1985; Willmann 1999). Another 'keystone' fossil that has at times been included here is *Triplosoba pulchella* from the Late Carboniferous of Commentry, France. *Triplosoba* had an unusual wing venation for mayflies such as the basal connection of MA with R and RP in the forewing, and MA remote from RP and basally fused with MP in the hindwing. Prokop and Nel (2009) transferred this taxon to Palaeodictyopterida as suggested by earlier authors (Forbes 1943; Willmann 1999). Nevertheless, doubt remains about the inclusion of *Triplosoba* in Palaeodictyopterida (Staniczek et al. 2011). Beckemeyer and Engel (2011) followed Prokop and Nel (2009) and excluded the Triplosobidae (their Triplosoboptera) from what they considered to represent a monophyletic Ephemeropterida, admitting that the former might be a stem group to Palaeodictyopterida or Metapterygota.

Odonatoptera, comprising the Recent dragonflies and damselflies along with the extinct griffenflies and others, are one of the most peculiar groups owing to their strikingly different wing articulation relative to other pterygotes. Odonatopterans are readily recognized by the presence of two large plates (costal plate and radio-anal plate), rather than an arrangement of multiple axillary sclerites, the former representing a unique synapomorphy for the clade. The lineage was abundant and diverse and their morphology stable over evolutionary time, with the earliest species known from the earliest Late Carboniferous (Namurian) (Riek and Kukalová-Peck 1984; Bechly et al. 2001; Ren et al. 2008). Wings of Odonatoptera have a strongly reduced anal area, especially in the forewings; ScP reaching the costal margin well before the wing apex, and an anal brace with a Z-like kink in CuP at the point of fusion with AA. Prothoracic winglets present in Eugeropteridae were probably articulated and movable, whereas in Erasipteridae, the articulation is not apparent (Bechly et al. 2001). Both of the latter groups represent the most basal lineages known from the Late Carboniferous. Geroptera, with the single family Eugeropteridae, is known only from Argentina and possess a rather coarse meshwork of simple crossveins and a short ScP reaching the costal margin at about wing mid-length, this being convergent with Erasipteridae, Paralogidae, and Odonatoclada (Riek and Kukalová-Peck 1984). Erasipteridae are known from the early Late Carboniferous of central Europe and retained a partially dense pattern of crossveins (the so-called archeodictyon) and a long median stem lacking the arculus and nodus (Pruvost 1933; Bechly 1996; Bechly et al. 2001). The Palaeozoic griffenflies of the Protodonata (Fig. 12.3d), also widely known as the Meganisoptera (a name which implicates their former common name as 'giant dragonflies' despite the fact that they are in no way 'dragonflies'), are famous for having given rise to the largest wingspans among insects. *Meganeuropsis permiana* from the Early Permian Wellington Formation of central Kansas and northcentral Oklahoma had a wing span around 710 mm, well exceeding that of any living insect (Carpenter 1939, 1947). Insect gigantism, known particularly from the Late Carboniferous, occurred also in other groups such as the stem-group mayfly *Bojophlebia prokopi*, with a wingspan reaching almost 500 mm (Kukalová-Peck 1985). One of the favoured hypotheses for the presence of such giant insects (and other massive arthropods) at this time is the corresponding hyperoxic atmospheres, thereby permitting the passive transfer of more oxygen via tracheae to metabolically active tissues (Graham et al. 1995). However, these insect giants coexisted together with even more diverse normal- to small-sized relatives (Nel et al. 2009). Another explanation of Palaeozoic insect gigantism

assumes that the increase in body size of some insects was a result of an evolutionary race in body size between aerial predators such as griffenflies and their putative prey among the Palaeodictyopterida (Hasenfuss 2008). While the latter is an enticing hypothesis, it is entirely ad hoc as there is no evidence for what griffenflies fed upon nor any phylogenetic evaluation of relationships for these lineages which demonstrate such an arms race. Similarly, Nel et al. (2008) also considered other factors such as the absence of flying vertebrate predators at this time, but again it is not immediately clear why this alone should lead to such a dramatic increase in size. A comprehensive and conclusive explanation for Palaeozoic insect gigantism remains to be seen.

The Protodonata, or Meganisoptera, are known entirely from the Late Palezoic and are currently considered as a stem group to true Odonata, differing mainly in wing venation by the absence of a nodus, discoidal cells, and a pterostigma (Nel et al. 2009). Wings of these species consist of hundreds to thousands of small polygonal cells, especially numerous in the Meganeuridae (Fig. 12.3d). A true odonatoid nodus with more or less oblique nodal and subnodal veinlets at about wing mid-length first appears in Nodialata, a clade comprising Protanisoptera and Discoidalia (Bechly 1996). The Upper Permian family Lapeyriidae has been considered to be the most basal group of Nodialata (Nel et al. 1999). The Protanisoptera were a widely distributed group of Permian Nodialata that had a partly developed nodus, the hindwings as long or even slightly longer than forewings, the brace formed by ScA uniquely oblique, and a special form of pterostigma crossed by RA, the latter considered as convergent with Dicliptera (Fig. 12.3l) (Palaeodictyopterida) and among modern Diptera.

The extinct superorder Palaeodictyopterida (=Dictyoneuridea) represents a widely diverse group of Palaeozoic insects ranging from the earliest Late Carboniferous to the Late Permian, with a peak in abundance in the Late Carboniferous (Sinitshenkova 2002). Triassic records of *Thuringopteryx gimmi* and *Paratitan reliquia* putatively suggesting the existence of palaeodictyopteridans after the Permian/Triassic mass extinction have been revised and unambiguously excluded from the superorder (Willmann 2008; Shcherbakov 2011). Palaeodictyopterida have a number of wing venation symplesiomorphies for pterygotes such as a pronounced CP and ScA and a convex ridge formed by stiffened membrane in conjunction with basal portions of AA (homologous to the anal brace in Ephemeroptera and Odonatoptera) (Kukalová-Peck 1991, 1997). Kukalová-Peck (2009) supposed palaeoptery as derived based on fusions between basivenalia and fulcalaria in the subcostal to jugal rows putatively visible in various palaeodictyopterid groups, but this requires confirmation. The Palaeodictyopterida were most remarkable for their haustellate mouthparts (Fig. 12.3b), their sucking beaks making them among the earliest of specialized herbivores. The classical divisions of Palaeodictyopterida recognize four main orders Diaphanopterodea, Palaeodictyoptera, Megasecoptera, and Dicliptera (Grimaldi and Engel 2005).

Palaeodictyoptera are the largest and most diverse group and are known from the earliest Late Carboniferous to the Late Permian, with a few giant species like *Mazothairos enormis* attributed to Homoiopteridae (Kukalová-Peck and Richardson 1983; Prokop et al. 2006). The fore- and hindwings were either similar in form, or in some families, the hindwings were distinctly broader, for example as in Spilapteridae. The wing venation had a complete set of main veins including MA and MP, a prominent corrugation of convex and concave veins, and usually lacked fusion between these systems. Intercalary veins were sometimes present, as in the families Calvertiellidae and Namuroningxianiidae (Prokop and Ren 2007), while the main longitudinal veins were frequently connected by numerous crossveins forming a dense pattern of irregular networks (='archeodictyon') and were well developed in families such as Dictyoneuridae. Articulated prothoracic winglets were putatively present in some members of Palaeodictyoptera such as in *Lithomantis carbonarius* (Lithomanteidae) and *Stenodictya pygmaea* (Dictyoneuridae) (Fig. 12.3b, c) (Kukalová-Peck

1978). However, most of the known lateral prothoracic extensions lacked any observable articulation with the prothorax, much like those in neopteran insects such as Lemmatophoridae (Fig. 12.3m) from the same deposits (Kukalová-Peck 1978, 1991). Nygmata-like structures were present in the wing membrane (Fig. 12.3e), observable as circular spots or punctures in distinctive positions principally alongside RP and the medial veins of various groups (e.g. Novák 1880, Carpenter 1963). However, the homology of these with similarly named structures present in different holometabolan orders like Neuroptera, Mecoptera, and Hymenoptera has never been elaborated (Forbes 1924, 1943), and they are assuredly of different evolutionary origins. Palaeodictyoptera are likely paraphyletic with respect to other palaeodictyopterid orders, lacking any distinctive synapomorphies and principally recognized by their exclusion from the other groups.

The Megasecoptera had homonomous wings that were typically slender and petiolate (Fig. 12.3j), with a complete set of main longitudinal veins and frequent coalescence of MA and MP (e.g. Mischopteridae), or more rarely fusion to partial connections between MA and RP and MP with CuA (e.g. Sphecopteridae or Corydaloididae). The costal margin was usually straight with closely parallel veins ScP and R and had a generally denser pattern of crossveins relative to the remaining orders. Overall, the wing venation was considerably similar to Palaeodictyoptera (Carpenter 1962; Sinitshenkova 1980), although these similarities are largely symplesiomorphic. It is not entirely clear what subgroup or families of the paraphyletic Palaeodictyoptera might be more closely allied to Megasecoptera.

As mentioned above, Diaphanopterodea had homonomous wings that could be held roof-like over the abdomen when at rest (Fig. 12.3h). In addition, species had markedly curved stems of R and M running closely parallel, with the stem of R subsequently diverging from the separation of MA and MP; otherwise, the pattern of wing venation frequently resembled that of Megasecoptera, likely symplesiomorphically.

The last clade of Palaeodicyopterida is the Dicliptera. This was a small group known strictly from the Permian, with the hindwings strongly reduced in Diathemidae or completely lost in Permothemistidae (Fig. 12.3l). The forewing had a well-developed and sclerotized pterostigma, a strong reduction in the crossveins to a single rs-m vein, a large anal area, and fusion near the wing base between M + Cu and AA + Cu + CuP as a double anal brace (Kukalová-Peck 1991; Grimaldi and Engel 2005).

Wootton and Kukalová-Peck (2000) utilized available morphological data from palaeopterous Palaeozoic insects to interpret the flight abilities and techniques for various groups. Their interpretations indicated that there existed marked differences in flight abilities among Carboniferous and Permian ephemeropteridans by comparison with Recent taxa, while Palaeozoic odonatopterans exhibited a similar wing construction and shape to modern dragonflies and damselflies as well as an apparent early adaptation to aerial predation. The palaeodictyopterid groups exhibited a broad spectrum of flight techniques and patterns, indicative of the diversity and various peculiar specializations within the lineage.

Lastly, the Neoptera are, of course, well known for their ability to fold the wings over the abdomen due to a unique organization of the axillary sclerites, particularly the Y-shaped third axillary, this suite representing one of the strongest synapomorphies. The Neoptera comprise the vast majority of all pterygote insects in the Recent fauna as well as the fossil record. The wings of neopterans are characterized by the separation of the anterior remigium from a posterior vannus by the claval furrow and the subsequent subdivision of the neala (jugum) from the vannus by the jugal furrow, particularly visible in the hindwing. The wing venation of Neoptera has lost the strong pattern of corrugation; the stem of M is basally concave; vein MA is not clearly convex and frequently hardly identifiable when fused with RP or completely suppressed. The course of MA and MP is the most controversial issue among authors. Forbes (1943) supposed that MA is connected to R or RP, and there is a free MP in all Neoptera,

while Sharov (1968) and others considered that MP was fused with CuA and only the basal part is retained as crossvein m-cua (='arculus'). The course of the medial and cubital veins close to the wing base and the orientation of the arculus play a role in the elasticity and function of the wing and have been utilized for phylogenetic interpretations of larger groupings among Pterygota. Neoptera is traditionally subdivided into three units: Polyneoptera, Paraneoptera, and Holometabola, the latter two united as the Eumetabola (Grimaldi and Engel 2005). Haas and Kukalová-Peck (2001) proposed a split between two major neopteran clades 'Pleconeoptera' + 'Orthoneoptera', both with a full anojugal lobe and 'Blattoneoptera' + (Paraneoptera + Holometabola) (although using the names Hemineoptera and Endopterygota for the latter two) with a partial anojugal lobe and reduced anterior anal sector (AA) as putative apomorphies. These divisions have not received support from more extensive morphological and molecular studies on insect relationships (e.g. Trautwein et al. 2012). The Neoptera are abundantly represented in the Palaeozoic, many of which were historically dumped into a wastebasket group called the 'Protorthoptera'. Handlirsch (1906) established the group for insects with orthopteroid affinities and 'Protoblattodea' for insects with blattoid affinities, although even he was unable to attribute several genera to either group (e.g. *Distasis*), and the groups overlapped in their characters as recognized. Martynov (1938), followed by Sharov (1961), attempted to separate Protoblattodea, Protorthoptera, and Paraplecoptera as stem groups to Blattaria, Orthoptera, and Plecoptera, respectively. Hennig (1981) supported the notion of a close relationship between Protorthoptera and Protoblattodea with the modern orders Orthoptera and Blattodea, respectively. By contrast, he noted difficulties with Paraplecoptera as it had been conceived, noting that it appeared to be based strictly on plesiomorphies. Carpenter (1966, 1992) took a conservative position and, in a retrograde classificatory scheme, merged Paraplecoptera and Protoblattodea with Protorthoptera pending future study. Sharov (1968) agreed that Protoblattodea and Paraplecoptera were inseparable and should be combined into one order. There are distinctive groups among these early Neoptera, such as the Paoliida (Fig. 12.3k) (a.k.a., Protoptera) and Caloneurodea, but retention of the remainder in 'Protorthoptera' obscures phylogenetic relationships, thereby serving no good purpose and is assuredly polyphyletic as constituted by Carpenter (1992) (e.g. Béthoux and Nel 2005; Béthoux 2007; Prokop and Nel 2007). While it is beyond the scope of the present work to summarize the entire geological history of insects, these taxa are of interest for wing origins as they factor into the reconstruction of basal character states for Neoptera. As such, a clarification of their relationships relative to extant neopteran orders, their implications for understanding the living clades, and certainly whether any may represent stem groups to Neoptera as a whole, is of vital importance to insect phylogenetics. For the moment, there is no robust phylogeny that comprehensively treats all 'protorthopteran' families alongside the full diversity of non-Eumetabolan insect orders, and this hinders any meaningful interpretation of primitive character states for Neoptera.

## 12.4 Conclusions

Mounting evolutionary developmental data is giving us a robust and greatly revised perspective on the homology of insect wings, thereby providing an immense leap towards answering the first of those questions posed in the introduction. There is a growing body of developmental evidence that the wing is largely a paranotal extension that integrated appendage patterning modules to develop a functional articulation incorporating portions of the upper pleuron. Unfortunately, there remains significant debate regarding the basal lineages of Pterygota, rendering it difficult to distinguish between competing interpretations of polarity relative to the form of the wing articulation. Palaeontological studies have advanced significantly during the last 25 years, particularly with a large number of critical reevaluations of taxa in a cladistic framework and by pushing back the timing of wing origins from the Early Carboniferous into the earliest Devonian, perhaps latest Silurian. Coupled with this has been the steady accumulation of new taxa from diverse time periods and deposits throughout the globe. While the recovery of an abundance of interpretable remains from the Devonian has not been forthcoming, work on clarifying the identity and relationships between Late Palaeozoic taxa has continued at a significant pace such that the principal lineages important for resolving basal relationships can be characterized and difficulties with particular taxa recognized (e.g. placement of *Triplosoba*, monophyly of Palaeodictyoptera). The currently expanding body of developmental work must unite with a newly invigorated study of insect palaeontology (including the reconstruction of life histories for immature Palaeozoic insects) and phylogeny. Once these elements are resolved and meaningfully united into a comprehensive picture of the early stages and ecologies of wing evolution, only then will we have a solid stance from which to build a consistent model for the origins of powered flight.

What we do know is that the ancestral pterygote lived in a seemingly barren world, quite foreign to anything we are familiar with today, and in which plant life was never far from a shoreline and the climate was generally warm with a moderately high $O_2$ level. Arborescence had not yet developed, and flight may have been a significant aid to reach nutritious sporangia at the apices of branches in early plants and/or for dispersal. Wings originated as paranotal extensions much like those observed in silverfish, suggesting that gliding may have been the initial stage in developing flight. This would have been followed by the integration of a hinge at the base and some early form of controlled flight. Beyond these few, overly simplified statements, we can say little else with certainty, and to speculate on elaborate adaptive scenarios is fruitless. Anything else about Nature's first flyer remains, for the moment, up in the air.

**Acknowledgments**  We are grateful to A. Minelli for the invitation to contribute this brief chapter concerning wing origins and evolution. Much appreciation is given to R. Ward for his comments on an earlier draft of the section on development, to I.A. Hinojosa-Díaz and A. Nawrocki for general discussion of hexapod morphology and homology, to N.P. Kristensen, A. Nel, and A. Minelli for their comments, and to T. Ohde and T. Niimi for sharing and discussing their recent findings. Illustrative figures were executed by S. Taliaferro under the direction of M.S.E. and S.R.D. and supported by the Engel Illustration Fund, University of Kansas College of Liberal Arts & Sciences. We thank the President and fellows of Harvard College for permission to use MCZ copyrighted material. We are grateful to P. Perkins for permission to study and take photographs of type specimens from F.M. Carpenter's collection housed in the Museum of Comparative Zoology, Harvard University; C. Mellish and A.J. Ross for access to collections and permission to take photographs of *Idoptilus onisciformis* and *Lithomantis carbonarius* in the Natural History Museum, London; A. Nel for access to collections and permission to take photographs of *Stenodictya pygmaea* in the Museum national d'Histoire naturelle, Paris; L. Schöllmann for access to collections and permission to take photographs of *Kemperala hagenensis* in the LWL-Museum für Naturkunde, Münster; A.P. Rasnitsyn for access to and permission to photograph specimens in the Laboratory of Paleoentomology, Russian Academy of Sciences, Moscow; and T. Nichterl for permission to take photographs of *Lithomanthis bohemica* in the Naturhistorisches Museum Wien, Austria. Partial support for this work was provided by US National Science Foundation grants DEB-0542909 (to M.S.E.) and DEB-1110590 (to M.S.E., P. Cartwright, and S.R.D.), and the Grant Agency of the Czech Republic No. P210/10/0633 (to J.P.). This is a contribution of the Division of Entomology, University of Kansas Natural History Museum.

# References

Abzhanov A, Kaufman TC (2000a) Embryonic expression patterns of the Hox genes of the crayfish *Procambarus clarkii* (Crustacea, Decapoda). Evol Dev 2:271–283

Abzhanov A, Kaufman TC (2000b) Crustacean (malacostracan) Hox genes and the evolution of the arthropod trunk. Development 127:2239–2249

Alexander DE (2002) Nature's flyers: birds, insects, and the biomechanics of flight. John's Hopkins University Press, Baltimore

Alonso J, Santarén JF (2005) Proteomic analysis of the wing imaginal discs of *Drosophila melanogaster*. Proteomics 5:474–489

Angelini DR, Kaufman TC (2005) Comparative developmental genetics and the evolution of arthropod body plans. Annu Rev Genet 39:95–119

Angelini DR, Liu PZ, Hughes CL, Kaufman TC (2005) Hox gene function and interaction in the milkweed bug *Oncopeltus fasciatus* (Hemiptera). Dev Biol 287:440–455

Averof M (1997) Arthropod evolution: same Hox genes, different body plans. Curr Biol 7:R634–R636

Averof M, Cohen SM (1997) Evolutionary origin of insect wings from ancestral gills. Nature 385:627–630

Azpiazu N, Morata G (2000) Function and regulation of *Homothorax* in the wing imaginal disc of *Drosophila*. Development 127:2685–2693

Bechly G (1996) Morphologische Untersuchungen am Flügelgeäder der rezenten Libellen und deren Stammgruppenvertreter (Insecta; Pterygota; Odonata), unter besonderer Berücksichtigung der Phylo-genetischen Systematik und des Grundplanes der Odonata. Petalura 2:1402

Bechly G, Brauckmann C, Zessin W, Gröning E (2001) New results concerning the morphology of the most ancient dragonflies (Insecta: Odonatoptera) from the Namurian of Hagen-Vorhalle (Germany). Z Zool Syst Evol 39:209–226

Beckemeyer RJ, Engel MS (2011) Upper Carboniferous insects from the Pottsville Formation of northern Alabama (Insecta: Ephemeropterida, Palaeodictyopterida, Odonatoptera). Sci Pap Nat Hist Mus Univ Kansas 44:1–19

Béthoux O (2007) Emptying the Paleozoic wastebasket for insects: members of a Carboniferous 'protorthopterous family' reassigned to natural groups. Alavesia 1:41–48

Béthoux O, Briggs DEG (2008) How *Gerarus* lost its head: stem-group Orthoptera and Paraneoptera revisited. Syst Entomol 33:529–547

Béthoux O, Nel A (2005) Some palaeozoic 'Protorthoptera' are 'ancestral' orthopteroids: major wing braces as clues to a new split among the 'Protorthoptera'. J Syst Palaeontol 2:285–309

Béthoux O, Nel A, Schneider JW, Gand G (2007) *Lodetiella magnifica* nov. gen. and nov. sp. (Insecta: Palaeodictyoptera; Permian), an extreme situation in wing morphology of palaeopterous insects. Geobios 40:181–189

Béthoux O, Kristensen NP, Engel MS (2008) Hennigian phylogenetic systematics and the 'groundplan' vs. 'post-groundplan' approaches: a reply to Kukalová-Peck. Evol Biol 35:317–323

Beutel RG, Gorb S (2006) A revised interpretation of the evolution of attachment structures in Hexapoda with special emphasis on Mantophasmatodea. Arthropod Syst Phyl 64:3–25

Blanke A, Wipfler B, Letsch H, Koch M, Beckmann F, Beutel R, Misof B (2012) Revival of Palaeoptera: head characters support a monophyletic origin of Odonata and Ephemeroptera (Insecta). Cladistics 28:560–581

Bolker JA, Raff RA (1996) Developmental genetics and traditional homology. BioEssays 18:489–494

Börner C (1904) Zur Systematik der Hexapoden. Zool Anz 27:511–533

Boudreaux HB (1979) Arthropod phylogeny, with special reference to insects. Wiley, New York

Bowsher JH, Nijhout HF (2007) Evolution of novel abdominal appendages in a sepsid fly from histoblasts, not imaginal discs. Evol Dev 9:347–354

Boxshall GA (2004) The evolution of arthropod limbs. Biol Rev 79:253–300

Bryant PJ (1975) Pattern formation in the imaginal wing disc of *Drosophila melanogaster*: fate map, regeneration and duplication. J Exp Zool 193:49–77

Buratovich MA, Wilder EL (2001) The role of wingless in the development of the proximal wing hinge of *Drosophila*. Drosophila Info Serv 84:145–157

Butler MJ, Jacobsen TL, Cain DM, Jarman MG, Hubank M, Whittle JR, Simcox A (2003) Discovery of genes with highly restricted expression patterns in the *Drosophila* wing disc using DNA oligonucleotide arrays. Development 130:659–670

Cameron SL, Beckenbach AT, Dowton MA, Whiting MF (2006) Evidence from mitochondrial genomics on interordinal relationships in insects. Arthropod Syst Phyl 64:27–34

Carpenter FM (1938) Two carboniferous insects from the vicinity of Mazon Creek, Illinois. Amer J Sci 36:445–452

Carpenter FM (1939) The lower permian insects of Kansas. Part 8. Additional Megasecoptera, Protodonata, Odonata, Homoptera, Psocoptera, Plecoptera, and Protoperlaria. Proc Amer Acad Arts Sci 73:29–70

Carpenter FM (1947) Lower permian insects from Oklahoma. Part 1. Introduction and the orders Megasecoptera, Protodonata, and Odonata. Proc Amer Acad Arts Sci 76:25–54

Carpenter FM (1962) A permian megasecopteron from Texas. Psyche 69:37–41

Carpenter FM (1963) Studies on Carboniferous insects of Commentry, France: Part V. The genus *Diaphanoptera* and the order Diaphanopterodea. Psyche 70:240–256

Carpenter FM (1966) The lower permian insects of Kansas. Part 11. The orders Protorthoptera and Orthoptera. Psyche 73:46–88

Carpenter FM (1987) Review of the extinct family Syntonopteridae (order uncertain). Psyche 94:373–388

Carpenter FM (1992) Superclass hexapoda. In: Kaesler RL (ed) Treatise on invertebrate paleontology (Part R, Arthropoda 4, Vol. 3–4). Geological Society of America, Boulder, pp. 1–655

Carroll SB, Weatherbee SD, Langeland JA (1995) Homeotic genes and the regulation of insect wing number. Nature 375:58–61

Casares F, Mann RS (2000) A dual role for *homothorax* in inhibiting wing blade development and specifying proximal wing identities in *Drosophila*. Development 127:1499–1508

Cavodeassi F, Rodríguez I, Modolell J (2002) Dpp signalling is a key effector of the wing-body wall subdivision of the *Drosophila* mesothorax. Development 129:3815–3823

de Celis JF (2003) Pattern formation in the *Drosophila* wing: the development of the veins. BioEssays 25:443–451

de Celis JF, Diaz-Benjumea FJ (2003) Developmental basis for vein pattern variations in insect wings. Int J Dev Biol 47:653–663

Chesebro J, Hrycaj S, Mahfooz N, Popadić A (2009) Diverging functions of *Scr* between embryonic and post-embryonic development in a hemimetabolous insect, *Oncopeltus fasciatus*. Dev Biol 329:142–151

Cho E, Irvine KD (2004) Action of *fat*, *four-jointed*, *dachsous*, and *dachs* in distal-to-proximal wing signaling. Development 131:4489–4500

Cohen B, Simcox AA, Cohen SM (1993) Allocation of the thoracic imaginal primordia in the *Drosophila* embryo. Development 117:597–608

Cohen B, Wimmer EA, Cohen SM (1991) Early development of leg and wing primordia in the *Drosophila* embryo. Mech Dev 33:229–240

Cohen SM (1993) Imaginal disc development. In: Bate M, Arias AM (eds) The development of *Drosophila melanogaster*, vol 2. Cold Spring Harbor Laboratory Press, Plainview, pp 747–841

Crampton GC (1916) The phylogenetic origin and the nature of the wings of insects according to the paranotal theory. J N Y Entomol Soc 24:1–39

Crampton GC (1924) The phylogeny and classification of insects. Pomona J Entomol Zool 16:33–47

Dixey FA (1931) Development of wings in Lepidoptera. Trans Entomol Soc London 79:365–393

Dudley R (2000) The biomechanics of insect flight: Form, function, evolution. Princeton University Press, Princeton

Dürken B (1907) Die Tracheenkiemenmuskulatur der Ephemeriden unter Berücksichtigung der Morphologie des Insektenflügels. Z wiss Zool 87:435–550

Dürken B (1923) Die postembryonale Entwicklung der Tracheenkiemen und ihrer Muskulatur bei *Ephemerella ignita*. Zool Jahrb Anat 44:439–614

Edmunds GF Jr (1972) Biogeography and evolution of Ephemeroptera. Annu Rev Entomol 17:21–42

Edmunds GF Jr, Travers JR (1954) The flight mechanics and evolution of the wings of Ephemeroptera, with notes on the archetype insect wing. J Wash Acad Sci 44:390–400

Edwards D, Selden PA, Richardson JB, Axe L (1995) Coprolites as evidence for plant-animal interactions in Siluro-Devonian terrestrial ecosystems. Nature 377:329–331

Engel MS, Grimaldi DA (2004) New light shed on the oldest insect. Nature 427:627–630

Forbes WTM (1924) The occurrence of nygmata in the wings of insecta: Holometabola. Entomol News 35:230–232

Forbes WTM (1943) The origin of wings and venational types in insects. Amer Midland Nat 29:381–405

Fristrom DK, Rickoll WL (1982) The morphogenesis of imaginal discs of Drosophila. In: King RC, Akai H (eds) Insect ultrastructure, vol 1. Plenum Press, New York, pp 247–277

Garrouste R, Clément G, Nel P, Engel MS, Grandcolas P, D'Haese C, Lagebro L, Denayer J, Gueriau P, Lafaite P, Olive S, Prestianni C, Nel A (2012) A complete insect from the late Devonian period. Nature 488:82–85

Gaunt MW, Miles MA (2002) An insect molecular clock dates the origin of the insects and accords with palaeontological and biogeographic landmarks. Mol Biol Evol 19:748–761

González-Crespo S, Morata G (1996) Genetic evidence for the subdivision of the arthropod limb into coxopodite and telopodite. Development 122:3921–3928

Graham JB, Dudley R, Aguilar NM, Gans C (1995) Implications of the later Palaeozoic oxygen pulse for physiology and evolution. Nature 375:117–120

Grenier JK, Garber TL, Warren R, Whitington PM, Carroll S (1997) Evolution of the entire arthropod Hox gene set predated the origin and radiation of the onychophoran/arthropod clade. Curr Biol 7:547–553

Grimaldi D, Engel MS (2005) Evolution of the insects. Cambridge University Press, Cambridge

Haas F, Kukalová-Peck J (2001) Dermaptera hindwing structure and folding: new evidence for familial, ordinal and superordinal relationships within Neoptera (Insecta). Eur J Entomol 98:445–509

Hall BK (1994) Homology: the hierarchical basis of comparative biology. Academic Press, San Diego

Hall BK (2007) Homoplasy and homology: dichotomy or continuum? J Hum Evol 52:473–479

Handlirsch A (1906–1908) Die fossilen Insekten und die Phylogenie der rezenten Formen: Ein Handbuch für Paläontologen und Zoologen. Engelmann, Leipzig

Hasenfuss I (2002) A possible evolutionary pathway to insect flight starting from lepismatid organization. J Zool Syst Evol Res 40:65–81

Hasenfuss I (2008) The evolutionary pathway to insect flight: a tentative reconstruction. Arthropod Syst Phyl 66:19–35

Held LI Jr (2002) Imaginal discs: the genetic and cellular logic of pattern formation. Cambridge University Press, Cambridge

Hennig W (1981) Insect phylogeny. Wiley, New York

Hovmöller R, Pape T, Källersjö M (2002) The Paleoptera problem: Basal pterygote phylogeny inferred from 18S and 28S rDNA sequences. Cladistics 18:313–323

Hubbard MD, Kukalová-Peck J (1980) Permian mayfly nymphs: new taxa and systematic characters. In: Flannagan JR, Marshall KE (eds) Advances in Ephemeroptera biology. Plenum Press, London, pp 19–31

Hughes CL, Kaufman TC (2002) Hox genes and the evolution of the arthropod body plan. Evol Dev 4:459–499

Ishiwata K, Sasaki G, Ogawa J, Miyata T, Su Z (2011) Phylogenetic relationships among insect orders based on three nuclear protein-coding gene sequences. Mol Phylogenet Evol 58:169–180

Jeram AJ, Selden PA, Edwards D (1990) Land animals in the silurian: arachnids and myriapods from shropshire, England. Science 250:658–661

Jockusch EL, Nagy LM (1997) Insect evolution: how did insect wings originate? Curr Biol 7:R358–R361

Jockusch EL, Ober KA (2004) Hypothesis testing in evolutionary developmental biology: a case study from insect wings. J Hered 95:382–396

Jockusch EL, Williams TA, Nagy LM (2004) The evolution of patterning of serially homologous appendages in insects. Dev Genes Evol 214:324–338

Kjer KM, Carle FL, Litman J, Ware J (2006) A molecular phylogeny of Hexapoda. Arthropod Syst Phyl 64:35–44

Klein T (2001) Wing disc development in the fly: the early stages. Curr Opin Genet Dev 11:470–475

Klein T, Martinez Arias A (1998) Different spatial and temporal interactions between *notch*, *wingless*, and *vestigial* specify proximal and distal pattern elements of the wing in *Drosophila*. Dev Biol 194:196–212

Kristensen NP (1991) Phylogeny of extant hexapods. In: CSIRO (ed) The insects of Australia 2nd ed, vol 1. Melbourne University Press, Melbourne, pp 125–140

Kukalová J (1968) Permian mayfly nymphs. Psyche 75:311–327

Kukalová-Peck J (1978) Origin and evolution of insect wings and their relation to metamorphosis, as documented by the fossil record. J Morphol 15:53–126

Kukalová-Peck J (1983) Origin of the insect wing and wing articulation from the arthropodan leg. Can J Zool 61:1618–1669

Kukalová-Peck J (1985) Ephemeroid wing venation based upon new gigantic carboniferous mayflies and basis morphological phylogeny and metamorphosis of pterygote insects (Insecta, Ephemerida). Can J Zool 63:933–955

Kukalová-Peck J (1991) Fossil history and the evolution of hexapod structures. In: CSIRO (ed) The insects of Australia 2nd ed, vol 1. Melbourne University Press, Melbourne, pp 141–179

Kukalová-Peck J (1997) Arthropod phylogeny and 'basal' morphological structures. In: Fortey RA, Thomas RH (eds) Arthropod relationships (Systematics Association Special), vol 55. Chapman & Hall, London, pp 249–268

Kukalová-Peck J (2009) Carboniferous protodonatoid dragonfly nymphs and the synapomorphies of Odonatoptera and Ephemeroptera (Insecta: Palaeoptera). Palaeodiversity 2:169–198

Kukalová-Peck J, Beutel RG (2012) Is the Carboniferous †Adiphlebia lacoana really the "oldest beetle"? Critical reassessment and description of a new Permian beetle family. Eur J Entomol 109:633–645

Kukalová-Peck J, Brauckmann C (1990) Wing folding in pterygote insects, and the oldest Diaphanopterodea from the early late carboniferous of West Germany. Can J Zool 68:1104–1111

Kukalová-Peck J, Peters JG, Soldán T (2009) Homologisation of the anterior articular plate in the wing base of Ephemeroptera and Odonatoptera. Aquat Insect 31:459–470

Kukalová-Peck J, Richardson ES Jr (1983) New Homoiopteridae (Insecta: Paleodictyoptera) with wing articulation from upper carboniferous strata of Mazon Creek, Illinois. Can J Zool 61:1670–1687

Labandeira CC, Beall BS, Hueber FM (1988) Early insect diversification: evidence from a lower devonian bristletail from Québec. Science 242:913–916

Labandeira CC, Sepkoski JJ Jr (1993) Insect diversity in the fossil record. Science 261:310–315

La Greca M (1980) Origin and evolution of wings and flight in insects. Boll Zool 47:65–82

Lameere A (1922) Sur la nervation alaire des insectes. Bull Class Sci Acad R Belg 8:138–149

Lemche H (1940) The origin of winged insects. Vidensk Medd Dansk Naturh Foren 104:127–168

Madhavan MM, Schneiderman HA (1977) Histological analysis of the dynamics of growth of imaginal discs and histoblast nests during the larval development of *Drosophila melanogaster*. Roux's Arch Dev Biol 183:269–305

Malpighi M (1687) Opera omnia, figuris elegantissimus in aes incisis illustrata. Tomis duobus, comprehensa. Quorum catalogum sequens pagina exhibit. Robertum Littlebury, Little Brittain

Martynov AV (1925) Über zwei Grundtypen der Flügel bei den Insekten und ihre Evolution. Z Morphol Ökol Tiere 4:465–501

Martynov AV (1938) Essays on the geological history and phylogeny of insect orders. 1. Palaeoptera and Neoptera-Polyneoptera. Trudy Paleontologicheskogo Instituta Akademii Nauk SSSR 7:1–150 (In Russian)

Matsuda R (1970) Morphology and evolution of the insect thorax. Mem Entomol Soc Can 102:1–431

Matsuda R (1981) The origin of insect wings (Arthropoda: Insecta). Int J Insect Morphol Embryol 10:387–398

Mercer WF (1900) The development of the wings in the Lepidoptera. J N Y Entomol Soc 8:1–20

Moczek AP, Rose DJ (2009) Differential recruitment of limb patterning genes during development and diversification of beetle horns. Proc Natl Acad Sci USA 106:8992–8997

Milner MJ, Bleasby AJ, Kelly SL (1984) The role of the peripodial membrane of leg and wing imaginal discs of *Drosophila melanogaster* during evagination and differentiation in vitro. Roux's Arch Dev Biol 193:180–186

Morata G (2001) How drosophila appendages develop. Nature Rev 2:89–97

Murray FV, Tiegs OW (1935) The metamorphosis of *Calandra oryzae*. Quart J Micr Sci 77:405–495

Nardi JB, Hardt TA, Magee-Adams SM, Osterbur DL (1985) Morphogenesis in wing imaginal discs: its relationship to changes in the extracellular matrix. Tissue Cell 17:473–490

Nel A, Fleck G, Garrouste R, Gand G (2008) The Odonatoptera of the Late Permian Lodève Basin (Insecta). J Iber Geol 34:115–122

Nel A, Fleck G, Garrouste R, Gand G, Lapeyrie J, Bybee SM, Prokop J (2009) Revision of Permo-Carboniferous griffenflies (Insecta: Odonatoptera: Meganisoptera) based upon new species and redescription of selected poorly known taxa from Eurasia. Palaeontogr Abt A 289:89–121

Nel A, Gand G, Garric J (1999) A new family of Odonatoptera from the continental upper permian: the Lapeyriidae (Lodève Basin, France). Geobios 32:63–72

Ng M, Diaz-Benjumea J, Cohen SM (1995) *Nubbin* encodes a POU-domain protein required for proximal-distal patterning in the *Drosophila* wing. Development 121:589–599

Ng M, Diaz-Benjumea J, Vincent J-P, Wu J, Cohen SM (1996) Specification of the wing by localized expression of *wingless* protein. Nature 381:316–318

Niitsu S (2003) Postembryonic development of the wing imaginal discs in the female wingless bagworm moth *Eumeta variegata* (Lepidoptera, Psychidae). J Morphol 257:164–170

Niitsu S, Kobayashi Y (2008) The developmental process during metamorphosis that results in wing reduction in females of three species of wingless-legged bagworm moths, *Taleporia trichopterella*, *Bacotia sakabei* and *Proutia* sp. (Lepidoptera: Psychidae). Eur J Entomol 105:697–706

Niitsu S, Lobbia S (2010) An improved method for the culture of wing discs of the wingless bagworm moth, *Eumeta variegata* (Lepidoptera: Psychidae). Eur J Entomol 107:687–690

Ninomiya T, Yoshizawa K (2009) A revised interpretation of the wing base structure in Odonata. Syst Entomol 34:334–345

Niwa N, Akimoto-Kato A, Miimi T, Tojo K, Machida R, Hayashi S (2010) Evolutionary origin of the insect wing via integration of two developmental modules. Evol Dev 12:168–176

Novák O (1880) Über *Gryllacris bohemica*, einen neuen Locustiden-Rest aus der Steinkohlenformation von Stradonitz in Böhmen. Jahrb KK Geol Reichsanstalt 30:69–74

Ogden TH, Whiting MF (2003) The problem with "the Paleoptera problem:" sense and sensitivity. Cladistics 19:432–442

Peterson MD, Rogers BT, Popadić A, Kaufman TC (1999) The embryonic expression pattern of *labial*, posterior homeotic complex genes and the *teashirt* homologue in an apterygote insect. Dev Genes Evol 209:77–90

Powell PB (1904) The development of wings of certain beetles, and some studies of the origin of the wings of insects. J N Y Entomol Soc 12:237–243

Powell PB (1905) The development of wings of certain beetles, and some studies of the origin of the wings of insects. J N Y Entomol Soc 13:5–22

Pratt HS (1900) The embryonic history of imaginal discs in *Melophagus ovinus* L., together with an account of the earlier stages in the development of the insect. Proc Boston Soc Nat Hist 29:241–272

Prokop J, Nel A (2007) An enigmatic Palaeozoic stem-group: Paoliida, designation of new taxa from the upper Carboniferous of the Czech Republic (Insecta: Paoliidae, Katerinkidae fam. n.). Afr Invertebr 48:77–86

Prokop J, Nel A (2009) Systematic position of *Triplosoba*, hitherto the oldest mayfly from upper Carboniferous of Commentry in Central France (Insecta: Palaeodictyopterida). Syst Entomol 34:610–615

Prokop J, Nel A, Hoch I (2005) Discovery of the oldest known Pterygota in the lower Carboniferous of the Upper Silesian Basin in the Czech Republic (Insecta: Archaeorthoptera). Geobios 38:383–387

Prokop J, Nel A, Tenny A (2010) On the phylogenetic position of the palaeopteran Syntonopteroidea (Insecta: Ephemeroptera), with a new species from the upper Carboniferous of England. Org Divers Evol 10:331–340

Prokop J, Ren D (2007) New significant fossil insects from the upper Carboniferous of Ningxia in northern China (Palaeodictyoptera, Archaeorthoptera). Eur J Entomol 104:267–275

Prokop J, Smith R, Jarzembowski E, Nel A (2006) New homoiopterids from the late Carboniferous of England (Insecta: Palaeodictyoptera). CR Palevol 5:867–873

Prud'homme B, Gompel N, Carroll SB (2007) Emerging principles of regulatory evolution. Proc Natl Acad Sci USA 104:8605–8612

Prud'homme B, Minervino C, Hocine M, Cande JD, Aouane A, Dufour HD, Kassner VA, Gompel N (2011) Body plan innovation in treehoppers through the evolution of an extra wing-like appendage. Nature 473:83–86

Pruvost P (1933) Un ancêtre des libellules dans le terrain houiller de Tchécoslovaquie. Ann Soc Geol du Nord 58:149–154

Quennedey A, Quennedey B (1990) Morphogenesis of the wing Anlagen in the mealworm beetle *Tenebrio molitor* during the last larval instar. Tissue Cell 22:721–740

Regier JC, Shultz JW, Zwick A, Hussey A, Ball B, Wetzler R, Martin JW, Cunningham CW (2010) Arthropod relationships revealed by phylogenomic analysis of nuclear protein-coding sequences. Nature 463:1079–1083

Rehm P, Borner J, Meusemann K, von Reumont BM, Simon S, Hadrys H, Misof B, Burmester T (2011) Dating the arthropod tree based on large-scale transcriptome data. Mol Phylogenet Evol 61:880–887

Ren D, Nel A, Prokop J (2008) New early griffenfly, *Sinomeganeura huangheensis* from the late Carboniferous of northern China (Meganisoptera: Meganeuridae). Insect Syst Evol 38:223–229

Riek EF, Kukalová-Peck J (1984) A new interpretation of dragonfly wing venation based upon early Carboniferous fossils from Argentina (Insecta: Odonatoidea) and basic character states in pterygote wings. Can J Zool 62:1150–1166

Rogers BT, Peterson MD, Kaufman TC (1997) Evolution of the insect body plan as revealed by the *Sex combs reduced* expression pattern. Development 124:149–157

Ross MH (1964) Pronotal wings in *Blattella germanica* (L.) and their possible evolutionary significance. Amer Midl Nat 71:161–180

Schmidt AR, Jancke S, Lindquist EE, Ragazzi E, Roghi G, Nascimbene PC, Schmidt K, Wappler T, Grimaldi DA (2012) Arthropods in amber from the Triassic period. Proc Natl Acad Sci USA 109:14796–14801

Schwanwitsch BN (1943) Subdivision of Insecta Pterygota into subordinate groups. Nature 152:727–728

Sharov AG (1961) On the system of the orthopterous insects. In: Strouhal H, Beier M (eds) Verhandlungen: XI. Internationaler Kongreß für Entomologie, Wien 1960, Band I (Sektion I bis VI). Organisationskomitee des XI Internationalen Kongresses für Entomologie, Vienna, pp 295–296

Sharov AG (1968) Phylogeny of the Orthopteroidea. Trudy Paleontologicheskogo Instituta Akademii Nauk SSSR 118:1–216 (In Russian)

Shcherbakov DE (2011) The alleged Triassic palaeodictyopteran is a member of Titanoptera. Zootaxa 3044:65–68

Shear WA, Bonamo PM, Grierson JD, Rolfe WDI, Smith EL, Norton RA (1984) Early land animals in North America: evidence from the Devonian age arthropods from Gilboa, New York. Science 224:492–494

Simcox AA, Hersperger E, Shearn A, Whittle JRS, Cohen SM (1991) Establishment of imaginal discs and histoblast nests in *Drosophila*. Mech Develop 34:11–20

Simon S, Strauss S, von Haeseler A, Hadrys H (2009) A phylogenomic approach to resolve the basal pterygote divergence. Mol Biol Evol 26:2719–2730

Sinitshenkova ND (1980) Order Dictyoneurida. Order Mischopterida. Order Permothemistida. Trudy Paleontologicheskogo Instituta Akademii Nauk SSSR 175:44–49 (In Russian)

Sinitshenkova ND (2002) Superorder Dictyoneurida Handlirsch, 1906 (=Palaeodictyopteroidea). In: Rasnitsyn AP, Quicke DLJ (eds) History of insects. Kluwer, Dordrecht, pp 115–124

Staniczek AH (2000) The mandible of silverfish (Insecta: Zygentoma) and mayflies (Ephemeroptera): its morphology and phylogenetic significance. Zool Anz 239:147–178

Staniczek AH, Bechly G, Godunko R (2011) Coxoplectoptera, a new fossil order of Palaeoptera (Arthropoda: Insecta), with comments on the phylogeny of the stem group of mayflies (Ephemeroptera). Insect Syst Evol 42:101–138

Šulc K (1927) Das Tracheensystem von *Lepisma* (Thysanura) und Phylogenie der Pterygogenea. Acta Soc Sci Nat Morav 4:227–344

Švácha P (1992) What are and what are not imaginal discs: reevaluation of some basic concepts (Insecta, Holometabola). Dev Biol 154:101–117

Tannreuther GW (1910) Origin and development of the wings of Lepidoptera. Arch Entwicklung Org 29:275–286

Tillyard RJ (1928) Some remarks on the Devonian fossil insects from the Rhynie Chert beds, Old Red Sandstone. Trans Entomol Soc London 76:65–71

Tomoyasu Y, Wheeler SR, Denell RE (2005) *Ultrabithorax* is required for membranous wing identity in the beetle *Tribolium castaneum*. Nature 433:643–647

Tower WL (1903) The origin and development of the wings of Coleoptera. Zool Jahrb Anat Ont 17:517–570

Trautwein MD, Wiegmann BM, Beutel R, Kjer KM, Yeates DK (2012) Advances in insect phylogeny at the dawn of the postgenomic era. Annu Rev Entomol 57:449–468

Truman JW, Riddiford LM (1999) The origins of insect metamorphosis. Nature 401:447–452

Vigoreaux JO (2005) Nature's versatile engine: insect flight muscle inside and out. Springer, Berlin

Waddington CH (1941) The genetic control of wing development in *Drosophila*. J Genet 41:75–139

Warren RW, Nagy L, Selegue J, Gates J, Carroll S (1994) Evolution of homeotic gene regulation and function in flies and butterflies. Nature 372:458–461

Wasik BR, Moczek AP (2011) *Decapentaplegic* (*dpp*) regulates the growth of a morphological novelty, beetle horns. Dev Genes Evol 221:17–27

Wasik BR, Rose DJ, Moczek AP (2010) Beetle horns are regulated by the *Hox* gene, *Sex combs reduced*, in a species- and sex-specific manner. Evol Dev 12:353–362

Wehman HJ (1969) Fine structure of *Drosophila* wing imaginal discs during early stages of metamorphosis. Dev Genes Evol 163:375–390

Weismann A (1864) Die Entwicklung der Dipteren: Ein Beitrag zur Entwicklungsgeschichte der Insecten. Engelmann, Leipzig

Whalley P, Jarzembowski EA (1981) A new assessment of *Rhyniella*, the earliest known insect, from the Devonian of Rhynie, Scotland. Nature 291:317

Whitfield JB, Kjer KM (2008) Ancient rapid radiations of insects: challenges for phylogenetic analysis. Annu Rev Entomol 53:449–472

Wigglesworth VB (1972) The principles of insect physiology, 7th edn. Chapman & Hall, London

Williams JA, Carroll SB (1993) The origin, patterning and evolution of insect appendages. BioEssays 15:567–577

Willkommen J (2009) The tergal and pleural wing base sclerites: homologous within the basal branches of Pterygota? Aquat Insect, Suppl 31:443–457

Willmann R (1999) The upper Carboniferous *Lithoneura lameerei* (Insecta, Ephemeroptera?). Palaeont Z 73:289–302

Willmann R (2008) *Thuringopteryx*: eine "Permische" Eintagsfliege im Buntsandstein (Insecta, Pterygota). Palaeont Z 82:95–99

Wootton RJ (1972) Nymphs of Palaeodictyoptera (Insecta) from the Westphalian of England. Palaeontology 15:662–675

Wootton RJ (1981) Palaeozoic insects. Annu Rev Entomol 26:319–344

Wootton RJ, Kukalová-Peck J (2000) Flight adaptations in Palaeozoic Palaeoptera (Insecta). Biol Rev 75:129–167

Yoshizawa K (2011) Monophyletic Polyneoptera recovered by wing base structure. Syst Entomol 36:377–394

Yoshizawa K (2012) The treehopper's helmet is not homologous with wings (Hemiptera: Membracidae). Syst Entomol 37:2–6

Zhang H, Shinmyo Y, Mito T, Miyawaki K, Sarashina I, Ohuchi H, Noji S (2005) Expression patterns of the homeotic genes *Scr*, *Antp*, *Ubx*, and *Abd-A* during embryogenesis of the cricket *Gryllus bimaculatus*. Gene Expr Patterns 5:491–502

Zhang J, Zhou C, Gai Y, Song D, Zhou K (2008) The complete mitochondrial genome of *Parafronurus youi* (Insecta: Ephemeroptera) and the phylogenetic position of Ephemeroptera. Gene 424:18–24

# Architectural Principles and Evolution of the Arthropod Central Nervous System

**13**

Rudolf Loesel, Harald Wolf, Matthes Kenning,
Steffen Harzsch and Andy Sombke

## Contents

R. Loesel
Institut für Biologie, RWTH Aachen, Lukasstr.
1, 52056, Aachen, Germany
e-mail: loesel@bio2.rwth-aachen.de

H. Wolf
Institute of Neurobiology, University of Ulm,
Albert-Einstein-Allee 11, 89081, Ulm/Donau,
Germany
e-mail: harald.wolf@uni-ulm.de

M. Kenning · S. Harzsch · A. Sombke (✉)
Zoological Institute and Museum, Department
of Cytology and Evolutionary Biology, Ernst Moritz
Arndt University of Greifswald, Soldmansstrasse
23, 17487, Greifswald, Germany
e-mail: andy.sombke@uni-greifswald.de

M. Kenning
e-mail: matthes.kenning@googlemail.com

S. Harzsch
e-mail: steffen.harzsch@uni-greifswald.de

This is an exciting time for arthropod neuro-anatomists! A wealth of reviews, special issues, book chapters, and entire book volumes published during the last 10 years shows the unbroken interest in and enthusiasm for the arthropod nervous system and for gaining insights into its architecture, physiology, and aspects of neuroethology (Barth and Schmid 2001; Wiese 2001, 2002; Barth 2002; North and Greenspan 2007; Breithaupt and Thiel 2011; Galizia et al. 2012; Land and Nilsson 2012; Strausfeld 2012). Numerous review articles and book chapters witness that neurobiology is one of the most active fields of arthropod research. Recently featured topics are, for example, the crustacean central nervous system (Schmidt and Mellon 2011; Harzsch et al. 2012; Sandeman et al. in press), structure and function of crustacean chemosensory sensilla (e.g. Hallberg and Skog 2011; Mellon and Reidenbach 2011),

A. Minelli et al. (eds.), *Arthropod Biology and Evolution*,
DOI: 10.1007/978-3-642-36160-9_13, © Springer-Verlag Berlin Heidelberg 2013

chelicerate strain detection systems (Barth 2012), and insect olfaction (Galizia and Szyska 2008; Hansson and Stensmyr 2011; Hansson et al. 2011; Sachse and Krieger 2011). Moreover, the central nervous system and visual organs of neglected taxa such as Myriapoda (Sombke et al. 2011a, 2012), Onychophora (Mayer 2006; Strausfeld et al. 2006a, b; Eriksson and Stollewerk 2010; Whitington and Mayer 2011), Trilobita (Clarkson et al. 2006), and Xiphosura (Battelle 2006) have been analyzed with contemporary techniques. Furthermore, detailed reviews have been provided on specific substructures of the arthropod brain such as the central complex (Loesel et al. 2002; Homberg 2008), mushroom bodies (MBs) (e.g. Farris 2005, 2011; Strausfeld et al. 2009; Loesel and Heuer 2010; Heuer et al. 2012), and the peripheral and central olfactory pathways (e.g. Sandeman and Mellon 2002; Schachtner et al. 2005; Mellon 2007; Masse et al. 2009; Galizia and Rössler 2010; Hansson and Stensmyr 2011; Rössler and Zube 2011). Functional anatomy, physiology, and development of arthropod eyes and the optic neuropils seem to be endlessly appealing for arthropod neurobiologists (e.g. Egelhaaf et al. 2009; Borst et al. 2010; Borst and Euler 2011).

The past decade has also seen the emergence of the discipline of 'neurophylogeny' that is the synthesis of neurobiological questions and evolutionary aspects (e.g. Harzsch et al. 2005a, b; Harzsch 2006, 2007; Loesel 2006, 2011; Strausfeld 2009; Strausfeld and Andrews 2011). Methods such as immunohistochemistry combined with confocal laser scan microscopy have facilitated the analysis of neuroanatomy of non-model arthropods. These comparative data have yielded new insights into arthropod phylogeny. Within the limitations, a book chapter imposes the following: (i) we will focus on the central nervous system only and for all aspects of sensory systems refer the reader to some of the literature mentioned above; (ii) as a systematic overview touching all anatomical structures of the nervous system in all major taxa is impossible, we will try to extract some common architectural principles of the arthropod ventral

nerve cord and brain and will highlight evolutionary trends of these structures.

## 13.1 The Ventral Nerve Cord

### 13.1.1 The Arthropod Ventral Nerve Cord is Segmentally Organized

As a basic scheme, segmentation of the ventral nerve cord matches body segmentation, in the form of segmental ganglia connected by a pair of connectives. This holds for tagmata such as head and thorax, although the fusion of the segmental ganglia does not always follow the fusion pattern of the visible cuticle segments. Often, ganglia shift along the longitudinal body axis to join other ganglia, thus lengthening the nerves attached to them. This may be the result of actual morphogenetic movements in the embryonic nervous system. The segmental ganglia receive sensory input from the corresponding body segment, and the motoneurons in that ganglion supply the segmental muscles (Fig. 13.1a). There are, however, many exceptions, for instance, as far as intersegmental muscles are concerned. These muscles may be supplied from motoneurons in either of the adjacent segmental ganglia. Sensory neurons often do not branch just in the segmental ganglion but ascend further, sometimes up to the brain. Commissures connect the two sides of the body, in many Mandibulata via two sets of pathways: the anterior and posterior commissures. The commissures consist primarily of axons, whereas dendrites do not usually cross the ganglion midline (anatomical details, e.g., in Tyrer and Gregory 1982; Elson 1996) (Fig. 13.2).

In annelids, on each side of a body segment, separate ganglia which are connected by distinct axon bundles as commissures are formed by neuronal somata and neuropil center (Denes et al. 2007). The latter is defined as a network of dendrites and axons where synapses are present and in which somata do not occur (Richter et al. 2010). In Arthropoda, these two ganglia are usually fused across the body midline (exceptions include many Branchiopoda), thus forming just a single segmental ganglion which consists of two hemiganglia, connected by the anterior and posterior commissures. The ganglia of adjacent body segments communicate via the connectives. Anastomoses of peripheral nerves are common and allow innervation across segment borders. The axons running in the connectives often do not terminate in the ganglia joined by the latter, but may extend along the ventral nerve cord for several neuromeres, or even the whole length. The latter is true for brain neurons descending all the way to the terminal ganglion and, vice versa, neurons from the terminal ganglion or from any of the more anteriorly located segmental ganglia that send axons into the brain. The axons in the connectives thus usually pass through the ganglia, giving off a few branches, and are joined by axons originating in the particular ganglion. The connectives do not pass the ganglion as a solid bundle but are arranged in separate longitudinal axon bundles that proceed through the ganglion's neuropil (Fig. 13.2).

In the arthropods, and actually in many invertebrates including molluscs and annelids, the somata of neurons are arranged around the periphery of the segmental ganglia. The soma layer may form a continuous rind, or cortex, coating the whole ganglion, particularly where the ganglion neuropil is relatively small and does not bulge and displace the soma cortex. A much larger number of somata and accordingly a thicker soma cortex invariably occur on the ventral side of the segmental ganglia, with a few soma groups extending towards the lateral and dorsal ganglion surfaces. Bundles of primary neurites extend from soma groups into the neuropil where they split up into dendritic and axonal fibers (Fig. 13.2b, d). Primary neurites of motoneurons perforate the ventral neuropil to reach the dorsal side of the ganglion where the motor neuropils are located.

Examples for neurons that occur near the dorsal midline of the ganglion are the so-called dorsal unpaired median neurons, or DUMs (Fig. 13.1a, light green). In Hexapoda, this group of neurons originates in development from special unpaired neuroblasts and forms

important neurosecretory cells that release octopamine (review in Pflüger and Stevenson 2005). This neuron type or its precursors may represent an apomorphy of Mandibulata (Linne et al. 2012). It is also suggested that unpaired midline precursors evolved from the bilateral median domain of the ventral neuroectoderm.

## 13.1.2 The Segmental Ganglia are Highly Structured

The pattern of the connectives branching into the tracts is quite stereotypic, at least within a given arthropod subtaxon but probably beyond. It appears that corresponding tract patterns are present even across the different arthropod groups, such as hexapods, malacostracan crustaceans (Fig. 13.2a, c) (Skinner 1985a, b; Leise et al. 1986, 1987; Elson 1996), and chelicerates (Wolf and Harzsch 2002a). The conservation of fasciculation patterns in the development of axon pathways in the arthropods examined so far (reviews Whitington 1996; 2004, 2006; Whitington and Bacon 1997; Whitington and Mayer 2011) lends support to such an idea as far as hexapods and malacostracan crustaceans are concerned. Similarly, the presence of an anterior and a posterior commissure per segmental ganglion is consistent across the Tetraconata at least (compare Fig. 13.3). The segmental neuropils, too, exhibit structural properties that are common amongst the arthropods, and beyond. Motor neuropils are located in the dorsal half of the segmental ganglion, and sensory neuropils in the ventral half (Fig. 13.2). Besides this general pattern, sensory projections are also present in intermediate areas, between the dorsal and ventral neuropils proper, and some afferents even synapse in dorsal and medial neuropil areas. In the latter cases, there are usually monosynaptic connections from sensory afferents to motoneurons that support fast reflexes, for instance, in the context of locomotor control (Burrows 1996).

Within the neuropil, different sensory modalities often segregate to different regions (compare sensory projections in the brain, Sect.

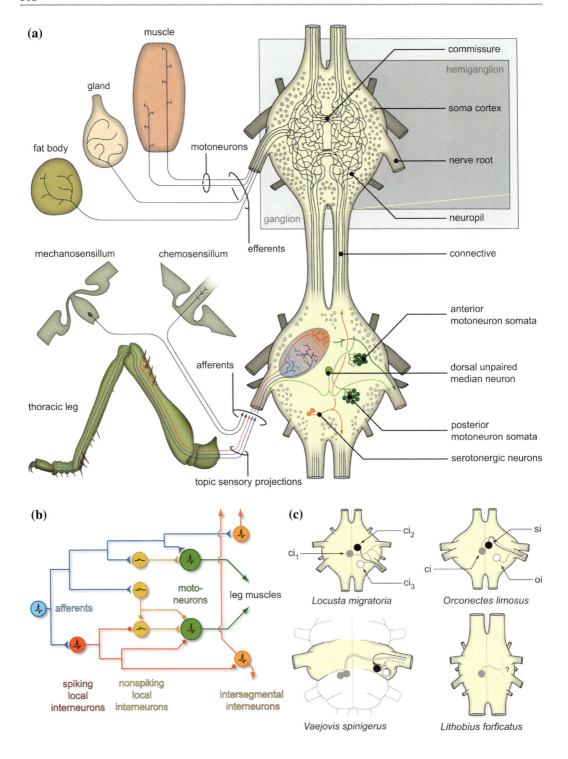

(a)

muscle

gland

fat body

motoneurons

commissure

hemiganglion

soma cortex

nerve root

ganglion

neuropil

efferents

connective

mechanosensillum

chemosensillum

anterior
motoneuron somata

dorsal unpaired
median neuron

thoracic leg

afferents

posterior
motoneuron somata

serotonergic neurons

topic sensory projections

(b)

afferents

spiking
local
interneurons

nonspiking
local
interneurons

moto-
neurons

leg muscles

intersegmental
interneurons

(c)

$ci_1$

$ci_2$

$ci_3$

*Locusta migratoria*

si

ci

oi

*Orconectes limosus*

*Vaejovis spinigerus*

?

*Lithobius forficatus*

◀ **Fig. 13.1** Architecture of the ventral nerve cord in an insect or malacostracan crustacean. **a** Two adjacent segmental ganglia are shown to illustrate major features and anatomical terms (*top ganglion*) and properties of selected neuron groups (*bottom ganglion*) of the ventral nerve cord; note color coding of topological sensory projections. Modified after Richter et al. (2010) and Burrows and Newland (1993). **b** Basic wiring diagram of the sensorimotor pathways in leg motor control. Modified after Burrows (1996). **c** Inhibitory motoneurons in four sample arthropods: hexapod *top left*, malacostracan *top right*, scorpion *bottom left*, chilopod *bottom right*. The three different, and probably homologous, types of common inhibitors are marked by different shading (*grey*: hexapod $ci_1$, *black*: hexapod $ci_2$, *white*: hexapod $ci_3$). No homologization is possible yet for chilopods. Modified after Wiens and Wolf (1993), Harzsch et al. (2005a). *ci* common inhibitor, *si* stretcher-closer inhibitor, *oi* opener inhibitor

13.2 ff.), although exceptions exist. One such exception is the parallel projection of mechanosensory and gustatory input from the locust tarsus. In their target region within the central nervous system, the input from mechanosensory *versus* gustatory sensilla of the same region of the tarsus does not segregate into separate neuropil regions according to the two sensory modalites but rather project into largely overlapping areas in a topologically organized pattern (Newland et al. 2000). In the thoracic ganglia of hexapods, mechanoafferent neurites project mainly to three regions of the neuropil: the most ventral and dorsolateral regions, and the medioventral level of the neuropil. Mechanosensory receptors from the legs exhibit mostly local projections, while receptors from sternites and chordotonal organs form intersegmental projections in addition to local ones (Bräuning et al. 1983). Within a given sensory modality, an ordered structure of neuropil areas is usually observed, in the form of arrangement of sensory projections along gradient axes. For example, mechanosensory input from appendages is usually arranged in a topologically organized pattern (Fig. 13.1a, lower ganglion). That is, the neighboring relationships of sensory input from the body surface are preserved, thus producing a topographic representation of body surface within the ganglion (Burrows 1996). Input from more distal areas, for instance, on an appendage, typically projects to more distal areas in the segmental ganglion. Similarly, the anterior–posterior axis is preserved in central nervous projections, although distortions occur as a result of differential growth in development.

Further sensorimotor processing is brought about by different groups of interneurons with specific properties (Fig. 13.1b). A coarse outline is as follows: worked out primarily in hexapods (Burrows 1996) such as locust, stick insect or cockroach, and in crustaceans such as crayfish and lobster. The ordered projections of *sensory afferents* facilitate the generation of receptive fields in the first group of interneurons, the *local spiking interneurons* (LSIs). The receptive fields may have the shape of particular small regions of body surface and may possess an inhibitory surrounding area that supports contrast enhancement (e.g. von Békésy 1967). The sensory afferents may make contacts to all other neuron groups downstream of the LSIs, however, including the motoneurons as mentioned above. This downstream connectivity holds for all the other groups of interneurons, in principle, although it is dependent on a neurons' function in detail. One important function of the LSIs is transport of sensory information from the ventral primary projection areas to the dorsal motor areas. Consequently, LSIs typically have axons that extend from ventral dendrites to dorsal axonal processes. The LSIs make connections to *local non-spiking interneurons* (NSIs). A major function of this group of interneurons is the organization of a coordinated motor output. This is achieved by connections to the appropriate sets of motoneurons and by inhibitory connections amongst the NSIs that prevent co-contraction of antagonistic muscles, for example. This is illustrated by the fact that intracellular stimulation of a particular NSI will often result in the execution of a well coordinated movement, such as leg extension or leg flexion

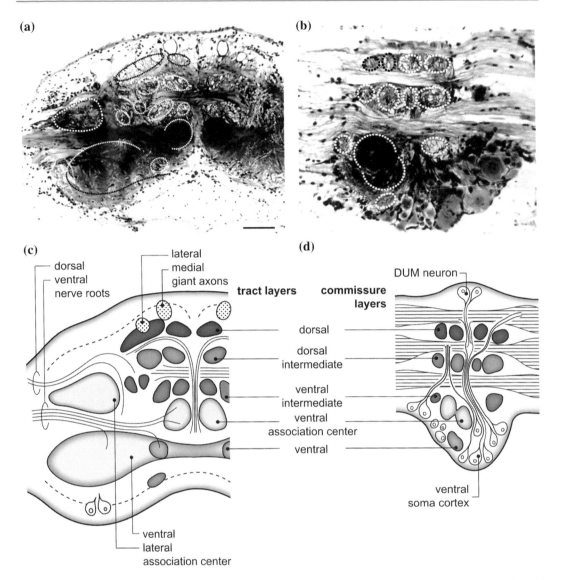

**Fig. 13.2** Anatomical features of ventral ganglia, exemplified in a crayfish. Modified after Elson 1996. **a** Histological cross section and **b** parasagittal section illustrate the main features of the segmental ganglion, indicated by *dotted outlines*. The drawings in **c** and **d** provide the corresponding labeling. Lateral and medial giant axons are particularities of crayfish used in reflex escape (review in Reichert 1988). Note dorsal DUM somata in **d**

involving all the appropriate joints and muscles (Burrows 1996). Signal propagation and transmitter release in the NSIs is via graded potentials, a mechanism that is possible due to the small length of the processes which are restricted to the particular ganglion or even hemiganglion (hence the term *local* interneurons). *Intersegmental interneurons* receive input from all the upstream neurons and convey signals into neighboring ganglia, and sometimes up to the brain or down to the terminal ganglion. These are spiking neurons, of course, since they have to transfer signals across large distances to support the coordination of movement across the different body segments. The *motoneurons*, finally, convey excitation to the muscles to

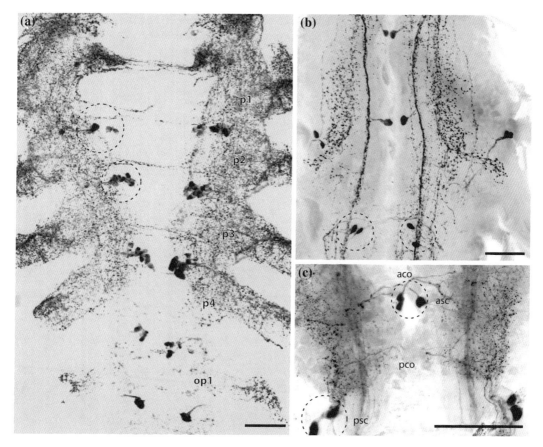

**Fig. 13.3** Serotonin immunoreactivity in the ventral nerve cord. The situation in the fused ganglion complex of *Limulus polyphemus* (**a**), is compared to that in unfused segmental ganglia of *Lithobius forficatus* (**b**), and *Triops cancriformis* (**c**). Note posterior groups of serotonergic cell bodies with primary neurites extending contralaterally through the posterior commissures. A similar, anterior soma group with neurites extending contralaterally through the anterior commissure is present in *Triops*. Selected soma groups are indicated by *dotted circles*. Further note larger number of somata per group in *Limulus*. Modified after Harzsch (2004a), Harzsch and Waloszek (2000). *aco* anterior commissure, *asc* anterior soma cluster, *op1* opisthosomal neuromere, *p1–p4* prosomal neuromeres, *pco* posterior commissure, *pp* pedipalp neuromere, *psc* posterior soma cluster. Anterior is to the top. *Scale bars*: 50 μm

produce muscle contraction and movement. In hexapods, motoneurons appear to be primarily output elements that do not usually make output connections within the central nervous system. In crustaceans, by contrast, motoneurons are often integral parts of motor control circuitry and thus make synapses to other motoneurons and interneurons. It should be noted, however, that the neural basis for sensorimotor processing in two other major arthropod groups—myriapods and chelicerates—has not been defined in

anywhere near the same detail as in hexapods and crustaceans.

Interesting examples with respect to the ordered arrangement of sensory projections are auditory receptor neurons that originate in tympanal organs. These have been studied in much detail in several hexapod groups (e.g. Oldfield 1988; Römer et al. 1988). Auditory input is usually represented in a tonotopic, or frequency-dependent manner. This tonotopic organization appears to be derived from the somatotopic organization of mechanosensory input. Different sound frequencies are received by different though adjacent groups of sensory cells within the tympanal organs. In this way, map-like representations of mechanosensory input in the central nervous system translate into tonotopic representations in auditory neuropils (e.g. Kämper and Murphey 1987) (compare ordered mechanosensory projections indicated in Fig. 13.1a, lower half).

Chemosensory inputs, by contrast, are typically organized according to the molecular identity of the chemosensory neurons. That is, chemosensory cells responding to a particular group of chemicals—odorants or gustatory substances—project to particular small delineated areas of neuropil (details see Sect. 13.2.6). These neuropil areas are typically organized as circular glomeruli, ensheathed by glia and the axons of interneurons. The glomeruli formed by all the different groups of chemosensory receptor neurons form the chemosensory neuropil in the ganglion. The pectine neuropils of scorpions are segmental chemosensory and mechanosensory neuropils with glomerular organization (Wolf and Harzsch 2002b, 2012; Wolf 2008). Such organization appears to be a common feature in chelicerate arthropods, although their primary chemosensors are located on very different appendages (Strausfeld 2012). Again, exceptions exist and chemosensory inputs may project in parallel with the mechanosensory inputs from the respective body regions, for instance, in the bimodal chemo- and mechanosensory sensilla of the locust leg (Newland et al. 2000).

A similar segregation as outlined for the sensory neuropils may exist in the motor neuropils. For example, the arborizations of flight motoneurons in pterygote hexapods occupy the dorsalmost layer of the motor neuropil, while leg motoneurons occupy the ventrally adjacent neuropil areas with their dendritic arborizations (e.g. Robertson et al. 1982; Tyrer and Gregory 1982). Study of a possible segregation of motor neuropils is, unfortunately, more difficult than for sensory neuropils and has received much less attention.

## 13.1.3 Common Features in Arthropod Ventral Nerve Cord Structure are Based on Developmental and Genetic Similarities

The similarities of ventral nerve cord organization amongst the arthropod groups extend to individually identified neurons. This is true in particular for pioneer neurons that lay down the basic scaffold of axonal pathways in the developing peripheral and central nervous systems. There are apparent homologies of pioneer neurons and other individually identified nerve cells in hexapods and malacostracan crustaceans (Patel et al. 1989a, b; Whitington and Bacon 1997; Whitington 1996, 2004, 2006; Duman-Scheel and Patel 1999). It is not surprising, thus, that some individually identifiable neurons, especially motoneurons, can be homologized across a number of arthropod groups, with hexapods and malacostracan crustaceans having received particular attention in this respect (Wiens and Wolf 1993; Kutsch and Breidbach 1994).

The soma cortex consists of sometimes rather distinct groups of somata which in some cases may not immediately be obvious in histology (Fig. 13.2) but which have an ontogenetic basis. It is thought that during development of hexapods, neurons are generated by stereotyped patterns of cell divisions of neuronal stem cells that are the progeny of the neuroectoderm. Each of these stem cells—neuroblasts in hexapods and malacostracan crustaceans—generates a group of neurons, the somata of which are located in close proximity in the soma cortex, due to their common origin from a particular neuroblast (reviews Harzsch 2003; Whitington 2004, 2006; Stollewerk and Simpson 2005; Stollewerk and Chipman 2006; Stollewerk 2008). In Myriapoda, stem cells apparently of the hexapod/malacostracan neuroblast type do not exist (Whitington et al. 1991; Whitington 2004, 2006). The identity and location of neuronal progenitor cells in myriapods and chelicerates have been discussed by Whitington and Mayer (2011) who also reviewed the possible homologies between neuron progenitor cells in the various arthropod groups.

For some insect neuroblasts, there is evidence that the progeny of one particular stem cell share physiological properties, for example, transmitter phenotype, and thus excitatory or inhibitory action on postsynaptic neurons. Or the progeny may be motoneurons or particular types of interneurons. However, in many cases, mixed lineages occur with the progeny even including glia cells (Bossing et al. 1996; Schmidt et al. 1997).

An obvious commonality across all arthropod groups is the arrangement of motoneuron somata which supply the leg muscles into two characteristic groups. These soma groups are located on the ventral side of the ganglion, one just anterior and the other just posterior to the entrance of the segmental leg nerve into the ganglion (Fig. 13.1a, dark green somata in lower ganglion). The respective motoneurons tend to innervate leg muscles that are located in the more anterior or the more posterior half of the appendage, respectively (Tyrer and Gregory 1982). By the same token, inhibitory interneurons occur in stereotyped groups that exhibit morphological and functional correspondences amongst the different arthropod groups (Watson 1986; Wolf and Harzsch 2002b) suggesting at least partial homology (Fig. 13.2c).

The structural properties outlined above for individual ganglia are maintained where several ganglia are fused. A typical example is the so-called subesophageal ganglia of scorpions—which comprises the neuromeres of the chelicerae, the pedipalp, and the four walking leg segments and two more posterior segments including that of the pectines (Wolf and Harzsch 2002a). Another example is the subesophageal ganglion of higher dipterans that represents the fusion product of all segmental ganglia posterior to the esophagus. These fused ganglia with their distinctly segmented structure exhibit almost all the characteristics outlined above for the individual ganglia within the respective neuromeres. The same is true for crustaceans, namely, the highly fused ventral nervous system of the crab, or the chelicerate *Limulus polyphemus* (shown in Fig. 13.3a, and compared to the

situation in *Triops cancriformis* and *Lithobius forficatus*).

## 13.1.4 Homologies Across the Arthropod Taxa

Considering the features outlined above, it is not surprising that several neurons, or groups of neurons, occur in more or less stereotyped fashion in most or all arthropods. Such neurons or neuron groups would appear to be homologous (Kutsch and Breidbach 1994). Correspondences occur not just between different arthropod groups but also in the ganglia along the ventral nerve cord of a given species. These so-called homonomies (serial homology) will vary, of course, depending on the segmental identity and the functional properties of that particular segment (e.g. Kutsch and Heckmann 1995a, b). For example, neurons relevant for the control of appendages, such as legs and wings, will be absent in neuromeres where the appendages have been reduced and are missing, or in species that lack the structures altogether. This is certainly true for the motoneurons supplying the appendage muscles, while the interneurons may be conserved and function in different contexts (e.g. Robertson et al. 1982).

Typical examples for homology across arthropods are the inhibitory motoneurons characteristic of arthropod motor control (Belanger 2005) (Fig. 13.1c). In hexapods and malacostracan crustaceans, the musculature of each walking leg is supplied by a set of three inhibitory motoneurons that adjust muscle performance in the time/velocity domain (Rathmayer 1990; Wolf 1990). It is not just the number of motoneurons but also soma location, anatomical characteristics, and muscle innervation patterns that support homology of the inhibitory leg motoneurons in the Tetraconata. Intriguingly, two of these inhibitory motoneurons serve different functions in hexapods and malacostracans. In hexapods, all three are common inhibitors, supplying partially different sets of muscles (the term common inhibitor alludes to the fact that it is common to several leg muscles). This function is fulfilled in the malacostracans by just one of the inhibitors innervating all leg muscles. The other inhibitors are used to uncouple two distal leg muscles that are innervated by a single (common) excitatory motoneuron (Wiens 1989). Inhibitory motoneurons or groups of inhibitory motoneurons that possess intriguingly similar characteristics concerning soma location, certain anatomical features, and innervation patterns of leg muscles also occur in scorpions and centipedes (Harzsch et al. 2005a) (Fig. 13.1c). Apparent similarities are that, (i) these neurons use gamma-aminobutyric acid (GABA) as neurotransmitter, (ii) physiological activity of the inhibitors induces hyperpolarization in the muscles that they target, (iii) the number of inhibitory leg motoneurons within one hemiganglion is always three, (iv) the somata share corresponding positions within the ganglionic framework, and (v) their axons show a specific pattern of exiting the ganglia via the anterior or posterior nerve roots.

Kutsch and Heckmann (1995a, b) analyzed the innervation of a group of body wall muscles, the dorsal longitudinal muscles (DLMs) in *Lithobius forficatus* (Chilopoda) and compared it with that in Hexapoda. Their study indicated that the set of motoneurons that innervate the DLMs of one segment is composed of two subgroups, the somata of which are arranged in two adjacent neuromeres. Kutsch and Heckmann (1995a, b) suggest that this situation is a plesiomorphic character state of Mandibulata. Considering morphological characteristics, several of the DLM motoneurons may be homologized across the hexapods. Further, the number of motoneurons that supply the DLMs in *L. forficatus* is close to that in the hexapods. However, the authors point out that the motoneurons' morphologies are dissimilar in hexapods and chilopods, a fact that argues against a homology of hexapodan and chilopodan longitudinal muscle motoneurons. The same appears to apply to the motoneurons supplying the intersegmental dorsoventral musculature (Kutsch and Heckmann 1995a, b). Not only the architecture of the motoneurons differs between hexapods and chilopods but also the pattern of axon exit through the ganglionic nerve roots. Once again,

these patterns share considerable similarities between malacostracan crustaceans and hexapods. Similar to the inhibitory leg motoneurons, more detailed analyzes of longitudinal muscle motoneuron architecture in a wider range of taxa will be necessary to fully appreciate and exploit the neurophylogenetic potential of these structures.

So far, similarities have been emphasized that unite the different arthropod taxa—suggesting homology—and similarities of the different segmental ganglia in any given species ('homonomy' sensu Kutsch and Heckmann 1995a, b). However, the partly different functions of inhibitory motoneurons in hexapods and malacostracans illustrate that idiosyncratic specializations may in fact be more interesting under physiological and evolutionary perspectives than the commonalities in basic structure. These differences are important since they may be used to delimit crown groups if they represent apomorphies. Moreover, such specializations may be of particular interest if they can be related to functional properties in physiology and ecology.

This holds true for serotonin-immunoreactive (5HT-ir) neuron groups in the different arthropod taxa. The segmental ganglia of virtually all arthropods investigated so far are characterized by the presence of a set of 5HT-ir cell bodies or small soma groups that possess a number of common features. This pattern is maintained if the segmental ganglia fuse into a larger complex (illustrated for *Limulus*, and compared with *Lithobius* and *Triops* in Fig. 13.3). A posterior group of 5HT-ir cell bodies with primary neurites that extend contralaterally through the posterior commissure is one such characteristic (indicated as orange neuron group in Fig. 13.1a). A similar, anterior soma group with neurites extending contralaterally through the anterior commissure is present in hexapods and malacostracan and other crustaceans, while it is absent in the chilopods. The situation in diplopods and chelicerates is less clear, although anterior and posterior 5HT-ir soma groups exist. The cell bodies are more numerous in the chelicerates, as appears to be typical of most or even all neuron types investigated so far, including the inhibitory motoneurons mentioned above (Wolf and Harzsch 2002a, b). The features of 5HT-ir soma groups have actually been used to reconstruct arthropod phylogeny by exploiting both common features to be interpreted as plesiomorphies and consistent differences amongst the groups that have to be interpreted as apomorphies (Harzsch 2004a).

## 13.2 The Brain

The arthropod brain is a *syncerebrum* formed by the close association and structural and functional transformation of segmental cephalic ganglia (Richter et al. 2010). It is considered to be composed of three neuromeres, the protocerebrum, deutocerebrum, and tritocerebrum (Scholtz and Edgecombe 2006; Bitsch and Bitsch 2007, 2010; see Scholz and Richter in this book (arthropod head)) and hence has been termed a *tripartite brain* (Lichtneckert and Reichert 2005) although it needs to be critically evaluated where the posterior limit is of what we term 'brain' (Harzsch 2004b). Each neuromere is usually compartmentalized to some degree into definable clusters of neurons in the periphery that surround central neuropils (Strausfeld 1976; Sandeman et al. 1993; Doeffinger et al. 2010; Richter et al. 2010). A neuropil is defined as a network of dendrites and axons where synapses are present but neural somata do not occur. However, glial cell somata, tracts, hemolymph vessels, and tracheae may be embedded within a neuropil. A neuropil itself can also be compartmentalized into units which are also termed neuropils (Richter et al. 2010). However, these compartments usually are given specific names such as, for example, olfactory glomeruli (OG) (Fig. 13.4). In some Mandibulata, for example, *Scutigera coleoptrata* (Chilopoda) or *Apis mellifera* (Hexapoda), the axis of brain neuromeres (neuraxis) is bent out of the anterior–posterior body axis resulting in, for example, a dorsal or even posteriodorsal location of the protocerebrum with regard to body axis (Sandeman et al. 1993; Burrows 1996).

Therefore, the ventral surface of the brain can face forward in the head (compare Fig. 13.4d).

The chelicerate brain has been described in few species, most detailed in *Cupiennius salei* (Fig. 13.4a). Here, the nervous system is supraesophageal into two fused masses: the dorsal supraoesophageal ganglion (brain) and the ventral subesophageal ganglion (VNC). The division of the three brain neuromeres in Chelicerata is, however, not easily identifiable. Traditionally, the neuromere associated with the chelicerae was considered to be homologous with the tritocerebrum of Mandibulata resulting in the absence of a deutocerebrum (Bitsch and Bitsch 2007). However, Mittmann and Scholtz (2003) and Harzsch et al. (2005b) showed similarities in the larval nervous system of *L. polyphemus* to that of Mandibulata which confirmed the assumption of a tripartite brain in Arthropoda. Recent comparisons of expression domains of the head *Hox* genes corroborate the assumption that a deutocerebrum is indeed present supporting the existence of a tripartite brain in the Chelicerata (Damen et al. 1998; Telford and Thomas 1998; Abzhanov and Kaufman 2004; Scholtz and Edgecombe 2006).

The *protocerebrum* is the anteriormost neuromere according to the neuraxis and receives input from the eyes (lateral compound eyes and/or median eyes) if present. Thus, the protocerebrum contains the optic neuropils and forms a prominent part of the brain (compare Fig. 13.4c, d, *Birgus latro* and *A. mellifera*). In *C. salei*, four pairs of optic nerves innervate the four first-order optic neuropils (anterior median, posterior median, posterior lateral, and anterior lateral; compare Fig. 13.4a). Besides the optic neuropils, the protocerebrum houses the mushroom bodies and the central body (see Sects. 13.2.8 and 13.2.9). In the Arthropoda, neurosecretory cells often form clusters whose axons leave the neuropil and project to neurohemal release sites and non-neuronal endocrine glands (Hartenstein 2006). The majority of neurosecretory cells are associated with the protocerebrum (pars intercerebralis and lateralis). Axons of neurosecretory cells project to neuroendocrine (or neurohemal) glands. In the brain of Arthropoda,

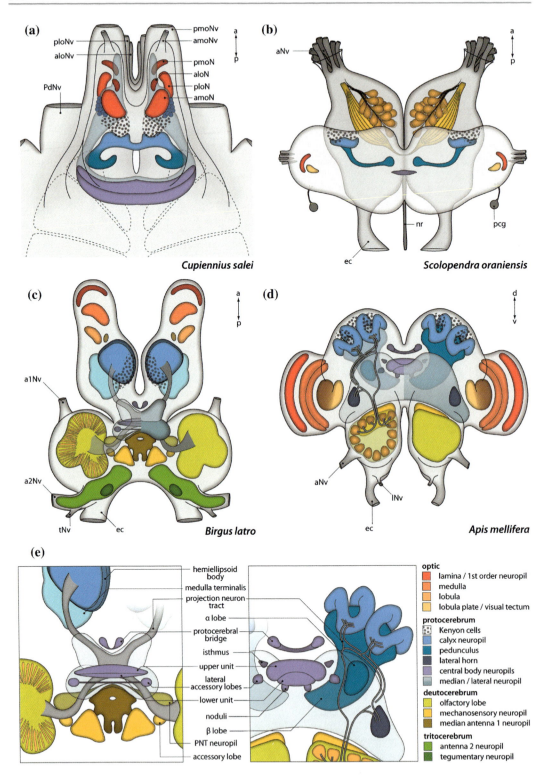

◄ **Fig. 13.4** Schematic representation of selected arthropod brains (**a–c** dorsal, **d** frontal view). Compiled after Barth (2001), Galizia and Rössler (2010), Krieger et al. (2010), Sombke et al. (2012). **a** *Cupiennius salei* (Chelicerata). The first-order optic neuropils (*red*) are associated with a group of optic glomeruli (*blue*). The optic tracts (*transparent blue*) project to the central body. The nerves of the chelicerae are obscured by the optic nerves. **b** *Scolopendra oraniensis* (Myriapoda). The protocerebrum is bent dorsoposteriorly, thus resulting in a dorsal position with regard to body axis. The protocerebral glands (*pcg*) are located posteriorly. The antennal nerve (*aNv*) innervates the olfactory lobe and the mechanosensory neuropil. The nervus recurrens (*nr*) projects caudally on top of the esophagus. **c** *Birgus latro* (Crustacea). The optic neuropils as well as the hemiellipsoid body with the medulla terminalis are located in the anteriormost lateral protocerebrum. The central body is embedded in the median protocerebrum (**e**, *left*). The accessory neuropil as well as the projection neuron tract neuropil (*PNT neuropil*) are located in the median deutocerebrum (**e**, *left*). Besides the antenna 2 nerve (*a2Nv*), the tegumentary nerve (*tNv*) innervates the tritocerebrum. **d** *Apis mellifera* (Hexapoda). The pedunculus of the mushroom body houses the lateral horn and extends into the $\alpha$ and $\beta$ lobes. The mechanosensory neuropil is located posteriorly of the olfactory lobe. The labral nerves (*lNv*) project ventrally (**e**, *right*). In all mandibulate taxa, the esophageal connectives (*ec*) link the tritocerebrum with the mandibular ganglion. **e** Detailed description of the proto- and deutocerebral neuropils of *Birgus latro* (*left*) and *Apis mellifera* (*right*) as well as the color code for all given structures. *a↔p* anterior↔posterior, *a1Nv* and *aNv* antenna 1 nerve, *a2Nv* antenna 2 nerve, *aloN* anterior lateral optic neuropil, *aloNv* anterior lateral optic nerve, *amoN* anterior median optic neuropil, *amoNv* anterior median optic nerve, *d↔v* dorsal↔ventral, *ec* esophageal connective, *lNv* labral nerve, *nr* nervus recurrens, *pcg* protocerebral gland, *PdNv* pedipalp nerve, *ploN* posterior lateral optic neuropil, *ploNv* posterior lateral optic nerve, *pmoN* posterior median optic neuropil, *pmoNv* posterior median optic nerve, *tNv* tegumentary nerve

they have different names like the Schneider's organ in Chelicerata, the protocerebral gland in Chilopoda (Fig. 13.4b), the corpora cardiaca and allata in Hexapoda, or the sinus gland in Crustacea (Tsuneki 1992; Hartenstein 2006; Sombke et al. 2011a).

In Chelicerata, the *deutocerebrum* is associated with the chelicerae while in the Mandibulata, it is associated with the first antennae. In the latter, it houses the olfactory lobes and the mechanosensory neuropils (see below). The antennal nerve contains axons of sensory receptor neurons (chemo- and/or mechanosensoric) and motor neurons innervating the antennal muscles. In hexapods, a tegumentary nerve (innervating parts of the head capsule) is deutocerebral while in Crustacea, this nerve is tritocerebral and innervates an associated neuropil (Fig. 13.4c).

The *tritocerebrum*, flanking the esophagus, links the brain with the subesophageal ganglia. Both hemispheres are connected by tritocerebral commissures that are always located postorally. It is assumed that the possession of two tritocerebral commissures (like in the trunk ganglia) is a plesiomorphic feature of arthropods (Harzsch 2004b). In Chelicerata, the tritocerebrum is associated with the pedipalps, yet it is not clearly

demarcated in the adult brain. In Crustacea, the second antenna innervates the prominent antenna 2 neuropil which processes mostly mechanosensory information. The reduction of the second antenna in Myriapoda and Hexapoda (intercalary, postantennal, or premandibular segment) results in the absence of primary processing neuropils.

In addition, the tritocerebrum links the brain with the *stomatogastric nervous system* which consists of ganglia and nerves supplying the foregut and the clypeolabral region of the head (Bullock and Horridge 1965; Harzsch and Glötzner 2002; Bitsch and Bitsch 2010; Sombke et al. 2012). The frontal ganglion is connected via a pair of frontal connectives with the tritocerebrum (the stomatogastric bridge) and gives rise to the posteriorly projecting unpaired nervus recurrens (Fig. 13.4b: nr). In Chelicerata, a loop-shaped stomatogastric bridge innervates also a so–called labrum in Xiphosura and Scorpiones (Barth 2001; Harzsch et al. 2005b). However, it is assumed that in the ground pattern of Arthropoda, the stomatogastric bridge is formed by fibers of the deuto- and tritocerebrum (Harzsch 2007).

In the Onychophora, the sister group to Arthropoda, the number of brain neuromeres is under debate (Mayer et al. 2010; Whitington and

Mayer 2011). Strausfeld et al. (2006b) proposed that the onychophoran brain is tripartite. However, what appears as a tritocerebrum could be part of the proto- or deutocerebrum or even the ventral nerve cord (Mayer et al. 2010; Whitington and Mayer 2011). The protocerebrum is innervated by the lateral eyes and antenna-like appendages that are regarded to be convergent to the mandibulate antennae (Mayer and Koch 2005; Scholtz and Edgecombe 2006). Within the protocerebrum, a distinct midline neuropil, antennal glomeruli, and MBs have been identified (Strausfeld et al. 2006a, b). The deutocerebrum is associated with the jaws. Backfills of the papillae suggest that the neural region supplying the appendages is part of the ventral nerve cord (Mayer et al. 2010). In conclusion, the brain architecture of Onychophora may represent plesiomorphic characters compared with arthropods, and the tritocerebrum represents an arthropod apomorphy (Whitington and Mayer 2011).

## 13.2.1 The Compound Eyes and Visual Neuropils

The facetted eyes of arthropods have fascinated arthropod neurobiologists for more than 100 years. Numerous book contributions were devoted to this topic and amongst the first and most important ones is probably Sigmund Exner's (1891) treatise on *Die Physiologie der facettierten Augen von Krebsen und Insekten* which was translated into English some 100 years later (Exner and Hardie 1989). Additional book volumes that are either exclusively devoted to arthropod eyes or contain significant chapters on arthropod visual systems are those by Wehner (1972), Horridge (1975), Autrum (1979), Eguchi and Tominaga (1999), as well as Stavenga and Hardie's (1989) *Facets of vision* and Warrant and Nilsson's (2006) *Invertebrate vision*. Evolutionary aspects of arthropod visual systems were dealt with in two special issues of *Arthropod Structure and Development* (Stavenga et al. 2006, 2007). The latest addition to this body of literature is the new edition of Land

and Nilsson's (2012) *Animal eyes*. Because the present chapter focuses on the central nervous system, sensory systems will not be treated here in any depth so that the reader who wants to newly engage in arthropod vision research is referred to the sources listed above.

It has long been known that the cellular architecture of the compound eye's ommatidia shows a strong correspondence between Crustacea and Insecta (Melzer et al. 1997, 2000; Paulus 2000; Dohle 2001; Richter 2002; Harzsch et al. 2005a) but the evolutionary relationships between the eyes of other Arthropoda is matter of debate (Nilsson and Osorio 1997; Paulus 2000; Müller et al. 2003; Spreitzer and Melzer 2003; Bitsch and Bitsch 2005; Harzsch et al. 2005a, b, 2007; Harzsch and Hafner 2006; Nilsson and Kelber 2007). Research on the architecture of the visual neuropils that process the retinal input has strongly focused on flies (Pterygota, Diptera; reviews Strausfeld et al. 2006c, Strausfeld 2012) and crayfish (Malacostraca, Decapoda; Nässel 1976, 1977; Nässel and Waterman 1977; Strausfeld and Nässel 1981) whereas the Chelicerata and Myriapoda have been unjustifiably neglected.

As for the ommatidial structure, a strong correspondence of the cellular components of the visual neuropils of crayfish and flies is obvious (Strausfeld and Nässel 1981; Nilsson and Osorio 1997; Strausfeld 2012). In most decapod crustaceans and pterygote insects, the visual input from the compound eyes is mapped onto four columnar optic neuropils, the lamina, medulla, and the lobula/lobula plate complex which are connected by two successive chiasms (Figs. 13.5a, 13.7a). The hexapod medulla is divided into two distinct layers that are transversed by an axonal projection called the Cuccati bundle or serpentine layer (Strausfeld and Nässel 1981). In the visual neuropils, typically a columnar arrangement of neuronal elements interacts with the neurites of interneurons arranged in a stratified or tangential pattern. One ommatidium of both insects and malacostracan crustaceans contains a group of eight photoreceptors R1–R8 with the same optic axis. Developmental data indicate a homology of the insect and crustacean photoreceptor cells (Melzer et al. 1997, 2000; Hafner and Tokarski 2001; Harzsch and Waloszek 2001). These photoreceptors together constitute the rhabdom where light is absorbed by the visual pigments (reviews Paulus 2000; Osorio 2007; Friedrich et al. 2011). The photoreceptor axons project the retinal mosaic topically onto the first optic neuropil, the lamina (Fig. 13.5a, b), and histamine seems to be the neurotransmitter of these photoreceptors (review Hardie 1989; Callaway and Stuart 1999). Ontogenetically, the R1–R6 develop in three pairs, R1/R6, R2/R5, R3/R4, both in crustaceans and flies (Melzer et al. 1997; Friedrich et al. 2011), and the axons from R1 to R6 ('short' photoreceptor axons) innervate distinct underlying columnar modules in the lamina and retain their neighborhood relationship amongst themselves between the retina and lamina (Strausfeld and Nässel 1981; Sanes and Zipursky 2010; Strausfeld 2012). This architecture gives rise to retinotopic processing units in the lamina, the 'optic cartridges' with an almost crystalline regularity (Fig. 13.5b). The projection pattern of the dipteran photoreceptors is more complex; these animals have an open rhabdom and use the neural apposition mechanism (Nilsson 1989). In these animals, seven rhabdomeres of each ommatidium have divergent optical axes but single receptors (of the R1–R6 type) in six neighboring ommatidia project into one common cartridge in the lamina (Fig. 13.5a; Strausfeld and Nässel 1981 and references therein). Hence, in taxa with neural superposition, a complex sorting of the retina-lamina projections takes place which is not the case in the taxa with apposition and optic superposition designs (Nilsson 1989). In these, the photoreceptor axons project into the lamina cartridge directly beneath their parent ommatidium. In flies, R1–R6 are achromatic and most sensitive to green light whereas in crayfish, they are characterized as yellow–green sensitive. R7 and R8 develop as single units, and in flies, their axons project through the lamina ('long' photoreceptor axons) to terminate in the second optic neuropil, the medulla (Fig. 13.5a, b). They have a narrow

spectral sensitivity with R7 being a UV receptor and R8 being sensitive for blue light. In crayfish, however, only the axons of the blue/violet receptor R8 project through the lamina to terminate in the medulla (Nässel 1976, 1977) whereas R7 has a short axon to the lamina only. The evolutionary correspondence of insect and crustacean R7 and R8 cells needs further clarification.

## 13.2.2 The Lamina

Within the crayfish lamina, which is subdivided into two horizontal strata, the centripetal input provided from the photoreceptor axons diverges greatly and is relayed to visual interneurons. Of these, ten distinct classes have been identified according to their characteristic dendritic or axonal domains as well as their cell body locations, and more cell classes await discovery: five types of monopolar cells (M1–M5), two types of tangential T-neurons, one type of small-field T-neuron, one type of centrifugal cell, and one type of amacrine (anaxonal) cell (Strausfeld and Nässel 1981; Meinertzhagen 1991). All these neurons, except the anaxonal amacrine cells, connect the lamina with the medulla via the outer optic chiasm that also contains the 'long' photoreceptor axons. In the outer optic chiasm, the linear order of the columns is reversed but their spatial relationships are retained. The crayfish lamina monopolar cells as well as the transmedullary cells associated with the medulla constitute the retinotopic columnar pathway whereas amacrine (anaxonal) neurons, wide-field, and tangential elements possess neurites arranged in horizontal layers and modulate the excitability of the columnar projections (Strausfeld and Nässel 1981). The somata of the lamina monopolar neurons are located distally to the neuropil whereas the amacrine cells and the T-neurons have their cell bodies proximal to the lamina neuropil.

There is a strong correspondence between crayfish and fly laminae not only concerning the general arrangement of neuronal elements but also at the level of single classes of visual

**(a)**

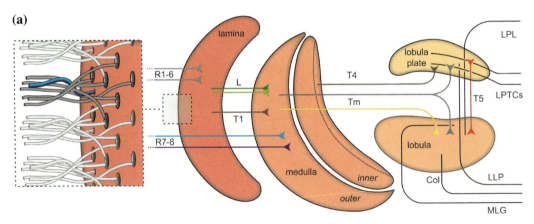

**(b)**

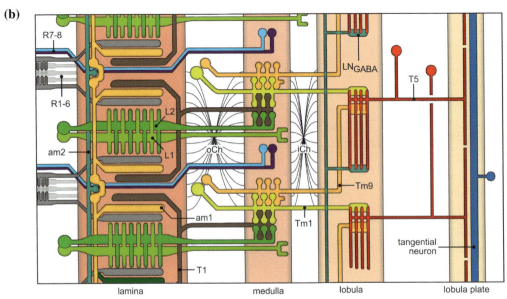

**(c)**

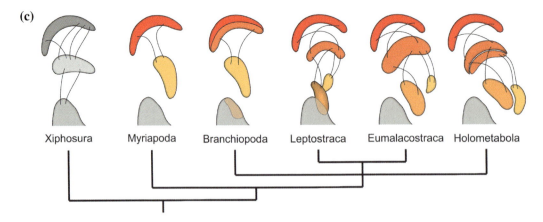

◀ **Fig. 13.5** **a** Schematic overview of the dipteran visual system with neural superposition showing some of the known classes of neuronal elements (compiled from Strausfeld and Nässel 1981; Strausfeld 1989; Douglas and Strausfeld 2003). The left box shows the complex sorting pattern of the R1–R6 photoreceptor axons (*grey*) from four rhabdoms of the retina that project to several neighboring lamina cartridges (*circles*). The axons of R7 and R8 (*blue* and *violet*) are not distributed to several cartridges but extend in tandem to pierce the lamina below their parent cartridge and to terminate in the medulla, which is divided into an inner and outer portion. Several types of lamina monopolar cells (L) are postsynaptic to the R1–R6 input and relay information to the medulla. Small-field T1-neurons also connect lamina and medulla. Transmedullary neurons (*Tm*) and T4 bushy T cells associated with the medulla relay information to the lobula and lobula plate, respectively. Wide-field lobula plate tangential cells (*LPTCs*) have dendrites in direction-specific layers of the lobula plate. Transmedullary neurons supply information about motion to directionally selective motion-sensitive neurons such as the male specific giant neurons (*MLG*) in the lobula. Small-field neurons associated with the lobula plate and lobula provide axonal outputs to the medial brain, and dendrites have their distalmost processes either in the lobula plate (*LPL*) or in the lobula. **b** Schematic representation of some identified neurons serving the achromatic photoreceptors in the fly visual system and successive levels of synaptic connections in the lamina, medulla, lobula, and lobula plate (figure and legend reproduced with modifications from Strausfeld et al. 2006a, b, c). Several known cell types are omitted for clarity. The axons of the color-sensitive R7 and R8 photoreceptors are also shown to pass through the lamina and terminate in the medulla. The inner and outer

chiasms (*iCh*, *oCh*) are indicated schematically. R1–R6 photoreceptors (*grey*) that use histamine as their transmitter provide inputs to type 1 amacrines (*am1*, *yellow*) and lamina monopolar cells Ll and L2 neurons (*green*; glutamatergic). The glutamate-immunoreactive type 1 amacrines are shown serially connected via NMDARl-immunopositive type 2 amacrines (*am2*). The basket dendrites of Tl cells (*brown*) interact with type 1 amacrines. Tl cells, accompanied by L2 of the same optic cartridge, terminate at the dendrites of ChAT-positive paired transmedullary neurons (*Tm1*, *yellow*), the dendrites of which are coincident with those of the GABA-immunoreactive Tm9 neurons (*orange*). The Tm9 axon from the neighboring retinotopic medulla column converges with terminals of Tml neurons at the aspartate immunopositive T5 layer (*red*) in the lobula. A GABA-immunoreactive local interneuron (*LN* GABA, *blue*) provides arborizations within the T5 ensemble. T5 neurons terminate on glutamate-immunoreactive directionally selective tangential neurons in the lobula plate. **c** Evolution of optic neuropils associated with the lateral eyes of Euarthropoda. Modified from Strausfeld (2005). *Red*: outer plexiform layer (lamina), *yellow*: visual tectum (lobula plate), *dark orange*: outer medulla, *light orange*: lobula Col columnar neurons, *iCh* outer chiasm, *L* lamina monopolar cells, *L1*, *L2* lamina monopolar cells type one and two, *LLP* Lobula-lobula plate neurons, *LPL* Lobula plate-lobula neurons, *LN*$_{GABA}$ GABA-immunoreactive local interneuron of the lobula, *LPTCs* wide-field lobula plate tangential cells, *MLG* male specific giant neurons, *oCh* outer chiasm, *R1–R8* axons of photoreceptors R1–R8, *T1* small-field T-neuron, *T4* bushy T cell, *T5* aspartate immunopositive bushy T cell, *Tm* transmedullary neurons, *Tm1* and *Tm9* transmedullary neurons types one and nine

interneurons (Strausfeld and Nässel 1981; Meinertzhagen 1991; Nilsson and Osorio 1997; Sinakevitch et al. 2003; Strausfeld et al. 2006c). Flies, like crayfish, possess five types of monopolar cells, termed L1–L5 (Fig. 13.5a, b). Three of these, the large monopolars (LMCs) L1–L3 are non-spiking neurons and directly postsynaptic to the R1–R6 afferents. L1 and L2 provide color-independent information by signaling changes of luminance. The L3 axons extend to the medulla alongside the long visual fibers of R7 and R8, together providing a trichromatic input to the medulla (Fig. 13.5a, b; Strausfeld 1989; Douglas and Strausfeld 2003; Strausfeld 2012). L4 and L5 are smaller cells that receive inputs from the LMCs. Based on physiological properties and architectural features, Strausfeld and Nässel (1981) and Nilsson and Osorio

(1997) suggested the fly LMCs to be equivalent to the crayfish monopolar neurons M1–M4 which are small-field elements, with their dendritic arbors restricted to the parent cartridge. The crayfish M5 represents a class of wide-field neurons the neurites of which spread through several (six or eight) cartridges and may correspond to the fly L4 or L5 monopolars. The lamina monopolar cells M1–M4 (crayfish) and L1–L3 (fly) in both cases are characteristically wired up to specific receptor terminal combinations by synapses arranged in triads (see Strausfeld and Nässel 1981).

The small-field T-neuron (T1) with dendritic fields in the lamina and a cell body located close to the medulla is another columnar neuron that is part of the optic cartridges (Fig. 13.5a, b). Fly and crayfish small-field T-neurons were suggested to be homologous (Nilsson and Osorio 1997). Tangential cells (Tan 1) of both crayfish and flies have dendritic fields whose arborizations invade both lamina strata and are not restricted to one optic cartridge but spread across several of these. The crayfish lamina has a second type of tangential neuron (Tan 2) with large vertically arranged branches beneath the lamina from which fibers ascend distally into the lamina's plexiform layer. Tan 2 lacks an obvious counterpart in the fly lamina. The axons of both types of tangential neurons project towards the medulla. The cell bodies of centrifugal neurons (C cells) are located between the medulla and lobula, and their axons project distally to invade the lamina and arborize diffusely over several cartridges. The architecture of these GABAergic centrifugal feedback neurons is very similar between insects and a malacostracan crustacean, an isopod in this case (Sinakevitch et al. 2003). Finally, anaxonal or amacrine neurons are associated with the lamina (Fig. 13.5a, b). Physiological and anatomical studies suggest a close correspondence of insect and crayfish amacrine cells (Nilsson and Osorio 1997). Their somata are located at the lamina's proximal surface and give rise to tangential branches from which numerous processes project through the plexiform layer, finally giving rise to lateral branchlets at the distal surface of the lamina. The amacrine neurons exert a presynaptic inhibitory action on the photoreceptor terminals and are thought to be part of the pathway that mediates lateral inhibition in the lamina (Glantz et al. 2000). All the aforementioned wide-field and tangential elements do not seem to be directly postsynaptic to receptor terminals but most probably interact with sets of other relay neurons in the lamina (Strausfeld and Nässel 1981).

### 13.2.3 The Medulla

As mentioned above, in crayfish and flies, the axons of M1–M5/L1–L5 and Tan1, Tan 2 travel towards the medulla via the first (outer) optic chiasm in which the fibers cross but retain the retinotopic organization. The chiasm also comprises the axons of the R8 (crayfish) or R7 and R8 (fly), T1, and centrifugal neurons (Strausfeld and Nässel 1981). The fly medulla is divided into an outer and an inner neuropil by a layer of thick tangential axons, the serpentine layer or Cuccati bundle, but such a bundle does not seem to be present in malacostracan crustaceans (Sinakevitch et al. 2003). However, fly and crayfish show strong correspondence in their medullae in that the distal three-quarters (outer layer) contain the terminals of the M2–M4 lamina monopolar cells, the endings of the long visual fibers (R8) and the arborizations of the lamina tangentials, Tan 1 and 2 (Fig. 13.5a, b). In addition, the dendrites of medulla columnar neurons (the transmedullary neurons), as well as amacrine arbors, are arranged within the outer layers of this region (Strausfeld and Nässel 1981). In flies (but not necessarily other insects), this input to the medulla comprises at least four information channels: two color-insensitive channels, one polychromatic channel, and one channel relaying information about the E-vector of polarized light. In both taxa, small-field transmedullary neurons (Tm1-6) are arranged periodically in association with the long visual fibers (R7/8) and the incoming axons from the lamina monopolar cells. These transmedullay neurons relay the incoming retinotopic picture through the medulla and project to the lobula via the second (inner) optic chiasma. In addition, three classes of amacrine cells (Am) are present in the medulla, the neurites of which are either restricted to a single column or a specific domain of medulla columns and project to different depths of the neuropil (Strausfeld and Nässel 1981). Once again, the amacrine cells are involved in processes of lateral inhibition (Glantz and Miller 2002). The neurochemical architecture of both lamina and medulla is diverse and covered in the following reviews: Hardie 1989; Homberg 1994; Sinakevitch et al. 2003; Harzsch et al. 2012.

## 13.2.4 The Deeper Neuropils and Image Analysis

Whereas in the crayfish and fly, the lamina and medulla receive a direct photoreceptor input, visual interneurons relay information from the lamina and medulla to the deeper neuropils, lobula, and lobula plate (Fig. 13.5a, b). The structure of these two secondary neuropils cannot be described in any depth in this section which focusses on primary processing units. Nevertheless, structural properties of lobula and lobula plate are quite well understood (e.g. Strausfeld and Nässel 1981; Strausfeld 1989, 2012; Strausfeld et al. 2006c). The functions of lobula and lobula plate have been primarily discussed so far in the context of motion detection and these neuropils in Tetraconata are considered to play an integral part in processing optokinetic information (Sztarker et al. 2005). The entire field of how the visual input is processed to extract meaningful information about the image is a research field of its own that cannot be touched here (reviews, e.g., Wiersma et al. 1982; Franceschini et al. 1989; Glantz and Miller 2002; Zeil and Layne 2002; Douglas and Strausfeld 2003; Egelhaaf 2006; Egelhaaf et al. 2009; Borst et al. 2010; Borst and Euler 2011).

In general, it appears that the visual systems of insects and malacostracan crustaceans are organized into parallel processor channels that encode information about contrast and intensity separately from information about color and shape (Douglas and Strausfeld 2003; Strausfeld 2012). Most of the visual field is simultaneously analyzed in a sophisticated parallel-distributed information pathway by multiple classes of interneurons associated with the optic neuropils. Contrast, polarity, polarization angle, and local and global motion are assessed across the visual space at multiple loci defined by the visual receptive field (Glantz and Miller 2002). These aspects are best understood in the fly visual system (Douglas and Strausfeld 2003) and identified parallel retinotopic pathways through the dipteran nervous system include an achromatic pathway with information about the orientation and direction of motion, three parallel channels that are achromatic and non-directional-sensitive, and a fifth channel that serves color vision.

## 13.2.5 Evolution of Visual Neuropils

There is little doubt about the homology of the ommatidia of insects and crustaceans (Melzer et al. 1997, 2000; Nilsson and Osorio 1997; Paulus 2000; Dohle 2001; Hafner and Tokarski 2001; Richter 2002; Bitsch and Bitsch 2005; Harzsch et al. 2005b, Harzsch and Hafner 2006; Nilsson and Kelber 2007), and the strong architectural correspondence of crayfish and fly laminae and medullae is unquestionable (Strausfeld and Nässel 1981; Meinertzhagen 1991; Nilsson and Osorio 1997; Harzsch 2002; Sinakevitch et al. 2003; Strausfeld et al. 2006c). However, it has long been noted that the visual neuropils of non-malacostracan crustaceans, especially studied in the branchiopod genera *Artemia*, *Triops*, *Branchinecta*, and *Daphnia* do not fit into this pattern because these taxa have only two visual neuropils, commonly termed lamina and medulla (reviewed in Strausfeld and Nässel 1981) that are linked by straight fibers without any chiasm. Whereas the neuroarchitecture of the branchiopod lamina resembles that of Malacostraca and Hexapoda even at the level of single cell types (Nässel et al. 1978; Elofsson and Hagberg 1986), the linking fibers take a different course in the two groups. More importantly, it is impossible to reconcile the neuroarchitecture of the branchiopod medulla with that of the other two taxa. Since the influential review by Elofsson and Dahl (1970) on this topic, several studies have readdressed this issue, either by collecting ontogenetic data on branchiopod taxa (Harzsch and Waloszek 2001; Harzsch 2002; Wildt and Harzsch 2002; reviewed in Harzsch and Hafner 2006) or by analysing the connectivity of the adult vision system of the taxa in question (Sinakevitch et al. 2003; Strausfeld 2005). This issue is far from settled and further complicated by the fact that we do not have a robust scenario about the evolutionary position of Branchiopoda with regard to Hexapoda and Malacostraca. Currently, three hypotheses have been put forward to account for the fundamental differences of the malacostracan/hexapod lamina on one side and that of Branchiopoda on the other:

(i)   There has been convergent evolution of the visual pathways associated with the compound eyes in Branchiopoda versus Malacostraca/Hexapoda (Nilsson and Osorio 1997).

(ii)  Evolutionary changes concerning the proliferative activity of stem cells that give rise to the optic anlagen are responsible for an axonal rewiring of the fibers between lamina and medulla (Elofsson and Dahl 1970; Harzsch 2002).

(iii) The branchiopod medulla does not correspond to the malacostracan/hexapod medulla but to a deeper optic neuropil (Strausfeld 2005).

In the light of the cellular similarities of the compound eyes and laminae in these three taxa, the first hypothesis seems unlikely. Strausfeld (2005) combined hypotheses (ii) and (iii) into a new scenario of optic neuropil evolution in Tetraconata with the fundamentally new idea that a mandibulate ancestor possessed only two visual neuropils, the plexiform layer and the visual tectum which correspond to the hexapod/malacostracan lamina and lobula plate, respectively (Fig. 13.5c). Both neuropils are connected by uncrossed fibers, an arrangement that characterizes Branchiopoda and Myriapoda (Melzer et al. 1996; Harzsch and Waloszek 2001; Harzsch 2002; Wildt and Harzsch 2002; Strausfeld 2005; Sombke et al. 2011a). The subsequent evolutionary scenario proposed by Strausfeld (2005) relies on the idea that Branchiopoda and Myriapoda represent a plesiomorphic character state from which the situation in Malacostraca and Hexapoda evolved. However, considering the unstable position of Branchiopoda in recent phylogenetic studies (Regier et al. 2010; Rota-Stabelli et al. 2011; Trautwein et al. 2012), we need to take into account that the architecture of the branchiopod visual system is derived and a

simplification from a more complex pattern. Furthermore, we know very little about the cellular architecture of the myriapod visual system beyond the simple facts that they have two visual neuropils and straight fibers, and therefore, we cannot claim that both share a similar neuroarchitecture representing an ancestral mandibulatan state.

Strausfeld (2005) proposed the following scenario for the evolution of the optic neuropils in the Tetraconata (Fig. 13.5c):

Step1: The malacostracan and hexapod medullae initially arose by a duplication of the outer optic anlagen, the proliferation zone of the lamina. This duplication led to a division of the ancestral plexiform layer into an outer and an inner stratum—the lamina and the nascent medulla, respectively. Due to the developmental organization of both layers, they are connected by means of a chiasm. The visual tectum now receives uncrossed projections from the inner layer.

Step2: The third optic neuropil, the lobula, is a protocerebral derivate and originated in a duplication event of the inner proliferation zone. It has been shown that this inner zone is separate from the outer one that generates the lamina (Nässel and Geiger 1983; Harzsch et al. 1999; Harzsch and Waloszek 2001). The lobula formed as an outgrowth of the lateral protocerebrum, as seen during development in some species. It is connected to the medulla via a chiasm, while the visual tectum is still linked by straight fibers. Based on structural similarities, the latter is regarded as the progenitor of the hexapodan and malacostracan lobula plate.

Step3: Within the hexapods, a reduplication of the inner optic anlagen gave rise to the proximal layer of the medulla.

In conclusion, Branchiopoda, Malacostraca, and Hexapoda are characterized by deep homologies of the cellular architecture of their compound eyes and laminae whereas strong differences of the deeper visual neuropils separate the Branchiopoda on the one side from

Malacostraca and Hexapoda on the other. It is very difficult to frame a simple evolutionary scenario that could transform the cellular architecture of the deeper branchiopod optic neuropil into that of Malacostraca/Hexapoda. This difficulty persists regardless of whether the branchiopod condition is plesiomorphic for Mandibulata or an apomorphy of Branchiopoda.

## 13.2.6 Olfactory Lobes

In the arthropod brain, the primary processing neuropils for chemosensory qualities are the olfactory lobes. In most bilaterians, olfactory receptor cells terminate in *glomerular neuropils* which are the subunits of the olfactory lobe (or olfactory bulb in Mammalia). In principle, a glomerulus is a spheroid synaptic complex that may be ensheathed by glia. Given their widespread phylogenetic distribution, glomeruli have either evolved once in a common ancestor or are a case of evolutionary convergence. The latter assumption points to a functional adaption related to processing olfactory information or a space-efficient architecture bringing together axons of similarly tuned receptor neurons (reviewed in Eisthen 2002). Olfactory glomeruli (OG) are also known in Mollusca (Wertz et al. 2006), Annelida (Heuer and Loesel 2009), Onychophora (Strausfeld et al. 2006b), and several Chelicerata (Brownell 1989) as well as Mammalia (Strotmann 2001). In general, olfactory receptor neurons (ORNs) are bipolar and project into a fluid medium within olfactory sensilla. In detail, however, there are striking differences between arthropod and vertebrate olfactory systems: (1) odorant binding proteins (OBPs) that mediate the transfer of ligands to receptors on the ORNs do not show any structural similarity in Hexapoda vs. Mammalia (Bianchet et al. 1996) and (2) odorant receptors (ORs) known from Hexapoda show no homology to the OR families of Mammalia and Nematoda (Hansson and Stensmyr 2011). This clearly points to a *convergent evolution of olfactory systems* in bilaterians (Strausfeld and Hildebrand 1999). Ionotropic receptors (IRs),

which occur in ORNs proposed to be the ancestral chemosensory receptor, are found only in protostomes and are absent in vertebrates (Croset et al. 2010). IRs are specifically divided into antennal IRs and divergent IRs which are expressed in peripheral and internal gustatory neurons.

Not all chemosensory input from antennae, walking appendages, and even wings is processed in the olfactory lobes of the brain. As a consequence, in arthropods the processing of chemosensory input is achieved in any neuromere that innervates chemosensory appendages. However, usually only specialized appendages lead to distinct olfactory lobes. In Mandibulata, these specialized appendages are the antennae associated with the deutocerebrum. Within several taxa of Chelicerata, olfactory lobes composed of OG are known in parts of the nervous system other than the deutocerebrum (Brownell 1989; Szlendak and Oliver 1992; van Wijk et al. 2006a, b; Wolf 2008; Strausfeld and Reisenman 2009). Here, OG occur, for example, in association with chemosensory walking appendages, like the first leg pair in Acari (Szlendak and Oliver 1992) or Solifugae (Strausfeld and Reisenman 2009) or the pectines in scorpions (see Sect. 13.1.2). In Onychophora, the antenna-like appendages supply chemosensory centers in the protocerebrum which are also composed of glomerular neuropils (Strausfeld et al. 2006b). However, the onychophoran antennae are not homologous to the mandibulate antennae (Scholtz and Edgecombe 2006).

The sensory deutocerebral antenna is an apomorphic character of Mandibulata (Scholtz and Edgecombe 2006). Grounded in a consistent architecture, the olfactory lobes within the deutocerebrum of Mandibulata have been suggested to be homologous structures (e.g. Schachtner et al. 2005; Strausfeld 2009; Sombke et al. 2012). The paired olfactory lobes of Mandibulata are usually located in the anterior or ventral deutocerebrum (Fig. 13.4). The array of OG in Hexapoda is thought to represent a chemotopic map, which forms the basis of the olfactory code (Galizia and Menzel 2000; 2001; Ignell and Hansson 2005; Galizia and Szyska

2008). The olfactory lobes or rather the OG are innervated by axons of ORNs from antennal olfactory and/or gustatory sensilla (Keil and Steinbrecht 1984; Tichy and Barth 1992; Hallberg and Skog 2011; Schmidt and Mellon 2011; Sombke et al. 2011b; Keil 2012). The fllowing architectural characteristics apply to both the olfactory system of insects and malacostracan crustaceans. Within the clearly demarcated dense OG, antennal ORNs terminate and form first synapses (Fig. 13.6). The input is integrated by local interneurons and then relayed to protocerebral neuropils via projection neurons (Schachtner et al. 2005). Local interneurons branch unilaterally within one, two, or even all OG resulting in connections of specific glomeruli. In addition, subclasses of interneurons can innervate certain regions of the OG (rim and core interneurons in Fig. 13.6a). Projection neurons connect single or several glomeruli with secondary processing centers such as the mushroom bodies via the projection neuron tract (PNT), also called antennocerebral tract in Hexapoda. In Malacostraca, the PNT (also called olfactory globular tract) targets the hemiellipsoid bodies (Galizia and Rössler 2010; Schmidt and Mellon 2011; Sandeman et al. in press; Strausfeld 2012; compare Fig. 13.6). In Tetraconata, the interconnection of primary and secondary processing centers is achieved by different pathways. While an ipsilateral connection is suggested to be plesiomorphic, in malacostracan Crustacea and Remipedia, a subset of neurons of the projection neuron tract projects to the contralateral hemiellipsoid body/medulla terminalis-complex (Fanenbruck and Harzsch 2005; Fig. 13.6a). In hexapods, several projection neuron tracts occur, the median, mediolateral, and lateral tracts (Galizia and Rössler 2010; compare Fig. 13.6b). In the honeybee, three different mediolateral tracts which target the lateral horn also branch in the lateral network (consisting of ring neuropil, triangle, and lateral bridge; compare Kirschner et al. 2006). The median and lateral tracts project either firstly into the MBs (lip- and basal ring region of the calyces) and secondly into the lateral horn, or vice versa (compare pathways in Fig. 13.6b).

Strausfeld (2012) listed a number of differences between hexapod and malacostracan OG. In most hexapods, each olfactory glomerulus gives rise to two or more uniglomerular projection neurons (with arborizations in only one glomerulus) whereas in malacostracan Crustacea, projection neurons are multiglomerular (with arborizations in several glomeruli). These multiglomerular projections might result in a higher discrimination capacity. Although in several tetraconate taxa (Crustacea + Hexapoda) olfactory lobes may be absent and structural differences occur, several shared characters are present that have been modified in many taxon-specific ways (Schachtner et al. 2005). The olfactory lobes of malacostracan Crustacea and neopteran Hexapoda share the following synapomorphies: (1) the OG are embedded in coarse neuropil, (2) ORNs are cholinergic, possess uniglomerular terminals, and penetrate the olfactory lobes in a radial manner from the periphery, (3) local interneurons are inhibitory, GABAergic or histaminergic, and contain neuropeptides as cotransmitters, (4) the olfactory lobe is innervated by at least one prominent serotonergic neuron (or dorsal giant neuron) with multiglomerular arborizations, (5) projection neurons (forming the projection neuron tract) pass the central body posteriorly and link the olfactory lobe with neuropils in the protocerebrum. Most of these characters are also present in representatives of the Myriapoda although projection neuron tracts linking the olfactory lobes with the MBs have not been demonstrated conclusively, most likely due to their diffuse arrangement of axons (Strausfeld et al. 1995). In this view, the absence of olfactory lobes in various Crustacea (for example in certain Branchipoda, Branchiura, and Thecostraca) and Hexapoda (Odonata, certain Hemiptera, and Coleoptera) can be interpreted as reductions (Sombke et al. 2012) within Tetraconata.

The shape and arrangement of OG are probably rather subjected to functional and/or physiological aspects than to phylogenetic constraints (Schachtner et al. 2005). Structural and physiological changes that lead to improved

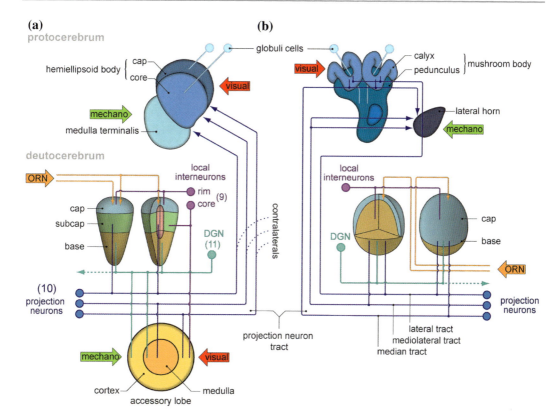

**Fig. 13.6** Overview of the central olfactory pathway in a malacostracan crustacean and a hexapod. The ORNs (*orange*) are the primary sensory input and innervate the cap of the olfactory glomeruli. Local interneurons (*purple*) and dorsal giant neurons (serotonergic, turquoise) are associated with the olfactory and the accessory lobe (in malacostracan crustaceans). Processed information is relayed from the olfactory lobe to the secondary computational centers via projection neuron tracts (*blue*). **a** *Cherax destructor* (Crustacea). Modified after Sandeman et al. in press. The olfactory glomeruli are compartmentalized into the cap, subcap, and base as well as the central rod (*red*). Local interneurons innervate specific compartments of the olfactory glomeruli, for example, the rim local interneurons. Core local interneurons relay information from the subcap to the cortex of the accessory lobe. The dorsal giant neuron (*DGN*) innervates the olfactory glomeruli as well as the accessory lobe. Olfactory information from the olfactory and accessory lobe is then relayed to the protocerebral medulla terminalis and the cap region of the hemiellipsoid body. In addition, the accessory lobe and the hemiellipsoid body's core region receive mechanosensory and visual input via interneurons. Furthermore, information from the accessory lobe and the core region of the hemiellipsoid body converges in the medulla terminalis. **b** *Apis mellifera* (Hexapoda). Compiled after Kirschner et al. (2006), Galizia and Szyska (2008). ORNs (*orange*) innervate the cap of the olfactory glomeruli from the periphery though they enter the core of the olfactory lobe and resurface between the glomeruli (as indicated). Local interneurons (*purple*) innervate the cap and base of the olfactory glomeruli. The dorsal giant neuron (serotonergic, turquoise) innervates multiple olfactory glomeruli. Different projection neuron populations (*blue*) relay information from the olfactory lobe to the mushroom body and the lateral horn. The lateral tract (multiglomerular) projects through the lateral horn into the calyces (with arborizations in the lip and basal ring). The median tract (uniglomerular) projects through the calyces (with arborizations in the lip and basal ring) into the lateral horn. The mediolateral tracts project into the lateral horn either through the lateral protocerebrum or with arborizations in the lateral network (not shown)

function drive phylogenetic change. This means that the shape of olfactory neuropils does not provide a stable phylogenetic signal as far as large-scale phylogeny of arthropods is concerned. However, a trend in transforming the shape of OG can be observed in interordinal relationships and thus could provide phylogenetically informative characters. This is for example the case when looking at decapod crustaceans. While a spheroid shape is present in

Onychophora and Chelicerata, tremendous diversity in shape and arrangements occurs within the Mandibulata. In Myriapoda, for example, the shape of OG ranges from elongated cylindrical in the Scutigeromorpha through drop-shaped to spheroid in the Geophilomorpha (Sombke et al. 2012). In centipedes, the glomeruli are arranged in a parallel or grape-like pattern (Fig. 13.4b). As in scutigeromorph Chilopoda, the olfactory lobe in Archaeognatha (Hexapoda) and Cephalocarida (Crustacea) is composed of elongated cylindrical glomeruli (Mißbach et al. 2011; Stegner and Richter 2011). In many pterygote hexapod species, the OG are spheroid and surround a coarse neuropil, for example, in Dictyoptera (Boeckh and Tolbert 1993), Hymenoptera (Galizia et al. 1999), Lepidoptera, and Diptera (reviewed in Schachtner et al. 2005) (Fig. 13.4d). In malacostracan Crustacea, the OG are arranged radially around the periphery of a loose core of neuronal processes. Interestingly, the trend of transforming OG seen in Chilopoda (elongated to spheroid) is found in the malacostracans as well, but according to the phylogenetic relationships in this taxon, it is reversed (spheroid to elongated). The shape ranges from spheroid in the Leptostraca (Strausfeld 2012), marine Isopoda, and Euphausiacea (Johansson and Hallberg 1992; Harzsch et al. 2011) across wedge-shaped in several reptantian Decapoda (Sandeman et al. 1992, 1993; Schmidt and Ache 1996a; Schachtner et al. 2005; Krieger et al. 2012) to markedly elongated columns which are aligned in parallel in eureptant Anomura (Harzsch and Hansson 2008; Krieger et al. 2010) (Fig. 13.4c). Moreover, in hermit crabs, the olfactory lobes can be enlarged by the presence of sublobes (Krieger et al. 2010). In Remipedia, the olfactory lobes are also divided into several sublobes, however, the shape of OG is roughly spheroid (Fanenbruck and Harzsch 2005).

Sexual dimorphism of olfactory lobes and OG is known in several neopteran Hexapoda, for example, cockroaches (Rospars 1988), moths (Rospars and Hildebrand 2000), or honeybees (Galizia et al. 1999) and have most likely occurred convergently (Schachtner et al. 2005).

Macroglomeruli (or macroglomerular complexes) are present in males and are innervated by specific sex-pheromone receptors on the antennae. The OG themselves can be compartmentalized. In honeybees (Hexapoda), OG have a layered organization (Pareto 1972; Arnold et al. 1985; Fonta et al. 1993; Sun et al. 1993; Galizia et al. 1999) where only the periphery (or cap) is innervated by sensory afferents (Fig. 13.6). Different populations of projection neurons and local interneurons innervate the central and peripheral areas (Fig. 13.6b). A longitudinal subdivision of the OG into cap, subcap, and base has been well documented in malacostracan crustaceans such as crayfish, clawed and clawless lobsters, hermit crabs, and brachyuran crabs (Sandeman and Luff 1973; Sandeman and Sandeman 1994; Langworthy et al. 1997; Schmidt and Ache 1997; Wachowiak et al. 1997; Harzsch and Hansson 2008; Krieger et al. 2010; 2012; compare Fig. 13.6a). In Archaeognatha and Chilopoda, the OG are not compartmentalized (Mißbach et al. 2011; Sombke et al. 2011c). The number of OG is thought to be species specific. In Chilopoda, the number per olfactory lobe ranges from 34 to 97 (Sombke et al. 2012), in Hexapoda from about 20 in Collembola to approx. 250 in ants (reviewed in Schachtner et al. 2005; Kollmann et al. 2011) and seems to be invariant within species (Chambille and Rospars 1981; Rospars 1983; Rospars and Hildebrand 1992; Galizia et al. 1999; Laissue et al. 1999; Berg et al. 2002; Huetteroth and Schachter 2005; Kirschner et al. 2006; Ghaninia et al. 2007; Zube et al. 2008; Dreyer et al. 2010). In Crustacea, the number of OG varies from approx. 150 to 1,300 (reviewed in Beltz et al. 2003; Schachtner et al. 2005; Krieger et al. 2010). It should be noted that crustaceans probably do not feature a species-specific constant number of OG (compare Blaustein et al. 1988; Beltz et al. 2003).

In addition to the olfactory lobes, several deutocerebral *accessory neuropils* occur in some tetraconate taxa. In eureptant Crustacea (e.g. Homarida, Brachyura and Achelata), large and complex accessory lobes occur (Figs. 13.4, 13.6a). In spiny lobsters, the accessory lobe

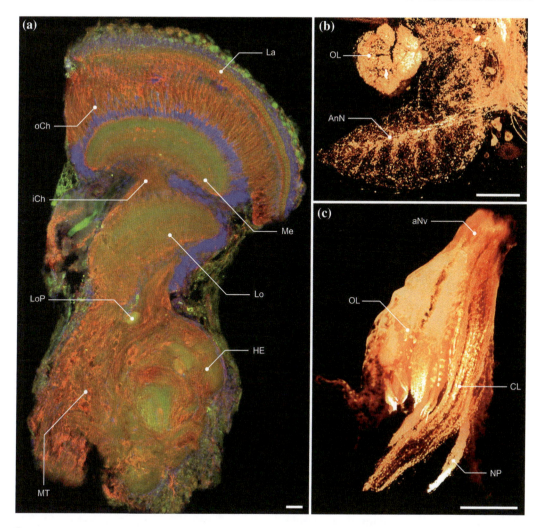

**Fig. 13.7 a** Lateral protocerebrum of *Carcinus maenas* (Crustacea: Decapoda) and optic neuropils (*red*: synapsin-like immunoreactivity, *green*: serotonin-like immunoreactivity, *blue*: nuclear stain). **b** Structural composition of the tritocerebral antenna 2 neuropil of *Idotea baltica* (Crustacea: Isopoda). The AnN is transversely divided into segment-like synaptic fields. The olfactory lobe is composed of spheroid olfactory glomeruli (3D reconstruction, RFamid-like immunoreactivity). **c** Antennal neurobiotin backfill of *Lithobius forficatus* (Myriapoda: Chilopoda) showing the structural composition in distinct lamellae of the deutocerebral *c. lamellosum*. Antennal afferents innervate the olfactory lobe as well as the subesophageal ganglia via antennal neurite projections (3D reconstruction). *AnN* antenna 2 neuropil, *aNv* antennal nerve, *CL c. lamellosum*, *HE* hemiellipsoid body, *iCh* inner optic chiasm, *La* lamina, *Lo* lobula, *LoP* lobula plate, *Me* medulla, *MT* medulla terminalis, *NP* neurite projections, *OL* olfactory lobe. *Scale bars*: 100 µm

(AcL) is composed of three neuropilar layers (Blaustein et al. 1988). The AcL receives input from the olfactory lobe via local interneurons (Sullivan and Beltz 2005). In *Cherax destructor* (Decapoda), it has been shown that the accessory lobe receives unilateral input from the proto- and deutocerebrum and bilateral input from the tritocerebrum. In addition, bilateral output to the contralateral olfactory and accessory lobe via the deutocerebral commissure occurs (Sandeman et al. 1995). Therefore, it is suggested that the AcL integrates mechano- and chemosensory information. In certain hemimetabolous Hexapoda, gustatory and probably olfactory input from the mouthparts is processed in the lobus glomerulatus (LG). The output is transferred via the tritocerebral tract to the mushroom bodies. In holometabolous hexapods, the LG is not present as a distinct neuropil but instead appears to be fused with the olfactory lobe (Farris 2008).

## 13.2.7 Mechanosensory Neuropils

Apart from their presence in the ventral nerve cord of Arthropoda (see above), mechanosensory neuropils are known from the brains of Mandibulata. The deutocerebrum is characterized by (at least) one bilaterally paired neuropil processing mechanosensory input from the first antennae (Fig. 13.4a–c). The deutocerebral mechanosensory neuropils have been called *dorsal lobe* in Hexapoda, *lateral antennular neuropil* in malacostracan Crustacea, and *corpus lamellosum* in Myriapoda. The general organization of the lateral antennular neuropil and the *c. lamellosum* in many respects matches the innervation and connections of the hexapod dorsal lobe. Therefore, these paired neuropils have been unified under the term deutocerebral mechanosensory neuropil (Sombke et al. 2012). In some mandibulate taxa, mechanosensory neuropils with a general striate or palisade shape are known, for example, in Zygentoma and Chilopoda (Tautz and Müller-Tautz 1983; Strausfeld 1998; Sombke et al. 2011a, 2012). In contrast to ORNs, mechanosensory neurons appear much thicker and possess several side branches.

In Chilopoda, the *c. lamellosum* is innervated by the posterior partition of the antennal nerve. The neuropil is composed of parallel neuropilar lamellae (Sombke et al. 2011a, b, 2012) including a contralateral connection. In malacostracan Crustacea, mechanosensory and non-olfactory input from the first antennae is processed in the lateral antennular neuropil (LAN) (Schmidt and Ache 1993, 1996b; Harzsch and Hansson 2008) which contains synaptic fields of the motor neurons that control the movements of the ipsilateral antennule (Sandeman et al. 1992; Schmidt et al. 1992). In Decapoda, contralateral connections between the LANs occur. In malacostracan Crustacea and Remipedia, an additional *median antennular*

*neuropil* (MAN) processes mechanosensory input (Sandeman et al. 1992, 1993; Schmidt and Ache 1996b, Fanenbruck and Harzsch 2005; Harzsch and Hansson 2008). In crabs and crayfish, it receives branches of interneurons related to input from the statocysts and mechanoreceptive input from the base of the antennae (Schmidt and Ache 1993; Schmidt et al. 1992). Whether the MAN of Malacostraca and Remipedia are homologous neuropils, is still debated. In pterygote Hexapoda, mechanosensory afferents from the scapus and pedicellus of the antennae project into the dorsal lobe (or AMMC = antennal mechanosensory and motor center). The dorsal lobe is also innervated by neurites of antennomuscular motoneurons. The flagellar sensilla whose neurons project into the olfactory lobe are mostly specialized for chemoreception (Rospars 1988; Homberg et al. 1989). Usually, the mechanosensory neuropil is located in the posterior region of the deutocerebrum, for example, in *Periplaneta americana* (Burdohan and Comer 1996; Nishino et al. 2005), *A. mellifera* (Kloppenburg 1995), *Gryllus bimaculatus* (Staudacher 1998; Staudacher and Schildberger 1999), and *Aedes aegypti* (Ignell and Hansson 2005; Ignell et al. 2005). In these organisms, presumptive tactile antennal afferents provide two pairs of long and several short branches which are orientated laterally and form a multilayered arrangement medially in the dorsal lobe.

In malacostracan Crustacea, the tritocerebral neuromere is characterized by the bilaterally paired *antenna 2 neuropil* (AnN), stretching posterolaterally to either side of the esophageal foramen (Fig 13.4c). Afferents ascending from the second antenna project into this neuropil which may have a specialized chemosensory function (reviewed in Schmalfuss 1998) in addition to its role in processing mechanosensory information (Sandeman and Luff 1973; Hoese 1989; Sandeman et al. 1992; Schmidt and Ache 1992; Schachtner et al. 2005). Moreover the tritocerebrum of Malacostraca and Remipedia is innervated by the tegumentary nerve which carries mechanosensory information from the carapace, and projects into the *tegumentary*

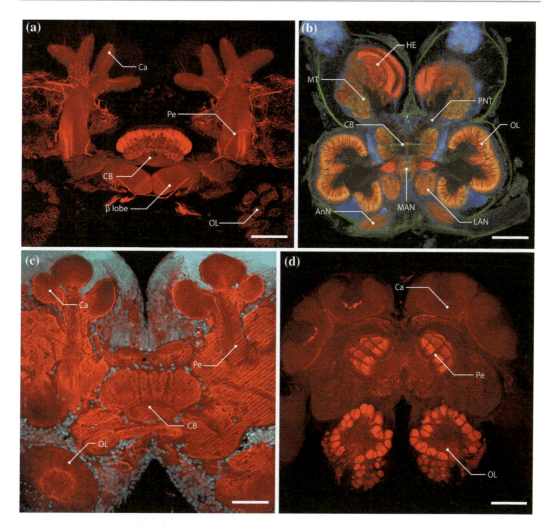

**Fig. 13.8** Sections through the brains of **a** the cockroach *Rhyparobia maderae* (allatostatin-like immunoreactivity) from Loesel and Heuer (2010), **b** *Birgus latro* modified from Krieger et al. (2010) (*red*: synapsin immunoreactivity, *green*: allatostatin-like immunoreactivity, *blue*: nuclear stain), **c** the desert locust *Locusta migratoria* (*red*: HRP-like immunoreactivity, *blue*: nuclear stain), and **d** the ant *Camponotus floridanus* (allatostatin-like immunoreactivity). *Ca* calyx of mushroom body, *CB* central body, *OL* olfactory lobe, *Pe* peduncle of mushroom body. Scale bars: **a, c** 200 μm, **d** 100 μm, **b** 500 μm

*neuropil*, located within the complex of the AnN (Sandeman et al. 1992). In some representatives of Decapoda and Isopoda, the AnN is transversely divided into repetitive synaptic fields (Tautz and Müller-Tautz 1983; Harzsch et al. 2011; Sandeman et al. in press; see Fig. 13.7). It has been suggested that this organization might be a somato- or spatiotopic representation of the mechanoreceptors along the length of the second antenna. Evidence for this comes from behavioral studies on blinded crayfish, which precisely touch the point of the antenna where they have been stimulated (Zeil et al. 1985; Sandeman and Varju 1988). For Remipedia, not much is known about the tritocerebral morphology, although in two species two pairs of tritocerebral nerves have been identified (i.e. tegumentary nerves and antenna 2 nerves; Fanenbruck and Harzsch 2005) suggesting the presence of associated neuropils (antenna 2 neuropil and tegumentary neuropil). Tritocerebral neuropils that are associated with the antenna 2 and tegumentary nerves, suggestive of a tegumentary neuropil, have also been reported for Branchiopoda (Fritsch and Richter 2010) and Cephalocarida (Stegner and Richter 2011).

In Hexapoda and Myriapoda, the tritocerebrum lacks tritocerebral appendages and consequently associated neuropils. In Onychophora and Chelicerata, distinct mechanosensory neuropils associated with antennae (Onychophora) and cheliceres/pedipalps (Chelicerata) have not been described satisfactorily. Whereas the pedipalp provides mechanosensory input into what can be regarded as the chelicerate tritocerebrum, nothing is known about the corresponding brain regions in Onychophora that are connected to the jaws (see Sect. 13.2).

## 13.2.8 The Mushroom Bodies

The mushroom bodies (MB) are the most prominent and conspicuous neuropils in the central brain of arthropods, onychophorans, and vagile polychaete annelids but have not been described in any other animal group with complex brain architecture. Due to a number of unique neuroanatomical characters, MBs can easily be identified and distinguished from other brain centers.

A good starting point for a morphological comparison of MB structures is the insects, where the literature on brain architecture and function is vast. In this group, the MBs are located in the protocerebrum (Figs. 13.4d, e, 13.6b, 13.8a, c, d). They act as centers for sensory integration (Gronenberg 2001), memory formation (Heisenberg 2003), and represent the neuronal basis for associative and flexible behaviors (Farris and Roberts 2005). With the exception of the archaeognathans (see below), the remaining insect taxa share a common ground plan in terms of mushroom body cellular architecture and connectivity. MBs consist of several thousand parallel fibers of intrinsic neurons, called Kenyon cells. Their perikarya are densely packed and surround the calyces which contain the dendritic arborizations of the Kenyon cells. The calyces represent the major synaptic input region to the MBs. The most

prominent inputs to the calyces originate in the antennal lobes through collaterals of olfactory interneurons that connect the antennal lobe with the protocerebrum via an projection neuron tract. MBs, however, are not merely higher order olfactory neuropils, but are present even in anosmic insects (Strausfeld et al. 1998). In a variety of social hymenopterans and in the cockroach *Periplaneta americana,* additional inputs originate in the optic lobes.

The axons of Kenyon cells project from the calyx into the peduncle. They then bifurcate and form the lobes (usually an α- and a β-lobe), the major output regions of the MBs (Laurent and Naraghi 1994). This basic neuroanatomical motif is highly conserved and has been described in all insect representatives investigated so far (Strausfeld 1998; Farris and Sinakevitch 2003; Strausfeld 2012).

Brain centers that adhere to the architectural designs of insect MBs are found in other arthropod groups (albeit not analyzed in as much detail) including myriapods (Holmgren 1916; Hanström 1928; Strausfeld et al. 1995; Loesel et al. 2002) and chelicerates (in the older literature referred to as 'corpora pedunculata'; Holmgren 1916; Hanström 1928; Strausfeld and Barth 1993; Wegerhoff and Breidbach 1995; Strausfeld et al. 2006a, b). In crustaceans, on the other hand, second-order olfactory neuropils connected to the antennal lobes differ in morphology from the insect MBs. Since crustaceans are the sister taxon to the hexapods, or even the group containing the hexapods (Dohle 2001; Giribet et al. 2001; Loesel et al. 2002; Regier et al. 2005; Ungerer and Scholtz 2008), these apparent differences have resulted in conflicting views on the evolutionary origin of insect MBs and their possible homology to their namesakes in myriapods and chelicerates.

In the Malacostraca (higher crustaceans) and in the homonomously segmented Remipedia, interneurons originating in the olfactory lobes innervate the so-called hemiellipsoid bodies (Figs. 13.4c, e, 13.6a, 13.8b) that either reside in the eyestalks or in the protocerebrum as demonstrated for several crab species. The hemiellipsoid bodies are associated with thousands of densely packed perikarya of globuli cells (the equivalent of Kenyon cells in insects), the ramifications of which contribute to the internal matrix of the hemiellipsoid bodies. This neuropil, however, does not comprise a peduncle and elongated lobes, thus being dissimilar to the external shape of the hexapod MB (Hanström 1928; Sandeman and Scholtz 1995; Strausfeld et al. 1995; Fanenbruck and Harzsch 2005). Since the neuronal organization of the hemiellipsoid body, however, is similar to the cellular architecture of the hexapod MB, hemiellipsoid bodies are today being interpreted by most specialists as modified MBs (Wolff et al. 2012). This view is supported by the fact that the brain of a basal hexapod taxon, the Archaeognatha, contains second-order olfactory neuropils which resemble hemiellipsoid bodies rather than MBs (Strausfeld 2012).

Investigations into myriapod brain anatomy (Fig. 13.4b) are scarce, and descriptions of MB neuropils are available only for a limited number of species. Diplopods (*Orthoporus ornatus, Julus scandinavius*) and chilopods (*Lithobius variegatus*) exhibit clusters of small-diameter globuli cells that supply ramifications to MBs which comprise a pedunculus and lobes and which are connected to the antennal lobes via a tract of interneurons (Hanström 1928; Strausfeld et al. 1995). These commonalities suggest close affinities with the insect MB. In *Lithobius variegatus,* the lobes have been described to represent spherical outswellings, a motif similar to the MB organization of the apterygote hexapod *Lepisma saccharina*, where the pedunculus provides several outswellings as well (Böttger 1910; Strausfeld et al. 1995).

Amongst the chelicerate taxa, the neuronal architecture of the MBs has probably been most thoroughly investigated in the xiphosuran *L. polyphemus* (Hanström 1926; Fahrenbach 1977, 1979; Chamberlain and Wyse 1986; Fahrenbach and Chamberlain 1987). Dwarfing the MBs of other arthropod and non-arthropod species alike, the MBs in adult horseshoe crabs are composed of an estimated 100 Mio globuli (=Kenyon) cells (Fahrenbach 1979) and account for approximately 80 % of the total brain volume (Hanström 1926). The aggregated somata of the globuli cells form a ventral hemisphere which enfolds the elaborately lobed neuropil in a fashion that has been likened to a cauliflower. The orientation of the MBs in *L. polyphemus* is clearly ventral—a unique condition that has not been observed in any other MBs investigate so far. Despite their highly deviant morphology, the xiphosuran MBs seem to be involved in the same tasks as their hexapod counterparts. Lacking antennae, the input to the MBs is provided by glomeruli that receive terminals from chemoreceptors located in the legs and gills (Fahrenbach 1979). In other chelicerates, chemoreceptors are located in specialized organs (pectines in scorpions, malleoli in solpugids) or modified walking limbs (uropygids and amblypygids). Similar to the condition observed in *L. polyphemus*, the OG in these taxa are not situated within the brain but in the segmental neuromeres associated with the olfactory appendages (Strausfeld et al. 1998; Wolf 2008; and Sect. 13.1.2). Ascending axons of projection neurons relay the information to the protocerebral MBs. In derived araneans such as the wandering spider *C. salei* (Fig. 13.4a), evidence for an association of the MBs with chemosensory pathways is lacking. The neuropils receive, however, direct input from a second-order visual neuropil (Strausfeld and Barth 1993). The changed connectivity suggests that the MBs in this group might have undergone major evolutionary changes with regard to their function.

Direct connections to a second-order visual neuropil are also evident in onychophoran MBs (Strausfeld et al. 2006a, b). As in arthropod taxa, onychophoran MBs are composed of a peduncle and several output lobes formed through parallel axons of a dense cluster of small-diameter perikarya of intrinsic neurons (globuli cells). One characteristic the MBs of onychophorans share with those of chelicerates but with no other arthropod group is the presence of a commissure that renders the MBs into one confluent structure.

Although architectural differences in the MBs of hexapods, myriapods, chelicerates, onychophorans, and the hemiellipsoid bodies of crustaceans still stimulate discussions about the homology of these brain centers, the assumption of a common phylogenetic origin nevertheless seems a plausible working hypothesis due to the many commonalities that have been described. Hence, we suggest that in the ground pattern of the common ancestor of arthopods, a basal computational circuit was present that included olfactory afferents, local olfactory interneurons, second-order olfactory projection neurons, and intrinsic MB cells (Kenyon- or globuli cells). This ancestral circuitry has been retained in all osmic panarthropods and was elaborated in different functional directions at least within Tetraconata, as the difference in the structure between the olfactory system in hexapods and crustaceans suggest.

This implies the possibility that arthropod MBs are symplesiomorphic and may be of a more ancient evolutionary origin than the arthropods themselves. This idea has recently been supported by the finding that the neuronal architecture of MB-like neuropils described in vagile polych-eates is almost identical to that of insects (Heuer and Loesel 2008) and that insect and annelid MBs express the same specific set of genes that orchestrate MB development (Tomer et al. 2010). Taken together these findings suggest a deep-time origin of MBs possibly dating back to the last common ancestor of protostomes, or beyond.

## 13.2.9 The Central Body

All major arthropod groups possess a brain midline neuropil called the central body. Several lines of evidence from behavioral and comparative studies suggest that the central body serves as a motor control centers that is involved in orchestrating limb actions. Based on its neuronal architecture, the central body of various arthropod groups is characterized by several distinct features that will be summarized below.

Again, the most thoroughly investigated taxa are insects and crustaceans. Here, the so-called 'central complex' is an assemblage of unpaired midline neuropils that comprises the central body and the protocerebral bridge (Figs. 13.4d, e,

13.8a–c). The central body itself is subdivided into several layers and columns. Connections of the central body to the protocerebral bridge are provided via columnar neurons that form a complicated but highly conserved pattern of chiasmata. The entire complex receives and provides axons from and to other protocerebral neuropils. The most prominent of these satellite neuropils are the paired noduli and the lateral accessory lobes. Presumably, the lateral accessory lobes are the sites where the central body connects to ascending and descending fibers from and to the thoracic ganglia. These features are common in all investigated neopteran insects and have been demonstrated in several decapod crustaceans. (For a synopsis of literature on central complex neuroarchitecture of individual species see Loesel et al. 2002). Especially in flies and locusts, the neuroarchitecture of the central complex has been analyzed to an extent that made it possible to construct a wiring diagram of this brain region. While the number and arrangement of columns is identical in insects and decapods (Utting et al. 2000), there are differences in the number and shape of layers of the central body which, especially in decapods, differs even between closely related species (Loesel et al. 2002). It will be an interesting challenge for future research to homologize individual layers between different species and correlate the findings with the behavioral repertoire of the animals.

Immunohistochemical studies demonstrated the presence of a wide variety of neuroactive substances in different parts of the central complex (Homberg 1994). In addition to biogenic amines such as octopamine, serotonin, dopamine, and allatostatin, to name just a few, neurotransmitters like histamine (Loesel and Homberg 1999) and GABA (Homberg et al. 1999) have been found. Immunocytochemistry proved to be a powerful phylogenetic tool that has helped to establish homologies between subcompartments of the central complex in different species. GABA, for example, has been shown to be present only in one central body layer, the so-called ellipsoid body (also termed lower division) in all investigated insect species (Homberg et al. 1999).

Despite our detailed knowledge about the central complex neuroarchitecture, its role in controlling the animal's behavior is only partially understood, especially when it comes to assigning certain functions to individual subunits of the central complex. In intracellular recordings from locusts and bees, a subset of neurons of the central body respond to multimodal stimuli and the e-vector of polarized light (Homberg 1985; Milde 1988; Vitzthum et al. 2002; Heinze and Homberg 2007). The central body of flies incorporated radioactive-labeled deoxyglucose during visual stimulation, indicating that the metabolic rate of central body neurons was elevated at this time (Bausenwein et al. 1994). Such studies suggest that the central complex receives visual input. It is, however, unlikely that its main function is visual information processing since eyeless workers of the ant *Mystrium* sp. (Gronenberg, personal communication) and the blind cave beetle *Neaphaenops tellkampfii* (Ghaffar et al. 1984) possess a well-developed central complex.

Functions ascribed to the central complex come mainly from studies of behavioral mutants of *Drosophila*, where specific deficits in limb coordination relate to structural defects in the central complex. Especially, the fly's ability to execute asymmetrical limb movements (e.g. in turning) is severely impaired in animals with a damaged central complex (Strauss and Heisenberg 1990, 1993; Strauss et al. 1992; reviewed by Strauss 2002). In these mutants, the overall locomotor activity and the ability to retain direction towards a landmark, which becomes invisible during approach, are also diminished (Strauss 2003). Comparative studies further support the notion that the central complex relates to leg coordination. In cell-building social insects that can perform complicated and heterolaterally independent limb movements, the cellular organization of the individual subunits of the central complex is elaborate. In nocturnal Lepidoptera, which mainly use their legs for grasping but not for walking, central body layers and the protocerebral bridge are reduced (Strausfeld 1999). In aquatic Hemiptera, which primarily perform bilaterally coupled swimming strokes, central complex elaboration is significantly impoverished (Strausfeld 1999).

Similar studies have been carried out in decapod crustaceans. Fiddler crabs of the genus *Uca* exhibit strong sexual dimorphism in the size of their claws. Male fiddler crabs have one large front claw and one small one, while females have two small claws. The male fiddler crab waves this claw and wrestles other males to mark his territory and attract mates. The small claw is needed for gathering food. Since males use their front claws for two different tasks, they have a higher ability to uncouple the movements of right and left front claws as compared to females. This difference in locomotor abilities possibly correlates with a pronounced dimorphism in the relative size and shape of the central body (Loesel 2004). Such a dimorphism has not been observed in any of the other investigated crustacean genera. Together, the available data suggest that the central complex is a higher brain center of insects and malacostracan crustaceans for navigational control and limb coordination that is especially involved in locomotor patterns that require heterolaterally independent leg movements.

In all investigated species of the Chelicerata (spiders, scorpions, horseshoe crabs), the central body (for historical reasons often referred to as 'arcuate body') is a crescent-shaped, unpaired neuropil extending across the entire width of the brain (Fig. 13.4a). In each representative, the central body was found to be the posterior-most cerebral neuropil. A protocerebral bridge and satellite neuropils such as noduli or ventral bodies have not been found. A variety of staining techniques reveal that the neuroarchitecture of the central body of chelicerates is characterized by a palisade-like arrangement of a large number of columnar fibers that project through consecutive layers. These layers are demarcated by successive strata of tangential neurons and by the density of presynaptic terminals. Unlike in insects and decapods, columnar neurons of chelicerates are not bundled into discrete columns but innervate the central body uniformly across its entire lateral extension. Immunocytochemical studies revealed that a large number of neuroactive substances such as allatostatin, proctolin, crustacean-cardioactive peptide, and GABA that have been detected in the insect central body are likewise present in the central body of spiders (Loesel et al. 2002, 2011). While the spider's central body receives direct visual input from second-order optic neuropils (Strausfeld et al. 1993), its main function seems to be locomotor control, too. The evidence is somewhat anecdotal, but comparative studies have demonstrated that chelicerate central body elaboration and relative size correlate to the animal's motoric repertoire (Strausfeld 2012). Besides striking similarities in neuroarchitecture and transmitter content, a further argument for a common origin of the central bodies in chelicerates and in the hexapod–crustacean clade is the presence of a central body in even the most basal chelicerate representatives such as the horseshoe crab *L. polyphemus* and the segmented spider *Heptathela kimurai*.

Onychophorans which are the sister group to arthropods have a well-developed central body, as well. A thorough analysis of the brain of *Euperipatoides rowellii* revealed that the basic neuroarchitecture of the central body of this velvet worm is practically indistinguishable from that of chelicerates. As is the case in spiders, allatostatin is a major neurotransmitter in certain layers of the central body in this onychophoran species, too (Strausfeld et al. 2006a, b, c).

From the morphological point of view, the central body of chilopods (Fig. 13.4b) is an intermediate between the central bodies of the Tetraconata and the Chelicerata/Onychophora. In several representatives of the Chilopoda the central body is a roughly hemiellipsoid midline neuropil that is situated between the proximal tips of the mushroom body's medial lobe. It consists of several horizontal layers and is innervated by columnar fibers. Two classes of columnar neurons can be distinguished. One (allatostatin-immunoreactive) class innervates the central body without crossing the trajectory of another allatostatin-ir fiber. The second subset of columnar neurons forms a system of interweaving fibers across the entire lateral extension

of the central body. A protocerebral bridge or any other satellite neuropils have not been identified. Columnar neurons of the chilopod central body, however, are in close spatial contact to commissures that connect both hemispheres of the brain and might retrieve information from there (Loesel et al. 2002). The anatomy of the central body of centipedes bears resemblance to the central bodies of chelicerates as well as of that of insects. The position, external shape, and relative size of the centipede central body are comparable to the conditions found in insects. On the other hand, the way the columnar neurons innervate consecutive layers of the central body is reminiscent of the situation found in chelicerates. A shared feature in all these taxa is the presence of allatostatin-immunoreactive columnar fibers.

Several staining techniques failed to identify a midline neuropil in the brain of the second major myriapod clade, the Diplopoda (Loesel et al. 2002). Here, the entire midbrain consists of numerous commissural tracts that connect both hemispheres. This would require a secondary loss of the central body in the diplopod brain. The hypothesis is at least plausible since diplopods do not change the locomotor pattern of their many legs when turning, as is the case in chilopods. Diplopods rather bend their body, presumably by contracting their lateral body muscles, and their legs just follow this curve. Hence, necessity for a brain control center that uncouples bilateral limb movements would not be there.

Taking all the available data together, it seems highly unlikely that the central body has evolved several times during arthropod evolution. Beside striking similarities in its neuroanatomical Leitmotif, the shared neurotransmitter equipment (e.g. allatostatin is always found in columnar fibers) argues for a one-time evolutionary event that brought about the progenitor of the central body. Another strong argument is the fact that a central body has been found even in the most basal representatives of crustaceans, hexapods, and chelicerates. In crustaceans, the central body neuroarchitecture of basal groups is not well investigated. However, in the branchiopod *Triops longicaudatus,* a central body is certainly present

(Loesel 2004) and it contains allatostatin and tachykinin-related peptide as is also the case in decapods and hexapods.

What did the ancestral central body look like? Many basal features may have been retained in the central body of chilopods. Its palisade-like arrangement of columnar neurons that possibly retrieve information form commissural fibers and subsequently innervate a small and layered midline neuropil may initially have served to compare neuronal information from both sides of the brain. To enable this, a subset of columnar neurons evolved into a chiasmatic pattern of interweaving fibers that has finally been organized into discrete columns. In the Tetraconata, the protocerebral bridge and satellite neuropils have been added. In chelicerates and onychophorans, the ground pattern that has been retained in the chilopod central body may have been conserved but elaborated with respect to the neuropil's relative size, width, and stratification.

The available data suggest that the neuroanatomical Leitmotif of the central body has been highly conserved during arthropod evolution and originated at least 600 million years ago, well before the first terrestrial arthropods emerged. Small central midline neuropils that are in some respects reminiscent of the arthropod central body have recently been described in predatory polychaetes that use their parapodia for rapid locomotion during hunting (Heuer et al. 2010). This might be another piece of evidence that the central body evolved together with the ability to coordinate limb movements.

## 13.3 Phylogenetic Overview and Outlook

Despite three decades of molecular phylogenetic analyzes of arthropod relationships using constantly increasing data sets and analysis methods (recent examples are Koenemann et al. 2010; Regier et al. 2010; Rota-Stabelli et al. 2011; Trautwein et al. 2012; von Reumont et al. 2012), we are still far from a reliable and robust hypothesis on arthropod main clade phylogeny (see Chap. 2). Arthropod neuroanatomists

realized early that the arthropod nervous system includes a wealth of structures that can be used both for analysing the phylogeny of arthropods and for describing evolutionary transformations within the arthropod brain (Holmgren 1916; Hanström 1928). In the past 25 years, structure and development of the arthropod nervous system have made a major contribution to the raging debate on arthropod phylogeny and the discipline of 'neurophylogeny' attempts to synthesize neurobiological questions and evolutionary aspects (e.g. Arbas et al. 1991; Breidbach and Kutsch 1995; Whitington 1996; Nilsson and Osorio 1997; Strausfeld 1998; Loesel 2004; Harzsch and Hafner 2006; Harzsch 2006, 2007; Loesel 2006, 2011; Strausfeld 2009; Strausfeld and Andrews 2011; Strausfeld 2012). In the following, we summarize architectural features of the nervous system that we consider to be part of the ground pattern of Arthropoda.

### 13.3.1 Ground pattern of the Arthropod Nervous System

(i) The three preoral neuromeres of the arthropod brain are the protocerebrum (ocular segment), deutocerebrum (chelicera segment in Chelicerata, first antennal segment in Mandibulata), and tritocerebrum (pedipalp segment in Chelicerata, second antennal segment in Crustacea, intercalary segment in Hexapoda and Myriapoda)

(ii) The axons of bilaterally symmetric median eyes project into a protocerebral neuropil (the median eye center). The median eyes can be paired or unpaired (ocellar ganglia in Xiphosura, nauplius eye center in Branchiopoda, ocellar plexus in Pterygota). The median eye center is innervated by interneurons with somata located anteriorly (dorsal median group in Xiphosura, cluster 6 in Crustacea, pars intercerebralis in Hexapoda)

(iii) The lateral eyes are associated with two optic neuropils which are linked by straight fibers and provide input to the protocerebrum

(iv) During growth of the lateral eye, new elements are added to the side of the existing eye field and elongate the rows of earlier generated units

(v) Photoreceptors of both median and lateral eyes use histamine as neurotransmitter

(vi) The central body is enwrapped by layers of neuronal somata and is also innervated by columnar neurons with somata located in the anterior soma cluster (see median eye center somata)

(vii) A preoral frontal commissure (e.g. stomatogastric bridge) is present which is composed of deuto- and tritocerebral fibers and gives rise to nerves innervating the hypostome, esophagus, and the anterior part of the gut

(viii) In the ventral nerve cord, an anterior and posterior group of serotonergic neurons with variable number is present in each hemiganglion, which is connected by transversely linked fibers to its contralateral unit

(ix) Excitatory motoneurons supplying appendages are arranged in an anterior and a posterior soma cluster (according to the muscles they innervate)

(x) In addition to excitatory motoneurons, single or small groups of inhibitory motoneurons are present that also innervate the appendage muscles.

A number of character complexes related to the nervous system promise to contribute meaningful data for future in-depth phylogenetic analyzes. Studying mechanisms of neurogenesis such as stem cell proliferation and growth of pioneer neurons has already provided important insights into arthropod relationships. At the cellular level, individually identifiable neurons in the ventral nerve cord and the brain which can be identified by their transmitter expression and the morphology of which can be mapped by confocal laser scan analysis have a high potential to unravel homologies between arthropod

taxa. At the level of brain neuropils and associated sensory systems, structures that are in the focus of comparative analyzes are the optic neuropils and optic chiasmata, the architecture and connectivity of the central complex, olfactory, and mechanosensory systems as well as the mushroom and hemiellipsoid bodies. It will be vital in the future to analyze these structures in more and more non-model organisms.

# References

Abzhanov A, Kaufman TC (2004) *Hox* genes and tagmatization of the higher Crustacea (Malacostraca). In: Scholtz G (ed) Evolutionary developmental biology of Crustacea (Crustacen issues 15). Balkema, Lisse, pp 41–74

Arbas EA, Meinertzhagen IA, Shaw SR (1991) Evolution in nervous systems. Annu Rev Neurosci 14:9–38

Arnold G, Masson C, Budharugsa S (1985) Comparative study of the antennal lobes and their afferent pathway in the worker bee and the drone (*Apis mellifera*). Cell Tiss Res 242:593–605

Autrum H (1979) Comparative physiology and evolution of vision in invertebrates. A: Invertebrate photoreceptors. Springer, Berlin

Barth FG (2001) Sinne und Verhalten—Aus dem Leben einer Spinne. Springer, Berlin

Barth FG (2002) A spider's world: senses and behavior. Springer, Berlin

Barth FG (2012) Spider strain detection. In: Barth FG, Humphrey JAC (eds) Frontiers in sensing. Springer, Wien, pp 251–273

Barth FG, Schmid A (2001) Ecology of sensing. Springer, Berlin

Battelle B-A (2006) The eyes of *Limulus polyphemus* (Xiphosura, Chelicerata) and their afferent and efferent projections. Arthropod Struct Dev 35:261–274

Bausenwein B, Müller NR, Heisenberg M (1994) Behavior-dependent activity labeling in the central complex of *Drosophila* during controlled visual stimulation. J Comp Neurol 340:255–268

Belanger JH (2005) Contrasting tactics in motor control by vertebrates and arthropods. Integr Comp Biol 45:672–678

Beltz BS, Kordas K, Lee MM, Long JB, Benton JL, Sandeman DC (2003) Ecological, evolutionary, and functional correlates of sensilla number and glomerular density in the olfactory system of decapod crustaceans. J Comp Neurol 455:260–269

Berg BG, Galizia CG, Brandt R, Mustaparta H (2002) Digital atlases of the antennal lobe in two species of tobacco budworm moths, the oriental *Helicoverpa assulta* (male) and the American *Heliothis virescens* (male and female). J Comp Neurol 446:123–134

Bianchet MA, Bains G, Pelosi P, Pevsner J, Snyder SH, Monaco HL, Amzel LM (1996) The three-dimensional structure of bovine odorant binding protein and its mechanism of odor recognition. Nature Struct Mol Biol 3:934–939

Bitsch C, Bitsch J (2005) Evolution of eye structure and arthropod phylogeny. In: Koenemann S, Jenner RA (eds) Crustacea and arthropod relationships. Taylor & Francis, New York, pp 81–111

Bitsch J, Bitsch C (2007) The segmental organization of the head region in Chelicerata: a critical review of recent studies and hypotheses. Acta Zool 88:317–335

Bitsch J, Bitsch C (2010) The tritocerebrum and the clypeolabrum in mandibulate arthropods: segmental interpretations. Acta Zool 91:249–266

Blaustein DN, Derby CD, Simmons RB, Beall AC (1988) Structure of the brain and medulla terminals of the spiny lobster *Panulirus argus* and the crayfish *Procambarus clarkii* with an emphasis on olfactory centers. J Crustac Biol 8:493–519

Boeckh J, Tolbert LP (1993) Synaptic organization and development of the antennal lobe in insects. Microsc Res Tech 24:260–280

Borst A, Euler T (2011) Seeing things in motion: models circuits, and mechanisms. Neuron 71:974–994

Borst A, Haag J, Reiff DF (2010) Fly motion vision. Annu Rev Neurosci 33:49–70

Bossing T, Udolph G, Doe CQ, Technau GM (1996) The embryonic central nervous system lineages of *Drosophila melanogaster*. I. Neuroblast lineages derived from the ventral half of the neuroectoderm. Dev Biol 179:41–64

Böttger O (1910) Das Gehirn eines niederen Insektes (*Lepisma saccharina* L.). Jenaer Ztschr Naturwiss 46:801–844

Bräuning P, Pflüger H-J, Hustert R (1983) The specificity of central nervous projections of locust mechanoreceptors. J Comp Neurol 218:197–207

Breidbach O, Kutsch W (1995) The nervous systems of invertebrates: an evolutionary and comparative approach. Birkhäuser Verlag, Basel

Breithaupt T, Thiel M (2011) Chemical communication in crustaceans. Springer, New York

Brownell PH (1989) Glomerular cytoarchitectures in chemosensory systems of Archnids. Ann N Y Acad Sci 855:502–507

Bullock TH, Horridge GA (1965) Structure and function in the nervous system of invertebrates, vol II. Freeman, San Francisco

Burdohan JA, Comer CM (1996) Cellular organization of an antennal mechanosensory pathway in the cockroach *Periplaneta americana*. J Neurosci 16: 5830–5843

Burrows M (1996) Neurobiology of an insect brain. Oxford University Press, Oxford

Burrows M, Newland PL (1993) Correlation between the receptive fields of locust interneurons, their dendritic

morphology, and the central projections of mechanosensory neurons. J Comp Neurol 329:412–426

Callaway JC, Stuart AE (1999) The distribution of histamine and serotonin in the barnacle's nervous system. Microsc Res Tech 44:94–104

Chamberlain SC, Wyse GA (1986) An atlas of the brain of the horseshoe crab *Limulus polyphemus*. J Morphol 187:363–386

Chambille I, Rospars JP (1981) Le deutocerebron de la blatte *Blaberus craniifer* Burm. (Dictyoptera: Blaberidae). Étude qualitative et identification visuelle des glomerules. Int J Insect Morphol Embryol 10:141–165

Clarkson ENK, Levi-Setti R, Horváth G (2006) The eyes of trilobites; the oldest preserved visual system. Arthropod Struct Dev 35:247–259

Croset V, Rytz R, Cummins SF, Budd A, Brawand D, Kaessmann H, Gibson TJ, Benton R (2010) Ancient protostome origin of chemosensory ionotropic glutamate receptors and the evolution of insect taste and olfaction. PLoS Genet 6(8):e1001064. doi: 10.1371/journal.pgen.1001064

Damen WGM, Hausdorf M, Seyfarth E-A, Tautz D (1998) A conserved mode of head segmentation in arthropods revealed by the expression pattern of *Hox* genes in a spider. Proc Natl Acad Sci U S A 95:10665–10670

Denes AS, Jékely G, Steinmetz PRH, Raible F, Snyman H, Prud'homme B, Ferrier DEK, Balavoine G, Arendt D (2007) Molecular architecture of annelid nerve cord supports common origin of nervous system centralization in Bilateria. Cell 129:277–288

Doeffinger C, Hartenstein V, Stollewerk A (2010) Compartmentalization of the prech, eliceral neuroectoderm in the spider *Cupiennius salei*: development of the arcuate body, optic ganglia, and mushroom body. J Comp Neurol 518:2612–2632

Dohle W (2001) Are the insects terrestrial crustaceans? a discussion of some new facts and arguments and the proposal of the proper name 'Tetraconata' for the monophyletic unit Crustacea + Hexapoda. Ann Soc Entom France (NS) 37:85–103

Douglas JK, Strausfeld NJ (2003) Anatomical organization of retinotopic motion-sensitive pathways in the optic lobes of flies. Microsc Res Tech 62:132–150

Dreyer D, Vitt H, Dippel S, Goetz B, el Jundi B, Kollmann M, Huetteroth W, Schachtner J (2010) 3D standard brain of the red flour beetle *Tribolium castaneum*: a tool to study metamorphic development and adult plasticity. Front Syst Neurosci 4:3. doi: 10.3389/neuro.06.003.2010

Duman-Scheel M, Patel NH (1999) Analysis of molecular marker expression reveals neuronal homology in distantly related arthropods. Development 126:2327–2334

Egelhaaf M (2006) The neural computation of visual motion information. In: Warrant E, Nilsson D-E (eds) Invertebrate vision. Cambridge University Press, Cambridge, pp 399–462

Egelhaaf M, Kern R, Lindemann J, Braun E, Geurten B (2009) Active vision in blowflies: strategies and mechanisms of spatial orientation. In: Floreano D, Zufferey J-C, Srinivasan MV, Ellington C (eds) Flying insects and robots. Springer, Heidelberg, pp 51–61

Eguchi E, Tominaga Y (1999) Atlas of arthropod sensory receptors: dynamic morphology in relation to function. Springer, Berlin

Eisthen HL (2002) Why are olfactory systems of different animals so similar? Brain Behav Evol 59:273–293

Elofsson R, Dahl E (1970) The optic neuropils and chiasmata of Crustacea. Z Zellforsch mikrosk Anat 107:343–360

Elofsson R, Hagberg M (1986) Evolutionary aspects on the construction of the first optic neuropil (lamina) in Crustacea. Zoomorphol 106:174–178

Elson RC (1996) Neuroanatomy of a crayfish thoracic ganglion: sensory and motor roots of the walking-leg nerves and possible homologies with insects. J Comp Neurol 365:1–17

Eriksson BJ, Stollewerk A (2010) The morphological and molecular processes of onychophoran brain development show unique features that are neither comparable to insects nor to chelicerates. Arthropod Struct Dev 39:478–490

Exner S (1891) Die Physiologie der facettirten Augen von Krebsen und Insekten. Deuticke, Leipzig

Exner S, Hardie RC (1989) The physiology of the compound eyes in insects and crustaceans. Springer, Berlin

Fahrenbach WH (1977) The brain of the horseshoe crab (*Limulus polyphemus*) II. Architecture of the corpora pedunculata. Tissue Cell 9:157–166

Fahrenbach WH (1979) The brain of the horseshoe crab (*Limulus polyphemus*) III. Cellular and synaptic organization of the corpora pedunculata. Tissue Cell 11:163–200

Fahrenbach WH, Chamberlain SC (1987) The brain of the horseshoe crab, *Limulus polyphemus*. In: Gupta AP (ed) Arthropod brain: its evolution, development, structure, and functions. Wiley, New York, pp 63–93

Fanenbruck M, Harzsch S (2005) A brain atlas of *Godzilliognomus frondosus* Yager, 1989 (Remipedia, Godzilliidae) and comparison with the brain of *Speleonectes tulumensis* Yager, 1987 (Remipedia, Speleonectidae): implications for arthropod relationships. Arthropod Struct Dev 34:343–378

Farris SM, Sinakevitch I (2003) Development and evolution of the insect mushroom bodies: towards the understanding of conserved developmental mechanisms in a higher brain center. Arthropod Struct Dev 32:79–101

Farris SM (2005) Evolution of insect mushroom bodies: old clues, new insights. Arthropod Struct Dev 34:211–234

Farris SM (2008) Tritocerebral tract input to the insect mushroom bodies. Arthropod Struct Dev 37:492–503

Farris SM (2011) Are mushroom bodies cerebellum-like structures? Arthropod Struct Dev 40:368–379

Farris SM, Roberts NS (2005) Coevolution of generalist feeding ecologies and gyrencephalic mushroom bodies in insects. Proc Natl Acad Sci U S A 102:17394–17399

Fonta C, Sun X-J, Masson C (1993) Morphology and spatial distribution of bee antennal lobe interneurons responsive to odours. Chemi Senses 18:101–119

Franceschini N, Riehle A, Le Nestour A (1989) Directional selective motion detection by insect neurons. In: Stavenga DG, Hardie RC (eds) Facets of vision. Springer, Berlin, pp 360–390

Friedrich M, Wood EJ, Wu M (2011) Developmental evolution of the insect retina: Insights from standardized numbering of homologous photoreceptors. J Exptl Zool 316:484–499

Fritsch M, Richter S (2010) The formation of the nervous system during larval development in *Triops cancriformis* (Bosc) (Crustacea, Branchiopoda): an immunohistochemical survey. J Morphol 271:1457–1481

Galizia CG, Eisenhardt D, Giurfa M (2012) Honeybee neurobiology and behavior: a tribute to Randolf Menzel. Springer, New York

Galizia CG, McIlwrath SL, Menzel R (1999) A digital three-dimensional atlas of the honey-bee antennal lobe based on optical sections acquired by confocal microscopy. Cell Tissue Res 395:383–394

Galizia CG, Menzel R (2000) Odour perception in honeybees: coding information in glomerular patterns. Curr Opin Neurobiol 10:504–510

Galizia CG, Menzel R (2001) The role of glomeruli in the neural representation of odours: results from optical recording studies. J Insect Physiol 47:115–129

Galizia CG, Rössler W (2010) Parallel olfactory systems in insects: anatomy and function. Annu Rev Entomol 55:399–420

Galizia CG, Szyska P (2008) Olfactory coding in the insect brain: molecular receptive ranges, spatial and temporal coding. Entom Exp Appl 128:81–92

Ghaffar H, Larsen JR, Booth GM, Perkes R (1984) General morphology of the brain of the blind cave beetle, *Neaphaenops tellkampfii* Erichson (Coleoptera: Carabidae). Int J Insect Morphol Embryol 13:357–371

Ghaninia M, Hansson BS, Ignell R (2007) The antennal lobe of the African malaria mosquito, *Anopheles gambiae*—innervations and three dimensional reconstruction. Arthropod Struct Dev 36:23–39

Giribet G, Edgecombe GD, Wheeler WC (2001) Arthropod phylogeny based on eight molecular loci and morphology. Nature 413:157–161

Glantz RM, Miller CS (2002) Signal processing in the crayfish optic lobe: contrast, motion and polarization vision. In: Wiese K (ed) The crustacean nervous system. Springer, Berlin, pp 486–498

Glantz RM, Miller CS, Nässel DR (2000) Tachykinin-related peptide and GABA-mediated presynaptic inhibition of crayfish photoreceptors. J Neurosci 20:1780–1790

Gronenberg W (2001) Subdivisions of hymenopteran mushroom body calyces by their afferent supply. J Comp Neurol 436:474–489

Hafner GS, Tokarski TR (2001) Retinal development in the lobster *Homarus americanus*: comparison with compound eyes of insects and other crustaceans. Cell Tissue Res 305:147–158

Hallberg E, Skog M (2011) Chemosensory sensilla in crustaceans. In: Breithaupt T, Thiel M (eds) Chemical communication in crustaceans. Springer, New York, pp 103–121

Hansson BS, Harzsch S, Knaden M, Stensmyr MC (2011) The neural and behavioral basis of chemical communication in terrestrial crustaceans. In: Breithaupt T, Thiel M (eds) Chemical communication in crustaceans. Springer, New York, pp 149–173

Hansson BS, Stensmyr MC (2011) Evolution of insect olfaction. Neuron 72:698–711

Hanström B (1926) Das Nervensystem und die Sinnesorgane von *Limulus* polyphemus. Lunds Univ Årsskr NF 22:1–79

Hanström B (1928) Vergleichende Anatomie des Nervensystems der wirbellosen Tiere unter Berücksichtigung seiner Funktion. Springer, Berlin

Hardie RC (1989) Neurotransmitters in compound eyes. In: Stavenga DG, Hardie RC (eds) Facets of vision. Springer, Berlin, pp 235–256

Hartenstein V (2006) The neuroendocrine system of invertebrates: a developmental and evolutionary perspective. J Endocrinol 190:555–570

Harzsch S (2002) The phylogenetic significance of crustacean optic neuropils and chiasmata: a re-examination. J Comp Neurol 453:10–21

Harzsch S (2003) Ontogeny of the ventral nerve cord in malacostracan crustaceans: a common plan for neuronal development in Crustacea and Hexapoda? Arthropod Struct Dev 32:17–38

Harzsch S (2004a) Phylogenetic comparison of serotonin-immunoreactive neurons in representatives of the Chilopoda, Diplopoda, and Chelicerata: implications for arthropod relationships. J Morphol 259:198–213

Harzsch S (2004b) The tritocerebrum of Euarthropoda: a "non-*drosophilo*centric" perspective. Evol Dev 6:303–309

Harzsch S (2006) Neurophylogeny: architecture of the nervous system and a fresh view on arthropod phylogeny. Integr Comp Biol 46:162–194

Harzsch S (2007) Architecture of the nervous system as a character for phylogenetic reconstructions: examples from the Arthropoda. Species Phylog Evol 1:33–57

Harzsch S, Benton J, Darwirs RR, Beltz B (1999) A new look at embryonic development of the visual system in decapod crustaceans: neuropil formation, neurogenesis and apoptotic cell death. J Neurobiol 39:294–306

Harzsch S, Glötzner J (2002) An immunhistochemical study of structure and development of the nervous system in the brine shrimp *Artemia salina* Linnaeus, 1758 (Branchiopoda, Anostraca) with remarks on the evolution of the arthropod brain. Arthropod Struct Dev 30:251–270

Harzsch S, Hafner G (2006) Evolution of eye development in arthropods: phylogenetic aspects. Arthropod Struct Dev 35:319–340

Harzsch S, Hansson BS (2008) Brain architecture in the terrestrial hermit crab *Coenobita clypeatus* (Anomura, Coenobitidae): neuroanatomical evidence for a superb aerial sense of smell. BMC Neurosci 9:1–35

Harzsch S, Melzer RR, Müller CHG (2007) Mechanisms of eye development and evolution of the arthropod visual systems: the lateral eyes of myriapoda are not modified insect ommatidia. Org Divers Evol 7:20–32

Harzsch S, Müller CHG, Wolf H (2005a) From variable to constant cell numbers: cellular characteristics of the arthropod nervous system argue against a sister-group relationship of Chelicerata and "Myriapoda" but favour the Mandibulata concept. Dev Gen Evol 215:53–68

Harzsch S, Rieger V, Krieger J, Seefluth F, Strausfeld NJ, Hansson BS (2011) Transition from marine to terrestrial ecologies: changes in olfactory and tritocerebral neuropils in land-living isopods. Arthropod Struct Dev 40:244–257

Harzsch S, Sandeman D, Chaigneau J (2012) Morphology and development of the central nervous system. In: Forest J, von Vaupel Klein JC (eds) Treatise on zoology—anatomy, taxonomy, biology. The Crustacea, vol. 3. Brill, Leiden, pp 9–236

Harzsch S, Waloszek D (2000) Serotonin-immunoreactive neurons in the ventral nerve cord of Crustacea: a character to study aspects of arthropod phylogeny. Arthropod Struct Dev 29:307–322

Harzsch S, Waloszek D (2001) Neurogenesis in the developing visual system of the branchiopod crustacean *Triops longicaudatus* (LeConte, 1846): corresponding patterns of compound-eye formation in Crustacea and Insecta? Dev Genes Evol 211:37–43

Harzsch S, Wildt M, Battelle B, Waloszek D (2005b) Immunohistochemical localization of neurotransmitters in the nervous system of larval *Limulus polyphemus* (Chelicerata, Xiphosura): evidence for a conserved protocerebral architecture in Euarthropoda. Arthropod Struct Dev 34:327–342

Heinze S, Homberg U (2007) Map-like representation of celestial E-vector orientations in the brain of an insect. Science 315:995–997

Heisenberg M (2003) Mushroom body memoir: from maps to models. Nature Rev Neurosci 4:266–275

Heuer CM, Kollmann M, Binzer M, Schachtner J (2012) Neuropeptides in insect mushroom bodies. Arthropod Struct Dev 41:199–226

Heuer CM, Loesel R (2008) Immunofluorescence analysis of the internal brain anatomy of *Nereis diversicolor* (Polychaeta, Annelida). Cell Tissue Res 331:713–724

Heuer CM, Loesel R (2009) Three-dimensional reconstruction of mushroom body neuropils in the polychaete species *Nereis diversicolor* and *Harmothoe areolata* (Phyllodocida, Annelida). Zoomorphol 128:219–226

Heuer CM, Müller CHG, Loesel R (2010) Comparative neuroanatomy suggests repeated reduction of neuro-architectural complexity in Annelida. Front Zool 7:13. doi:10.1186/1742-9994-7-13

Hoese B (1989) Morphological and comparative studies on the second antennae of terrestrial isopods. Monitore Zoologico Italiano (N.S.) Monografia 4:127–152

Holmgren N (1916) Zur vergleichenden Anatomie des Gehirns von Polychaeten, Onychophoren, Xiphosuren, Arachniden, Crustaceen, Myriapoden und Insekten. Kongl Svensk Vetenskap Akad Handl 56:1–303

Homberg U (1985) Interneurons in the central complex in the bee brain (*Apis mellifera*, L.). J Insect Physiol 31:251–264

Homberg U (1994) Distribution of neurotransmitters in the insect brain. Fischer, Stuttgart

Homberg U (2008) Evolution of the central complex in the arthropod brain with respect to the visual system. Arthropod Struct Dev 37:347–362

Homberg U, Christensen TA, Hildebrand JG (1989) Structure and function of the deutocerebrum in insects. Annu Rev Entomol 34:477–501

Homberg U, Vitzthum H, Müller M, Binkle U (1999) Immunocytochemistry of GABA in the central complex of the locust *Schistocerca gregaria*: identification of immunoreactive neurons and colocalization with neuropeptides. J Comp Neurol 409:495–507

Horridge GA (1975) The compound eye and vision of insects. Clarendon Press, Oxford

Huetteroth W, Schachter J (2005) Standard three-dimensional glomeruli of the *Manduca sexta* antennal lobe: a tool to study both developmental and adult neuronal plasticity. Cell Tissue Res 319:513–524

Ignell R, Dekker T, Ghaninia M, Hansson BS (2005) Neuronal architecture of the mosquito deutocerebrum. J Comp Neurol 493:207–240

Ignell R, Hansson BS (2005) Projection patterns of gustatory neurons in the suboesophageal ganglion and tritocerebrum of mosquitoes. J Comp Neurol 492:214–233

Johansson KUI, Hallberg E (1992) The organization of the olfactory lobes in Euphausiacea and Mysidacea (Crustacea, Malacostraca). Zoomorphol 112:81–90

Kämper G, Murphey RK (1987) Synapse formation by sensory neurons after cross-species transplantation in crickets: the role of positional information. Dev Biol 122:492–502

Keil TA, Steinbrecht RA (1984) Mechanosensitive and olfactory sensilla of insects. In: King RC, Akai H (eds) Insect ultrastructure, vol 1. Plenum, New York, pp 402–433

Keil TA (2012) Sensory cilia in arthropods. ASD 41:515–534

Kirschner S, Kleineidam CJ, Zube C, Rybak J, Grünewald B, Rössler W (2006) Dual olfactory pathway in the honeybee, *Apis mellifera*. J Comp Neurol 499:933–952

Kloppenburg P (1995) Anatomy of the antennal motor neurons in the brain of the honeybee (*Apis mellifera*). J Comp Neurol 363:333–343

Koenemann S, Hoenemann M, Stemme T, Jenner RA, v Reumont BM (2010) Arthropod phylogeny revisited, with a focus on crustacean relationships. Arthropod Struct Dev 39:88–110

Kollmann M, Huetteroth W, Schachtner J (2011) Brain organization in Collembola (springtails). Arthropod Struct Dev 40:304–316

Krieger J, Sandeman RE, Sandeman DC, Hansson BS, Harzsch S (2010) Brain architecture of the largest living land arthropod, the giant robber crab *Birgus latro* (Crustacea, Anomura, Coenobitidae): evidence for a prominent central olfactory pathway? Front Zool 7:25. doi:10.1186/1742-9994-7-25

Krieger J, Sombke A, Seefluth F, Kenning M, Hansson BS, Harzsch S (2012) Comparative brain architecture of the European shore crab *Carcinus maenas* (Brachyura) and the common hermit crab *Pagurus bernhardus* (Anomura). Cell Tissue Res 348:47–69

Kutsch W, Breidbach O (1994) Homologous structures in the nervous system of Arthropoda. Adv Insect Physiol 24:1–113

Kutsch W, Heckmann R (1995a) Motor supply of the dorsal longitudinal muscles I: homonomy and ontogeny of the motoneurones in locusts (Insecta, Caelifera). Zoomorphol 115:179–195

Kutsch W, Heckmann R (1995b) Motor supply of the dorsal longitudinal muscles II: comparison of motoneurone sets in Tracheata. Zoomorphol 115:197–211

Laissue PP, Reiter C, Hiesinger PR, Halter S, Fischbach KF, Stocker RF (1999) Three-dimensional reconstruction of the antennal lobe in *Drosophila melanogaster*. J Comp Neurol 405:543–552

Land MF, Nilsson D-E (2012) Animal eyes. Oxford University Press, Oxford

Langworthy K, Helluy S, Benton J, Beltz B (1997) Amines and peptides in the brain of the American lobster: immunocytochemical localization patterns and implications for brain function. Cell Tissue Res 288:191–206

Laurent G, Naraghi M (1994) Odorant-induced oscillations in the mushroom bodies of the locust. J Neurosci 14:2993–3004

Leise EM, Hall W, Mulloney B (1986) Functional organization of crayfish abdominal ganglia: I. The flexor systems. J Comp Neurol 253:25–45

Leise EM, Hall WM, Mulloney B (1987) Functional organization of crayfish abdominal ganglia. II: sensory afferents and extensor motor neurons. J Comp Neurol 266:495–518

Lichtneckert R, Reichert H (2005) Insights into the urbilaterian brain: conserved genetic patterning mechanisms in insect and vertebrate brain development. Heredity 94:465–477

Linne V, Eriksson BJ, Stollewerk A (2012) *Single-minded* and the evolution of the ventral midline in arthropods. Dev Biol 364:66–76

Loesel R (2004) Comparative morphology of central neuropils in the brain of arthropods and its evolutionary and functional implications. Acta Biol Hung 55:39–51

Loesel R (2006) Can brain structures help to resolve interordinal relationships in insects? Arthropod Syst Phylog 64:101–106

Loesel R (2011) Neurophylogeny—retracing early metazoan brain evolution. In: Pontarotti P (ed) Evolutionary biology: concepts, biodiversity, macroevolution, and genome evolution. Springer, Heidelberg, pp 169–191

Loesel R, Heuer CM (2010) The mushroom bodies—prominent brain centers of arthropods and annelids with enigmatic evolutionary origin. Acta Zool 91:29–34

Loesel R, Homberg U (1999) Histamine-immunoreactive neurons in the brain of the cockroach Leucophaea maderae. Brain Res 842:408–418

Loesel R, Nässel DR, Strausfeld NJ (2002) Common design in a unique midline neuropil in the brains of arthropods. Arthropod Struct Dev 31:77–91

Loesel R, Seyfarth EA, Bräunig P, Agricola HJ (2011) Neuroarchitecture of the arcuate body in the brain of the spider Cupiennius salei (Araneae, Chelicerata) revealed by allatostatin-, proctolin-, and CCAP-immunocytochemistry. Arthropod Struct Dev 40:210–220

Masse NY, Turner GC, Jefferis GS (2009) Olfactory information processing in Drosophila. Curr Biol 19:R700–R713

Mayer G (2006) Structure and development of onychophoran eyes—what is the ancestral visual organ in arthropods? Arthropod Struct Dev 35:231–245

Mayer G, Koch M (2005) Ultrastructure and fate of the nephridial anlagen in the antennal segment of Epiperipatus biolleyi (Onychophora, Peripatidae)—evidence for the onychophoran antennae being modified legs. Arthropod Struct Dev 34:471–480

Mayer G, Whitington PM, Sunnucks P, Pflüger H-J (2010) A revision of brain composition in Onychophora (velvet worms) suggests that the tritocerebrum evolved in arthropods. BMC Evol Biol 10:255. doi: 10.1186/1471-2148-10-255

Meinertzhagen IA (1991) Evolution of the cellular organization of the arthropod compound eye and optic lobe. In: Cronly-Dillon JR, Gregory RL (eds) Vision and visual dysfunction, vol 2., Evolution of the eye and visual systemMacmillan, London, pp 341–363

Mellon Jr DeF (2007) Combining dissimilar senses: central processing of hydrodynamic and chemosensory inputs in aquatic crustaceans. Biol Bull 213:1–11

DeF Mellon Jr, Reidenbach MA (2011) Fluid mechanical problems in crustacean active chemoreception. In: Barth F, Humphrey JAC, Srinivasan M (eds) Frontiers in sensing. Springer, New York, pp 159–170

Melzer RR, Diersch R, Nicastro D, Smola U (1997) Compound eye evolution: highly conserved retinula and cone cell patterns indicate a common origin of the insect and crustacean ommatidium. Naturwiss 84:542–544

Melzer R, Michalke C, Smola U (2000) Walking on insect paths: early ommatidial development in the compound eye of the ancestral crustacean Triops cancriformis. Naturwiss 87:308–311

Melzer RR, Petyko Z, Smola U (1996) Photoreceptor axons and optic neuropils in Lithobius forficatus (Linnaeus, 1758) (Chilopoda, Lithobiidae). Zool Anz 235:177–182

Milde JJ (1988) Visual responses of interneurons in the posterior median protocerebrum and the central complex of the honeybee Apis mellifera. J Insect Physiol 34:427–436

Mißbach C, Harzsch S, Hansson B (2011) New insights into an ancient insect nose: the olfactory pathway of Lepismachilis y-signata (Archaeognatha: Machilidae). Arthropod Struct Dev 40:317–333

Mittmann B, Scholtz G (2003) Development of the nervous system in the 'head' of Limulus polyphemus (Chelicerata: Xiphosura): morphological evidence for a correspondence between the segments of the chelicerae and of the (first) antennae of Mandibulata. Dev Genes Evol 213:9–17

Müller CHG, Rosenberg J, Richter S, Meyer-Rochow VB (2003) The compound eye of Scutigera coleoptrata (Linnaeus, 1758) (Chilopoda: Notostigmophora): an ultrastructural reinvestigation that adds support to the mandibulata concept. Zoomorphol 122:191–209

Nässel DR (1976) The retina and retinal projection on the lamina ganglionaris of the crayfish Pacifastacus leniusculus (Dana). J Comp Neurol 167:341–360

Nässel DR (1977) Types and arrangement of neurons in the crayfish optic lamina. Cell Tissue Res 179:45–75

Nässel DR, Elofsson R, Odselius R (1978) Neuronal connectivity patterns in the compound eyes of Artemia salina and Daphnia magna (Crustacea: Branchiopoda). Cell Tissue Res 190:435–457

Nässel DR, Geiger G (1983) Neuronal organization in fly optic lobes altered by laser ablations early in development or by mutations of the eye. J Comp Neurol 217:86–102

Nässel DR, Waterman TH (1977) Massive diurnally modulated photoreceptor membrane turnover in crab light and dark adaptation. J Comp Physiol A 131:205–216

Newland PL, Rogers SM, Gaaboub I, Matheson T (2000) Parallel somatotopic maps of gustatory and mechanosensory neurons in the central nervous system of an insect. J Comp Neurol 425:82–96

Nilsson D-E (1989) Optics and evolution of the compound eye. In: Stavenga DG, Hardie RC (eds) Facets of vision. Springer, Berlin, pp 30–73

Nilsson D-E, Kelber A (2007) A functional analysis of compound eye evolution. Arthropod Struct Dev 36:373–385

Nilsson D-E, Osorio D (1997) Homology and parallelism in arthropod sensory processing. In: Fortey RA, Thomas RH (eds) Arthropod relationships. Chapman and Hall, London, pp 333–347

Nishino H, Nishikawa M, Yokohari F, Mizunami M (2005) Dual, multilayered somatosensory maps formed by antennal tactile and contact chemosensory afferents in an insect brain. J Comp Neurol 493:291–308

North G, Greenspan RJ (2007) Invertebrate neurobiology. Cold Spring Harbor Laboratory Press, New York

Oldfield BP (1988) Tonotopic organization of the insect auditory pathway. Trends Neurosci 11:267–270

Osorio D (2007) *Spam* and the evolution of the fly's eye. BioEssays 29:111–115

Pareto A (1972) Die zentrale Verteilung der Fühlerafferenz bei Arbeiterinnen der Honigbiene, *Apis mellifera*, L. Ztschr Zellforsch mikr Anat 131:109–140

Patel NH, Kornberg TB, Goodman CS (1989a) Expression of *engrailed* during segmentation in grasshopper and crayfish. Development 107:201–212

Patel NH, Martin-Blanco E, Coleman KG, Poole SJ, Ellis MC, Kornberg TB, Goodman CS (1989b) Expression of *engrailed* proteins in arthropods, annelids, and chordates. Cell 58:955–968

Paulus HF (2000) Phylogeny of the Myriapoda—Crustacea—Insecta: a new attempt using photoreceptor structure. J Zool Syst Evol Res 38:189–208

Pflüger HJ, Stevenson P (2005) Evolutionary aspects of octopaminergic systems with emphasis on arthropods. Arthropod Struct Dev 34:379–396

Rathmayer W (1990) Inhibition through neurons of the common inhibitory type (CI-neurons) in crab muscles. In: Wiese K (ed) Frontiers in crustacean neurobiology. Birkhäuser, Basel, pp 271–278

Regier JC, Shultz JW, Kambic RE (2005) Pancrustacean phylogeny: hexapods are terrestrial crustaceans and maxillopods are not monophyletic. Proc R Soc B 272:395–401

Regier JC, Shultz JW, Zwick A, Hussey A, Ball B, Wetzer R, Martin JW, Cunningham CW (2010) Arthropod relationships revealed by phylogenomic analysis of nuclear protein-coding sequences. Nature 463:1079–1083

Reichert H (1988) Control of sequences of movements in crayfish escape behavior. Experientia 44:395–401

Richter S (2002) The Tetraconata concept: hexapod-crustacean relationships and the phylogeny of Crustacea. Org Divers Evol 2:217–237

Richter S, Loesel R, Purschke G, Schmidt-Rhaesa A, Scholtz G, Stach T, Vogt L, Wanninger A, Brenneis G, Doring C, Faller S, Fritsch M, Grobe P, Heuer CM, Kaul S, Moeller OS, Müller CHG, Rieger V, Rothe BG, Stegner MEJ, Harzsch S (2010) Invertebrate neurophylogeny: suggested terms and definitions for a neuroanatomical glossary. Front Zool 7:29. doi: 10.1186/1742-9994-7-29

Roberston RM, Pearson KG, Reichert H (1982) Flight interneurons in the locust and the origin of insect wings. Science 217:177–179

Römer H, Marquart V, Hardt M (1988) Organization of a sensory neuropile in the auditory pathway of two groups of orthoptera. J Comp Neurol 275:201–215

Rospars JP (1983) Invariance and sex-specific variations of the glomerular organization in the antennal lobes of a moth, *Mamestra brassicae* and a butterfly, *Pieris brassicae*. J Comp Neurol 220:80–96

Rospars JP (1988) Structure and development of the insect antennodeutocerebral system. Int J Insect Morphol Embryol 17:243–294

Rospars JP, Hildebrand JG (1992) Anatomical identification of glomeruli in the antennal lobes of the male sphinx moth *Manduca sexta*. Cell Tissue Res 270:205–227

Rospars JP, Hildebrand JG (2000) Sexually dimorphic and isomorphic glomeruli in the antennal lobes of the sphinx moth *Manduca sexta*. Chem Senses 25:119–129

Rössler W, Zube C (2011) Dual olfactory pathway in Hymenoptera: evolutionary insights from comparative studies. Arthropod Struct Dev 40:349–357

Rota-Stabelli O, Campbell L, Brinkmann H, Edgecombe GD, Longhorn SJ, Peterson KJ, Pisani D, Philippe H, Telford MJ (2011) A congruent solution to arthropod phylogeny: Phylogenomics, microRNAs and morphology support monophyletic Mandibulata. Proc R Soc B 278:298–306

Sachse S, Krieger J (2011) Olfaction in insects—the primary processes of odor recognition and coding. E-Neuroforum 2:49–60

Sandeman D, Kenning M, Harzsch S (in press) Adaptive trends in malacostracan brain form and function related to behaviour. In: Derby C, Thiel M (eds) Crustacean nervous system and their control of behaviour. The natural history of the Crustacea, vol. 3

Sandeman DC, Luff SE (1973) The structural organization of glomerular neuropile in the olfactory and accessory lobes of an australian freshwater crayfish, *Cherax destructor*. Ztschr Zellforsch mikr Anat 142:37–61

Sandeman DC, Beltz BS, Sandeman RE (1995) Crayfish brain interneurons that converge with serotonin giant cells in accessory lobe glomeruli. J Comp Neurol 352:263–279

Sandeman D, DeF Mellon Jr (2002) Olfactory centers in the brain of freshwater crayfish. In: Wiese K (ed) The crustacean nervous system. Springer, Berlin, pp 386–404

Sandeman DC, Sandeman RE (1994) Electrical responses and synaptic connections of giant serotonin-immunoreactive neurons in crayfish olfactory and accessory lobes. J Comp Neurol 341:130–144

Sandeman DC, Sandeman RE, Derby C, Schmidt M (1992) Morphology of the brain of crayfish, crabs, and spiny lobsters: a common nomenclature for homologous structures. Biol Bull 183:304–326

Sandeman DC, Scholtz G (1995) Ground plans, evolutionary changes and homologies in decapod crustacean brains. In: Breidbach O, Kutsch W (eds) The nervous systems of invertebrates: an evolutionary and comparative approach. Birkhäuser, Basel, pp 329–347

Sandeman DC, Scholtz G, Sandeman RE (1993) Brain evolution in decapod crustacea. J Exptl Zool 265:112–133

Sandeman DC, Varju D (1988) A behavioral study of tactile localization in the crayfish *Cherax destructor*. J Comp Physiol A 163:525–536

Sanes JR, Zipursky SL (2010) Design principles of insect and vertebrate visual systems. Neuron 66:15–36

Schachtner J, Schmidt M, Homberg U (2005) Organization and evolutionary trends of primary olfactory brain centers in Tetraconata (Crustacea + Hexapoda). Arthropod Struct Dev 35:257–299

Schmalfuss H (1998) Evolutionary strategies of the antennae in terrestrial isopods. J Crustac Biol 18:10–24

Schmidt H, Rickert C, Bossing T, Vef O, Urban J, Technau GM (1997) The embryonic central nervous system lineages of *Drosophila melanogaster*. II. Neuroblast lineages derived from the dorsal part of the neuroectoderm. Dev Biol 189:186–204

Schmidt M, Ache BW (1992) Antennular projections to the midbrain of the spiny lobster. II. Sensory innervation of the olfactory lobe. J Comp Neurol 318:291–303

Schmidt M, Ache BW (1993) Antennular projections to the midbrain of the spiny lobster. III. Central arborizations of motoneurons. J Comp Neurol 336:583–594

Schmidt M, Ache BW (1996a) Processing of antennular input in the brain of the spiny lobster, *Panulirus argus*. II. The olfactory pathway. J Comp Physiol A 178:605–628

Schmidt M, Ache BW (1996b) Processing of antennular input in the brain of the spiny lobster, *Panulirus argus*. I. Non-olfactory chemosensory and mechanosensory pathway of the lateral and median antennular neuropils. J Comp Physiol A 178:579–604

Schmidt M, Ache BW (1997) Immunocytochemical analysis of glomerular regionalization and neuronal diversity in the olfactory deutocerebrum of the spiny lobster. Cell Tissue Res 287:541–563

Schmidt M, DeF Mellon Jr (2011) Neuronal processing of chemical information in crustaceans. In: Breithaupt T, Thiel M (eds) Chemical communication in crustaceans. Springer, New York, pp 123–147

Schmidt M, van Ekeris L, Ache BW (1992) Antennular projections to the midbrain of the spiny lobster. I. Sensory innervation of the lateral and medial antennular neuropils. J Comp Neurol 318:277–290

Scholtz G, Edgecombe GD (2006) The evolution of arthropod heads: reconciling morphological, developmental and palaeontological evidence. Dev Genes Evol 216:395–415

Sinakevitch I, Douglass JK, Scholtz G, Loesel R, Strausfeld NJ (2003) Conserved and convergent organization in the optic lobes of insects and isopods, with reference to other crustacean taxa. J Comp Neurol 467:150–172

Skinner K (1985a) The structure of the fourth abdominal ganglion of the crayfish, *Procambarus clarki* (Girad).

I. Tracts in the ganglionic core. J Comp Neurol 234:168–181

Skinner K (1985b) The structure of the fourth abdominal ganglion of the crayfish, *Procambarus clarki* (Girad). II. Synaptic neuropils. J Comp Neurol 234:182–191

Sombke A, Harzsch S, Hansson BS (2011a) Organization of deutocerebral neuropils and olfactory behavior in the centipede *Scutigera coleoptrata* (Linnaeus, 1758) (Myriapoda: Chilopoda). Chem Senses 36:43–61

Sombke A, Lipke E, Kenning M, Müller CHG, Hansson BS, Harzsch S (2012) Comparative analysis of deutocerebral neuropils in Chilopoda (Myriapoda): implications for the evolution of the arthropod olfactory system and support for the Mandibulata concept. BMC Neurosci 13:1. doi:10.1186/1471-2202-13-1

Sombke A, Rosenberg J, Hilken G (2011b) Chilopoda—the nervous system. In: Minelli A (ed) Treatise on zoology—anatomy, taxonomy, biology—the Myriapoda, vol I. Brill, Leiden, pp 217–234

Sombke A, Rosenberg J, Hilken G, Westermann M, Ernst A (2011c) The source of chilopod sensory information: external structure and distribution of antennal sensilla in *Scutigera coleoptrata* (Chilopoda, Scutigeromorpha). J Morphol 272:1376–1387

Spreitzer A, Melzer RR (2003) The nymphal eyes of Parabuthus transvaalicus Purcell, 1899 (Buthidae): an accessory lateral eye in a scorpion. Zool Anz 242:137–143

Staudacher E (1998) Distribution and morphology of descending brain neurons in the cricket. Cell Tissue Res 294:187–202

Staudacher E, Schildberger K (1999) A newly described neuropile in the deutocerebrum of the cricket: antennal afferents and descending interneurons. Zoology 102:212–226

Stavenga DG, Hardie RC (1989) Facets of vision. Springer, Berlin

Stavenga DG, Melzer RR, Harzsch S (eds) (2006, 2007) Origin and evolution of arthropod visual systems. Arthropod Struct Dev 35(4) (2006), 36(4) (2007)

Stegner MEJ, Richter S (2011) Morphology of the brain in *Hutchinsoniella macracantha* (Cephalocarida, Crustacea). Arthropod Struct Dev 40:221–243

Stollewerk A (2008) Evolution of neurogenesis in arthropods. In: Minelli A, Fusco G (eds) Evolving pathways. Cambridge University Press, Cambridge, pp 359–380

Stollewerk A, Chipman AD (2006) Neurogenesis in myriapods and chelicerates and its importance for understanding arthropod relationships. Integr Comp Biol 46:195–206

Stollewerk A, Simpson P (2005) Evolution of early development of the nervous system: a comparison between arthropods. BioEssays 27:874–883

Strausfeld NJ (1976) Atlas of an insect brain. Springer, Berlin

Strausfeld NJ (1989) Beneath the compound eye: neuroanatomical analysis and physiological correlates in the study of insect vision. In: Stavenga DG, Hardie

RC (eds) Facets of vision. Springer, Berlin, pp 317–359

Strausfeld NJ (1998) Crustacean–insect relationships: the use of brain characters to derive phylogeny amongst segmented invertebrates. Brain Behav Evol 52:186–206

Strausfeld NJ (1999) A brain region in insects that supervises walking. Progr Brain Res 123:273–284

Strausfeld NJ (2005) The evolution of crustacean and insect optic lobes and the origins of chiasmata. Arthropod Struct Dev 34:235–256

Strausfeld NJ (2009) Brain organization and the origin of insects: an assessment. Proc R Soc B 276:1929–1937

Strausfeld NJ (2012) Arthropod brains. Evolution, functional elegance, and historical significance. The Belknap Press of Harvard University Press, Cambridge, MA

Strausfeld NJ, Andrews DR (2011) A new view of insect–crustacean relationships I. Inferences from neural cladistics and comparative neuroanatomy. Arthropod Struct Dev 40:276–288

Strausfeld NJ, Barth FG (1993) Two visual systems in one brain: neuropils serving the secondary eyes of the spider—Cupiennius salei. J Comp Neurol 328:43–62

Strausfeld NJ, Buschbeck E, Gomez RS (1995) The arthropod mushroom body: its functional roles, evolutionary enigmas and mistaken identities. In: Breidbach O, Kutsch W (eds) The nervous system of invertebrates—an evolutionary and comparative approach. Birkhäuser, Basel, pp 349–381

Strausfeld NJ, Douglas J, Campbell H, Higgins C (2006a) Parallel processing in the optic lobes of flies and the occurence of motion computing circuits. In: Warrant E, Nilsson D-E (eds) Invertebrate vision. Cambridge University Press, Cambridge, pp 349–399

Strausfeld NJ, Hansen L, Li Y, Gomez RS, Ito K (1998) Evolution, discovery, and interpretation of arthropod mushroom bodies. Learn Mem 5:11–37

Strausfeld NJ, Hildebrand JG (1999) Olfactory systems: common design, uncommon origins? Curr Opin Neurobiol 9:634–939

Strausfeld NJ, Nässel DR (1981) Neuroarchitecture of brain regions that subserve the compound eyes of Crustacea and insects. In: Autrum H (ed) Handbook of sensory physiology, vol VII/6B. Invertebrate visual center and behaviors I. Springer, Berlin, pp 1–132

Strausfeld NJ, Reisenman CE (2009) Dimorphic olfactory lobes in the Arthropoda. Ann NY Acad Sci 1170:487–496

Strausfeld NJ, Sinakevitch I, Brown SM, Farris SM (2009) Ground plan of the insect mushroom body: functional and evolutionary implications. J Comp Neurol 513:265–291

Strausfeld NJ, Strausfeld CM, Loesel R, Rowell D, Stowe S (2006b) Arthropod phylogeny: onychophoran brain organization suggests an archaic relationship with a chelicerate stem linage. Proc R Soc B 273:1857–1866

Strausfeld NJ, Strausfeld CM, Stowe S, Rowell D, Loesel R (2006c) The organization and evolutionary implications of neuropils and their neurons in the brain of the onychophoran Euperipatoides rowelli. Arthropod Struct Dev 35:169–196

Strausfeld NJ, Weltzien P, Barth FG (1993) Two visual systems in one brain: neuropils serving the principal eyes of the spider Cupiennius salei. J Comp Neurol 328:63–75

Strauss R (2002) The central complex and the genetic dissection of locomotor behaviour. Curr Opini Neurobiol 12:633–638

Strauss R (2003) Control of Drosophila walking and orientation behavior by functional subunits localized in different neuropils in the central brain. In: Elsner N, Zimmermann H (eds) Proceedings of the 29th Göttingen neurobiology conference. Thieme, Stuttgart, p 206

Strauss R, Hanesch U, Kinkelin M, Wolf R, Heisenberg M (1992) No-bridge of Drosophila melanogaster—portrait of a structural mutant of the central complex. J Neurogenet 8:125–155

Strauss R, Heisenberg M (1990) Coordination of legs during straight walk and turning in Drosophila melanogaster. J Comp Physiol A 167:403–412

Strauss R, Heisenberg M (1993) Higher control center of locomotor behavior in the Drosophila brain. J Neurosci 13:1852–1861

Strotmann J (2001) Targeting of olfactory neurons. Cell Mol Life Sci 58:531–537

Sullivan JM, Beltz BS (2005) Integration and segregation of inputs to higher-order neuropils of the crayfish brain. J Comp Neurol 481:118–126

Sun X-J, Fonta C, Masson C (1993) Odour quality processing by bee antennal lobe interneurons. Chem Senses 18:355–377

Szlendak E, Oliver JH (1992) Anatomy of synganglia, including their neurosecretory regions, in unfed, virgin female Ixodes scapularis Say (Acari: Ixodidae). J Morphol 213:349–364

Sztarker J, Strausfeld NJ, Tomsic D (2005) Organization of optic lobes that support motion detection in a semiterrestrial crab. J Comp Neurol 493:396–411

Tautz J, Müller-Tautz R (1983) Antennal neuropile in the brain of the crayfish: morphology of neurons. J Comp Neurol 218:415–425

Telford MJ, Thomas RH (1998) Expression of homeobox genes shows chelicerate arthropods retain their deutocerebral segment. Proc Natl Acad Sci U S A 95:10671–10675

Tichy H, Barth FG (1992) Fine structure of olfactory sensilla in myriapods and arachnids. Microsc Res Tech 22:372–391

Tomer R, Denes A, Tessmar-Raible K, Arendt D (2010) Profiling by image registration reveals common origin of annelid mushroom bodies and vertebrate pallium. Cell 241:800–809

Trautwein MD, Wiegmann BM, Beutel R, Kjer KM, Yeates DK (2012) Advances in insect phylogeny at the dawn of the postgenomic era. Annu Rev Entomol 57:449–468

Tsuneki K (1992) Endocrine System of arthropods other than crustaceans and insects. In: Matsumoto A, Ishii S (eds) Atlas of endocrine organs. Springer, Berlin, pp 227–229

Tyrer NM, Gregory GE (1982) A guide to the neuroanatomy of locust suboesophageal and thoracic ganglia. Phil Trans R Soc B 297:91–123

Ungerer P, Scholtz G (2008) Filling the gap between identified neuroblasts and neurons in crustaceans adds new support for Tetraconata. Proc R Soc B 275: 369–376

Utting M, Agricola H, Sandeman RE, Sandeman DC (2000) Central complex in the brain of crayfish and its possible homology with that of insects. J Comp Neurol 416:245–261

van Wijk M, Wadman WJ, Sabelis MW (2006a) Gross morphology of the central nervous system of a phytoseiid mite. Exptl Appl Acarol 40:205–216

van Wijk M, Wadman WJ, Sabelis MW (2006b) Morphology of the olfactory system in the predatory mite *Phytoseiulus persimilis*. Exptl Appl Acarol 40:217–229

Vitzthum H, Müller M, Homberg U (2002) Neurons of the central complex of the locust *Schistocerca gregaria* are sensitive to polarized light. J Neurosci 22:1114–1125

von Békésy G (1967) Sensory inhibition. Princeton University Press, Princeton, NJ

von Reumont BM, Jenner RA, Wills MA, Dell'Ampio E, Pass G, Ebersberger I, Meusemann K, Meyer B, Koenemann S, Iliffe TM, Stamatakis A, Niehuis O, Misof B (2012) Pancrustacean phylogeny in the light of new phylogenomic data: support for Remipedia as the possible sister group of Hexapoda. Mol Biol Evol 29:1031–1045

Wachowiak M, Diebel CE, Ache BW (1997) Local interneurons define functionally distinct regions within olfactory glomeruli. J Exptl Biol 200: 989–1001

Warrant E, Nilsson D-E (2006) Invertebrate vision. Cambridge University Press, Cambridge

Watson AHD (1986) The distribution of GABA-like immunoreactivity in the thoracic nervous system of the locust *Schistocerca gregaria*. Cell Tissue Res 246:331–341

Wegerhoff R, Breidbach O (1995) Comparative aspects of the chelicerate nervous system. In: Breidbach O, Kutsch W (eds) The nervous systems of invertebrates: an evolutionary and comparative approach. Birkhäuser, Basel, pp 159–179

Wehner R (1972) Information processing in the visual system of arthropods. Springer, Berlin

Wertz A, Rössler W, Obermayer M, Bickmeyer U (2006) Functional anatomy of the rhinophore of *Aplysia punctata*. Front Zool 3:11. doi:10.1186/1742-9994-3-6

Whitington PM (1996) Evolution of neuronal development in arthropods. Semin Cell Dev Biol 7:605–614

Whitington PM (2004) The development of the crustacean nervous system. In: Scholts G (ed) Evolutionary developmental biology of Crustacea, Crustacean Issues, vol 15. Balkema, Lisse, pp 135–167

Whitington PM (2006) The evolution of arthropod nervous systems: insights from neural development in the Onychophora and Myriapoda. In: Striedler GF, Rubenstein JLR, Kaas JH (eds) Theories, development, invertebrates. Academic, Oxford, pp 317–336

Whitington PM, Bacon JP (1997) The organization and development of the arthropod ventral nerve cord: insights into arthropod relationships. In: Fortey RA, Thomas RH (eds) Arthropod relationships. Chapman and Hall, London, pp 295–304

Whitington PM, Mayer G (2011) The origins of the arthropod nervous system: Insights from the Onychophora. Arthropod Struct Dev 40:193–209

Whitington PM, Meier T, King P (1991) Segmentation, neurogenesis and formation of early axonal pathways in the centipede, *Ethmostigmus rubipes* (Brandt). Roux's Arch Dev Biol 199:349–363

Wiens TJ (1989) Common and specific inhibition in leg muscles of decapods: sharpened distinctions. J Neurobiol 20:458–469

Wiens TJ, Wolf H (1993) The inhibitory motoneurons of crayfish thoracic limbs: structures and phylogenetic comparisons. J Comp Neurol 336:61–278

Wiersma CAG, Roach J, Glantz RM (1982) Neuronal integration in the optic system. In: Sandeman DC, Atwood HL (eds) The biology of Crustacea, vol 4., Neuronal integration and behavior. Academic Press, New York, pp 1–31

Wiese K (2001) The crustacean nervous system. Springer, Berlin

Wiese K (2002) Crustacean experimental systems in neurobiology. Springer, Berlin

Wildt M, Harzsch S (2002) A new look at an old visual system: structure and development of the compound eyes and optic ganglia of the brine shrimp *Artemia salina* Linnaeus, 1758 (Branchiopoda, Anostraca). J Neurobiol 52:117–132

Wolf H (1990) Activity patterns of inhibitory motoneurons and their impact on leg movement in tethered walking locusts. J Exptl Biol 152:281–304

Wolf H (2008) The pectine organs of the scorpion, *Vaejovis spinigerus*: structure and central (glomerular) projections. Arthropod Struct Dev 37:67–80

Wolf H, Harzsch S (2002a) The neuromuscular system in the walking legs of a scorpion. 1. Arrangement of muscles and innervation in the walking legs of a scorpion: *Vaejovis spinigerus* (Wood, 1863) Vaejovidae, Scorpiones, Arachnida. Arthropod Struct Dev 31:185–202

Wolf H, Harzsch S (2002b) The neuromuscular system in the walking legs of a scorpion. 2. Inhibitory innervation of the walking legs of a scorpion: *Vaejovis spinigerus* (Wood, 1863), Vaejovidae, Scorpiones, Arachnida. Arthropod Struct Dev 31:203–215

Wolf H, Harzsch S (2012) Serotonin-immunoreactive neurons in the scorpions' pectine neuropils:

similarities to insect and crustacean olfactory centers? Zoology 115:151–159

Wolff G, Harzsch S, Hansson BS, Brown S, Strausfeld NJ (2012) Neuronal organization of the hemiellipsoid body of the land hermit crab *Coenobita clypeatus*: correspondence with the mushroom body ground pattern. J Comp Neurol 520:2824–2846

Zeil J, Layne J (2002) Path integration in fiddler crabs and its relation to habitat and social life. In: Wiese K (ed) Crustacean experimental systems in neurobiology. Springer, Heidelberg, pp 227–246

Zeil J, Sandeman RE, Sandeman DC (1985) Tactile localisation: the function of active antennal movements in the crayfish *Cherax destructor*. J Comp Physiology A 157:607–617

Zube C, Kleineidam CJ, Kirschner S, Neef J, Rössler W (2008) Organization of the olfactory pathway and odor processing in the antennal lobe of the ant *Camponotus floridanus*. J Comp Neurol 506:425–441

# The Arthropod Circulatory System

**14**

## Christian S. Wirkner, Markus Tögel and Günther Pass

## Contents

C. S. Wirkner (✉)
Allgemeine und Spezielle Zoologie, Universität Rostock, Universitätsplatz 2, 18055, Rostock, Germany
e-mail: christian.wirkner@web.de

M. Tögel · G. Pass
Department of Integrative Zoology, University of Vienna, Althanstraße 14, 1090, Vienna, Austria
e-mail: m.toegel@aon.at

G. Pass
e-mail: guenther.pass@univie.ac.at

## 14.1 Introduction

Arthropods have a genuine circulatory system. Their exoskeleton encloses a liquid-filled body cavity, the *haemocoel*. Thereby all organs and tissues are permanently exposed to a liquid medium, the *haemolymph*, which consists of plasma and suspended haemocytes. The circulation of haemolymph is actively forced by special pumping organs referred to as hearts. The flow of haemolymph may additionally be facilitated by movements of other organs and body parts. Emanating from the hearts, arteries deliver the haemolymph to the various body regions and compartments. These arterial systems are developed to differing extents in arthropods and together with the heart constitute the *cardiovascular system* or *haemolymph*

A. Minelli et al. (eds.), *Arthropod Biology and Evolution*,
DOI: 10.1007/978-3-642-36160-9_14, © Springer-Verlag Berlin Heidelberg 2013

343

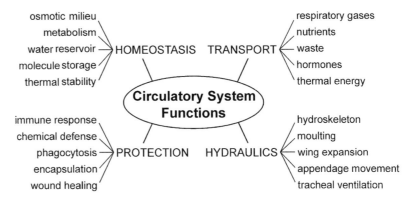

**Fig. 14.1** The circulatory system of arthropods is involved in a wide spectrum of functional areas

*vascular system.* As haemolymph leaves the vascular system, it empties into the *lacunar system*, where it follows distinct routes that are determined by the anatomy of the internal organs and by channels or diaphragms formed of connective tissue.

The physiological functions of the circulatory system of arthropods are astoundingly diverse and can be assigned to four main areas: (1) homeostasis, (2) transport, (3) hydraulics and (4) protection (Fig. 14.1). (1) Maintaining the internal organs in a state of *homeostatic equilibrium* is the most basic function of haemolymph. This comprises the regulation of pH and inorganic ion values, as well as maintaining proper levels of amino acids, proteins, nucleic acids, carbohydrates and lipids. In terrestrial arthropods, it is crucial in the storage of water for use by tissues in case of desiccation. (2) Another main task of the circulatory system is the *transport* and circulation of nutrients, metabolites and wastes, as well as neuroactive substances and hormones. Oxygen is primarily transported in haemolymph by means of the respiratory pigment haemocyanin. This function was lost in some arthropod lineages in the course of the evolution of tracheae, which supply oxygen directly to the tissues. In insects

capable of *thermoregulation*, the circulatory system is responsible for the transport of thermal energy. (3) In arthropods with a soft cuticle, such as many insect larvae, haemolymph is the supporting medium for the hydroskeleton. Moreover, *haemolymph pressure* is responsible for the extension of appendages and legs in various arthropods. Pressure changes also play an important role in supporting tracheal ventilation in high-performance fliers. (4) Finally, haemolymph has numerous *protective functions*. It contains cryoprotective agents which are responsible for cold-hardiness and carries the various components of the immune system. Attacks by pathogens and parasites can be defended against by means of humoral factors, as well as by phagocytosis and encapsulation. In some species, haemolymph contains toxic substances which fend off predators. Finally, haemolymph *seals injuries* and contributes to wound healing through the clotting of haemocytes.

This review focuses on the structure and evolution of the cardiovascular system in arthropods. Although this organ system has been reviewed for all major subgroups separately (see below), no detailed and comparative overview of arthropods as a whole has been carried out so far. The aim of

**Fig. 14.2** Schematic representations of the circulatory organs in various arthropods. **a** Generalised spider. The cardiovascular system consists of a well-developed heart, which extends through the opisthosoma and a complex arterial system. Pulmo-pericardial sinuses channel oxygenated haemolymph (coloured in *light blue* in the diagram) from the book lungs into the pericard from where it enters the cardiovascular system to be distributed throughout the body. **b** Generalised crayfish.

Complex cardiovascular system as found in most▶ malacostracans. Branchio-pericardial sinuses channel oxygenated haemolymph from the gills into the pericard from where it enters the cardiovascular system to be distributed throughout the body. **c** Generalised insect. The cardiovascular system consists of only a tubular dorsal vessel. Appendages are supplied by accessory circulatory organs

**(a)**

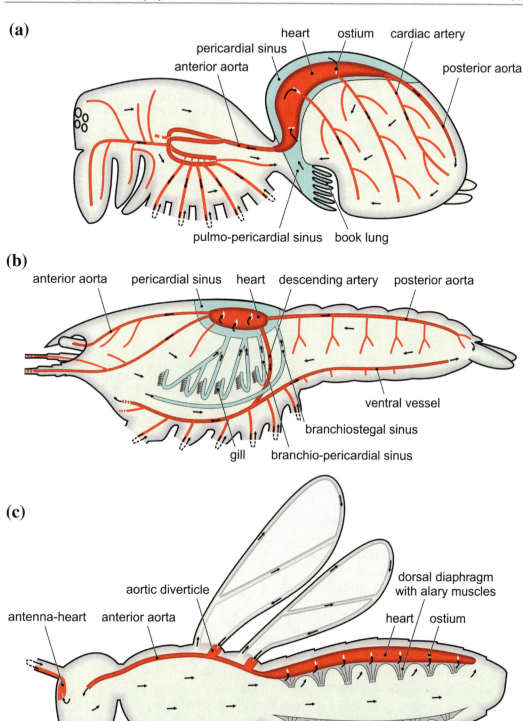

**(b)**

**(c)**

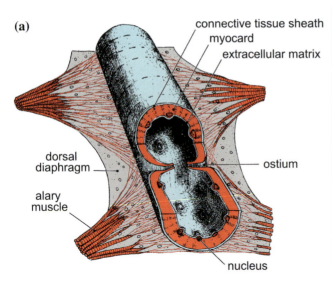
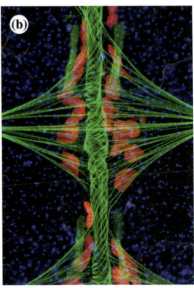

**Fig. 14.3** Anatomy of arthropod hearts. **a** Schematic representation of a section of the heart with incurrent ostia. The wall consists of a thick myocard, which is covered on the outside by a layer of connective tissue and on the luminal side by a layer of extracellular matrix. Dorsal diaphragm and alary muscles attach to the heart (modified after Weber 1954). **b** Fluorescence micrograph of a dissected whole mount of the mosquito *Anopheles gambiae* showing heart and alary muscles; note spirally arranged cardiomyocytes. Pericardial cells (red) flank the heart (from King and Hillyer 2012)

this contribution is to provide a comprehensive picture of the functional and evolutionary morphology of the circulatory organs, their disparity and evolutionary transformations. Particular attention will be paid to the changes this organ system has undergone in the context of major environmental transitions, to the demands associated with such transitions and to the evolutionary novelties which appeared as a result.

## 14.2 General Anatomy of the Circulatory Organs

The circulatory system of arthropods can be divided into a *vascular part* and a *lacunar part*. The central element of the vascular system is the *heart*, which together with the anterior and posterior aorta is often referred to as the *dorsal vessel* (Figs. 14.2, 14.3). It is situated in the midline just beneath the tergal cuticle. Additionally, arthropods may have a varying number of so-called *accessory pulsatile organs*. Some are understood to facilitate a rise in haemolymph

pressure in different body parts, and others are clearly associated with vascular structures and reinforce the pumping action of the main heart. Moreover, some accessory pulsatile organs are completely independent of the vascular system and operate as autonomous organs, particularly to supply haemolymph to body appendages. In many arthropods, a number of vessels emanate from the heart which can collectively be considered peripheral vessels. In analogy to vertebrates, they are termed *arteries* since they channel haemolymph away from the heart. Distally, they have terminal openings through which haemolymph empties into the haemocoel. True veins, that is, vessels that channel haemolymph back in to the heart are absent in arthropods. Instead, after passage through the lacunar system, the haemolymph flows directly from the haemocoel into the dorsal vessel through ostia.

In many arthropods, the body cavity is divided by septa of connective tissue into various compartments, that is, the *sinuses*, which mainly serve to channel haemolymph. In almost all arthropods, one sinus encloses the dorsal vessel and is therefore designated the pericardial sinus. Sometimes a ventral sinus is also present: the perineural sinus, which accommodates the ventral nerve cord. The intermediate compartment is termed the perivisceral sinus. It represents the most voluminous compartment containing all other internal organs. Further, there are sinuses limited by delicate septa which form wide channels, such as those which guide haemolymph from the respiratory organs to the pericardial sinus (in the literature sometimes confusingly referred to as veins, Fig. 14.2a, b). The narrow spaces between the various internal organs and muscles are termed *lacunae*. They form a highly complex system of spaces through which the haemolymph circulates along distinct routes. Movements of organs, such as the gut or muscles, may enhance the flow generated by pumping of the heart. In general, haemolymph flows through the body of most arthropods in much more strictly defined pathways than commonly conceived for an open circulatory system, a fallacy that has been propagated by the extremely simplified diagrams of textbooks.

## 14.2.1 Hearts

Xavier-Neto et al. (2010) propose that the term 'heart' should be restricted to the chambered heart of molluscs and vertebrates, but we do not follow this view (cf. also the critical discussion in McMahon 2012) and use it in the broad sense for any discrete pumping organ that promotes the circulation of blood or haemolymph throughout the body of an animal. A heart is present in almost all arthropods lacking only in minute forms of mites and insects, as well as pauropods and some small crustaceans. Hearts are usually the section of the dorsal vessel which are particularly well developed and which are responsible for the major pumping action; moreover, they bear mostly segmentally organised, paired openings with valves referred to as *ostia*. The shape and performance of hearts varies extensively in the various groups and can be in general correlated with the functional type

of the respiratory system. For example, arthropods that use the circulatory system for oxygen transport generally have hearts that are robust and well developed.

*Histology*

The anatomy of arthropod hearts follows a general scheme. They consist of a muscular layer, the *myocard,* and an outer covering of connective tissue, as well as an inner lining with an extracellular matrix (Fig. 14.3a). In many arthropods, the myocard consists of a single layer of cardiomyocytes; however, in large species with a robust heart, it is formed by several thick layers. Generally, the arrangement of myofibrils in the myocard is described as circular or spiral (Fig. 14.3b). In this manner, the myofibrils form a narrow tube which enables contractions (Fig. 14.3). In some cases, the cardiomyocytes form quite different arrangements, for example irregular meshworks (e.g. Fig. 14.11e; compare also Fig. 14.11b). The cardiomyocytes resemble vertebrate cardiac muscle cells in their ultrastructure and dimensions. Their sarcomeres are very short (about 2 μm), and intercalated discs have been reported in a number of arthropods (Leyton and Sonnenblick 1971; Tjønneland et al. 1985a, b; Midttun 1977; Økland et al. 1982). The mitochondria are relatively small but numerous and located mainly in peripheral pockets of the cardiomyocytes.

The outer covering of hearts consists of a layer of connective tissue which may vary in thickness. In fully developed hearts, the fibrocytes are often reduced, and the outer layer consists only of extracellular material that contains bundles of collagenous fibrils. These fibrils provide the organ with both stiffness and elasticity. In the literature, the outer layer is often called adventitia, but since it does not form a true epithelium, this term should be avoided (see Seifert and Rosenberg 1978). The inner side of the heart is lined with an extracellular matrix (sometimes misleadingly called endocard). It is exceptionally thick and presumably renders the inner surface a smoothness that reduces frictional forces caused by the passing haemolymph stream, a condition also found in vertebrates

(Wagenseil and Mecham 2009). The parts of the heart, which are made of connective tissue, work together as antagonists to achieve the dilation of the dorsal vessel after relaxation of the cardiomyocytes. Dilation is supported by elastic suspensory strands and, in some cases, additionally by muscles associated with the dorsal diaphragm i.e. alary muscles. In the diastolic phase, the haemolymph enters the lumen of the heart through the ostia. In the succeeding systolic phase, haemolymph is further transported due to the contraction of the myocard.

*Ostia and Valves*

Hearts of arthropods are equipped with pairs of *incurrent ostia* that are mostly segmentally arranged. The ostia are slit-like openings (Fig. 14.3a) in the lateral wall of the heart and are oriented either vertically or obliquely (Fig. 14.5b). Often, they have a pair of lips which form a funnel, allowing haemolymph to enter the lumen of the heart. The lips consist of special cells which differ in their development from ordinary cardiomyocytes. In some cases, the lips are elongated to join those on the opposing side of the lumen; together, they form a valve, which prevents backflow into the lumen of the heart during diastole. In addition, internal valves may be present which are independent of the ostia. Aside from such incurrent ostia, some hexapods and crustaceans exhibit *excurrent ostia* with sphincter-like muscular valves. In some hexapods which perform heartbeat reversal, *two-way ostia* are found enabling both inflow and outflow depending on the flow direction within the dorsal vessel.

## 14.2.2 Arterial Systems

As stated above, dorsal vessels are usually not uniform along their longitudinal extension but differentiated into distinct sections. The anterior part of the dorsal vessel is referred to as the anterior aorta, and in some arthropods, there is additionally a posterior aorta. In chelicerates, myriapods and crustaceans, the border of these sections is marked by valves; however, in hexapods, it cannot be always clearly delineated. In contrast to most other peripheral vessels, the

walls of the aortae can contain myocytes and may thus be contractile to some extent. The number of arteries which emanate from the heart varies considerably. Four major arterial systems occur: (1) anterior aorta, (2) posterior aorta, (3) cardiac arteries and (4) ventral vessels. (1) The *anterior aorta* is in many groups the only artery. It is either a simple, unbranched tube or has a high number of secondary branches which supply various appendages, organs and tissues in the anterior body region. (2) The *posterior aorta* is likewise either a simple tube or shows more or less complex branching patterns which supply musculature and other organs along their course. (3) In many arthropods, there are pairs of segmentally arranged *lateral cardiac arteries* branching off from the heart in close vicinity to the ostia. At their origin, they have sphincter-like muscular valves which ensure a one-way flow of haemolymph out of the heart. These valves are innervated and can be induced by neural and/or neurohormonal stimulation to contract progressively to restrict cardiac outflow into a particular vessel (Alexandrowicz 1932; Kihara and Kuwasawa 1984; Kuramoto and Ebara 1984). Since the valves of the different arterial systems are innervated separately, this provides a mechanism by which the outflow of the heart can be diverted into particular arterial systems and thus to particular body regions (McMahon 2001). (4) In addition, some arthropods possess a *longitudinal ventral vessel*, which may lie dorsal or ventral of the ventral nerve cord (*supraneural* or *subneural vessel*). Ventral vessels are connected to the dorsal vessel by a vessel ring, a number of segmental vessels or by a so-called descending artery. Frequently, peripheral arteries branch off the longitudinal vessels and extend into various body appendages. In most cases, the walls of the peripheral vessels contain no contractile elements.

## 14.2.3 Pericardial Sinuses and Associated Structures

The pericardial sinus harbours the dorsal vessel, or at least the heart, and is more or less separated from the remaining body cavity by the pericardial septum which consists of connective tissue. The pericardial sinus is dorsally confined by the cuticle. Especially in arthropods which transport oxygen by means of haemolymph, the pericardial sinus is closed except for channel-like connections to the respiratory organs. This enclosure allows for a direct and rapid transport of oxygenated haemolymph to the heart. Evidently, the pericardial septum has a high elasticity, and some muscles are associated with the septum. Contraction of these muscles leads to enlargement of the lumen of the pericardial sinus, and thereby, haemolymph is drawn from the general body cavity into the pericardial sinus. Following muscle relaxation, the pericardial septum returns to its original state due to its elasticity. The increase in pressure of the haemolymph inside the pericard supports the influx of haemolymph into the lumen of the dorsal vessel through the ostia. The pericardial sinus can thus be functionally considered as an additional heart chamber analogous to the atrium of vertebrates. It is the receiving chamber, while the heart itself represents the discharging chamber.

In some arthropods, the pericardial sinus is not strictly separated from the remaining body cavity. In these cases, the pericardial septum may be fenestrated and is traditionally termed the dorsal diaphragm. Associated muscle cells may be integrated into the septum and span over the entire width of the body. Sometimes they are directly attached to the dorsal vessel and extend as triangular plates of muscles from both sides of the dorsal vessel to the lateral body wall (Fig. 14.3b). They are denoted alary muscles, because they evoked the far-fetched notion that wings are attached to the dorsal vessel. Especially in small arthropods, the dorsal diaphragm is commonly extremely small and may even be entirely absent.

## 14.2.4 Heart Function and Circulation

The haemolymph flow in the dorsal vessel is generated in the various arthropods by three different modes. (1) *Unidirectional flow*: the myocard contracts metachronously such that the contraction waves begin at the posterior end and press haemolymph through the dorsal vessel in the anterior direction. (2) *Bidirectional flow*: the contraction wave begins at a particular point in the dorsal vessel and proceeds along both directions. (3) *Heartbeat reversals*: anterograde pumping phases alternate with retrograde phases.

The contractions of the dorsal vessel may occur quite differently. Thus, in some species, the whole tube contracts more or less simultaneously, while in others, there are distinct waves of contraction which run along the organ. The heartbeat rates cover an enormous range. In insects, for example, the highest measured rates are 7 Hz in *Drosophila* larvae and about 3 Hz in the adults (Sláma and Farkas 2005). The hearts of larger insects beat much more slowly, for example, about 1 Hz in resting locusts and cockroaches (up to 3 Hz during periods of high activity: Miller 1997) and only 4–9 beats per min in adults of the aquatic moth *Acentropus niveus* (Nigmann 1908).

*Heartbeat control* is achieved in arthropods by a wide range of physiological modes. Generally, cardiac cycles are triggered by pacemaker cells which spontaneously depolarise slowly. When a certain threshold potential is reached, an action potential occurs which causes the whole myocard to contract. In many arthropods, the pacemaker cells are special cardiomyocytes; this kind of heart control is termed *myogenic automatism*. It stands in contrast to hearts with a *neurogenic automatism* in which the pacemakers are special cardiac neurons. These two kinds of heart control, however, represent only the extremes. Myogenic hearts are often intensely innervated, and the heartbeat may be modulated by neuronal activity or hormones. This cardioregulatory system can be both excitatory, as well as inhibitory, and both the heartbeat rate and the amplitude can be affected. Regulatory neurons may be located in the ventral nerve cord or the stomatogastric nervous system. Intrinsic cardiac neurons also exist, which in some species are concentrated to form their own cardiac ganglion. In many instances, neurosecretory cells are associated with the heart. They often form extensive neurohaemal sites, which release hormones directly into the lumen of the pumping organ, thereby reaching their destinations very quickly and efficiently. Furthermore, numerous studies have dealt with the impact of cardioactive substances, in particular, neurotransmitters. The overall picture, however, presents no uniform condition for all arthropods.

The *pattern and velocity of circulation* through the body of arthropods is still a largely unexplored field due to methodological problems. Detailed analyses of the dynamics and flow patterns first became available with the implementation of new methods, such as ultrasonic Doppler velocimetry (Hetz et al. 1999) and synchrotron X-ray phase-contrast imaging (Lee and Socha 2009) in combination with tracer particles. Further and comparative investigations with these new methods are required to allow for a holistic approach to the entire circulation process in arthropods.

## 14.3 Circulatory Systems in Major Arthropod Groups

The following section gives an overview of the variation in structure and function of the circulatory system among arthropods. It is organised along the four major groups, chelicerates, myriapods, crustaceans and hexapods. Some of these taxa may be paraphyletic assemblages (for a discussion of arthropod phylogeny, see Chap. 2); however, the traditional subdivision of arthropods was chosen as a manageable framework for this synopsis. The terminology used follows functional and descriptive comparative anatomy. It should be evident that the terms do not always imply homology of structures.

### 14.3.1 Chelicerates

The chelicerates is a diverse group of arthropods; it contains some of the largest arthropods (horseshoe crabs) and some of the smallest (mites). Since size has a major impact on the design of circulatory organs, it should be clear that large chelicerates have some of the most complex vascular systems (Fig. 14.4a), while

the small and minute species usually rely exclusively on diffusion and organ movement for the transport of necessary substances.

The Arachnida is the largest group within the chelicerates and may be divided into two major groups with regard to respiratory organs, i.e. pulmonates (Fig. 14.4b) and apulmonates (Fig. 14.4c). Each group has largely the same circulatory organ equipment respectively, probably due to the functional linkage between respiration and circulation (see Sect. 14.6.1).

Most comparative investigations of the chelicerate circulatory system date back several decades. A gross overview was given by Firstman (1973) for all chelicerates and by Gerhardt & Kästner (1938) for the major subgroups.

#### 14.3.1.1 Hearts

Hearts in chelicerates are located mainly in the opisthosoma (Fig. 14.5); however, they extend into the prosoma in pycnogonids, xiphosurans and certain arachnids (solifugids and uropygids). Some small mites are devoid of any heart. The thickness of the hearts and the arrangement of cardiomyocytes vary considerably. They are extraordinarily thin, and only their lateral parts contain cardiomyocytes in pycnogonids (Tjønneland et al. 1985a), while in *Limulus polyphemus*, the myocard is a thick meshwork of cardiomyocytes giving it a sponge-like appearance (Meek 1909). Nonetheless, in all chelicerates except pycnogonids, we can speak of tubular hearts since the myocard concentrically surrounds the lumen (Fig. 14.5). In spiders, scorpions, amblypygids and uropygids, the cardiomyocytes are arranged in parallel and form semilunar lamellae that protrude into the lumen of the heart (Fig. 14.5a). Apart from the transversely arranged cardiomyocytes, longitudinal ones occur, at least, in Xiphosura, Araneae and Scorpiones.

Hearts are suspended in the body cavity by partly muscular *ligaments* (Fig. 14.5d,e). Presumably, these function as antagonists to expand the lumen of the myocard after relaxation. By this action, haemolymph is sucked into the heart through the paired incurrent ostia. Whether the ligaments or parts of them are homologous to alary muscles awaits clarification.

Heartbeat in chelicerates is triggered by a neurogenic automatism and has been investigated in xiphosurans (e.g. Watson and Groome 1989), scorpions (e.g. Farley 1985, 1987) and spiders (e.g. Sherman et al. 1969; Sherman 1985). According to McMahon et al. (1997), the discovery of a myogenic automatism in juvenile *Limulus* would be an exception and needs further verification. In chelicerates, hearts are innervated by a dorsal cardiac ganglion running the length of the heart tube. Two different neurons were described for *Limulus*: pacemaker neurons and follower neurons. The former generate action potentials which are distributed throughout the myocard by the latter (Watson and Groome 1989).

#### 14.3.1.2 Arterial Systems

A wide range of arterial systems are present in chelicerates. All taxa possessing a heart exhibit an *anterior aorta*. It is the only artery in pycnogonids, palpigrads and pseudoscorpions. Where the anterior aorta merges into the heart, there is an internal valve. In Xiphosura, Pycnogonida and apulmonate arachnids, the anterior aorta is a rather short vessel that connects to the perineural sinus (see below).

In pulmonate arachnids, the anterior aorta runs to the posterior end of the dorsal part of the prosomal ganglion where it splits into two trunks-often referred to as sinuses-which bend ventrally and run backwards (Fig. 14.4c). At the posterior part of the prosomal ganglion, the two aortic trunks merge in the body midline and continue as an unpaired ventral vessel in posterior direction. The arteries for all six pairs of prosomal appendages emanate successively from the two aortic trunks. The arteries which supply the chelicerae are usually the largest. They also supply the brain, eyes and further tissues of the prosoma via a number of side branches. The other arteries supply muscles and tissues in the respective limb and likewise have numerous small side branches. All major

prosomal arteries (aortic trunks and arteries for appendages) are closely associated with the prosomal ganglion since they lie directly on the nervous tissue. In spiders and scorpions, the nerves for the legs show a longitudinal groove in which the leg arteries are embedded.

Many small arteries emerge from the major prosomal arteries and run directly into the prosomal ganglion penetrating the thick perineurium (Fig. 14.6b, c; Huckstorf et al. 2013; Klussmann-Fricke et al. 2013). They are true vessels, in possessing a wall made of connective tissue and therefore represent a true blood–brain barrier (Lane et al. 1981). The vessels inside the nervous tissue form networks to supply those regions having high physiological capacities (Huckstorf et al. 2013). The networks surround the surface of these areas and send vessel loops directly into the dense neuropile. The major vessels which emerge from the anterior aorta resemble afferent vessels based on studies of the networks in scorpions (Klussmann-Fricke et al. 2013). Other vessels collect the haemolymph out of the network and channel it out of the nervous tissue. It is thus justifiable to speak of a *capillarisation* between an afferent and an efferent system of vessels.

In most arachnids, hearts are extended rearward by a *posterior aorta*. In Xiphosura, Pycnogonida, Pseudoscorpiones and Palpigradi, the heart terminates blindly. Posterior aortae run to the rear of the opisthosoma; in spiders they split and supplies the region of the spinnerets. Apart from these two unpaired extensions of the heart i.e. anterior and posterior aortae, pairs of *lateral cardiac arteries* can be present. In spiders, a reduction from five pairs (Mesothelae) to three occurred in different lineages and, like the reduction in pairs of ostia, is probably linked to the development of tracheal respiratory systems. Cardiac arteries emanate from the heart lateroventrally to the ostia (Fig. 14.5b) and supply the viscera in the opisthosoma. They may exhibit a complex branching pattern and have been well studied in spiders, particularly *Cupiennius salei* (Fig. 14.6a; Huckstorf et al. 2013). In that species, three pairs of cardiac arteries are present which supply the anterior, middle and posterior

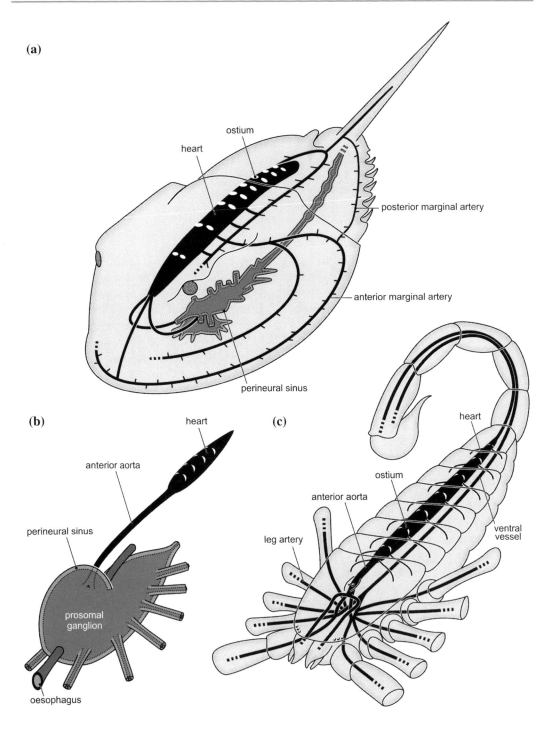

**(a)**

ostium

heart

posterior marginal artery

anterior marginal artery

perineural sinus

**(b)**          heart                    **(c)**          heart

anterior aorta                                                      ostium

perineural sinus                         anterior aorta

leg artery                          ventral
                                    vessel

prosomal
ganglion

oesophagus

parts of the opisthosoma, respectively (Fig. 14.6a). In Xiphosura, the lateral cardiac arteries merge shortly after their origin into a lateral artery running parallel to the heart (Fig. 14.4a; Milne-Edwards 1872; Redmond et al. 1982). This represents one of the many special features of the circulatory system in xiphosurans.

◄ **Fig. 14.4** Schematic representations of circulatory organs in chelicerates. Only major vessels are shown. **a** Horseshoe crab (*Limulus polyphemus*) with a highly complex cardiovascular system. Four pairs of lateral cardiac arteries connect to an artery running parallel to the heart on each side. From the anterior part of the heart, a pair of arteries arches ventrally and connects to the perineural sinus, enveloping the entire central nervous system up to the distal endings of the peripheral nerves. **b** Generalised diagram of the circulatory organs in apulmonate chelicerates. The anterior aorta merges into the perineural sinus which envelopes the complete nervous system. Lateral cardiac arteries are absent (modified after Firstman 1973). **c** Generalised scorpion representing the cardiovascular condition in pulmonate arachnid groups. A complex vascular system branches off from the anterior aorta with numerous secondary arteries that supply the whole prosoma (details shown in Fig. 14.6b, c) (modified after Wirkner and Prendini 2007)

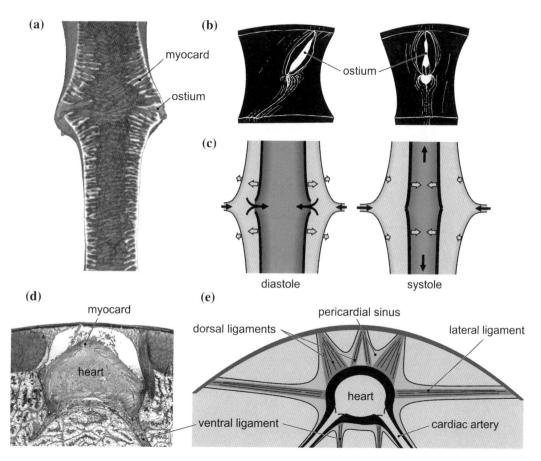

**Fig. 14.5** Hearts in chelicerates. **a** Tomographic radiograph of the heart in the spider *Cupiennius salei*. Dorsal part omitted to allow view of the semilunar cardiomyocytes protruding into the heart's lumen. Upper side is anterior (from Huckstorf et al. 2013, with permission from Urban & Fischer Verlag). **b** Ostia in the scorpions *Uroctonus mordax* (*right*) and *Centruroides sculpturatus* (*left*) representing two different orientations of ostia in scorpions. White areas underneath ostia represent origins of cardiac arteries (modified after Randall 1966). **c** Schematic diagram of the functional linkage of heart and pericard acting as suction-pressure pump (explanation in the text). *Black arrows* represent haemolymph flow, and *grey arrows* represent movements of heart wall and pericard. Note that during both diastole and systole, haemolymph is sucked out of the pulmo-pericardial sinuses (modified after Paul et al. 1994). **d** Cross semithin section through the heart in the scorpion *Buthacus arenicola* showing the well-developed myocard and ventral ligaments (from Wirkner and Prendini 2007, with permission from Wiley and Sons). **e** Schematic representation of a spider heart in cross section depicting arrangement of ligaments which act as antagonists to the myocard (modified after Gerhardt and Kästner 1938)

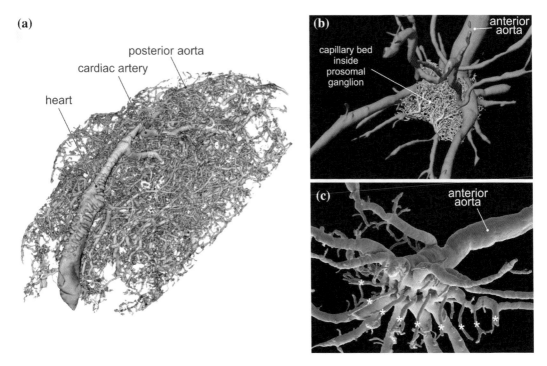

**Fig. 14.6** Arterial systems in chelicerates. **a** Tomographic radiograph of a corrosion cast of the cardiovascular system in the spider *Cupiennius salei*, showing complexity of the cardiac arteries. Three pairs of cardiac arteries branch off the heart and ramify between the extensive midgut gland and other organs in the opisthosoma (from Huckstorf et al. 2013, with permission from Urban & Fischer Verlag). **b** Tomographic radiograph of a cast of the complete anterior aorta system in the scorpion *Brotheas granulatus*. Note complex arterial network supplying the prosomal ganglion (courtesy Bastian-Jesper Klussmann-Fricke). **c** Scanning electron micrograph of an incomplete cast of the anterior aorta system in the scorpion *Centruroides exilicauda*. Main vessels supply the prosomal ganglion. *Asterisks* mark trans-ganglionic arteries channelling haemolymph through midline of the prosomal ganglion (from Wirkner and Prendini 2007, with permission from Wiley and Sons)

### 14.3.1.3 Diaphragms and Sinuses

The lacunar system in chelicerates seems to be highly organised, yet detailed investigations are scarce. In xiphosurans, the route of the haemolymph follows an astoundingly distinct pattern and has been made visible by injection methods. The so-called sinuses have the shape of vessels and lead haemolymph towards the book gills (Fig. 14.7a). In scorpions, two ventral, longitudinal sinuses lead haemolymph towards the book lungs (partly visible in Fig. 14.7b). *Branchio-* or *pulmo-pericardial sinuses* lead from the centralised respiratory organs directly into the pericard. These sinuses have often been termed 'veins' although they only connect to the pericard and not to the heart. The *pericard* itself encloses the heart (Figs. 14.5c, e, 14.7b). Heart and pericardial sinus have been described to act together as a *suction-pressure pump* (Paul et al. 1994; Paul and Bihlmayer 1995). After this model, the contraction of the myocard (systole), results in a change in volume not only in the heart but also in the pericardial sinus where it produces a negative pressure (Fig. 14.5c). The negative pressure in turn causes haemolymph to be sucked out of the pulmo-pericardial sinuses at the same time as it is pumped through the heart. Even during diastole of the heart, the pressure remains negative in the pericardial sinus (Fig. 14.5c) since the pericardial septum is widened by action of the ligaments. This means that haemolymph continues to be sucked out of the pulmo-pericardial sinuses even when it is being sucked into the heart.

In xiphosurans, pycnogonids and apulmonate chelicerates, the central nervous system is

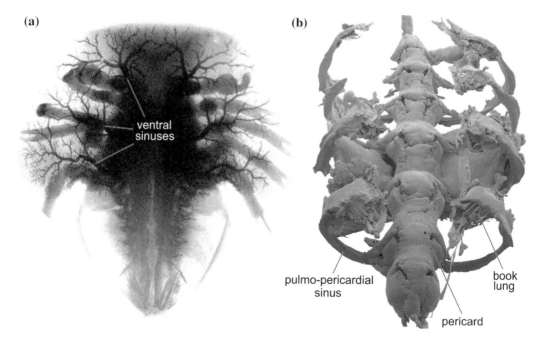

**(a)**

ventral
sinuses

**(b)**

pulmo-pericardial
sinus

book
lung

pericard

**Fig. 14.7** Lacunar systems in chelicerates. **a** X-ray photograph of the ventral lacunar system in the horseshoe crab *Limulus polyphemus*. Complex sinuses channel haemolymph out of the prosoma towards the book gills (from Makioka 1988, permission from Science House pending). **b** Electron micrograph showing a cast of lacunar system in the scorpion *Centruroides exilicauda*. Ventral aspect. Four pairs of pulmo-pericardial sinuses connect to the book lungs (top left two book lungs not completely filled with resin). The entire pericard filled with resin (compare Fig. 14.5d, e; from Wirkner and Prendini 2007, with permission from Wiley and Sons)

embedded into a vascular septum which is connected to the vascular system via the anterior aorta (Fig. 14.4a, b). The haemolymph space enclosed by this septum is called *perineural sinus*. In the studied taxa, the anterior aorta widens close to the posterior dorsal wall of the prosomal ganglion to surround this organ completely. Nerves that emanate from the prosomal ganglion are also surrounded by this septum (Dumont et al. 1965). The sinuses surrounding the nerves apparently become thinner distally and finally disappear near the terminations of the nerves (Milne-Edwards 1872). In xiphosurans and pycnogonids, the intestine is likewise surrounded by a sinus which branches off the perineural sinus.

#### 14.3.1.4 Hydraulic Functions

A special function of the circulatory system is the production of *hydraulic pressure for the movement of limbs* or more precisely particular joints (Fig. 14.1). This has been best studied in spiders which heavily depend on hydraulics for the movement of different appendages. Some of their leg joints are not provided with extensor muscles, and leg extension is accomplished by hydraulic pressure (Ellis 1944; Frank 1957; Parry and Brown 1959). Pedipalps that serve as secondary copulatory structures in male spiders are also moved by changes in hydraulic pressure (Grasshoff 1968, 1973; Lamoral 1973; Huber 1993; Eberhard and Huber 1998; Huber 2004). The pressure that is functionally responsible for the hydraulic movement of prosomal appendages is generated by contraction of dorso-ventral muscles in the prosoma (Kropf 2013).

### 14.3.2 Myriapods

Myriapods show relatively uniform circulatory systems with few small branched arteries. Apart from the minute Pauropoda, which lack any circulatory organs, we find a tubular heart extending

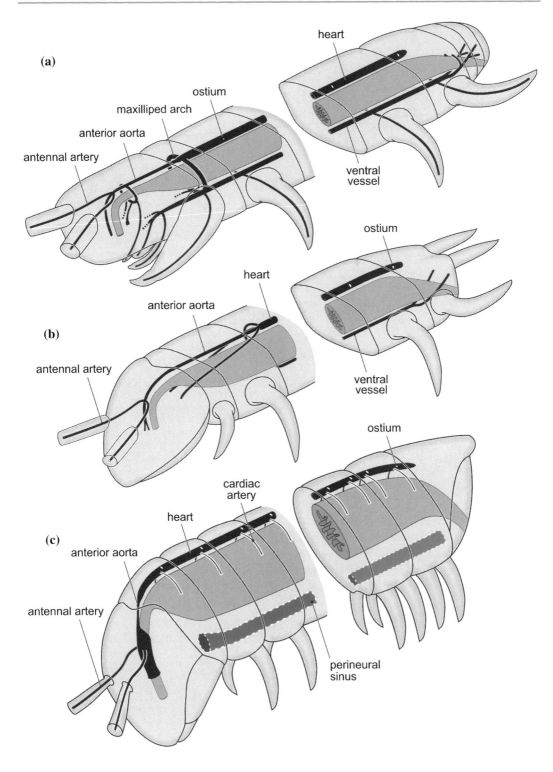

◄ **Fig. 14.8** Schematic representations in circulatory organs in myriapods. **a** Generalised lithobiomorph representing the condition in centipedes with two central longitudinal vessels extending along the entire body which are interconnected by the maxilliped arch (modified after Wirkner and Pass 2002). **b** Generalised

symphylan. Both a dorsal vessel and a supraneural ventral vessel are present. **c** Generalised juliform representing the condition in diplopods. Dorsal vessel extends along the entire body. Each diplosegment of the heart is equipped with two pairs of ostia and two pairs of cardiac arteries (modified after Hertel 2009)

more or less through the length of the trunk. In Diplopoda (Fig. 14.8c), a heart together with lateral cardiac arteries and an anterior aorta constitutes the entire vascular system. In Chilopoda and Symphyla (Fig. 14.8a, b), we additionally find a ventral vessel system which supplies the ventral nerve cord and, in chilopods, also the legs.

A comparative account of all myriapod taxa has yet to be written. Only a handful of papers deal with Symphyla (Tiegs 1940, 1945) and Diplopoda (Seifert 1932; Leiber 1935). More attention has been dedicated to centipedes (Herbst 1891; Fahlander 1938; Wirkner and Pass 2002; Wirkner et al. 2011).

### 14.3.2.1 Hearts

Aside from Pauropoda, all studied myriapod species have *tubular hearts* which extend along the greater part of the trunk. The myocard is single-layered with circularly arranged cardiomyocytes and segmental pairs of incurrent ostia (Seifert and Rosenberg 1978; Økland et al. 1982). Outer and inner sides of the hearts are covered by a thick layer of extracellular matrix. Hearts are attached to the dorsal integument by a number of irregularly arranged suspending strands and laterally by alary muscles which are embedded into a horizontal, dorsal diaphragm (Wirkner and Pass 2002).

In Chilopoda, *Scutigera coleoptrata* the myocard is considerably thicker (Fig. 14.9b) than in other chilopods, since it consists of extraordinarily large cardiomyocytes which extend shingle-like into the lumen of the heart (Wirkner and Pass 2002). In many species, the lips of the ostia project deeply into the heart lumen, thus forming valves which prevent the backflow of haemolymph into the heart. In most myriapods, the posterior end of the heart is closed. Only in some centipedes, hearts are extended posteriorly by one or two arteries. In Diplopoda, each diplosegment

has two pairs of ostia, two pairs of cardiac arteries and two pairs of alary muscles (Fig. 14.9c; Seifert 1932; Leiber 1935).

Few studies have focused on the cardiac physiology in myriapods. In Chilopoda, heartbeat is triggered by a neurogenic automatism which is superimposed over a basic myogenic rhythm, occurring under experimental conditions (Hertel et al. 2002). In Diplopoda, however, it seems that heartbeat generation is achieved by a myogenic automatism (Rajulu 1967; Hertel 2009).

### 14.3.2.2 Arterial Systems

An *anterior aorta* extends the hearts in all myriapod species. In centipedes and symphylans, two pairs of arteries branch off the anterior aorta. In centipedes, there is a pair of arteries supplying the mandibles and a further pair running into the antennae. In Lithobiomorpha, the two mandibular arteries unite to form the so-called *mandibular arch*. In Symphyla, the transition of the heart into the anterior aorta lies relatively posteriorly in the third trunk segment (Fig. 14.8b; Tiegs 1940). Shortly before this transition, a pair of arteries branches off the aorta and runs into the head where the open ends envelop the sacculi of the maxillary glands (Juberthie-Jupeau 1971; Haupt 1976). In the head, a pair of antennal arteries branches off. The anterior aorta in Diplopoda fuses with the dorsal wall of the oesophagus, and antennal arteries emanate from this widening (Fig. 14.8c; Pass 1991).

All chilopod species studied so far possess a pair of cardiac arteries which emanates from the anterior part of the heart and connects to a ventral vessel, thus forming a further vessel arch termed *maxilliped* or *forcipular arch*. The part of the ventral vessel anterior to the maxilliped arch runs into the head and supplies the mouthparts. The posterior part, the *supraneural vessel*, runs above

**(a)**

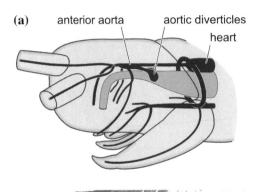

anterior aorta    aortic diverticles    heart

**(b)**

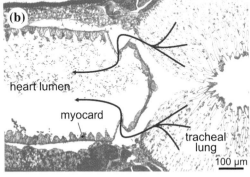

heart lumen

myocard

tracheal lung

100 µm

**(c)**

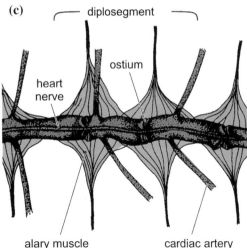

diplosegment

ostium

heart nerve

alary muscle    cardiac artery

**Fig. 14.9** **a** Schematic drawing of head of the centiped *Scutigera coleoptrata* showing position of aortic diverticles. **b** Horizontal section through the heart in the centiped *Scutigera coleoptrata*. Thickness of myocard and close vicinity of ostia to tracheal lungs are interpreted as adaptations to linkage of circulatory and respiratory systems. Arrows mark route of oxygenated haemolymph into heart. **c** Schematic drawing of the heart, cardiac arteries and alary muscles in the diplopod *Strongylosoma pallipes* (from Seifert 1932)

the ventral nerve cord and gives rise to a pair of arteries for each pair of walking legs. A pair of arteries also branches off the maxilliped arch to supply the maxillipeds. In Chilopoda, a complete reduction in the remaining lateral cardiac arteries has taken place in two lineages, i.e. Craterostigmomorpha and Geophilomorpha. In Scutigeromorpha and Scolopendromorpha, a pair of cardiac arteries can be assigned to each leg-bearing segment, while in Lithobiomorpha, only two pairs occur at the 11th and 12th pair of ostia (Rilling 1968; Wirkner and Pass 2002). In Symphyla, an unpaired artery runs laterally around the gut and eventually in a posterior direction. As in chilopods, this supraneural vessel continues in a posterior direction above the ventral nerve cord (Fig. 14.8b). Further, cardiac arteries are absent, while Diplopoda have two pairs in each diplosegment.

### 14.3.2.3 Accessory Pulsatile Organs (Aortic Diverticles)

In Scutigeromorpha, a short unpaired artery branches off the anterior aorta ventrally at the level of the mandibles. This vessel divides, and each branch widens into a blind-ending sack termed aortic diverticle (Fig. 14.9a; Hilken and Rosenberg 2005; Hilken et al. 2006). The diverticles are made of a single-layered myoepithelium (thickness ~15–20 µm) covered on both sides by a layer of extracellular matrix. The myofibrils are circularly and longitudinally arranged. With regard to the function of this enigmatic structure, only speculations can be made. The fact that the aortic diverticles possess a myoepithelium clearly speaks in favour of some sort of pumping capacity, as proposed by Herbst (1891) and Wirkner and Pass (2002). For this to occur, some sort of antagonist, such as elastic connective tissue, would be necessary. However, no such structures were observed in a detailed ultrastructural study (Hilken et al. 2006).

### 14.3.3 Crustaceans

Due to the great disparity of crustaceans, we find a great variety of circulatory organ equipment. It ranges from only short sack-like hearts to complex arrays of structures, such as in decapods which exhibit the most complex circulatory systems within arthropods from both morphological and functional points of view. With respect to the major subgroups, only Malacostraca show a rich evolution of arterial systems, while in the remaining crustaceans, mostly an unbranched dorsal vessel prevails. In cirripeds, a secondary circulatory system has evolved consisting mainly of channels between the organs and a pumping

structure that enables a distinct haemolymph circulation together with the movement of various other organs, such as cirri and penis (Cannon 1947; Burnett 1972; Walker 1991).

A number of reviews exist that provide a detailed physiological and morphological overview on the circulatory system in crustaceans (e.g. Maynard 1960; Mayrat et al. 2006; Wirkner and Richter 2013). With respect to major groups, malacostracans are the best studied from morphological (e.g. Siewing 1956; Wirkner and Richter 2003, 2004, 2007a, b, c, 2008, 2009, 2010) and physiological (e.g. Wilkens et al. 1997a, b; McMahon 2001) points of view.

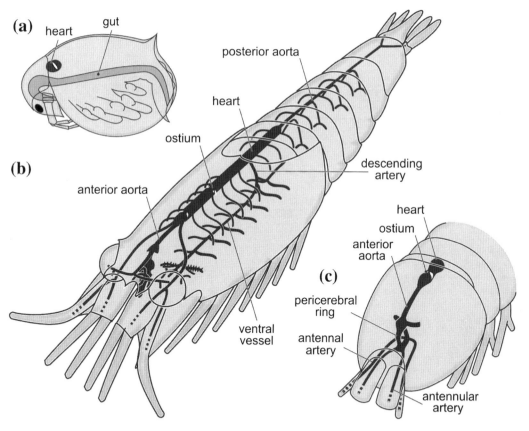

**Fig. 14.10** Schematic representations of circulatory organs in crustaceans. **a** Generalised water flea (*Daphnia pulex*) representing the condition with a completely reduced arterial system (Compare Fig. 14.11d; modified after Pirow et al. 1999). **b** Generalised lophogastrid malacostracan (*Lophogaster typicus*) representing the condition with a complex cardiovascular system. The

tubular heart (compare Fig. 14.2b) is connected to a complex vascular system (modified after Wirkner and Richter 2007a). **c** Generalised thermosbaenacean malacostracan (*Tethysbaena argentarii*) representing the condition of a tubular heart extended by an anterior aorta to supply the anterior cephalothorax (modified after Wirkner and Richter 2009)

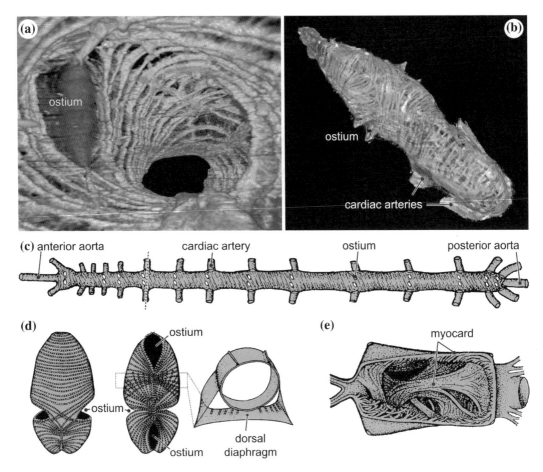

**Fig. 14.11** Hearts in crustaceans. **a** 3D reconstruction of heart in the tanaidacean *Tanais dulongii* from a serial thin-section series. View into heart from posterior. Anterior part of heart omitted (black central area). Cardiomyocytes are arranged circularly to form tubular heart (from Wirkner and Richter 2008, with permission from Pergamon Press). **b** 3D reconstruction from a semithin-section series of heart in the cumacean *Leucon nasica*. The tubular myocard formed by cardiomyocytes not arranged strictly circularly but running into different directions (from Wirkner and Richter 2008, with permission from Pergamon Press). **c** Schematic drawing of heart in the mantis shrimp *Squilla mantis* representing the ancestral condition in malacostracans with a tubular heart extending the entire trunk, a posterior and an anterior aorta, segmentally arranged pairs of ostia and paired cardiac arteries. Broken line marks plane of section depicted in Fig. 14.12a (modified after Alexandrowicz 1932). **d** Schematic drawing of heart in the copepod *Calanus finmarchicus* with a short tubular heart; in contrast to the hearts in decapod (Fig. 14.11e) and euphausiacean malacostracans in which the myocard is arranged in a complex fashion (modified after Lowe 1935). **e** Schematic drawing of heart in the crayfish *Astacus astacus* representing the derived condition of a globular heart made up of a complex three-dimensional and multilayered myocard (modified after Baumann 1921)

## 14.3.3.1 Hearts

In crustaceans, some groups, such as most copepods, lack a heart completely. In those taxa exhibiting a heart, however, the form and structure vary immensely. Two distinct heart forms are observable. In most taxa, a contractile tube is formed by a spiral or circular arrangement of the myocard. These hearts are therefore termed *tubular hearts* (Figs. 14.10, 14.11a–d, 14.12a). In Decapoda and Euphausiacea, the myocard is compact and composed of a meshwork of muscular strands that run in different directions through the heart (Fig. 14.11e; Balss et al. 1940–61; Huckstorf and Wirkner

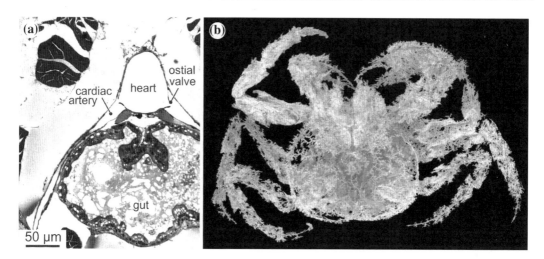

**Fig. 14.12** Arterial systems in crustaceans. **a** Cross section through heart and pair of cardiac arteries in the mantis shrimp *Gonodactylaceus falcatus*. Arteries are equipped with valve preventing haemolymph backflow during diastole (from Wirkner and Richter 2013, with permission from Oxford University Press). **b** Corrosion cast of cardiovascular system of the king crab *Paralomis granulosa* showing the complexity of arterial systems in decapods (from Wirkner and Richter 2013, with permission from Oxford University Press)

2011). Theses hearts are therefore termed *globular hearts*. In addition to these different arrangements of the cardiomyocytes, hearts can vary in thickness and distinctness of the myocard and connective tissue covering; furthermore, they may vary in length, number of pairs of ostia and pairs of cardiac arteries.

The length of tubular hearts ranges from spanning one or few segments to extending through most of the body (Fig. 14.10a, b, c). Globular hearts in euphausiaceans and decapods lie in the posterior thorax and have an extension of just few segments. The sack-like hearts in copepods and cladocerans are sometimes erroneously termed globular hearts (Figs. 14.10a, 14.11d, 14.13a; e.g. Mayrat et al. 2006). However, while the decapod and euphausicean hearts consist of complex myocards (Fig. 14.11e) with a complex physiology, the short tubular hearts in cladocerans and copepods resemble reduced and simple remains of a cardiovascular system.

Haemolymph enters the heart through a number of mostly paired incurrent ostia. Arrangement, size and shape of ostia vary immensely in the different groups.

Haemolymph flow generation is also not uniform. Though empirical data on crustaceans are scarce, hearts with both ends open necessitate a bidirectional flow, as seen at least in lophogastrids (Belman and Childress 1976), where contractions start in the middle of the heart and proceed in both directions. A similar flow is generated through a simultaneous contraction of the complete myocard in branchiopods. In globular hearts, myocardial contraction forces haemolymph centrifugally into the various arterial systems.

While the heart in branchiopods seems to be *myogenic* (Yamagishi et al. 1997), it is *neurogenic* at least in adult isopods and decapods (Mayrat et al. 2006). In the latter, cardiac physiology is studied in detail at least for some species of lobsters and crayfish (detailed reviews by Cooke 2002, McMahon 2012). In malacostracans, three components of the nervous system are associated with the vascular system (Mayrat et al. 2006): (1) the cardiac ganglion, (2) nerves regulating the cardiac ganglion and (3) nerves innervating the cardioarterial valves and the pericard. The cardiac ganglion initiates the contraction of the heart and innervates the entire myocard, as well as the lips of the ostia. As in xiphosurans (see above), pacemaker and follower neurons can be distinguished.

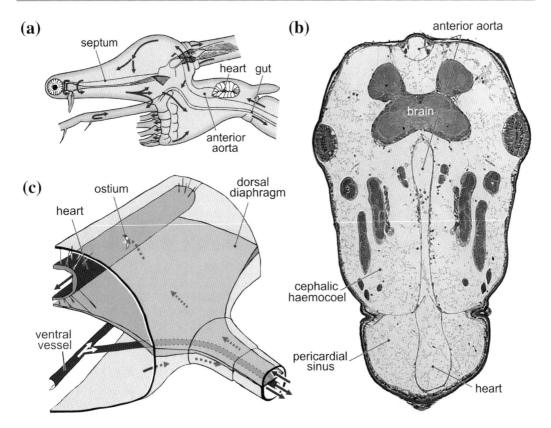

**Fig. 14.13** Lacunar systems in crustaceans. **a** Generalised water flea (*Leptodora kindtii*) depicting complex flow currents in the haemocoel. As in most cladocerans, *L. kindtii* is equipped with a short tubular heart and a short unbranched anterior aorta (modified after Saalfeld 1936). **b** Horizontal section through cephalothorax in the amphipod *Caprella mutica* showing the vast cephalic haemolymph spaces between organs (from Wirkner and Richter 2007a, with permission from Blackwell Publishing). **c** Schematic representation of a thoracic segment in the anaspidacean *Anaspides tasmaniae*. Dorsal diaphragm is extended laterally into the legs, thus dividing the leg haemocoel in an afferent and an efferent sinus (modified after Siewing 1959)

The regulatory nerves act either to accelerate and decelerate heartbeat or to modulate the strength of heartbeat (Wilkens and Walker 1992). The third component is more or less segmentally arranged and affects the distribution of haemolymph into different arterial systems by control of the arterial valves.

### 14.3.3.2  Arterial Systems

Generally, at least one artery, the anterior aorta, emanates from the heart. In ostracods and some cladocerans, however, there is only an anterior excurrent ostium (Fig. 14.10a). Only, malacostracans have extensive arterial systems, some of which reach levels of complexity comparable with those in vertebrates. Arterial ultrastructure has rarely been studied. Three layers are described for decapod arteries. Both external and internal layers are made of connective tissue, while the middle layer consists of an aggregation of cells which contain actin, myosin and tropomyosin; this gives the arteries an unexpected elasticity and contractibility (Chan et al. 2006; Wilkens et al. 2008). The posterior aorta is the only artery containing myocytes (Burnett 1984; Chan et al. 2006). This and the fact that all lateral arteries emanating from the posterior aorta have innervated valves at their origins lead to the hypothesis that the posterior aorta in Decapoda resembles a transformed part of a formerly tubular heart (see Wilkens et al. 1997a).

The *anterior aorta* is unbranched, for example, in branchiopods (Fig. 14.13a), while in malacostracans, it always shows a number of secondary branches that supply at least antennae and the brain (Fig. 14.10b, c). In some cases, these branches form vessel rings around the brain. In euphausiaceans and decapods, complex networks of arteries branch off the anterior aorta to supply the brain and optic lobes. In most malacostracans, various visceral muscles are internalised in the aorta and are thought to supplement the pumping action of the heart, thus resembling accessory pulsatile organs (see below).

Again, a *posterior aorta* occurs only in Malacostraca. In most cases, the posterior aorta is an unbranched artery, although in krill (Euphausiacea) two posterior aortae are present (Huckstorf and Wirkner 2011). In the remaining groups, it is either an unbranched tube running to the posterior body segments, or it has segmentally arranged side branches to supply the tissues and appendages in the pleon.

*Lateral cardiac arteries* are only present in Malacostraca. The complexity of these arterial systems varies greatly. Some taxa lack lateral cardiac arteries entirely (Fig. 14.10c), and others possess few slightly branched pairs of arteries (e.g. Amphipoda and Tanaidacea). Decapoda have few pairs of arteries that display highly complex branching patterns reminiscent of vertebrate vascular systems (Fig. 14.12b). Stomatopoda, Leptostraca and Anaspidacea exhibit to some extent the plesiomorphic condition of one pair of cardiac arteries per trunk segment. The cardiac arteries lead to two main destinations. They either supply visceral organs, such as the gut and the gonads, or the main stems of these arteries run into the thoracopods. However, in the latter instance, branches also supply the viscera. A special case has been described for the Anaspidacea, in which few anterior cardiac arteries supply viscera, while a greater number of posterior cardiac arteries supply the pleopods (Siewing 1954). In Tanaidacea, Cumacea and Isopoda, the last pair of cardiac arteries runs into the pleon, most probably compensating for the lack of a posterior aorta.

A *descending artery* (often confusingly referred to as a sternal artery) is described in Decapoda, Euphausiacea, Anaspidacea and Mysidacea. Although some details vary intra- and/or interspecifically (Vogt et al. 2009; Wirkner and Richter 2013), it can be described as a mostly unpaired artery connecting the heart with the ventral vessel. Despite the variation, it is likely that the descending artery is homologous within Malacostraca.

A longitudinal *ventral vessel* is described for several malacostracan taxa. In Stomatopoda, Decapoda, Euphausiacea, Lophogastrida (Fig. 14.10b), Mysida and some Isopoda, it runs along the ventral side of the nerve cord (*subneural vessel*), while in Anaspidacea, the anterior part extends above the nerve cord (*supraneural vessel*). In contrast to taxa possessing a descending artery (see above), the stomatopod ventral vessel is connected to the dorsal vessel via a number of shunts (rami communicantes) branching off from the lateral cardiac arteries. They function to supply the ventral nerve cord (Siewing 1956). The ventral vessel described for some isopods (Silen 1954) shows some similarities to that of stomatopods but is a structure acquired independently within Isopoda (see Wirkner and Richter 2003). In the remaining taxa, the ventral vessel supplies the thoracopods (and mouthparts) through lateral branches. The part which extends anteriorly from the descending artery is termed the sternal artery, while the part which extends posteriorly is termed the caudal artery.

### 14.3.3.3 Accessory Pulsatile Organs (Myoarterial Formations)

In some crustaceans, parts of somatic muscles such as oesophageal dilators lie within the lumen in sections of the anterior aorta. Such so-called *myoarterial formations* have been described in Ostracoda, Copepoda and Malacostraca (cf. Maynard 1960), although data on the first two taxa are scarce (Wirkner and Richter 2013). These units are functionally interpreted as accessory pulsatile organs or *frontal hearts* since associated muscles rhythmically dilate the vessel wall, thereby reinforcing the haemolymph flow (Steinacker 1978, 1979; Huber 1992; Wirkner and Richter 2007a). According to the anatomy of the associated muscles, three different forms of myoarterial formations have been described (Wirkner and Richter 2010).

#### 14.3.3.4 Lacunar Systems

In crustaceans, the most definitively enclosed sinuses are the *pericardial sinus*, the *branchio-pericardial* and/or the *podo-pericardial* sinuses (often wrongly referred to as veins). The pericardial sinus is confined by the dorsal diaphragm (or pericardial septum) and the dorsal cuticle. Using the analogy of the vertebrate heart, it has been interpreted as a second heart chamber (e.g. McMahon and Burnett 1990). The podo- and branchio-pericardial sinuses channel haemolymph out of the legs (and gills) into the pericardial sinus and therefore play a major role in respiration (Wirkner and Richter 2013). They are confined by a connective tissue septum that is attached to the lateral cuticle. In the cirripeds, a complex channel system is often developed, but these spaces are nevertheless usually considered to be haemolymph sinuses rather than vessels. Their walls are apparently formed from parenchymatous connective tissue cells and interlacing fibres (Cannon 1947).

Generally, lacunae are complex three-dimensional systems of haemolymph spaces, but they are often divided by septa for a greater or lesser portion of their length. Such septa can be found in trunk appendages (legs or gills). They run most of the length of these appendages and thus divide the haemolymph space of the limb into two channels (Fig. 14.13c). They receive haemolymph from the ventral lacunae which are connected to the limbs. Only in a few taxa are they additionally supplied by arteries (Fig. 14.13c). One of the channels usually opens into the lower-pressure region of the pericardial sinus and the other into the ventral sinus, while the pressure drop in the limb ensures the flow of

fluids. In certain amphipods, the appendage septum may be a direct continuation of the pericardial septum, and in decapod malacostracan gills can become quite complex. In flattened gill plates, the essentially continuous septum is replaced by a lacunar network in which oxygenation of haemolymph takes place.

### 14.3.4 Hexapods

The circulatory system of hexapods is often portrayed as uniform and simply organised. This is largely due to the fact that the vascular system appears greatly reduced and is restricted in almost all species to the dorsal vessel. However, insects have numerous supplementary circulatory organs, such as the dorsal and ventral diaphragms as well as great number of so-called accessory pulsatile organs (Fig. 14.14b). The latter are evolutionary novelties of pterygote insects and present an astounding diversity of functional construction, a fact which has become appreciated only in the recent decades. Since insects have undergone immense radiation, it may not be surprising to find all sorts of special adaptations of the circulatory system, particularly in context with the evolution of flight (see Sect. 14.6.2.).

The literature on the structure and function of the circulatory system of hexapods was thoroughly collected and treated in detail in a book by Jones (1977). More recent reviews are given in Miller (1985, 1997), Chapman (1998), Hertel and Pass (2002), Wasserthal (2003a), Pass et al. (2006) and Miller and Pass (2009).

#### 14.3.4.1 Dorsal Vessel

The dorsal vessel is a more or less uniform tube in some basal hexapods only. In the remaining groups, it is differentiated into an anterior aorta and a posterior part, the heart. The heart section bears the ostia and has a much thicker musculature since it does most of the pumping. In the plesiomorphic condition, the entire dorsal vessel lies directly under the tergal cuticle, whereas in many pterygotes, the aorta is often detached, looped or coiled, and in some species, it runs in a straight line through the middle of the thorax (Hessel 1969; Krenn and Pass 1994, 1995).

The anterior end of the dorsal vessel differs among hexapods. The plesiomorphic condition is characterised by antennal arteries which occur only in Diplura (Gereben-Krenn and Pass 1999; Fig. 14.14a). In most other insects, the dorsal vessel terminates in the head just behind the brain where it is fused with the connective tissue sheath of the brain. Haemolymph passes further to the frontal region of the head through the channel-like sinus between the brain and oesophagus. The posterior end of the dorsal vessel, likewise, differs in anatomy. In Diplura, it continues into a caudal chamber that has emanating arteries which extend into the cercal appendages (Gereben-Krenn and Pass 1999). Remarkably, the flow of haemolymph in the dorsal vessel of these animals is bidirectional and regulated by intracardiac valves. Haemolymph thus flows constantly not only in the anterior direction but also posteriorly. This *bidirectional flow* apparently represents the plesiomorphic condition in hexapods and is essential for the supply of haemolymph to long abdominal appendages (Gereben-Krenn and Pass 2000). In most exopterygotes and larvae of endopterygotes, the posterior end of the dorsal vessel is closed and a peristaltic wave of contraction is possible only in the forward direction (*unidirectional anterograde flow*) (Fig. 14.14b). In the adults of Coleoptera, Lepidoptera, Diptera and probably many other insects, a periodic reversal of the flow takes place in the dorsal vessel that is referred to as *heartbeat reversal* (Gerould 1933; Jones 1977; Wasserthal 1980, 2007, 2012; Angioy and Pietra 1995; Dulcis and Levine 2005; Sláma and Farkás 2005; Glenn et al. 2010; Sláma 2010). Several special anatomical adaptations are related to this and have been well studied in some higher Diptera (Angioy et al. 1999; Curtis et al. 1999; Wasserthal 1999, 2007; Glenn et al. 2010; Lehmacher et al. 2012) (Fig. 14.15a, b) and Lepidoptera (Wasserthal 1980, 2003c). The physiological function

of heartbeat reversal can be manifold: on the one hand, it supports a proper mixing of molecules throughout the insect body; on the other hand, it may be involved in thermoregulation and tracheal ventilation (see Sect. 14.6.2).

The ostia exist in a broad variety among the various hexapods, and different kinds of ostia may even occur in the same animal (Fig. 14.15c; see Pass et al. 2006 for a summary). Most common are incurrent ostia. In addition, excurrent ostia with sphincter-like valves occur in some exopterygotes (Nutting 1951; Pass et al. 2006). In insects known to undergo heartbeat reversal, paired or unpaired excurrent ostia occur at the caudal end of the dorsal vessel. In Lepidoptera, two-way ostia are also present (Wasserthal 1980, 2003c). They have only a singular valve and allow in- or outflow depending on the flow direction.

Dorsal vessel contractions are based on myogenic pacemakers which are modulated by neuronal activities and eventually by hormonal factors. The innervation can attain quite a complexity, but no consistent pattern has been found that holds true for all insects. It may be comprised of several components, such as segmental nerves from the ventral nerve cord, nerves coming from the stomatogastric nervous system or even peripheral cardiac neurones (reviews: Miller 1985, 1997). Of special interest is the innervation of the dorsal vessel in insects with heartbeat reversal. In some larvae, the dorsal vessel has no obvious innervation, and the anterograde pumping is driven myogenically. The innervation develops later in the pupal stage together with the beginning of heartbeat reversal (Davis et al. 2001; Dulcis and Levine 2003). Both the anterior and the posterior myogenic pacemakers in their dorsal vessel are subject to a complex neuronal control which is still poorly understood (Ai and Kuwasawa 1995; Kuwasawa et al. 1999; Dulcis et al. 2001; Dulcis and Levine 2003, 2005).

### 14.3.4.2  Other Vessels

A so-called circum-oesophageal vessel ring occurs only in some basal hexapods. It arises from the dorsal vessel just behind the brain,

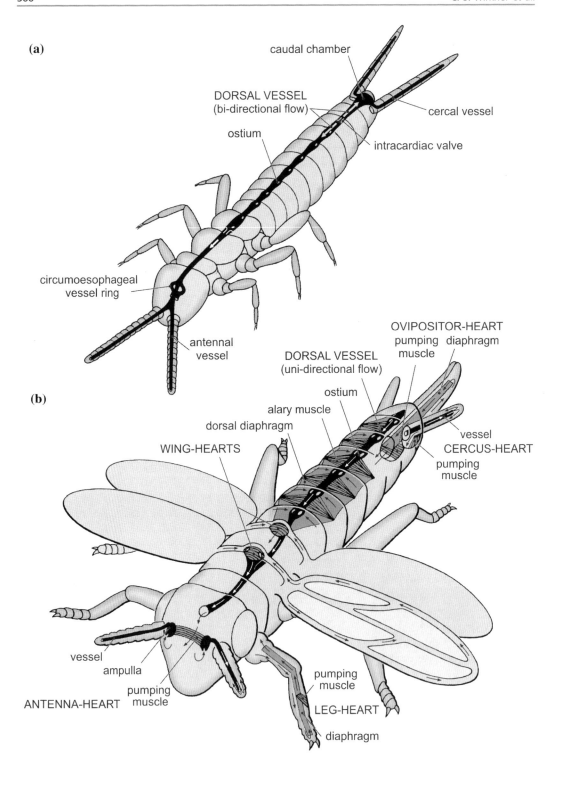

◀ **Fig. 14.14** Schematic representations in the circulatory organs in hexapods. **a** Dipluran *Campodea* representing an ancestral condition in hexapods having antennal and cercal arteries, circum-oesophageal vessel ring in head and bidirectional flow in the dorsal vessel (modified after Gereben-Krenn and Pass 1999). **b** Generalised pterygote insect representing a derived condition with all potentially occurring accessory pulsatile organs and an unidirectional anterograde flow in the dorsal vessel. Accessory pulsatile organs evolved with functional mechanisms differing among insects even for a given appendage. Vessels in *solid black*, diaphragms and pumping muscles in *grey*; *arrows* indicate direction of haemolymph flow (modified after Pass 2000)

encompasses the oesophagus and has a large ventral opening that is anteriorly directed (Gereben-Krenn and Pass 1999; Pass et al. 2006) (Fig. 14.14a). Additional vessel structures are the lateral segmental arteries in some mantids and cockroaches (Nutting 1951) (Fig. 14.15d). At the vessel base are sphincter-like muscular valves which only permit an outflow of haemolymph from the dorsal vessel (Fig. 14.15e). In the cockroach *Periplaneta*, the valves open and close independently from each other, whereby their physiological control has not yet been elucidated (Miller 1997). From an evolutionary point of view, it is remarkable that these segmental arteries resemble those of other arthropods. However, since they are absent in all ancestral groups of hexapods, they are mostly considered as evolutionary novelties (e.g. Kristensen 1975). Finally, it should be noted that no hexapod has a ventral longitudinal vessel.

### 14.3.4.3  Diaphragms and Sinuses

Hexapods have up to two diaphragms, which extend horizontally in the body cavity. The dorsal diaphragm is formed in most species by a fenestrated septum of connective tissue and associated alary muscles. These muscle plates are limited to the abdominal section, and their number varies (e.g. 12 in *Periplaneta*, 4 in *Apis*). In most cases, their myocytes join together beneath the heart. When they contract, the dorsally arched diaphragm is flattened, thereby causing haemolymph to be sucked from the central body cavity into the pericardial sinus. However, the dorsal diaphragm may have an entirely different organisation, for example, in Lepidoptera, it is completely flat and the muscles are arranged in a loose network (Wasserthal 2003b), or in the fly *Calliphora*, the muscle fibres run in the longitudinal direction and surround the dorsal vessel like a basket meshwork (Wasserthal 1999). The functional significance of these arrangements is not clear. The ventral diaphragm is differently formed in the various insects and often completely absent as in basal hexapods (Richards 1963). It apparently plays an important role in the perfusion of the ventral nerve cord in that it enforces the haemolymph flow with undulating movements.

### 14.3.4.4  Accessory Pulsatile Organs

The pumping activity of the dorsal vessel and the diaphragms results in the circulation of haemolymph in the central body cavity but is insufficient for circulation to outlying dead-end structures, such as long body appendages. Arteries, which fulfil this task, are common in many other arthropods, but in hexapods, they are only found in few basal taxa. In the pterygotes, special circulation pumps exist for this task which are independent of the dorsal vessel. They are autonomous pumping organs with their own beat rates and are therefore appropriately referred to as auxiliary hearts. Such organs may be associated with the antennae, some mouthparts, legs, wings and long abdominal appendages, such as cerci and ovipositors (Fig. 14.14b; reviews: Pass 1998, 2000).

*Antennal Circulatory Organs*
Aside from the Diplura, which possess antennal arteries, almost all other insects have autonomous pulsatile organs (Pass 1991, 1998). Of these, the best investigated with respect to both morphology and physiology is the antenna-heart of the

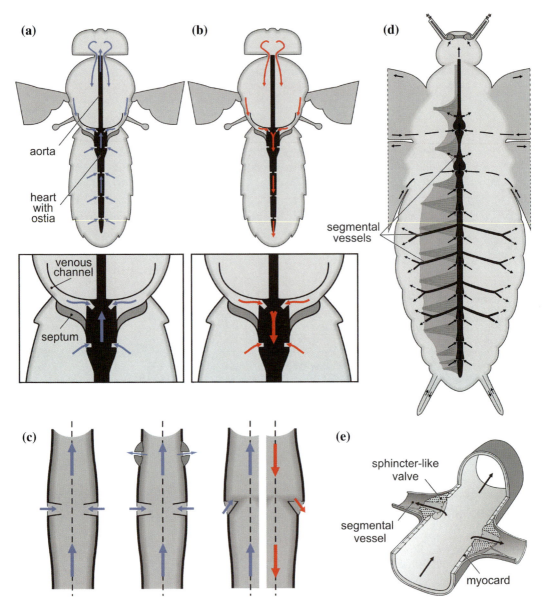

**Fig. 14.15** **a–b** Haemolymph circulation in *Drosophila* during forward (**a**) and backward (**b**) heartbeats. Overview in upper diagrams; magnified details in lower diagrams showing enlarged anterior part of abdominal heart and connection to venous thoracic channels. Haemolymph from lateral thorax flows through these channels directly to first ostia pair. Note that lateral flow is maintained independently of heartbeat direction (modified after Wasserthal 2007). **c** Various kinds of ostia in hexapods. *Left* incurrent ostia with paired lips; *middle*: incurrent ostia plus excurrent ostia with sphincter-like valves; *Right* two-way ostia with a single lip, *left half* showing incurrent flow during anterograde phase, *right half* showing excurrent flow during retrograde phase (modified after Pass et al. 2006). **d** Cardiovascular system in the cockroach *Blaberus*. Dorsal vessel with two short thoracic and four long abdominal segmental arteries. Haemolymph flow from posterior wing veins into thoracic aorta indicated by *broken lines*. In right half of diagram, dorsal diaphragm and alary muscles omitted for reasons of clarity (modified after Nutting 1951). **e** Portion of dorsal vessel with branching of segmental vessels in the cockroach *Periplaneta*. At the base of segmental vessels, sphincter-like valve controlling outflow into segmental vessels (valves shown open) (modified after Pass et al. 2006). *Coloured arrows* in the various diagrams indicate direction of haemolymph flow: anterograde in *blue* and retrograde in *red*

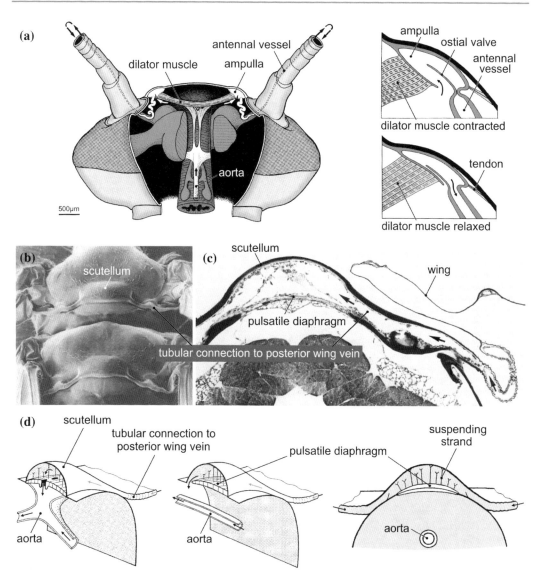

**Fig. 14.16** Accessory pulsatile organs of antennae and wings of insects. **a** Antenna-heart of the cockroach *Periplaneta americana* consisting of an ampulla at the base of each antennal vessel and associated dilator muscles (right half of brain removed). Diastole in upper scheme: lowered pressure in widened ampulla enables haemolymph to enter and causes constriction of the antennal vessel base hindering backflow. Systole in lower scheme: ampulla flattened due to elasticity of wall and of an associated tendon (modified after Pass 1985). **b–d** Wing circulatory organs of insects. **b** SEM of dorsal thorax in the scorpionfly *Panorpa communis*. Wing-hearts are located below the scutellum elevation; small lumen beneath connected by cuticular tubes to posterior wing veins (from Krenn and Pass 1993, with permission from Pergamon Press). **c** Semi-thin cross section through dorsal metathorax of the scorpionfly *Nannochorista neotropica* showing pulsatile diaphragm and haemolymph pathway (*arrows*) from wings to scutellum (from Krenn and Pass 1993, with permission from Pergamon Press). **d** Schemes of different wing circulatory organs. Winged thorax segment with scutellum elevation and associated pulsatile structures (left and middle diagram viewed from an angle behind, right in cross section). Contraction of the muscular parts induce haemolymph flow from posterior wing veins to scutellum and further into dorsal vessel or thoracic cavity. Left: aortic diverticle with thickened dorsal wall and paired incurrent ostia. Middle and right: separate dorsal diaphragm with anterior valve allowing haemolymph flow directly into thorax (modified after Pass 1998)

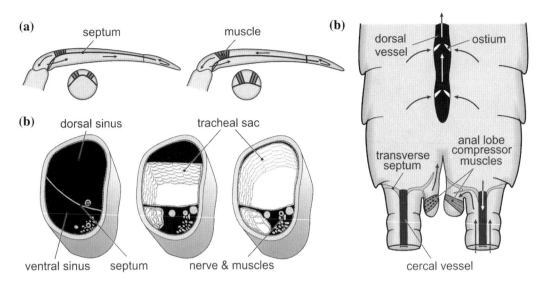

**Fig. 14.17 a** Leg-heart in the waterbug *Ranatra*. The leg haemocoel is divided by a thin septum into two sinuses. Contraction of the associated muscle narrows the dorsal sinus and forces haemolymph towards the thorax. At the same time, haemolymph is sucked from the thorax into the ventral sinus. The diaphragm is lacking at the apex, enabling turn of flow (modified after Brocher 1909). **b** Cross section of the tibia in the hawkmoth *Deilephila*. Pupal leg (*left diagram*): haemocoel divided by a thin septum into two sinuses with counter-current haemolymph flow. Adult leg (*middle and right diagram*): haemocoel divided into two sinuses by a large tracheal sac whose volume changes in correlation with heartbeat reversal; during retrograde beating phases (*right diagram*), haemolymph is removed, and the elastic tracheal sac compensatorily expands (modified after Wasserthal 1998). **c** Cercus-heart in Plecoptera. Cercal vessels are separate from the dorsal vessel and originate at a transverse septum at the cercus base. Pulsatile part in anal lobe depicted in different phases of action. *Left body half* contraction of the pumping muscle compresses the anal lobe and forces haemolymph through a slit valve into body cavity. *Right body half* relaxation of muscle allows the anal lobe to return to its original shape; thereby, haemolymph is drawn from cercus haemocoel into anal lobe and from body cavity into cercal vessel (modified after Pass 1987)

cockroach *Periplaneta americana* (Hertel et al. 1985, 1997, 2012; Pass 1985; Pass et al. 1988a, b) (Fig. 14.16a). Remarkably, it also functions as a neurohaemal organ which probably releases hormones that may be involved in the control of the antennal sensory system of the cockroach (Pass et al. 1988a, b). An endocrine function is also assumed for a tissue mass associated with the antennal circulatory organs in some Lepidoptera (Schneider and Kaissling 1959; Raina et al. 2003). The pumping muscles associated with the antenna-hearts exhibit a widely different anatomy and function either as dilators or as compressors; the antagonists to these muscles are the elastic wall of the ampullae or associated elastic suspensory structures (Clements 1956; Selman 1965; Pass 1980, 1988, 1991; Sun and Schmidt 1997; Matus and Pass 1999; Wasserthal 2003b; Baum et al. 2007).

## Wing Circulatory Organs

The wings of insects are not simply dead cuticular structures. Their veins are hollow tubes filled with haemolymph to supply living cells, such as nerves which innervate sensilla on the wing surfaces. The haemolymph circulates in most insects according to a typical pattern: efferent in the anterior veins and afferent in the posterior (Arnold 1964). In some Lepidoptera, an exchange of the haemolymph occurs simultaneously in all wing veins by an oscillatory mode (Wasserthal 1980). The motor for haemolymph movement is located in the thorax just below the scutellum which is evident as a small elevation of the tergal cuticle (Fig. 14.16b–d). The scutellum is the pump housing and is connected to the posterior wings via cuticular tubes (Krenn and Pass 1993). The anatomy of the pump differs (Fig. 14.16d); it may consist of specially modified parts of the dorsal vessel (aortic diverticles) or organs completely independent of the dorsal vessel which may thus be labelled wing-hearts (Whedon 1938; Selman 1965; Krenn and Pass 1994, 1995). All wing circulatory organs function according to the same principle: haemolymph is sucked out of the posterior wing veins into the dorsal vessel or directly into the thoracic haemocoel (Fig. 14.16d).

*Leg Circulatory Organs*
The legs of hexapods contain no vessels for circulation. Their haemocoel is usually divided by a longitudinal septum of connective tissue into two sinuses (Fig. 14.17a, b). Only at the tip of the appendage is the septum lacking, allowing for a counter-current flow of haemolymph. How this flow is generated remains unclear in most cases and may be different among species (Pass et al. 2006). It can be induced by periodic abdominal pressure pulses (Ichikawa 2009). In some species of Heteroptera, however, a rhythmically contracting muscle is associated with the leg septum. Thereby, haemolymph is actively forced through the appendages, and these organs can thus be designated as leg-hearts (Debaisieux 1936; Hantschk 1991) (Fig. 14.17a). A leg-heart with a completely different functional morphology was described in the locust (Hustert 1999).The anatomy of the leg circulatory organs may even change during development, as in the hawkmoth *Deilephila*, where a thin septum is replaced by a large elastic tracheal sac during metamorphosis (Wasserthal 2003b) (Fig. 14.17b).

*Circulatory Organs of Abdominal Appendages*
Some basal hexapods, such as Archaeognatha, Zygentoma and Ephemeroptera, exhibit a long terminal filum and long cerci. These insects are characterised by a bidirectional flow in their dorsal vessel, whereby the caudal end forms a strong muscular chamber with a vessel extending into the terminal filum. Curiously, the chamber contracts with a different frequency from that of the anterior part of the dorsal vessel (Meyer 1931; Gereben-Krenn and Pass 2000). Autonomous cercus-hearts have been detected in Plecoptera (Pass 1987) (Fig. 14.17c). In most other insects, longitudinal septa are present in the cerci, but no specific pulsatile apparatus has been found (Murray 1967; Pass et al. 2006). Recently, in the

cricket *Acheta*, autonomous ovipositor-hearts were discovered (Hustert and Pass in prep).

Finally, it should be noted that accessory hearts may also contribute to the hydraulic-based movements of body appendages. For example, the long proboscis of Lepidoptera is uncoiled by special pumping organs in the head at the base of the proboscis (reviewed in Krenn 2010), and the lamellate antennae in scarabaeid beetles are spread out by increased haemolymph pressure produced by action of the antenna-heart (Pass 1980).

### 14.3.4.5 Extracardiac Pressure Pulses
In addition to the action of the dorsal vessel, various regular abdominal pressure pulses occur which clearly influence flow conditions, for example, the very pronounced abdominal breathing movements. Beyond that, there are pressure pulses which are barely discernible from the outside. The functional significance of these extracardiac pulses, sometimes called coelopulses, is hotly debated (Kuusik et al. 1996; Sláma 1999, 2008; Tartes et al. 2002). In the mosquito *Anopheles*, it was conclusively demonstrated that there are slight but regular ventral abdominal contractions which are correlated with heartbeat reversal. The contractions obviously promote extracardiac haemolymph propulsion in the abdominal haemocoel during periods of anterograde heart flow (Andereck et al. 2010).

## 14.4    Development of the Circulatory System

### 14.4.1 General Development of the Circulatory Organs

All major components of the circulatory system such as vessels and muscles arise as derivatives of the mesoderm [reviewed in Johannsen and Butt (1941), Siewing (1969), Anderson (1973)]. The *heart* arises from the cardiac primordium which is located at the dorsal-most edge of the

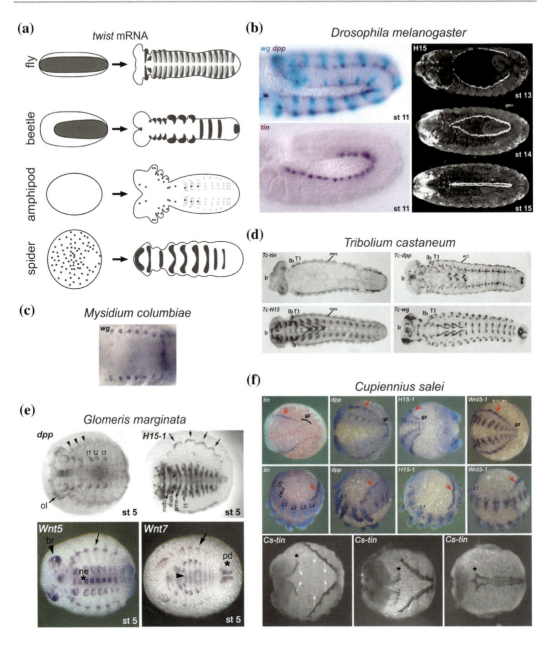

somites. Formation of the tubular vessel is accomplished by the upward migration of the mesoderm and the subsequent joining of opposite cardioblasts in the dorsal midline (Fig. 14.18b, f). This occurs mostly in conjunction with a process called dorsal closure during which the lateral margins of the embryo extend dorsally and enclose the yolk. The dorsal mesoderm adjacent to the heart primordium gives rise to the pericardial septum and to the alary muscles. While the formation of the heart appears similar among arthropods, the formation of the vessels varies greatly.

The *anterior aorta* develops independently of the heart in many arthropods either from the precheliceral, antennal, preantennulary or antennulary somite. The initial gap between anterior aorta and heart is spanned in these species later during development by a posterior extension of the anterior aorta (Manton 1928, 1934;

◀ **Fig. 14.18** Genes involved in heart development.
**a** Schematic representation of *twist* expression at early
stages in different arthropods (modified after Price and
Patel 2008). **b** *Drosophila melanogaster*, *left panel*
shows lateral views of stage 11 embryos: ectodermal
*wg* (*turquoise*) and *dpp* (*purple*) expressions (*upper left
panel*) restrict *tin* to dorsal mesoderm (*lower left panel*;
from Bodmer and Frasch 2010, with permission from
Academic Press). *Right panel* shows heart formation
from opposite rows of cardioblasts visualised by *H15*
expression in heart precursors (dorsal view, numerals
indicate stages; from Griffin et al. 2000, with permission
from Academic Press). **c** *Mysidium columbiae*, ventral
view: *wg* expressed in patches at the dorsal margin of the
embryo (from Duman-Scheel et al. 2002, with permission from Springer Verlag). **d** *Tribolium castaneum*,
ventral views of germ band, retraction stage embryos: *tin*

and *H15* are expressed in the dorsal mesoderm, while
*dpp* and *wg* (in patches) are found in the overlying dorsal
ectoderm (from Janssen and Damen 2008, with permission from Blackwell Publishing). **e** *Glomeris marginata*,
ventral views of stage five embryos: expressions of *dpp*,
*Wnt5*, *Wnt7* and *H15-1* in heart primordium (from Prpic
2004, with permission from BioMed Central; Janssen
et al. 2004, with permission from Academic Press; Prpic
et al. 2005, with permission from Blackwell Publishing).
**f** *Cupiennius salei*: formation of heart from opposite
rows of *tin*-positive cardioblasts (lower panel). Note that
the anterior part is formed from a separate population of
*tin*-positive cells (*asterisks*). *tin*, *dpp*, *Wnt5* and *H15-1*
are expressed at the dorso-lateral margin of the germ
band, which will give rise to the heart (upper panel; from
Janssen and Damen 2008, with permission from Blackwell Publishing)

Hickman 1936; Johannsen and Butt 1941; Nair
1956; Weygoldt 1961, 1975; Scholl 1963, 1977;
Anderson 1973; Janssen and Damen 2008). In
many other arthropods, for example, in *Scolopendra* (Anderson 1973) or in *Drosophila* (e.g.
Bodmer and Frasch 2010), the anterior aorta and
heart forms *en bloc* from the mesoderm of successive segments.

The development of the *lateral arteries* differs among arthropods. In chelicerates and
myriapods, they are formed similarly to the heart
by joining of cells from the lateral margins of
adjacent somites (Johannsen and Butt 1941;
Weygoldt 1975; Scholl 1977), while in malacostracan crustaceans, it is described that they
grow out from the heart. Initially, all crustacean
lateral arteries arise as pairs including the aorta
descendens of which one artery later degenerates
(Weygoldt 1961; Vogt et al. 2009).

The *ventral vessel* of myriapods can be
formed in two different ways. In the chilopod
*Scolopendra*, it resembles the formation of the
heart as it develops by the joining of the ventral
edges of the somites in the ventral midline (Johannsen and Butt 1941; Anderson 1973). In
contrast, in the symphylan *Hanseniella*, it arises
from cells of the median mesoderm anlage
(Anderson 1973). The subneural artery of
decapod crustaceans is assumed to originate
from the aorta descendens (Weygoldt 1961).
Similarly, the neural artery of *Limulus polyphemus* is formed by the outgrowth of the aortal
trunks which extend posteriorly as two separate

tubes above the nerve cord. During later stages
in development, these arteries wrap around the
nerve cord and fuse to form the perineural sinus
(Kingsley 1893).

The origin of the *arteries supplying the
appendages* was studied in a whip spider
(Weygoldt 1975). Initially, the mesoderm
extending into the appendages forms only one-
half of the blood channel. The completion of the
artery occurs later during development, presumably by cells from the mesodermal sheath of
the accompanying nerve. The formation of the
fine *peripheral vascular system*, especially the
highly branched vessels in malacostracans and
arachnids, is poorly investigated. In the shrimp
*Metapenaeus ensis*, it was shown that the complexity of the vascular system increases with
each larval stage. Starting with a bulbous heart,
one pair of ostia and a single anterior aorta in the
protozoëa, the vascular system progressively
changes until the juvenile stage to a heart with
three pairs of ostia, a posterior aorta, lateral
branched arteries and loss of the anterior aorta
anterior to the first pair of lateral arteries
(McMahon et al. 2002).

## 14.4.2 Genetic Control of Mesoderm and Dorsal Vessel Formation

With respect to developmental genetics, *Drosophila melanogaster* is by far the most intensively studied arthropod, and almost our entire

knowledge of heart formation is derived from research on this model organism. To date, a large number of transcription factors and signalling pathways are known which play crucial roles in cardiogenesis and have been extensively reviewed (e.g. Bodmer and Frasch 2010; Cripps and Olson 2002; Zaffran and Frasch 2002; Tao and Schulz 2007; Bryantsev and Cripps 2009; Reim and Frasch 2010 and references therein). In *Drosophila*, the ventral third of the blastoderm is specified as presumptive mesoderm by activation of the transcription factor *twist* (Fig. 14.18a). Twist is a master regulator of mesoderm development and subsequently activates genes required for gastrulation, patterning and differentiation of the mesoderm. One of its target genes is the transcription factor *tinman* which is initially expressed throughout the whole mesoderm. Following invagination, the mesoderm spreads out dorsally along the ectoderm, forming a monolayer on either side of the embryo. This migration brings the dorsal region of the mesoderm into close contact with ectodermal cells that express *decapentaplegic* (*dpp*). Dpp signalling maintains *tinman* expression only in the dorsal mesoderm which subsequently differentiates into cardiac and visceral mesoderm. A second signal encoded by the segment polarity gene *wingless* (a member of the *Wnt* family) which is primarily expressed in the ectoderm in transverse stripes perpendicular to the *dpp* domain leads to the restriction of *tinman* expression to the cardiac precursors (Fig. 14.18b). Activation of GATA and T-box genes (e.g. *H15*) within these precursors results in the functional diversification of the cardioblasts into cardiomyocytes and ostia cells. While cardiomyocytes remain *tinman* positive, ostia cells express the transcription factor *seven-up* which probably acts as a repressor of *tinman* in these cells. The subdivision of the dorsal vessel into a thin aorta and a muscular heart is established by the intrinsic and partly overlapping expression of homeotic selector (Hox) genes (Lovato et al. 2002; Ponzielli et al. 2002; Perrin et al. 2004; Monier et al. 2005; Ryan et al. 2005; Monier et al. 2007).

Cloning of orthologs and subsequent in situ hybridisation showed that some of the genes known to play a role in mesoderm and dorsal vessel development in *Drosophila* are also expressed in other arthropods (Fig. 14.18c–f). *Twist*: in the beetle *Tribolium castaneum* (Handel et al. 2005) and the spider *Achaearanea tepidariorum* (Yamazaki et al. 2005), *twist* expression starts in the presumptive mesoderm before gastrulation as in *Drosophila*. In contrast, in the amphipod *Parhyale hawaiensis*, *twist* is first activated in the mesoderm after gastrulation, indicating that the role of *twist* in mesoderm specification is not conserved among arthropods. However, its expression during the subsequent development suggests a conserved role in mesoderm patterning and differentiation (Price and Patel 2008). *Dpp*: like in *Drosophila*, *dpp* is expressed in the ectoderm overlying the heart precursors in *Tribolium* (Janssen and Damen 2008). In the millipede *Glomeris marginata* (Prpic 2004) and the spider *Cupiennius salei* (Janssen and Damen 2008), expression was detected at the dorso-lateral edge of the germ band where the heart precursors form. *Wnts*: *wingless* was found in the ectoderm in segmental spots at the dorsal edge of the embryo in *Tribolium* (Janssen and Damen 2008) and in a similar position in the mesoderm of the crustacean *Mysidium columbiae* (Duman-Scheel et al. 2002). In *Glomeris*, *Wnt5* and *Wnt7* are expressed in the heart primordium (Janssen et al. 2004), and in *Cupiennius*, *Wnt5-1* was detected in the dorsal edge of the embryo (Janssen and Damen 2008). A recent study in *Drosophila* showed that *Wnt4* also plays a role in heart formation, indicating that various Wnt family members are involved in cardiogenesis within arthropods (Tauc et al. 2012). *Tinman*: in *Tribolium* and *Cupiennius*, *tinman* is expressed in the cardiac mesoderm and later in the dorsal vessel (Janssen and Damen 2008). In the spider, *tinman* expression showed that the anterior part of the dorsal vessel is formed by a separate population of *tinman*-positive cells, which were reported to originate from the ridge of the dorsal cephalic lobe (Janssen and Damen 2008) and may correspond to the anterior aorta described as developing separately from the heart (Fig. 14.18f). *H15*: in *Tribolium*, *H15* expression was detected in heart precursors (Janssen and Damen 2008). In *Glomeris*, *H15-1* but not *H15-2* is expressed in the developing dorsal vessel (Prpic et al. 2005; Janssen et al. 2006, 2008), whereas both were found in the dorsal vessel in *Cupiennius* (Janssen and Damen 2008).

Although remarkable differences in the temporal onset and expression pattern of some of the genes exist, the occurrence of *tinman* expression in the dorsal vessel of so distantly related groups as flies and spiders suggests a largely conserved regulatory network of dorsal vessel formation in arthropods. In this hypothetical pathway, *twist* activates *tinman* in the mesoderm, whose expression is then maintained by *dpp*. *Wnt* genes specify heart precursors, and T-box genes like *H15* maintain *tinman* in the heart precursors after *dpp* expression has ceased, leading to the functional diversification of cells within the dorsal vessel.

### 14.4.3 Accessory Pulsatile Organs

These organs deserve special interest from the developmental point of view since they are clearly evolutionary novelties. In the decapod crustacean *Palaemonetes varians*, the pumping muscles associated with the aortic wall that constitute the *frontal heart* are derived from adjacent parts of the mesoderm and not from the cardiac mesoderm (Weygoldt 1961). Reports on the development of *wing circulatory* organs of insects are known from two species. While in *Locusta migratoria*, a modified part of the heart forms the pulsatile apparatus during postembryonic development (Krenn 1993), in *Drosophila*, there are heart-independent progenitors which give rise to the wing-hearts (Tögel et al. 2008). The progenitors develop close to the heart during embryogenesis and are specified by downregulation of *tinman* expression. Remarkably, during their development, they express

several genes typical for cardiomyogenesis. The mature wing-heart muscles, nonetheless, resemble adult somatic muscles (Lehmacher et al. 2009).

## 14.5 Major Evolutionary Trends

### 14.5.1 Ancestral Condition

The vascular system of all arthropods is organised in a way which is clearly segmental. The individual elements are found in serial configuration in a number of segments. Even in unrelated lineages, the same vascular elements occur in a similar arrangement. It is therefore plausible that a distinct segmental set of circulatory organ structures are part of the ground pattern of arthropods. This set may comprise a portion of the dorsal vessel plus one pair of ostia, one pair of cardiac arteries and possibly also a portion of the ventral vessel.

How is the circulatory system organised in groups related to the arthropods? Conditions vary significantly in the two other taxa of the panarthropod clade. Tardigrades have no circulatory organs, while onychophorans, in contrast, have a dorsal vessel with paired ostia, a dorsal diaphragm and antennal arteries (Pass 1991; Storch and Ruhberg 1993). We may, then, tentatively conclude that the dorsal vessel and the dorsal diaphragm at least belong to the ground pattern of the panarthropods, while the condition found in tardigrades may be due to their minute body size and therefore represent a derived condition.

Evolutionary scenarios for the circulatory system in the sister group to the Panarthropoda differ according to phylogenetic hypothesis. In the traditional Articulata hypothesis, the open circulatory system of arthropods is hypothesised to be derived from the closed circulatory system of annelids. The differences in the body cavities of these two groups are held to be a plausible explanation for the evolutionary changes that are thought to have occurred. In annelids, the vascular system bridges a metameric arrangement of separate coelomic cavities. In arthropods, in contrast, a single body cavity, i.e. the mixocoel, developed in the context of the evolution of the cuticular exoskeleton. In this manner, the closed system could have transformed into an open one. Although the origin of the mixocoel is still a matter of debate, transitory coelomic cavities occur at least in Onychophora (Mayer et al. 2004) and according to Bartholomaeus et al. (2009) in some arthropods too. In the light of the now widely accepted Ecdysozoa hypothesis, however, completely different scenarios must be developed. All closely related ecdysozoan sister groups of panarthropods are completely devoid of circulatory organs. That leaves two possible explanations: either the circulatory system as seen in arthropods is a completely new acquisition, or circulatory organs were lost several times. Although the second possibility is not parsimonious, deep homology (sensu Shubin et al. 2009) of circulatory organ developmental pathways between arthropods and other bilaterian taxa (Harvey 1996; Xavier-Neto et al. 2010) possessing circulatory systems nonetheless speaks in favour of the latter scenario.

## 14.5.2 Evolutionary Trends and Factors

When we look at arthropods from the viewpoint of the circulatory system, the degree of variation in structural and functional complexity is striking. Parts of the cardiovascular system in some of the major taxa are highly sophisticated, as is the case in most chelicerates and malacostracan crustaceans, while in other groups, such as copepods, branchiopods and—last but not least—insects, the cardiovascular system is confined to the dorsal vessel. This structural and functional diversity contrasts starkly with that of other major organ systems such as the central nervous system and the digestive system. Respiratory systems, on the other hand, also display great variation between air and water breathing organs, and this seems, at least in some cases, to be correlated with evolutionary changes in circulatory organ design (see below).

It seems appropriate in a first step, to describe what actually changed during the evolution of circulatory organs in arthropods, before in a second step, we address the potential causes of these changes.

The greatest changes took place in arterial systems. On the one hand, *reductions* are apparent in a number of lineages. First, we see a reduction in the complexity of the branching patterns of lateral cardiac arteries (e.g. within malacostracans), and secondly, a reduction in the number of pairs of cardiac arteries (e.g. in spiders and malacostracans). This was often accompanied by the loss of a posterior aorta. The anterior aorta, in contrast, is reduced only in very few taxa (e.g. copepods), probably due to its functional significance in supplying the cephalic region.

On the other hand, an *increase in the structural complexity* of the vascular system also occurred in different lineages independently. In pulmonate arachnids, the arterial supply of the central nervous system reached a degree of complexity (Fig. 14.6b, c) that is not seen in other chelicerates (Wirkner and Prendini 2007; Klussmann-Fricke et al. 2013). The same is true of the supply of the central nervous system in decapods (e.g. Sandeman 1967). In the latter taxon, the degree of complexity of the lateral cardiac arteries also increased (Fig. 14.12b).

The hearts, too, were subject to both reductions and increases in complexity. In branchiopods, the plesiomorphic state of a tubular heart extending right the way through the trunk (anostracans) was reduced in a more or less stepwise manner to a simple, short tubular heart with no arteries in some cladocerans. In decapods and euphausiaceans, on the other hand, we find high-performance globular hearts made up of a multilayered myocard and a complicated regulatory system which evolved from the tubular hearts found in other malacostracans (e.g. Wilkens et al. 1997a).

Can some of the major factors be identified that triggered this range of changes? Here, two processes are suggested to have affected each other reciprocally: changes in body size and changes in respiratory modes. In various groups, a decrease in body size is clearly correlated with a decrease in arterial complexity. Changes in respiratory modes will be dealt with in detail later, but it is obvious that body size had a greater effect since even in groups with coupled respiratory and circulatory systems (e.g. various crustaceans), small species exhibit a reduced vascular system. Another factor shaping changes in the circulatory system has been the need to supply haemolymph to the central nervous system. In the majority of species, the brain (in mandibulates) or prosomal ganglion (in chelicerates) is supplied indirectly via the anterior aorta, though a system of direct supply via a complex arrangement of arteries has evolved at least twice, that is, in pulmonate arachnids and decapods. In a number of malacostracans, chilopods, symphylans and various arachnids, the ventral nerve cord is supplied via a ventral vessel, and this mode of supply too has evolved several times. An alternative way of supplying haemolymph to the central nervous system, namely via a perineural sinus, has also evolved at least twice, in chelicerates and diplopods. The central nervous system is on the one hand the most consumptive tissue found in arthropods but on the other hand also the most dense, which may go some way to explaining why the need to supply it with haemolymph has had such a central influence on the evolution of the circulatory system. There are also various developmental constraints which may have affected circulatory design. It is known from vertebrates that the tubulogenesis of arteries proceeds along oxygen gradients (e.g. Fraisl et al. 2009). Similar processes are imaginable in the brain of decapods or the prosomal ganglion of pulmonate arachnids. Furthermore, in chelicerate taxa, main arteries to appendages are found directly on top of the nerves, supplying the same appendages (Huckstorf et al. 2013; Klussmann-Fricke et al. 2013). A developmental correlation between the nerves and the arteries similar to that between oxygen gradients and arteries is therefore feasible.

### 14.5.3 Transformations and Evolutionary Novelties

The combination of moderate complexity and high variability makes the arthropod circulatory system eminently suitable for the study of evolutionary transformations and novelties. A comprehensible and well-studied example is haemolymph supply to body appendages. While most arthropods have arteries for this task, pterygote insects evolved autonomous circulatory organs which are independent of the vascular system (reviews: Pass 1998, 2000). These auxiliary hearts cannot be homologised with any organ in their ancestors and are therefore considered evolutionary novelties (see West-Eberhard 2003; Pigliucci and Müller 2010; Brigandt and Love 2012). The basic elements needed for the formation of a new accessory heart are pumping muscles and an elastic antagonist of connective tissue or cuticle. These elements were recruited from various organ systems by decoupling, individualisation and local displacement to form a novel functional unit. Comparative analyses and developmental studies of the pumping muscles in several accessory hearts have led to the conclusion that they are derived from various nearby muscle systems (Pass 2000; Tögel et al. 2008). Results pertaining to the development of the muscles associated with the frontal heart in decapods fit well with this hypothesis (Weygoldt 1961; see also Sect. 14.4.3).

### 14.6 Adaptations to Major Environmental Transitions

### 14.6.1 From Water to Land: New Modes of Respiration and the Effects on the Cardiovascular System

When we think about the most influential environmental changes and transitions that occurred during evolution, those involving the change between the two major habitats water and land come to mind. Although this transition can be interpreted to have occurred in both directions, the one of interest here is *terrestrialisation*: the change from an aquatic life style to a terrestrial one. For an overview of these key events in the evolution of arthropods, see Chap. 16.

Terrestrialisation processes have had a major influence on the structure and function of almost all organ systems. A main challenge for the circulatory system in this regard was maintaining water balance and homeostatic conditions in the haemolymph. Even more important, however, was the switch between the two respiratory media water and air. But did the major changes in respiratory modes in the different groups also have a strong impact on the cardiovascular system?

The question of how often terrestrialisation took place within arachnids is currently under debate (see e.g. Scholtz and Kamenz 2006). This means that it is still unclear how often book lungs and tracheal systems developed in this clade. Regardless of the answer, it has been suggested that the xiphosuran book gills were the prerequisite in the evolution of book lungs (Lankester 1881; Kingsley 1885; Farley 2010). In our context, it is unimportant whether book lungs evolved once or twice. The point is that the haemolymph supply to book gills (Xiphosura) and book lungs (Tetrapulmonata, Scorpiones, i.e. pulmonate arachnids) is more or less the same, since they both represent *centralised respiratory organs*. Centralised respiratory organs must be irrigated with haemolymph and require an oxygen transporter (haemocyanin in the arthropods) to maintain a constant supply of oxygen to the organs and tissues. This is reflected in the circulatory design in xiphosurans and pulmonate arachnids (see above). Book gills and lungs are well connected to the pericardial sinus via sinuses which channel oxygenated haemolymph towards the heart, from where it is distributed to the tissues via highly complex arterial systems.

All apulmonate arachnids display a reduction in their arterial systems, following a transition from *centralised* to *decentralised respiratory organs*: they breathe through a system of tubular tracheae which directly supply the tissues with

oxygen (Gruner 1993). In these groups, a dorsal tubular heart is connected via an anterior aorta to an extensive perineural and in some cases also perivisceral sinus (Firstman 1973; Gruner 1993). It is interesting to note that the xiphosuran circulatory system (see above) can probably be seen as a prerequisite for that found in both pulmonate and apulmonate arachnids since xiphosurans possess both a well-defined arterial system and a system of perineural and perivisceral sinuses (see Fig. 14.4; Milne-Edwards 1872; Redmond et al. 1982).

A similar example of terrestrialisation is found in myriapods. All recent representatives are obligate air breathers, that is, they all possess tracheal systems, but when it comes to oxygen transport, there are distinct differences. Within chilopods, scutigeromorphs possess centralised tracheal lungs (Dubuisson 1928; Dohle 1985; Hilken 1997, 1998) and haemocyanin serves as the oxygen transporter (Rajulu 1969; Mangum et al. 1985). Wirkner and Pass (2002) therefore interpreted some distinctive features of the circulatory organs in scutigeromorphs as being related to this coupling of the respiratory and circulatory systems. These features include a voluminous heart with an unusually thick myocard for high-performance haemolymph circulation, the close proximity of the tracheal lungs to the heart (Fig. 14.9b), which shortens the transport route of oxygen-enriched haemolymph, sinuses which channel a centripetal haemolymph flow towards the heart via the tracheal lung sinus, and lastly, aortic diverticles in the head which act as accessory pulsatile organs (Fig. 14.9a). In all other chilopods, which have tracheal systems that bring the oxygen directly to the tissues, none of these specialised circulatory system features are found.

In decapod crustaceans, structures in the gill chamber were recruited for air breathing purposes, meaning that the respiratory organs remained centralised. As a result, the arterial systems in this group did not undergo any major changes, though the lacunar system found in the respiratory organs either altered slightly or became more extensively developed.

In Isopoda, we find all sorts of degrees of terrestrialisation, from amphibious species to species that are completely independent of water as a habitat. The pleopods act as gills in all species, but in the terrestrial species, the pleopodal gills are kept wet through a sophisticated system of water canals that channel the products of the maxillary glands to the pleopods (Wägele 1992). On the way back, ammoniac evaporates and oxygen is taken up. In strictly terrestrial species, the gills are supplemented by pleopodal lungs (see Chap. 16). Silen (1954) states that the arterial system in terrestrial isopods is somewhat reduced in comparison with that in aquatic forms. However, this relates to one pair of cardiac arteries and only some species. Apart from the shift of the margins of the heart into the pleon (see Wirkner and Richter 2003), no drastic rearrangement of the arterial system took place.

The simple vascular system of hexapods has also often been interpreted to be the result of a reduction caused by the evolution of the tracheal system. This interpretation implies that the hexapod ancestor had a complex vascular system. However, according to the now widely accepted Tetraconata/Pancrustacea hypothesis, the potential sister groups of hexapods are aquatic crustaceans such as Branchiopoda, Cephalocarida and/or Remipedia which are all relatively small in terms of body size and possess a very simple cardiovascular system, comprising only a tubular heart with an anterior aorta. In this scenario, it seems obvious to assume that the last common ancestor of hexapods and their crustacean sister group already had this simple cardiovascular system.

This morphological analysis of the cardiovascular system shows that it was not primarily terrestrialisation that affected the restructuring of circulatory organs but rather the transition from centralised to decentralised air breathing systems. Once the cardiovascular system was no longer needed for oxygen transport, it underwent a reduction in both extent and performance.

## 14.6.2 From Land to Air: Lightweight Body Construction and New Tasks for the Circulatory System

The acquisition of the ability to fly was a key event in the evolution of insects, requiring not only the development of flight organs but affecting nearly all the other body parts and organ systems (reviews: Brodsky 1994; Dudley 2000). The circulatory system played an important role in facilitating flight, but most reviewers have paid little attention to this fact. The ways in which the circulatory system contributed to flight are wide-ranging and include aspects such as wing formation, body-weight reduction, tracheal ventilation and thermoregulation.

The importance of the circulatory system for the *development and maintenance of the wings* is vital (Fig. 14.19a). Increased haemolymph pressure in the thorax is necessary for eclosion and wing inflation and is achieved through muscular contractions, especially in the abdomen (Moreau 1974; Moreau and Lavenseau 1975). Heartbeat reversal may also contribute to the required pressure changes (Wasserthal 1975, 2003c). In wing maturation, the accessory wing circulatory organs play an important role by sucking out dissolved epidermal cells to enable the dorsal and ventral cuticular lamellae of the wing to merge together (Tögel et al. 2008). Proper haemolymph circulation through the veins of mature wings is then necessary to maintain them in a healthy and functional state. Wings deprived of circulating haemolymph by the elimination of the wing circulatory organs become dry, brittle and tear easily.

An important selective factor in the evolution of flight is the *reduction in body weight*. The greatest contribution to weight loss is achieved by a radical reduction in the amount of haemolymph in the body. About 50 % of the body weight of lepidopteran caterpillars, for instance, is haemolymph, while in adults, it is only 12–18 % (Wasserthal 1996, 2003c). The reduction occurs immediately after the moult to adult,

**(a)**

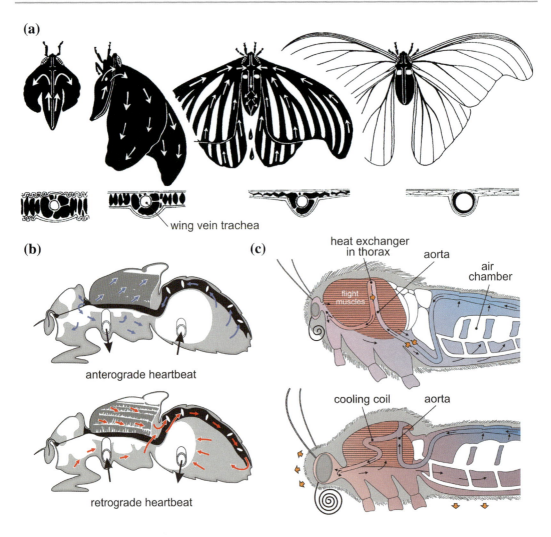

wing vein trachea

**(b)**

anterograde heartbeat

retrograde heartbeat

**(c)**

heat exchanger in thorax    aorta    air chamber

flight muscles

cooling coil    aorta

with haemolymph becoming highly concentrated via diuresis and excess water being excreted through the anus (Nicolson 1976). To compensate for this loss of haemolymph volume, the tracheal air sacs inflate to extend throughout the body.

A further vital functional area in which the circulatory system plays an important role in winged insects is breathing. Generally, the circulatory system in insects is not considered to be used to deliver respiratory gases to the various tissues since this is usually accomplished by the tracheal system. However, there are numerous reports that the circulatory system actually plays an *active role in the ventilation of the*

*respiratory gases*, or at least, it does when metabolic needs are high. During flight, the demand for oxygen increases enormously, sometimes up to 200 times the resting metabolic rate (Harrison and Roberts 2000). This means that fresh air must be delivered to the flight muscles quickly, which is only possible if the tracheal system is efficiently ventilated (reviews: Dudley 2000; Harrison et al. 2012). The tracheal system of winged insects is particularly well developed and contains, beside tubular components, large air sacs which can be inflated or compressed. The ventilation of these voluminous structures occurs in connection with changes in hydraulic pressure, for which there are several

◀ **Fig. 14.19** From land to air: newly acquired functions of circulatory systems. **a** Wing expansion and wing blade formation in giant silkmoth *Attacus*. *Upper row* of diagrams: different phases of wing expansion; *lower row* of diagrams: small portion of wing with a vein in cross section. The voluminous space between dorsal and ventral lamella of freshly expanded wings is strongly reduced by sucking out haemolymph and remnants of autolytically resolved epidermis through action of wing-hearts and backward pumping of the dorsal vessel. At completion of this process, dorsal and ventral cuticles merge together leaving only wing vein tubes, which contain a small amount of haemolymph, and a distended trachea. *Arrows* indicate direction of haemolymph flow (from Wasserthal 1996, permission from Elsevier pending). **b** Discontinuous tracheal ventilation associated with heartbeat reversal in the fly *Calliphora. Upper* diagram: during forward periods, haemolymph is drawn from the abdomen and pushed into thorax and head. Thereby abdominal parts of the tracheal system expanded and those in head and thorax compressed. In return air is drawn through the abdominal spiracles and forced out through the thoracic spiracles. *Lower* diagram: during backward periods, the events in thorax and abdomen are reversed. *Coloured arrows* indicate direction of haemolymph flow: anterograde in *blue* and retrograde in *red*. *Black arrows* indicate air flow through spiracles (modified after Wasserthal 1996). **c** Thermoregulation and relevant anatomy of two moths living under different temperature conditions. *Upper* diagram: winter-active noctuid moth *Eupsilia*. Thorax thermally insulated by numerous cuticular hairs and large tracheal sacs; two counter-current heat exchangers help to keep thermal energy in the thorax: the anterior portion of the heart runs downward into an area where warm haemolymph flows back into the abdomen; in the thorax, the aorta itself functions as heat exchanger by forming a loop in which ascending and descending portions are placed tightly together. *Lower* diagram: summer-active hawk-moth *Manduca*. Aorta seen in loops whereby ascending and descending portions are far from each other; the aorta takes up excessive thermal energy on the way through thorax to head where it is dissipated; another site for heat radiation is the naked ventral abdomen. During flight, thoracic temperature is kept largely constant by modulation of heartbeat frequency and intensity which affects speed of haemolymph circulation in the thorax. *Red arrows* indicate heat radiation (modified after Heinrich 1987b)

mechanisms. The most common is abdominal pumping, which is effected by the contraction of segmental muscles. Another is heartbeat reversal, which in addition requires the insect body to be separated into an anterior and an abdominal compartment by a narrow waist and/or by internal organs which function as valves. Haemolymph can then be periodically shifted between the two body parts, resulting in alternating pressure changes which effect discontinuous tracheal ventilation (Wasserthal 1996). The rate at which haemolymph oscillates between the two compartments may be quite high. In a *Calliphora* fly, for example, the pumping direction of the heart is regularly reversed, with a sequence of forward and backward pulses lasting approximately 30s (Wasserthal 2012) (Fig. 14.19b). Extracardiac pressure pulses, which cannot be detected from the outside, are considered to act as an additional ventilation mechanism. At least in the pupal stage of some endopterygotes, periodically repeated extracardiac pressure pulses are clearly coordinated with brief openings of the spiracle valves to expel carbon dioxide and draw in fresh, well-oxygenated respiratory air (Sláma et al. 1979; Sláma 2000; Sláma and

Neven 2001). The abdomen functions in these events as a superordinate pressure pump (Tartes et al. 2002). To summarise, there are complicated interactions between the pumping of the circulatory organs and extracardiac pressure changes on the one hand, and between the tracheal system and its regulatory spiracles on the other, which together guarantee efficient ventilation distribution of respiratory air.

An additional adaptation to flight in many advanced flying insects is the capacity for *thermoregulation*, which has emancipated the species in question from ambient temperatures and provided them with advantages in several important aspects of their lives, including the search for food, mates and the avoidance of predators (reviews: Heinrich 1993, 1996). To achieve the physiological feat of thermoregulation, a great number of sophisticated mechanisms have evolved in which the circulatory system is often an essential component. Body temperature is the result of a balance between the rates of heat gain and heat loss. Heat gain can be achieved by two different means: the utilisation of external heat, for example, basking in the sun, or the internal production of heat through muscle contraction. In the latter method, known as endothermy, heat is produced by the more or less synchronous contractions of the flight muscles, as observable as shivering of the thorax and wings. Prior to flight, and when temperatures are low, insects first warm themselves to the operating temperature of the flight apparatus. During extended flight, however, excessive amounts of heat are produced by the contraction of the flight muscles, and a cooling mechanism is necessary to prevent the thorax from overheating. The circulatory system controls thermoregulation in that haemolymph is the carrier of the thermal energy, and the circulatory pumps regulate its flow by modulating both the frequency and the amplitude of the heartbeats.

The role of the circulatory system in endothermic thermoregulation was convincingly demonstrated in nocturnally active hawkmoths (Heinrich 1971; Heinrich and Bartolomew 1971) and winter-flying endothermic moths (Heinrich 1987a) (Fig. 14.19c). A precondition for effective thermoregulation in these species is the thermal insulation of the thorax by a thick covering of hairs which form a kind of fur and the thermal isolation of the thorax from the abdomen via a narrow waist or specific valves. During the warm-up phase, a temporary reduction or pause in heartbeat slows down heat loss from the thorax, while during flight the thorax is kept cool by two means. On the one hand, cold haemolymph is pumped by the heart from the abdomen into the thorax where the aorta loops between the heated flight muscles serve as cooling coils, and on the other hand, warm blood is forced out of the thorax into the ventral area of the abdomen. There are large areas of cuticle, which lack thermal insulation by hairs or tracheal sacs functioning as heat radiators. In this manner, the temperature in the thorax of these insects can be kept constant during flight.

A more subtle mechanism of thermoregulation by the circulatory system is found in certain hymenopterans. In honeybee workers, the aorta in the petiole, the narrow waist between thorax and abdomen, is convoluted into loops. The function of this construction was puzzling to early anatomists (Freudenstein 1928) but becomes understandable when viewed in connection with thermoregulation. The coiling and lengthening of the aorta in the petiole are alleged to permit heat exchange between the countercurrent haemolymph flows (Heinrich 1979). Within this structure, the cold haemolymph of the abdomen absorbs heat, allowing warm haemolymph to pass into the thorax. Curiously, the radiation of excessive heat in these animals occurs through the head and can be further increased by evaporative cooling by using regurgitated nectar (Heinrich 1979, 1980a, b; Heinrich and Esch 1994).

In some insects, thermoregulation may serve other purposes than flight. In bumblebees, for example, heat produced in the thorax by shivering is used to warm the brood (Heinrich 1976). Warm haemolymph from the thorax passes through the perineural sinus into the abdomen, which is ventrally not well insulated. Bumblebees transfers heat from the body to the brood cells by pressing their abdomen against them. Perhaps, the most spectacular use of thermoregulation is the 'thermo-execution' of hornets by Japanese honeybees (*Apis cerana japonica*). Intruding hornets are engulfed by hundreds of bees that heat their thoraces up to 47 °C. The resulting heat coat formed around the hornet's body exceeds the upper lethal temperature of the hornet, which dies within minutes (Ono et al. 1995). The heat is probably released from the haemolymph in distinct areas of the body, but this has not yet been investigated.

## 14.7 Concluding Remarks

The vascular system of arthropods spans a broad spectrum of complexity. In some groups, it is confined to a compact heart, and in others, it is an extensive system of vessels and peripheral capillarisation which has been described as 'partially closed' (McGaw 2005; Reiber and McGaw 2009). However, even in this latter case, the fundamental difference between closed circulatory systems, such as those in vertebrates, and the open system in arthropods is not obscured: in an open system, no vessels lead directly back into the heart. In all arthropods, haemolymph is collected in the pericardial sinus before it enters the heart via the ostia. The functional morphology of the pericardial sinus may differ from species to species. In many arthropods, the separation of the pericardial sinus from the rest of body haemocoel is incomplete, and in others, there is a sack-like pericard which rhythmically dilates. From a functional point of view, the latter arrangement, which comprises an inflow and an outflow chamber, is comparable with the more developed hearts of molluscs and vertebrates (cf. Xavier-Neto et al. 2010).

The range of physiological functions fulfilled by the circulatory system in arthropods are enormous. There is hardly a functional area in which this organ system is not involved in some form. The most important factors behind the evolution of an effective circulation system are probably body size and the improvement of oxygen transport. Within the latter process, the regional concentration of the respiratory organs combined with the needs to follow the shortest possible pathway to the heart required the greatest architectural transformations. This can be seen clearly in the circulatory organs of xiphosurans and pulmonate arachnids and those of malacostracan crustaceans. In arthropods with tracheal systems, the circulatory system has lost the function of oxygen transportation. These groups are characterised in general by relatively simple vascular systems. The reduction in the vascular system, however, may have occurred prior to the acquisition of the tracheal system, for example, in hexapods, which are probably derived from relatively small-sized aquatic ancestors. In insects, the circulatory system acquired completely new tasks and features, such as tracheal ventilation and thermoregulation, in connection with the ability to fly.

The ancestral design of arthropods is probably relatively complex, and the vascular system is based on a segmental arrangement (but see Chap. 9 for a different perspective). The individual modules consist of a portion of the dorsal longitudinal vessel plus a pair of ostia and segmental arteries. These elements were eventually joined by a ventral longitudinal vessel with segmental arteries extending into the appendages and by vessels connecting the dorsal and the ventral vessels. Many examples of vessel reduction can be found in arthropods, but so too can lineages which possess complex cardiovascular systems, even though they were completely lost in many of their close relatives. Novelties which appeared in insects include the numerous auxiliary hearts which facilitate the supply of haemolymph to the various body appendages. Their evolutionary origin can be traced to the recruitment of construction elements from other organs systems.

Studies of the embryonic development of the vascular system in the various arthropods have shown that despite the formation of the heart being relatively conserved, there is astounding variation in the development of the arterial system. The discovery of strikingly similar gene networks and signalling pathways in the development of the hearts of arthropods and vertebrates indicates extensive deep homologies inbetween the circulatory organs of bilaterians (Hartenstein and Mandal 2006; Xavier-Neto et al. 2010; Grigorian et al. 2011). This fact has become the basis for an applied approach in research into arthropod circulatory systems in recent years. Due to its highly conserved molecular and developmental features, *Drosophila* has proved to be a powerful model in unravelling the genetic mechanisms underlying cardiac ageing and certain cardiac diseases in humans (Bier and Bodmer 2004; Wolf et al. 2006; Ocorr et al. 2007; Nishimura et al. 2011; Iliadi et al. 2012).

**Acknowledgments** Cordial thanks to John Plant and Lucy Cathrow for the linguistic revision of the manuscript, and Torben Göpel for technical assistance. We thank Heidemarie Grillitsch for drawing of some graphs and Bastian-Jesper Klussmann-Fricke (University Rostock) for a photograph. CSW thanks all members of his lab for discussions and technical help. The work was supported by the German Research Foundation (DFG) 3334/1-2, 3334/3-1, 3334/4-1 (C.S.W) and the Austrian Science Fund P23251 (G.P. and M.T.).

# References

Ai H, Kuwasawa K (1995) Neural pathways for cardiac reflexes triggered by external mechanical stimuli in larvae of *Bombyx mori*. J Insect Physiol 41:1119–1131

Alexandrowicz JS (1932) The innervation of the heart of Crustacea I. Decapoda. Q J Microsc Sci 75:182–247

Andereck JW, King JKG, Hillyer JF (2010) Contraction of the ventral abdomen potentiates extracardiac retrograde hemolymph propulsion in the mosquito hemocoel. PLoS ONE 5(9):e12943. doi:10.1371/journal.pone.0012943

Anderson DT (1973) Embryology and phylogeny in annelids and arthropods. Pergamon Press, Oxford

Angioy AM, Boassa D, Dulcis D (1999) Functional morphology of the dorsal vessel in the adult fly *Protophormia terraenovae* (Diptera, Calliphoridae). J Morphol 240:15–31

Angioy AM, Pietra P (1995) Mechanism of beat reversal in semi-intact heart preparations of the blowfly *Phormia regina* (Meigen). J Comp Physiol B 165:165–170

Arnold JW (1964) Blood circulation in insect wings. Mem Entomol Soc Can 38:5–60

Balss H, Buddenbrock Wv, Gruner HE, Korschelt E (1940–61) Decapoda. In: Schellenberg A, Gruner HE (eds) Bronn's Klassen und Ordnungen des Tierreichs. Akademische Verlagsgesellschaft Geest and Portig, Leipzig

Bartolomaeus T, Quast B, Koch M (2009) Nephridial development and body cavity formation in *Artemia salina* (Crustacea: Branchiopoda): no evidence for any transitory coelom. Zoomorphology 128:247–262

Baumann H (1921) Das Gefäßsystem von *Astacus fluviatilis* (*Potamobius* L.). Z wiss Zool 118:246–312

Baum E, Hertel W, Beutel RG (2007) Head capsule, cephalic central nervous system and head circulatory system of an aberrant orthopteran, *Prosarthria teretirostris* (Caelifera, Hexapoda). Zoology 110:147–160

Belman BW, Childress JJ (1976) Circulatory adaptations to the oxygen minimum layer in the bathypelagic mysid *Gnathophausia ingens*. Biol Bull 150:15–37

Bier E, Bodmer R (2004) *Drosophila*, an emerging model for cardiac disease. Gene 342:1–11

Bodmer R, Frasch M (2010) Development and aging of the *Drosophila* heart. In: Rosenthal N, Harvey RP (eds) Heart development and regeneration. Academic, San Diego, pp 47–86

Brigandt I, Love AC (2012) Conceptualizing evolutionary novelty: moving beyond definitional debates. J Exp Zool B Mol Dev Evol 318:417–427

Brocher F (1909) Sur l'organe pulsatile observé dans les pattes des hémiptères aquatiques. Ann Biol Lacustre 4:33–41

Brodsky AK (1994) The evolution of insect flight. Oxford University Press, New York

Bryantsev AL, Cripps RM (2009) Cardiac gene regulatory networks in *Drosophila*. Acta Biochim Biophys 1789:343–353

Burnett BR (1972) Aspects of the circulatory system of *Pollicipes polymerus* J. B. Sowerby (Cirripedia: Thoracica). J Morphol 136:79–107

Burnett BR (1984) Striated muscle in the wall of the dorsal abdominal aorta of the California spiny lobster *Panulirus interruptus*. J Crustac Biol 4:560–566

Cannon HG (1947) On the anatomy of the pedunculate barnacle *Lithotrya*. Phil Trans R Soc B 595:89–136

Chan KS, Cavey MJ, Wilkens JL (2006) Microscopic anatomy of the thin-walled vessels leaving the heart of the lobster *Homarus americanus*: anterior lateral arteries. Invertebr Biol 125:70–82

Chapman RF (1998) The insects: structure and function, 4th edn. Cambridge University Press, Cambridge

Clements AN (1956) The antennal pulsatile organs of mosquitoes and other Diptera. Q J Microsc Sci 97:429–435

Cooke IM (2002) Reliable, responsive pacemaking and pattern generation with minimal cell numbers: the crustacean cardiac ganglion. Biol Bull 202:108–136

Cripps RM, Olson EN (2002) Control of cardiac development by an evolutionarily conserved transcriptional network. Dev Biol 246:14–28

Curtis NJ, Ringo JM, Dowse HB (1999) Morphology of the pupal heart, adult heart, and associated tissues in the fruit fly, *Drosophila melanogaster*. J Morphol 240:225–235

Davis N-T, Dulcis D, Hildebrand JG (2001) Innervation of the heart and aorta of *Manduca sexta*. J Comp Neurol 440:245–260

Debaisieux P (1936) Organes pulsatiles des tibias de Notonectes. Ann Soc Sci Brux B 56:77–87

Dohle W (1985) Phylogenetic pathways in the Chilopoda. Bijdr Dierk 55:55–66

Dubuisson M (1928) Recherches sur la circulation du sang chez les Crustacés. 1er note: Amphipodes. Circulation chez les Gammariens; synchronisme des mouvements respiratoires et des pulsations cardiaque. Arch Zool Exp Gen 67:93–104

Dudley R (2000) The biomechanics of insect flight. Princeton University Press, Princeton

Dulcis D, Davis NT, Hildebrand JG (2001) Neuronal control of heart reversal in the hawkmoth *Manduca sexta*. J Comp Physiol A 187:837–849

Dulcis D, Levine RB (2003) Innervation of the heart of the adult fruit fly, *Drosophila melanogaster*. J Comp Neurol 465:560–578

Dulcis D, Levine RB (2005) Glutamatergic innervation of the heart initiates retrograde contractions in adult *Drosophila melanogaster*. J Neurosci 25:271–280

Duman-Scheel M, Pirkl N, Patel NH (2002) Analysis of the expression pattern of *Mysidium columbiae wingless* provides evidence for conserved mesodermal and retinal patterning processes among insects and crustaceans. Dev Genes Evol 212:114–123

Dumont JN, Anderson E, Chomyn E (1965) The anatomy of the peripheral nerve and its ensheathing artery in the horseshoe crab, *Xiphosura* (*Limulus*) *polyphemus*. J Ultrastruct Res 13:38–64

Eberhard WG, Huber BA (1998) Courtship, copulation, and sperm transfer in *Leucauge mariana* (Araneae, Tetragnathidae) with implications for higher classification. J Arachnol 26:342–368

Ellis CH (1944) The mechanism of extension in the legs of spiders. Biol Bull 86:41–50

Fahlander K (1938) Beiträge zur Anatomie und systematischen Einteilung der Chilopoden. Zool Bidr Uppsala 17:1–148

Farley RD (1985) Cardioregulation in the desert scorpion, *Paruroctonus mesaensis*. Comp Biochem Physiol 82C:377–387

Farley RD (1987) Postsynaptic potentials and contraction pattern in the heart of the desert scorpion, *Paruroctonus mesaensis*. Comp Biochem Physiol 86A:121–131

Farley RD (2010) Book gill development in embryos and first and second instars of the horseshoe crab *Limulus polyphemus* L. (Chelicerata, Xiphosura). Arthropod Struct Dev 39:369–381

Firstman B (1973) The relationship of the chelicerate arterial system to the evolution of the endosternite. J Arachnol 1:1–54

Fraisl P, Mazzone M, Schmidt T, Carmeliet P (2009) Regulation of angiogenesis by oxygen and metabolism. Dev Cell 16:167–179

Frank H (1957) Untersuchungen zur funktionellen Anatomie der lokomotorischen Extremitäten von *Zygiella x-notata*, einer Radnetzspinne. Eberhard-Karls-Universität zu Tübingen, Dissertation

Freudenstein K (1928) Das Herz und das Circulationssystem der Honigbiene (*Apis mellifica* L.). Z Wiss Zool 132:404–475

Gereben-Krenn B-A, Pass G (1999) Circulatory organs of Diplura (Hexapoda): the basic design in Hexapoda? Int J Insect Morphol 28:71–79

Gereben-Krenn B-A, Pass G (2000) Circulatory organs of abdominal appendages in primitive insects (Hexapoda: Archaeognatha, Zygentoma and Ephemeroptera). Acta Zool 81:285–292

Gerhardt U, Kästner A (1938) Araneae. In: Kükenthal WG, Krumbach T (eds) Handbuch der Zoologie, vol 3. De Gruyter, Berlin, pp 394–656

Gerould JH (1933) Orders of insects with heart-beat reversal. Biol Bull 64:424–431

Glenn JD, King JG, Hillyer JF (2010) Structural mechanics of the mosquito heart and its function in bidirectional hemolymph transport. J Exp Biol 213:541–550

Grasshoff M (1968) Morphologische Kriterien als Ausdruck von Artgrenzen bei Radnetzspinnen der Subfamilie Araneinae (Arachnida, Araneae, Araneidae). Abh Senckenberg Naturforsch Ges 516:1–100

Grasshoff M (1973) Bau und Mechanik der Kopulationsorgane der Radnetzspinne *Mangora acalypha* (Arachnida, Araneae). Z Morphol Tiere 74:241–252

Griffin KJ, Stoller J, Gibson M, Chen S, Yelon D, Stainier DY, Kimelman D (2000) A conserved role for H15-related T-box transcription factors in zebrafish and *Drosophila* heart formation. Dev Biol 218:235–247

Grigorian M, Mandal L, Hakimi M, Ortiz I, Hartenstein V (2011) The convergence of Notch and MAPK signaling specifies the blood progenitor fate in the *Drosophila* mesoderm. Dev Biol 353:105–118

Gruner HE (ed) (1993) Arthropoda. In: Lehrbuch der Speziellen Zoologie. Vol. I/4. Gustav Fischer Verlag, Jena

Handel K, Basal A, Fan X, Roth S (2005) *Tribolium castaneum twist*: gastrulation and mesoderm formation in a short-germ beetle. Dev Genes Evol 215:13–31

Hantschk A (1991) Functional morphology of accessory circulatory organs in the legs of Hemiptera. Int J Insect Morphol 20:259–274

Harrison JF, Roberts SP (2000) Flight respiration and energetics. Annu Rev Physiol 62:179–205

Harrison JF, Woods HA, Roberts SP (2012) Ecological and environmental physiology of insects. Oxford University Press, Oxford

Hartenstein V, Mandal L (2006) The blood/vascular system in a phylogenetic perspective. BioEssays 28:1203–1210

Harvey RP (1996) NK-2 homeobox genes and heart development. Dev Biol 178:203–216

Haupt J (1976) Die segmentalen Kopfdrüsen von *Scutigerella* (Symphyla, Myriapoda). Zool Beitr 22:19–37

Heinrich B (1971) Temperature regulation of the sphinx moth, *Manduca sexta*. II. Regulation of heat loss by control of blood circulation. J Exp Biol 54:153–166

Heinrich B (1976) Heat exchange in relation to blood flow between thorax and abdomen. J Exp Biol 64:561–585

Heinrich B (1979) Keeping a cool head: honeybee thermoregulation. Science 205:1269–1271

Heinrich B (1980a) Mechanisms of body-temperature regulation in honeybees, *Apis mellifera* I. Regulation of head temperature. J Exp Biol 85:61–73

Heinrich B (1980b) Mechanisms of body-temperature regulation in honeybees, *Apis mellifera*. II. Regulation of thoracic temperatures at high air temperatures. J Exp Biol 85:73–87

Heinrich B (1987a) Thermoregulation by winter-flying endothermic moths. J Exp Biol 127:313–332

Heinrich B (1987b) Thermoregulation in winter moths. Sci Amer 256:104–111

Heinrich B (1993) The hot-blooded insects: mechanisms and evolution of thermoregulation. Harvard University Press, Cambridge

Heinrich B (1996) The thermal warriors: strategies of insect survival. Harvard University Press, Cambridge

Heinrich B, Bartholomew GA (1971) An analysis of preflight warm-up in the sphinx moth, *Manduca sexta*. J Exp Biol 55:223–239

Heinrich B, Esch H (1994) Thermoregulation in bees. Am Sci 82:164–170

Herbst C (1891) Beiträge zur Kenntnis der Chilopoden. Bibliotheca Zoologica 9:1–42

Hertel W (2009) Contributions on cardiac physiology in Diplopoda (Myriapoda). Physiol Entomol 34:296–299

Hertel W, Neupert S, Eckert M (2012) Proctolin in the antennal circulatory system of lower Neoptera: a comparative pharmacological and immunohistochemical study. Physiol Entomol 37:160–170

Hertel W, Pass G (2002) An evolutionary treatment of the morphology and physiology of circulatory organs in insects. Comp Biochem Physiol A 133:555–575

Hertel W, Pass G, Penzlin H (1985) Electrophysiological investigation of the antennal heart of *Periplaneta americana* and its reactions to proctolin. J Insect Physiol 31:563–572

Hertel W, Rapus J, Richter M, Eckert M, Vettermann S, Penzlin H (1997) The proctolinergic control of the antenna-heart in *Periplaneta americana* (L.). Zoology 100:70–79

Hertel W, Wirkner CS, Pass G (2002) Studies on the cardiac physiology of Onychophora and Chilopoda. Comp Biochem Physiol A 133:605–609

Hessel JH (1969) The comparative morphology of the dorsal vessel and accessory structures of the Lepidoptera and its phylogenetic implications. Ann Entomol Soc Am 62:353–370

Hetz SK, Psota E, Wasserthal LT (1999) Roles of aorta, ostia and tracheae in heartbeat and respiratory gas exchange in pupae of *Troides rhadamantus* Staudinger 1888 and *Ornithoptera priamus* L. 1758 (Lepidoptera, Papilionidae). Int J Insect Morphol 28:131–144

Hickman V (1936) The embryology of the syncarid crustacean *Anaspides tasmaniae*. Pap Proc R Soc Tas 1–35

Hilken G (1997) Tracheal systems in Chilopoda: a comparison under phylogenetic aspects. Entomol Scand Suppl 51:49–60

Hilken G (1998) Vergleich von Tracheensystemen unter phylogenetischem Aspekt. Verh Naturwiss Vereins Hamburg (NF) 37:5–94

Hilken G, Rosenberg J (2005) A new cell type associated with haemolymph vessels in the centipede *Scutigera coleoptrata* (Chilopoda, Notostigmophora). Entomol Gen 27:201–210

Hilken G, Wirkner CS, Rosenberg J (2006) On the structure and function of the aortic diverticles in *Scutigera coleoptrata* (Chilopoda, Scutigeromorpha). Norw J Entomol 53:107–112

Huber BA (1992) Frontal heart and arterial system in the head of Isopoda. Crustaceana 63:57–69

Huber BA (1993) Genital mechanics and sexual selection in the spider *Nesticus cellulanus* (Araneae: Nesticidae). Can J Zool 71:2437–2447

Huber BA (2004) Evolutionary transformation from muscular to hydraulic movements in spider (Arachnida, Araneae) genitalia: a study based on histological serial sections. J Morphol 261:364–376

Huckstorf K, Kosok G, Seyfarth EA, Wirkner CS (2013) The hemolymph vascular system in *Cupiennius salei* (Araneae: Ctenidae). Zool Anz (in press)

Huckstorf K, Wirkner CS (2011) Comparative morphology of the hemolymph vascular system in krill (Euphausiacea; Crustacea). Arthropod Struc Dev 40:39–53

Hustert R (1999) Accessory hemolymph pump in the mesothoracic legs of locusts (*Schistocerca gregaria* Forskal) (Orthoptera, Acrididae). Int J Insect Morphol 28:91–96

Hustert R, Pass G (in prep) New accessory pumps for circulation in insect appendages: mechanism and neuronal drive of the cricket ovipositor-hearts

Ichikawa T (2009) Mechanism of hemolymph circulation in the pupal leg of tenebrionid beetle *Zophobas atratus*. Comp Biochem Physiol A 153:174–180

Iliadi KG, Knight D, Boulianne GL (2012) Healthy aging—insights from *Drosophila*. Front Physiol 3:106. doi:10.3389/fphys.2012.00106

Janssen R, Budd GE, Damen WG, Prpic NM (2008) Evidence for Wg-independent tergite boundary formation in the millipede *Glomeris marginata*. Dev Genes Evol 218:361–370

Janssen R, Damen WG (2008) Diverged and conserved aspects of heart formation in a spider. Evol Dev 10:155–165

Janssen R, Prpic NM, Damen WG (2004) Gene expression suggests decoupled dorsal and ventral segmentation in the millipede *Glomeris marginata* (Myriapoda: Diplopoda). Dev Biol 268:89–104

Janssen R, Prpic NM, Damen WG (2006) Dorso-ventral differences in gene expression in *Glomeris marginata* (Villers 1789) (Myriapoda: Diplopoda). Norw J Entomol 53:129–137

Johannsen OA, Butt FH (1941) Embryology of insects and myriapods. McGraw-Hill, New York

Jones JC (1977) The circulatory system of insects. Thomas, Springfield

Juberthie-Jupeau L (1971) Glandes à sécrétion externe de la tête des Symphyles. Rev Écol Biol Sol 8:617–629

Kihara A, Kuwasawa K (1984) A neuroanatomical and electrophysiological analysis of nervous regulation in the heart of an isopod crustacean, *Bathynomus doederleini*: excitatory and inhibitory junction potentials. J Comp Physiol 154:883–894

Kingsley JS (1885) Notes on the embryology of *Limulus*. Q J Microsc Sci 25:521–576

King JG, Hillyer JF (2012) Infection-induced interaction between the mosquito circulatory and immune systems. PLoS Pathogens. 8(11): e1003058

Kingsley JS (1893) The embryology of *Limulus*, part II. J Morphol 8:195–268

Klussmann-Fricke B-J, Prendini L, Wirkner CS (2013) Evolutionary morphology of the hemolymph vascular system in scorpions: a character analysis. Arthropod Struct Dev 41:545–560

Krenn HW (1993) Postembryonic development of accessory wing circulatory organs in *Locusta migratoria* (Orthoptera: Acrididae). Zool Anz 230:227–236

Krenn HW (2010) Feeding mechanisms of adult Lepidoptera: structure, function, and evolution of the mouthparts. Annu Rev Entomol 55:307–327

Krenn HW, Pass G (1993) Wing-hearts in Mecoptera (Insecta). Int J Insect Morphol 22:63–76

Krenn HW, Pass G (1994) Morphological diversity and phylogenetic analysis of wing circulatory organs in insects, part I: non-Holometabola. Zoology 98:7–22

Krenn HW, Pass G (1995) Morphological diversity and phylogenetic analysis of wing circulatory organs in insects, part II: Holometabola. Zoology 98:147–164

Kristensen NP (1975) The phylogeny of hexapod "orders" a critical review of recent accounts. Z Zool Syst Evol 13:1–44

Kropf C (2013) Hydraulic system of locomotion. In: Nentwig W (ed) Spider ecophysiology. Springer, Heidelberg (in press)

Kuramoto T, Ebara A (1984) Neurohormonal modulation of the cardiac outflow flow through the cardioarterial valve in the lobster Homarus americanus. J Exp Biol 111:123–130

Kuusik A, Harak M, Hiiesaar K, Metspalu L, Tartes U (1996) Different types of external gas exchange found in pupae of greater wax moth Galleria mellonella (Lepidoptera: Pyralidae). Eur J Entomol 93:23–35

Kuwasawa K, Ai H, Matsushita T (1999) Cardiac reflexes and their neural pathways in lepidopterous insects. Comp Biochem Physiol A 124:581–586

Lamoral BH (1973) On the morphology, anatomy, histology and function of the tarsal organ on the pedipalpi of Palystes castaneus (Sparassidae, Araneidae). Ann Natal Mus 21:609–648

Lane NJ, Harrison JB, Bowermann RF (1981) A vertebrate-like blood brain barrier, with intraganglionic blood channels and occluding junctions in the scorpion. Tissue Cell 13:557–576

Lankester ER (1881) Limulus an arachnid. Q J Microsc Sci 21:504–548

Lee W-K, Socha JJ (2009) Direct visualization of hemolymph flow in the heart of a grasshopper (Schistocerca americana). BMC Physiol 9:2. doi: 10.1186/1472-6793-9-2

Lehmacher C, Abeln B, Paululat A (2012) The ultrastructure of Drosophila heart cells. Arthropod Struct Dev 41:459–474

Lehmacher C, Tögel M, Pass G, Paululat A (2009) The Drosophila wing hearts consist of syncytial muscle cells that resemble adult somatic muscles. Arthropod Struct Dev 38:111–123

Leiber G (1935) Beiträge zur vergleichenden Anatomie des Gefäßsystems der Diplopoden. Zool Jahrb Anat Ontog Tiere 59:333–354

Leyton RA, Sonnenblick EH (1971) Cardiac muscle of the horseshoe crab, Limulus polyphemus I. Ultrastructure. J Cell Biol 48:101–119

Lovato TL, Nguyen TP, Molina MR, Cripps RM (2002) The Hox gene abdominal-a specifies heart cell fate in the Drosophila dorsal vessel. Development 129:5019–5027

Lowe E (1935) On the anatomy of a marine copepod, Calanus finmarchicus (Gunner). Trans R Soc Edinb 58:561–603

Makioka T (1988) Internal morphology. In: Sekiguchi K (ed) Biology of horseshoe crabs. Science House, Tokyo, pp 104–132

Mangum CP, Scott JL, Black REL, Miller KI, Van Holde KE (1985) Centipedal haemocyanin: its structure and its implication for arthropod phylogeny. Proc Natl Acad Sci USA 82:3721–3725

Manton SM (1928) On the embryology of a mysid crustacean, Hemimysis lamornae. Phil Trans R Soc B 216:363–463

Manton SM (1934) On the embryology of the crustacean Nebalia bipes. Phil Trans R Soc B 223:163–238

Matus S, Pass G (1999) Antennal circulatory organ of Apis mellifera L. (Hymenoptera: Apidae) and other Hymenoptera: functional morphology and phylogenetic aspects. Int J Insect Morphol 28:97–109

Mayer G, Ruhberg H, Bartolomaeus T (2004) When the epithelium ceases to exist - an ultrastructural study on the fate of the embryonic coelom in Epiperipatus biolleyi (Onychophora, Peripatidae). Acta Zool 85:163–170

Maynard DM (1960) Circulation and heart function. In: Waterman TH (ed) The physiology of Crustacea. Academic, New York, pp 161–226

Mayrat A, McMahon BR, Tanaka K (2006) The circulatory system. In: Forest J, von Vaupel Klein JC, Schram FR (eds) Treatise on zoology—Anatomy, taxonomy, biology. The Crustacea revised and updated from the Traité de Zoologie, vol 2. Brill, Leiden, pp 3–84

McGaw IJ (2005) The decapod crustacean circulatory system: a case that is neither open nor closed. Microsc Microanal 11:18–36

McMahon BR (2001) Control of cardiovascular function and its evolution in Crustacea. J Exp Biol 204:923–932

McMahon BR (2012) Comparative evolution and design in non-vertebrate cardiovascular systems. In: Sedmera D, Wang T (eds) Ontogeny and phylogeny of the vertebrate heart. Springer, New York, pp 1–33

McMahon BR, Burnett BR (1990) The crustacean open circulatory system: a reexamination. Physiol Zool 63:35–71

McMahon BR, Smith PJS, Wilkens JL (1997) Invertebrate circulatory systems. In: Danzler WH (ed) Handbook of physiology. Comparative physiology, Sect 13. American Physiological Society, New York, pp 931–1008

McMahon BR, Tanaka K, Doyle JE, Chu KH (2002) A change of heart: cardiovascular development in the shrimp Metapenaeus ensis. Comp Biochem Physiol A: Mol Integr Physiol 133:577–587

Meek WJ (1909) Structure of Limulus heart muscle. Amer J Physiol 20:403–412

Meyer E (1931) Über den Blutkreislauf der Ephemeriden. Zeitschr Morphol Ökol Tiere 22:1–52

Midttun B (1977) Ultrastructure of cardiac muscle of Trochosa terricola Thor, Pardosa amentata Clerck, P. pullata Clerck, and Pisaura mirabilis Clerck (Araneae: Lycosidae, Pisauridae). Cell Tissue Res 181:299–310

Miller TA (1985) Structure and physiology of the circulatory system. In: Kerkut GA, Gilbert LI (eds)

Comprehensive insect physiology, biochemistry and pharmacology, vol 3. Pergamon Press, Oxford, pp 289–353

Miller TA (1997) Control of circulation in insects. Gen Pharmacol 29:23–38

Miller T, Pass G (2009) Circulatory system In: Resh VH, Cardé RT (eds) Encyclopedia of insects, 2nd ed. Elsevier, Burlington, pp 169–173

Milne-Edwards A (1872) Recherches sur l'anatomie des Limules. Ann Sci Nat Zool 17:1–67

Monier B, Astier M, Semeriva M, Perrin L (2005) Steroid-dependent modification of Hox function drives myocyte reprogramming in the Drosophila heart. Development 132:5283–5293

Monier B, Tevy MF, Perrin L, Capovilla M, Semeriva M (2007) Downstream of homeotic genes: in the heart of hox function. Fly 1:59–67

Moreau R (1974) Variations de la pression interne au cours de l'émergence et de l'expansion des ailes chez Bombyx mori et Pieris brassicae. J Insect Physiol 20:1475–1480

Moreau R, Lavenseau L (1975) Rôle des organes pulsatiles thoraciques et du coeur pendant l'émergence et l'expansion des ailes des Lépidoptères. J Insect Physiol 21:1531–1534

Murray JA (1967) Morphology of the cercus in Blattella germanica (Blattaria: Pseudomopinae). Ann Entomol Soc Am 60:10–16

Nair SG (1956) On the embryology of the isopod Irona. J Embryol Exp Morphol 4:1–33

Nicolson SW (1976) Diuresis in the cabbage white butterfly Pieris brassicae: fluid excretion by the malpighian tubules. J Insect Physiol 22:1347–1356

Nigmann M (1908) Anatomie und Biologie von Acentropus niveus Oliv. Zool Jahrb Syst 26:489–560

Nishimura M, Ocorr K, Bodmer R, Cartry J (2011) Drosophila as a model to study cardiac aging. Exp Gerontol 46:326–330

Nutting WL (1951) A comparative anatomical study of the heart and accessory structures of the orthopteroid insects. J Morphol 89:501–597

Ocorr K, Akasaka T, Bodmer R (2007) Age-related cardiac disease model of Drosophila. Mech Ageing Dev 128:112–116

Økland S, Tjønneland A, Nylund A, Larsen LN, Christ I (1982) The membrane systems and the sarcomere in the heart of Lithobius forficatus L. (Arthropoda, Chilopoda). Zool Anz 208:124–131

Ono M, Igarashi T, Ohno E, Sasaki M (1995) Unusual thermal defense by a honeybee against mass attack by hornets. Nature 377:334–336

Parry DA, Brown RHJ (1959) The jumping mechanism of salticid spiders. J Exp Biol 36:654–664

Pass G (1980) The anatomy and ultrastructure of the antennal circulatory organs in the cockchafer beetle Melolontha melolontha L. (Coleoptera, Scarabaeidae). Zoomorphology 96:77–89

Pass G (1985) Gross and fine structure of the antennal circulatory organ in cockroaches (Blattodea, Insecta). J Morphol 185:255–268

Pass G (1987) The "cercus heart" in stoneflies—a new type of accessory circulatory organ in insects. Naturwiss 74:440–441

Pass G (1988) Functional morphology and evolutionary aspects of unusual antennal circulatory organs in Labidura riparia Pallas (Labiduridae), Forficula auricularia L. and Chelidurella acanthopygia Géné (Forficulidae) (Dermaptera: Insecta). Int J Insect Morphol 17:103–112

Pass G (1991) Antennal circulatory organs in Onychophora, Myriapoda and Hexapoda: functional morphology and evolutionary implications. Zoomorphology 110:145–164

Pass G (1998) Accessory pulsatile organs. In: Harrison F, Locke M (eds) Microscopic anatomy of invertebrates, vol 11B. Wiley-Liss, New York, pp 621–640

Pass G (2000) Accessory pulsatile organs: evolutionary innovations in insects. Annu Rev Entomol 45:495–518

Pass G, Agricola H, Birkenbeil H, Penzlin H (1988a) Morphology of neurones associated with the antennal heart of Periplaneta americana (Blattodea, Insecta). Cell Tissue Res 253:319–326

Pass G, Gereben-Krenn B-A, Merl M, Plant J, Szucsich NU, Tögel M (2006) Phylogenetic relationships of the orders of Hexapoda: Contributions from the circulatory organs for a morphological data matrix. Arthr Syst Phyl 64:165–203

Pass G, Sperk G, Agricola H, Baumann E, Penzlin H (1988b) Octopamine in a neurohaemal area within the antennal heart of the American cockroach. J Exp Biol 135:495–498

Paul R, Bihlmayer S (1995) Circulatory physiology of a tarantula (Eurypelma californicum). Zoology 98:69–81

Paul R, Bihlmayer S, Colmorgan M, Zahler S (1994) The open circulatory system of spiders (Eurypelma californicum, Pholcus phalangoides): A survey of functional morphology and physiology. Physiol Zool 67:1360–1382

Perrin L, Monier B, Ponzielli R, Astier M, Semeriva M (2004) Drosophila cardiac tube organogenesis requires multiple phases of Hox activity. Dev Biol 272:419–431

Pigliucci M, Müller GB (2010) Evolution: the extended synthesis. MIT Press, Cambridge

Pirow R, Wollinger F, Paul RJ (1999) The sites of respiratory gas exchange in the planktonic crustacean Daphnia magna: an in vivo study employing blood haemoglobin as an internal oxygen probe. J Exp Biol 202:3089–3099

Ponzielli R, Astier M, Chartier A, Gallet A, Therond P, Semeriva M (2002) Heart tube patterning in Drosophila requires integration of axial and segmental information provided by the Bithorax complex genes and hedgehog signaling. Development 129:4509–4521

Price AL, Patel NH (2008) Investigating divergent mechanisms of mesoderm development in arthropods: the expression of Ph-twist and Ph-mef2 in Parhyale hawaiensis. J Exp Zool B 310:24–40

Prpic NM (2004) Homologs of wingless and decapentaplegic display a complex and dynamic expression profile during appendage development in the milli-pede *Glomeris marginata* (Myriapoda: Diplopoda). Front Zool 1:6. doi:10.1186/1742-9994-1-6

Prpic NM, Janssen R, Damen WG, Tautz D (2005) Evolution of dorsal-ventral axis formation in arthropod appendages: H15 and optomotor-blind/bifid-type T-box genes in the millipede *Glomeris marginata* (Myriapoda: Diplopoda). Evol Dev 7:51–57

Raina A, Meola S, Wergin W, Blackburn M, Bali G (2003) Antennal ampullary glands of *Helicoverpa zea* (Lepidoptera: Noctuidae). Cell Tissue Res 312: 127–134

Rajulu G (1967) Physiology of the heart of *Cingalobolus bugnioni* (Diplopoda: Myriapoda). Cell Mol Life Sci 23:388

Rajulu GS (1969) Presence of haemocyanin in the blood of a centipede *Scutigera longicornis* (Chilopoda: Myriapoda). Curr Sci India 38:168–169

Randall WC (1966) Microanatomy of the heart and associated structures of two scorpions, *Centruroides sculpturatus* Ewing and *Uroctonus mordax* Thorell. J Morphol 119:161–180

Redmond JR, Jorgensen DD, Bourne GB (1982) Circulatory physiology of *Limulus*. Prog Clin Biol Res 81:133–146

Reiber CL, McGaw IJ (2009) A review of the "open" and "closed" circulatory systems: new terminology for complex invertebrate circulatory systems in light of current findings. Int J Zool. doi:10.1155/2009/301284

Reim I, Frasch M (2010) Genetic and genomic dissection of cardiogenesis in the *Drosophila* model. Pediatr Cardiol 31:325–334

Richards AG (1963) The ventral diaphragm of insects. J Morphol 113:17–47

Rilling G (1968) Großes zoologisches Praktikum. Teil 13b: *Lithobius forficatus*. Gustav Fischer, Stuttgart

Ryan KM, Hoshizaki DK, Cripps RM (2005) Homeotic selector genes control the patterning of seven-up expressing cells in the *Drosophila* dorsal vessel. Mech Dev 122:1023–1033

Saalfeld E (1936) Untersuchungen über den Blutkreislauf bei *Leptodora hyalina*. Z vergl Physiol 24:58–70

Sandeman DC (1967) The vascular circulation in the brain, optic lobes and thoracic ganglia of the crab *Carcinus*. Proc R Soc B 168:82–90

Schneider D, Kaissling KE (1959) Der Bau der Antenne des Seidenspinners *Bombyx mori* L. III. Das Bindegewebe und das Blutgefäß. Zool Jahrb Anat Ontog 77:111–132

Scholl G (1963) Embryologische Untersuchungen an Tanaidaceen (*Heterotanais oerstedi* Kroyer). Zool Jahrb Anat Ontog 80:500–554

Scholl G (1977) Beitrage zur Embryonalentwicklung von *Limulus polyphemus* L. (Chelicerata, Xiphosura). Zoomorphologie 86:99–154

Scholtz G, Kamenz C (2006) The book lungs of Scorpiones and Tetrapulmonata (Chelicerata, Arachnida): evidence for homology and a single

terrestrialisation event of a common arachnid ancestor. Zoology 109:2–13

Seifert B (1932) Anatomie und Biologie des Diplopoden *Strongylosoma pallipes* Oliv. Zoomorphology 25:362–507

Seifert G, Rosenberg J (1978) Feinstruktur der Herzwand des Doppelfüßers *Oxidus gracilis* (Diplopoda: Paradoxosomatidae) und allgemeine Betrachtungen zum Aufbau der Gefäße von Tracheata und Onychophora. Entomol Gener 4:224–233

Selman BJ (1965) The circulatory system of the alder fly *Sialis lutaria*. Proc Zool Soc Lond 144:487–535

Sherman RG (1985) Neural control of the heartbeat and skeletal muscle in spiders and scorpions. In: Barth FG (ed) Neurobiology of arachnids. Springer, New York, pp 319–336

Sherman RG, Bursey CR, Fourtner CR, Pax RA (1969) Cardiac ganglia in spiders (Arachnida, Araneae). Experientia 25:438

Shubin N, Tabin C, Carroll S (2009) Deep homology and the origins of evolutionary novelty. Nature 457:818–823

Siewing R (1954) Über die Verwandtschaftsbeziehungen der Anaspidaceen. Verh Deutsch Zool Ges 210–252

Siewing R (1956) Untersuchungen zur Morphologie der Malacostraca (Crustacea). Zool Jahrb Anat Ontog Tiere 75:39–176

Siewing R (1959) Syncarida. In: Gruner H-E (ed) Bronn's Klassen und Ordnungen des Tierreichs, vol 5., 1. Abteilung, Akademische Verlagsgesellschaft Geest & Portig, Leipzig, pp 1–121

Siewing R (1969) Lehrbuch der vergleichenden Entwicklungsgeschichte der Tiere. Paul Parey, Hamburg

Silen L (1954) On the circulatory system of the Isopoda Oniscoidea. Acta Zool 35:11–70

Sláma K (1999) Active regulation of insect respiration. Ann Entomol Soc Am 92:916–929

Sláma K (2000) Extracardiac versus cardiac haemocoelic pulsations in pupae of the mealworm (*Tenebrio molitor* L.). J Insect Physiol 46:977–992

Sláma K (2008) Extracardiac haemocoelic pulsations and the autonomic neuroendocrine system (coelopulse) of terrestrial insects. Terr Arthrop Rev 1:39–80

Sláma K (2010) Physiology of heartbeat reversal in adult *Drosophila melanogaster* (Diptera: Drosophilidae). Eur J Entomol 107:13–31

Sláma K, Baudry-Partiaogolou N, Provansal-Baudez A (1979) Control of extracardiac haemolymph pressure pulses in *Tenebrio molitor*. J Insect Physiol 25:825–831

Sláma K, Farkas R (2005) Heartbeat patterns during the postembryonic development of *Drosophila melanogaster*. J Insect Physiol 51:489–503

Sláma K, Neven L (2001) Active regulation of respiration and circulation in pupae of the codling moth (*Cydia pomonella*). J Insect Physiol 47:1321–1336

Steinacker A (1978) The anatomy of the decapod crustacean auxiliary heart. Biol Bull 154:497–507

Steinacker A (1979) Neural and neurosecretory control of the decapod crustacean auxiliary heart. Am Zool 19:67–75

Storch V, Ruhberg H (1993) Onychophora. In: Harrison FW, Rice ME (eds) Microscopic anatomy of invertebrates, vol 12. Wiley-Liss, New York, pp 11–56

Sun BD, Schmidt JM (1997) The structure of the antennal heart of *Aedes aegypti* (Linnaeus). Can J Zool 75:444–458

Tao Y, Schulz RA (2007) Heart development in *Drosophila*. Semin Cell Dev Biol 18:3–15

Tartes U, Vanatoa A, Kuusik A (2002) The insect abdomen: a heartbeat manager in insects? Comp Biochem Physiol A 133:611–623

Tauc HM, Mann T, Werner K, Pandur P (2012) A role for *Drosophila* Wnt-4 in heart development. Genesis 50:466–481

Tiegs OW (1940) The embryology and affinities of the Symphyla, based on a study of *Hanseniella agilis*. Q J Microsc Sci 82:1–225

Tiegs OW (1945) The post-embryonic development of *Hanseniella agilis* (Symphyla). Q J Microsc Sci 85:191–328

Tjønneland A, Kryvi H, Ostnes JP, Økland S (1985a) The heart ultrastructure in two species of pycnogonids and its phylogenetic implications. Zool Scr 14:215–219

Tjønneland A, Økland S, Midttun B (1985b) Myocardial ultrastructure in five species of scorpions (Chelicerata). Zool Anz 214:7–17

Tögel M, Pass G, Paululat A (2008) The *Drosophila* wing hearts originate from pericardial cells and are essential for wing maturation. Dev Biol 318:29–37

Vogt G, Wirkner CS, Richter S (2009) Symmetry variation in the heart-descending artery system of the parthenogenetic marbled crayfish. J Morphol 207:221–226

Wägele JW (1992) Isopoda. In: Harrison FW, Humes AG (eds) Microscopic anatomy of invertebrates, vol 9. Wiley-Liss, New York, pp 529–617

Wagenseil JE, Mecham RP (2009) Vascular extracellular matrix and arterial mechanics. Physiol Rev 89:957–989

Walker G (1991) Cirripedia. In: Harrison FW, Humes AG (eds) Microscopic anatomy of invertebrates, vol 9. Wiley-Liss, New York, pp 225–311

Wasserthal LT (1975) Periodische Herzschlag-Umkehr bei Insekten. Umschau 75:93–94

Wasserthal LT (1980) Oscillating haemolymph "circulation" in the butterfly *Papilio machaon* L. revealed by contact thermography and photocell measurements. J Comp Physiol 139:145–163

Wasserthal LT (1996) Interaction of circulation and tracheal ventilation in holometabolous insects. Adv Insect Physiol 26:297–351

Wasserthal LT (1998) The open hemolymph system of Holometabola and its relation to the tracheal space. In: Harrison F, Locke M (eds) Microscopic anatomy of invertebrates, vol 11B. Wiley-Liss, New York, pp 583–620

Wasserthal LT (1999) Functional morphology of the heart and of a new cephalic pulsatile organ in the blowfly *Calliphora vicina* (Diptera: Calliphoridae) and their roles in hemolymph transport and tracheal ventilation. Int J Insect Morphol 28:111–129

Wasserthal LT (2003a) Hämolymphe und Hämolymphtransport. In: Dettner K, Peters W (eds) Lehrbuch der Entomologie, 2nd edn. Spektrum, Heidelberg, pp 185–203

Wasserthal LT (2003b) Respiration. In: Kristensen NP (ed) Handbook of zoology, vol 4/36–2. Lepidoptera. de Gruyter, Berlin, pp 189–203

Wasserthal LT (2003c) Circulation and thermoregulation. In: Kristensen NP (ed) Handbook of zoology, vol 4/36-2. Lepidoptera. de Gruyter, Berlin, pp 205–228

Wasserthal LT (2007) *Drosophila* flies combine periodic heartbeat reversal with a circulation in the anterior body mediated by a newly discovered anterior pair of ostial valves and 'venous' channels. J Exp Biol 210:3707–3719

Wasserthal LT (2012) Influence of periodic heartbeat reversal and abdominal movements on hemocoelic and tracheal pressure in resting blowflies *Calliphora vicina*. J Exp Biol 215:362–373

Watson WH, Groome JR (1989) Modulation of *Limulus* heart. Am Zool 29:1287–1303

Weber H (1954) Grundriß der Insektenkunde. 3. Aufl. Gustav Fischer, Stuttgart

West-Eberhard MJ (2003) Developmental plasticity and evolution. Oxford University Press, New York

Weygoldt P (1961) Beitrag zur Kenntnis der Ontogenie der Dekapoden: embryologische Untersuchungen an *Palaemonetes varians* (Leach). Zool Jahrb Anat Ontog 79:223–294

Weygoldt P (1975) Untersuchungen zur Embryologie und Morphologie der Geißelspinne *Tarantula marginemaculata* C.L.Koch (Arachnida, Amblypygi, Tarantulidae). Zoomorphologie 82:137–199

Whedon A (1938) The aortic diverticula of the Odonata. J Morphol 63:229–261

Wilkens JL, Cavey MJ, Shovkivska I, Zhang ML, ter Keurs HEDJ (2008) Elasticity, unexpected contractility and the identification of actin and myosin in lobster arteries. J Exp Biol 21:766–772

Wilkens JL, Davidson GW, Cavey MJ (1997a) Vascular peripheral resistance and compliance in the lobster *Homarus americanus*. J Exp Biol 200:477–485

Wilkens JL, Walker RL (1992) Nervous control of crayfish cardiac hemodynamics. Comp Physiol 11:115–122

Wilkens JL, Yazawa T, Cavey MJ (1997b) Evolutionary derivation of the American lobster cardiovascular system: an hypothesis based on morphological and physiological evidence. Invertebr Biol 116:30–38

Wirkner CS, Hilken G, Rosenberg J (2011) The circulatory system. In: Minelli A (ed) Treatise on zoology—anatomy, taxonomy, biology, vol 1., Myriapoda. Brill, Leiden, pp 157–176

Wirkner CS, Pass G (2002) The circulatory system in Chilopoda: functional morphology and phylogenetic aspects. Acta Zool 83:193–202

Wirkner CS, Prendini L (2007) Comparative morphology of the hemolymph vascular system in scorpions—a survey using corrosion casting, MicroCT, and 3D-reconstruction. J Morphol 268:401–413

Wirkner CS, Richter S (2003) The circulatory system in Phreatoicidea: implications for the isopod ground pattern and peracarid phylogeny. Arthropod Struct Dev 32:337–347

Wirkner CS, Richter S (2004) Improvement of microanatomical research by combining corrosion casts with MicroCT and 3D reconstruction, exemplified in the circulatory organs of the woodlouse. Microsc Res Tech 64:250–254

Wirkner CS, Richter S (2007a) Comparative analysis of the circulatory system in Amphipoda (Malacostraca, Crustacea). Acta Zool 88:159–171

Wirkner CS, Richter S (2007b) The circulatory system and its spatial relations to other major organ systems in Spelaeogriphacea and Mictacea (Malacostraca, Crustacea)—a three-dimensional analysis. Zool J Linn Soc 149:629–642

Wirkner CS, Richter S (2007c) The circulatory system in Mysidacea—Implications for the phylogenetic position of Lophogastrida and Mysida (Malacostraca, Crustacea). J Morphol 268:311–328

Wirkner CS, Richter S (2008) Morphology of the haemolymph vascular system in Tanaidacea and Cumacea: Implications for the relationships of "core group" Peracarida (Malacostraca; Crustacea). Arthropod Struct Dev 37:141–154

Wirkner CS, Richter S (2009) The hemolymph vascular system in *Tethysbaena argentarii* (Thermosbaenacea, Monodellidae) as revealed by 3D reconstruction of semi-thin sections. J Crust Biol 29:13–17

Wirkner CS, Richter S (2010) Evolutionary morphology of the circulatory system in Peracarida (Malacostraca; Crustacea). Cladistics 26:143–167

Wirkner CS, Richter S (2013) Circulatory system and respiration. In: Thiel M, Watling L (eds) The natural history of the Crustacea, vol 1. Oxford University Press, Oxford, pp 376–412

Wolf MJ, Amrein H, Izatt JA, Choma MA, Reedy MC, Rockman HA (2006) *Drosophila* as a model for the identification of genes causing adult human heart disease. Proc Natl Acad Sci USA 103:1394–1399

Xavier-Neto J, Davidson B, Simoes-Costa MS, Castro RA, Castillo HA, Sampaio AC (2010) Evolutionary origins of hearts. In: Rosenthal N, Harvey RP (eds) Heart development and regeneration. Elsevier, London, pp 3–46

Yamagishi H, Ando H, Makioka T (1997) Myogenic heartbeat in the primitive crustacean *Triops longicaudatus*. Biol Bull 193:350–358

Yamazaki K, Akiyama-Oda Y, Oda H (2005) Expression patterns of a twist-related gene in embryos of the spider *Achaearanea tepidariorum* reveal divergent aspects of mesoderm development in the fly and spider. Zool Sci 22:177–185

Zaffran S, Frasch M (2002) Early signals in cardiac development. Circ Res 91:457–469

# The Arthropod Fossil Record

Gregory D. Edgecombe and David A. Legg

## Contents

G. D. Edgecombe (✉) · D. A. Legg
Department of Earth Sciences, The Natural History
Museum, Cromwell Road, London, SW7 5BD, UK
e-mail: g.edgecombe@nhm.ac.uk

D. A. Legg
e-mail: d.legg10@imperial.ac.uk

D. A. Legg
Department of Earth Science and Engineering,
Imperial College London, Prince Consort Road,
London, SW7 2AZ, UK

## 15.1 Introduction

With respect to animal life, we inhabit a planet dominated by arthropods. The Recent is not geologically anomalous in this respect—arthropods have been megadiverse for some 520 million years, since the early Cambrian. The trace fossil and body fossil records of Arthropoda extend to the main burst of the Cambrian explosion, alongside the earliest fossils of most other animal phyla. Because their calcite exoskeleton confers a high fossilisation potential, trilobites are the most species-rich fossil group in the Cambrian, but even when the fossil record is extended to non-biomineralised groups, arthropods are the most numerically abundant and diverse phylum in Early Palaeozoic sites of exceptional preservation (Briggs et al. 1994; Hou et al. 2004a; Van Roy et al. 2010; Zhao et al. 2010).

Fossils afford glimpses of vanished (that is, extinct) morphologies. These may reveal novel patterns of morphospace occupation (e.g. gigantism in groups that are now diminished in size) or novel ecologies or physiology (e.g. the ability to see the world through lenses composed of prisms of calcite, as in trilobites). Fossils contribute unique character combinations to phylogenetic analysis and, when their temporal occurrence lies near to the time of splitting events, they can alter the inferred relationships between extant taxa (Edgecombe 2010). Even when fossils do not change the relationships of living groups, they can still provide otherwise inaccessible

A. Minelli et al. (eds.), *Arthropod Biology and Evolution*,
DOI: 10.1007/978-3-642-36160-9_15, © Springer-Verlag Berlin Heidelberg 2013

information about the sequence of character evolution in stem groups (sensu Hennig 1966), and they can increase the accuracy of root positions for a tree by breaking up long branches that separate extant taxa from each other. Arthropod origins provide an obvious example. The sister group relationship between Onychophora and Arthropoda indicated by anatomy and phylogenomics (Campbell et al. 2011) may withstand the addition of fossils to datasets, but fossils are critical to understanding that arthropods evolved from a grade of lobopodian animals, some of which possessed character states now lost (e.g. the armoured frontal appendages of *Megadictyon* or the defensive spines of *Hallucigenia*) (Ma et al. 2009).

The temporal information provided by fossils is vital for inferring divergence dates, fossils being the usual source of minimal divergence dates for calibrating nodes in molecular trees, and modern methods of molecular dating use relaxed clocks and probabilistic calibrations that incorporate uncertainties in the fossil record. The deepest splits between the extant lineages of Arthropoda are inferred to have occurred in the late Neoproterozoic, in the Ediacaran Period. Although this predates palaeontological estimates based on fossils (no Ediacaran organisms are convincingly attributed to Arthropoda), molecular dating of arthropod phylogeny is increasingly compatible with the fossil record (Sanders and Lee 2010; Rehm et al. 2011).

This chapter highlights some styles of fossil preservation that have provided the most anatomically informative arthropod material in the fossil record. Because of the enormity of literature on fossil arthropods, citations are weighted in favour of the most recent descriptions or revisions of species named in earlier studies.

## 15.2 Burgess Shale-Type Faunas and Cambrian Stem Groups

The term "Burgess Shale-type" biota (abbreviated as BST) refers to preservation of non-biomineralised fossils as more or less two-dimensional carbonaceous compressions

(Fig. 15.1). In the Cambrian, dozens of such sites are now known across an expanding geographic range. Rapid burial and inhibited microbial decay due to oxidant starvation, thereby allowing organic tissues to preserve, are the signatures of BSTs (Gaines et al. 2012). Some taxa are shared between several of the better known BSTs, despite occurrences in different palaeogeographic settings and a range of time. For example, the arthropods *Isoxys, Naraoia, Leanchoilia, Canadaspis,* and *Tuzoia,* and the anomalocaridids *Anomalocaris* and *Amplectobelua,* among others, occur in both the Burgess Shale in Cambrian Stage 5 in Canada (on the palaeocontinent Laurentia) and in the Chengjiang biota in Cambrian Stage 3 in China (South China Plate).

### 15.2.1 The Chengjiang Lagerstätte

The discovery in 1984 of naraoiids—non-biomineralised trilobite relatives known from the Burgess Shale (Whittington 1977; Zhang et al. 2007b)—at Maotianshan in Yunnan Province, China, initiated a burst of intense taxonomic description. Hou et al. (2004a, b) summarised knowledge of arthropods from the Chengjiang Lagerstätte to that time, and several species have since been redescribed and new taxa documented. Of the 228 species known from Chengjiang by 2010, 37 % were arthropods (this increases when stem-group arthropods such as anomalocaridids are added) and they make up about 52 % of the individuals collected (Zhao et al. 2010). The importance of the Chengjiang Lagerstätte results from its diversity and quality of preservation, but also its age, being some 10–15 million years older than the Burgess Shale, and thus closer to deep divergences during the Cambrian Explosion.

#### 15.2.1.1 Naraoiids
The first arthropods to be described from Chengjiang were two species of naraoiid (Lamellipedia: Artiopoda: Nektaspida), assigned to *Naraoia,* a genus previously documented from the Burgess Shale and coeval sites in Utah.

**Fig. 15.1** Arthropods from Cambrian Burgess Shale-type biotas (BSTs). **a-d, g** Chengjiang fauna (Cambrian Stage 3), Yunnan Province, China: **a** the naraoiid *Naraoia spinosa*, showing gut (*gt*) and gut diverticula (*gd*) in cephalon; **b** the naraoiid *Misszhouia longicaudata*, showing an exopod (*ex*) bearing lamellar setae (*lm*); **c** the fuxianhuiid *Liangwangshania biloba* (photo courtesy of Chen Ai-lin); **d** the bradoriid *Kunmingella douvillei* with left (*lv*) and right (*rv*) valves agape, exposing posterior-most appendages (*a9, a10*) (photo courtesy of Derek Siveter); **g**, the trilobite *Eoredlichia intermedia*, ventral view showing antennae (*an*) emerging from sides of hypostome (photo courtesy of Xiaoya Ma). **e, f, h, i,** Burgess Shale (Cambrian Stage 5), British Columbia, Canada: **e** the bivalved arthropod *Perspicaris dictynna* (image courtesy of Allison Daley); **f** the marrellomorph *Marrella splendens* (image courtesy of Diego García-Bellido); **h** the megacheiran *Leanchoilia superlata* (image courtesy of Allison Daley). Cephalon in lateral view, showing "great appendages" with spinose projections bearing flagellae (*fg*); **i** the possible stem-group chelicerate *Sanctacaris uncata*, cephalic appendages in dorsal view. Other abbreviations: (*as*) anterior eye-bearing sclerite; (*cp*) carapace; (*gs*) genal spine: (*le*) left eye; (*st*) sternite

Subsequent documentation of "soft" (i.e. non-biomineralised) anatomy of the Chengjiang species led to a proposal that two genera were represented (*Naraoia* and *Misszhouia*: Chen et al. 1997), and recent monographic revision has clarified their morphology and systematics (Zhang et al. 2007b).

The naraoiid exoskeleton (Fig. 15.1a) consists of a head shield and a single trunk tergite (i.e. lacking articulated thoracic segments). The head covers the antennae and four biramous appendage pairs (Zhang et al. 2007b) that resemble those on the trunk. The morphology of head and trunk appendages is similar to those of trilobites. The endopod consists of seven articles distal to the protopodite, and the exopod is subdivided into two parts, the proximal of which bears long, imbricated setae ("lamellar setae": Hou and Bergström 1997) and the distal of which bears short setae (Fig. 15.1b). Different naraoiids show substantial differences in the extent of the gut diverticula (Fig. 15.1a), but a consensus has emerged that they were epibenthic predators and scavengers (Vannier and Chen 2002).

### 15.2.1.2 Fuxianhuiids
Fuxianhuiids [Class Yunnanata Hou and Bergström (1997): Order Fuxianhuiida Bousfield (1995)] are known from five species— *Fuxianhuia protensa*, *Chengjiangocaris longiformis*, *Shankouia zhenghei*, *Liangwangshania biloba* (Fig. 15.1c), and *Guangweicaris spinatus*—all from Chengjiang and the early Cambrian Guanshan Lagerstätte in Yunnan. An additional undescribed species occurs in the slightly younger Kaili Biota in Guizhou (Zhu et al. 2004). Investigations of the head composition and appendage structure of *Fuxianhuia* led to a hypothesis that this taxon occupied a pivotal, early-diverging position in the arthropod stem group (Chen et al. 1995; Hou and Bergström 1997). Anatomical data have since been added for *Shankouia* (Waloszek et al. 2005), and *Fuxianhuia* has been revised based on new collections (Bergström et al. 2008), including specimens that preserve a tripartite brain and neural tissue in the optic lobes (Ma et al. 2012).

Fuxianhuiids have a pair of stalked, moveable eyes associated with an anterior head sclerite (Fig. 15.1c). The head bears one pair of antennae and curved structures that have variously been identified as gut diverticula (Waloszek et al. 2005; Bergström et al. 2008) or a uniramous appendage pair (Budd 2008). The trunk consists of numerous tergites; in some members it is differentiated into a thoracic tagma with wide paratergal folds and a narrower abdominal tagma. In at least some taxa (such as *Fuxianhuia*) the trunk appendages are decoupled from the tergites, the appendages being more numerous. The limb stem is pediform, composed of ca 20 articles, with a blunt distal tip, and the exopods are simple rounded flaps. Although fuxianhuiids had been interpreted as deposit feeders on the basis of a sediment-filled gut (Bergström and Hou 2005), at least some taxa are observed to have trilobite fragments in the gut and must have been durophagous predators (Zhu et al. 2004).

### 15.2.1.3 Bradoriida
Bradoriids are bivalved arthropods ranging from early Cambrian to the Early Ordovician, long considered to be members of the Ostracoda based on their shell form. The ostracod affinities of bradoriids have been refuted by several investigations of the softpart morphology of these fossils (Shu et al. 1999; Hou et al. 2010). The best preserved appendage morphology comes from *Kunmingella douvillei* (=*K. maotianshensis* of earlier descriptions), from Chengjiang (Fig. 15.1d). The body consists of 10 limb-bearing segments, apparently five of them in the head and five in the trunk. The antenna is long and uniramous, limbs 2–8 are biramous, and limbs 9–10 are again uniramous. The biramous limbs form a homonomous series, all having a leaf-shaped exopod and a five-segmented endopod. The number of head segments may vary within the group, a species of *Kunyangella* apparently having four segments rather than five (Hou et al. 2010). Though arguments have been made that bradoriids are stem-group crustaceans (Shu et al. 1999; Hou

et al. 2010), few strong characters place them there or indeed within Mandibulata.

#### 15.2.1.4 Bivalved Miscellanea

The lesson from bradoriids described above (misleading resemblance of the bivalved carapace to that of ostracods) is reinforced by many other bivalved arthropods from Chengjiang and other BSTs (Fig. 15.1e). Many of these groups had been assigned to crustacean clades that have bivalved carapaces (e.g. Phyllocarida or some Branchiopoda) but discovery of soft-parts has negated membership of the crustacean crown-group or, in other cases, even the euarthropod crown group.

*Canadaspis perfecta* is a bivalved arthropod from the Burgess Shale (Briggs 1978), but a Chengjiang species, *C. laevigata* (Hou and Bergström 1997), has also been referred to the genus. The interpretation of paired antennae, mandibles and a five-segmented head, and more particularly the number of somites in its trunk tagmata, led to *Canadaspis* being identified as a malacostracan crustacean (Briggs 1978, 1992). This identification was contested by carcinologists (Dahl 1984; Boxshall 1997) based on its limb structure, and the Chengjiang *C. laevigata* possesses just a single pair of antennae, followed by uniform biramous post-antennal limbs, without clear head-trunk tagmosis (Hou and Bergström 1997). *Canadaspis* has accordingly been viewed as a stem-group arthropod (Budd 2002; Legg et al. 2012b) or at least being outside the pancrustacean crown-group (Bergström and Hou 2005) rather than being a phyllocarid malacostracan.

Another bivalved arthropod, *Isoxys*, is represented by three species from Chengjiang (Vannier et al. 2009) and other Lagerstätten in South China, two from each of the Burgess Shale (García-Bellido et al. 2009b) and the Emu Bay Shale, Australia (García-Bellido et al. 2009a), and others are distributed in the western USA, Greenland, Spain, France and Siberia. *Isoxys* has a large pair of eyes, a spiniferous, prehensile anterior appendage pair, and a uniform series of biramous limbs on the remaining segments. The body was largely covered by the paired valves, terminating in a flap-like telson. The form of the anterior appendage has been used as evidence that *Isoxys* is related to "great appendage" arthropods or to anomalocaridids (Vannier et al. 2009), though the homologies have not been strongly established; it is likely that raptorial frontal appendages have multiple origins in Cambrian arthropods (Fu et al. 2011).

The bivalved Chengjiang species *Pectocaris spatiosa* has been identified as a crustacean for having two pairs of antennae and mandibles, and more specifically as a branchiopod based on trunk limbs with a blade-like exopod and numerous (ca. 40) spinose endites on the endopod (Hou et al. 2004b). The gnathal edge of its purported mandible does not show the characteristic differentiation of a crown-group pancrustacean, and it has alternatively been relegated to the arthropod stem-group (Harvey and Butterfield 2008). A recently documented Chengjiang arthropod, *Jugatacaris agilis*, resembles *Pectocaris* in having a highly polypodous trunk (55–65 segments), a comparably large number of podomeres in the legs, and paddle-like furcal rami. It too has been called a "crustaceanomorph" (Fu and Zhang 2011), a rather vaguely-delimited assemblage for which resemblance to Crustacea may be superficial. Phylogentic analyses resolve *Pectocaris*, *Jugatocaris*, *Canadaspis* and several other Cambrian bivalved forms as a grade at the base of the arthropod stem group (Legg et al. 2012b).

### 15.2.2 The Burgess Shale

Charles Walcott's discovery of the Burgess Shale in what is now Yoho National Park in British Columbia, Canada, in 1909, provided the first insights into the composition of a predominantly non-biomineralised Cambrian marine community (Briggs et al. 1994). The initial descriptions of arthropods were made by Walcott himself, but a concerted program of investigation was not resumed until the late 1960s, when Harry Whittington led the Geological Survey of Canada Burgess Shale project (Whittington 1985).

Several arthropod taxa from the Burgess Shale that were described by Whittington's group in the 1970s and 1980s have been revisited, applying new techniques of study or using new collections. These include new works on *Marrella* (García-Bellido and Collins 2006), *Leanchoilia* (García-Bellido and Collins 2007; Haug et al. 2012a), *Sarotrocercus* (Haug et al. 2011), *Emeraldella* (Stein and Selden 2012), and *Yohoia* (Haug et al. 2012b).

### 15.2.2.1 Soft Anatomy of Trilobites

Trilobita is a monophyletic group that persisted through some 270 million years, from the early Cambrian to the end of the Permian. The known diversity of the clade consists of more than 19,000 species (Adrain in Zhang 2011), rendering trilobites perhaps the most familiar of all invertebrate fossils. In spite of their ubiquity, appendages are preserved in just 20 species, and of the dozen or so trilobite species in the Burgess Shale, only three preserve their soft parts. By far the most informative of these is *Olenoides serratus*, for which the anatomy and functional morphology were treated in detail by Whittington (1975, 1980).

*Olenoides serratus* has a pair of antennae and three pairs of biramous head appendages, the latter all similar in detail to the trunk appendages (a situation that, as far as is known, pertains to all trilobites). The trunk has an appendage pair under each of its seven thoracic segments and four to six pairs under the pygidium. Endopods of both the head and trunk limbs have seven podomeres; these and the protopodite are strongly spinose. The exopods are divided into a proximal and distal lobe, the former with lamellar setae (as in naraoiids and other "lamellipedians"). *Olenoides* is the only trilobite known to have a pair of long, antenniform, terminal cerci that project far behind the pygidium. The appendage morphology of *Olenoides* corresponds closely to that of another Cambrian

trilobite, *Eoredlichia intermedia* (Hou et al. 2009) from Chengjiang (Fig. 15.1g), the two belonging to different orders that diverged from each other near the root of Trilobita and thus providing a reasonable estimate for plesiomorphic characters for all trilobites.

Since the mid-20th century, trilobites have generally been regarded as part of an "arachnomorph" assemblage (Bergström 1992; Cotton and Braddy 2004) that includes chelicerates and their stem group. The status of the putative apomorphies of Arachnomorpha/Arachnata has been challenged (Scholtz and Edgecombe 2005), and an assignment of Trilobita to the mandibulate stem group has advanced as an alternative.

### 15.2.2.2 Marrellomorpha

*Marrella splendens* is the most abundant Burgess Shale arthropod (Fig. 15.1f), known from tens of thousands of specimens, and was the first to be revised in the modern era (Whittington 1971). *Marrella*'s most distinguishing feature is its head shield, which is dominated by two pairs of spines, one lateral and one medial (Fig. 15.1f). The head appendages are a pair of elongate antennae and a more robust second appendage that has been ascribed a natatory role (García-Bellido and Collins 2006). The body has 26 post-cephalic segments in the largest individuals (García-Bellido and Collins 2006). The trunk segments are ring-like, without pleurae, and each bears a pair of biramous appendages in which the exopods bear slender filaments. *Marrella* belongs to a clade named Marrellomorpha [reviewed by Rak et al. (2013)] that persisted until at least the Early Devonian. Its best known members include the Ordovician *Furca* from Bohemia and Morocco (Van Roy et al. 2010; Rak et al. 2013), the British Silurian *Xylokorys chledophilia* (Siveter et al. 2007a) (Fig. 15.5c), and the German Devonian *Mimetaster hexagonalis* (Kühl and Rust 2010).

### 15.2.2.3 Megacheira: "Great Appendage Arthropods"

The Burgess Shale and the other main BSTs preserve the remains of arthropods popularly known as great appendage arthropods. The proper name Megacheira has been applied to this assemblage (Hou and Bergström 1997), but whether megacheirans are monophyletic or are paraphyletic is debated. If paraphyletic, they are variously regarded as either stem-group Chelicerata based on a hypothesis that the frontal appendage is a precursor of a chelifore or chelicera (Chen et al. 2004; Cotton and Braddy 2004), or stem-group Euarthropoda (Budd 2002; Legg et al. 2012b). The most anatomically complete megacheiran is *Leanchoilia* (Fig. 15.1h), a genus best known from two species in the Burgess Shale (García-Bellido and Collins 2007; Haug et al. 2012a) and another from Chengjiang (Liu et al. 2007b). The frontal ("great") appendage of *Leanchoilia* consists of a two-segmented peduncle and four distal articles, three of which have a long spinose projection bearing a flagellum. The head shield covers three pairs of biramous appendages (the first reduced in size) that are similar to those of the 11 trunk segments. The endopods of *L. superlata* are composed of no more than seven podomeres (Haug et al. 2012a), fewer than in previous descriptions of the genus (García-Bellido and Collins 2007; Liu et al. 2007a, b). The exopods are two-segmented flaps fringed with spines. The telson is triangular, bearing a fringe of marginal spines. Gut diverticula are preserved as a series of paired reniform organs in the head and anterior part of the trunk (Butterfield 2002; García-Bellido and Collins 2007).

Other megacheirans from the Burgess Shale are *Alalcomenaeus* (Briggs and Collins 1999), a genus that is closely allied to *Leanchoilia*, and *Yohoia* (Haug et al. 2012b). Chengjiang is a source of additional megacheiran diversity, including *Jianfengia* and *Haikoucaris* (Chen et al. 2004), and the Emu Bay Shale in Australia has another leanchoiliid, *Oestokerkus* (Edgecombe et al. 2011).

### 15.2.2.4 *Sanctacaris*: A Chelicerate?

Nicknamed "Santa Claws" prior to its official description (Briggs and Collins 1988), *Sanctacaris uncata* sports an impressive set of raptorial claws under its cephalic shield (Fig. 15.1i). This taxon was originally considered the oldest representative of the chelicerates based on the supposed presence of (1) six pairs of head appendages, (2) a cardiac lobe resembling that of extant horseshoe crabs, (3) the division of the body into a prosoma and opisthosoma, and (4) the location of the anus on the last trunk segment. It does, however, lack a pivotal chelicerate synapomorphy—chelicerae—and cladistic analyses (e.g. Wills et al. 1998) have failed to resolve this taxon close to chelicerates. Bousfield (1995) suggested that chelicerae were definitely absent, the limbs instead representing a single "great appendage"-like limb basket. This interpretation was followed by Budd (2002), who resolved *Sanctacaris* amongst the "great appendage" arthropods within the euarthropod stem-lineage. Analyses that resolve "great appendage" arthropods as stem-lineage chelicerates (Cotton and Braddy 2004) reignite the hypothesis that *Sanctacaris* may indeed be close to Chelicerata. Boxshall's (2004) interpretation, in which the raptorial limbs correspond to those of a prosoma (but are biramous, with a slender exite), is most consistent with a chelicerate identity for *Sanctacaris*.

### 15.2.2.5 *Anomalocaris* and Other Radiodonta: Stem-, Crown- or Non-Arthropods?

The history of the piecing together of *Anomalocaris* has been told many times, the end result being the largest known Cambrian animal (Whittington and Briggs 1985; Collins 1996). The diversity of anomalocaridids in the Burgess Shale has increased as a result of new species being discovered from their heavily sclerotised, spinose frontal appendages (Daley and Budd 2010), as well as better understanding of the articulated specimens that contribute the bulk of data bearing on their systematic affinities. Notably, a formerly unrecognised anomalocaridid with a three-part carapace-like structure, *Hurdia*, has been shown to be a common element in the Burgess Shale (Daley et al. 2009), the characteristic oral cone of the group is more

**Fig. 15.2** Reconstruction of the anomalocaridid (Radiodonta) *Amplectobelua* [based on specimens figured by Hou et al. (1995), Daley and Budd (2010)]

variable than previously suspected (Daley and Bergström 2012), and new genera have been assigned to the group from the Chengjiang Lagerstätte (Hou et al. 1995).

The anomalocaridid body (Fig. 15.2) can attain a length of over 90 cm (Van Roy and Briggs 2011). The mouth is situated ventrally on the head, opening at the centre of a circlet of overlapping plates. The only appendages are the frontal appendages, attaching anterolaterally on the head, their segmentation and spinosity providing many of the main taxonomic characters within the group. Stalked compound eyes (Paterson et al. 2011) attach dorsolaterally. The trunk is composed of at least seven pairs of imbricated lateral lobes ["swimming flaps" fide Hou and Bergström (2006)], each of which is associated with blade-like lamellae. A posterior tagma is variably developed as a three-segmented tail fan, and some species have a pair of elongate cerci.

Anomalocaridids have been united under the proper name Radiodonta (Collins 1996), with reference to the oral cone. Their affinities to Arthropoda are controversial, the diversity of current perspectives ranging from them being stem-group arthropods (Budd 1997; Daley et al. 2009), part of the arthropod crown group (Chen et al. 2004; Haug et al. 2012b), or part of a clade of Early Palaeozoic predatory ecdysozoans with convergent similarity to arthropods (Hou and

Bergström 2006). Perhaps the least plausible of these alternatives is the crown group proposal, in which anomalocaridids are viewed as a paraphyletic group from which chelicerates evolved via a megacheiran grade. This hypothesis is based only on a few characters of one appendage pair (anomalocaridid frontal appendages, megacheiran "great appendages" and chelicerae), and it employs character states for which no evidence exists (e.g. a two-segmented peduncle in the anomalocaridid frontal appendage).

### 15.2.3 Sirius Passet

Early Cambrian fossils from the Buen Formation at Sirius Passet in North Greenland (Peel and Ineson 2011) came to widespread attention in the 1980s with the discovery of articulated specimens of halkieriids, stem-group molluscs with an elaborate scleritome. This biota is approximately coeval with the Chengjiang biota ("early Cambrian"/Cambrian Stage 3). To date, some 40 species have been discovered, of which nine arthropods and four lobopodians or stem-group arthropods have been formally described.

*Kiisortoqia soperi* is a common arthropod from Sirius Passet, reaching a body length in excess of 50 cm (Stein 2010). The head is covered by a simple shield, and bears a pair of elongate appendages composed of at least 15 articles, followed by three biramous limb pairs that resemble those on the trunk. The trunk has 16 segments, each of which has a tergite with paratergal folds that covers a pair of biramous appendages. The first appendage has been compared to an anomalocaridid frontal appendage (Stein 2010) and this has been claimed to solve the segmental identity of the latter (by identifying the *Kiisortoqia* appendage as an "antennule", i.e. deutocerebral). The comparison with anomalocaridids is not so clear-cut. Stein drew heavily on *Kiisortoqia* and anomalocaridids having a shared number of articles ("about 15") in their frontal appendages, but the variability in numbers of articles in anomalocaridids [e.g. 9 in *Hurdia*, 11 in *Peytoia*; Daley et al. (2009)] weakens the argument. As well, the spines on the antennule are not rigid processes oriented perpendicular to the axis of the appendage as in anomalocaridids but rather resemble movable setae similar to those of other arthropods.

*Campanamuta mantonae* is even more abundant than *Kiisortoqia*, its 1,700 specimens representing one-fifth of the known Sirius Passet collections (Budd 2011). This taxon has tagmosis resembling that of many trilobitomorphs (cephalon, a trilobed thorax of eight segments, pygidum of similar size to the cephalon). The cephalon bears antennae and possibly five appendage pairs, but the post-antennal appendages are not at all well known, and its exact position within Artiopoda [arthropods with lamellar setae on their exopods and undifferentiated biramous post-antennal limbs, like trilobites: Hou and Bergström (1997)] is uncertain.

Sirius Passet is the source of *Kerygmachela kierkegaardi* [revised by Budd (1999)] and *Pambdelurion whittingtoni* (Budd 1997), animals that have been described as "gilled lobopodians" and which have played a major role in discussions about the affinities of anomalocaridids and character evolution in the arthropod stem group. Both have a pair of large, spine-bearing frontal appendages that are annulated rather than arthropodised, and their bodies are composed of 11 pairs of imbricated flaps (homologised with the "swimming flaps" of anomalocaridids) that are each associated with a pair of lobopodial appendages. *Pambdelurion* at least has a radial oral cone composed of plates with dentate inner margins, situated on the ventral side of the head as in anomalocaridids (Budd 1997 and GDE pers. obs.). Debate continues about whether the similarities of the "gilled lobopodians" and anomalocaridids are due to unique common ancestry (i.e. membership in a clade of giant Cambrian predators called Dinocaridida: Chen et al. 1994; Hou et al. 2006) or whether *Kerygmachela*, *Pambdelurion* and anomalocaridids branched successively from the arthropod stem lineage (Budd 1997, 1999; Liu et al. 2007a; Daley et al. 2009).

## 15.2.4 Emu Bay Shale

The early Cambrian (Stage 4) Emu Bay Shale Konservat-Lagerstätte, Kangaroo Island, South Australia, constitutes the only diverse BST known from Australia (Gehling et al. 2011). This biota is presently known from over 50 species, more than half of which are stem- or crown-group arthropods. It includes elements known from other BSTs, such as the large bivalved arthropods *Isoxys* and *Tuzoia* (García-Bellido et al. 2009a), the only Australian records of nektaspidids (Paterson et al. 2010) and leanchoiliids (Edgecombe et al. 2011), and some genera otherwise known only from Chengjiang, such as *Squamacula* (Paterson et al. 2012). Compound eyes are well preserved by phosphate replication, and demonstrate that predatory arthropods had evolved visual specialisations such as a bright zone by 515 mya (Lee et al. 2011).

## 15.2.5 Fezouata: An Ordovician BST

Early Ordovician faunas from the Lower and Upper Fezouata Formations in Morocco (Van Roy et al. 2010) demonstrate the survival of several taxa otherwise known only from Cambrian BSTs into the Ordovician. Examples include the only known Ordovician anomalocaridid (Van Roy and Briggs 2011) and species apparently related to the Burgess Shale arthropods *Thelxiope* and *Skania* (Van Roy et al. 2010). These assemblages also provide the earliest records of some lineages that proliferated later in the Palaeozoic, such as Xiphosura.

## 15.3  Origins of Crown Groups

### 15.3.1 Orsten

The term "Orsten" refers to small fossils preserved by calcium phosphate replacement of cuticle, concentrated in a temporal window from the early Cambrian to the Early Ordovician. A review of the taphonomy, occurrences and the

most informative fossils from Orsten was provided by Maas et al. (2006). This material is
important because of the fidelity of preservation
of soft anatomy, especially appendages, its
undistorted three-dimensionality, and because it
is concentrated around a size fraction (the fossils
being no larger than 2 mm) that allows larval
stages to be documented. Even some internal
anatomical structures, such as muscles, tendons
and parts of the gut, can be imaged in Orsten
arthropods by microtomography (Eriksson et al.
2012).

Although Orsten fossils are known from
Poland, Siberia, the UK (Fig. 15.3), Canada,
Australia, and China (see below), the most
thoroughly documented material is that from the
Alum Shale in Sweden, where Orsten preservation spans the interval from Cambrian Stage 7
(ca 500 mya) to the terminal Cambrian stage,
the Furongian (ca 490 mya).

### 15.3.1.1 *Rehbachiella*: A Branchiopod

*Rehbachiella kinnekullensis* is one of the best
understood Cambrian arthropods in terms of its
morphology and ontogeny, owing to the detailed
description by Walossek (1993). This species is
known from 30 ontogenetic stages from the
nauplius through a long larval series. An
assignment of *Rehbachiella* to the branchiopod
crown-group emphasised characters of the filter

feeding apparatus, with a more precise alliance
to Anostraca (Walossek 1993). An alternative
position in the branchiopod stem-group rather
than in the anostracan lineage has subsequently
found favour (Schram and Koenemann 2001;
Olesen 2009).

### 15.3.1.2 *Agnostus*

Agnostids are small, blind Cambrian-Ordovician
arthropods that are often classified as trilobites
based on their similarity to a securely-placed
group of trilobites, the Eodiscina (Cotton and
Fortey 2005). In both cases the cephalic and
pygidial tagmata are of similar proportions, the
thorax bears two or three segments, and the
exoskeleton is trilobed. Appendage morphology
of an agnostid, *Agnostus pisiformis*, is magnifi-
cently preserved in Swedish Orsten (Müller and
Walossek 1987). This description underpins
arguments that the resemblance between agnostids and eodiscids is superficial, and that
agnostids are instead an early diverging lineage
in the crustacean stem group (Bergström and
Hou 2005; Haug et al. 2010a). The limbs of
*Agnostus* have a hanging, rather laterally-
splayed stance, and the first three pairs of
appendages are structurally differentiated from
the posterior most head appendage and the trunk
appendages (as in the naupliar limbs of crustaceans). The exopod setae have round sections,

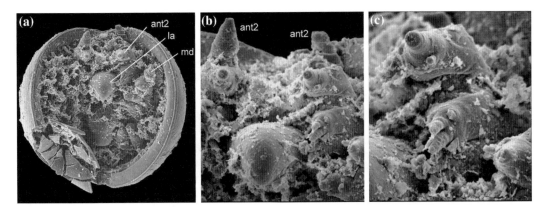

**Fig. 15.3** "Orsten" preservation of *Klausmuelleria sa-
lopiensis*, a phosphatocopid with soft parts from the
Lower Cambrian, Shropshire, UK (Siveter et al. 2001,
2003a, b). All in ventral view. **a** Open shield with soft
parts (see **b** for labels); **b** labrum (*la*), second antennae
(*ant2*), and medial tip of basipod of left mandible (*md*);
**c** proximal parts of left second antenna, bearing setulate
spines. SEM images courtesy of David Siveter

and are not arranged in the comb-like pattern as are the flattened setae of trilobites.

### 15.3.1.3 *Cambropycnogon*: A Pycnogonid

Pycnogonids (sea spiders) are rare in the fossil record, limited to a handful of species from the upper Cambrian, Silurian, Devonian and Jurassic (Charbonnier et al. 2007). *Cambropycnogon klausmuelleri* is the oldest unequivocal pycnogonid (Waloszek and Dunlop 2002). Known exclusively from protonymphs resembling those of extant pygnogonids, this fossil demonstrates both how early the lineage diverged from other arthropod groups and how little their morphology has changed since this time. This taxon differs from extant pygnogonids in the retention of gnathobasic post-cheliceral limbs and the possession of a pair of pre-cheliceral appendages. The exact identity of these appendages is unknown but their position has been used as evidence that primitive arthropods possessed an appendage pair innervated from the protocerebrum (Scholtz and Edgecombe 2006). These appendages were originally described as deutocerebral (Waloszek and Dunlop 2002), identified as homologous with the antenna of other arthropods, and used as evidence that the chelicerae were innervated from the tritocerebrum, as in the traditional view of brain morphology (Bitsch and Bitsch 2007). This has been countered by neuroantomical and, especially, gene expression studies that instead show the antenna of mandibulates and the chelicera of chelicerates to be homologous, both originating from the deuterocerebral somite [reviewed by Scholtz and Edgecombe (2006)]. The identity of the small anterior appendages in *Cambropycnogon* remains unsettled.

### 15.3.1.4 "Stem-Group Crustaceans"

Perhaps the most significant contribution of Swedish Orsten fossils has been the insights they provide into the early evolution of Tetraconata (=pancrustaceans). Walossek and Müller (1990) first outlined a hypothesis in which a series of Orsten taxa constitute a grade in the crustacean stem group. Current understanding of this group resolves the recently-revised *Oelandocaris* (Stein et al. 2008) and *Henningsmoenicaris* (Haug et al. 2010a ), *Goticaris* and *Cambropachycope* (Haug et al. 2009), and *Martinssonia* (Haug et al. 2010b) in progressively more crownward positions in the crustacean stem group. The stem group hypothesis draws heavily on the development of a proximal endite on the limb bases. At the base of Crustacea s.l. (Haug et al. 2009; Zhang et al. 2012), this endite is present only on the third appendage (the positional homologue of the mandible), as exemplified by *Oelandocaris*. Species positioned more crown ward have a proximal endite on at least the second appendage as well.

Phosphatocopina consists of ca 30 known species with a bivalved carapace that, like Bradoriida, were traditionally classified with ostracods. The discovery of their soft-anatomy in Swedish Orsten (Müller 1979) and subsequently in even older, early Cambrian rocks in England (Siveter et al. 2001, 2003b) (Fig. 15.3), provided evidence that they are neither ostracods nor crown-group crustaceans. A monographic treatment of Orsten Phosphatocopina (Maas et al. 2003) elaborated a hypothesis that phosphatocopines are sister group to Eucrustacea in a clade named Labrophora, the name deriving from the shared presence of a labrum with setulate sides and paragnath humps. They fall outside Eucrustacea based on their unspecialised post-mandibular appendages (lacking differentiation of maxillae) and their four- (rather than three) segmented larva.

The Labrophora scheme was developed in the light of crustacean monophyly, i.e. inferring that Eucrustacea is a clade than includes all extant "crustaceans" but not hexapods. Zhang et al. (2007a, their Fig. 3) note that the group would correspond to Mandibulata if hexapods and myriapods were included. Accommodating the morphologies of these fossils in light of the well-supported hypothesis supported by neuroanatomy and molecular sequences—"Eucrustacea" as a grade that includes hexapods rather than a clade—remains to be undertaken.

### 15.3.1.5 Chinese Orsten

Important Orsten fossils have recently been documented from sites in Hunan and Yunnan Provinces, South China. Furongian strata in Hunan are the source of *Skara hunanensis* (Liu and Dong 2007), a congener of the two Swedish Orsten species that formed the basis for the maxillopodan crustacean order Skaracarida (Müller and Walossek 1985). Late Cambrian Orsten from Hunan has also yielded phosphatocopines assigned to genera based on Swedish Orsten type species, *Hesslandona* and *Vestrogothia* (Zhang and Dong 2009; Zhang et al. 2012).

Even older Orsten fossils from Yunnan provide evidence for crown-group crustaceans in the early Cambrian, from strata that represent the same trilobite zone as the Chengjiang Lagerstätte. *Yicaris dianensis* Zhang et al. (2007a), was originally assigned to the Entomostraca (a grouping that is consistently resolved in molecular phylogenies as paraphyletic: Regier et al. 2010; von Reumont et al. 2012) and compared to branchiopods and cephalocarids based on similarities in the endites on its protopodites (Zhang et al. 2007a). *Yicaris* prompted a debate over whether outgrowths from its legs are (Maas et al. 2009) or are not (Boxshall 2007; Boxshall and Jaume 2009) epipodites. Another Yunnan Orsten taxon, *Wujicaris muelleri*, is based on metanauplius larvae that are closely comparable to those of copepods and barnacles (Zhang et al. 2010). The presence of crown-group crustaceans in early Cambrian Orsten is corroborated by another kind of fossil preservation, discussed in the following section.

### 15.3.2 Small Carbonaceous Fossils

Organic preservation in the form of cuticle fragments extracted from shales by dissolution in hydrofluoric acid ("small carbonaceous fossils": Butterfield and Harvey 2012) is proving to be especially informative for understanding the early history of crustaceans. Fragments of this kind

attributed to Crustacea were first illustrated by Butterfield (1994) from rocks of late early Cambrian age in Canada. Subsequently, Harvey and Butterfield (2008) documented mouthparts, in particular mandibular gnathal edges, that indicated the megascopic crustaceans had evolved by Cambrian Stage 4. More recently investigated material from the middle and late Cambrian (Fig. 15.4) allows more precise systematic attributions of some of these crustaceans (Harvey et al. 2012), with the result that at least three crustacean lineages are known to have originated considerably earlier than previously known. Mandibles indicate the presence of crown-group Branchiopoda (based on distinctive asymmetries characteristic of Anostraca; Fig. 15.4b, c), as well as stem- or crown-group Copepoda (Fig. 15.4d–f) and Ostracoda (Fig. 15.4g). In addition, distinctive filter plates in these residues have details of their setae that assign them to the Branchiopoda, or at least its stem-group (Harvey et al. 2012). This material (Fig. 15.4a) shows such detailed similarity with the geologically older Mount Cap filter plates described by Harvey and Butterfield (2008) that the presence of total-group Branchiopoda in the early Cambrian (Stage 4) can be defended. This carries with it the important consequence that the clades that include Branchiopoda (e.g. Vericrustacea, Altocrustacea, Pancrustacea and Mandibulata, in the phylogeny and classification of Regier et al. 2010) had evolved by at least Cambrian Stage 4.

### 15.3.3 Virtual Fossils

The Herefordshire Lagerstätte, from the Silurian (Wenlock Series, 525 Mya) of western England, involves three-dimensional soft tissue preservation of small fossils in concretions (Briggs et al. 2008). The specimens are sparry calcite fill of the void space left after decay of the animal, the sample being serially ground and then reconstructed with computer software to generate a 3-D model (Fig. 15.5). This technique has allowed

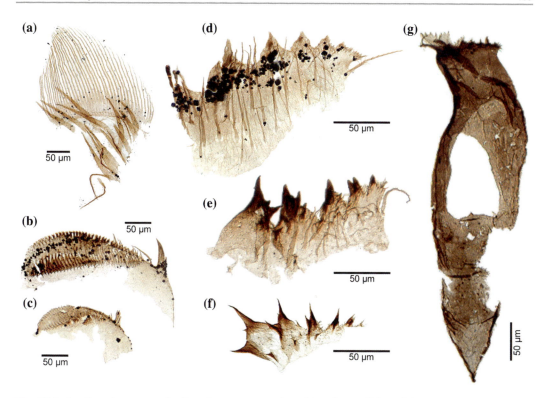

**Fig. 15.4** Small carbonaceous fossils of crustaceans from the late Cambrian (Furongian Stage), Saskatchewan, Canada (Harvey and Butterfield 2012). **a** Branchiopod filter plate and associated setae; **b**, **c** anostracan-type branchiopod mandibles; **d–f** copepod-type mandible; **g** ostracod-type mandible. Photos courtesy of Tom Harvey and Nick Butterfield

the detailed morphology of many arthropods to be reconstructed, including several phylogenetically important species.

*Offacolus kingi* (Sutton et al. 2002) is a chelicerate (Fig. 15.5a), its tagmosis in particular inspiring comparisons with Synziphosurina, a probably paraphyletic assemblage of Palaeozoic stem-group Xiphosura (Moore et al. 2011). Unexpectedly, however, *Offacolus* is reconstructed with five pairs of biramous prosomal appendages (the chelicera alone being uniramous), in contrast to the restriction of biramy in Xiphosura to walking leg 5 (the flabellum on that limb being homologous with either an exopod or an epipodite; see Boxshall 2004 for evidence for each view). This character has been used to argue that *Offacolus* may be sister to all other chelicerates (Dunlop 2006).

*Haliestes dasos* (Siveter et al. 2004) is a pycnogonid, the earliest known adult sea spider

(the older *Cambropycnogon* is known only from larval stages). Even compared to species described from the Early Devonian of Germany, *Haliestes* is decidedly modern in its combination of characters, and when included in morphological cladistic analyses of Pycnogonida, it is resolved in the crown-group (Arango and Wheeler 2007).

Ostracods are the most abundant arthropods in the fossil record (in terms of both number of specimens and number of species), but the example of bradoriids recounted above exposes the challenge of identifying and classifying ostracods based on carapaces alone. The Herefordshire Lagerstätte provided the earliest well-preserved appendages that allowed a Palaeozoic ostracod to be classified with confidence in an extant order (Siveter et al. 2003a) (Fig. 15.5b). Subsequent discoveries of soft parts from Herefordshire myodocopids, including eggs

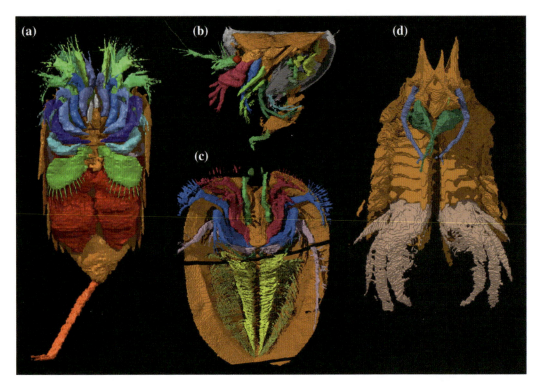

**Fig. 15.5** Computer reconstructions of Silurian arthropods from the Herefordshire Konservat-Lagerstätte, UK. **a** The chelicerate *Offacolus kingi*, ventral view (from Sutton et al. 2002); **b** the myodocopid ostracod *Nymphatelina gravida*, lateral view (from Siveter et al. 2007b); **c** the marrellomorph *Xylokorys chledophilia*, ventral view (from Siveter et al. 2007a); **d** the stem group mandibulate *Tanazios dokeron*, ventral view of head and anterior part of trunk (from Siveter et al. 2007c)

brooded within the carapace have shown that the carapace morphology alone is not a reliable indicator of affinity (Siveter et al. 2007b).

*Tanazios dokeron* (Siveter et al. 2007c), has a trunk composed of at least 64 segments with some 60 appendage pairs (Fig. 15.5d). The head is mandibulate, but the post-mandibular head appendages resemble the trunk limbs rather than being differentiated as maxillae. The pre-mandibular head appendages were originally interpreted as two pairs of antennae, and the species assigned to the crustacean stem lineage (Siveter et al. 2007c), but Boxshall (2007) proposed an alternative in which the "first antennae" are instead a pair of frontal filaments (as in remipedes), such that the species has only one pair of antennae. Applying Boxshall's coding of the head segments shifts *Tanazios* to the stem

lineage of Mandibulata rather than "Crustacea"/Tetraconata (Rota-Stabelli et al. 2011).

## 15.4 The Early Terrestrial Record

The terrestrial body fossil record of arthropods dates to at least the mid Silurian (Wenlock Series), constrained by the first appearance of millipedes (Wilson and Anderson 2004; Wilson 2005) in rocks some 425 million years old in Scotland. Although the older Silurian material is preserved as moulds or compressions in shales, knowledge of Silurian and Devonian terrestrial arthropods is biased by a few sites in Euramerica that have exceptional preservation, two of which are treated below.

The body fossil record for terrestrial arthropods is substantially preceded by trackways that record subaerial locomotion by arthropods. Trace fossils assigned to the ichnogenera *Diplopodichnus* and *Diplichnites* in Upper Ordovician rocks in the English Lake District (Johnson et al. 1994) are likely millipede-produced (Wilson 2006). If this is correct, these trackways pre-date the earliest body fossil record of millipedes by some 25 million years. The subaerial trackway record of arthropods extends back into the Cambrian (Hagadorn et al. 2011), and in at least some cases the taxonomic identify of the tracemaker can be determined, such as traces ascribed to euthycarcinoids by Collette and Hagadorn (2010). This does not allow terrestriality in extant lineages to be constrained because the phylogenetic affinities of euthycarcinoids within Mandibulata remain imprecisely determined (Racheboeuf et al. 2008), and the early trackways likely reflect amphibious rather than fully terrestrial habits.

Marine stem-group hexapods and myriapods remain to be identified from the Palaeozoic fossil record, even though these lineages are predicted to have diverged from crustaceans by the late Cambrian at the latest (Rehm et al. 2011). The proposal that the Lower Devonian marine fossil *Devonohexapodus* was a stem-group hexapod (Haas et al. 2003) has been dismissed; the fossil belongs to a previously documented species, *Wingertshellicus backesi*, and lacks the diagnostic characters of Hexapoda, or Tetraconata, or possibly even the mandibulate crown group (Kühl and Rust 2009).

### 15.4.1 Rhynie

The Rhynie chert in Aberdeenshire, Scotland, together with nearby sites known as the Windyfield chert, is interpreted as sinters that preserve a hot spring system of Early Devonian age, the fossils being silicified in three dimensions. Radiometric dating indicates an age of $411.5 \pm 1.3$ million years (Parry et al. 2011).

Rhynie includes the first records of several major terrestrial arthropod lineages. These include Acari, Opiliones (that exhibit preservation of the genitalia: Dunlop et al. 2004), and the earliest Hexapoda, in the form of at least three taxa that sample both Enthognatha and Insecta. The springtail (Collembola) *Rhyniella* may belong to an extant family (Greenslade and Whalley 1986). The recently described *Leverhulmia* has been reinterpreted as a primitively-flightless insect (Fayers and Trewin 2005), and *Rhyniognatha* is known from a mandible that is certainly a member of the insect clade Dicondylia (i.e. having an anterior mandibular articulation) and may even be a pterygote (Engel and Grimaldi 2004). *Rhyniognatha* (Fig. 15.6) extends the range of winged insects downwards from the Carboniferous. The detail of soft anatomical preservation in the Rhynie chert is adequate for resolution of, e.g. the book lung structure of trigonotarbid arachnids (Kamenz et al. 2008), showing microanatomy that is conserved in arachnids over more than 400 million years.

### 15.4.2 Gilboa

Dissolution of mudstones in hydrofluoric acid to liberate insoluble organic remains (as in Sect. 3.2) provides a wealth of data about terrestrial arthropods from the Middle Devonian at

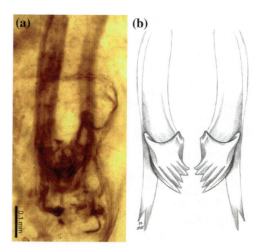

**Fig. 15.6** Mandible of the earliest pterygote insect, *Rhyniognatha hirsti* from the Lower Devonian. **a** Light photograph of mandibles and apodemes (image copyright The Natural History Museum); **b** reconstruction

Gilboa and South Mountain (ca 386 mya and 374 mya, respectively), New York State.

Chelicerates from Gilboa include mites (Norton et al. 1988), the earliest known pseudoscorpion (Judson 2012), several species of trigonotarbids (Shear et al. 1987), and one of two species assigned to the extinct arachnid order Uraraneida (Selden et al. 2008). This last, *Attercopus fimbriunguis*, was first identified as a trigonotarbid but subsequently reinterpreted as the earliest known spider (Selden et al. 1991). Reinvestigation of the material revealed the presence of spigots arranged on the edges of plates rather than on spinnerets (Selden et al. 2008), prompting its assignment to Uraraneida.

Myriapoda at Gilboa and South Mountain include representatives of both Diplopoda and Chilopoda. The arthropleurid-like *Microdecemplex* (Wilson and Shear 2000), assigned to an extinct Order Microdecemplicida, is classified either within Arthropleuridea (Wilson and Shear 2000) or is instead placed more closely to the chilognathan millipedes (Kraus and Brauckmann 2003). In either case, arthropleurideans are now regarded as ingroup Diplopoda, following a long history of being treated simply as enigmatic myriapods. The centipedes at Gilboa are the stem-group scutigeromorph *Crussolum* (Shear et al. 1998) and *Devonobius delta* (Shear and Bonamo 1988), which is assigned to an extinct monotypic order.

Organic preservation of terrestrial arthropods extends back into the Silurian, the best example being the Ludlow Bone Bed, of latest Silurian age in western England (Jeram et al. 1990). This deposit is the source of the earliest centipedes (Scutigeromorpha: Shear et al. 1998) and trigonotarbids (Dunlop 1996).

### 15.4.3 Coal Measures Nodules

Numerous arthropod lineages, e.g. many orders of insects, have their earliest records in Upper Carboniferous deposits that represent widespread coal swamps in Euramerica. Fossils preserved in siderite nodules in the Carboniferous Coal Measures of the UK, Montceau-les-Mines in France,

and the Mazon Creek area of Illinois, USA, often preserve complete articulated exoskeletons, including the appendages. Non-invasive imaging of this material by micro-computer tomography permits three-dimensional models of specimens to be reconstructed, including body parts that are otherwise concealed in the rock, such as the distal parts of appendages. This technique has been successfully applied to Carboniferous arachnids, including opilionids, scorpions (Fig. 15.7a; Legg et al. 2012a) and trigonotarbids (Garwood et al. 2009, 2011), as well as to insects (Garwood and Sutton 2010).

The detailed knowledge provided by the Upper Carboniferous sites contrasts with a generally patchy record for many groups in the Early Carboniferous, a lacuna that is known as Romer's Gap (for the vertebrate palaeontologist Romer, who detected it in fossil vertebrates). Early Late Carboniferous (Namurian) strata from Ningxia, China, extend the record of some groups of insects back earlier than the apparent evolutionary burst suggested by first appearances in Coal Measures-equivalent strata. Examples from Ningxia include recent discoveries of the earliest Dictyoptera (Zhang et al. 2013) and stem-group Orthoptera (Gu et al. 2011). Millipedes and scorpions are coming to light from Early Carboniferous sites (Scottish material in Smithson et al. 2012), suggesting that the dearth of terrestrial fossils from this time interval stems from collection failure rather than requiring a biological explanation.

## 15.5 Amber Fossils

Fossilised tree resin—amber—provides enormous insights into the origins of modern arthropod diversity, not least because of the quality of anatomical preservation. Fossiliferous amber is known from hundreds of deposits that span the world, ranging as far back as the Late Triassic (Schmidt et al. 2012; see Penney 2010a for accounts of all major amber deposits, including comprehensive lists of all arthropods). New techniques of microscopy and imaging, such as phase contrast X-ray computed tomography

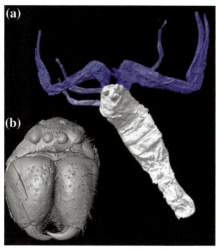

**Fig. 15.7** Amber and micro-tomographic reconstructions of terrestrial arthropod fossils. **a** Carboniferous scorpion *Compsoscorpius buthiformis* (from Legg et al. 2012a), reconstructed from micro-CT; **b** *Eusparassus crassipes*, a huntsman spider from Baltic amber (Eocene), reconstructed by phase contrast-enhanced X-ray CT (Dunlop et al. 2011), anterior view of chelicerae and carapace. Image by Andrew McNeil; **c** Baltic amber pseudoscorpion *Geogarypus* sp. (copyright The Natural History Museum)

(Dunlop et al. 2011) and X-ray synchrotron imaging and microCT (Lak et al. 2008; Pohl et al. 2010), are being applied to amber-preserved fossils and allowing even more anatomical detail to be extracted and depicted (Fig. 15.7b). This can include non-destructive documentation of soft anatomy rarely tractable from the fossil record, e.g. the brain, subesophageal ganglion, and flight muscles of a Baltic amber strepsipteran insect (Pohl et al. 2010).

Lebanese amber is the oldest major source, with most collections dating to the Lower Cretaceous (Azar et al. 2010). It contains a greater number of extinct insect families than does geologically later amber, and provides the earliest records of many extant families. Other major Cretaceous arthropod faunas are known from ambers from France, Spain, western Canada, Burma, and Ethiopia. Owing to its vast geographic extent across northern Europe and an especially long history of study, the Eocene (44–49 mya) Baltic amber (Fig. 15.7c) is the source of more species than any other. Over 3,000 arthropod species are known, with enormous numbers in some insect orders (800 species of Diptera); arthropods comprise 98 % of all Baltic amber inclusions (Weitschat and Wichard 2010).

Especially pristine preservation is seen in the Miocene (ca 16 mya) amber from the Dominican Republic, from which more than 1,000 arthropod species have been documented, including 100+ species of each of Diptera, Hymenoptera, Coleoptera and Araneae (Penney 2010b).

**Acknowledgments** For providing photographs used to illustrate this chapter, we thank Chen Ai-lin (Chengjiang Fauna National Geological Park), Allison Daley (The Natural History Museum), Jason Dunlop (Museum für Naturkunde, Berlin), Diego García-Bellido (CSIC, Universidad Complutense Madrid), Tom Harvey and Nick Butterfield (University of Cambridge), Xiaoya Ma (The Natural History Museum/Yunnan University), David Siveter (University of Leicester), and Derek Siveter (University of Oxford). Comments on the manuscript by Allison Daley, John Paterson, and a referee are appreciated.

# References

Adrain JM (2011) Class Trilobita Walch, 1771. In: Zhang Z-Q (ed) Animal biodiversity: An outline of higher-level classification and survey of taxonomic richness. Zootaxa 3148:104–109

Arango CP, Wheeler WC (2007) Phylogeny of sea spiders (Arthropoda: Pycnogonida) based on direct optimization of six loci and morphology. Cladistics 23:253–293

Azar D, Géze R, Acra F (2010) Lebanese amber. In: Penney D (ed) Biodiversity of fossils in amber from the major world deposits. Siri Sci Press, Manchester, pp 271–298

Bergström J (1992) The oldest arthropods and the origin of Crustacea. Acta Zool 73:287–291

Bergström J, Hou X-G (2005) Early Palaeozoic non-lamellipedian arthropods. In: Koenemann S, Jenner RA (eds) Crustacean issues 16: Crustacea and arthropod relationships. Taylor and Francis, Boca Raton, pp 73–93

Bergström J, Hou X, Zhang X, Clausen S (2008) A new view of the Cambrian arthropod Fuxianhuia. GFF 130:189–201

Bitsch J, Bitsch C (2007) The segmental organization of the head region in Chelicerata: a critical review of recent studies and hypotheses. Acta Zool 88:317–335

Bousfield EL (1995) A contribution to the natural classification of Lower and Middle Cambrian arthropods: food gathering and feeding mechanisms. Amphipacifica 2:3–34

Boxshall G (1997) Comparative limb morphology in major crustacean groups: the coxa-basis joint in postmandibular limbs. In: Fortey RA, Thomas RH (eds) Arthropod relationships. Chapman and Hall, London, pp 155–167

Boxshall GA (2004) The evolution of arthropod limbs. Biol Rev 79:253–300

Boxshall GA (2007) Crustacean classification: on-going controversies and unresolved problems. Zootaxa 1668:313–325

Boxshall GA, Jaume D (2009) Exopodites, epipodites and gills in crustacean. Arthropod Syst Phylogeny 67:229–254

Briggs DEG (1978) The morphology, mode of life, and affinities of Canadaspis perfecta (Crustacea: Phyllocarida), Middle Cambrian, Burgess Shale, British Columbia. Phil Trans R Soc B 281:439–487

Briggs DEG (1992) Phylogenetic significance of the Burgess Shale crustacean Canadaspis. Acta Zool 73:293–300

Briggs DEG, Collins D (1988) A middle Cambrian chelicerate from Mount Stephen, British Columbia. Palaeontology 31:779–798

Briggs DEG, Collins D (1999) The arthropod Alalcomenaeus cambricus Simonetta, from the Middle Cambrian Burgess Shale of British Columbia. Palaeontology 42:953–977

Briggs DEG, Erwin DH, Collier FJ (1994) The fossils of the Burgess Shale. Smithsonian, Washington

Briggs DEG, Siveter DJ, Siveter DJ, Sutton MD (2008) Virtual fossils from a 425 million-year-old volcanic ash. Amer Sci 96:474–481

Budd GE (1997) Stem group arthropods from the Lower Cambrian Sirius Passet fauna of North Greenland. In: Fortey RA, Thomas RH (eds) Arthropod relationships. Chapman and Hall, London, pp 125–138

Budd GE (1999) The morphology and phylogenetic significance of Kerygmachela kierkegaardi Budd (Buen Formation, Lower Cambrian, N Greenland). Trans R Soc Edinburgh Earth Sci 89:249–290

Budd GE (2002) A palaeontological solution to the arthropod head problem. Nature 417:271–275

Budd GE (2008) Head structure in upper stem-group arthropods. Palaeontology 51:1–13

Budd GE (2011) Campanamuta mantonae gen. et. sp. nov., an exceptionally preserved arthropod from the Sirius Passet Fauna (Buen Formation, lower Cambrian, North Greenland). J Syst Palaeontol 9:217–260

Butterfield NJ (1994) Burgess Shale-type fossils from a Lower Cambrian shallow-shelf sequence in northwestern Canada. Nature 369:477–479

Butterfield NJ (2002) Leanchoilia guts and the interpretation of three-dimensional structures in Burgess Shale-type fossils. Paleobiology 28:155–171

Butterfield NJ, Harvey THP (2012) Small carbonaceous fossils (SCFs): a new measure of early Paleozoic paleobiology. Geology 40:71–74

Campbell LI, Rota-Stabelli O, Edgecombe GD, Marchioro T, Longhorn SJ, Telford MJ, Rebecchi L, Peterson KJ, Pisani D (2011) MicroRNAs and phylogenomics resolve the phylogenetic relationships of the Tardigrada, and suggest the velvet worms as the sister group of Arthropoda. Proc Natl Acad Sci USA 108:15920–15924

Charbonnier S, Vannier J, Riou B (2007) New sea spiders from the Jurassic La Voulte-sur-Rhône Lagerstätte. Proc R Soc B 274:2555–2561

Chen J, Ramsköld L, Zhou G (1994) Evidence for monophyly and arthropod affinity of Cambrian giant predators. Science 264:1304–1308

Chen J, Edgecombe GD, Ramsköld L, Zhou G (1995) Head segmentation in Early Cambrian Fuxianhuia: implications for arthropod evolution. Science 268:1339–1343

Chen J, Edgecombe GD, Ramsköld L (1997) Morphological and ecological disparity in naraoiids (Arthropoda) from the Early Cambrian Chengjiang fauna, China. Rec Aust Mus 49:1–24

Chen J, Waloszek D, Maas A (2004) A new 'great-appendage' arthropod from the Lower Cambrian of China and homology of chelicerate chelicerae and raptorial antero-ventral appendages. Lethaia 37:3–20

Collette JH, Hagadorn JW (2010) Three-dimensionally preserved arthropods from Cambrian Lagerstätten of Quebec and Wisconsin. J Paleontol 84:646–667

Collins D (1996) The "evolution" of Anomalocaris and its classification in the arthropod Class Dinocarida (nov.) and Order Radiodonta (nov.). J Paleontol 70:280–293

Cotton TJ, Braddy SJ (2004) The phylogeny of arachnomorph arthropods and the origin of the Chelicerata. Trans R Soc Edinburgh Earth Sci 94:169–193

Cotton TJ, Fortey RA (2005) Comparative morphology and relationships of the Agnostida. In: Koenemann S, Jenner RA (eds) Crustacean Issues 16: Crustacea and arthropod relationships. Taylor and Francis, Boca Raton, pp 95–136

Dahl E (1984) The subclass Phyllocarida (Crustacea) and the status of some early fossils; a neontologist's view. Vidensk Meddel Dansk naturhist For 145:61–76

Daley AC, Bergström J (2012) The oral cone of *Anomalocaris* is not a classic "peytoia". Naturwiss 99:501–504

Daley AC, Budd GE (2010) New anomalocaridid appendages from the Burgess Shale. Palaeontology 53:721–738

Daley AC, Budd GE, Caron J-B, Edgecombe GD, Collins D (2009) The Burgess Shale anomalocaridid *Hurdia* and its significance for early euarthropod evolution. Science 323:1597–1600

Dunlop JA (1996) A trigonotarbid arachnid from the Upper Silurian of Shropshire. Palaeontology 39:605–614

Dunlop JA (2006) New ideas about the euchelicerate stem lineage. In: Deltshev C, Stoev P (ed), European Arachnology 2005. Acta Zool Bulg Suppl 1:9–23

Dunlop JA, Anderson LI, Kerp H, Hass H (2004) A harvestman (Arachnida: Opiliones) from the early Devonian Rhynie cherts, Aberdeenshire, Scotland. Trans R Soc Edinburgh Earth Sci 94:341–354

Dunlop JA, Penney D, Dalüge N, Jäger P, McNeil A, Bradley RS, Withers PJ, Preziosi RF (2011) Computed tomography recovers data from historical amber: an example from huntsman spiders. Naturwiss 98:519–527

Edgecombe GD (2010) Palaeomorphology: fossils and the inference of cladistic relationships. Acta Zool 91:72–80

Edgecombe GD, García-Bellido DC, Paterson JR (2011) A new leanchoiliid megacheiran arthropod from the lower Cambrian Emu Bay Shale, South Australia. Acta Palaeontol Pol 56:373–388

Engel MS, Grimaldi DA (2004) New light shed on the oldest insect. Nature 427:627–630

Eriksson ME, Terfelt F, Elofsson R, Marone F (2012) Internal soft-tissue anatomy of Cambrian 'Orsten' arthropods as revealed by synchrotron X-ray tomographic microscopy. PLoS ONE 7:e42582

Fayers SR, Trewin NH (2005) A hexapod from the Early Devonian Windyfield Chert, Rhynie, Scotland. Palaeontology 48:1117–1130

Fu D, Zhang X (2011) A new arthropod *Jugatocaris agilis* n. gen. n. sp. from the Early Cambrian Chengjiang biota South China. J Paleontol 85:567–586

Fu D, Zhang X, Shu D (2011) Soft anatomy of the Early Cambrian arthropod *Isoxys curvirostratus* from the Chengjiang biota of South China with a discussion on the origination of great appendages. Acta Palaeontol Pol 56:843–852

Gaines RR, Hammarlund EU, Hou X, Qi C, Gabbott SE, Zhao Y, Peng J, Canfield DE (2012) Mechanism for Burgess Shale-type preservation. Proc Natl Acad Sci USA 109:5180–5184

García-Bellido DC, Collins DH (2006) A new study of *Marrella splendens* (Arthropoda, Marrellomorpha) from the Middle Cambrian Burgess Shale, British Columbia, Canada. Can J Earth Sci 43:721–742

García-Bellido DC, Collins D (2007) Reassessment of the genus *Leanchoilia* (Arthropoda, Arachnomorpha) from the Middle Cambrian Burgess Shale, British Columbia, Canada. Palaeontology 50:693–709

García-Bellido DC, Paterson JR, Edgecombe GD, Jago JB, Gehling JG, Lee MSY (2009a) The bivalved arthropods *Isoxys* and *Tuzoia* with soft-part preservation from the lower Cambrian Emu Bay Shale Lagerstätte (Kangaroo Island, Australia). Palaeontology 52:1221–1241

García-Bellido DC, Vannier J, Collins D (2009b) Soft-part preservation in two species of the arthropod *Isoxys* from the middle Cambrian Burgess Shale of British Columbia, Canada. Acta Palaeontol Pol 54:69–712

Garwood R, Sutton M (2010) X-ray micro-tomography of Carboniferous stem-Dictyoptera: new insights into early insects. Biol Lett 6:699–702

Garwood RJ, Dunlop JA, Sutton MD (2009) High-fidelity X-ray micro-tomography reconstruction of siderite-hosted carboniferous arachnids. Biol Letters 5:841–844

Garwood RJ, Dunlop JA, Giribet G, Sutton MD (2011) Anatomically modern Carboniferous harvestmen demonstrate early cladogenesis and stasis in Opiliones. Nature Comm 2:444

Gehling JG, Jago JB, Paterson JR, García-Bellido DC, Edgecombe GD (2011) The geological context of the lower Cambrian (Series 2) Emu Bay Shale Lagerstätte and adjacent stratigraphic units, Kangaroo Island, South Australia. Aust J Earth Sci 58:243–257

Greenslade P, Whalley PES (1986) The systematic position of *Rhyniella praecursor* Hirst and Maulik (Collembola), the earliest known hexapod. In: Dallai R (ed) Second international seminar on Apterygota. Univ Siena, Siena, pp 319–323

Gu J-J, Béthoux O, Ren D (2011) *Longzhua loculata* n. gen. n. sp., one of the most completely documented Pennsylvanian Archaeorthoptera (Insecta; Ningxia, China). J Paleontol 85:303–314

Haas F, Waloszek D, Hartenberger R (2003) *Devonohexapodus bockbergensis*, a new marine hexapod from the Lower Devonian Hunsrück Slates, and the origin of the Atelocerata and Hexapoda. Org Divers Evol 3:39–54

Hagadorn JW, Collette JH, Belt ES (2011) Eolian-aquatic deposits and faunas of the Middle Cambrian Potsdam group. Palaios 26:314–334

Harvey THP, Butterfield NJ (2008) Sophisticated particle-feeding in a large Early Cambrian crustacean. Nature 452:868–871

Harvey THP, Vélez MI, Butterfield NJ (2012) Exceptionally preserved crustaceans from western Canada reveal a cryptic Cambrian radiation. Proc Natl Acad Sci USA 109:1589–1594

Haug JT, Maas A, Waloszek D (2009) Ontogeny of two Cambrian stem crustaceans, †*Goticaris longispinosa* and †*Cambropachycope clarksoni.* Palaeontographica A 289:1–43

Haug JT, Maas A, Waloszek D (2010a) †*Henningsmoenicaris scutula,* †*Sandtorpia vestrogothieni*s gen. et sp. nov. and heterochronic events in early crustacean evolution. Trans R Soc Edinburgh Earth Sci 100:311–350

Haug JT, Waloszek D, Haug C, Maas A (2010b) High-level phylogenetic analysis using developmental sequences: The Cambrian †*Martinssonia elongata,* † *Muscacaris gerdgeyeri* gen. et sp. nov., and their position in early crustacean evolution. Arthropod Struct Dev 39:154–174

Haug JT, Maas A, Haug C, Waloszek D (2011) *Sarotrocercus obletus*—small arthropod with great impact on the understanding of arthropod evolution? Bull Geosci 86:725–736

Haug JT, Briggs DEG, Haug c (2012a) Morphology and function in the Cambrian Burgess Shale arthropod *Leanchoilia superlata* and the application of a descriptive matrix. BMC Evol Biol 12:162

Haug JT, Waloszek D, Maas A, Liu Y, Haug C (2012b) Functional morphology, ontogeny and evolution of mantis shrimp-like predators in the Cambrian. Palaeontology 55:369–399

Hennig W (1966) Phylogenetic systematics. University Illinois Press, Urbana

Hou X-G, Bergström J (1997) Arthropods of the Lower Cambrian Chengjiang fauna, southwest China. Fossils Strata 45:1–116

Hou X, Bergström J (2006) Dinocarids—anomalous arthropods or arthropod-like worms. In: Rong J, Fang Z, Zhou Z, Zhan R, Wang X,Yuan X (eds), Originations, radiations and biodiversity changes–evidences from the Chinese fossil record. Science Press, Beijing, pp. 139–158, 847–850

Hou X, Bergström J, Ahlberg P (1995) *Anomalocaris* and other large animals in the Lower Cambrian Chengjiang fauna of southwest China. GFF 117:163–183

Hou X, Aldridge RJ, Bergström J, Siveter DJ, Siveter DJ, Feng XH (2004a) The Cambrian fossils of Chengjiang, China. Blackwell, Oxford

Hou X-G, Bergström J, Xu G-X (2004b) The Lower Cambrian crustacean *Pectocaris* from the Chengjiang biota, Yunnan, China. J Paleontol 78:700–708

Hou X-G, Bergström J, Yang J (2006) Distinguishing anomalocaridids from arthropods and priapulids. Geol J 41:259–269

Hou X, Clarkson ENK, Yang J, Zhang X, Wu G, Yuan Z (2009) Appendages of early Cambrian *Eoredlichia* (Trilobita) from the Chengjiang biota, Yunnan, China. Trans R Soc Edinburgh Earth Sci 99:213–223

Hou X, Williams M, Siveter DJ, Siveter DJ, Aldridge RJ, Sansom RS (2010) Soft-part anatomy of the Early Cambrian bivalved arthropods *Kunyangella* and *Kunmingella*: significance for the phylogenetic relationships of Bradoriida. Proc R Soc B 277:1835–1841

Jeram AJ, Selden PA, Edwards D (1990) Land animals in the Silurian: arachnids and myriapods from Shropshire, England. Science 250:659–661

Johnson EW, Briggs DEG, Suthren RJ, Wright JL, Tunnicliff SP (1994) Non-marine arthropod traces from the subaerial Ordovician Borrowdale Volcanic Group, English Lake District. Geol Mag 131:395–406

Judson MLI (2012) Reinterpretation of *Drachochela deprehendor* (Arachnida: Pseudoscorpiones) as a stem-group pseudoscorpion. Palaeontology 55:261–283

Kamenz C, Dunlop JA, Scholtz G, Kerp H, Hass H (2008) Microanatomy of Early Devonian book lungs. Biol Lett 4:212–215

Kraus O, Brauckmann C (2003) Fossil giants and surviving dwarfs. Arthropleurida and Pselaphognatha (Atelocerata, Diplopoda): Characters, phylogenetic relationships and construction. Verh Naturwiss Ver Hamburg NF 40:5–50

Kühl G, Rust J (2009) *Devonohexapodus bocksbergensis* is a synonym of *Wingertshellicus backesi* (Euarthropoda)—no evidence for marine hexapods living in the Devonian Hunsrück Sea. Org Divers Evol 9:215–231

Kühl G, Rust J (2010) Re-investigation of *Mimetaster hexagonalis*: a marrellomorph arthropod from the Lower Devonian Hunsrück Slate (Germany). Palaeontol Z 84:397–411

Lak M, Azar D, Nel A, Néraudeau D, Tafforeau P (2008) The oldest representative of the Trichomyiinae (Diptera: Psychodidae) from the Lower Cenomanian French amber studied with phase-contrast synchrotron X-ray imaging. Invert Syst 22:471–478

Lee MSY, Jago JG, García-Bellido DC, Edgecombe GD, Gehling JG, Paterson JR (2011) Modern optics in exceptionally preserved eyes of Early Cambrian arthropods from Australia. Nature 474:631–634

Legg DA, Garwood RJ, Dunlop JA, Sutton MA (2012a) A revision of orthosternous scorpions from the English Coal Measures aided by x-ray micro-tomography (XMT). Palaeontol Electronica 15.2. 14A:1–16

Legg DA, Sutton MD, Edgecombe GD, Caron J-B (2012b) Cambrian bivalved arthropod reveals origin of arthrodisation. Proc R Soc B (Biol Sci) 279: 4699–4704

Liu J, Dong XP (2007) *Skara hunanensis* a new species of Skaracarida (Crustacea) from Upper Cambrian (Furongian) of Hunan, South China. Chin Sci Bull 17:934–942

Liu J, Shu D, Han J, Zhang Z, Zhang X (2007a) Morphoanatomy of the lobopod *Megadictyon* cf. *haikouensis* from the Early Cambrian Chengjiang Lagerstätte, South China. Acta Zool 88:279–288

Liu Y, Hou X-G, Bergström J (2007b) Chengjiang arthropod *Leanchoilia illecebrosa* (Hou 1987) reconsidered. GFF 129:263–272

Ma XY, Hou XG, Bergström J (2009) Morphology of *Luolishania longicruris* (Lower Cambrian, Chengjiang Lagerstätte, SW China) and the phylogenetic relationships within lobopodians. Arthropod Struct Dev 38:271–291

Ma XY, Hou XG, Edgecombe GD, Strausfeld NJ (2012) Complex brain and optic lobes in an early Cambrian arthropod. Nature 490:258–261

Maas A, Walossek D, Müller K (2003) Morphology, ontogeny and phylogeny of the Phosphatocopina (Crustacea) from the Upper Cambrian 'Orsten' of Sweden. Fossils Strata 49:1–238

Maas A, Braun A, Dong X-P, Donoghue PCJ, Müller KJ, Olempska E, Repetski JE, Siveter DJ, Stein M, Waloszek D (2006) The 'Orsten'—More than a Cambrian Konservat-Lagerstätte yielding exceptional preservation. Palaeoworld 15:266–282

Maas A, Haug C, Haug JT, Olesen J, Zhang X, Waloszek D (2009) Early crustacean evolution and the appearance of epipodites and gills. Arthropod Syst Phylog 67:255–273

Moore RA, Briggs DEG, Braddy SJ, Shultz JW (2011) Synziphosurines (Xiphosura: Chelicerata) from the Silurian of Iowa. J Paleontol 85:83–91

Müller KJ (1979) Phosphatocopine ostracodes with preserved appendages from the Cambrian of Sweden. Lethaia 12:1–27

Müller KJ, Walossek D (1985) Skaracarida, a new order of Crustacea from the Upper Cambrian of Västergötland, Sweden. Fossils Strata 17:1–65

Müller KJ, Walossek D (1987) Morphology, ontogeny, and life habit of Agnostus pisiformis from the Upper Cambrian of Sweden. Fossils Strata 19:1–124

Norton RA, Bonamo PM, Grierson JD, Shear WA (1988) Oribatid mite fossils from a terrestrial Devonian deposit near Gilboa, New York. J Paleontol 62:259–269

Olesen J (2009) Phylogeny of Branchiopoda (Crustacea)—Character evolution and contribution of uniquely preserved fossils. Arthropod Syst Phylog 67:3–39

Parry SF, Noble SR, Crowley QG, Wellman CH (2011) A high-precision U-Pb age constraint on the Rhynie Chert Konservat-Lagerstätte: Time scale and other implications. J Geol Soc 168:863–872

Paterson JR, Edgecombe GD, García-Bellido DC, Jago JB, Gehling JG (2010) Nektaspid arthropods from the Lower Cambrian Emu Bay Shale Lagerstätte, South Australia, with a reassessment of lamellipedian relationships. Palaeontology 53:377–402

Paterson JR, García-Bellido DC, Lee MSY, Brock GA, Jago JB, Edgecombe GD (2011) Acute vision in the giant Cambrian predator Anomalocaris and the origin of compound eyes. Nature 480:237–240

Paterson JR, García-Bellido DC, Edgecombe GD (2012) New artiopodan arthropods from the early Cambrian Emu Bay Shale Konservat-Lagerstätte, South Australia. J Paleontol 86:340–357

Peel JS, Ineson JR (2011) The Sirius Passet Lagerstätte (early Cambrian) of North Greenland. Palaeontographica Canadiana 39:109–118

Penney D (2010a) Biodiversity of fossils in amber from the major world deposits. Siri Sci Press, Manchester

Penney D (2010b) Dominican amber. In: Penney D (ed) Biodiversity of fossils in amber from the major world deposits. Siri Sci Press, Manchester, pp 22–41

Pohl H, Wipfler B, Grimaldi D, Beckmann F, Beutel RG (2010) Reconstructing the anatomy of the 42-million-year-old fossil †Mengea tertiaria (Insecta, Strepsiptera). Naturwiss 97:855–859

Racheboeuf P, Vannier J, Schram FR, Chabard D, Sotty D (2008) The euthycarcinoid arthropods from Montceau-les-Mines, France: Functional morphology and affinities. Trans R Soc Edinburgh Earth Sci 99:11–25

Rak S, Ortega-Hernández J, Legg DA (2013) A revision of the Late Ordovician marrellomorph arthropod Furca bohemica from Czech Republic. Acta Palaeontol Pol. doi:10.4202/app.2011.0038

Regier JC, Shultz JW, Zwick A, Hussey A, Ball B, Wetzer R, Martin JW, Cunningham CW (2010) Arthropod relationships revealed by phylogenomic analysis of nuclear protein-coding sequences. Nature 463:1079–1083

Rehm P, Borner J, Meusemann K, von Reumont BM, Simon S, Hadrys H, Misof B, Burmester T (2011) Dating the arthropod tree based on large-scale transcriptome data. Mol Phylogenet Evol 61:880–887

Rota-Stabelli O, Campbell L, Brinkmann H, Edgecombe GD, Longhorn SJ, Peterson KJ, Pisani D, Philippe H, Telford MJ (2011) A congruent solution to arthropod phylogeny: Phylogenomics, microRNAs and morphology support monophyletic Mandibulata. Proc R Soc B 278:298–306

Sanders KL, Lee MSY (2010) Arthropod molecular divergence times and the Cambrian origin of pentastomids. Syst Biodiv 8:63–74

Schmidt AR, Jancke S, Lindquist EE, Ragazzi E, Roghi G, Nascimbene PC, Schmidt K, Wappler T, Grimaldi DA (2012) Arthropods in amber from the Triassic Period. Proc Natl Acad Sci USA 109:14796–14801

Scholtz G, Edgecombe GD (2005) Heads, Hox and the phylogenetic position of trilobites. In: Koenemann S, Jenner RA (eds) Crustacea and arthropod relationships (Crustacean Issues 16). Taylor and Francis, Boca Raton, pp 139–165

Scholtz G, Edgecombe GD (2006) The evolution of arthropod heads: Reconciling morphological, developmental and palaeontological evidence. Dev Genes Evol 216:395–415

Schram FR, Koenemann S (2001) Developmental genetics and arthropod evolution: Part I, on legs. Evol Dev 3:243–354

Selden PA, Shear WA, Bonamo PM (1991) A spider and other arachnids from the Devonian of New York, and reinterpretations of Devonian Araneae. Palaeontology 34:241–281

Selden PA, Shear WA, Sutton MD (2008) Fossil evidence for the origin of spider spinnerets, and a proposed arachnid order. Proc Natl Acad Sci USA 105:20781–20785

Shear WA, Bonamo PM (1988) Devonobiomorpha, a new order of centipeds (Chilopoda) from the Middle Devonian of Gilboa, New York State, USA, and the phylogeny of centiped orders. Amer Mus Novitates 2927:1–30

Shear WA, Selden PA, Rolfe WDI, Bonamo PM, Grierson JD (1987) New terrestrial arachnids from the Devonian of Gilboa, New York (Arachnida, Trigonotarbida). Amer Mus Novitates 2901:1–74

Shear WA, Jeram AJ, Selden PA (1998) Centiped legs (Arthropoda, Chilopoda, Scutigeromorpha) from the Silurian and Devonian of Britain and the Devonian of North America. Amer Mus Novitates 3231:1–16

Shu D, Vannier J, Luo H, Chen L, Zhang X, Hu S (1999) Anatomy and lifestyle of *Kunmingella* (Arthropoda, Bradoriida) from the Chengjiang fossil Lagerstätte (lower Cambrian; Southwest China). Lethaia 32:279–298

Siveter DJ, Williams M, Waloszek D (2001) A phosphatocopine crustacean with appendages from the Lower Cambrian. Science 293:479–481

Siveter DJ, Sutton MD, Briggs DEG, Siveter DJ (2003a) An ostracode crustacean with soft parts from the Lower Silurian. Science 300:1749–1751

Siveter DJ, Waloszek D, Williams M (2003b) An Early Cambrian phosphatocopid crustacean with three-dimensionally preserved soft-parts from Shropshire, England. Spec Pap Palaeontol 70:9–30

Siveter DJ, Sutton MD, Briggs DEG, Siveter DJ (2004) A Silurian sea spider. Nature 431:978–980

Siveter DJ, Fortey RA, Sutton MD, Briggs DEG, Siveter DJ (2007a) A Silurian 'marrellomorph' arthropod. Proc R Soc B 274:2223–2229

Siveter DJ, Siveter DJ, Sutton MD, Briggs DEG (2007b) Brood care in a Silurian ostracod. Proc R Soc B 274:465–469

Siveter DJ, Sutton MD, Briggs DEG, Siveter DJ (2007c) A new probable stem lineage crustacean with three-dimensionally preserved soft parts from the Herefordshire (Silurian) Lagerstätte, UK. Proc R Soc B 274:2099–2107

Smithson TR, Wood SP, Marshall JEA, Clack JA (2012) Earliest Carboniferous tetrapod and arthropod faunas from Scotland populate Romer's Gap. Proc Natl Acad Sci USA 109:4532–4537

Stein M (2010) A new arthropod from the Early Cambrian of North Greenland, with a 'great append-age'-like antennula. Zool J Linn Soc 158:477–500

Stein M, Selden PA (2012) A restudy of the Burgess Shale (Cambrian) *Emeraldella brocki* and reassessment of its affinities. J Syst Palaeontol 10:361–383

Stein M, Waloszek D, Mass A, Haug JT, Müller KJ (2008) The stem crustacean *Oelandocaris oelanica* re-visited. Acta Palaeontol Pol 53:461–484

Sutton MD, Briggs DEG, Siveter DJ, Siveter DJ, Orr PJ (2002) The arthropod *Offacolus kingi* (Chelicerata) from the Silurian of Herefordshire, England: computer based morphological reconstructions and phylogenetic affinities. Proc R Soc B 269:1195–1203

Van Roy P, Briggs DEG (2011) A giant Ordovician anomalocaridid. Nature 473:510–513

Van Roy P, Orr PJ, Botting JP, Muir LA, Vinther J, Lefebvre B, el Hariri K, Briggs DEG (2010) Ordovician faunas of Burgess Shale type. Nature 465:215–218

Vannier J, Chen J-Y (2002) Digestive system and feeding mode in Cambrian naraoiid arthropods. Lethaia 35:107–120

Vannier J, García-Bellido GC, Hu S-X, Chen A-L (2009) Arthropod visual predators in the early pelagic ecosystem: evidence from the Burgess Shale and Chengjiang biotas. Proc R Soc B 276:2567–2574

von Reumont BJ, Jenner RA, Wills MA, Dell-Ampio E, Pass G, Ebersberger I, Meyer B, Koenemann S, Iliffe TM, Stamatakis A, Niehuis O, Meusemann K, Misof B (2012) Pancrustacean phylogeny in the light of new phylogenomic data: support for Remipedia as the possible sister group of Hexapoda. Mol Phylogenet Evol 29:1031–1045

Walossek D (1993) The Upper Cambrian *Rehbachiella* and the phylogeny of Branchiopoda and Crustacea. Fossils Strata 32:1–202

Walossek D, Müller KJ (1990) Upper Cambrian stem-lineage crustaceans and their bearing upon the monophyletic origin of Crustacea and the position of *Agnostus*. Lethaia 24:409–427

Waloszek D, Dunlop JA (2002) A larval sea spider (Arthropoda: Pycnogonida) from the Upper Cambrian 'Orsten' of Sweden, and the phylogenetic position of pycnogonids. Palaeontology 45:421–446

Waloszek D, Chen J, Maas A, Wang X (2005) Early Cambrian arthropods—new insights into arthropod head and structural evolution. Arthropod Struct Dev 34:189–205

Weitschat W, Wichard W (2010) Baltic amber. In: Penney D (ed) Biodiversity of fossils in amber from the major world deposits. Siri Sci Press, Manchester, pp 80–115

Whittington HB (1971) Redescription of *Marrella splendens* (Trilobitoidea) from the Burgess Shale, Middle Cambrian, British Columbia. Geol Soc Canada Bull 209:1–24

Whittington HB (1975) Trilobites with appendages from the Middle Cambrian, Burgess Shale, British Columbia. Fossils Strata 4:97–136

Whittington HB (1977) The Middle Cambrian trilobite *Naraoia*, Burgess Shale, British Columbia. Phil Trans R Soc B 280:409–443

Whittington HB (1980) Exoskeleton, moult stage, appendage morphology, and habits of the Middle Cambrian trilobite *Olenoides serratus*. Palaeontology 23:171–204

Whittington HB (1985) The Burgess Shale. Yale Univ Press, New Haven

Whittington HB, Briggs DEG (1985) The largest Cambrian animals, *Anomalocaris*, Burgess Shale, British Columbia. Phil Trans R Soc B 309:569–609

Wills MA, Briggs DEG, Fortey RA, Wilkinson M, Sneath PHA (1998) An arthropod phylogeny based on fossil and recent taxa. In: Edgecombe GD (ed) Arthropod fossils and phylogeny. Columbia University Press, New York, pp 33–105

Wilson HM (2005) Zosterogrammida, a new order of millipedes from the Middle Silurian of Scotland and the Upper Carboniferous of Euramerica. Palaeontology 48:1101–1110

Wilson HM (2006) Juliformian millipedes from the Lower Devonian of Euramerica: Implications for the timing of millipede cladogenesis in the Paleozoic. J Paleontol 80:638–649

Wilson HM, Anderson LI (2004) Morphology and taxonomy of Paleozoic millipedes (Diplopoda: Chilognatha: Archipolypoda) from Scotland. J Paleontol 78:169–184

Wilson HM, Shear WA (2000) Microdecemplicida, a new order of minute arthropleurideans (Arthropoda: Myriapoda) from the Devonian of New York State, U.S.A. Trans R Soc Edinburgh Earth Sci 90:351–375

Zhang HQ, Dong XP (2009) Two new species of *Vestrogothia* (Phosphatocopina, Crustacea) of Orsten-type preservation from the Upper Cambrian in western Hunan, South China. Sci China Earth Sci 52:784–796

Zhang X, Siveter DJ, Waloszek D, Maas A (2007a) An epipodite-bearing crown-group crustacean from the Lower Cambrian. Nature 449:595–598

Zhang X, Shu D, Erwin DH (2007b) Cambrian naraoiids (Arthropoda): morphology, ontogeny, systematics, and evolutionary relationships. Paleontol Soc Mem 68:1–52

Zhang X, Maas A, Haug JT, Siveter DJ, Waloszek D (2010) A crustacean metanauplius from the Lower Cambrian. Curr Biol 20:1075–1079

Zhang HQ, Dong XP, Xiao S (2012) Three head-larvae of *Hesslandona angustata* (Phosphatocopida, Crustacea) from the upper Cambrian of western Hunan, South China and the phylogeny of Crustacea. Gondwana Res 21:1115–1127

Zhang Z, Schneider JW, Hong Y (2013) The most ancient roach (Blattida): a new genus and species from the earliest Late Carboniferous (Namurian) of China, with discussion on the phylomorphogeny of early blattids. J Syst Palaeontol 11(1)

Zhao FC, Zhu MY, Hu SX (2010) Community structure and composition of the Cambrian Chengjiang biota. Sci China Earth Sci 53:1784–1799

Zhu M-Y, Vannier J, Van Iten H, Zhao Y-L (2004) Direct evidence for predation on trilobites in the Cambrian. Proc R Soc B Suppl 271:S277–S280

# Water-to-Land Transitions

16

## Jason A. Dunlop, Gerhard Scholtz and Paul A. Selden

## Contents

J. A. Dunlop (✉)
Museum für Naturkunde, Leibniz Institute
for Research on Evolution and Biodiversity
at the Humboldt University Berlin, Invalidenstr.
43, 10115 Berlin, Germany
e-mail: jason.dunlop@mfn-berlin.de

G. Scholtz
Humboldt-Universität zu Berlin, Institut für
Biologie, Vergleichende Zoologie, Philippstr. 13,
10115 Berlin, Germany
e-mail: gerhard.scholtz@rz.hu-berlin.de

P. A. Selden
Paleontological Institute, University of Kansas,
1475 Jayhawk Boulevard, Lawrence, Kansas 66045,
USA
e-mail: selden@ku.edu

## 16.1 Introduction

Arthropods are, by a considerable margin, the most species-rich group of animals alive today and have long been a major component of the Earth's biodiversity. Exact counts of the total number of species are not easy to come by, but Zhang (2011) offered a recent summary. Together the ca. 1,023,559 described living species of hexapods, 11,885 myriapods and 110,615 arachnids—most of which live on land—massively outnumber the ca. 66,914 recorded crustaceans, 1,322 sea spiders and the four species of horseshoe crab. Put bluntly, in terms of raw species numbers the primarily terrestrial lineages (Hexapoda, Myriapoda, Arachnida) outnumber the primarily aquatic ones ('Crustacea', Pycnogonida, Xiphosura) by a factor of almost seventeen to one. In fairness, there is a degree of bias in these figures. Some arachnids, such as water mites, are secondarily aquatic, but by the same token, some crustaceans such as woodlice and a number of crab species are also to a greater or lesser extent terrestrial. Despite the common opinion that terrestrial arthropods are easier to collect than (deep) marine ones, which may influence the total number of described taxa, new species of terrestrial insects, arachnids and myriapods are still being regularly described, particularly from the tropics. Aquatic arthropods seem unlikely to approach the

A. Minelli et al. (eds.), *Arthropod Biology and Evolution*,
DOI: 10.1007/978-3-642-36160-9_16, © Springer-Verlag Berlin Heidelberg 2013

diversity levels seen in megadiverse insect orders such as beetles, butterflies, true flies, or the bees, wasps and ants. This begs the questions *how*, *when*, *why* and *how often* did arthropods become successful in the terrestrial environment. We should also consider whether they came onto land via salt or fresh water.

### 16.1.1 The Significance of the Sister Group

Phylogeny has a crucial impact on evolutionary scenarios. If the sister group of a given terrestrial arthropod taxon also lives on land, we can infer that the last common ancestor of both groups was already terrestrial too. For instance, the traditional view of a close, or even a sister-group, relationship between hexapods and myriapods (the Antennata, Atelocerata or Tracheata concept) implied that the shift onto land took place in the common stem lineage of Antennata (see Kraus and Kraus 1994). This in turn biased the search for, and interpretation of, fossils as putative stem-lineage representatives of Antennata, further influencing hypotheses about the water–land transition within this group (see Haas et al. 2003; Kühl and Rust 2009). Today, the prevailing view is that Hexapoda is deeply nested within Crustacea (e.g. Regier et al. 2010; Rehm et al. 2011), although the hexapod sister group among the crustacean subgroups remains ambiguous (Glenner et al. 2006; von Reumont and Burmester 2010; von Reumont et al. 2012). Depending on the analysis, Remipedia, Branchiopoda or Malacostraca emerge as the closest relatives of the hexapods. This leads to the next problem. In contrast to remipedes and malacostracans, recent branchiopods are—with a few evidently derived exceptions—freshwater animals (e.g. fairy shrimps, tadpole shrimps, water fleas). If their sister-group relationship to hexapods were to be corroborated by further data, then we have to consider that the water-to-land transition in the hexapod lineage began in a freshwater habitat. This scenario might change if fossil branchiopods are included, since these animals were most likely marine (Olesen 2009).

Where monophyletic taxa are exclusively terrestrial, it is plausible to assume that the last common ancestor of the clade was terrestrial too. Nevertheless, given the frequent water-to-land transitions among arthropods—and animals in general—one has to ask whether parsimony considerations alone are enough to make a clear case for a single-terrestrialisation event, even in monophyletic groups. The case becomes stronger if structural apomorphies are also present as adaptations to a terrestrial life style; the book lungs of arachnids being a good example here (Scholtz and Kamenz 2006). This may seem trivial, but for the major terrestrial lineages of arthropods, there are surprisingly few unambiguous examples of anatomical terrestrial adaptations defining monophyletic groups. For instance, the tracheae of myriapods are so diverse in their position and structure that their homology has been seriously doubted (Ripper 1931; Dohle 1988; Hilken 1997). The same might apply to hexapod tracheae, and functionally similar tracheal tubes have also evolved more than once in the arachnids; at least twice just within the spiders where they can occur as tube or sieve tracheae (cf. Levi 1967) whose origins may be independent.

## 16.2 Secrets of Success

It seems self-evident from modern phylogenies (e.g. Regier et al. 2010) that land-living clades evolved independently within the overall arthropod tree. On the face of it, there must have been at least seven separate terrestrialisation events, enacted by: hexapods, myriapods, arachnids and at least four groups of crustaceans, namely isopods, amphipods, ostracods and decapods. Some authors have inferred multiple events within the arachnids too, but see Scholtz and Kamenz (2006) for counterarguments. Recall as well that among the arthropods' closest relatives (see Chap. 2), both Onychophora (velvet worms) and Tardigrada (water bears) are also now wholly or partially terrestrial. What these groups all share in common is a body plan with legs. While terms such as 'preadaptation' have fallen out of favour, it is fair to say that the arthropod ground pattern—which presumably originated in the early Palaeozoic seas (see Chap. 15)—possessed much which would later prove very useful in animals attempting to make the transition onto land. In addition to their jointed legs, which required only minimal modifications for locomotion on land, arthropods possess a tough external cuticle over the whole body which could later be waterproofed with a waxy epicuticle layer. In most of these cases, the respiratory organs are essentially invaginations of the body wall with thin cuticle.

### 16.2.1 Brave New Worlds?

We can only speculate about the evolutionary pressures that drove arthropods to exploit terrestrial ecosystems. Avoidance of predators would be one possibility. Horseshoe crabs were present by at least the early Ordovician (Van Roy et al. 2010), and it is conceivable that their modern mating behaviour—emerging onto shorelines to lay their eggs—is a relict of a distant time when the land was a safe and predator-free environment. In general, it is hard to envisage a terrestrial community of animals existing without at least some degree of plant cover and/or soil habitat. When this first became available remains a topic of debate, but there are microfossils of tough, desiccation-resistant spores as far back as the mid-Cambrian (Strother et al. 2004). Whether these indicate land plants is controversial (cf. Kenrick et al. 2012). By the mid-Ordovician, there are dispersed spores, often called cryptospores, consistent with belonging to terrestrial plants (Strother et al. 1996), and these presumably evolved into the small, branching *Cooksonia* type of vegetation recorded as body fossils by the mid-Silurian. Once plants became established as primary producers, there was clearly an opportunity for herbivores—or more likely at first the decomposers and detritivores—to exploit this new niche. Predators would then be able to follow too. The sclerotised head limbs of arthropods may also have played a role here, being easily

modified into a variety of mouthpart structures suitable for different feeding ecologies.

Although the focus here is on crown-group arthropods, we should not forget that soft-bodied organisms, from protozoans to tardigrades to various platyhelminth, nematode and oligochaete worms, may also have played a role in the earliest soil habitats (see also comments in Pisani et al. 2004). The chance of such animals being preserved as fossils is unfortunately very small. There is a Devonian plant-associated nematode (Poinar et al. 2008), and creatures like this would have been a potential food source for at least the smaller early terrestrial arthropods.

### 16.3  What is 'Terrestrial'?

Before discussing the timing and mechanisms of terrestrialisation further, we need to be clear about what we mean by a terrestrial animal. At what point does an arthropod become fully terrestrial? A water/land dichotomy is too simplistic, since a whole range of intermediate habitats can be envisaged. Examples would include regularly inundated algal strandlines on beaches. These are a typical feeding habitat today for sandhoppers (Amphipoda: Talitridae). Another would be the wet interstitial spaces between soil or sand particles. Marine interstitial environments play host to certain collembolans (Thibaud 2007), halacarid and occasionally oribatid mites (Bartsch 1989; Bayartogtokh and Chatterjee 2010) and to the rare palpigrade arachnids (Condé 1965). We must caution against assuming that such modern arthropods are 'primitive', or relicts of the first semi-terrestrial fauna. Today, we may be looking at secondary colonisations of beaches or river banks. For example some water mites (Acari: Hydrachnida) effectively live in a water/land transition zone, but phylogenies do not recover them as a particularly basal mite clade.

The marine interstitial route onto land has been widely suggested for early arthropods (Little 1990) and would have allowed a gradual accumulation of terrestrial adaptations.

However, in some cases, a transition from fresh water to terrestrial habitats is more likely (De Deckker 1983; Diesel et al. 2000). The typical micro- and meiofauna of a sandy beach today were summarised by Armonies and Reise (2000), who recorded tiny arthropods such as copepods, ostracods and mites, as well as the arthropod relatives, the tardigrades. Microarthropods living in soil and/or sand can effectively remain in an aquatic habitat by living in water films around the sediment particles; see, for example, Villani et al. (1999) for a review of the implications of edaphic habitats for terrestrialisation. As we will argue for crustaceans in particular, there are gradations of increasingly terrestrial habitats and lifestyles and corresponding gradations of anatomical adaptations. As is often the case in biology, the boundaries of what can be defined as terrestrial are not very sharp and we prefer to leave it somewhat fuzzy. Is a crab terrestrial when it lives completely on land and only larval spawning and development take place in seawater? This means the crab feeds on leaf litter, breathes air, excretes purine, mates in the forests and carries the eggs on land for a certain phase of development. If we define terrestrial arthropods by their life history, then reproduction and development (see also Cannicci et al. 2011) are among the strongest limiting factors hindering transitions onto land. For the purposes of this chapter, we consider an arthropod to be fully terrestrial if it does not need to return to water in order to complete its life cycle.

## 16.4 A Time Framework for the Transition

For a broad perspective on the origins of animal biodiversity on land—and the associated key events in Earth history—see Benton (2010). We cannot say for certain which arthropod group(s) first placed their feet on the shore, or exactly when they achieved it. But we can make inferences by combining direct evidence, in the form of body fossils, with indirect evidence drawn from trace fossils and molecular clock data. Arthropod terrestrialisation has been reviewed in its wider context by Størmer (1976), Rolfe (1985), Selden and Edwards (1989), Shear and Kukalová-Peck (1990), Shear (1991), Selden (2001, 2012), Shear and Selden (2001), Garwood and Edgecombe (2011) and Kenrick et al. (2012). We refer to these studies for further details and additional literature.

As outlined in Fig. 16.1, from the Cambrian–Ordovician boundary onwards (ca. 488 Ma), there are strong hints that arthropods of some description were able to walk, if only briefly, across terrestrial sediments. By the Silurian (ca. 416–443 Ma), myriapods and arachnids were unequivocally living on land and hexapods appear soon afterwards in the early Devonian (ca. 398–416 Ma). Today's land-living crustaceans do not appear to have been part of this early radiation. The oldest fossils implicit of terrestrial crustacean clades are Mesozoic. It is nevertheless important to remember that the early terrestrial fossil record of arthropods remains fragmentary. Much of our present knowledge is based on only a handful of 'windows' of opportunity. Key localities include the Silurian of Ludford Lane in England (Jeram et al. 1990), the Early Devonian Rhynie and Windyfield cherts of Scotland (reviewed in Anderson and Trewin 2003) and Alken an der Mosel and some adjacent sites in Germany (e.g. Størmer 1976), as well as the Middle Devonian of Gilboa in the USA (Shear et al. 1984, 1987). The more important discoveries are outlined below, and new fossils from even older localities would undoubtedly change the overall picture.

### 16.4.1 Trackways

Trace fossils (ichnofossils) cover a broad spectrum of fossilised animal activity, including faeces, burrows and nests, as well as locomotion traces from individual footprints through to fully developed trackways left by animals walking over the substrate. A review of the ichnological

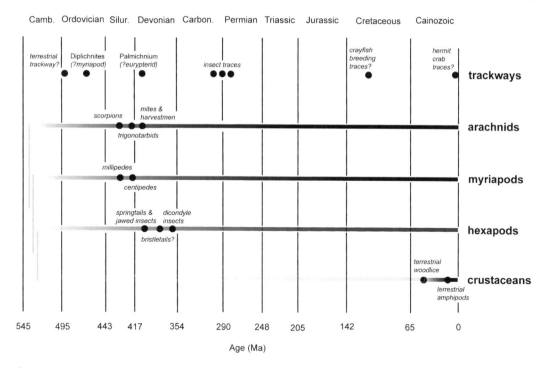

Camb. Ordovician Silur. Devonian Carbon. Permian Triassic Jurassic Cretaceous Cainozoic

**Fig. 16.1** A time framework for terrestrialisation based on trace fossils and the body fossil record of the four major arthropod lineages. Note that, hexapods are now conventionally regarded as having evolved from within the crustaceans

evidence for early life on land can be found in Braddy (2004), and a recent comprehensive case study based on the Siluro–Devonian Old Red Sandstone of Britain was published by Morrissey et al. (2012). Ichnologists refer to trackways made in a terrestrial environment (Fig. 16.2) as 'subaerial', differentiating them from those left in sediments under water. Subaerial trackways can be recognised by, for example, the footprints of the animal crossing desiccation cracks, which imply that the original animal (named the producer) walked across mud which was already exposed to the air and was in the process of drying out. Other examples can involve animals walking over ash falls. However, in both cases, we must exclude the possibility that the trackway (or ash fall) was made in water and was later exposed as the mud or ash dried out.

The oldest putative record of an arthropod walking across land comes from the Cambrian–Ordovician Nepean Formation in Ontario, Canada (MacNaughton et al. 2002). Dating to around 488 Ma, these trackways assigned to the ichnogenera *Diplichnites* and *Protichnites* were interpreted as having been made in a near-shore environment and as having possibly been produced by an enigmatic group of extinct arthropods known as the Euthycarcinoidea (see also Chap. 15). Collette et al. (2012) provided further model-based experimental evidence that Cambrian *Protichnites* traces could have been made by euthycarcinoids. These extinct (Cambrian–Triassic) mandibulate arthropods have multiple pairs of uniramous trunk limbs, but in the absence of unequivocal respiratory organs, it is unclear whether euthycarcinoids were aquatic, amphibious or terrestrial creatures.

Considerably younger trackways from the Ordovician Borrowdale Volcanic Group of England (Johnson et al. 1994) were again assigned to *Diplichnites*, and to another ichnogenus *Diplopodichnus*. These trackways were interpreted as non-marine, with millipedes tentatively suggested as a possible producer. Again, it is unclear whether they were left by fully terrestrial animals. Retallack and Feakes (1987)

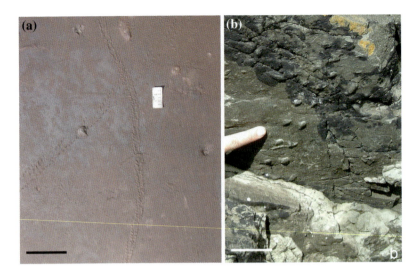

**Fig. 16.2** Arthropod trackways can imply life on land. **a** *Diplichnites*, a terrestrial trail probably made by a millipede-like animal, from the Early Devonian of Wales; **b** *Palmnichnium*, thought to be a semi-terrestrial eurypterid trackway, from the late Silurian of Wales. Scale bars equal 50 mm. Images courtesy of Rob Hillier and Lance Morrissey

documented supposed terrestrial burrows in fossil soils from late Ordovician Juniata Formation of central Pennsylvania, USA. Millipedes were inferred as the possible producers of these impressions, but this was disputed by Wilson (2006) and others, and the terrestrial palaeoenvironment of the Juniata Formation has now been comprehensively disproved by sedimentological analysis (Davies et al. 2010).

From the Silurian onwards, there are further records of terrestrial (millipede?) trackways (e.g. Wright et al. 1995; Morrissey et al. 2012). The latter authors also noted examples of a putative scorpion trace named *Paleohelcura* in the Old Red Sandstone of South Wales. Both these localities and the similarly aged Alken an der Mosel in Germany yield another ichnogenus called *Palmichnium*, recovered from sediments probably lain down in the intertidal zone. These impressions have been ascribed to the walking activities of sea scorpions (Eurypterida). The scenario implied here is that at least some eurypterids might have been semi-terrestrial (see also below) and perhaps clambered onto beaches to mate in a similar fashion to the modern horseshoe crabs (Poschmann and Braddy 2010).

Trackways unequivocally produced by insects first appear rather later in the fossil record, first being picked up in the Permo-Carboniferous (e.g. Braddy and Briggs 2002). Trace fossils assignable to terrestrial crustaceans are also very rare and so far do not belong to the Cambrian–Devonian phase of radiation either. Genise et al. (2008) described putative terrestrial breeding traces from the mid-Cretaceous (ca. 125 Ma?) of Patagonia in Argentina which they believed to be consistent with the activities of crayfish. Barely counting as fossils, Walker et al. (2003) described trackways from the Holocene (i.e. less than 12,000 years old) of the West Indies which they interpreted as the activities of a terrestrial hermit crab (Decapoda: Paguroidea).

There are two general problems with at least the older trace fossil discoveries. First, we cannot be exactly sure which arthropod produced a given trackway. Inferences can be made, for example three pairs of impressions would suggest a hexapod, four an arachnid and multiple pairs or groove-like furrows a myriapod. However, it is extremely rare to find a trackway with the producer (quite literally) stopped dead in its tracks at the end, such as in the aquatic 'death marches' of horseshoe crabs from the Jurassic of Solnhofen (Malz 1964). PreSilurian trackways have sometimes been ascribed to a myriapod-like animal, but it would be premature to accept them as explicit evidence for crown-group Myriapoda. As with the euthycarcinoids mentioned above, it is conceivable that there were other extinct arthropods around at this time with multiple limbs capable of leaving such impressions. Second, as discussed by Johnson et al. (1994), we cannot be certain that an arthropod walking across dry land in the mid-Palaeozoic was habitually terrestrial. These may represent (semi)-aquatic animals capable of brief excursions onto land, but who were trying to cross from one body of water to another, or were trying to escape a drying pool.

To recap, from the end of the Cambrian through to the Ordovician and Silurian, it appears that some arthropods—possibly including myriapods and eventually also arachnids—could walk across land, but it is unclear from the trace fossils alone whether they lived in this environment on a long-term basis. Apart from some possibly semi-aquatic crustacean burrows from the Devonian (Morrissey et al. 2012), there is no convincing trace fossil evidence for terrestrial locomotion by insects or crustaceans prior to the Carboniferous and Cretaceous, respectively.

## 16.4.2 Body Fossils

A general overview of the arthropod fossil record can be found in Chap. 15. Although there is a dubious record of an Ordovician mite, the oldest arthropod fossil which can be assigned with confidence to a terrestrial habitat is the millipede (Diplopoda) *Pneumodesmus newmani* from the Silurian (ca. 428 Ma) of Scotland (Wilson and Anderson 2004). Significantly, this fossil (Fig. 16.3d, arrow) reveals the putative openings of spiracles which imply a tracheal system (see also 16.5.5). There is also an enigmatic group called Kampecarida—probably millipedes of some description—known from a range of Siluro–Devonian localities. They were

last reviewed by Almond (1985) and would merit further study. A small number of scorpions (Scorpiones) also occur in strata dating from the mid-to-late Silurian. Laurie's (1899) species *Dolichophonus loudonensis* is stratigraphically the oldest known arachnid and is approximately the same age as the oldest terrestrial millipede *Pneumodesmus*; both fossils incidentally coming from Scotland. Note that there has been a long debate about whether the early scorpions (Fig. 16.3a) were terrestrial or aquatic, the latter hypothesis championed by Kjellesvig-Waering (1986) in particular. However, as critiqued by Scholtz and Kamenz (2006) and Kühl et al. (2012), these views have rarely been supported by convincing morphological features. The trend seems to be shifting towards interpreting all fossil scorpions as potentially terrestrial animals.

### 16.4.2.1 Ludford Lane
Late Silurian fossils from the ca. 419 Ma Ludford Lane consist of cuticle remains acid-macerated out of the sediment. These include fragments of the oldest centipede (Chilopoda), which can be provisionally assigned to the Scutigeromorpha (Jeram et al. 1990; Shear et al. 1998). Scutigeromorphs (Fig. 16.3e) are widely perceived as sister group of all other centipedes. There is also a millipede belonging to the extinct Arthropleurida group (Shear and Selden 1995). A further interesting Ludford Lane find is the oldest non-scorpion arachnid which belongs to an extinct spider-like order called Trigonotarbida (Fig. 16.3b). Since younger trigonotarbids are demonstrably terrestrial (see below), this habitat has been assumed for the Ludford Lane fossil too. Finally, there are also examples of fossil faeces, or coprolites (Fig. 16.3c), which may have been produced by a detritivore such as a millipede (Edwards et al. 1995).

### 16.4.2.2 Rhynie
The next oldest localities are from the Early Devonian. Prominent among these are the Rhynie and adjacent Windyfield cherts of northwest

**Fig. 16.3** Terrestrial arthopod life in the Silurian. **a** the scorpion *Proscorpius osborni* from the 'Bertie Waterlime' of the USA (note that, some authors have interpreted early scorpions as aquatic); **b** the trigonotarbid arachnid *Palaeotarbus jerami* from Ludford Lane, UK; **c** scanning electron micrograph of a coprolite, possibly from a myriapod, from Ludford Lane, UK; **d** the millipede *Pneumodesmus newmani* from Cowie Harbour, Scotland with slit-like spiracles arrowed (used with permission by the Paleontological Society); **e** femur–tibia articulation of a scutigeromorph centipede, *Crussolum* sp., from Ludford Lane, UK

Scotland which are dated to ca. 410 Ma and preserve an entire terrestrial ecosystem of early plants and animals with extraordinary three-dimensional fidelity. Rhynie has yielded more trigonotarbid arachnids (Hirst 1923), as well as the oldest unequivocal mites (Acari), such as *Protacarus crani* and some further species named later from among Hirst's original specimens (Fig. 16.4a). From the same locality, Dunlop et al. (2004) described the oldest harvestman (Opiliones) as *Eophalangium sheari*. Further significant finds at Rhynie are the oldest terrestrial hexapods. These include a springtail (Collembola) *Rhyniella praecursor* described by Hirst and Maulik (1926) and Scourfield (1940) (Fig. 16.4b). Perhaps, even more significant is *Rhyniognatha hirsti*, a fossil primarily known

from its mandibles (Tillyard 1928). It was later reinterpreted as the oldest true insect (Engel and Grimaldi 2004); the authors even speculating that this animal may have borne wings. Other Rhynie/Windyfield records include euthycarcinoids, centipedes and an additional hexapod of uncertain affinity (Anderson and Trewin 2003; Fayers and Trewin 2005).

### 16.4.2.3 Alken and Other Sites

Marginally, younger than Rhynie are a number of localities in the German Rhineland, the most famous of which is Alken an der Mosel (cf. Størmer 1970, 1976). As well as semi-terrestrial eurypterid trace fossils (see above), these sites have yielded early terrestrial arthropods

**Fig. 16.4** Terrestrial arthropod life in the Devonian. **a** a mite, possibly *Protospeleorchestes pseudoprotacarus*, from the Rhynie chert, Scotland; **b** drawing of the collembolan *Rhyniella praecursor* (after Scourfield 1940), also from the Rhynie chert; **c** the trigonotarbid arachnid *Alkenia mirabilis* from Alken an der Mosel, Germany; **d** the oldest pseudoscorpion *Dracochela deprehendor* from Gilboa, New York, USA; **e** the centipede *Devonobius delta*, also from Gilboa

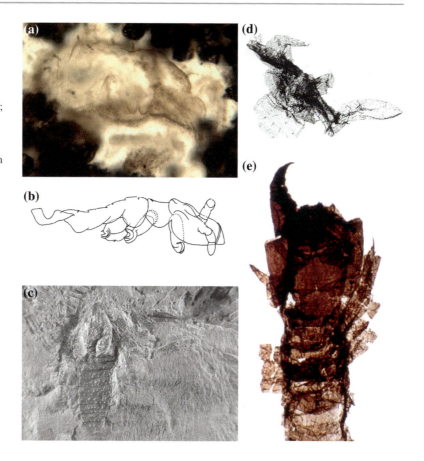

such as trigonotarbids (Fig. 16.4c), scorpions and arthropleurids, as well the oldest example of another extinct arachnid order named Phalangiotarbida (Poschmann et al. 2005). More generally, a number of early Devonian sites have yielded records of scorpions reviewed by Kjellesvig-Waering (1986), but as noted above, his interpretations must be treated with caution. Kjellesvig-Waering assumed that almost all Palaeozoic scorpions were aquatic and at least one set of scorpion 'gills' later turned out to be part of an arthropleurid millipede (Shear and Selden 1995). In general, millipedes are also quite well represented at this time. Numerous Devonian examples belonging to the juliform group—the classic, long-bodied millipedes which burrow through soil and litter—were critically reviewed by Wilson (2006).

### 16.4.2.4  Gilboa

The Middle Devonian, in particular the ca. 390 Ma Gilboa locality near New York, has produced more trigonotarbid arachnids, as well as both oribatid and alicorhagiid mites (Norton et al. 1988; Kethley et al. 1989). Other notable finds at Gilboa include the oldest pseudoscorpion (Schawaller et al. 1991, revised as a stem-group taxon by Judson 2012) (Fig. 16.4d). Also significant is *Attercopus fimbriunguis*. This fossil was first thought to be a trigonotarbid (Shear et al. 1987), but was later reinterpreted as the oldest spider (Araneae). More recently, it has been shown to be an example of an extinct, spider-like order called Uraraneida (Selden et al. 2008). These animals probably resembled spiders, albeit with a flagelliform tail like that of a whip scorpion (Uropygi). The well-preserved cuticle fragments suggest that *Attercopus* could produce silk, but lacked the discrete spinnerets which define the true spiders. Gilboa also hosts centipedes (Shear and Bonamo 1988; Shear et al. 1998) including another scutigeromorph, together with a further record assigned to an extinct group (Fig. 16.4e).

### 16.4.2.5 Other Ancient Insects and Crustaceans

A contemporary find (Labandeira et al. 1988) from the mid-Devonian of Gaspé in Québec, Canada, is a bristletail (Insecta: Archaeognatha). After the Rhynie mandibles (see above), this fossil would be only the second oldest example of a true insect. However, the fossiliferous nature of this find was seriously questioned by Jeram et al. (1990) shortly after its publication and unless more specimens or evidence are forthcoming, it would be unwise to accept this as an unequivocal insect record. Recently, Garrouste et al. (2012) reported the discovery of a putative terrestrial insect from the Late Devonian (ca. 365 Ma) of Belgium. *Strudiella devonica* was assigned to the Dicondylia clade and noted for having 'orthopteroid' mouthparts suitable for an omnivorous diet.

What should by now be apparent is the preponderance of arachnid and myriapod fossils making up these Siluro–Devonian terrestrial assemblages, as compared to the relative paucity of hexapods/insects and the complete absence of any demonstrably terrestrial crustaceans. The hexapods—and in particular the pterygote (winged) insects—only really seem come into their own from the Carboniferous onwards (e.g. Prokop et al. 2005) by which time land-based communities of plants and animals were already well established.

Body fossils of unequivocally terrestrial crustaceans are much younger. The oldest terrestrial woodlice (Isopoda: Oniscoidea) are currently known as fossils from Eocene (ca. 49 Ma) Baltic amber (reviewed by Schmidt 2008), although the larger clade to which they belong—the Scutocoxifera—can be traced back

to the Jurassic. The oldest known Amphipoda referable to the (semi-)terrestrial family Talitridae are even younger, being first recorded from the Miocene (ca. 16 Ma) Chiapas or Mexican amber (Bousfield and Poinar 1994). It has also been suggested (Bousfield 1983) that since modern Talitridae live in, and feed on, angiosperm litter then the group is unlikely to be older than the flowering plants which themselves radiated in the mid-Mesozoic.

Fossils of a number of land crabs (Decapoda, Brachyura) can be found in the Quaternary (ca 2–3 Ma). These include fossils assignable to Gecarcinidae from the Caribbean (Donovan and Dixon 1998), Potamidae from Japan (Naruse et al. 2004) and Grapsidae from Hawaii (Paulay and Starmer 2011). The last of these is an interesting case study in which the species was probably driven extinct by human activity. However, for all these remains, we do not know whether they are indicative for a fully terrestrial lifestyle according to our definition. There is currently no body fossil record of the terrestrial hermit crabs (Coenobitidae), although a possible trace fossil of this group was mentioned above.

### 16.4.3 Molecular Clocks

An alternative way to infer the age of terrestrial crown-group clades is to use the molecular clock. If we assume—and it is an assumption—that arachnids, hexapods and myriapods each had a common terrestrial ancestor, then determining when each of these groups separated off from their nearest relatives (i.e. the time of cladogenesis) would yield an approximate date by which the clade may have come onto land. The problem here comes if the split occurred far back in time in an aquatic environment, and if the crown-group terrestrial arthropods had a stem group of aquatic ancestors which continued living for millions of years in the water and about which we know little or nothing from the fossil record. In essence, how tightly is cladogenesis coupled to terrestrialisation? How soon did the last common ancestor make it onto land?

The principal value of the fossils here is in imposing constraints on these models. For example, terrestrial millipedes must have evolved by the mid-Silurian at the very latest (Wilson and Anderson 2004). In another case study, Dunlop and Selden (2009) pointed out that, at 428 Ma, fossil scorpions are older than a published estimate of $393 \pm 23$ Ma for the split between spiders and scorpions based on a mitochondrial phylogeny. In this particular example, the fossil showed that the split based on molecular data was an underestimate. In this context, well-preserved fossils whose systematic position is robust can act as calibration points, helping to improve the overall reliability of molecular-based phylogenies. Further discussion of the strengths and limitations of molecular dating—such as the risks of treating stem-group fossils as calibration points for crown-group organisms—can be found in Kenrick et al. (2012).

More usually, molecular methods suggest older dates (sometimes substantially so) for life on land, as compared to the direct evidence of the fossil record. This is unsurprising as it is unlikely that we will ever find a fossil of the very first terrestrial animal in a given clade. Problems come when there is large discrepancy between the fossil and molecular dates. A prime example would be the study of Schaefer et al. (2010) who calibrated their tree using data for oribatid mites; a group with a reasonable fossil record thanks to their often quite tough and resilient bodies. Their data suggested that oribatid mites—and by inference interstitial soil microarthropods in general—originated and began moving onto land as early as the late Precambrian ($571 \pm 37$ Ma). By contrast, the oldest fossil oribatids from the Gilboa locality are considerably younger, being Devonian (390 Ma) in age. Realistically, a Precambrian date for land arthropods seems much too early. As noted by Kenrick et al. (2012), most Precambrian fossils are barely recognisable as animals, let alone members of crown-group arthropod clades, and these authors suggested that the discrepancy in this case may be due to an analysis which relied on only a single gene.

Authors such as Pisani et al. (2004) recovered somewhat younger dates of $475 \pm 53$ Ma for the split between xiphosurans and arachnids, and $442 \pm 50$ Ma for a split between millipedes and centipedes. These dates (Fig. 16.5) are more consistent with the fossil record since the oldest arachnids and myriapods are Silurian (and thus about 430 Ma). However, for the crustacean–hexapod split, a very old (i.e. Precambrian) date of $666 \pm 58$ Ma was recovered in this paper. Alternatively, Regier et al. (2004) published a younger date of ca. 488–461 Ma for crown-group hexapods—this is still about 50 million years before the first body fossil—while Sanders and Lee (2010) found an older date of ca. 504 Ma. In another study, Rehm et al. (2011) recovered a ca. 555 Ma split for myriapods and chelicerates—with a ca. 500 Ma split for millipedes and centipedes—and a ca. 480 Ma split for spiders and horseshoe crabs; in their scheme, mites were their sister clade coming off at ca. 495 Ma. The hexapods and crustaceans were dated to a split of ca. 520 Ma (Rehm et al. 2011, Fig. 16.1). As noted above, elucidating the sister group of hexapods among the crustaceans is crucial to dating their origins. Future developments in this field will hopefully refine the methods further and reduce the gap between the inferred (molecular) and observed (fossil) data.

## 16.5 Challenges and Solutions

As we have argued above, the major arthropod groups almost certainly moved onto land independent of one another, but once they did they all faced an identical set of problems. These animals developed similar, sometime even identical, responses to these challenges. The example of tracheae arising in parallel in insects, myriapods and arachnids has already been discussed. It is a prime example of how moving onto land automatically creates homoplastic characters among different groups of arthropods exposed to the same selective pressures. The physiological challenges faced by animals moving from an aquatic to a terrestrial environment have been summarised in some detail

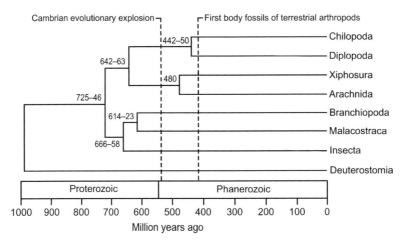

**Fig. 16.5** A fossil-calibrated timescale of arthropod evolution based on two nuclear genes, reproduced from (Pisani et al. 2004, Fig. 2). Numbers associated with nodes are divergence times (Ma) and their standard errors. Molecular clocks offer an alternative approach for estimating when terrestrial arthropod clades first appeared (see text for details)

by Little (1990). In brief, these relate primarily to gas exchange, reproduction, osmoregulation and exposure to ultraviolet radiation. Likewise, sensory structures have to be modified for life on land (e.g. Stensmyr et al. 2005). Changes in locomotory biology would also have been important and—on a related point—it is interesting to speculate whether smaller or larger arthropods were better able to make the initial transition onto land?

## 16.5.1 Body Size

If terrestrialisation did take place via the interstitial route, then tiny microarthropods would have been better adapted to exploit such environments. They could easily occupy wet spaces between soil or sand particles and, like many such organisms in modern soil ecosystems (Villani et al. 1999), they could burrow deeper into the substrate in search of moisture when the upper layers dried out. However, minute organisms are highly dependent on local conditions—they are less able to walk far or fast—and if there is a major change in the environment, they run the risk of drying out more quickly. Larger animals, with a lower surface area to body size ratio, would have been more resilient against

such fluctuations in the environment, but would presumably have been more exposed to such changes.

Palaeontology does not, as yet, yield a clear picture about which of these scenarios is correct. Smaller, less strongly sclerotised arthropods are, in principle, less likely to be preserved as fossils, and so key components of the early land fauna may simply not be visible. In any case, different taxa may have adopted different strategies. The oldest (Silurian) trigonotarbid arachnid is only ca. 1.5 mm long, and most of the Rhynie arachnids described so far have body lengths from less than 1 mm (the mites) up to about 4 mm (trigonotarbids). By contrast, scorpions seem to have been at least a few centimetres long throughout their known geological history (cf. Kjellesvig-Waering 1986). Indeed, in the Devonian a few became huge, with body lengths approaching a metre!

The known Silurian juliform millipedes preserve body lengths in the 35–45 mm range (Wilson and Anderson 2004), comparable to modern temperate species found in soils today. However, the Devonian also yields tiny arthropleuridean millipedes less than 5 mm long (Wilson and Shear 2000). This is interesting given that Carboniferous arthropleurids achieved enormous body lengths of two metres or more (Kraus and Brauckmann 2003). The earliest hexapods also tend towards a small body size, whereas (Shear et al. 1989) documented some fairly large early terrestrial arachnids and myriapods from the early Devonian of Canada.

An origin from small-sized aquatic ancestors is also likely for terrestrial ostracods (De Deckker 1983). The various malocostracan terrestrial crustacean lineages reveal, however, a different pattern of land colonisation. For instance, it is obvious that among true land crabs and the various 'amphibious' hermit crabs and brachyurans, it was relatively large animals which gave rise to terrestrial populations starting from a variety of marine and freshwater origins. In the case of the proper land brachyurans, it seems apparent that they colonised land from a freshwater habitat (Diesel et al. 2000). In contrast to this, the monophyletic oniscoid isopods invaded land only once and most likely from the sea, since the *Ligia* species—which serve as model for the transition—mostly occur on sea shores (see Schmalfuss 1978; Carefoot and Taylor 1995; Schmidt 2008). Again, it is apparent that the first terrestrial isopods were animals in the centimetre range (Schmidt 2008).

## 16.5.2 Locomotion

We have already argued that the presence of jointed legs in ancestral arthropods was probably an important factor in facilitating a smooth transition from water onto land. Many aquatic arthropods live on (or in) the substrate, and it was presumably fairly straightforward to adapt such a body plan to a terrestrial environment. Although quantitative data are lacking, when we compare the walking legs of purely aquatic arthropods with their (semi-)terrestrial relatives (insects, spiders and certain crabs), there seems to have been a tendency for the legs to become larger and often thicker in proportion to the rest of the body. Terrestrial animals are no longer supported by the buoyancy of water, and larger, thicker legs can accommodate a more extensive musculature to

support the animals and overcome the effects of gravity on land; see, for example, discussion in (Dalingwater 1985).

Related to this, authors such as Selden and Jeram (1989)—based on the studies of Manton (1977)—proposed the presence of a plantigrade foot in scorpions as one of the criteria for recognising terrestrial animals. The idea here is that marine animals, supported by the buoyancy of water, can effectively walk on the tips of their toes (digitigrade stance), but terrestrial animals would have little purchase on the substrate, and abrade their toes, without the larger surface area of a foot on the ground. Manton (1977) showed how rocking joints in the leg allowed the foot to remain stable while the body and leg moved forwards during locomotion. Terrestrial arthropods hang from their legs, which have rocking joints at the bases of the leg and the tarsus. Note that this applies better to insects and arachnids, which have relatively few legs, than to the multilimbed millipedes and centipedes who can still use a digitigrade stance on land today as their weight is distributed across many more individual appendages. Another interesting point of convergence between arachnids and insects is the tendency (there are exceptions in both groups) to have two large claws, or ungues, at the end of the leg. In some arachnids, a smaller third claw may be present and both insects and arachnids may have a fleshy (adhesive) pad between the claws. It is interesting to speculate whether this two-clawed pattern evolved in parallel as an advantageous feature for gripping the substrate and/or clambering through the early vegetation.

## 16.5.3 Osmoregulation

Water balance—and specifically water loss—is one of the main challenges facing primarily terrestrial animals. The presence of a cuticular exoskeleton in arthropods (cf. Chap. 8) was undoubtedly a major advantage here, and most terrestrial groups have a waxy epicuticle layer which reduces water loss directly over the cuticle. Behavioural adaptations should also be mentioned here, and numerous arthropods avoid desiccation by favouring damp, humid habitats or by burrowing deeper into the substrate. Another option is to avoid activity during daylight; groups like scorpions are predominantly nocturnal. In tracking the evolution of relevant features, the fossil record is less helpful. Even the best preservation (e.g. Rhynie) does not yield structures such as Malphigian tubules. Other osmoregulatory organs, such as coxal glands, are also hard to detect in fossils.

Osmoregulation is also highly relevant to the question about the route taken onto land, specifically did the ancestors of a given terrestrial arthropod group come directly from a marine environment or did they go first via fresh water? Little (1990) concluded that because intertidal animals could tolerate a greater range of salinities—as well as being better adapted to variable water supply, temperature, etc.—they were thus better adapted to move onto land than those which became highly adapted to freshwater or interstitial habitats.

## 16.5.4 Reproduction and Development

As discussed above, we suggest here that the ability to complete the life cycle without having to return to water is a reasonable definition of a fully terrestrial arthropod. Achieving this imposes major constraints upon the organism. Sperm can no longer simply be released over the eggs as in an aquatic environment, but must be delivered directly to the eggs or the female genital opening. This can either be done directly, for example, via an intromittent organ such as a penis or the palpal organ of spiders, or indirectly via a sperm package (Witte and Döring 1999). These packages, or spermatophores, can either be deposited on the substrate to be discovered later by the female, or they can be deposited as part of a controlled mating ritual in which the female is usually led directly over the sperm package by the male (Witte and Döring 1999). Similarly, the eggs now have to be provided

with a protective layer to prevent them from drying out (see below). A few groups such as scorpions (reviewed by Warburg 2012) have adopted live-birth strategies instead.

### 16.5.4.1 Mating Chelicerates and Myriapods

Tracing the evolution of reproductive systems in the fossil record can be challenging, but under optimal conditions of preservation, key developments can still be identified. Kamenz et al. (2011) argued that at least some Silurian eurypterids had arachnid-like spermatophores, the precursors of which are occasionally fossilised as the eurypterid 'horn-organs' (Fig. 16.6a). While eurypterids are thought to have been primarily aquatic—with some speculation about trends towards an amphibious mode of life (see Sect.16.4.1)—the evolution of spermatophores in the common ancestor of eurypterids and arachnids would have provided a useful way for the first arachnids to transfer their sperm on land. It is tempting to speculate about eurypterids practising some sort of scorpion-like mating dance. Other arachnids have taken indirect sperm transfer further and developed direct techniques. The Rhynie chert harvestman preserves both a male penis and female ovipositor (Dunlop et al. 2004), which implies that by the early Devonian, these animals had already developed a mechanism for impregnating females directly and then laying eggs (Figs. 16.6b, c) without the need for water.

Among the myriapods, the oldest evidence for direct copulation is the millipede *Cowiedesmus eroticopodus* from the Silurian of Scotland (Wilson and Anderson 2004). Most modern millipedes belong to the derived clade Helminthomorpha, which is characterised by the presence in the male of modified appendages a short distance behind the head called gonopods. These are actively used in sperm transfer, and their presence in a Silurian fossil (Fig. 16.6d) shows that the helminthomorph mating strategy of direct insemination was already present in some of the oldest millipede fossils.

**Fig. 16.6** Fossil reproductive organs. **a** the genital operculum of a eurypterid (*Eurypterus* sp. from the Silurian of Sareema in Estonia) with its genital appendage, whereby the associated horn organ was recently interpreted as a precursor of a (male) spermatophore; **b–c** the penis and ovipositor of male and female harvestmen, respectively, from the Devonian Rhynie chert of Scotland (after Dunlop et al. 2004); **d** the millipede *Cowiedesmus eroticopodus*, which preserves limbs modified into gonopods, from Cowie Harbour, Scotland (used with permission by the Paleontological Society)

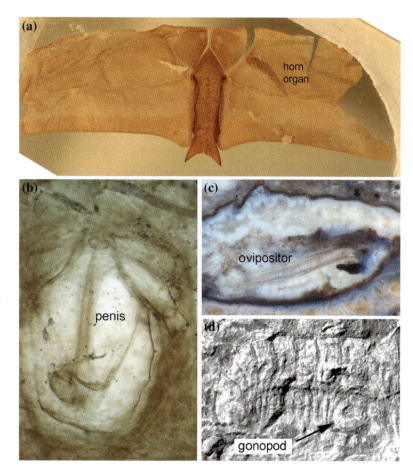

### 16.5.4.2 Brooding Crustaceans

Among crustaceans, there are no fossil data for changes in reproductive strategies, but the developmental biology and physiology of today's semi-terrestrial species offers some interesting insights into a shift onto land which is still in progress (e.g. Greenaway 2003). The study by Cannicci et al. (2011) stressed the role of the embryo and maternal care in the terrestrialisation of crabs. In particular, they focussed on the fact that developing embryos in amphibious and shallow water species can be bimodal in their respiration, with oxygen uptake possible both from water and air. This can affect brooding strategies since the relative ease of extracting oxygen from air compared to water can reduce the level of parental care required—such as the mother having to create water currents to ventilate the eggs. The authors stressed

that these crab embryos are still not independent of water, and parental effort is still needed to maintain a suitable microhabitat. For example in some genera, brooding mothers tend to remain in their burrows, which limits their offspring's exposure to desiccation.

### 16.5.4.3 Arthropod Eggs

The major leap forward for vertebrates getting onto land was the evolution of the amniote egg, which freed previously amphibious tetrapods from their reproductive link to the water by providing a miniature pond (the amniotic cavity and fluid) within the egg itself in which the embryo develops. To protect the egg and embryo from evaporation either leathery or calcified shells enclose the egg, or as a secondary evolutionary step, the eggs develop inside the mother. Similar solutions have been realised in arthropods, as reviewed by Zeh et al. (1989) who argued that adaptations for life on land affecting the egg stage were key to freeing up a wider range of niches for these animals to exploit. In particular, in the fully terrestrial arachnids, myriapods and hexapods, the eggs are generally enclosed in thick outer envelopes, and in some cases, such as scorpions, eggs undergo development within the maternal uterus. Additional brood care devices such as silken cocoons in spiders or maternal brood chambers (e.g. pseudoscorpions, whip spiders) have also evolved. The more derived centipedes enrol themselves around their eggs and millipedes produce a protective coat for the eggs, the nematomorphans specifically using silk to build a nesting chamber.

As in other aspects of their biology, terrestrialisation in relation to egg protection in crustaceans took a different pathway. Most 'terrestrial' crustaceans, such as the decapod land crabs and land hermit crabs including the robber crab *Birgus latro*, are effectively still at the 'amphibian' stage in that their larvae develop in water (Türkay 1987). The only fully terrestrial crustaceans are found among decapods and peracarids, namely within brachyurans and within isopods and amphipods. All these taxa are characterised by complex maternal brood care structures which already evolved partly or completely in the marine environment.

In the pleocyematan decapods, the eggs are attached to the maternal pleopods and, in particular, in brachyuran crabs, the ventrally folded pleon forms a protecting structure which encapsulates a tightly closed brood chamber. This is the structural prerequisite for terrestrialisation, since only a slightly increased degree of tightness between the pleon margin and the sternites allows for the generation of a humid or water-filled chamber for the embryos. In combination with increasingly embryonised larval stages—that is, a direct development, which evolved in freshwater crabs—this leads to a complete terrestrial life cycle in the few true land crab species (see e.g. Diesel et al. 2000).

A comparable but convergent solution to the problem of egg protection occurs within the peracarid crustaceans. Terrestrial isopods and amphipods maintain the eggs in a ventral brood pouch—the marsupium (Fig. 16.7)—which effectively acts as a mobile pond (Hoese and Janssen 1989). This peracarid marsupium is formed by a number of plates, the oostegites, originating from the coxae of a number of thoracopods. Interestingly, the marsupium as such is not an adaptation to a terrestrial life style, but was already present in the marine peracarid stem species (Richter and Scholtz 2001). Hence, terrestrial isopods and amphipods inherited this structure form their marine ancestors. Likewise, the direct development and the absence of distinct larval stages that we observe in terrestrial isopods and amphipods were also already present in their marine ancestor. Again, they cannot be regarded as unique terrestrial adaptations. Nevertheless, the marsupium of terrestrial isopods in particular is a very effective structure for brood protection and even allowed the colonisation of desert habitats.

### 16.5.5 Gas Exchange

Terrestrial arthropods presumably evolved from aquatic, gill-bearing forebears. In the standard arthropod Bauplan, the trunk limbs originally comprised a leg-like branch for locomotion and a flap-like branch with numerous blade-like lamellae for gas exchange and/or swimming. External gills of this form are impractical in air, where they would collapse under their own weight and/or dry out far too quickly. Terrestrial arthropods were thus faced with two options: adapt or innovate. An example of adaptation would be to internalise an existing system, which appears to have been the case in the book lungs of the pulmonate arachnids.

#### 16.5.5.1 Book Lungs
Arachnid book lungs are widely regarded as homologous with the book-gills of horseshoe

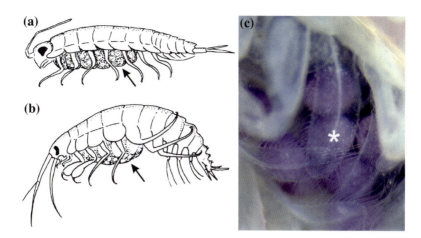

**Fig. 16.7** The marsupium of peracarid crustaceans. **a** schematised isopod showing the position of the marsupium (*arrow*) in the pereon region; **b** schematised amphipod showing the position of the marsupium (*arrow*) in the pereon region; **c** photograph of ventral aspect of the marsupium of the (semi) terrestrial amphipod *Orchestia cavimana*. The embryos are enclosed and protected by plate-like coxal structures, the oostegites (*asterisk*), which form the marsupium. Line drawings modified from Dohle (1976)

crabs (e.g. Kingsley 1885). For a recent study, comparing and contrasting lung and gill development see Farley (2012). In essence, one can argue that arachnids have largely retained the leg branch in the prosoma for locomotion, but in groups like spiders and scorpions, they have lost the legs and retained the gill branch on the opisthosoma for gas exchange. Fossil book lungs (Fig. 16.8b) can be observed in the Devonian Rhynie chert trigonotarbids (Claridge and Lyon 1961; Kamenz et al. 2008). The quality of preservation even allows us to identify the small cuticular struts keeping adjacent lung lamellae apart. It is worth reiterating that these fossilised respiratory systems are anatomically indistinguishable from the lungs and tracheae of living arachnids. Shear et al. (1989) and Kühl et al. (2012) also documented putative book lung material preserved in Devonian scorpions from Canada and Germany, respectively.

The Devonian also throws up some unusual morphologies. Fossil scorpions in the genus *Waeringoscorpio* from both Alken an der Mosel and a nearby locality have unusual projections from the sides of the opisthosoma (Størmer 1970; Poschmann et al. 2008). These are associated with the lung-bearing body segments in living scorpions. Poschmann et al. (2008) likened these scorpion structures to the gills seen today in some (secondarily) aquatic insect larvae. It raises the intriguing possibility that these early terrestrial faunas included animals with unique respiratory systems, and perhaps even animals which secondarily re-entered water.

### 16.5.5.2  Modified Branchial Chamber Walls and Gills

As noted above, some authors have speculated that the extinct eurypterids were to a certain extent amphibious. One line of evidence in favour of this is some well-preserved fossils expressing paired structures on the opisthosoma referred to as 'gill tracts' or Kiemenplatten (Fig. 16.8a). These appear to have been modified areas of spongy tissue with a fine microornament of conical projections occupying the upper walls of the gill chambers (Selden 1985; Manning and Dunlop 1995). In this scenario, eurypterids are hypothesised to have relied on these gill tracts during brief excursions onto land. Analogies have been drawn with the branchial lungs of certain (semi-) terrestrial brachyuran crabs and hermit crabs, whereby a folded gill chamber wall creates an increased surface area for gas exchange (e.g. Farrelly and

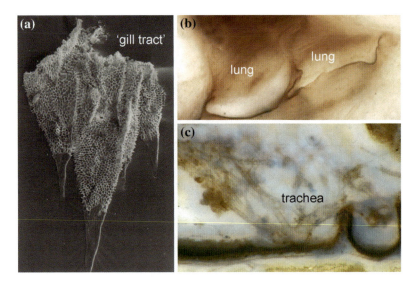

**Fig. 16.8** Fossil respiratory organs. **a** scanning electron micrograph of 'gill tract' tissue from a eurypterid (after Manning and Dunlop 1995); modified regions of the gill chamber wall interpreted as potentially analogous to the branchial lungs of certain modern terrestrial crabs; **b** the oldest book lungs (after Kamenz et al. 2008), from a Devonian Rhynie chert trigonotarbid arachnid; **c** the oldest tracheal tubes (after Dunlop et al. 2004), from a Rhynie chert harvestman

Greenaway 1993; 2005). This is often accompanied by adaptive modifications of gill structures such as stiffening and the presence of nodules, which prevent the gill lamellae from collapsing, and a functional shift from a predominant role in gas exchange to more ion-regulation (Farrelly and Greenaway 1992).

Terrestrial oniscoid isopods still use their gills for respiration. These are, uniquely in malacostracans, formed by the pleopodal endopods (Gruner 1993). The use of gills in a terrestrial environment was made possible due to a complex water-conducting system, which constantly supplies the ventral side of the pleon with a water flow (Hoese 1981). In addition, within some oniscoidean groups, the anterior exopods are equipped with invaginations forming lung-like structures (Hoese 1981, 1983; Gruner 1993). These are additional examples of animals modifying an existing structure for terrestrial respiration.

### 16.5.5.3 Tracheae

Alternatively, arthropods could innovate and evolve an entirely new respiratory system. The best example here would be the branching tracheae which evolved independently in multiple arthropod groups (Hilken 1997) and supply the tissues with oxygen directly. Both book lungs and tracheae open through small rounded or slit-like spiracles which reduce the amount of water lost over the respiratory organs via evaporation. The specific discontinuous gas exchange cycle used by many tracheate arthropods—essentially the cycle in which the spiracles are opened and closed—was investigated by Klok et al. (2002), who again concluded that this essentially identical physiological process must have evolved in different groups simultaneously. The oldest direct evidence for unequivocally terrestrial respiration (Sect. 16.4.2) is the spiracle openings preserved in a Silurian millipede (Fig. 16.3d). The oldest example of the actual tracheal tubes themselves (Fig. 16.8c) comes from the Devonian Rhynie chert harvestman (Dunlop et al. 2004). The branching pattern observed here is almost identical to that seen in modern harvestmen today. Thus, both lung-based and tracheal systems for gas exchange were already clearly established by the Devonian.

## 16.6    Concluding Remarks

If we step back and take a broad view of the arthropod fossil record (see also Chap. 15), it is notable that crustaceans appear quite early as a recognisable group, alongside trilobites and various other Cambrian marine arthropods. By contrast, contemporary fossils assignable to today's primarily terrestrial groups (arachnids, myriapods and hexapods) are unknown. A major challenge has been to identify convincing stem-group representatives of all three primarily terrestrial clades back in their original marine environment in the early to mid-Palaeozoic. Arthropods called megacheirans may have given rise to the chelicerates, but fossils proposed as potential hexapod or myriapod ancestors have invariably proved to be highly controversial; see Haas et al.'s (2003) supposed marine Devonian hexapod with its reinterpretation by Kühl and Rust (2009). At this point, it is worth remembering that, according to current phylogenies, the hexapods were merely the first—and by a considerable margin the most successful—of a number of attempts by crustaceans throughout their geological history to colonise the land.

Despite the success of the insects as a megadiverse branch of now largely land-living crustaceans, much of the focus in the present chapter has instead been on early fossils of myriapods and arachnids. Both groups were likely to have been on land from about 430 million years ago and have been recorded, often with quite modern-looking body plans, from an increasing number of Siluro–Devonian fossil assemblages. If the fossil record does reflect the composition of the original ecosystem, it is tempting to envisage them together as part of an early detritivore/carnivore association. Arachnids and centipedes today do not commonly eat millipedes and one could speculate whether millipedes evolved defence mechanisms such as calcified cuticle and noxious secretions in response to predator attacks back in these early terrestrial ecosystems.

Although hexapods are numerically the most significant terrestrial arthropod group today, the fossil record suggests that their transition onto land may have begun slightly later. They were present both as collembolans and as early jawed insects by at least 410 million years ago, but their early evolution is less well understood, being based on only a handful of (sometimes controversial) precarboniferous records. By the time of the Carboniferous Coal Measures, about 300 million years ago, a truly diverse fauna of winged insects begins to appear. Yet, the origins—or at least the principal radiations—of the most diverse modern clades among the holometabolus insects (beetles, flies, wasps, etc.) did not take place until the early part of the Mesozoic. Some authors (e.g. Grimaldi 1999) point to the rise of the flowering plants, as part of a so-called Cretaceous Terrestrial Revolution about 100 million years ago. This may have created new ecological niches (e.g. as pollinators) which further facilitated insect diversification. Thus, one hypothesis could be that insects came late, but gradually came to dominate terrestrial ecosystems from the end of the Palaeozoic onwards, supplanting the myriapods and arachnids in both diversity and abundance.

Finally, the non-hexapod crustaceans have also attempted to invade the land on more than one occasion. Precise dates for the origins of groups like land crabs and (terrestrial) woodlice are unfortunately not well constrained, but we should reiterate that there is no evidence for any of these groups living on land during the Palaeozoic. Physiologically, none of these terrestrial crustaceans is as well adapted for life on land as the insects, myriapods and arachnids. Thus, it is tempting to see today's amphibious crustaceans as living models for the problems—and solutions—faced by all terrestrial arthropods having to evolve the mechanisms which they need to free themselves completely from their aquatic ancestry.

**Acknowledgments** We thank C. Oliver Coleman for advice on fossil amphipods; and Lyall Anderson, Wolfgang Dohle, Diane Edwards, Hagen Hass, Rob Hillier, Carsten Kamenz, Andrew Jeram and Lance Morrissey for making images used here available. The editors and a reviewer provided valuable comments on an earlier version. PAS's visit to Berlin was funded by the Alexander von Humboldt Foundation.

# References

Almond JE (1985) The Silurian-Devonian fossil record of the Myriapoda. Phil Trans R Soc B 309:227–237

Anderson LI, Trewin N (2003) An Early Devonian arthropod fauna from the Windyfield cherts, Aberdeenshire, Scotland. Palaeontology 46:467–509

Armonies W, Reise K (2000) Faunal diversity across a sandy shore. Mar Ecol Prog Ser 196:49–57

Bartsch I (1989) Marine mites (Halacaroidea: Acari): a geographical and ecological survey. Hydrobiologia 178:21–42

Bayartogtokh B, Chatterjee T (2010) Oribatid mites from marine littoral and freshwater habitats in India with remarks on world species of *Thalassozetes* (Acari: Oribatida). Zool Stud 49:839–854

Benton MJ (2010) The origins of modern biodiversity on land. Phil Trans R Soc B 365:3667–3679

Bousfield EL (1983) An updated phyletic classification and palaeohistory of the Amphipoda. In: Schram F (ed) Crustacean Phylogeny (Crustacean Issues 1). Balkema, Rotterdam, pp 257–277

Bousfield EL, Poinar GO Jr. (1994) A new terrestrial amphipod from Tertiary amber deposits of Chiapas province, southern Mexico. Hist Biol 7:105–114

Braddy SJ (2004) Ichnological evidence for the arthropod invasion of land. Fossils Strata 51:136–140

Braddy SJ, Briggs DEG (2002) New Lower Permian nonmarine arthropod trace fossils from New Mexico and South Africa. J Palaeont 76:546–557

Cannicci S, Simoni R, Giomi F (2011) Role of the embryo in crab terrestrialization: an ontogenetic approach. Mar Ecol Prog Ser 430:121–131

Carefoot T, Taylor B (1995) *Ligia*: a prototypal terrestrial isopod. In: Alikhan MM (ed) Terrestrial Isopod Biology (Crustacean Issues 9). Balkema, Rotterdam, pp 47–60

Claridge MF, Lyon AG (1961) Lung-books in the Devonian Palaeocharinidae (Arachnida). Nature 191:1190–1191

Collette JH, Gass KC, Hagadorn JW (2012) *Protichnites eremita* unshelled? experimental model-based neoichnology and new evidence for a euthycarcinoid affinity for this ichnospecies. J Palaeont 86:442–454

Condé B (1965) Présence de Palpigrades dans le milieu interstitial littoral. C r Acad Sci Paris 261:1898–1900

Dalingwater JE (1985) Biomechanical approaches to eurypterid cuticles and chelicerate exoskeletons. Trans R Soc Edinb Earth Sci 76:359–364

Davies NS, Rygel MC, Gibling MR (2010) Marine influence in the Upper Ordovician Juniata Formation (Potters Mills, Pennsylvania): implications for the history of life on land. Palaios 25:527–539

De Deckker P (1983) Terrestrial ostracods in Australia. Austr Mus Mem 18:87–100

Diesel R, Schubart CD, Schuh M (2000) A reconstruction of the invasion of land by Jamaican crabs (Grapsidae: Sesarminae). J Zool 250:141–160

Dohle W (1976) Zur Frage des Nachweises von Homologien durch die komplexen Zell- und Teilungsmuster in der embryonalen Entwicklung höherer Krebse (Crustacea, Malacostraca, Peracarida). Sitzungsber Ges Naturf Freunde Berlin 16:125–144

Dohle W (1988) Myriapoda and the ancestry of insects. Manchester Polytechnic, Manchester

Donovan SK, Dixon HL (1998) A fossil land crab from the late Quaternary of Jamaica (Decapoda, Brachyura, Gecarcinidae). Crustaceana 71:824–826

Dunlop JA, Anderson LI, Kerp H, Hass H (2004) A harvestman (Arachnida: Opiliones) from the Early Devonian Rhynie cherts, Aberdeenshire, Scotland. Trans R Soc Edinb Earth Sci 94:341–354

Dunlop JA, Selden PA (2009) Calibrating the chelicerate clock: a paleontological reply to Jeyaprakash and Hoy. Exp Appl Acarol 48:183–197

Edwards D, Selden PA, Richardson JB, Axe L (1995) Coprolites as evidence for plant–animal interaction in Siluro-Devonian terrestrial ecosystems. Nature 377:329–331

Engel MS, Grimaldi D (2004) New light shed on the oldest insect. Nature 427:627–630

Farley R (2012) Ultrastructure of book gill development in embryos and first instars of the horseshoe crab *Limulus polyphemus* L (Chelicerata, Xiphosura). Front Zool 9:4. doi:10.1186/1742-9994-9-4

Farrelly CA, Greenaway P (1992) Morphology and ultrastructure of the gills of terrestrial crabs (Crustacea, Gecarcinidae and Grapsidae): adaptations for air breathing. Zoomorphology 112:39–49

Farrelly CA, Greenaway P (1993) Land crabs with smooth lungs: Grapsidae, Gecarcinidae, and Sundathelphusidae ultrastructure and vasculature. J Morph 215:245–260

Farrelly CA, Greenaway P (2005) The morphology and vasculature of the respiratory organs of terrestrial hermit crabs (*Coenobita* and *Birgus*): gills, branchiostegal lungs and abdominal lungs. Arthropod Struct Dev 34:63–87

Fayers SR, Trewin NH (2005) A hexapod from the Early Devonian Windyfield Chert, Rhynie, Scotland. Palaeontology 48:1117–1130

Garrouste R, Clément G, Nel P, Engel MS, Grandcolas P, D'Haese C, Lagebro L, Denayer J, Gueriau P, Lafaite P, Olive S, Prestianni C, Nel A (2012) A complete insect from the Late Devonian period. Nature 488:82–85

Garwood RJ, Edgecombe GD (2011) Early terrestrial animals, evolution and uncertainty. Evol Edu Outreach 4:489–501

Genise JF, Bedatou E, Melchor RN (2008) Terrestrial crustacean breeding trace fossils from the Cretaceous of Patagonia (Argentina): palaeobiological and evolutionary significance. Palaeogeo Paleoclim Palaeoecol 264:128–139

Glenner H, Thomsen PF, Hebsgaard MB, Sørensen MV, Willerslev E (2006) The origin of insects. Science 314:1883–1884

Greenaway P (2003) Terrestrial adaptations in the Anomura (Crustacea: Decapoda). Mem Mus Vict 60:13–26

Grimaldi D (1999) The co-radiations of pollinating insects and angiosperms in the Cretaceous. Ann Miss Bot Gard 86:373–406

Gruner H-E (1993) Klasse Crustacea, Krebse. In: Gruner H-E (ed) Lehrbuch der speziellen Zoologie, Band I, 4.Teil Arthropoda. Gustav Fischer, Jena, pp 448–1030

Haas F, Waloszek D, Hartenberger R (2003) Devonohexapodus bocksbergensis, a new marine hexapod from the Lower Devonian Hunsrück Slates, and the origin of Atelocerata and Hexapoda. Organ Divers Evol 3:39–54

Hilken G (1997) Vergleich von Tracheensystemen unter phylogenetischem Aspekt. Verh Naturwiss Ver Hamburg 37:5–94

Hirst S (1923) On some arachnid remains from the Old Red Sandstone (Rhynie Chert bed, Aberdeenshire). Ann Mag Nat Hist 12(70):455–474

Hirst S, Maulik S (1926) On some arthropod remains from the Rhynie Chert (Old Red Sandstone). Geol Mag 63:69–71

Hoese B (1981) Morphologie und Funktion des Wasserleitungssystems der terrestrischen Isopoden (Crustacea, Isopoda, Oniscoidea). Zoomorphology 98:135–167

Hoese B (1983) Struktur und Entwicklung der Lungen der Tylidae (Crustacea, Isopoda, Oniscoidea). Zool Jb Anat 109:487–501

Hoese B, Janssen HH (1989) Morphological and physiological studies on the marsupium of terrestrial isopods. Monit zool ital (NS) 4:153–173

Jeram AJ, Selden PA, Edwards D (1990) Land animals in the Silurian: arachnids and myriapods from Shropshire, England. Science 250:658–661

Johnson EW, Briggs DEG, Suthren RJ, Wright JL, Tunnicliff SP (1994) Non-marine arthropod traces from the subaerial Ordovician Borrowdale Volcanic Group, English Lake District. Geol Mag 131:395–406

Judson MLI (2012) Reinterpretation of Dracochela deprehendor (Arachnida: Pseudoscorpiones) as a stem-group pseudoscorpion. Palaeont 55:261–283

Kamenz C, Dunlop JA, Scholtz G, Kerp H, Hass H (2008) Microanatomy of Early Devonian book lungs. Biol Lett 4:212–215

Kamenz C, Staude A, Dunlop JA (2011) Sperm carriers in Silurian sea scorpions. Naturwissenschaften 98:889–896

Kenrick P, Wellman CH, Schneider H, Edgecombe GD (2012) A timeline for terretrialization: consequences for the carbon cycle in the Palaeozoic. Phil Trans R Soc B 367:519–536

Kethley JB, Norton RA, Bonamo PM, Shear WA (1989) A terrestrial alicorhagiid mite (Acari: Acariformes) from the Devonian of New York. Micropaleontology 35:367–373

Kingsley JS (1885) Notes on the embryology of Limulus. Q J Microsc Sci 25:521–576

Kjellesvig-Waering EN (1986) A restudy of the fossil Scorpionida of the world. Palaeontographica Am 55:1–287

Klok CJ, Mercer RD, Chown SL (2002) Discontinuous gas-exchange and its convergent evolution in tracheated arthropods. J Exp Biol 205:1019–1029

Kraus O, Brauckmann C (2003) Fossil giants and surviving dwarfs. Arthropleurida and Pselaphognatha (Atelocerata, Diplopoda): characters, phylogenetic relationships and construction. Verh naturwiss Ver Hamburg (NF) 40:5–50

Kraus O, Kraus M (1994) Phylogenetic system of the Tracheata (Mandibulata): on "Myriapoda"–Insecta interrelationships, phylogenetic age and primary ecological niches. Verh naturwiss Ver Hamburg (NF) 34:5–31

Kühl G, Bergmann A, Dunlop JA, Garwood RJ, Rust J (2012) Redescription and palaeobiology of Palaeoscorpius devonicus Lehmann, 1944 from the Lower Devonian Hunsrück Slate of Germany. Palaeontology 55:775–787

Kühl G, Rust J (2009) Devonohexapodus bocksbergensis is a synonym of Wingertshellicus backesi (Euarthropoda)—no evidence for marine hexapods living in the Devonian Hunsrück Sea. Org Divers Evol 9: 215–231

Labandeira CC, Beall BS, Hueber FM (1988) Early insect diversification: evidence from a Lower Devonian bristletail from Québec. Science 242:913–916

Laurie M (1899) On a Silurian scorpion and some additional eurypterid remain from the Pentland Hills. Trans R Soc Edinb 39:575–590

Levi HW (1967) Adaptations of respiratory systems of spiders. Evolution 21:571–583

Little C (1990) The Terrestrial Invasion—an Ecophysiological Approach to the Origins of Land Animals. Cambridge University Press, Cambridge

Macnaughton RB, Cole JM, Dalrymple RW, Braddy SJ, Briggs DEG, Lukie TD (2002) First steps on land: arthropod trackways in Cambrian–Ordovician eolian sandstone, southeastern Ontario, Canada. Geology 30:391–394

Malz H (1964) Kouphichnium walchi. Die Geschichte einer Fährte und ihres Tieres. Nat Mus 94:81–97

Manning PL, Dunlop JA (1995) The respiratory organs of eurypterids. Palaeontology 38:287–297

Manton SM (1977) The Arthropoda: Habits, Functional Morphology, and Evolution. Oxford University Press, Oxford

Morrissey LB, Hiller RD, Marriott SB (2012) Late Silurian and early Devonian terrestrialization: ichnological insights from the Lower Old Red Sandstone of the Anglo-Welsh Basin, U.K. Palaeogeog Palaeoclim Palaeoecol 337–338:194–215

Naruse T, Karasawa H, Shokita S, Tanaka T, Moriguchi M (2004) A first fossil record of the terrestrial crab, *Geothelphusa tenuimanus* (Miyake and Minei 1965) (Decapoda, Brachyura, Potamidae) from Okinawa Island, Central Ryukyus, Japan. Crustaceana 76:121–1218

Norton RA, Bonamo PN, Grierson JD, Shear WA (1988) Oribatid mite fossils from a terrestrial Devonian deposit near Gilboa, New York. J Paleontol 62:259–269

Olesen J (2009) Phylogeny of Branchiopoda (Crustacea) – character evolution and contribution of uniquely preserved fossils. Arthrop Syst Phylog 67:3–39

Paulay G, Starmer J (2011) Evolution, insular restriction, and extinction of oceanic land crabs, exemplified by the loss of an endemic *Geograpsus* in the Hawaiian Islands. PLoS ONE 6(5):e19916. doi:10.1371/journal.pone.0019916

Pisani D, Poling LL, Lyons-Weiler M, Hedges SB (2004) The colonization of land by animals: Molecular phylogeny and divergence times among arthropods. BMC Biology 2:1; doi:10.1186/1741-7007-2-1

Poinar GO Jr., Kerp H, Hass H (2008) *Palaeonema phyticum* gen. n., sp. n. (Nematoda: Palaeonematidae fam. n.), a Devonian nematode associated with early land plants. Nematology 10:9–14

Poschmann M, Anderson LI, Dunlop JA (2005) Chelicerate arthropods, including the oldest phalangiotarbid arachnid, from the Early Devonian (Siegenian) of the Rhenish Massif, Germany. J Paleontol 79:110–124

Poschmann M, Braddy SJ (2010) Eurypterid trackways from early Devonian tidal facies of Alken an der Mosel (Rheinisches Schiefergebirge, Germany) Palaeobiodivers Palaeoenvir 90:111–124

Poschmann M, Dunlop JA, Kamenz C, Scholtz G (2008) The Lower Devonian scorpion *Waeringoscorpio* and the respiratory nature of its filamentous structures, with a description of a new species from the Westerwald area, Germany. Paläont Ztschr 82:418–436

Prokop J, Nel A, Hoch I (2005) Discovery of the oldest known Pterygota in the Lower Carboniferous of the Upper Silesian Basin in the Czech Republic (Insecta: Archaeorthoptera). Geobios 38:383–387

Regier JC, Shultz JW, Kambic RE (2004) Phylogeny of basal hexapod lineages and estimates of divergence times. Ann Entomol Soc Am 97:411–419

Regier JC, Shultz JW, Zwick A, Hussey A, Ball B, Wetzer R, Martin JW, Cunningham CW (2010) Arthropod relationships revealed by phylogenomic analysis of nuclear protein-coding sequences. Nature 463:1079–1083

Rehm P, Borner J, Meusemann K, von Reumont BM, Simon S, Hadrys H, Misof B, Burmester T (2011) Dating the arthropod tree based on large-scale transcriptome. Mol Phyl Evol 61:880–887

Retallack GJ, Feakes CR (1987) Trace fossil evidence for Late Ordovician animals on land. Science 235:61–63

Richter S, Scholtz G (2001) Phylogenetic analysis of the Malacostraca (Crustacea). J Zool Syst Evol Res 39:113–136

Ripper W (1931) Versuch einer Kritik der Homologiefrage der Arthropodentracheen. Z wiss Zool 138:303–369

Rolfe WDI (1985) Early terrestrial arthropods: A fragmentary record. Phil Trans R Soc B 309:207–218

Sanders K, Lee MSY (2010) Arthropod molecular divergence times and the Cambrian origin of pentastomids. Syst Biodivers 8:63–74

Schaefer I, Norton RA, Scheu S, Maraun M (2010) Arthropod colonization of land – linking fossils in oribartid mites (Acari, Oribatida). Mol Phyl Evol 57:113–121

Schawaller W, Shear WA, Bonamo PM (1991) The first Paleozoic pseudoscorpions (Arachnida, Pseudoscorpionida). Amer Mus Novit 3009:1–17

Schmalfuss H (1978) *Ligia simony*: a model for the evolution of terrestrial isopods. Stuttg Beitr Naturk A 317:1–5

Schmidt C (2008) Phylogeny of the terrestrial Isopoda (Oniscidea): a review. Arthropod Syst Phyl 66:191–226

Scholtz G, Kamenz C (2006) The book lungs of Scorpiones and Tetrapulmonata (Chelicerata, Arachnida): Evidence for homology and a single terrestrialisation event of a common arachnid ancestor. Zoology 109:2–13

Scourfield DJ (1940) The oldest known fossil insect (*Rhyniella praecursor* Hirst and Maulik). Further details from additional specimens. Proc Linn Soc 152:113–131

Selden PA (1985) Eurypterid respiration. Phil Trans R Soc B 309:209–226

Selden PA (2001) Terrestrialization (invertebrates). In:Briggs DEG, Crowther PR (eds). Palaeobiology II. Blackwell, Oxford, 71–74

Selden PA (2012) Terrestrialisation (Precambrian–Devonian) (Version 3.0).In: Encyclopedia of Life Sciences. Wiley, Chichester.doi: 10.1002/97804 70015902.a0001641.pub3

Selden PA, Edwards D (1989) Chapter 6. Colonisation of the land. In: Allen KC, Briggs DEG (eds) Evolution and the fossil record. Belhaven, London, pp 122–152

Selden PA, Jeram AJ (1989) Palaeophysiology of terrestrialisation in the Chelicerata. Trans R Soc Edinb Earth Sci 80:303–310

Selden PA, Shear WA, Sutton MD (2008) Fossil evidence for the origin of spider spinnerets, and a proposed arachnid order. Proc Natl Acad Sci USA 105:20781–20785

Shear WA (1991) The early development of terrestrial ecosystems. Nature 351:283–289

Shear WA, Bonamo PM (1988) Devonobiomorpha, a new order of centipeds (Chilopoda) from the Middle

Devonian of Gilboa, New York State USA, and the phylogeny of centiped orders. Amer Mus Novitat 2927:1–30

Shear WA, Bonamo PM, Grierson JD, Rolfe WDI, Smith EL, Norton RA (1984) Early land animals in North America: evidence from Devonian age arthropods from Gilboa, New York. Science 224:492–494

Shear WA, Gensel PG, Jeram AJ (1989) Fossils of large terrestrial arthropods from the Lower Devonian of Canada. Nature 384:555–557

Shear WA, Jeram AJ, Selden PA (1998) Centiped legs (Arthropoda, Chilopoda, Scutigeromorpha) from the Silurian and Devonian of Britain and North America. Amer Mus Novitat 3231:1–16

Shear WA, Kukalová-Peck J (1990) The ecology of Palaeozoic terrestrial arthropods: the fossil evidence. Can J Zool 68:1807–1834

Shear WA, Selden PA (1995) Eoarthropleura (Arthropoda, Arthropleurida) from the Silurian of Britain and the Devonian of North America. N Jahrb Geol Paläont Abh 196:347–375

Shear WA, Selden PA (2001) Rustling in the undergrowth: animals in early terrestrial ecosystems. In: Gensel PG, Edwards D (eds) Plants invade the land: Evolutionary and environmental perspectives. Columbia University Press, New York, pp 29–51

Shear WA, Selden PA, Rolfe WDI, Bonamo PM, Grierson JD (1987) New terrestrial arachnids from the Devonian of Gilboa, New York. Amer Mus Novitat 2901:1–74

Stensmyr MC, Erland S, Hallberg E, Wallén R, Greenaway P, Hansson BS (2005) Insect-like olfactory adaptations in the terrestrial giant robber crab. Curr Biol 15:116–121

Størmer L (1970) Arthropods from the Lower Devonian (Lower Emsian) of Alken an der Mosel, Germany. Part 1: Arachnida. Senckenberg Leth 51:335–369

Størmer L (1976) Arthropods from the Lower Devonian (Lower Emsian) of Alken an der Mosel, Germany. Part 5: Myriapoda and additional forms, with general remarks on the fauna and problems regarding invasion of land by arthropods. Senckenberg Leth 57:87–183

Strother PK, Al-Hajri S, Traverse A (1996) New evidence for land plants from the lower Middle Ordovician of Saudi Arabia. Geology 24:55–58

Strother PK, Wood GD, Taylor WA, Beck JH (2004) Middle Cambrian cryptospores and the origin of land plants. Mem Ass Australas Palaeontol 29:99–113

Thibaud J-M (2007) Recent advances and synthesis in biodiversity and biogeography of arenicolous Collembola. Ann soc Entomol France NS 43:181–185

Tillyard RJ (1928) Some remarks on the Devonian fossil insects from the Rhynie chert beds, Old Red Sandstone. Trans Ent Soc Lond 76:65–71

Türkay M (1987) Landkrabben. Nat Mus 117:143–150

Van Roy P, Orr PJ, Botting JP, Muir LA, Vinter J, Lefebvre B, el Hariri K, Briggs DEG (2010) Ordovician faunas of Burgess Shale type. Nature 465:215–218

Villani MG, Allee LL, Díaz A, Robbins PS (1999) Adaptive strategies of edaphic arthropods. Ann Rev Entomol 44:233–256

von Reumont B, Burmester T (2010) Remipedia and the evolution of hexapods. In: Encyclopedia of Life Sciences (ELS). Wiley, Chichester. doi/10.1002/9780470015902.a0022862

von Reumont BM, Jenner RA, Wills MA, Dell'Ampio E, Pass G, Ebersberger I, Meyer B, Koenemann S, Iliffe TM, Stamatakis A, Niehuis O, Meusemann K, Misof B (2012) Pancrustacean phylogeny in the light of new phylogenomic data: support for Remipedia as the possible sister group of Hexapoda. Mol Biol Evol 29:1031–1045

Walker SE, Holland SM, Gardiner L (2003) Coenobichnus currani (new ichnogenus and ichnospecies): fossil trackway of a land hermit crab, early Holocene, San Salvador, Bahamas. J Paleontol 77:576–582

Warburg MR (2012) Pre-and post-parturial aspects of scorpion reproduction: a review. Eur J Entomol 109:139–146

Wilson HM (2006) Juliformian millipedes from the Lower Devonian of Euramerica: implications for the timing of millipede cladogenesis in the Paleozoic. J Paleontol 80:638–649

Wilson HM, Anderson LI (2004) Morphology and taxonomy of Paleozoic millipedes (Diplopoda: Chilognatha: Archipolypoda) from Scotland. J Paleontol 78:169–184

Wilson HM, Shear WA (2000) Microdecemplicida, a new order of minute arthropleurideans (Arthropoda, Myriapoda) from the Devonian of New York State, U.S.A. Trans R Soc Edinb Earth Sci 90:351–375

Witte H, Döring D (1999) Canalized pathways of change and constraints in the evolution of reproductive modes of microarthropods. Exp Appl Acarol 23:181–216

Wright JL, Quinn L, Briggs DEG, Williams SH (1995) A subaerial arthropod trackway from the upper Silurian Clam Bank formation of Newfoundland. Can J Earth Sci 32:304–313

Zeh DW, Zeh JA, Smith RL (1989) Ovipositors, amnions and eggshell architecture in the diversification of terrestrial arthropods. Quart Rev Biol 64:147–168

Zhang Z-Q (2011) Phylum Arthropoda von Siebold, 1848. In: Zhang Z-Q (ed) Animal biodiversity: an outline of higher-level classification and survey of taxonomic richness. Zootaxa 4138:99–103

# Arthropod Endosymbiosis and Evolution

# 17

Jennifer A. White, Massimo Giorgini, Michael R. Strand and Francesco Pennacchio

## Contents

J. A. White
Department of Entomology, University of Kentucky, S-225 Agricultural Science Center N, Lexington, KY 40546-0091, USA

M. Giorgini
Istituto per la Protezione delle Piante, CNR, Via Università 133, 80055, Portici, Naples, Italy

M. R. Strand
Department of Entomology, University of Georgia, 120 Cedar Street, 413 Biological Sciences Building, Athens, GA 30602, USA

F. Pennacchio (✉)
Dipartimento di Agraria Laboratorio di Entomologia "E. Tremblay", Università di Napoli "Federico II", Via Università 100, 80055, Portici, Naples, Italy
e-mail: f.pennacchio@unina.it

## 17.1 Introduction

The association of "two species that live on or in one another" was first described in the nineteenth century, and the word symbiosis was proposed to denote this biological phenomenon (Sapp 1994). The discovery that lichens are organisms generated by the integration of a fungus and blue-green algae, that is, cyanobacteria, was followed by a number of other studies that have shown how the association of different species is widespread in nature and characterized by different degrees of benefit-sharing. Symbiosis encompasses both antagonistic relationships, in which one organism takes advantage of the other, and mutualistic relationships, where both partners gain advantage from their association. There are also cases where no clear benefit or harm is evident for both interacting species, which are then, in some cases, considered commensals. The term symbiosis applies to all these type of species associations, and not only to mutualism, as is sometimes erroneously done (Sapp 1994).

A. Minelli et al. (eds.), *Arthropod Biology and Evolution*,
DOI: 10.1007/978-3-642-36160-9_17, © Springer-Verlag Berlin Heidelberg 2013

The Darwinian model of gradual evolution, based on competitive selection acting on random mutations accumulated over time, was considered for a long time the major evolutionary pattern driving the diversification of all living organisms. However, gradualism does not account for evident leaps revealed from fossil records and molecular data. These evolutionary leaps have stimulated two hypotheses, not mutually exclusive (Ryan 2002). The punctuated equilibrium theory states that evolutionary leaps, denoted as saltations, are generated by drastic changes in the selection scenario, for example, the intense colonization of new ecological niches becoming available after catastrophic mass extinctions. The symbiosis theory provides an alternative interpretation of these evolutionary leaps, which are considered the final result of the fusion of different biological entities, to give rise to a new taxon (Ryan 2002). Therefore, symbiosis can be viewed not only as a peculiar mode of life, but also as a biological phenomenon that has important evolutionary implications, significantly contributing to the evolution of life on Earth (Sagan 1967; Margulis 1993). This possibility has been for a long time underestimated, but an increasing amount of molecular and functional data corroborate this hypothesis (Margulis 2009).

The most species-rich taxon of multicellular organisms is the arthropods, which have successfully colonized virtually every habitat and niche on Earth. Recent surveys reveal an equally rich diversity of symbiotic associations between different types of arthropods and micro-organisms. A comprehensive overview of these interactions is beyond the scope of a single review chapter. Thus, the purpose of this contribution is to summarize current knowledge about the interactions between primarily insects and certain groups of bacteria and viruses, which have been studied in more detail and allow the best appreciation of the considerable impact of endosymbiosis on the evolution of arthropods. In the first section of this chapter, we discuss the range of beneficial endosymbiotic associations that have evolved between insects and bacteria. In the second, we discuss the role of intracellular

bacteria in manipulating the reproduction of insects and other arthropods, while our third section discusses the role selected taxa of viruses play as beneficial symbionts of parasitoid wasps and other insects.

## 17.2 Bacteria as Obligate and Facultative Symbionts of Insects

A comprehensive understanding of the microbial diversity associated with arthropod populations is far from being defined. The information currently available clearly indicates that bacteria, in particular those in the groups of α- and γ-Proteobacteria, are among the major players, as they are more prone to establish tight interactions with arthropod tissues, either as pathogens or as mutualists. The mechanistic bases of these latter associations, and the details of how the interacting symbionts share the emerging benefit, have not always been fully elucidated. In some cases, the symbionts are mutually obligate, due to strong functional ties, such as nutritional complementation of poor diets, which do not allow them to live independently of one another. There are many other forms of symbiosis that are facultative associations, in which the arthropod can survive in the absence of the associated micro-organism. A large variety of evolutionary novelties are generated by these facultative symbioses, which are not always obvious to interpret from a mechanistic point of view and often have more than one function. Currently, our best understanding of both obligate and facultative symbioses is derived from the study of aphids and other plant feeding Hemiptera, which will offer the large majority of the case studies presented hereafter.

### 17.2.1 Obligate Nutritional Endosymbionts

Obligate microbial symbionts are a common feature among arthropods that have nutritionally poor or imbalanced diets (Buchner 1965; Douglas

1989; Moran et al. 2008). The microbial partners are highly diverse, representing a wide array of bacterial and fungal lineages acquired independently by a variety of ancestral arthropods (Moran et al. 2008; Gibson and Hunter 2010). The vast majority of research on these obligate associations has focused on bacterial partners; hence, our discussion here will also centre upon bacteria.

Like most animals, arthropods including insects are incapable of synthesizing essential amino acids and are generally dependent on gaining these protein building blocks through consumption (Chapman 1998). In contrast, many microbes are competent to synthesize all amino acids. Likewise, many vitamins and cofactors can be synthesized by microbes, but not by arthropods (Chapman 1998). For arthropods that feed upon diets that are deficient in amino acids (plant sap) or vitamins (animal blood), the inadequately available components must be provided by other means. Through light microscopy, early researchers found that arthropods feeding on such poor dietary sources often housed microbes in specialized cells (interchangeably referred to as mycetocytes or bacteriocytes) sometimes grouped together in organ-like structures (mycetomes or bacteriomes) and that these microbes were transmitted vertically from mother to offspring (reviewed in Buchner 1965). Early hypotheses that these mycetocyte-associated microbes play vital nutritional roles have now been validated by empirical, molecular, and genomic analyses (Buchner 1965; Moran et al. 2008; Shigenobu and Wilson 2011).

The pea aphid, *Acyrthosiphon pisum*, and its obligate bacterial symbiont, *Buchnera aphidicola*, present a case study. Aphids feed exclusively on plant phloem, which is carbohydrate rich but very low in essential amino acids (Gündüz and Douglas 2009 and references therein). Analyses of the *Buchnera* genome have found it to be highly reduced compared with free-living bacterial relatives, having lost many critical functions including synthetic pathways for many non-essential amino acids (e.g. Shigenobu et al. 2000). However, despite the overall genome erosion exhibited by the symbiont, the essential amino acid synthetic pathways remain largely intact, suggesting that the essential amino acids are provided to the host by the bacterial symbiont. Analyses of the recently published pea aphid genome, along with expression studies, have confirmed the perfect complementarity of aphid and bacterial metabolisms; the aphid generally provides the non-essential amino acids, and the bacterium synthesizes the essential amino acids (Hansen and Moran 2011; Shigenobu and Wilson 2011). Furthermore, synthesis of some amino acids (e.g. valine, leucine) requires metabolites contributed by both partners: neither host nor bacterium would be capable of synthesizing these amino acids on their own (Hansen and Moran 2011).

These intricate metabolic interdependencies reflect millions of years of coevolution between aphids and bacteria. Almost all aphids contain *Buchnera* symbionts, and phylogenetic analyses have indicated parallel evolution between bacteria and host, with divergences among the bacterial lineages corresponding to divergences among the aphids (Baumann 2005). This pattern of cocladogenesis is consistent with an initial infection of the ancestor of all aphids, estimated to have occurred more than 180 million years ago (mya) (Moran et al. 2008).

Similar stories can be told with other arthropod hosts and their obligate symbiotic lineages, as summarized in Table 17.1. Many of these symbioses are ancient in origin, although some more recent associations have been identified (e.g. Lamelas et al. 2008). Despite the diverse origins of the microbes, some common themes are evident in their evolutionary histories. First, it is very common for the symbiont genome to be extremely reduced, sometimes approaching an order of magnitude smaller than free-living bacterial relatives (Table 17.1; Moran et al. 2008; Toft and Anderson 2010). Factors contributing to this process of genome shrinkage include vertical transmission and insufficient purifying selection of small populations leading to fixation of deleterious alleles, high mutational rates due to loss of DNA repair machinery, and/or mutational bias towards adenine and thymine leading to transcription slippage (Moran et al. 2008; Tamas et al. 2008; Allen et al. 2009;

**Table 17.1** Some obligate bacterial nutritional endosymbionts of arthropods

| Host taxa[a] | Bacterial symbiont | Bacterial phylum | Est. age of association | Location of bacteriocytes | Bacterial Genome size | GC content (%) | References |
|---|---|---|---|---|---|---|---|
| **Blattodea** | | | | | | | |
| Blattidae + Mastotermes darwiniensis | Blattabacterium spp. | Bacteroidetes | >150 my | Fat body | 637 kB | 27 | Moran et al. (2008); Lopez-Sanchez et al. (2009) |
| **Phthiraptera** | | | | | | | |
| Pediculus spp. (human lice) | Riesia pediculicola | γ-Proteobacteria | 13–25 my | Stomach disc bacteriome | 575 kB | | Allen et al. (2009); Kirkness et al. (2010) |
| Pedicinus obtusus (old world monkey louse) | Puchtella pedicinophila | γ-Proteobacteria | | Midgut epithelium + ovaries | | | Fukatsu et al. (2009) |
| **Hemiptera** | | | | | | | |
| "**Auchenorrhyncha**" | Sulcia spp. | Bacteroidetes | 260–280 my | 2 Bacteriomes in haemocoel | 246–277 kB | 21–22.6 | Moran et al. (2005a); McCutcheon and Moran (2010) |
| Cicadellidae (sharpshooters) | + Baumannia cicadellinicola[b] | γ-Proteobacteria | >100 my | Same bacteriomes as Sulcia | 686 kB | 33 | Moran et al. (2008) |
| Pentastirini (Cixiidae planthoppers) | + Purcelliella pentastirinorum | γ-Proteobacteria | 25–120 my | Separate bacteriome from Sulcia | | | Bressan et al. (2009) |
| Clastoptera arizonana (spittlebug) | + Zinderia insecticola | β-Proteobacteria | | Same bacteriomes as Sulcia | 208 kB | 13.5 | McCutcheon and Moran (2010) |
| Cicadidae (cicadas) | + Hodgkinia cicadicola | α-Proteobacteria | | Same bacteriocytes as Sulcia | 144 kB | 58.4 | McCutcheon et al. (2009) |
| **Sternorrhyncha** | | | | | | | |
| Pseudococcidae (scales) | Tremblaya spp. | β-Proteobacteria | 100–200 my | 1 Bacteriome in haemocoel | 139 kB | 58.8 | Baumann (2005); McCutcheon and von Dohlen (2011) |
| | + Moranella endobia[c] | γ-Proteobacteria | | Within Tremblaya cells | 538 kB | 43.5 | McCutcheon and von Dohlen (2011) |
| Rhizoecini | Brownia rhizoecola | Bacteroidetes | | 1 Bacteriome in haemocoel | | | Gruwell et al. (2010) |
| Diaspididae (armoured scales) | Uzinura diaspidicola | Bacteroidetes | >100 my | Dispersed throughout haemocoel | | | Moran et al. (2008) |
| Psylloidea (psyllids) | Carsonella ruddii | γ-Proteobacteria | 100–250 my | 1 Bacteriome in haemocoel | 160 kB | 17 | Baumann (2005); Nakabachi et al. (2006) |

(continued)

**Table 17.1** (continued)

| Host taxa[a] | Bacterial symbiont | Bacterial phylum | Est. age of association | Location of bacteriocytes | Bacterial Genome size | GC content (%) | References |
|---|---|---|---|---|---|---|---|
| Aleyrodoidea (whiteflies) | *Portiera aleyrodidarum* | γ-Proteobacteria | 100–200 my | 2 Bacteriomes in haemocoel | | | Baumann (2005) |
| Aphidoidea (aphids) | *Buchnera aphidicola* | γ-Proteobacteria | >180 my | 1 Bacteriome in haemocoel | 416–641 kB | 20–26 | Moran et al. (2008) |
| *Cinara cedri* | + *Serratia symbiotica* | γ-Proteobacteria | | Same bacteriome as *Buchnera* | 1.76 Mb | 29 | Lamelas et al. (2011) |
| **Heteroptera** | | | | | | | |
| Cimicidae (bed bugs) | *Wolbachia* sp. | α-Proteobacteria | | 2 gonad-associated bacteriomes | ~1.3 Mb | | Hosokawa (2010) |
| *Chilacis typhae* (a lygaeid) | *Rohrkolberia cinguli* | γ-Proteobacteria | | Midgut epithelium | | | Kuechler et al. (2011) |
| *Kleidocerys resedae* (a lygaeid) | *Kleidoceria schneideri* | γ-Proteobacteria | | 1 Bacteriome near midgut | | | Kuechler et al. (2010) |
| **Coleoptera** | | | | | | | |
| Dryophthoridae + Molytinae weevils | *Nardonella* spp. | γ-Proteobacteria | >125 my | 1 Larval bacteriome surrounding fore/midgut | | | Conord et al. (2008) |
| *Sitophilus* spp. (grain weevils) | SPE | γ-Proteobacteria | ~20 my | Ditto | ~3 Mb | | (Charles et al. 1997; Conord et al. (2008) |
| *Curculio* spp. (seed weevils) | *Curculioniphilus buchneri* | γ-Proteobacteria | | 1 Bacteriome at larval midgut | | | Toju et al. (2010) |
| **Hymenoptera** | | | | | | | |
| Camponotini (carpenter ants) | *Blochmannia* spp. | γ-Proteobacteria | >50 my | Midgut epithelium | 706–792 kB | 27–29 | Moran et al. (2008) |
| **Diptera** | | | | | | | |
| *Glossina* spp. (tsetse flies) | *Wigglesworthia glossinidia* | γ-Proteobacteria | >40 my | 1 Bacteriome at gut | 698 kB | 22 | Moran et al. (2008) |

[a] Inferred taxonomic distribution. In many cases, some lineages within the taxon have subsequently lost or replaced the symbiont. The absence of a taxonomic group from this list does not necessarily indicate a lack of primary symbionts. Additional taxa (e.g. Membracidae) have primary symbionts (Buchner 1965; Douglas 1989) but have not received modern molecular attention

[b] Symbionts with a plus sign are cosymbionts with the preceding symbiont that lacks a plus sign

[c] This symbiont is contained within the *Tremblaya* cosymbiont

Moran et al. 2009). Genome shrinkage is an ongoing process (Moran et al. 2009), but evidence suggests that the rate of erosion decreases with the age of the symbiotic association (Allen et al. 2009). Second, the genomes of obligate microbial symbionts are very stable, showing no rearrangement over millions of years (Tamas et al. 2002). This stability is partially explained by a lack of mobile genetic elements (Shigenobu et al. 2000; Moran et al. 2008) which, in combination with the isolation of these symbionts, also means there is little opportunity for gene acquisition through horizontal transfer. Finally, there is a correlation between the estimated age of the host/microbe association and symbiont genome size, with microbes in the oldest associations (e.g. *Sulcia*) having the smallest and least functional genomes. In many of these ancient symbioses, additional symbionts (cosymbionts) occur within the same host; however, in some cases (e.g. *Carsonella*), essential symbiont metabolic functions have been lost without compensation by other symbionts (Baumann 2005; Moran et al. 2005b; Nakabachi et al. 2006). Understanding the continued viability of these puzzling minimalistic symbionts will likely await detailed genomic analysis of their hosts.

Concurrent with all these evolutionary changes in symbiotic microbes, host arthropods have also evolved in many ways to accommodate their inhabitants. As previously indicated, many lineages house their symbionts in host-derived membranes, within specialized cells (bacteriocytes), which in turn may be clustered into epithelial-bound bacteriomes (Table 17.1; Buchner 1965; Douglas 1989). Multiple symbionts within the same host may occupy different bacteriomes, the same bacteriome or even the same bacteriocytes (Buchner 1965). Ensuring transmission of these obligate symbionts to subsequent generations is key to the continued existence of their hosts. Because many obligate nutritional symbionts are housed in bacteriomes that are physically separated from the germline, quite complicated and variable pathways of symbiont transmission have evolved among host taxa (Buchner 1965; Douglas 1989). To date, little is known regarding the mechanistic bases for transmission.

Progress is being made in developing an understanding of the general regulation of obligate symbionts by the host, and how an immune response by the host to the symbiont is avoided. Among aphids, *Buchnera* has lost many regulatory genes and has static transcriptional dynamics (Shigenobu and Wilson 2011 and references therein). The pea aphid host, in contrast, has a greater diversity of regulatory genes than any arthropod sequenced thus far and is thought to be largely responsible for regulation of its domesticated microbe (Shigenobu and Wilson 2011). With respect to immunity, aphids have lost many immune genes and pathways that are highly conserved among animals, including other insects (Shigenobu et al. 2000). Whether such loss occurred prior to initial colonization by *Buchnera* (thus facilitating development of a symbiotic interaction) or as an evolutionary consequence of the obligate bacterial association is unclear. Regardless, the absence of a strong immune response by the aphid likely facilitates further symbiotic interactions and probably contributes to the plethora of facultative bacterial symbioses that are also present in these hosts.

In *Sitobion* weevils, a different immune dynamic is evident between host and symbiont. The antibacterial host protein coleoptericin A (ColA) is strongly expressed in symbiont-bearing tissues and apparently contributes to the characteristic bacterial gigantism of the symbionts through inhibition of cytokinesis (Login et al. 2011). Moreover, when transcription of ColA was reduced using RNAi, the bacterial symbiont was subsequently found to have escaped from the bacteriome into other larval tissues, indicating that ColA plays an important role in controlling both symbiont location and number (Login et al. 2011). Interestingly, the genus *Sitobion* has undergone a relatively recent symbiont replacement (Conord et al. 2008), but ColA may have similar effects on the ancestral weevil symbiont *Nardonella* (Login et al. 2011), suggesting that regulatory mechanisms for one symbiosis may indeed facilitate subsequent symbiotic interactions.

These examples make it clear that arthropods and their symbionts have had profound evolutionary effects upon one another. The critical role these nutritional endosymbionts play in allowing their hosts to use otherwise inadequate diets suggests that initial symbiont acquisition was an evolutionary novelty that allowed expansion into un- or under-occupied ecological niches. Thereafter, however, genetic variation between obligate symbionts likely has not been directly responsible for the radiation and diversification of their host taxa (Clark et al. 2010). The evolutionary processes experienced by these symbionts have usually consisted of stable maintenance of genomic content or genome shrinkage, rather than recombination and innovation. In fact, loss of function in endosymbionts may act as a constraint upon the host. For example, at least some of the variation between pea aphid clones in amino acid requirements can be traced to deleterious mutations in the *Buchnera* genome (Vogel and Moran 2011). Such limitations on the part of nutritional symbionts would act as one (of presumably numerous) constraints on dietary breadth of their hosts.

In general, it is probably fair to conclude that obligate nutritional endosymbionts are not an ongoing source of evolutionary innovation for their hosts. The evolutionary forces that result in genome reduction, combined with a lack of recombination, winnow the genetic complement of obligate endosymbionts down to the bare minimum, or even below (e.g. *Carsonella*; Nakabachi et al. 2006). In such instances, compensation is often provided by other endosymbionts (Moran et al. 2008) that likely originated as facultative "guests" (Buchner 1965) within the host. Facultatively mutualistic endosymbionts have different genomic properties from their obligate counterparts (see below), and it seems most plausible that symbiont-derived evolutionary innovation among hosts is associated with facultative, rather than obligate, endosymbionts.

## 17.2.2 Facultative Endosymbionts

In contrast to obligate symbionts, "facultative" symbionts are not a requisite from the host's perspective: these bacteria often do not infect every member of a host species and can be experimentally removed (through heat or antibiotic curing) without ill effects on the host. From the perspective of the bacteria, association with an arthropod host is usually obligate. Without considering pathogenic bacteria, we focus on facultative endosymbionts that maintain themselves in host populations through one of two routes: reproductive manipulation or mutualism. Bacteria in the former category are parasites that manipulate host reproduction to promote their own spread and maintenance in the host population, whereas bacteria in the latter category provide their host with fitness benefits, resulting in a selective advantage for infected hosts (Moran et al. 2008). Historically, endosymbiont taxa were considered to fall exclusively into one category or the other, but this distinction has become blurred. A number of examples now have been described, wherein "reproductive manipulators" have been found to provide fitness benefits to their hosts under some circumstances (summarized in White 2011). Nevertheless, many of the evolutionary consequences associated with facultative symbiont infection are tied to the phenotypes elicited by these symbionts, so it is appropriate to consider the broad categories of reproductive manipulators and facultative mutualists separately; it should be recognized, however, that the two categories are not mutually exclusive (Himler et al. 2011) and that any particular bacterial taxon (e.g. *Wolbachia*) might be acting as a reproductive manipulator in some hosts, but as a facultative (Weeks et al. 2007) or obligate mutualist (Hosokawa et al. 2010) in others.

Facultative endosymbionts typically have reduced genomes relative to free-living bacteria, if not as massively reduced as obligate nutritional symbionts (Toft and Anderson 2010). They have usually lost some critical functions and are incapable of resuming a host-independent lifestyle (Degnan et al. 2009; Darby et al. 2010). Additionally, while transmission of facultative endosymbionts is primarily vertical, they lack the pattern of cocladogenesis that characterizes obligate symbionts and their hosts, indicating that horizontal transfer among host taxa has occurred over evolutionary time (Russell et al. 2003; Werren et al. 2008). Indeed, the genomes of facultative symbionts are typically characterized by evidence of genetic flexibility (e.g. mobile DNA; Newton and Bordenstein 2011), suggesting that these symbionts retain much greater capacity to adapt to new host environments than obligate nutritional symbionts that are irrevocably committed to a particular host lineage.

### 17.2.2.1 Facultative Symbionts as Mutualists

Facultative mutualists tend to affect their hosts in a manner that is conditionally, rather than universally, beneficial (White 2011). This distinction is partially tautological: symbionts that confer traits that are beneficial under all environmental conditions would likely be categorized as obligate rather than as facultative, because they are probably fixed at 100 % prevalence in host populations and cause a decrease in host fitness if removed. However, evolutionary pressures to ensure vertical transmission of such beneficial associates would presumably render these associations obligate in relatively short order. In contrast, facultative mutualists that provide conditional benefits will be selected for in some environments and selected against in others. These symbionts may, therefore, persist at less than fixed levels due to balancing selection and can be large contributors to the phenotypic variability expressed by the host (White 2011). To date, the host phenotypes that have been shown to be affected by facultative symbionts include (1) defence against natural enemies, (2) interaction with host plants, and (3) environmental tolerances.

### Facultative Symbionts and Defence

In aphids, the three most common facultative symbionts are *Hamiltonella defensa*, *Regiella insecticola*, and *Serratia symbiotica*. Each is present in ~15 % of aphid species (Russell et al. 2003; Oliver et al. 2010), and all have been implicated in defence of at least one host aphid species. *Hamiltonella* and *Serratia* have both been shown to protect the pea aphid from

parasitism by braconid parasitoids in the genus *Aphidius* (Oliver et al. 2003), and *Regiella* has been shown to have the same effect in the peach-potato aphid, *Myzus persicae* (Vorburger et al. 2010). *Regiella* also defends the pea aphid against the fungus *Pandora* (Scarborough et al. 2005). Outside of aphids, some symbionts previously considered to be solely reproductive manipulators have also been shown to have defensive properties. For example, *Spiroplasma*, which is a male-killer in multiple host taxa (Anbutsu and Fukatsu 2011), protects *Drosophila neotestacea* from *Howardula* nematodes (Jaenike et al. 2010). Similarly, *Wolbachia* in some populations of *D. melanogaster* protects the host against viruses (e.g. Hedges et al. 2008). The relative prevalence of mutualistic versus manipulative strains of these symbionts remains unclear, but given the widespread occurrence of *Wolbachia*, it is certainly possible that many more arthropods benefit from defensive bacterial symbionts than previously realized.

Clearly, defensive symbionts are only beneficial in environments in which the targeted natural enemies are present. For example, in laboratory population studies of the pea aphid, *Hamiltonella*-infected aphids were selectively favoured over uninfected aphids in the presence of the parasitoid *Aphidius ervi*, but were at a disadvantage and decreased in frequency when the parasitoids were absent (Oliver et al. 2008). This suggests that there is a fitness cost associated with maintaining *Hamiltonella* infection. Furthermore, natural enemies are not necessarily just passive victims of symbiotic defensive measures, but can themselves evolve resistance to host defence (Dion et al. 2011). The selective pressures exerted on defensive symbionts in natural populations are likely to depend on the prevalence, identity, and coevolutionary history of natural enemies that are present in a particular locale. It is, therefore, little surprise that the observed frequency of symbiont infection can be highly variable among host populations (e.g. Ferrari et al. 2012).

This dynamic selective environment is matched by symbionts that apparently have a much more dynamic and versatile genetic makeup than observed in obligate nutritional symbionts. Of the facultative mutualistic symbionts whose genome has received attention thus far, most have ample mobile DNA, often exceeding the amount found in free-living bacteria (Newton and Bordenstein 2011). For *Hamiltonella* in the pea aphid, the mechanism for defence appears to be directly mediated through a bacteriophage, known as APSE phage (Moran et al. 2005a; Oliver et al. 2009). Different variants of this phage are present in different host populations (Degnan and Moran 2008) and encode different toxins (Oliver et al. 2010), which presumably have a range of effectiveness against different natural enemies. Other facultative symbionts also have phages (e.g. Darby et al. 2010), suggesting the possibility of interspecific exchange of genetic material among co-occurring facultative mutualists (Degnan and Moran 2008), which can in turn be horizontally transmitted within and among host taxa (Russell et al. 2003; Moran and Dunbar 2006). While interspecific horizontal gene transfer is rampant in free-living bacteria (Ochman et al. 2000) and may also be common among *Wolbachia* strains (Klasson et al. 2009), evidence for phage exchange among different lineages of facultative mutualists remains elusive (Degnan et al. 2010).

## Facultative Symbionts as Mediators of Host Plant Specialization

One of the earliest patterns that became evident with respect to facultative symbionts was that for symbionts of polyphagous herbivores, symbiont prevalence could vary depending upon host plant (Leonardo and Muiru 2003). In the pea aphid, *Regiella* is much more common in aphid clones that are specialized on clover, whereas other symbionts (e.g. *Hamiltonella*, *Serratia*) are less common in aphids on clover (Ferrari et al. 2012). Likewise, a recent study of facultative symbionts of the weevil *Curculio sikkimensis* has also found host-associated differences in symbiont community composition (Toju and Fukatsu 2011).

Such correlative patterns are suggestive that facultative symbionts may play a role in host plant specialization or even the generation of host races or subspecies (Tsuchida et al. 2004). However, subsequent studies that have experimentally manipulated symbiont composition indicate that the relationship between symbiont and host plant utilization may be complex. Tsuchida et al. (2004) found that curing a pea aphid clone of *Regiella* substantially decreased the aphid's performance on clover, whereas Leonardo (2004) found no effect of *Regiella* removal on the performance of multiple aphid clones. Ferrari et al. (2007) found host genotype by symbiont interactions in aphid performance on clover, whereas McLean et al. (2011) found that *Regiella* removal generally decreased aphid fitness, regardless of host plant. The balance of evidence therefore does not support a direct role for *Regiella* in host plant specialization in pea aphid (McLean et al. 2011). It remains to be seen whether facultative symbionts of other polyphagous herbivores are more directly involved in host plant utilization.

Phytophagous arthropods may also benefit from facultative symbionts that influence plant physiology. Larvae of the leafminer *Phyllonorycter blancardella* that develop in senescent apple leaves have a distinctive "green island phenotype" in which the surrounding leaf material remains photosynthetically active long after the rest of the leaf, due to a high concentration of cytokinins within the mine (Giron et al. 2007). It was recently demonstrated that this physiological effect is bacterially mediated, presumably by the endosymbiont *Wolbachia* (Kaiser et al. 2010). Endophagy is a widespread feeding habit among phytophagous insects that encourages intimate and specialized interactions between the insect and the plant. While the *Wolbachia*/*Phyllonorycter*/*Malus* interaction is currently an isolated example, it seems likely that bacterial endosymbionts might play a role in other endophagous insect/plant interactions.

## Facultative Symbionts and Environmental Tolerance

Facultative symbionts can also modify the environmental tolerances of their hosts. Once again returning to the well-documented pea aphid system, the facultative symbiont *S. symbiotica* protects infected hosts from heat shock (Russell and Moran 2006). Survival of *Serratia*-infected pea aphids was greater, following brief exposure to high temperatures than aphids uninfected by *Serratia* (Russell and Moran 2006). Correlative evidence suggests that *Rickettsia* endosymbionts in the whitefly *Bemisia tabaci* may play a similar role (Brumin et al. 2011). Facultative symbionts have also been tested for influence on frost resistance in the aphid *Sitobion avenae*, but not found to have an effect (Lukasik et al. 2011). It is conceivable that facultative symbionts may mediate many other stress responses for their hosts (e.g. toxins such as insecticides, UV, salinity), but to date, no evidence has been presented to support these possibilities.

## The Role of Facultative Endosymbionts in Host Evolution

Given the major phenotypes generated by facultative endosymbionts, and the variability in symbiont prevalence among host populations, it is perhaps unsurprising that these bacteria can drive rapid evolutionary shifts in their hosts. Himler et al. (2011) recently demonstrated that the symbiont *Rickettsia* provides major fitness benefits to the whitefly *B. tabaci*. This selective advantage provided to symbiont-bearing whiteflies drove a "symbiont sweep" through whitefly populations in the south-western US: the symbiont was virtually absent from host populations prior to the year 2000, but was near fixation in multiple populations over hundreds of miles by 2006. Similarly, Jaenike et al. (2010) have shown a geographical gradient in *Spiroplasma* infection of *D. neotestacea* across North America, suggesting that the symbiont is spreading because symbiont-bearing flies enjoy protection against invasive nematodes. In aphids, field-cage studies have demonstrated that aphid populations can evolve quickly in response to altered climate conditions, with increased frequency of symbionts that protect against heat shock (Harmon et al. 2009). Latitudinal clines in facultative symbiont prevalence (e.g. Tsuchida et al. 2002) also suggest that symbionts are involved in the climatic adaptation of their hosts. As climate change and invasive species continue to modify prevailing environmental conditions, facultative symbionts may play an important role in the resilience of their hosts.

Facultative symbionts may, on occasion, also provide traits that lead their hosts to a new evolutionary trajectory. For example, it is suggestive that sharpshooters (leafhoppers in the tribes Proconiini and Cicadellini within the family Cicadellidae), one of the very few kinds of insects to feed on xylem, have the necessary vitamins provided to them by the cosymbiont *Baumannia* (McCutcheon and Moran 2007). *Baumannia* likely began its association with an ancestral sharpshooter as a facultative symbiont in conjunction with the more ancient obligate symbiont *Sulcia*. *Sulcia*, which infects the larger Auchenorryncha clade of hemipterans, does not provide vitamins to the hosts in any lineages examined thus far (McCutcheon and Moran 2007, 2010; McCutcheon et al. 2009). It is therefore tempting to conclude that acquisition of *Baumannia* by sharpshooters facilitated a shift in host ecology and evolution. More recent transitions from facultative to obligate symbiosis have been documented in some systems (e.g. *Serratia* in *Cinara cedri*; Lamelas et al. 2011) and are suspected in others (e.g. *Hamiltonella* in a group of *Uroleucon* aphids; Degnan and Moran 2008). In the latter case, the functional basis for the obligate nature of the symbiosis remains unclear. Given that *Hamiltonella* in pea aphid has lost much of its biosynthetic machinery (Degnan et al. 2009), it will be interesting to learn what role *Hamiltonella* might be playing in this clade of aphids. The presence of phage and other mobile DNA in facultative symbionts suggests that acquisition of new traits through horizontal gene transfer remains a possibility for this versatile group of symbionts.

## 17.2.2.2 Bacteria as Reproductive Parasites of Insects and Other Arthropods

Most known reproductive parasites of arthropods are heritable, maternally transmitted intracellular bacteria that alter the reproduction of their hosts in ways that promote their own fitness. To ensure their own vertical transmission to the host progeny, reproductive manipulators have evolved mechanisms that favour a female-biased host sex ratio and are detrimental to the non-transmitting sex (the male), including thelytokous parthenogenesis, feminization, and male-killing (MK) (Table 17.2). Alternatively, by inducing cytoplasmic incompatibility, they inhibit the reproduction of uninfected or differently infected individuals and can spread without skewing the sex ratio of the host population. Such manipulations can increase the number of infected hosts within a population even where they reduce the fitness of the host (Werren and O'Neill 1997; Engelstädter and Hurst 2009).

## Diversity and Transmission of Reproductive Parasites

The ability to manipulate arthropod reproduction has evolved frequently in phylogenetically diverse bacterial taxa including *Wolbachia* and *Rickettsia* (α-Proteobacteria), *Arsenophonus* (γ-Proteobacteria), *Cardinium*, and *Flavobacterium* (Bacteroidetes), and *Spiroplasma* (Mollicutes) (Duron et al. 2008). *Wolbachia* is the most abundant endosymbiont of insects, with 66 % of species estimated to be infected (Hilgenboeker et al. 2008). Similarly, its prevalence in isopods was estimated at 47 % (Bouchon et al. 2009). In contrast, other bacteria are less pervasive (Duron et al. 2008), with *Cardinium* species, for example, being found in 6–7 % of the arthropod species screened to date (Zchori-Fein and Perlman 2004), but reaching higher prevalence in arachnids (Perlman et al. 2010). In addition, sex ratio distortion phenotypes have also been found in crustacean amphipods infected by representatives of the eukaryotic lineage of the Microsporidia (Terry et al. 2004).

Bacterial reproductive parasites have evolved sophisticated adaptations to move in the cellular environment and infect host reproductive tissues. *Wolbachia*, for example, relies on the host cell cytoskeleton and molecular motors, like dynein and kinesin-1, to move inside and between host cells (Serbus et al. 2008). Reproductive parasites in general are distributed in the host's ovary and infect the developing oocytes.

Within the egg, most symbionts localize to the germ pole, a mechanism for increasing the probability that bacteria persist in germ cells and are transmitted to host progeny (Veneti et al. 2004; Giorgini et al. 2010). However, although reproductive tissues of germline origin are the main target tissue, reproductive parasites have also been detected in different somatic tissues in many hosts (Dobson et al. 1999; Ijichi et al. 2002) where they can diversely affect the host biology. For example, a *Wolbachia* strain, *w*MelPop, proliferates massively in adult *Drosophila*'s brain, retina, and muscles, causing tissue degeneration and early death of hosts (Min and Benzer 1997). Further, *Wolbachia* infect haemocytes in isopods and have been implicated in reducing the host immunocompetence and longevity of infected individuals (Chevalier et al. 2012). From an evolutionary perspective, *Wolbachia*'s ability to infect cells of the immune system is very intriguing as regulation of the host immune system can be regarded as a strategy that reproductive parasites use to form long-term symbiotic relationships with their hosts (Siozios et al. 2008). Finally, in some *Drosophila* species, *Wolbachia* are highly abundant in the somatic stem cell niche in the germarium and from there are able to reach the germline, implying that infection of somatic stem cell niche may contribute to efficient vertical transmission (Frydman et al. 2006).

While most bacterial reproductive parasites persist intracellularly, a few exceptions are known. For example, *Arsenophonus nasoniae*, the MK agent of the parasitoid wasp *Nasonia vitripennis*, establishes a persistent intercellular infection that is maternally inherited without infecting the egg cytoplasm. Larval progeny instead acquires *A. nasoniae* by feeding. Because many wasp larvae develop in a single fly pupa, horizontal transmission can also occur between matrilines and different species of *Nasonia* (Duron et al. 2010).

Studies indicate that strictly transovarially transmitted reproductive manipulators must also occasionally be transmitted horizontally as evidenced by the incongruence between reproductive symbionts and host phylogenies. That is, closely related bacterial strains infect evolutionary distant host species, indicating that horizontal transfer between host species has occurred multiple times (Werren et al. 1995, 2009). Reproductive parasites with strong ability to infect somatic tissues and circulate in the haemolymph have been thought to be more prone to horizontal transfer (Dobson et al. 1999; Caspi-Fluger et al. 2012). At an ecological timescale, possible mechanisms of horizontal transmission in a given ecological community include predation and parasitism (Huigens et al. 2004a; Dedeine et al. 2005; Jaenike et al. 2007), plant-mediated transmission (Caspi-Fluger et al. 2012), and passage of haemolymph between infected and uninfected individuals (Rigaud and Juchault 1995). However, incongruence between symbiont and host phylogenies suggests that the interaction between symbiont and host is rarely permanent and that arthropods often lose an infection over time.

## Cytoplasmic Incompatibility

The most common reproductive manipulation is cytoplasmic incompatibility (CI). CI is induced by both *Wolbachia* and *Cardinium* and has been reported to occur in many taxa of insects, mites, and isopods (Bourtzis et al. 2003; Ros and Breeuwer 2009). CI occurs when uninfected female hosts are reproductively incompatible with infected males (unidirectional CI), while all other crosses are compatible (Table 17.2). Because uninfected females do not produce offspring in incompatible crosses, they suffer a fitness cost compared with infected females that produce viable offspring. As a result, the infection will spread in the host population. In addition, for *Wolbachia*, bidirectional CI can occur when males and females are infected by different symbiont strains (O'Neill and Karr 1990). In general, the expression of CI is the mortality of the developing embryo due to the loss of the paternal set of chromosomes, but in haplodiploid Hymenoptera incompatible eggs may also develop as normal males (Breeuwer and Werren 1990; Perrot-Minnot et al. 2002). Sometimes *Wolbachia* and *Cardinium* stably infect a common host that expresses CI, but only one of the

two is the reproductive manipulator (Ros and Breeuwer 2009; White et al. 2009).

Although not transmitted through the male germline, *Wolbachia* are present in developing sperm and are eliminated only during the final stages of sperm maturation. CI is due to sperm modification occurring during spermatogenesis (Clark et al. 2008; Serbus et al. 2008), possibly through changes in the expression of genes associated with spermatogenesis (Zheng et al. 2011a, b). However, little is known about the molecular mechanism of symbiont-induced CI. A most striking hypothesis has been postulated for *Wolbachia* and is based on a two-component "modification–rescue" model according to which symbionts induce modifications of sperm during spermatogenesis in infected males and rescue of this modification happens if the egg is infected with the same strain. If the modified sperm do not meet the appropriate symbiont in the egg, embryonic development will be arrested (Werren 1997; Poinsot et al. 2003; Bossan et al. 2011).

Evidence on the cytological mechanism of CI for *Wolbachia* (nothing is yet known about *Cardinium*) suggests that asynchronous development of male and female pronuclei caused by disruption of the cell cycle in early embryonic mitosis prevents karyogamy in incompatible crosses. While the female chromosomes separate normally during anaphase, the paternal chromosomes either fail to segregate or exhibit extensive bridging and fragmentation during segregation. This results in an embryo with a complete maternal chromosome complement but with a reduced or absent paternal chromosome complement. In compatible crosses, *Wolbachia* present in the female reproductive tissues restore coordination between the male and female pronuclei. For extensive description of cytological mechanisms of *Wolbachia*-induced CI, see Tram and Sullivan (2002), Serbus et al. (2008), and Landmann et al. (2009).

## Thelytokous Parthenogenesis

Thelytokous parthenogenesis is a form of reproduction where unmated females produce only female offspring through restoration of diploidy in unfertilized eggs (Table 19.2).

**Table 17.2** Reproductive manipulations induced by microbial symbionts

| Reproductive manipulations | Micro-organisms involved | Type of reproduction | Offspring produced |
|---|---|---|---|
| Cytoplasmic incompatibility | *Wolbachia* *Cardinium* | Compatible crosses: | |
| | | i♀ × u♂[a] | i♀ + i♂ |
| | | u♀ × u♂ | u♀ + u♂ |
| | | i♀ × i♂ | i♀ + i♂ |
| | | Incompatible crosses: | No offspring or i♂ in haplodiploids |
| | | u♀ × i♂ (unidirectional) | |
| | | $i_a$♀ × $i_b$♂ (bidirectional)[b] $i_b$♀ × $i_a$♂ | |
| Male-killing | Many bacteria | Biparental reproduction: | |
| | Microsporidia an RNA virus | i♀ × u♂ | i♀ |
| Parthenogenesis in haplodiploid arthropods | *Wolbachia* | Thelytokous reproduction: | |
| | *Rickettsia* | i♀ (2n) → infected eggs (n) → i♀ embryos (2n)[c] | i♀ (2n) |
| | *Cardinium* | | |
| Feminization in diploid arthropods | *Wolbachia* | Biparental reproduction: | |
| | *Microsporidia* | i♀ ZW × u♂ ZZ | i♀ ZW + i♀ ZZ |
| | | i♀ ZZ × u♂ ZZ | i♀ ZZ |
| Feminization in haplodiploid arthropods | *Cardinium* | Thelytokous reproduction: | |
| | | *Brevipalpus phoenicis mites* | |
| | | i♀(n) → i♂ eggs (n) → i♀ embryo (n) | i♀ (n) |
| | | *Encarsia hispida wasps* | |
| | | i♀ (2n) → i♂ eggs (2n) → i♀ embryos (2n) | i♀ (2n) |

[a] i♀ and u♀, and i♂ and u♂ are infected and uninfected female, and infected and uninfected male, respectively
[b] $i_a$ and $i_b$ means infected by two different bacterial strains
[c] (n) and (2n) mean haploid and diploid, respectively

In haplodiploid arthropods, thelytokous reproduction is common and has evolved independently in many different lineages. However, in most cases, thelytoky is associated with the occurrence of an endosymbiotic micro-organism (Stouthamer 1997). Feeding females antibiotics restores the production of males in a number of thelytokous species, showing that micro-organisms are the causal agents of the reproductive phenotype. At least three intracellular bacteria, including *Cardinium*, *Rickettsia* and *Wolbachia*, induce thelytoky, especially in the Hymenoptera (Huigens and Stouthamer 2003; Hunter and Zchori-Fein 2006; Giorgini et al. 2010). Parthenogenesis-inducing (PI) *Wolbachia* and *Cardinium* also occur in other haplodiploid arthropods, such as mites, scale insects, and thrips.

In hymenopterans infected by PI-*Wolbachia*, thelytokous parthenogenesis is automictic and occurs by disruption of the cell cycle during early embryogenesis, followed by gamete duplication (Gottlieb et al. 2002; Pannebakker et al. 2004). The haploid nuclei fail to separate and result in a single diploid nucleus containing two identical sets of chromosomes. As this mechanism produces completely homozygous females, it should result in strong inbreeding depression and should not be expected in outbreeding species like most diploid organisms (Stouthamer 1997). Indeed, gamete duplication has only been found in hymenopteran species that tolerate high rates of inbreeding. However, in the parasitoid wasp *Neochrysocharis formosa*, *Rickettsia*-induced parthenogenesis occurs by an apomictic cloning mechanism with the absence of meiotic recombination and reduction, and final development of heterozygous females (Adachi-Hagimori et al. 2008). A functionally apomictic parthenogenesis is also induced by *Wolbachia* in the mite *Bryobia pretiosa* (Weeks and Breeuwer 2001). Apomixis is the most common form of parthenogenesis within diplo-diploid arthropods (Suomalainen et al. 1987) and occurs in some uninfected Hymenoptera (Vavre et al. 2004). Parthenogenesis mechanisms that maintain heterozygosity keep open the chance that endosymbiotic bacteria could be involved in the evolution of thelytokous reproduction in outbreeding species as well (Adachi-Hagimori et al. 2008; Rodriguero et al. 2010).

## Feminization

Feminization is the development of genetic males into functional females. A well-known example occurs in isopods with female heterogametic sex determination, where *Wolbachia*-infected ZZ males are morphologically, anatomically, and functionally identical to ZW females (Table 17.2). In ZZ males, *Wolbachia* inhibits androgenic gland differentiation and the synthesis of the androgenic hormone, which promotes the differentiation of male gonads and secondary characters (Bouchon et al. 2009). Similarly, microsporidia induce feminization in amphipods. In insects, however, sexual differentiation is not under hormonal control; consequently, for the full expression of feminization, symbionts have to infect all somatic cells and interact with the genes involved in sex determination. In diploid insects, feminizing *Wolbachia* is known in a ZZ/ZW butterfly species (Narita et al. 2007) and in a leafhopper with XX/X0 sex determination system (Negri et al. 2006). Within haplodiploid arthropods, feminization has been reported only for *Cardinium*. It induces obligate thelytokous reproduction in two host species using different mechanisms (Table 17.2). In the mite *Brevipalpus*, *Cardinium* feminizes unfertilized haploid eggs that develop into functional haploid females (Weeks et al. 2001). In the parasitoid wasp *Encarsia hispida*, diploid males are the by-product of diploidy restoration in unfertilized eggs and *Cardinium* is required to feminize diploid male embryos and guarantee female offspring production (Giorgini et al. 2009).

Because thelytokous females can produce progeny without males, PI and feminizing endosymbionts can reach fixation in haplodiploid species without causing population extinction (Huigens and Stouthamer 2003; Giorgini et al. 2009). In some exceptional cases, infected thelytokous females retain the ability to mate and produce infected daughters from both fertilized and unfertilized eggs. Consequently, infected females can coexist in the field with individuals of uninfected bisexual populations, as in *Wolbachia*-infected populations of *Trichogramma* wasps (Stouthamer et al. 2001). However, most natural parthenogenetic populations have lost the ability to reproduce sexually, and reproduction relies on infection by endosymbiotic bacteria that have now become obligate symbionts (Huigens and Stouthamer 2003; Russell and Stouthamer 2011).

## Male-Killing

MK endosymbionts selectively kill male offspring of their arthropod hosts (Table 17.2). A diversity of male-killers, from the bacterial genera *Wolbachia*, *Rickettsia*, *Spiroplasma*, and *Arsenophonus*, and undescribed Flavobacteria and $\gamma$-Proteobacteria has been reported in many insect orders and in pseudoscorpions (Hurst et al. 2003; Zeh and Zeh 2006; Majerus and Majerus 2010). Infectious male-killers in insects also include microsporidia and an RNA virus (Hurst et al. 2003; Nakanishi et al. 2008). The MK phenotype, because it favours the transmitting female sex, is not selected against in the bacteria and becomes an advantageous trait for the symbionts if female offspring benefit from the death of their brothers. Infected females gain an advantage over uninfected females through fitness compensation originating from reduced competition between siblings, resource reallocation obtained through the consumption of dead males, or reduced rates of inbreeding. In general, species that lay eggs in clutches, exhibit cannibalism behaviour or aggregated distributions in breeding sites and use temporary resources may be particularly susceptible to invasion by male-killers (Jaenike et al. 2003; Majerus 2003). Ladybird beetles are a classical example. MK bacteria have to interact with components of the sex determination system of their hosts to express selectively their phenotype in the two sexes, but the mechanism is not fully understood (Bentley et al. 2007). As there is a diversity of MK agents, different mechanisms are expected (Veneti et al. 2005; Ferree et al. 2008; Riparbelli et al. 2012).

All of the aforementioned forms of reproductive manipulation depend on bacterial density within the reproductive tissues of the host. Efficiency of symbiont transmission through the host germline, penetrance of the reproductive phenotype, and infection prevalence in the host population are all strictly correlated with bacterial density (Jaenike 2009). Bacterial density is regulated by genetic factors of the host and the symbiont itself and is strongly influenced by environmental factors, like temperature, antibiotics, and host age (Jaenike 2009; Bordenstein and Bordenstein 2011). The general variation in bacterial density in response to temperatures indicates that there can be large spatial, temporal, and seasonal differences in endosymbiont densities and functions in natural populations.

## Interactions Between Reproductive Manipulators and the Host Immune System

To establish successful symbiotic associations with diverse hosts and be able to infect both reproductive and somatic tissues, reproductive parasites must cope with the immune system of their hosts (Siozios et al. 2008), but the mechanisms that endosymbionts use to escape the cellular and humoral host defences are still unclear. Regulation of the host immune system can be regarded as a strategy that reproductive parasites use to form long-term symbiotic relationships with their hosts. *Wolbachia* up-regulation of the host immune genes leading to symbiont-mediated protection against pathogens or predators (Brennan et al. 2008; Moreira et al. 2009; Kambris et al. 2010) may be an effective way by which vertically transmitted symbionts

can invade a host population, possibly explaining the high prevalence of weak reproductive parasites in field populations (Brownlie and Johnson 2009). To date, however, the hypothesis that *Wolbachia* interferes with pathogens by preactivating the immune response of its host is based only on studies of immune genes expression in transinfected hosts (naturally uninfected hosts infected by *Wolbachia* in the laboratory). In contrast, no differences in the up-regulation of immune genes have been found between hosts naturally infected by *Wolbachia* and uninfected insects with identical genetic background (Rancés et al. 2012; Wong et al. 2011). It has been found, however, that in the case of viral pathogens, *Wolbachia* reduces virus replication in both naturally infected and transinfected hosts, suggesting that immune priming by *Wolbachia* might not be the only mechanism responsible for viral interference (Rancés et al. 2012).

*Wolbachia* can also reduce the immunocompetence of hosts by reducing the efficiency of the cellular immune response (for example, preventing the encapsulation of parasitoid wasp eggs; Fytrou et al. 2006) and by down-regulating immune genes (Chevalier et al. 2012), leading to a reduced lifespan of infected individuals (Braquart-Varnier et al. 2008; Sicard et al. 2010). Stable infections of such costly symbionts, like feminizing *Wolbachia* in isopods, can be maintained in natural populations as a by-product of the genomic conflict between symbionts and their hosts. For example, in natural populations of *Armadillidium vulgare*, the frequencies of infected feminized individuals are generally lower than what would be predicted based on feminizing effects alone, possibly due to the lower fitness of immunodepressed feminized individuals (Braquart-Varnier et al. 2008).

Apoptosis of infected cells is an effective immune barrier that intracellular bacteria have to overcome in order to survive and to establish stable associations with host tissues (Batut et al. 2004). To this end, *Wolbachia* shows antiapoptotic pathways which, in the parasitoid wasp *Asobara tabida,* appear to have also a positive impact on host oogenesis by regulating the apoptosis of nurse cells (Dedeine et al. 2001;

Pannebakker et al. 2007). It is reasonable to assume that, at least in *Wolbachia*, the immunomodulating function of antiapoptotic factors may significantly contribute to the regulation of host reproduction, thus driving the evolutionary shift from facultative parasitism towards obligate mutualism (Miller et al. 2010).

## Evolution of Host Resistance Genes, Sex Determination Mechanisms, and Genetic Systems

Reproductive parasites, and maternally inherited symbionts in general, have conflicting interests with their hosts. Within infected host populations, microbial genes are selected to favour a female-biased host sex ratio, which increases the prevalence of the symbionts, whereas host genes, generally biparentally inherited, are selected to prevent the action of the symbionts and restore an unbiased sex ratio (Werren and Beukeboom 1998; Caubet et al. 2000). As a consequence of the genetic conflict occurring between microbial genes and host nuclear genes, changes in the host sex determination system may evolve or resistance genes that prevent transmission of the symbiont to host germline or suppress the symbiont activity can be selected.

Occurrence of genetic conflict in response to feminizing *Wolbachia* has been observed in the isopod female heterogametic *A. vulgare* (reviewed by Bouchon et al. 2009). In this species, genetic ZZ males are converted to phenotypic functional females and the female-determining W chromosome is lost in the infected populations. However, a polygenic system of resistance genes involved in reducing the symbiont transmission rate compensates for the absence of males in infected populations. Furthermore, there are uninfected populations with ZZ individuals reversed to females by a feminizing (*f*) element thought to be a mobile genetic element acquired by the host nuclear genome via lateral transfer from *Wolbachia*. The *f* element can also be stabilized on a Z male chromosome, originating a new W-like chromosome. An autosomal dominant masculinizing (*M*) gene, which restores maleness in the presence of the *f* element but is ineffective for

feminizing *Wolbachia*, has been found in some populations and interpreted as an effect of the genomic conflict between the selfish *f* element and the host genome. The autosome carrying the *M* gene behaves as a new sex chromosome originating a male heterogametic system of sex determination. These findings have suggested a dynamic evolution of sex determination in *A. vulgare* driven by *Wolbachia* infections and by the occurrence of intragenomic conflicts between different sex ratio distorters and the autosomal genes that promote the selection of new autosomal masculinizing genes. This would explain the low or null morphological differentiation of sex chromosomes in isopods and the occurrence of female and male heterogametic systems in closely related species of isopods.

In a different example, the *Wolbachia* strain *wSca* manipulates the sex of the moth *Ostrinia scapulalis* by interfering with the sex-specific splicing of *Osdsx* gene (Sugimoto and Ishikawa 2012), a homologue of *doublesex* (*dsx*) working at the bottom of the sex determination cascade, which is transcribed into either a male or female isoform by sex-specific splicing and regulates the sex-specific gene expression in somatic cells of insects (Gempe and Beye 2011). *wSca* causes feminization of ZZ genetic males early in development (infected male embryos express the female-type $Osdsx^{FL}$) and subsequently kills the same individuals. However, the male-type $Osdsx^{M}$ is expressed in all individuals cured from infection irrespective of the genetic sex. This indicates that elimination of *wSca* causes the masculinization of ZW females, and consequently, a factor in the female-determining cascade is degraded in *wSca*-infected hosts (Sugimoto and Ishikawa 2012).

Genes that suppress MK have been identified in some insects (Majerus and Majerus 2010), and their spread can occur very quickly in the field, for example, taking only few generations to change the 99 % female sex ratio of some infected populations of the butterfly *Hypolimnas bolina* to a sex ratio near parity (Charlat et al. 2007a).

In the case of CI bacteria, being infected is beneficial for females as their eggs are saved from the deleterious effect of CI and is detrimental for males as they suffer a fitness cost in mating with uninfected females (Turelli 1994; Snook et al. 2000). Even if selection would favour infection to spread to fixation, it is expected that uninfected individuals are always produced due to inefficiency of endosymbiont transmission. So, nuclear genes reducing levels of CI can be selected for. CI-*Wolbachia* can produce a physiological cost on infected males by significantly reducing the production of sperm (Snook et al. 2000; Lewis et al. 2011); thus, it is expected that endosymbionts can promote evolutionary changes in the functioning of the male germline. Host resistance genes that prevent the entry of *Wolbachia* into testes have been suspected (Poinsot et al. 1998).

In addition to their role in the evolution of sex determination systems, bacterial endosymbionts have been hypothesized to be a driving factor in the evolution of genetic systems (Ross et al. 2010). In particular, haplodiploidy could have originated in diplodiploid arthropods, following the spread of MK endosymbionts that caused the destruction of the paternal chromosome set in diploid males (Normark 2004). Under such circumstances, coevolutionary responses by the host would be predicted and genes that save viability and fertility of haploid males can be selected. If this should be the case, models predict the evolution of a paternal genome elimination-based haplodiploid system if haplodizing endosymbionts become beneficial for female hosts and the infection fixed (Kuijper and Pen 2010).

In the thelytokous parasitoid wasp *E. hispida*, *Cardinium* is required to feminize diploid male embryos and thus must interact with elements of the host sex determination system (Giorgini et al. 2009). Diploid males are produced by antibiotic-fed females. These findings suggest a possible route for the collapse of haplodiploidy into a diplodiploid genetic system. Hosts may contribute to or take over the process of asexual diploidy restoration from symbionts if this reduces mortality of parthenogenetic daughters. Reversion to diplodiploidy from haplodiploidy is quite rare, but in one of two examples of scale insects highlighted by Normark (2004),

symbionts appear to play a role. Although relatives are haplodiploid, Buchner (1965) noted that female embryos in the family Stictococcidae are diploid and infected with a bacterium, whereas males are also diploid, but free of bacteria.

## Host Population Genetics

Vertically inherited reproductive parasites influence the evolutionary dynamics of host population genetics dramatically. Symbionts and mitochondria are simultaneously inherited through the egg cytoplasm, and because infected individuals have a reproductive advantage over the uninfected ones, the spread of a reproductive parasite will sweep from the infected host populations the mitochondrial haplotypes not associated with infection. The final result will be the reduction in host mtDNA diversity (Johnstone and Hurst 1996). Less frequently, reproductive parasites can also alter the frequency of host nuclear genes. Theoretical models suggest that infections with early MK bacteria impede the spread of beneficial alleles, facilitate the spread of deleterious alleles, and reduce nuclear genetic variation in infected host populations. The reason for this is the strongly reduced fitness of infected females combined with no or very limited gene flow from infected females to uninfected individuals. Most mutations originating in infected individuals are therefore lost, and the effective population size for nuclear genes is reduced almost to the number of uninfected individuals. The impact of reproductive parasites on host population genetics is reviewed by Engelstädter and Hurst (2009).

## Reproductive Manipulators as Drivers of Host Reproductive Isolation and Speciation

One effect of CI is the reproductive isolation between differently infected hosts. As a result, CI-inducing endosymbionts could have a role in driving speciation processes in their hosts (Werren 1998; Bordenstein 2003; Telschow et al. 2007). However, because the penetrance of CI is frequently incomplete, vertical transmission of symbionts is not always perfect, and gene flow can occur in compatible cross-directions, it is unlikely that CI alone drives speciation. Instead, theoretical and empirical works on *Wolbachia*-induced CI suggest a complementary role in species formation along with other genetic and/or geographical mechanisms that restrict gene flow between diverging populations (Telschow et al. 2005). For example, between two closely related parasitoid wasps, *N. vitripennis* and *N. giraulti*, reproductive isolation is maintained both by complete bidirectional CI and by nuclear incompatibilities, leading to hybrid inviability and hybrid sterility (Breeuwer and Werren 1995). However, bidirectional CI was found to be the principal contributor to reproductive isolation between the sibling species *N. giraulti* and *N. longicornis*, each fixed for infection by a specific *Wolbachia* strain; here, *Wolbachia*-induced reproductive isolation has occurred in the early stages of speciation, because other postmating isolating mechanisms, like hybrid inviability and hybrid sterility, are still not present (Bordenstein et al. 2001).

CI causing bacteria can also promote speciation by promoting premating isolation (Telschow et al. 2005). For example, asymmetrical reinforcement has been observed in the field in uninfected *Drosophila. subquinaria* as a consequence of secondary contact with *D. recens*, which is infected near fixation (98 % infection prevalence) with *Wolbachia* causing strong intraspecific and interspecific CI. However, hybrid inviability is not manifested in matings between infected *D. recens* females and uninfected *D. subquinaria* males. Females of *D. subquinaria* from the zone of sympatry exhibit stronger levels of mate discrimination against *D. recens* males than do females from allopatric populations. Furthermore, there was substantial behavioural isolation within *D. subquinaria*, because females sympatric with *D. recens* discriminate against allopatric conspecific males, whereas females allopatric with *D. recens* show no discrimination against any conspecific males. These findings show that interspecific CI may contribute not only to postmating isolation but also to reinforcement, particularly in the uninfected species. The resulting reproductive character displacement not only increases behavioural

isolation from the *Wolbachia*-infected species, but may also lead to behavioural isolation between populations of the uninfected species (Jaenike et al. 2006).

Coevolution of reproductive parasites with their host towards a mutualistic association may also play a role in diversifying and separating host populations and eventually driving speciation (Miller et al. 2010). As an example, *Wolbachia* has been implicated in driving sexual isolation between six semispecies of *D. paulistorum* that occur sympatrically in Middle and South America. Each semispecies harbours a specific *Wolbachia* strain that provides a fitness benefit to its host, being essential for oogenesis and development. *Wolbachia* are ancestrally fixed, obligate mutualists of all *D. paulistorum* semispecies, perfectly transmitted by the mother and causing strong bidirectional CI and hybrid male sterility in the laboratory. In nature, however, incompatible matings between semispecies are avoided by female mating choice and courtship behaviour. In their native *D. paulistorum* hosts, *Wolbachia* manipulate sexual behaviour by triggering premating isolation via selective mate avoidance, that is, avoiding mates harbouring another, incompatible symbiont variant. It was assumed that symbiont-directed mate recognition could have evolved in order to prevent strong bidirectional CI and reduced sexual success of potential hybrids, thereby ensuring continuing vertical transmission of the symbiont (Miller et al. 2010).

In asexual populations of haplodiploid arthropods, continuous thelytokous reproduction caused by PI or feminizing symbionts can lead to degradation of genes involved in sexual reproduction, for example, in genes involved in male mating behaviour and fertility or encoding female sexual traits, because these gene are not maintained by selection anymore (Pijls et al. 1996; Arakaki et al. 2000; Gottlieb and Zchori-Fein 2001). Furthermore, such mutations may be selected for if they improve the fitness of asexual females. For example, degradation of costly genes involved in female behaviour or sperm usage could reallocate resources in favour of oogenesis or other fitness traits. As infection by a PI symbiont spreads, degradation of sexual traits would accumulate, leading to prezygotic isolation between infected asexual populations and uninfected sexual ones (Pannebaker et al. 2005). Sexual degradation will make thelytokous reproduction irreversible in infected populations even if the symbiont is lost, eventually resulting in a speciation event (Bordenstein 2003; Adachi-Hagimori et al. 2011). Consequently, if the symbiont is lost and the host does not come up with an alternative mechanism, this will result in extinction of the infected host.

## Gene Acquisition from Reproductive Parasites

Reproductive parasites have also been a source of new genes for hosts via lateral transfer (Werren et al. 2008). It has been found that one-third of sequenced invertebrate genomes contain *Wolbachia* gene insertions and that 70 % of *Wolbachia*-infected arthropod and nematode hosts might have a nuclear insert (Dunning Hotopp 2011). The largest lateral transfer has been found in *D. ananassae* where almost the entire genome of *Wolbachia* ($\sim 1.4$ Mb) has been integrated into an insect chromosome. However, although some inserted *Wolbachia* genes are transcribed, their biological functions are still unknown. Recently, it was found that the genome of *N. vitripennis* encodes 13 ankyrin repeat proteins with a C-terminal domain (PRANC), and these proteins are found in diverse *Wolbachia* strains (Werren et al. 2010). Phylogenetic analysis of the PRANC domain reveals that *Nasonia* wasps acquired one or more of these proteins from *Wolbachia* with subsequent gene duplication and divergence. Most of the genes are transcribed in both males and females and in different life stages, suggesting that in some cases, lateral gene transfer can be an effective source of new functional genes.

## The Role of Reproductive Parasites in Altering Host Behaviour

Biases in the sex ratio of a population are expected to alter which sex competes for mates (Emlen and Oring 1977). Reproductive parasites that skew sex ratio towards females and then decrease the frequency of males in a population are expected to reduce both the intensity of male–male competition and the opportunity for female choice between males. As a consequence, alterations of the mating system and reproductive strategy should occur in favour of female–female competition and male choice (Charlat et al. 2003). As an example, a sex-role reversal has been reported in some populations of the butterfly *Acraea encedon* characterized by high frequency of MK-*Wolbachia* infection and female-biased sex ratio (Jiggins et al. 2000). However, in female-biased populations of the butterfly *H. bolina* infected by a MK-*Wolbachia*, contrary to expectation, female mating frequency increases rather than decreases along with infection prevalence, until male mating capacity becomes limiting (Charlat et al. 2007b). This increasing female promiscuity has been explained as a facultative response to the increasing fatigue and reduced mating resource of males, which produce smaller spermatophores as mating frequency increases. Reduced investment (sperm transfer) by males when paired with infected individuals, potentially leading to variation in host mate preferences, has also been found in crustaceans infected by feminizing *Wolbachia* (Rigaud and Moreau 2004) or microsporidia (Dunn et al. 2006). This is advantageous for males as they are severely sperm limited, and feminized males have lower fecundity than uninfected females.

Reproductive behaviours can also evolve in arthropods infected by reproductive parasites to limit the spread of costly infections. For example, some infections have been found to negatively influence host body size, fecundity, survival, larval competitiveness, male fertility, and sperm production (Snook et al. 2000; Huigens et al. 2004b; Rigaud and Moreau 2004). In the mite *Tetranychus urticae*, *Wolbachia*-associated unidirectional CI can be avoided by females at the premating level through both precopulatory and ovipositional behaviours that increase chances of successful compatible matings; infected females aggregate their offspring, thereby promoting sib mating, while uninfected females preferably mate with uninfected males and, in doing so, directly reduce opportunities for CI expression (Vala et al. 2004). In *D. melanogaster*, *Wolbachia* plays a role in mate discrimination between infected and uninfected populations with identical genetic background and it has been suggested that *Wolbachia* might have evolved the capacity to modulate host pheromone expression and/or perception (Koukou et al. 2006). In a different study, however, neither male nor female *D. melanogaster* nor *D. simulans* exhibit significant *Wolbachia*-associated precopulatory mate preferences (Champion de Crespigny and Wedell 2007).

CI-*Wolbachia* infection in *D. simulans* negatively affects sperm competition in infected males, suggesting that polyandrous females can utilize differential sperm competitive ability to bias the paternity of the progeny and reduce the penetrance of reproductive manipulators (Champion de Crespigny and Wedell 2006). Reduced success in sperm competition associated with infection of CI-*Wolbachia* in *Drosophila* could play a role in the evolution of host reproductive strategies, like the selection for polyandry in species with CI-inducing endosymbionts, to avoid the fitness cost associated with infections. However, this hypothesis has not been supported by theoretical models (Champion de Crespigny et al. 2008).

PI-bacteria can change the female's host selection behaviour to successfully invade a host population. For example, PI-*Cardinium* manipulates the oviposition choice of its parasitoid host *Encarsia pergandiella*, causing the female wasp to lay unfertilized infected eggs into hosts that are competent for female but not male development (Kenyon and Hunter 2007).

## 17.3 Viruses as Beneficial Symbionts of Insects

In contrast to bacteria, which are well recognized to form beneficial symbiotic associations, viruses are almost always considered parasites whose life cycles reduce host fitness while benefiting their own (Villarreal 2005; Moreira and Lopez-Garcia 2009). In the case of insects

and other arthropods, nearly all studies focus on the role of viruses in causing severe or chronic disease (Bonning 2005). However, recent studies reveal that some viruses have evolved to become obligate beneficial symbionts of parasitoid wasps. The most elegant example of this is for all members of the *Polydnaviridae*, which are exclusively beneficial symbionts of parasitoid wasps. However, selected poxviruses, ascoviruses, and phages that infect bacterial symbionts of insects also exhibit features that suggest they too have evolved into obligate or facultative beneficial symbionts.

## 17.3.1 Polydnaviruses as Beneficial Symbionts

By far the best example of viruses evolving into beneficial symbionts is the family *Polydnaviridae* (reviewed in Webb and Strand 2005; Strand 2010). These large, double-stranded (ds) DNA viruses are exclusively associated with approximately 40,000 species of parasitoid wasps in the families Braconidae and Ichneumonidae and are divided into two genera called the *Bracovirus* (BV) and Ichnovirus (IV). Most polydnavirus (PDV)-carrying wasps parasitize larval-stage hosts in the order Lepidoptera (moths and butterflies). Each wasp species carries a genetically unique PDV that exists in two forms. The proviral form is integrated into the genome of every cell in wasps of both sexes, and transmission to offspring is strictly vertical through the germline (Fig. 17.1). The encapsidated form of the genome that is packaged into virions consists of multiple, circular dsDNAs, which have aggregate sizes that range from 190 to more than 600 kbp. This makes PDVs the only known dsDNA viruses with multipartite genomes and also underlies the naming of the family. PDVs only replicate in pupal- and adult-stage female wasps in specialized cells that form a region of the ovary called the calyx (Fig. 17.1). Virions accumulate to high density in the lumen of the calyx to form "calyx fluid", and females inject a quantity of calyx fluid together with eggs into each host they parasitize. Virions rapidly infect

**Fig. 17.1** Polydnavirus life cycle. The infection of host tissues and expression of virulence factors largely contribute to the disruption of the immune response and the alteration of development and reproduction in parasitized hosts

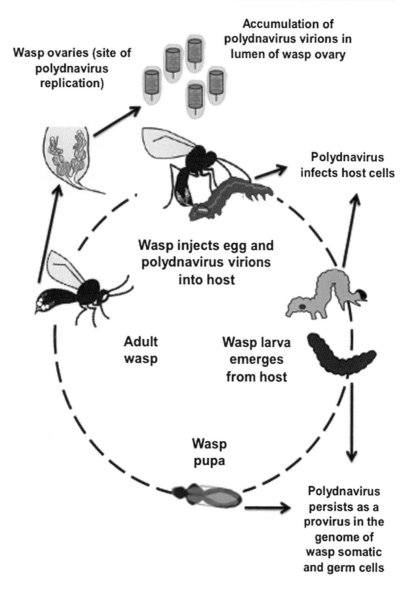

Wasp ovaries (site of polydnavirus replication)

Accumulation of polydnavirus virions in lumen of wasp ovary

Polydnavirus infects host cells

Wasp injects egg and polydnavirus virions into host

Adult wasp

Wasp larva emerges from host

Wasp pupa

Polydnavirus persists as a provirus in the genome of wasp somatic and germ cells

host cells, but they never replicate. They do, however, express a number of genes whose products alter host physiology in ways that are essential for survival of wasp offspring (Fig. 17.1). PDVs are thus beneficial symbionts because they fully depend on wasps for persistence, while wasps fully depend on the virulence genes that replication-defective PDV virions deliver to caterpillar hosts.

### 17.3.1.1 BVs and IVs Have Different Evolutionary Origins

BVs and IVs were originally placed into one family because of their multipartite dsDNA genomes, similar life cycles, and strict association with parasitoid wasps. However, three lines of evidence indicate that BVs and IVs have different evolutionary origins. First, phylogenetic studies indicate that BV-carrying braconids form a monophyletic assemblage called the micro-gastroid complex. This group evolved an estimated 100 mya and is relatively distant from the taxon of Ichneumonidae that carry IVs (Whitfield 2002; Murphy et al. 2008). Since no PDVs are known from a common ancestor of these wasp lineages, these data strongly suggest the association of BVs with braconids and IVs with ichneumonids arose independently. Second, partial or complete sequencing of 11 encapsidated genomes (6 BVs, 5 IVs) reveal that BVs from different wasps encode several homologous genes as do IVs. However, BVs share few genes with IVs. Third, morphological studies show that BV virions have barrel-shaped capsids surrounded by one envelope, while IV virions have fusiform capsids and two envelopes. Although each assembles in the nuclei of calyx cells, BV virions are released by cell lysis, while IV virions bud through the plasma membrane. Overall, these findings suggest that BVs evolved from a virus that interacted with the common ancestor of microgastroids. Whether the IV-ichneumonid association also arose from a single virus and wasp ancestor remains unclear, but these progenitors differ from the ancestors of BV-carrying braconids. Finally, the traits that BVs and IVs share most likely arose through convergent evolution and the similar roles each plays in parasitism.

BV and IV virions unquestionably look like viruses, but their encapsidated genomes yielded the surprising finding that no viral genes with predicted roles in genome replication, transcription, or virion formation are present (Burke and Strand 2012; Drezen et al. 2012). The encapsidated genomes of PDVs also have eukaryotic architectural features that include low coding densities and many intron-containing genes. These findings explain why PDVs do not replicate in the hosts that wasps parasitize, but also raise the spectre that PDVs are not of viral origin but products of wasp genes that generate virus-like structures. Recent studies from three braconids, however, provide strong evidence that BVs evolved from a viral ancestor (Bezier et al. 2009; Wetterwald et al. 2010). Specifically, analysis of transcripts expressed in ovaries

during BV replication identified homologues in each wasp that are core genes of nudiviruses. Nudiviruses themselves are relatively poorly studied, but they are a sister taxon of the *Baculoviridae*, which are well-known pathogens of insects (Wang and Jehle 2009; Jehle 2010).

Nudiviruses and baculoviruses infect insects or other arthropods, have large circular dsDNA genomes (>100 kbp), and like BVs produce single-enveloped virions. Most baculoviruses are virulent pathogens whose life cycles are distinguished by the coordinated expression of several core genes, high-level replication of the genome in host cell nuclei, and the release of large numbers of virions by cell lysis. Many nudiviruses also establish lytic infections, but some selectively infect the reproductive organs of their insect host (HzNV-2) and establish persistent infections (HzNV-1) that are characterized by a shutdown of most genes expressed during a productive infection and maintenance of the genome as both an episome and an integrated provirus. One route for the evolution of BVs, therefore, is that a nudivirus established a latent infection in the reproductive tract of the ancestral microgastroid. Studies of one ichneumonid also identify features that suggest a viral origin for IVs. However, a lack of homology with genes from other known viruses suggests IVs either evolved from an undiscovered or extinct virus group or have diverged, so greatly that it is not possible to detect sequence similarities with other viruses (Volkoff et al. 2010).

### 17.3.1.2 Roles of Polydnaviruses in Parasitism of Hosts

PDVs are essential for wasp survival because they deliver virulence genes to hosts that have two broad functions: (1) they prevent the host's immune system from killing wasp offspring and (2) they alter the growth, development, and metabolism of hosts in ways that facilitate wasp development while leading to host death. PDV-carrying wasps are highly specialized organisms with each species parasitizing only one or a few species of host insects. The immunological and developmental interactions between wasps and hosts also differ, with wasps from distantly related taxa generally exhibiting greater differences than wasps from the same or closely related genera. The encapsidated genomes of PDVs are thus likely to change in response to the selective pressures that act on each wasp in its coevolutionary interactions with hosts. Strict Mendelian inheritance, however, would also suggest that PDV isolates from wasps in distantly related taxa are likely to differ more than isolates from closely related wasps. Although relatively few isolates have been sequenced, current data strongly support these predictions by showing that aggregate genome size, genomic segment numbers, and gene content are most similar among isolates from closely related wasps (Webb and Strand 2005; Pennacchio and Strand 2006; Strand 2010). The encapsidated genomes of BVs from distantly related braconids in the subfamilies Microgastrinae and Cheloninae in contrast share no genes even though their proviral genomes encode the same nudivirus-like core gene set. The encapsidated genomes of BVs and IVs also, as expected, largely encode different genes.

Approximately half of the genes in PDV encapsidated genomes form multimember families that often reside on multiple genomic segments (Strand 2010). Several of these families show signatures of evolving by duplication from a single ancestral gene (Huguet et al. 2012). Several families also show evidence of evolving under diversifying or positive selection in response to alterations in a given host or shifts in the host range of a given wasp. The ankyrin repeat (*ank*) and protein tyrosine phosphatase (*ptp*) families of BVs, and *cys-motif* family of IVs are present in isolates from diverse taxa, suggesting that each have ancient origins and produce products of broad importance in parasitism. Other gene families are only known in isolates from a particular taxon of wasps, which suggests they were acquired more recently and have more specialized roles in parasitism.

Most of the genes in the encapsidated genomes of PDVs are homologues of known genes from eukaryotes, which in turn suggests that many were acquired from wasps or the hosts they parasitize (Burke and Strand 2012; Drezen et al. 2012). This is clearly the case for some genes like a family of predicted sugar transporters present in BVs from the wasp genus *Glyptapanteles*. However, the high divergence rates of more conserved gene families like *anks* and *ptps* make it impossible to discern whether the ancestral gene was acquired from an insect or another eukaryote. Recent studies also suggest that the *ank* gene family present in IVs was acquired by horizontal gene transfer (HGT) from a BV. Lastly, a few genes in the encapsidated genomes of PDVs are not of eukaryotic origin and have also been acquired by HGT through unknown mechanisms.

Understanding of PDV gene function in subversion of host immune defences or growth is restricted to a small number of BV isolates (reviewed in Strand 2010, 2012). The primary immune defence against parasitoids is encapsulation, which occurs when pattern recognition receptors bind to the surface of wasp eggs. This stimulates particular types of immune cells (haemocytes) to adhere to the parasitoid and form a multicellular sheath. Several pattern recognition receptors, cytokines, and adhesion molecules regulate haemocyte adhesion, while antimicrobial peptides (AMPs) and other genes regulated by NF-kB transcription factors of the Toll and/or Imd pathways are induced during the early phases of capsule formation. Capsules also often melanize due to activation of the phenoloxidase (PO) cascade. In turn, parasitoids die in capsules from asphyxiation, cytotoxic molecules generated by the PO cascade, and/or the activity of AMPs and other effector molecules. Members of the conserved members of viral *ank* family function as inhibitor kB (IkB) mimics that negatively regulate host NF-kBs, which are implicated in disabling haemocyte adhesion and phagocytosis. Two unique gene families (*glc, egf*) from *Microplitis demolitor* bracovirus (MdBV) have also been shown to block encapsulation and activation of the PO cascade, while studies of *Chelonus inanitus* bracovirus (CiBV) implicate three genes in altered host development. Physiological studies clearly show that IVs also produce gene products that disable encapsulation and host growth, but the specific genes involved largely remain uncharacterized.

## 17.3.2 Entomopoxviruses as Beneficial Symbionts

Whereas all members of the Polydnaviridae are beneficial symbionts, a few isolates from other virus families have also formed similar associations with parasitoid wasps. One such case is a member of the Poxiviridae, which are also large DNA viruses that have linear dsDNA genomes. Members of the family that infect insects in several orders are referred to as entomopoxviruses (Perera et al. 2010). However, one putative entomopox isolate named DlEPV is associated with the braconid wasp *Diachasmimorpha longicaudata*, which parasitizes the larval stage of the Caribbean fruit fly, *Anastrepha suspensa* and related tephritids (Lawrence 2005). Similar to PDV-carrying wasps, *D. longicaudata* injects DlEPV into its host when ovipositing. This wasp also injects another virus named DlRhV, because it morphologically resembles rhabdoviruses, which are negative-sense single-stranded (ss) RNA viruses. DlEPV and DlRhV both appear to replicate in the accessory glands of the wasp, and both entities also appear to replicate in parasitized hosts (Lawrence and Akin 1990; Lawrence and Matos 2005). The function of DlRhV remains unknown, but studies show that DlEPV infects host haemocytes and induces cytopathic effects that disable encapsulation, in order to allow the wasp's offspring survival (Lawrence 2005). Unlike PDVs, however, relatively little is known about the transmission and replication of either DlEPV or DlRhV in the wasp or its host. In turn, it is also unclear whether persistence of either virus totally depends on the wasp or whether these viruses are capable of persisting independently in tephritids or other insects.

## 17.3.3 Ascoviruses as Parasitoid-Vectored Pathogens and Potential Beneficial Symbionts

The Ascoviridae is a family whose members have large, circular dsDNA genomes and whose hosts are exclusively larval- and pupal-stage Lepidoptera in the family Noctuidae (Federici et al. 1991; Bideshi et al. 2010). These viruses also appear to be exclusively transmitted by parasitoids that acquire ascoviruses when they insert their ovipositor into an infected host. Most ascoviruses are then horizontally transmitted to a new host when the wasp parasitizes another host (Stasiak et al. 2005). Most ascoviruses cause distinct pathology that ultimately results in death of the host larva or pupa and the progeny of the parasitoid (Stasiak et al. 2005). However, one ascovirus isolate associated with the ichneumonid wasp species *Diadromus pulchellus*, DpAV4, has evolved to become an essential immunosuppressive symbiont, which persists without apparent replication as an episome in all cells of wasps and which is vertically transmitted to offspring (Bigot et al. 1997). However, DpAV4 replicates, asymptomatically, in the reproductive tract of female wasps, and DpAV4 virions are injected along with the egg at oviposition. Unlike most ascoviruses, when the pupae of the lepidopteran species *Acrolepiopsis assectella* are parasitized, DpAV4 replication does not immediately occur but is instead synchronized with the development of parasitoid juvenile offspring (Bigot et al. 1997). When this virus is transmitted by other *Diadromus* species, its mutualistic role remains unaltered, but it rapidly replicates and functions as a pathogen when transmitted by other ichneumonids (Bigot et al. 1997; Stasiak et al. 2005). Therefore, it seems that the success of ascovirus/wasp relationship is modulated by unknown factors in wasps from the genus *Diadromus* that control virus replication in the host. Thus, certain ascoviruses can potentially have a pathogenic, mutualistic, or non-pathogenic relationship with a specific wasp vector, depending upon the species system in which the relationship evolved. In particular, the presence of other interacting viruses may play an important role in determining the final outcome of this viral symbiosis, as explained in detail in the next section.

### 17.3.4 Cypoviruses as Modulators of Ascovirus Function in Parasitoids

The Reoviridae are segmented dsRNA viruses that infect a diversity of animals. One genus that specifically infects insects is the genus *Cypovirus*. Similar to ascoviruses like DpAV4, several cypoviruses have been identified from *Diodromus* ichneumonids (Renault et al. 2003, 2005). The best studied forms are DpRV1 and DpRV2, from *D. puchellus*. DpRV1 is always found in association with DpAV4 and appears to play a key role in modulating DpAV4 replication, so that the parasitoid's offspring develops before the host dies. This regulatory process appears to be controlled by an additional RNA of wasp origin, which is packaged in the DpRV1 virions (Renault et al. 2005; Stasiak et al. 2005). If true, this represents an alternative way to deliver wasp genetic material using viruses as vectors. Other cypoviruses, as well as other RNA viruses identified from parasitic wasps, have no known role in disabling host immune defences or promoting successful parasitism (Renault 2012).

### 17.3.5 Viruses that Manipulate Parasitoid Behaviour

The association between viruses and parasitic wasps goes beyond what is described above and may have a considerable influence not only on host physiological regulation and immune suppression, but also on other phenotypic traits, sometimes not very easy to define. For example, the figitid wasp *Leptopilina boulardi*, which parasitizes *Drosophila* spp., carries a virus, named LbFV, which promotes superparasitism (i.e. solitary parasitoids laying more than one

egg per host, which results in egg wastage as only one individual will complete development), a phenomenon that is rarely observed in uninfected *L. boulardi* wasps (Varaldi et al. 2005). LbFV is a filamentous virus of unclear taxonomic status that appears to be both maternally and horizontally transmitted (Varaldi et al. 2009). Superparasitism favours horizontal transmission of the virus but whether infection benefits the wasp is not fully clear. A population modelling approach suggests LbFV spread, and prevalence could positively influence parasitism rates of *L. boulardi* (Patot et al. 2010; Varaldi et al. 2012). Empirical studies also indicate that infected wasps have higher egg loads (i.e. fecundity) due possibly to superparasitism creating selection pressure that favours increased investment by wasps in egg production.

### 17.3.6 Viruses and Aphid Polyphenism

Aphids have complex life cycles, which usually include a sexual generation that consists of winged (alatae) adults, followed by a number of parthenogenetic generations, comprised of wingless (apterae) forms (Zera and Denno 1997). Parthenogenesis promotes rapid colony expansion when host plant resources are abundant and/or environmental conditions are favourable, while alatae formation promotes dispersal when resources become scarce and/or poor as a consequence of crowding. The cues involved in formation of alatae are vary and include a number of environmental factors, such as temperature, population density, nutrition quality of the plants, interactions with natural enemies and ants (Dixon 1998; Muller et al. 2001).

Many factors have been implicated in regulating aphid development (Fereres and Moreno 2009). Very recently, however, studies with rosy apple aphid, *Dysaphis plantaginea*, implicate infection by a densovirus named *D. plantaginea densovirus* (DplDNV) in alate and apterae development. DplDNV infection also often co-occurs with *Rosy apple aphid virus* (RAAV), a taxonomically unassigned RNA virus (Ryabov et al. 2009). Densoviruses have very small (4–6 kb) ssDNA genomes, and many isolates have been identified that infect insects. DplDNV-infected aphids produce winged aphids in response to crowding or poor plant quality, while uninfected aphids or aphids infected by RAAV fail to produce winged forms. DplDNV infection also reduces aphid reproduction, but by promoting dispersal through formation of winged forms, DplDNV is thought to benefit aphid carriers, on balance.

## 17.3.7 Aphid Bacterial Symbionts and their Viral Phages

As previously discussed, nearly all aphids depend on primary symbionts like *B. aphidicola* for survival and reproduction (Baumann 2005), while some facultative symbionts (*H. defensa, R. insecticola, S. symbiotica*) provide benefits to aphids through enhanced defence against parasitoids (Oliver et al. 2010). In the case of the pea aphid, *A. pisum*, resistance traits against parasitoids are associated with lysogenic lambdoid bacteriophages that infect the facultative symbiont (Moran et al. 2005a; Oliver et al. 2009). While not fully defined, some of these phage-associated factors disrupt development of parasitoid eggs while others affect the development of wasp larvae by possibly interfering with function of specialized wasp cells called teratocytes (Li et al. 2002; Falabella et al. 2005, 2009). The precise role of phages in the association that exists between aphids and their bacterial symbionts remains poorly understood. However, it is possible that the phages infecting symbionts may function as beneficial symbionts of the bacteria, the aphid, or both.

## 17.3.8 Plant Viruses and Insect Vectors

Several studies also report interactions between plant viruses and aphids, related homopterans, or other insects that serve as vectors (Stout et al. 2006). The most detailed information is available on aphids, for which alterations attributed to virus infection include increased attraction to infected host plants and increased fecundity

when feeding on virus-infected plants and subsequent increased production of alates which promotes virus spread (reviewed in Kluth et al. 2002; Colvin et al. 2006; Stout et al. 2006). The better performance of aphids on virus-infected plants has been poorly investigated at functional and molecular levels. Only recently, it has been demonstrated that *Cucumber mosaic virus* (CMV) encodes a protein that disrupts plant antiviral mechanisms, by enhancing the effect of salicylic acid (SA) on certain defence genes against pathogens and, at the same time, by inhibiting changes in the expression level of 90 % of the genes regulated by jasmonic acid (JA) (Lewsey et al. 2010), notably active against insects. Indeed, this reinforces the plant colonization strategy adopted by stylet feeders, such as aphids and whiteflies, which promote SA-induced gene expression, to down-regulate JA-mediated defence responses against insects, via JA/SA cross-talk (Thomma et al. 2001; Gatehouse 2002; De Vos et al. 2005; Zarate et al. 2007). The study of the salivary secretome of aphids shows the occurrence in the saliva of plant effector molecules similar to those used by different plant pathogens (Carolan et al. 2011), which may partly account for this interaction strategy. Collectively, all this evidence indicates that sucking insects may weaken plant barriers against insects by activating, with the help of plant pathogens, defence responses to which they are not sensitive but that redirect in their favour overall plant defence metabolism.

These complex tripartite interactions also may have a profound impact on biological invasions by alien species or populations, which may displace the local ones with the help of associated plant pathogens they transmit. The best studied case is offered by the biotype B of the whitefly *B. tabaci*, which has replaced the indigenous population (biotype ZHJ1) with the help of two whitefly-transmitted begomoviruses (Jiu et al. 2007). Host plants infected by these viruses enhance the fecundity and longevity of the biotype B individuals while have a negligible effect on the indigenous ZHJ1 population. This is considered one of the mechanisms accounting for the impressive invasive ability of biotype B (Jiu et al. 2007).

### 17.3.9 Viruses Helping Mosquitoes in Taking Their Blood Meal

Mosquitoes feeding on their hosts must complete their blood meal rapidly, to reduce the risk of being killed by an annoyed host. *Aedes aegypti* seems to do better in locating and using a blood vessel when host hamsters are infected by *Rift Valley fever virus* (RVFV), as this pathogen is supposed to disrupt haemostasis and facilitate mosquito feeding (Rossignol et al. 1985). Therefore, RVFV may provide a benefit to the vector insect, which enhances its own acquisition and transmission. The use of host pathogens by ectoparasitic arthropods to facilitate their feeding seems to be not an uncommon evolutionary pathway. It has been recently observed that *Varroa destructor*, a parasitic mite of honeybees, actively transmits the *Deformed wing virus* (DWV), which seem to disrupt host immune response, with likely negative effects on haemolymph clotting and melanization (Nazzi et al. 2012); this is an important functional requirement for feeding and successful development of mites and other ectoparasitic arthropods (Pennacchio and Strand 2006).

### 17.3.10 Evolutionary Patterns in Insect–Virus Associations

The examples provided above indicate that similar to bacteria, some viruses have evolved obligate or facultatively beneficial associations with insects. The most common theme is the domestication of viral pathogens by parasitic arthropods for suppressing the immune defences of hosts. The adaptive value of these symbiotic associations inextricably links life cycles and allows the colonization of new ecological niches. This generates fast evolution and speciation rates, both for parasitic organisms and for their viral symbionts, which effectively promote the introduction of new genomic traits favouring a rapid adaptation to new environments (Roossinck 2005). Current data also suggest

such symbioses begin as a loose association with a viral pathogen but can culminate, in the case of polydnaviruses, with entities that can no longer persist independently of one another.

The "alliance" of parasitic organisms with viral pathogens of the host seems to be an effective strategy also for insects attacking plants. The tight association between stylet feeding insects and viral plant pathogens provides a good example of how the latter can be used for suppressing plant defence responses. If and how some of the effector molecules present in aphid saliva, resembling those of pathogenic origin, may originate from HGT remains an intriguing question, which is certainly worthy of future research.

## 17.4 Conclusions

The study of symbiosis in arthropods demonstrates how the microbial diversity associated with these animals is an astonishing source of evolutionary novelty, which goes far beyond the simple complementation of nutritionally poor diets, required for the exploitation of difficult ecological niches. Even though our understanding of the biological bases of many specific symbiotic associations still remains incomplete, it is evident that the many impacts micro-organisms may have on host physiology, reproduction, and development promote the appearance of novel variants exposed to natural selection. Current molecular technologies offer the opportunity to unravel the intimate functional mechanisms underlying the establishment and maintenance of symbiotic associations and to shed light on some basic research issues, such as how chronic mutualistic interactions are discriminated by the host from pathogenic invasions and how these different categories of micro-organisms may be part of complex interactions affecting host immunity. These studies in arthropods offer new tools for investigating important aspects across the related fields of symbiosis and immunity, which have attracted increasing attention in the last few years

(Silverman and Paquette 2008; Gross et al. 2009; Ryu et al. 2010).

The Darwinian evolution model and that of symbiotic evolution are both based on the concept that evolution is driven by positive selection of the more fit variants, even though they propose different theories on how these variants are generated. In the Darwinian model, the gradual changes are the final outcome of random mutation and selection processes. The symbiogenic theory builds upon this model, by stating that genetic novelties and recombination occur among different biological entities (i.e. interacting symbionts) and not exclusively within the same species (Margulis 1993; Sapp 2009; Carrapiço 2010). These changes generated by cross-species association and recombination have the potential to generate much faster evolutionary rates, ultimately driven by natural selection. The unparalleled number and variety of host–symbiont associations in arthropods can offer a unique contribution to the active debate on the model best accounting for the evolution of the biological diversity we observe in nature. Based on the limited number of case studies presented in this chapter, we can reasonably conclude that symbiosis is one of the major drivers of diversification and evolution of natural populations of arthropods; the in-depth molecular and functional analysis of symbiotic interactions will disclose new basic information in evolutionary biology and will offer new tools for the development of innovative technologies for pest control. A cursory inspection of the available literature on arthropod symbiosis convincingly corroborates the model of symbiogenic evolution in multiple arthropod lineages. There is no doubt that selection, acting on horizontal mergers among different symbionts, has caused permanent and irreversible changes that result in new taxonomic entities. This can be interpreted as a consequence of a more general trend in biology: alliance of coexisting forms of life and integration of simpler functions generate new emergent properties.

# References

Adachi-Hagimori T, Miura K, Abe Y (2011) Gene flow between sexual and asexual strains of parasitic wasps: a possible case of sympatric speciation caused by a parthenogenesis-inducing bacterium. J Evol Biol 24:1254–1262

Adachi-Hagimori T, Miura K, Stouthamer R (2008) A new cytogenetic mechanism for bacterial endosymbiont-induced parthenogenesis in Hymenoptera. Proc R Soc B 275:2667–2673

Allen JM, Light JE, Perotti MA, Braig HR, Reed DL (2009) Mutational meltdown in primary endosymbionts: selection limits Muller's ratchet. PLoS ONE 4(3):e4969. doi:10.1371/journal.pone.0004969

Anbutsu H, Fukatsu T (2011) *Spiroplasma* as a model insect endosymbiont. Environ Microbiol Rep 3:144–153

Arakaki N, Noda H, Yamagishi K (2000) *Wolbachia*-induced parthenogenesis in the egg parasitoid *Telenomus nawai*. Entom Exper Appl 96:177–184

Batut J, Andersson SG, O'Callaghan D (2004) The evolution of chronic infection strategies in the alphaproteobacteria. Nat Rev Microbiol 2:933–945

Baumann P (2005) Biology of bacteriocyte-associated endosymbionts of plant sap-sucking insects. Annu Rev Microbiol 59:155–189

Bentley JK, Veneti Z, Heraty J, Hurst GD (2007) The pathology of embryo death caused by the male-killing *Spiroplasma* bacterium in *Drosophila nebulosa*. BMC Biol 5:9. doi:10.1186/1741-7007-5-9

Bézier A, Annaheim M, Herbiniere J, Wetterwald C, Gyapay G, Bernard-Samain S, Wincker P, Roditi I, Heller M, Belghazi M, Pfister-Wilhem R, Periquet G, Dupuy C, Huguet E, Volkoff A-N, Lanzrein B, Drezen J-M (2009) Polydnaviruses of braconid wasps derive from an ancestral nudivirus. Science 323:926–930

Bideshi DK, Bigot Y, Federici BA, Spears T (2010) Ascoviruses. In: Asgari S, Johnson K (eds) Insect virology. Caister Academic, Norfolk, pp 3–34

Bigot Y, Rabouille A, Doury G, Sizaret P-Y, Delbost F, Hamelin M-H, Periquet G (1997) Biological and molecular features of the relationships between *Diadromus pulchellus* ascovirus, a parasitoid hymenopteran wasp (*Diadromus pulchellus*) and its lepidopteran host, *Acrolepiopsis assectella*. J Gen Virol 78:1149–1163

Bonning BC (2005) Baculoviruses: biology, biochemistry, and molecular biology. In: Gilbert LI, Iatrou K, Gill SS (eds) Comprehensive molecular insect science, vol 6. Elsevier, Amsterdam, pp 233–270

Bordenstein SR (2003) Symbiosis and the origin of species. In: Bourtzis K, Miller TA (eds) Insect symbiosis, vol 1. CRC Press, Boca Raton, pp 283–304

Bordenstein SR, Bordenstein SR (2011) Temperature affects the tripartite interactions between bacteriophage WO, *Wolbachia*, and cytoplasmic incompatibility. PLoS One 6(12):e29106. doi:10.1371/journal.pone.0029106

Bordenstein SR, O'Hara FP, Werren JH (2001) *Wolbachia*-induced incompatibility precedes other hybrid incompatibilities in *Nasonia*. Nature 409:707–710

Bossan B, Koehncke A, Hammerstein P (2011) A new model and method for understanding *Wolbachia*-induced cytoplasmic incompatibility. PLoS ONE 6(5):e19757. doi:10.1371/journal.pone.0019757

Bouchon D, Cordaux R, Grève P (2009) Feminizing *Wolbachia* and the evolution of sex determination in isopods. In: Bourtzis K, Miller TA (eds) Insect symbiosis, vol 3., CRCBoca Raton, FL, pp 273–294

Bourtzis K, Braig HR, Karr TL (2003) Cytoplasmic incompatibility. In: Bourtzis K, Miller TA (eds) Insect symbiosis, vol 1. CRC, Boca Raton, pp 217–246

Braquart-Varnier C, Lachat M, Herbiniere J, Johnson M, Caubet Y, Bouchon D, Sicard M (2008) *Wolbachia* mediate variation of host immunocompetence. PLoS One 3 (9):e3286. doi:10.1371/journal.pone.0003286

Breeuwer JA, Werren JH (1990) Microorganisms associated with chromosome destruction and reproductive isolation between two insect species. Nature 346: 558–560

Breeuwer JAJ, Werren JH (1995) Hybrid breakdown between two haplodiploid species: the role of nuclear and cytoplasmic genes. Evolution 49:705–717

Brennan LJ, Keddie BA, Braig HR, Harris HL (2008) The endosymbiont *Wolbachia pipientis* induces the expression of host antioxidant proteins in an *Aedes albopictus* cell line. PLoS ONE 3(5):e2083. doi:10.1371/journal.pone.0002083

Bressan A, Arneodo J, Simonato M, Haines WP, Boudon-Padieu E (2009) Characterization and evolution of two bacteriome-inhabiting symbionts in cixiid planthoppers (Hemiptera: Fulgoromorpha: Pentastirini). Environ Microbiol 11:3265–3279

Brownlie JC, Johnson KN (2009) Symbiont-mediated protection in insect hosts. Trends Microbiol 17:348–354

Brumin M, Kontsedalov S, Ghanim M (2011) *Rickettsia* influences thermotolerance in the whitefly *Bemisia tabaci* B biotype. Insect Sci 18:57–66

Buchner P (1965) Endosymbiosis of animals with plant microorganisms. Interscience, New York

Burke GR, Strand MR (2012) Deep sequencing identifies viral and wasp genes with potential roles in replication of *Microplitis demolitor* bracovirus. J Virol 86:3293–3306

Carolan JC, Caragea D, Reardon KT, Mutti NS, Dittmer N, Pappan K, Cui F, Castanet M, Poulain J, Dossat C, Tagu D, Reese JC, Reeck GR, Wilkinson TL, Edwards OR (2011) Predicted effector molecules in the salivary secretome of the pea aphid (*Acyrthosiphon pisum*): a dual transcriptomic/proteomic approach. J Proteome Res 10:1505–1518

Carrapiço F (2010) How symbiogenic is evolution? Theory Biosci 129:135–139

Caspi-Fluger A, Inbar M, Mozes-Daube N, Katzir N, Portnoy V, Belausov E, Hunter MS, Zchori-Fein E (2012) Horizontal transmission of the insect symbiont *Rickettsia* is plant-mediated. Proc R Soc B 279:1791–1796

Caubet Y, Hatcher MJ, Mocquard JP, Rigaud T (2000) Genetic conflict and changes in heterogametic mechanisms of sex determination. J Evol Biol 13:766–777

Champion de Crespigny FE, Hurst LD, Wedell N (2008) Do *Wolbachia*-associated incompatibilities promote polyandry? Evolution 62:107–122

Champion de Crespigny FE, Wedell N (2006) *Wolbachia* infection reduces sperm competitive ability in an insect. Proc R Soc B 273:1455–1458

Champion de Crespigny FE, Wedell N (2007) Mate preferences in *Drosophila* infected with *Wolbachia*? Behav Ecol Sociobiol 61:1229–1235

Chapman RF (1998) The insects: structure and function, 4th edn. Cambridge University Press, New York

Charlat S, Hornett EA, Fullard JH, Davies N, Roderick GK, Wedell N, Hurst GD (2007a) Extraordinary flux in sex ratio. Science 317:214

Charlat S, Hurst GDD, Merçot H (2003) Evolutionary consequences of *Wolbachia* infections. Trends Genet 19:217–223

Charlat S, Reuter M, Dyson EA, Hornett EA, Duplouy A, Davies N, Roderick GK, Wedell N, Hurst GD (2007b) Male-killing bacteria trigger a cycle of increasing male fatigue and female promiscuity. Curr Biol 17:273–277

Charles H, Condemine G, Nardon C, Nardon P (1997) Genome size characterization of the principal endocellular symbiotic bacteria of the weevil *Sitophilus oryzae*, using pulsed field gel electrophoresis. Insect Biochem Mol Biol 27:345–350

Chevalier F, Herbiniere-Gaboreau J, Charif D, Mitta G, Gavory F, Wincker P, Greve P, Braquart-Varnier C, Bouchon D (2012) Feminizing *Wolbachia*: a transcriptomics approach with insights on the immune response genes in *Armadillidium vulgare*. BMC Microbiol 12(1):S1. doi:1471-2180-12-S1-S1 [pii] 10.1186/1471-2180-12-S1-S1

Clark EL, Karley AJ, Hubbard SF (2010) Insect endosymbionts: manipulators of insect herbivore trophic interactions? Protoplasma 244:25–51

Clark ME, Bailey-Jourdain C, Ferree PM, England SJ, Sullivan W, Windsor DM, Werren JH (2008) *Wolbachia* modification of sperm does not always require residence within developing sperm. Heredity 101:420–428

Colvin J, Omongo CA, Govindappa MR, Stevenson PC, Maruthi MN et al (2006) Host-plant viral infection effects on arthropod-vector population growth, development and behaviour: management and epidemiological implications. Adv Virol Res 67:419–452

Conord C, Despres L, Vallier A, Balmand S, Miquel C, Zundel S, Lemperiere G, Heddi A (2008) Long-term evolutionary stability of bacterial endosymbiosis in Curculionoidea: additional evidence of symbiont replacement in the Dryophthoridae family. Mol Biol Evol 25:859–868

Darby AC, Choi JH, Wilkes T, Hughes MA, Werren JH, Hurst GDD, Colbourne JK (2010) Characteristics of the genome of *Arsenophonus nasoniae*, son-killer bacterium of the wasp *Nasonia*. Insect Mol Biol 19:75–89

Dedeine F, Ahrens M, Calcaterra L, Shoemaker DD (2005) Social parasitism in fire ants (*Solenopsis* spp.): a potential mechanism for interspecies transfer of *Wolbachia*. Mol Ecol 14:1543–1548

Dedeine F, Vavre F, Fleury F, Loppin B, Hochberg ME, Bouletreau M (2001) Removing symbiotic *Wolbachia* bacteria specifically inhibits oogenesis in a parasitic wasp. Proc Natl Acad Sci USA 98:6247–6252

Degnan PH, Leonardo TE, Cass BN, Hurwitz B, Stern D, Gibbs RA, Richards S, Moran NA (2010) Dynamics of genome evolution in facultative symbionts of aphids. Environ Microbiol 12:2060–2069

Degnan PH, Moran NA (2008) Evolutionary genetics of a defensive facultative symbiont of insects: exchange of toxin-encoding bacteriophage. Mol Ecol 17:916–929

Degnan PH, Yu Y, Sisneros N, Wing RA, Moran NA (2009) *Hamiltonella defensa*, genome evolution of protective bacterial endosymbiont from pathogenic ancestors. Proc Natl Acad Sci USA 106:9063–9068

De Vos M, Van Oosten VR, Van Poecke RMP, Van Pelt JA, Pozo MJ, Mueller MJ, Buchala AJ, Métraux J-P, Van Loon LC, Dicke M, Pieterse CMJ (2005) Signal signature and transcriptome changes of *Arabidopsis* during pathogen and insect attack. Mol Plant-Microbe Interact 18:923–937

Dion E, Zele F, Simon JC, Outreman Y (2011) Rapid evolution of parasitoids when faced with the symbiont-mediated resistance of their hosts. J Evol Biol 24:741–750

Dixon AFG (1998) Aphid ecology. Chapman and Hall, London

Dobson SL, Bourtzis K, Braig HR, Jones BF, Zhou W, Rousset F, O'Neill SL (1999) *Wolbachia* infections are distributed throughout insect somatic and germ line tissue. Insect Biochem Mol Biol 29:153–160

Douglas AE (1989) Mycetocyte symbiosis in insects. Biol Rev 64:409–434

Drezen JM, Herniou EA, Bézier A (2012) Evolutionary progenitors of bracoviruses. In: Beckage NE, Drezen JM (eds) Parasitoid viruses. Elsevier, Amsterdam, pp 15–31

Dunn AM, Andrews T, Ingrey H, Riley J, Wedell N (2006) Strategic sperm allocation under parasitic sex-ratio distortion. Biol Lett 2:78–80

Dunning Hotopp JC (2011) Horizontal gene transfer between bacteria and animals. Trends Genet 27:157–163

Duron O, Bouchon D, Boutin S, Bellamy L, Zhou L, Engelstadter J, Hurst GD (2008) The diversity of reproductive parasites among arthropods: *Wolbachia* do not walk alone. BMC Biol 6:27. doi:10.1186/1741-7007-6-27

Duron O, Wilkes TE, Hurst GD (2010) Interspecific transmission of a male-killing bacterium on an ecological timescale. Ecol Lett 13:1139–1148

Emlen ST, Oring LW (1977) Ecology, sexual selection, and the evolution of mating systems. Science 197:215–223

Engelstädter J, Hurst GDD (2009) The ecology and evolution of microbes that manipulate host reproduction. Annu Rev Ecol Evol Syst 40:127–149

Falabella P, Perugino G, Caccialupi P, Riviello L, Varricchio P, Tranfaglia A, Rossi M, Malva C, Graziani F, Moracci M, Pennacchio F (2005) A novel fatty acid binding protein produced by teratocytes of the aphid parasitoid *Aphidius ervi*. Insect Mol Biol 14:195–205

Falabella P, Riviello L, De Stradis ML, Stigliano C, Varricchio P, Grimaldi A, de Eguileor M, Graziani F, Gigliotti S, Pennacchio F (2009) *Aphidius ervi* teratocytes release an extracellular enolase. Insect Biochem Mol Biol 39:801–813

Federici BA (1991) Viewing polydnaviruses as gene vectors of endoparasitic Hymenoptera. Redia 74:387–392

Fereres A, Moreno A (2009) Behavioural aspects influencing plant virus transmission by homopteran insects. Virus Res 14:158–168

Ferrari J, Scarborough CL, Godfray HCJ (2007) Genetic variation in the effect of a facultative symbiont on host-plant use by pea aphids. Oecologia 153:323–329

Ferrari J, West JA, Via S, Godfray HCJ (2012) Population genetic structure and secondary symbionts in host-associated populations of the pea aphid complex. Evolution 66:375–390

Ferree PM, Avery A, Azpurua J, Wilkes T, Werren JH (2008) A bacterium targets maternally inherited centrosomes to kill males in *Nasonia*. Curr Biol 18:1409–1414

Frydman HM, Li JM, Robson DN, Wieschaus E (2006) Somatic stem cell niche tropism in *Wolbachia*. Nature 441:509–512

Fukatsu T, Hosokawa T, Koga R, Nikoh N, Kato T, Hayama S, Takefushi H, Tanaka I (2009) Intestinal endocellular symbiotic bacterium of the macaque louse *Pedicinus obtusus*: distinct endosymbiont origins in anthropoid primate lice and the old world monkey louse. Appl Environ Microbiol 75:3796–3799

Fytrou A, Schofield PG, Kraaijeveld AR, Hubbard SF (2006) *Wolbachia* infection suppresses both host defence and parasitoid counter-defence. Proc R Soc B 273:791–796

Gatehouse JA (2002) Plant resistance towards insect herbivores: a dynamic interaction. New Phytol 156:145–169

Gempe T, Beye M (2011) Function and evolution of sex determination mechanisms, genes and pathways in insects. BioEssays 33:52–60

Gibson CM, Hunter MS (2010) Extraordinarily widespread and fantastically complex: comparative biology of endosymbiotic bacterial and fungal mutualists of insects. Ecol Lett 13:223–234

Giorgini M, Bernardo U, Monti MM, Nappo AG, Gebiola M (2010) *Rickettsia* symbionts cause parthenogenetic reproduction in the parasitoid wasp *Pnigalio soemius* (Hymenoptera: Eulophidae). Appl Environ Microbiol 76:2589–2599

Giorgini M, Monti MM, Caprio E, Stouthamer R, Hunter MS (2009) Feminization and the collapse of haplodiploidy in an asexual parasitoid wasp harboring the bacterial symbiont *Cardinium*. Heredity 102:365–371

Giron D, Kaiser W, Imbault N, Casas J (2007) Cytokinin-mediated leaf manipulation by a leafminer caterpillar. Biol Lett 3:340–343

Gottlieb Y, Zchori-Fein E (2001) Irreversible thelytokous reproduction in *Muscidifurax uniraptor*. Entom Exper Appl 100:271–278

Gottlieb Y, Zchori-Fein E, Werren JH, Karr TL (2002) Diploidy restoration in *Wolbachia*-infected *Muscidifurax uniraptor* (Hymenoptera: Pteromalidae). J Invert Pathol 81:166–174

Gross R, Vavre F, Heddi A, Hurst GDD, Zchori-Fein E, Bourtzis K (2009) Immunity and symbiosis. Mol Microbiol 73:751–759

Gruwell ME, Hardy NB, Gullan PJ, Dittmar K (2010) Evolutionary relationships among primary endosymbionts of the mealybug subfamily Phenacoccinae (Hemiptera: Coccoidea: Pseudococcidae). Appl Environ Microbiol 76:7521–7525

Gündüz EA, Douglas AE (2009) Symbiotic bacteria enable insect to use a nutritionally inadequate diet. Proc R Soc B 276:987–991

Hansen AK, Moran NA (2011) Aphid genome expression reveals host-symbiont cooperation in the production of amino acids. Proc Natl Acad Sci USA 108:2849–2854

Harmon JP, Moran NA, Ives AR (2009) Species response to environmental change: impacts of food web interactions and evolution. Science 323:1347–1350

Hedges LM, Brownlie JC, O'Neill SL, Johnson KN (2008) *Wolbachia* and virus protection in insects. Science 322:702

Hilgenboecker K, Hammerstein P, Schlattmann P, Telschow A, Werren JH (2008) How many species are infected with *Wolbachia*? A statistical analysis of current data. FEMS Microbiol Lett 281:215–220

Himler AG, Adachi-Hagimori T, Bergen JE, Kozuch A, Kelly SE, Tabashnik BE, Chiel E, Duckworth VE, Dennehy TJ, Zchori-Fein E, Hunter MS (2011) Rapid spread of a bacterial symbiont in an invasive whitefly is driven by fitness benefits and female bias. Science 332:254–256

Hosokawa T, Koga R, Kikuchi Y, Meng XY, Fukatsu T (2010) *Wolbachia* as a bacteriocyte-associated nutritional mutualist. Proc Natl Acad Sci USA 107:769–774

Huguet E, Serbielle C, Moreau SJM (2012) Evolution and origin of polydnavirus virulence genes. In: Beckage NE, Drezen JM (eds) Parasitoid viruses. Elsevier, Amsterdam, pp 63–78

Huigens ME, de Almeida RP, Boons PA, Luck RF, Stouthamer R (2004a) Natural interspecific and intraspecific horizontal transfer of parthenogenesis-inducing *Wolbachia* in *Trichogramma* wasps. Proc R Soc B 271:509–515

Huigens ME, Hohmann CL, Luck RF, Gort G, Stouthamer R (2004b) Reduced competitive ability due to *Wolbachia* infection in the parasitoid wasp *Trichogramma kaykai*. Entom Exper Appl 110:115–123

Huigens ME, Stouthamer R (2003) Parthenogenesis associated with *Wolbachia*. In: Bourtzis K, Miller TA (eds) Insect symbiosis, vol 2., CRCBoca Raton, FL, pp 247–266

Hunter MS, Zchori-Fein E (2006) Inherited bacteroidetes symbionts in arthropods. In: Bourtzis K, Miller TA (eds) Insect symbiosis, vol 2., CRCBoca Raton, FL, pp 39–56

Hurst GDD, Jiggins FM, Majerus MEN (2003) Inherited microorganisms that selectively kill male hosts: the hidden players of insect evolution? In: Bourtzis K, Miller TA (eds) Insect symbiosis, vol 2., CRCBoca Raton, FL, pp 177–198

Ijichi N, Kondo N, Matsumoto R, Shimada M, Ishikawa H, Fukatsu T (2002) Internal spatio-temporal population dynamics of infection with three *Wolbachia* strains in the adzuki bean beetle, *Callosobruchus chinensis* (Coleoptera: Bruchidae). Appl Environ Microbiol 68:4074–4080

Jaenike J (2009) Coupled population dynamics of endosymbionts within and between hosts. Oikos 118:353–362

Jaenike J, Dyer KA, Cornish C, Minhas MS (2006) Asymmetrical reinforcement and *Wolbachia* infection in *Drosophila*. PLoS Biol 4(10):e325. doi: 10.1371/journal.pbio.0040325

Jaenike J, Dyer KA, Reed LK (2003) Within-population structure of competition and the dynamics of male-killing *Wolbachia*. Evol Ecol Res 5:1023–1036

Jaenike J, Polak M, Fiskin A, Helou M, Minhas M (2007) Interspecific transmission of endosymbiotic *Spiroplasma* by mites. Biol Lett 3:23–25

Jaenike J, Unckless R, Cockburn SN, Boelio LM, Perlman SJ (2010) Adaptation via symbiosis: recent spread of a *Drosophila* defensive symbiont. Science 329:212–215

Jehle JA (2010) Nudiviruses: their biology and genetics. In: Asgari S, Johnson K (eds) Insect virology. Caister Academic, Norfolk, pp 153–170

Jiggins FM, Hurst GDD, Majerus MEN (2000) Sex-ratio-distorting *Wolbachia* causes sex-role reversal in its butterfly host. Proc R Soc B 267:69–73

Jiu M, Zhou XP, Tong L, Xu J, Yang X, Wan F-H, LiuS-S (2007) Vector-virus mutualism accelerates population increase of an invasive whitefly. PLoS ONE 2(1):e182. doi:10.1371/journal.pone.0000182

Johnstone RA, Hurst GDD (1996) Maternally inherited male-killing microorganisms may confound interpretation of mitochondrial DNA variability. Biol J Linn Soc 58:453–470

Kaiser W, Huguet E, Casas J, Commin C, Giron D (2010) Plant green-island phenotype induced by leaf-miners is mediated by bacterial symbionts. Proc R Soc B 277:2311–2319

Kambris Z, Blagborough AM, Pinto SB, Blagrove MS, Godfray HC, Sinden RE, Sinkins SP (2010)

*Wolbachia* stimulates immune gene expression and inhibits plasmodium development in *Anopheles gambiae*. PLoS Pathog 6(10):e1001143. doi:10.1371/journal.ppat.1001143

Kenyon SG, Hunter MS (2007) Manipulation of oviposition choice of the parasitoid wasp, *Encarsia pergandiella*, by the endosymbiotic bacterium *Cardinium*. J Evol Biol 20:707–716

Kirkness EF, Haas BJ, Sun W, Braig HR, Perotti MA, Clark JM, Lee SH, Robertson HM, Kennedy RC, Elhaik E, Gerlach D, Kriventseva EV, Elsik CG, Graur D, Hill CA, Veenstra JA, Walenz B, Tubío JMC, Ribeiro JMC, Rozas J, Johnston JS, Reese JT, Popadic A, Tojo M, Raoult D, Reed DL, Tomoyasu Y, Krause E, Mittapalli O, Margam VM, Li H-M, Meyer JM, Johnson RM, Romero-Severson J, Pagel VanZee J, Alvarez-Ponce D, Vieira FG, Aguadé M, Guirao-Rico S, Anzola JM, Yoon KS, Strycharz JP, Unger MF, Christley S, Lobo NF, Seufferheld MJ, Wang NK, Dasch GA, Struchiner CJ, Madey G, Hannick LI, Bidwell S, Joardar V, Caler E, Shao R, Barker SC, Cameron S, Bruggner RV, Regier A, Johnson J, Viswanathan L, Utterback TR, Sutton GG, Lawson D, Waterhouse RM, Venter JC, Strausberg RL, Berenbaum MR, Collins FH, Zdobnov EM, Pittendrigh BR (2010) Genome sequences of the human body louse and its primary endosymbiont provide insights into the permanent parasitic lifestyle. Proc Natl Acad Sci USA 107:12168–12173

Klasson L, Westberg J, Sapountzis P, Nasiund K, Lutnaes Y, Darby AC, Veneti Z, Chen LM, Braig HR, Garrett R, Bourtzis K, Andersson SGE (2009) The mosaic genome structure of the *Wolbachia* wRi strain infecting *Drosophila simulans*. Proc Natl Acad Sci USA 106:5725–5730

Kluth S, Kruess A, Tscharntke T (2002) Insects as vectors of plant pathogens: mutualistic and antagonistic interactions. Oecologia 133:193–199

Koukou K, Pavlikaki H, Kilias G, Werren JH, Bourtzis K, Alahiotis SN (2006) Influence of antibiotic treatment and *Wolbachia* curing on sexual isolation among *Drosophila melanogaster* cage populations. Evolution 60:87–96

Kuechler SM, Dettner K, Kehl S (2010) Molecular characterization and localization of the obligate endosymbiotic bacterium in the birch catkin bug *Kleidocerys resedae* (Heteroptera: Lygaeidae, Ischnorhynchinae). FEMS Microbiol Ecol 73:408–418

Kuechler SM, Dettner K, Kehl S (2011) Characterization of an obligate intracellular bacterium in the midgut epithelium of the bulrush bug *Chilacis typhae* (Heteroptera, Lygaeidae, Artheneinae). Appl Environ Microbiol 77:2869–2876

Kuijper B, Pen I (2010) The evolution of haplodiploidy by male-killing endosymbionts: importance of population structure and endosymbiont mutualisms. J Evol Biol 23:40–52

Lamelas A, Gosalbes MJ, Manzano-Marin A, Pereto J, Moya A, Latorre A (2011) *Serratia symbiotica* from the aphid *Cinara cedri*: a missing link from facultative to obligate insect endosymbiont. PLoS Genet 7(11):e1002357. doi:10.1371/journal.pgen.1002357

Lamelas A, Perez-Brocal V, Gomez-Valero L, Gosalbes MJ, Moya A, Latorre A (2008) Evolution of the secondary symbiont "Candidatus *Serratia symbiotica*" in aphid species of the subfamily Lachninae. Appl Environ Microbiol 74:4236–4240

Landmann F, Orsi GA, Loppin B, Sullivan W (2009) *Wolbachia*-mediated cytoplasmic incompatibility is associated with impaired histone deposition in the male pronucleus. PLoS Pathog 5(3):e1000343. doi:10.1371/journal.ppat.1000343

Lawrence PO (2005) Morphogenesis and cytopathic effects of the *Diachasmimorpha longicaudata* entomopoxvirus in host haemocytes. J Insect Physiol 51:221–233

Lawrence PO, Akin D (1990) Virus-like particles in the accessory glands of *Biosteres longicaudatus*. Can J Zool 68:539–546

Lawrence PO, Matos LF (2005) Transmission of the *Diachasmimorpha longicaudata* rhabdovirus (DlRhV) to wasp offspring: an ultrastructural analysis. J Insect Physiol 51:235–241

Leonardo TE (2004) Removal of a specialization-associated symbiont does not affect aphid fitness. Ecol Lett 7:461–468

Leonardo TE, Muiru GT (2003) Facultative symbionts are associated with host plant specialization in pea aphid populations. Proc R Soc B 270:S209–S212

Lewis Z, Champion de Crespigny FE, Sait SM, Tregenza T, Wedell N (2011) *Wolbachia* infection lowers fertile sperm transfer in a moth. Biol Lett 7:187–189

Lewsey MG, Murphy AM, MacLean D, Dalchau N, Westwood JH, Macaulay K, Bennet MH, Moulin M, Hanke DE, Powell G, Smith AG, Carr JP (2010) Disruption of two defensive signaling pathways by a viral RNA silencing suppressor. Mol Plant-Microbe Interact 23:835–845

Li S, Falabella P, Giannantonio S, Fanti P, Battaglia D, Digilio MC, Völkl W, Sloggett JJ, Weisser W, Pennacchio F (2002) Pea aphid clonal resistance to the endophagous parasitoid *Aphidius ervi*. J Insect Physiol 48:971–980

Login FH, Balmand S, Vallier A, Vincent-Monegat C, Vigneron A, Weiss-Gayet M, Rochat D, Heddi A (2011) Antimicrobial peptides keep insect endosymbionts under control. Science 334:362–365

Lopez-Sanchez MJ, Neef A, Pereto J, Patino-Navarrete R, Pignatelli M, Latorre A, Moya A (2009) Evolutionary convergence and nitrogen metabolism in *Blattabacterium* strain Bge, primary endosymbiont of the cockroach *Blattella germanica*. PLoS Genet 5(11):e1000721. doi:10.1371/journal.pgen.1000721

Lukasik P, Hancock EL, Ferrari J, Godfray HCJ (2011) Grain aphid clones vary in frost resistance, but this trait is not influenced by facultative endosymbionts. Ecol Entomol 36:790–793

Majerus MEN (2003) Sex wars: genes, bacteria and biased sex ratios. Princeton University Press, Princeton

Majerus TM, Majerus ME (2010) Intergenomic arms races: detection of a nuclear rescue gene of male-killing in a ladybird. PLoS Pathog 6(7):e1000987. doi:10.1371/journal.ppat.1000987

Margulis L (1993) Origins of species: acquired genomes and individuality. BioSystems 31(2–3):121–125

Margulis L (2009) Genome acquisition in horizontal gene transfer: symbiogenesis and macromolecular sequence analysis. Methods Mol Biol 532:181–191

McCutcheon JP, McDonald BR, Moran NA (2009) Convergent evolution of metabolic roles in bacterial co-symbionts of insects. Proc Natl Acad Sci USA 106:15394–15399

McCutcheon JP, Moran NA (2007) Parallel genomic evolution and metabolic interdependence in an ancient symbiosis. Proc Natl Acad Sci USA 104:19392–19397

McCutcheon JP, Moran NA (2010) Functional convergence in reduced genomes of bacterial symbionts spanning 200 my of evolution. Genome Biol Evol 2:708–718

McCutcheon JP, von Dohlen CD (2011) An interdependent metabolic patchwork in the nested symbiosis of mealybugs. Curr Biol 21:1366–1372

McLean AHC, van Asch M, Ferrari J, Godfray HCJ (2011) Effects of bacterial secondary symbionts on host plant use in pea aphids. Proc R Soc B 278:760–766

Miller WJ, Ehrman L, Schneider D (2010) Infectious speciation revisited: impact of symbiont-depletion on female fitness and mating behavior of Drosophila paulistorum. PLoS Pathog 6(12):e1001214. doi: 10.1371/journal.ppat.1001214

Min K-T, Benzer S (1997) Wolbachia, normally a symbiont of Drosophila, can be virulent, causing degeneration and death. Proc Nat Acad Sci USA 94:10792–10796

Moran NA, Degnan PH, Santos SR, Dunbar HE, Ochman H (2005a) The players in a mutualistic symbiosis: insects, bacteria, viruses, and virulence genes. Proc Natl Acad Sci USA 102:16919–16926

Moran NA, Dunbar HE (2006) Sexual acquisition of beneficial symbionts in aphids. Proc Natl Acad Sci USA 103:12803–12806

Moran NA, McCutcheon JP, Nakabachi A (2008) Genomics and evolution of heritable bacterial symbionts. Annu Rev Genet 42:165–190

Moran NA, McLaughlin HJ, Sorek R (2009) The dynamics and time scale of ongoing genomic erosion in symbiotic bacteria. Science 323:379–382

Moran NA, Tran P, Gerardo NM (2005b) Symbiosis and insect diversification: an ancient symbiont of sap-feeding insects from the bacterial phylum bacteroidetes. Appl Environ Microbiol 71:8802–8810

Moreira D, Lòpez Garcia P (2009) Ten reasons to exclude viruses from tree of life. Nat Rev Microbiol 7:306–311

Moreira LA, Iturbe-Ormaetxe I, Jeffery JA, Lu G, Pyke AT, Hedges LM, Rocha BC, Hall-Mendelin S, Day A, Riegler M, Hugo LE, Johnson KN, Kay BH, McGraw EA, van den Hurk AF, Ryan PA, O'Neill SL (2009) A Wolbachia symbiont in Aedes aegypti limits infection with dengue, Chikungunya, and Plasmodium. Cell 139:1268–1278

Müller CB, Williams IS, Hardie J (2001) The role of nutrition, crowding and interspecific interactions in the development of winged aphids. Ecol Entomol 26:330–340

Murphy N, Banks JC, Whitfield JB, Austin AD (2008) Phylogeny of the parasitic microgastroid subfamilies (Hymenoptera: Braconidae) based on sequence data from seven genes, with an improved time estimate of the origin of the lineage. Mol Phylogenet Evol 47:378–395

Nakabachi A, Yamashita A, Toh H, Ishikawa H, Dunbar HE, Moran NA, Hattori M (2006) The 160-kilobase genome of the bacterial endosymbiont Carsonella. Science 314:267

Nakanishi K, Hoshino M, Nakai M, Kunimi Y (2008) Novel RNA sequences associated with late male killing in Homona magnanima. Proc R Soc B 275:1249–1254

Narita S, Kageyama D, Nomura M, Fukatsu T (2007) Unexpected mechanism of symbiont-induced reversal of insect sex: feminizing Wolbachia continuously acts on the butterfly Eurema hecabe during larval development. Appl Environ Microbiol 73:4332–4341

Nazzi F, Brown SP, Annoscia D, Del Piccolo F, Di Prisco G, Varricchio P, Della Vedova G, Cattonaro F, Caprio E, Pennacchio F (2012) Synergistic parasite-pathogen interactions mediated by host immunity can drive the collapse of honeybee colonies. PLoS Pathog 8(6):e1002735. doi:10.1371/journal.ppat.1002735

Negri I, Pellecchia M, Mazzoglio PJ, Patetta A, Alma A (2006) Feminizing Wolbachia in Zyginidia pullula (Insecta, Hemiptera), a leafhopper with an XX/X0 sex-determination system. Proc Biol Sci 273:2409–2416

Newton ILG, Bordenstein SR (2011) Correlations between bacterial ecology and mobile DNA. Curr Microbiol 62:198–208

Normark BB (2004) Haplodiploidy as an outcome of coevolution between male-killing cytoplasmic elements and their hosts. Evolution 58:790–798

Ochman H, Lawrence JG, Groisman EA (2000) Lateral gene transfer and the nature of bacterial innovation. Nature 405:299–304

Oliver KM, Campos J, Moran NA, Hunter MS (2008) Population dynamics of defensive symbionts in aphids. Proc R Soc B 275:293–299

Oliver KM, Degnan PH, Burke GR, Moran NA (2010) Facultative symbionts of aphids and the horizontal transfer of ecologically important traits. Annu Rev Entomol 55:247–266

Oliver KM, Degnan PH, Hunter MS, Moran NA (2009) Bacteriophages encode factors required for protection in a symbiotic mutualism. Science 325:992–994

Oliver KM, Russell JA, Moran NA, Hunter MS (2003) Facultative bacterial symbionts in aphids confer resistance to parasitic wasps. Proc Natl Acad Sci USA 100:1803–1807

O'Neill SL, Karr TL (1990) Bidirectional incompatibility between conspecific populations of *Drosophila simulans*. Nature 348:178–180

Pannebakker BA, Loppin B, Elemans CP, Humblot L, Vavre F (2007) Parasitic inhibition of cell death facilitates symbiosis. Proc Natl Acad Sci USA 104:213–215

Pannebakker BA, Pijnacker LP, Zwaan BJ, Beukeboom LW (2004) Cytology of *Wolbachia*-induced parthenogenesis in *Leptopilina clavipes* (Hymenoptera: Figitidae). Genome 47:299–303

Pannebakker BA, Schidlo NS, Boskamp GJ, Dekker L, van Dooren TJ, Beukeboom LW, Zwaan BJ, Brakefield PM, van Alphen JJ (2005) Sexual functionality of *Leptopilina clavipes* (Hymenoptera: Figitidae) after reversing *Wolbachia*-induced parthenogenesis. J Evol Biol 18:1019–1028

Patot S, Martinez J, Allemand R, Gandon S, Varaldi J, Fleury F (2010) Prevalence of a virus inducing behavioural manipulation near species range border. Mol Ecol 19:2995–3007

Pennacchio F, Strand MR (2006) Evolution of developmental strategies in parasitic Hymenoptera. Annu Rev Entomol 51:233–258

Perera S, Li Z, Pavlik L, Arif B (2010) Entomopoxviruses. In: Asgari S, Johnson K (eds) Insect virology. Caister Academic, Norfolk, pp 83–102

Perlman SJ, Magnus SA, Copley CR (2010) Pervasive associations between *Cybaeus* spiders and the bacterial symbiont *Cardinium*. J Invert Pathol 103:150–155

Perrot-Minnot MJ, Cheval B, Migeon A, Navajas M (2002) Contrasting effects of *Wolbachia* on cytoplasmic incompatibility and fecundity in the haplodiploid mite *Tetranychus urticae*. J Evol Biol 15:808–817

Pijls JWAM, van Steenbergen HJ, van Alphen JJM (1996) Asexuality cured: the relations and differences between sexual and asexual *Apoanagyrus diversicornis*. Heredity 76:506–513

Poinsot D, Bourtzis K, Markakis G, Savakis C, Mercot H (1998) *Wolbachia* transfer from *Drosophila melanogaster* into *D. simulans*: host effect and cytoplasmic incompatibility relationships. Genetics 150:227–237

Poinsot D, Charlat S, Mercot H (2003) On the mechanism of *Wolbachia*-induced cytoplasmic incompatibility: confronting the models with the facts. BioEssays 25:259–265

Rancés E, Ye YH, Woolfit M, McGraw EA, O'Neill SL (2012) The relative importance of innate immune priming in *Wolbachia*-mediated dengue interference. PLoS Pathog 8(2):e1002548. doi:10.1371/journal.ppat.1002548

Renault S (2012) RNA viruses in parasitoid wasps. In: Beckage NE, Drezen JM (eds) Parasitoid viruses. Elsevier, Amsterdam, pp 193–201

Renault S, Bigot S, Lemesle M, Sizaret P-Y, Bigot Y (2003) The cypovirus *Diadromus pulchellus* DpRV-2 is sporadically associated with the endoparasitoid wasp *D. pulchellus* and modulates the defence mechanisms of pupae of the parasitized leek-moth *Acrolepiopsis assectella*. J Gen Virol 84:1799–1807

Renault S, Stasiak K, Federici BA, Bigot Y (2005) Commensal and mutualistic relationships of reoviruses with their parasitoid wasp hosts. J Insect Physiol 51:137–146

Rigaud T, Juchault P (1995) Success and failure of horizontal transfers of feminizing *Wolbachia* endosymbionts in woodlice. J Evol Biol 8:249–255

Rigaud T, Moreau J (2004) A cost of *Wolbachia*-induced sex reversal and female-biased sex ratios: decrease in female fertility after sperm depletion in a terrestrial isopod. Proc R Soc B 271:1941–1946

Riparbelli MG, Giordano R, Ueyama M, Callaini G (2012) *Wolbachia*-mediated male killing is associated with defective chromatin remodeling. PLoS One 7(1):e30045. doi:10.1371/journal.pone.0030045

Rodriguero MS, Confalonieri VA, Guedes JV, Lanteri AA (2010) *Wolbachia* infection in the tribe Naupactini (Coleoptera, Curculionidae): association between thelytokous parthenogenesis and infection status. Insect Mol Biol 19:631–640

Roossinck MJ (2005) Symbiosis versus competition in plant virus evolution. Nature Rev Microbiol 3:917–924

Ros VI, Breeuwer JA (2009) The effects of, and interactions between, *Cardinium* and *Wolbachia* in the doubly infected spider mite *Bryobia sarothamni*. Heredity 102:413–422

Ross L, Pen I, Shuker DM (2010) Genomic conflict in scale insects: the causes and consequences of bizarre genetic systems. Biol Rev 85:807–828

Rossignol PA et al (1985) Enhanced mosquito bloodfinding success on parasitemic host: evidence for vector-parasite mutualism. Proc Natl Acad Sci USA 82:7725–7727

Russell JA, Latorre A, Sabater-Munoz B, Moya A, Moran NA (2003) Side-stepping secondary symbionts: widespread horizontal transfer across and beyond the Aphidoidea. Mol Ecol 12:1061–1075

Russell JA, Moran NA (2006) Costs and benefits of symbiont infection in aphids: variation among symbionts and across temperatures. Proc R Soc B 273:603–610

Russell JE, Stouthamer R (2011) The genetics and evolution of obligate reproductive parasitism in *Trichogramma pretiosum* infected with parthenogenesis-inducing *Wolbachia*. Heredity 106:58–67

Ryabov EV, Keane G, Naish N, Evered C, Winstanley D (2009) Densovirus induces winged morphs in asexual clones of the rosy apple aphid, *Dysaphis plantaginea*. Proc Natl Acad Sci USA 21:8465–8470

Ryan F (2002) Darwin's blind spot. Houghton Miffin, Boston

Ryu JH, Ha EM, Lee WJ (2010) Innate immunity and gut-microbe mutualism in *Drosophila*. Dev Comp Immunol 34:369–376

Sagan L (1967) On the origin of mitosing cells. J Theor Biol 14(3):225–274

Sapp J (1994) Evolution by association. A history of symbiosis. Oxford University Press, New York

Sapp J (2009) The new foundations of evolution. On the tree of life. Oxford University Press, New York

Scarborough CL, Ferrari J, Godfray HCJ (2005) Aphid protected from pathogen by endosymbiont. Science 310:1781

Serbus LR, Casper-Lindley C, Landmann F, Sullivan W (2008) The genetics and cell biology of *Wolbachia*-host interactions. Annu Rev Genet 42:683–707

Shigenobu S, Watanabe H, Hattori M, Sakaki Y, Ishikawa H (2000) Genome sequence of the endocellular bacterial symbiont of aphids *Buchnera* sp APS. Nature 407:81–86

Shigenobu S, Wilson ACC (2011) Genomic revelations of a mutualism: the pea aphid and its obligate bacterial symbiont. Cell Mol Life Sci 68:1297–1309

Sicard M, Chevalier F, De Vlechouver M, Bouchon D, Greve P, Braquart-Varnier C (2010) Variations of immune parameters in terrestrial isopods: a matter of gender, aging and *Wolbachia*. Naturwissenschaften 97:819–826

Silverman N, Paquette N (2008) The right resident bugs. Science 319:734–735

Siozios S, Sapountzis P, Ioannidis P, Bourtzis K (2008) *Wolbachia* symbiosis and insect immune response. Insect Sci 15:89–100

Snook RR, Cleland SY, Wolfner MF, Karr TL (2000) Offsetting effects of *Wolbachia* infection and heat shock on sperm production in *Drosophila simulans*: analyses of fecundity, fertility and accessory gland proteins. Genetics 155:167–178

Stasiak K, Renault S, Federici BA, Bigot Y (2005) Characteristics of pathogenic and mutualistic relationships of ascoviruses in field populations of parasitoid wasps. J Insect Physiol 51:103–115

Stout MJ, Thaler JS, Thomma BPHJ (2006) Plant-mediated interactions between pathogenic microorganisms and herbivorous arthropods. Annu Rev Entomol 51:663–689

Stouthamer R (1997) *Wolbachia*-induced parthenogenesis. In: O'Neill SL, Hoffmann AA, Werren JH (eds) Influential passengers: inherited microorganisms and arthropod reproduction. Oxford University Press, Oxford, pp 102–122

Stouthamer R, van Tilborg M, de Jong JH, Nunney L, Luck RF (2001) Selfish element maintains sex in natural populations of a parasitoid wasp. Proc R Soc B 268:617–622

Strand MR (2010) Polydnaviruses. In: Asgari S, Johnson K (eds) Insect virology. Caister Academic, Norfolk, pp 171–197

Strand MR (2012) Polydnavirus gene products that interact with the host immune system. In: Beckage NE, Drezen JM (eds) Parasitoid viruses. Elsevier, Amsterdam, pp 149–161

Sugimoto TN, Ishikawa Y (2012) A male-killing *Wolbachia* carries a feminizing factor and is associated with degradation of the sex-determining system of its host. Biol Lett 8:412–415

Suomalainen E, Saura A, Lokki J (1987) Cytology and evolution in parthenogenesis. CRC, Boca Raton

Tamas I, Klasson L, Canback B, Naslund AK, Eriksson AS, Wernegreen JJ, Sandstrom JP, Moran NA, Andersson SGE (2002) 50 million years of genomic stasis in endosymbiotic bacteria. Science 296:2376–2379

Tamas I, Wernegreen JJ, Nystedt B, Kauppinen SN, Darby AC, Gomez-Valero L, Lundin D, Poole AM, Andersson SGE (2008) Endosymbiont gene functions impaired and rescued by polymerase infidelity at poly(A) tracts. Proc Natl Acad Sci USA 105:14934–14939

Telschow A, Flor M, Kobayashi Y, Hammerstein P, Werren JH (2007) *Wolbachia*-induced unidirectional cytoplasmic incompatibility and speciation: mainland-island model. PLoS One 2(8):e701. doi:10.1371/journal.pone.0000701

Telschow A, Hammerstein P, Werren JH (2005) The effect of *Wolbachia* versus genetic incompatibilities on reinforcement and speciation. Evolution 59:1607–1619

Terry RS, Smith JE, Sharpe RG, Rigaud T, Littlewood DT, Ironside JE, Rollinson D, Bouchon D, MacNeil C, Dick JT, Dunn AM (2004) Widespread vertical transmission and associated host sex-ratio distortion within the eukaryotic phylum Microspora. Proc R Soc B 271:1783–1789

Thomma BPHJ, Penninckx IAMA, Broekaert WF, Cammue BPA (2001) The complexity of disease signaling in *Arabidopsis*. Curr Opin Immunol 13:63–68

Toft C, Andersson SGE (2010) Evolutionary microbial genomics: insights into bacterial host adaptation. Nat Rev Genet 11:465–475

Toju H, Fukatsu T (2011) Diversity and infection prevalence of endosymbionts in natural populations of the chestnut weevil: relevance of local climate and host plants. Mol Ecol 20:853–868

Toju H, Hosokawa T, Koga R, Nikoh N, Meng XY, Kimura N, Fukatsu T (2010) "Candidatus *Curculioniphilus buchneri*", a novel clade of bacterial endocellular symbionts from weevils of the genus *Curculio*. Appl Environ Microbiol 76:275–282

Tram U, Sullivan W (2002) Role of delayed nuclear envelope breakdown and mitosis in *Wolbachia* induced cytoplasmic incompatibility. Science 296:1124–1126

Tsuchida T, Koga R, Fukatsu T (2004) Host plant specialization governed by facultative symbiont. Science 303:1989

Tsuchida T, Koga R, Shibao H, Matsumoto T, Fukatsu T (2002) Diversity and geographic distribution of secondary endosymbiotic bacteria in natural populations of the pea aphid, *Acyrthosiphon pisum*. Mol Ecol 11:2123–2135

Turelli M (1994) Evolution of incompatibility-inducing microbes and their hosts. Evolution 48:1500–1513

Vala F, Egas M, Breeuwer JA, Sabelis MW (2004) *Wolbachia* affects oviposition and mating behaviour of its spider mite host. J Evol Biol 17:692–700

Varaldi J, Boulétreau M, Fleury F (2005) Cost induced by viral particles manipulating superparasitism behaviour in the parasitoid *Leptopilina boulardi*. Parasitology 131:161–168

Varaldi J, Patot S, Nardin M, Gandon S (2009) A virus-shaping reproductive strategy in a *Drosophila* parasitoid. Adv Parasitol 70:333–363

Varaldi J, Martinez J, Patot S, Lepetit D, Fleury F, Gandon S (2012) An inherited virus manipulating the behavior of its parasitoid host: epidemiology and evolutionary consequences. In: Beckage NE, Drezen JM (eds) Parasitoid viruses. Elsevier, Amsterdam, pp 203–214

Vavre F, de Jong JH, Stouthamer R (2004) Cytogenetic mechanism and genetic consequences of thelytoky in the wasp *Trichogramma cacoeciae*. Heredity 93: 592–596

Veneti Z, Bentley JK, Koana T, Braig HR, Hurst GD (2005) A functional dosage compensation complex required for male killing in *Drosophila*. Science 307: 1461–1463

Veneti Z, Clark ME, Karr TL, Savakis C, Bourtzis K (2004) Heads or tails: host-parasite interactions in the *Drosophila-Wolbachia* system. Appl Environ Microbiol 70:5366–5372

Villarreal LP (2005) Viruses and the evolution of life. ASM, Washington DC

Vogel KJ, Moran NA (2011) Sources of variation in dietary requirements in an obligate nutritional symbiosis. Proc R Soc B 278:115–121

Volkoff AN, Jouan V, Urbach S, Samain S, Bergoin M, Wincker P, et al (2010) Analysis of virion structural components reveals vestiges of the ancestral ichnovirus genome. PLoS Pathog 6, e1000923. doi: 10.1371/journal.ppat.1000923

Vorburger C, Gehrer L, Rodriguez P (2010) A strain of the bacterial symbiont *Regiella insecticola* protects aphids against parasitoids. Biol Lett 6:109–111

Wang Y, Jehle JA (2009) Nudiviruses and other large, double-stranded circular DNA viruses of invertebrates: new insights on an old topic. J Invertebr Pathol 101:187–193

Webb B, Strand MR (2005) The biology and genomics of polydnaviruses. In: Gilbert LI, Iatrou K, Gill SS (eds) Comprehensive molecular insect science, vol 6. Elsevier, Amsterdam, pp 323–360

Weeks AR, Breeuwer JA (2001) *Wolbachia*-induced parthenogenesis in a genus of phytophagous mites. Proc Biol Sci 268:2245–2251

Weeks AR, Marec F, Breeuwer JAJ (2001) A mite species that consists entirely of haploid females. Science 292:2479–2482

Weeks AR, Turelli M, Harcombe WR, Reynolds KT, Hoffmann AA (2007) From parasite to mutualist: rapid evolution of *Wolbachia* in natural populations of *Drosophila*. PLoS Biol 5(5):e114. doi: 10.1371/journal.pbio.0050114

Weinert LA, Werren JH, Aebi A, Stone GN, Jiggins FM (2009) Evolution and diversity of *Rickettsia* bacteria. BMC Biol 7:6. doi:10.1186/1741-7007-7-6

Werren JH (1997) Biology of *Wolbachia*. Annu Rev Entomol 42:587–609

Werren JH (1998) *Wolbachia* and speciation. In: Howard DJ, Berlocher SH (eds) Endless forms: species and speciation. Oxford University Press, Oxford, pp 245–260

Werren JH, Baldo L, Clark ME (2008) *Wolbachia*: master manipulators of invertebrate biology. Nat Rev Microbiol 6:741–751

Werren JH, Beukeboom LW (1998) Sex determination, sex ratios and genetic conflict. Annu Rev Ecol Syst 29:233–261

Werren JH, O'Neil S (1997) The evolution of heritable symbionts. In: O'Neil S, Hoffmann AA, Werren JH (eds) Influential passengers: inherited microorganisms and arthropod reproduction. Oxford University Press, New York, pp 1–41

Werren JH, Richards S, Desjardins CA et al (2010) Functional and evolutionary insights from the genomes of three parasitoid *Nasonia* species. Science 327:343–348

Werren JH, Zhang W, Guo LR (1995) Evolution and phylogeny of *Wolbachia*: reproductive parasites of arthropods. Proc R Soc Lond B Biol Sci 261:55–63

Wetterwald C, Roth T, Kaeslin M, Annaheim M, Wespi G, Heller M, Mäser P, Roditi I, Pfister-Wilhelm R, Bézier A, Gyapay G, Drezen JM, Lanzrein B (2010) Identification of bracovirus particle proteins and analysis of their transcript levels at the stage of virion formation. J Gen Virol 91:2610–2619

White JA (2011) Caught in the act: rapid, symbiont-driven evolution. BioEssays 33:823–829

White JA, Kelly SE, Perlman SJ, Hunter MS (2009) Cytoplasmic incompatibility in the parasitic wasp *Encarsia inaron*: disentangling the roles of *Cardinium* and *Wolbachia* symbionts. Heredity 102:483–489

Whitfield JB (2002) Estimating the age of the polydnavirus/braconid wasp symbiosis. Proc Natl Acad Sci USA 99:7508–7513

Wong ZS, Hedges LM, Brownlie JC, Johnson KN (2011) *Wolbachia*-mediated antibacterial protection and immune gene regulation in *Drosophila*. PLoS ONE 6(9):e25430. doi:10.1371/journal.pone.0025430

Zarate SI, Kempema LA, Walling LL (2007) Silverleaf whitefly induces salicylic acid defenses and suppresses effectual jasmonic acid defenses. Plant Physiol 143:866–875

Zchori-Fein E, Perlman SJ (2004) Distribution of the bacterial symbiont *Cardinium* in arthropods. Mol Ecol 13:2009–2016

Zeh JA, Zeh DW (2006) Male-killing *Wolbachia* in a live-bearing arthropod: brood abortion as a constraint on the spread of a selfish microbe. J Invert Pathol 92:33–38

Zera AJ, Denno RF (1997) Physiology and ecology of dispersal polymorphism in insects. Annu Rev Entomol 42:207–230

Zheng Y, Ren PP, Wang JL, Wang YF (2011a) *Wolbachia*-induced cytoplasmic incompatibility is associated with decreased Hira expression in male *Drosophila*. PLoS One 6(4):e19512. doi: 10.1371/journal.pone.0019512

Zheng Y, Wang JL, Liu C, Wang CP, Walker T, Wang YF (2011b) Differentially expressed profiles in the larval testes of *Wolbachia* infected and uninfected *Drosophila*. BMC Genomics 12:595. doi: 10.1186/1471-2164-12-595

# The Evolvability of Arthropods

## Matthew S. Stansbury and Armin P. Moczek

# 18

## Contents

M. S. Stansbury (✉)
Center for Insect Science, University of Arizona,
1007 E. Lowell Street, Tucson, 85721-0106, USA
e-mail: mstansbury@email.arizona.edu

A. P. Moczek
Department of Biology, Indiana University, 915 E.
Third Street, Myers Hall 150, Bloomington, IN
47405-7107, USA

## 18.1 Introduction

By many metrics, arthropods constitute one of the most successful animal phyla on our planet, manifest in extreme species richness, enormous diversity in morphologies and developmental modes, and successful radiation into nearly every inhabitable ecological niche available to multicellular organisms (Storch and Welch 1991; Brusca and Brusca 2002; Ødegaard 2000; Valentine 2004; Gullan and Cranston 2004; Grimaldi and Engel 2005). In this chapter, we will explore some of the causes and mechanisms that have enabled arthropod diversification. We define evolvability broadly as a lineage's capacity to generate phenotypic diversity over evolutionary time. We begin by exploring two prominent axes of diversification in the arthropods: evolvability in (1) developmental space and in (2) developmental time, and their respective contributions to facilitating innovation, diversification, and radiation within the Arthropoda. We end our chapter by examining the role of (3) developmental plasticity in arthropod evolution. In each context, we explore the genetic, developmental, and ecological mechanisms that may have allowed arthropods to diversify more than any other group of animals, the interactions among these mechanisms, and the emergent properties of these interactions. Throughout, we highlight key questions for future research, in particular as created by

A. Minelli et al. (eds.), *Arthropod Biology and Evolution*,
DOI: 10.1007/978-3-642-36160-9_18, © Springer-Verlag Berlin Heidelberg 2013

the increased integration of evolution and ecology with developmental biology and genomics.

## 18.2 Evolvability in Developmental Space

The first major axis of arthropod evolution examined in this chapter concerns the diversification of body regions, segments, appendages, and other morphological "units". The early Cambrian arthropods already exhibited the characteristics that have come to define the group (Conway Morris et al. 1987; Hou et al. 2004) among these being meristic subdivision of the exoskeleton into distinct appendage-bearing segments. Among the most important themes in arthropod evolution is the specialization of these individual segments and segment groups and of the corresponding outgrowths they bear.

### 18.2.1 Redundancy in Arthropod Body Architecture

The ancestral arthropod is often represented as possessing a trunk composed of externally homonomous segments, each bearing an undifferentiated pair of appendages (Akam et al. 1988). The precise organization of the proto-arthropod remains to be fully elucidated, but what is clear is that it represented an evolutionary ground state endowed with vast potential for diversification (=evolvability), realized in extraordinarily varied arthropod morphologies. This potential appears rooted, at least in part, in the compartmentalization of repeating morphological units and in the redundancy inherent in such a body plan. Redundancy is integral to diversification across levels of biological organization (reviewed in Galis and Metz 2007). For instance, the evolution of new genes and gene functions is thought to be greatly facilitated by gene duplication events, which allow one copy to undergo modification while the other retains the ancestral function (Ohno 1970; Force et al. 1999). Similarly, an ancestral organism

composed of multiple morphological units of similar function may be deconstrained evolutionarily to a degree roughly proportional to the level of redundancy present in the system *provided one critical condition is met: that reiterated units can be developmentally decoupled*. In such a system, individual segments are afforded some measure of low-risk mutational and developmental exploration because neighbouring units continue to carry out crucial locomotory or food manipulation functions. Arthropod evolution is replete with examples of differential segment evolution giving rise to dramatic divisions of labour between neighbouring body regions. Segmental redundancy may thus have deconstrained the diversification of individual segments at a functional/anatomical level, facilitated by the genetic decoupling of segments into quasi-independent developmental units.

### 18.2.2 Compartmentalization of Arthropod Development: Genetically Decoupled Units

The evolutionary independence of morphological units is, of course, reliant upon a developmental system that is subdivided and decoupled in space. Because the diversification of segments and the appendages they bear is perhaps the most important theme of arthropod evolution, we focus our attention initially on the Hox genes. Hox genes encode highly conserved transcription factors that regulate segment identity along the anterior–posterior axis (Lewis 1978; Akam 1989). In other words, regions expressing unique suites of Hox genes (and other transcription factors) define quasi-independent developmental/evolutionary units, allowing downstream genetic programs to be activated or deactivated differentially based on spatial position.

Morphological units—in this case, segments or segment groups and their corresponding outgrowths—can thus diverge to a level permitted by the underlying genetic architecture (Fig. 18.1), such as Hox gene expression (Angelini and

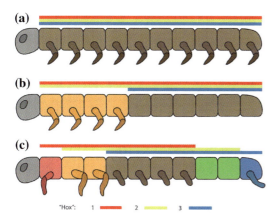

**Fig. 18.1** The principle of developmental decoupling and resolution of segment identity mediated by three Hox-like protein domains in three hypothetical arthropod-like embryos (note that these examples are not meant to represent actual species or hypothetical ancestral states). In "Species **a**", all three proteins are expressed over the entire length of the trunk defining this entire region as a single morphological unit. Appendage modifications are expected to affect all segments more or less equally. "Species **b**" has undergone a shift in the expression of "Hox 3", delineating two broad "tagmata" in the trunk. Note that homologous Hox expression profiles need not specify similar phenotypic states across lineages (overlapping domains of Hox1, 2, and 3 interact to repress appendage formation in "Species **b**" in contrast to "Species **a**" where they merely modify limb identity). The Hox domains in "Species **c**" are maximally offset relative to each other and therefore allow the greatest number of distinct segment-group identities possible in a 3-gene system with continuous Hox expression domains [2n − 1 distinct regions possible where n = # of Hox genes (Angelini and Kaufman 2005a)]. Note that the Hox expression profile represented in "Species **c**" is not common in nature and that real arthropods generally exhibit far fewer distinct segment identities than are theoretically allowed by simple Hox combinatorics (Minelli 2003)

Kaufman 2005a). For example, insects and myriapods differ markedly in the degree to which they have undergone morphological specialization of locomotory appendages. While myriapod walking legs have remained essentially uniform throughout their history, diverse insect orders have independently recruited individual pairs of thoracic legs for highly specialized roles such as food acquisition (mantids), jumping (orthopterans), digging (mole crickets), swimming (water boatmen), etc. This discrepancy is likely explained, in part, by the differing degrees of developmental resolution imparted by Hox gene expression in the two groups. Insects have three segments that bear locomotory appendages, each of which lies within the domain of a unique combination of Hox genes—prothorax, *Sex combs reduced* (*Scr*)/*Antennapedia* (*Antp*); mesothorax, *Antp*; and metathorax, *Antp*/*Ultrabithorax* (*Ubx*) (reviewed in Hughes and Kaufman 2002a). This bestows a distinct developmental identity and corresponding evolutionary degree of freedom to each insect thoracic segment as expression of downstream patterning genes can be activated, inhibited or modified within a given segment without affecting expression in other domains. In contrast, the locomotory segments of myriapods are regulated by a pair of broadly expressed Hox genes, *Ubx* and *abdominal-A* (*abd-A*) (Hughes and Kaufman 2002b). This genetic unity may underlie an evolutionary developmental indivisibility that precludes morphological and functional differentiation among walking legs of the myriapod lineages relative to those of the insects.

As Hox genes define segment identities and delineate domains of tagmatization, marked evolution of arthropod body organization may also be enabled through relatively simple shifts in the domains of Hox expression. Averof and Patel (1997) showed that, similar to myriapods, the locomotory tagma in some crustaceans is induced by the broad expression of posterior Hox genes. In some lineages, a variable number of the thoracic appendages have been modified for feeding—termed maxillipeds. The researchers found that the anterior reach of the *Ubx*/*abd-A* domain had shifted 1, 2, or 3 segments posteriorly, corresponding to the number of maxilliped pairs displayed by each group, respectively. This demonstrates the flexibility of the arthropod body plan as abrupt changes in the character of units—and therefore in the organization of the body plan and ecological strategies—are possible through relatively simple genetic modifications. These results may also suggest that the character of morphological units may be deconstrained when placed in a new regulatory context.

### 18.2.3 Compartmentalization of Arthropod Development: Semi-Autonomy of Gene Networks

It is important to point out that while Hox genes, through segment-specific activation or repression of various effector genes, can modify the character of the resultant appendage (or, in many cases, interrupt its expression entirely), the machinery associated with appendage induction *per se* is independent of Hox control in arthropods. This adds to the evolutionary flexibility of the system because domain-specific development does not require the evolution of novel pathways or genes for each domain. Instead, only patterns of activation, inhibition, and integration must be segment specific, whereas the genes and their products whose expression is modified in a domain-specific manner can themselves remain conserved. As a consequence, diversification is facilitated through changes in assembly, rather than changes in component parts.

The core components underlying the patterning of appendages are largely conserved in arthropods (reviewed in Angelini and Kaufman 2005b) and, to a lesser degree, throughout the animal kingdom (Panganiban et al. 1997). Much research in appendage induction has focused on the orthologs of the homeodomain-containing transcription factor, *Distal-less* (*Dll*), which activates the formation of the proximo-distal appendage axis (Cohen et al. 1989). *Dll*, in turn, is activated by signals that are present on each trunk segment, while Hox-mediated signals act in parallel, transmitting instructions to modify the appendage in a domain-specific manner (reviewed in Morata 2001). In many arthropod groups, Hox genes have evolved the capacity to inhibit the expression of *Dll* entirely—such as *abd-A* in insects (Vachon et al. 1992; Lewis et al. 2000) and *Antp* in arachnids (Khadjeh et al. 2012)—resulting in the characteristically legless abdomen (or opisthosoma). However, because the developmental machinery of appendage patterning is maintained independent of these repressive signals, trait recurrence is possible in body regions where the expression of that trait has previously been dormant. This potential has been demonstrated experimentally, where inhibiting the repressive effects of a single gene results in the ectopic recurrence of appendages on previously limbless body regions (Lewis et al. 2000), and also in nature, as in the case of caterpillar prolegs. Lepidopteran caterpillars exhibit functional pairs of appendages on five of their abdominal segments. Warren et al. (1994) showed that this feat is accomplished developmentally through segment-specific inactivation of the appendage-repressive *abd-A* gene. This secondary lifting of inhibition allows the ancestrally dormant yet conserved appendage-patterning network to be reactivated, resulting in the formation of abdominal limbs. In some extreme cases, reappearance of traits may be associated with such profound phenotypic modification that they may be considered true evolutionary novelties. For example, Prud'homme et al. (2011) presented intriguing developmental and anatomical evidence that the pronotal helmets of membracid treehoppers may be derived from the long-repressed T1 wing homologue (but see Miko et al. 2012; Kazunori 2012).

While the reactivation of conserved gene networks within their ancestral context can contribute to the reappearance of dormant phenotypes, the "reuse" of these signals outside of the confines of strict homology has the potential to create truly novel traits. Because the networks underlying various traits are integrated, self-contained units that may be induced by relatively few signals, they can theoretically be wired readily into other regulatory circuits and thus expressed in novel developmental and anatomical contexts. A well-known experimental example is the ectopic expression of the Drosophila *eyeless* gene. The *eyeless* induction in foreign domains such as legs or wings is sufficient to result in the ectopic formation of well-organized ommatidia in these body regions (Halder et al. 1995), illustrating the potential ease with which entire developmental-genetic modules can, in principle, become functionally co-opted into novel developmental contexts.

Such co-option may underlie much innovation in arthropod evolution. For instance, gene networks traditionally associated with appendage formation appear to have been co-opted repeatedly into novel contexts, facilitating for instance the evolution of beetle horns (Moczek and Nagy 2005; Moczek and Rose 2009; Wasik et al. 2010), which, though not homologous to appendages, at least share many of the same properties, or the foci of butterfly wing eyespots, whose similarities to conventional appendages appear far more remote (Carroll et al. 1994). A high degree of evolvability may thus be expected in biological systems that possess (a) traits underlain by modular gene networks under relatively simple regulatory control, and (b) a flexible regulatory scaffold into which these conserved networks can be wired in diverse ways.

In summary, arthropods are endowed with anatomical and developmental qualities that may make them particularly amenable to morphological change. The early organization of the arthropod body plan into a series of morphologically similar and functionally redundant units was likely a key step in predisposing the Arthropoda to evolutionary malleability. The extent to which exploration of morphological space was possible within segments and segment groups relied critically upon the degree of spatial decoupling present in the underlying genetic architecture. The flexibility of the system was further facilitated by the modular nature of gene networks under relatively simple regulatory control, enabling their transfer across a flexible regulatory scaffold by means of modest developmental-genetic modifications.

## 18.3 Evolvability in Developmental Time

The second major axis of diversification we examine concerns the evolvability of arthropods in developmental time, along the life cycle of individuals. Immature and mature stages, with or without distinct transitional forms, have evolved to varying degrees in different groups of arthropods (Storch and Welch 1991; Mente 2008; Grimaldi and Engel 2005). In many ways, these are best understood from studies on hexapods (Nijhout 1994, 1999a). Here, developmental modes range from ametabolous (continued moulting throughout adult life with no metamorphosis except for the addition of genitalia, e.g. silverfish) to hemimetabolous (terminal adult moult; more or less gradual transformation of immature into adult; metamorphic addition of wings and genitalia, e.g. cockroaches and grasshoppers) to holometabolous development (terminal adult moult; complete transformation of immature into adult via the intercalated pupal stage; e.g. beetles and butterflies). A similar diversity of developmental modes is also observed, though much less well understood, in the crustaceans, ranging from direct, largely ametamorphic development seen in groups such as the ostracods or cladocerans to highly disparate larval and transitional stages found in the life cycles of many Eucarida (Brusca and Brusca 2002; Mente 2008).

In all of these cases, parts of the life cycle have evolved more or less distinct identities, enabling them to diversify and specialize to varying degrees independently from other parts. Thus, like segments along the body axis, different stages in the life cycle have evolved increased modularity, with important consequences for the evolutionary and ecological success of many arthropod lineages, such as the origin of true larval stages with distinct ecologies and the evolution of metamorphosis in the Holometabola. Below, we discuss some of the mechanisms underlying this ontogenetic modularity, their origins, interactions and emergent properties, and their ability—by themselves, as well as in interaction with the mechanisms underlying spatial modularity discussed above— to foster innovation and diversification.

Before doing so, however, we would like to emphasize that the developmental decoupling of different life stage as discussed below, just like the developmental decoupling of adjacent segments or groups thereof discussed above, is of course not absolute, instead it is relative (for insightful discussions of these points see Minelli 2003, 2009;

Scholz 2004; Minelli et al. 2006, and references therein). Pleiotropic constraints may be reduced, but certainly not eliminated. Rather than pearls on a string that can be exchanged, added or lost, stages and body regions remain complex phenotypes whose development and existence *enable, and as such constrain*, subsequent stages and adjacent body regions, respectively. This is particularly important if we seek to homologise temporal and spatial developmental modules in phylogenetic studies (Minelli et al. 2006), or seek to infer causes or consequences of developmental evolution between associations and constellations of modules (Scholz 2004). Lastly, where one stage (or body region) ends and another begins is a non-trivial issue to consider when studying, comparing, and interpreting arthropod development (Minelli et al. 2006). For example, while the moulting cycle provides a convenient periodization of arthropod development, developmental, and physiological processes may vary greatly in the degree to which this periodization matters to their actions during each intermoult. Similarly, much developmental disparity may occur *within* traditional stages, as in late holometabolous larvae entering the prepupal stage (Nijhout 1994) or the induction of diapause during portions of larval or pupal development (Denlinger 2002). As such, boundaries between modules may or may not coincide with our preconceived notions. With these caveats in mind, however, we believe that thinking of arthropod development as being composed of, at least in part, temporal and spatial modules that can develop to varying degrees independent of each other, provides a valuable starting point for investigating how spatial and temporal modularity, by themselves and in interaction, may delineate the evolutionary degrees of freedom exploitable by an evolving lineage.

## 18.3.1 Mechanisms

The expression of, and transition between, distinct life stages requires mechanisms that specify life stage identity and order. Here, endocrine mechanisms play a key role in communicating throughout the body of a developing arthropod

what kind of stage in the life cycle to express, and when to transition to the next stage. A detailed presentation and discussion of arthropod endocrine mechanisms is given in Chap. 6 of this volume (Nijhout 2013). Here, we would like to briefly highlight and expand on a subset of issues, best understood through the study of insect development and metamorphosis.

In holometabolous insects, that is, insects that possess a distinct larval stage that transforms into the final adult via a larval-to-pupal and pupal-to-adult moult, the interplay between ecdysteroids and juvenile hormone orchestrates whether moults maintain the current developmental *status quo* (as in a larval-to-larval moult) or lead to the transition to a new stage (as in the larval-to-pupal and pupal-to-adult moults; reviewed in Nijhout 1994, 1999a; Truman and Riddiford 2002; Wheeler and Nijhout 2003). As such, the endocrine control of moulting and metamorphosis effectively subdivides the developing organisms into distinct temporal domains free to utilize, inhibit, or differentially integrate developmental pathways independent of other temporal domains. As a consequence, stage-specific gene expression and modulation of pathway activity are ubiquitous, enabling the promotion of larval-specific features during larval development (such as abdominal prolegs or feeding mandibles in caterpillars), their destruction (prolegs) or transformation into adult structures (feeding mandibles to proboscis) during the pupal stage, as well as the origin of adult-specific structures in late larval and pupal development (e.g. wings and genitalia; Chapman 1998; Heming 2003). Similarly, stage-specific activation of developmental and physiological processes underlie many ontogenetic diet shifts observed across holometabolous life stages, such as in mosquitoes (which switch from detritus-feeding in larvae to blood-(females) or pollen-feeding (males) in adults; Marinotti et al. 2006; Koutsos et al. 2007), or butterflies, (which switch from leaf feeding in caterpillars to nectar feeding in adults; Chapman 1998; Heming 2003; see also Rabossi et al. 2000). Much like spatial modularity discussed above enables adjacent segments to express very different morphologies or produce highly disparate appendages, stage-specific modularity in gene expression and pathway activation facilitates niche-specific adaptation while reducing pleiotropic constraints. Furthermore, as with segment-specific development, stage-specific development does not require the evolution of new genes or pathways: instead, only patterns of activation, inhibition and integration must be stage-specific whereas the genes and their products whose expression is modified can themselves remain conserved. Again, diversity is facilitated through changes in assembly, rather than changes in component parts.

## 18.3.2 Ontogenetic Modularity and Speciation

If ontogenetic modularity facilitates diversification, we would predict that lineages with relatively more modular development should diversify more readily. The most rigorous examination of this basic hypothesis comes from a study by Yang (2001), which compared rates of diversification and extinction at the family level across hemi- and holometabolous insect orders. Insects represent as close to an ideal set of taxa for this purpose as hemi- and holometabolous insect orders differ predominantly in developmental modes, that is, the absence/presence of an elaborate larval and distinct pupal stage, but not in tagmatization or other confounding issues that may complicate comparisons among many other arthropod taxa. One important complication nevertheless remains: hemimetabolous insects constitute a paraphyletic group, with the hemimetabolous Eumetabola (thrips, true bugs, lice, and book lice) being more closely related to the Holometabola than to the remaining Hemimetabola. Taking this factor into account, Yang (2001) calculated family-level rates of diversification from the fossil record and found that Holometabola exhibited a significantly and characteristically higher rate of diversification compared to the less modular Hemimetabola as a whole, or Eumetabola if

analysed separately. Importantly, analyses of survivorship curves for families of the Hemi- and Holometabola found no differences in extinction rates, suggesting that differential diversification, not extinction, underlies the relative taxonomic success of the Holometabola (Yang 2001). Compatible analyses have yet to be conducted in other arthropod lineages, though similar patterns may emerge there as well. For instance, crustacean lineages differ widely in developmental modes, with the most extreme degree of disparity among life stages seen in the Malacostraca, which also happens to represent one of the most species-rich crustacean lineages (Mente 2008; Regier et al. 2010).

Taken together, existing data clearly support the hypothesis that intrinsic differences in ontogenetic modularity influence the long-term diversification rates of lineages. Intriguingly, the same hypothetical framework makes an additional prediction, namely that characters in more ontogenetically modular clades should exhibit greater levels of variation due to their enhanced temporal independence. To date, this key prediction remains untested.

## 18.3.3 Ontogenetic and Spatial Modularity, Diversification and Innovation

Stage-specific modularity interacts with spatial modularity discussed above, allowing not only different body regions to develop independently of each other, but the "same" body region to develop very differently during different stages of the same life cycle. It is a characteristic feature of the holometabolous insects that hardly any body region or appendage looks remotely similar when larval (think maggot, caterpillar, grub) and adult (think fly, butterfly, beetle) stages of the *same* individual are compared. But the contributions of spatial and ontogenetic modularity to evolvability likely go even further, for instance when a formerly stage-restricted trait becomes expressed in a different stage in the same or different location. When this occurs, truly novel traits may originate, but may do so at least initially with modest developmental-genetic modifications (Fig. 18.2).

For example, all pupae in the extremely species-rich scarab beetle genus *Onthophagus* express more or less conspicuous thoracic horns (Fig. 18.2 top panel). A combination of histological and functional studies shows that these horns function as moulting devices, enabling the shedding of the highly sclerotized larval head capsule during the larval-to-pupal moult (Moczek et al. 2006). In the majority of species, these thoracic horns are resorbed during the pupal stage through programmed cell death (Moczek 2006; Kijimoto et al. 2010). Males of some species, however, convert this pupal outgrowth into a corresponding adult horn, which is then used as a weapon in male combat over access to females. Phylogenetic analyses strongly suggest that the moulting function of pupal horns predates the weapon function of adult horns (Moczek et al. 2006). This raises the possibility that adult thoracic horns, a novel trait lacking obvious homology to other insect structures, may have originated through the simple failure to remove a pupa-specific structure. Intriguingly, similar resorption failures occur in natural populations of thoracic-hornless species at a low but detectable frequency (Moczek et al. 2006; Kijimoto et al. 2010). More generally, this example illustrates how ontogenetic modularity enabled the evolution of an originally pupal-specific trait, which, once transferred into a new developmental stage, facilitated the rapid evolution and diversification of a novel trait and function, in this case that of a weapon of sexual selection.

A second example is illustrated by the bioluminescent photic organs in fireflies (beetle family Lampyridae, Fig. 18.2 bottom panel), which are thought to have originally evolved as a larval-specific trait (Branham and Wenzel 2003) likely used to generate aposematic signals to predators (De Cock and Matthysen 1999). While all known larval lampyrids develop photic organs, only a subset of lampyrid lineages also develops the more derived adult organ (Branham and Wenzel 2003). Although adult organs are similar to larval organs in the sense that both

emit light, they are not strictly homologous. Adult organs are more intricately organized, develop in different abdominal segments, and do so even when the larval organ is ablated experimentally (Harvey and Hall 1929). However, both organs derive from the same cell population, the fat body (Hess 1922), and utilize many of the same biochemical processes. Here, ontogenetic and spatial modularity appear to have facilitated the partial carry-over of a larval-specific trait into a different developmental stage, where it now functions in a completely different and novel context, the attraction of mates and, occasionally, of prey (Lloyd 1965).

In summary, ontogenetic modularity allows different life stages of the same life cycle to develop and evolve, partly independently of each other, thereby elevating the long-term diversification rates of lineages. Through its interactions with spatial modularity, it enables the "same" trait to develop very differently in different stages of the same individual, adding evolutionary degrees of freedom to evolving lineages. Lastly, by itself as well as in combination with spatial modularity, ontogenetic modularity can result in the transfer of stage-specific traits to new stages within the same life cycle, thereby creating complex novel traits with modest developmental-genetic means.

## 18.4  Evolvability through Developmental Plasticity

In this last section, we would like to step back from the two major axes of diversification examined above—developmental space and time—and towards a more universal property of all development—plasticity—and examine its contribution to arthropod evolvability. Developmental plasticity can be defined as a single individual's ability to adjust patterns of phenotype expression in response to changes in environmental conditions. Virtually all organisms as well as biological processes exhibit some degree of plasticity (West-Eberhard 2003; Whitman and Ananthakrishnan 2009). On one extreme, such effects may arise simply from the biochemical

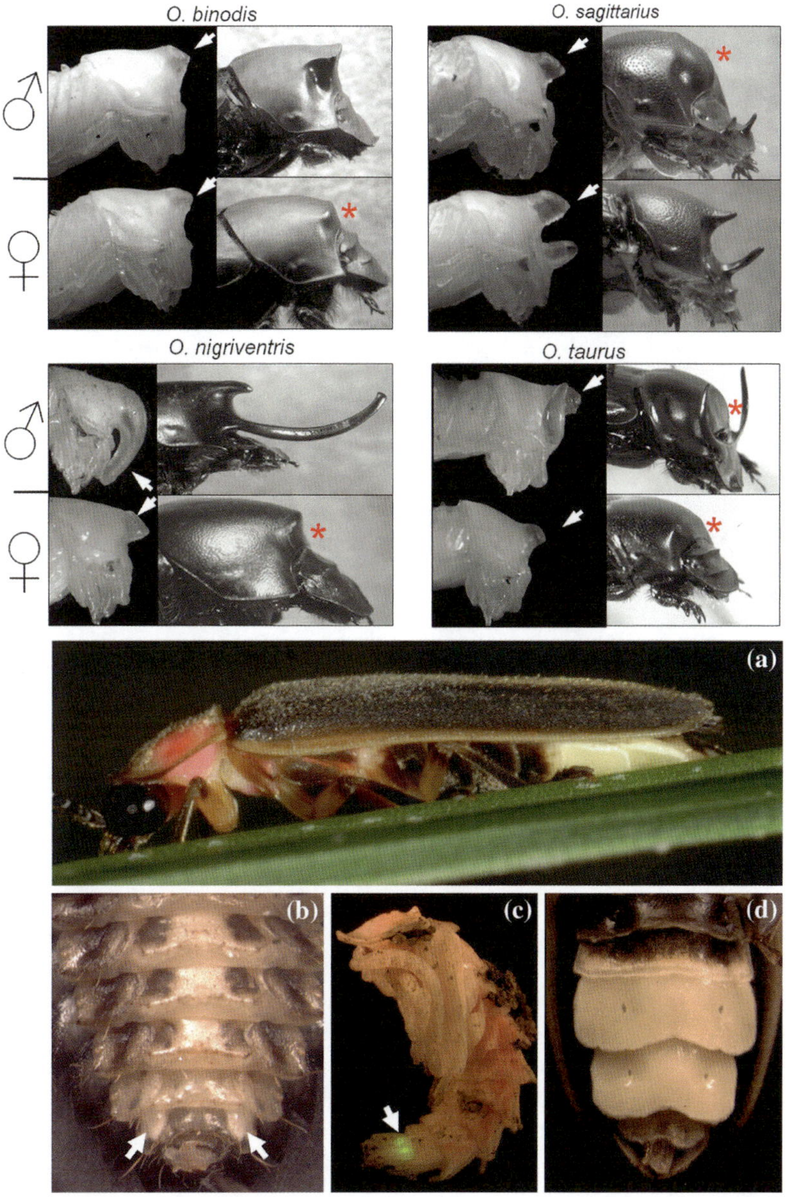

**Fig. 18.2** Two examples of innovation and diversification enabled through the interplay of spatial and ontogenetic modularity. (*Top panel*) Thoracic horn in the beetle genus *Onthophagus*. Shown are pupal and adult morphologies of males and females of four species. All pupae in the genus express conspicuous thoracic horns (marked by *arrow*), which play a critical role in the shedding of the highly sclerotized larval head capsule during the larval-to-pupal moult (Moczek et al. 2006). In the majority of species, these thoracic horns are resorbed during the pupal stage through programmed cell death (marked by *asterisks*) and regardless of sex, as exemplified by *O. taurus* (Moczek 2006a; Kijimoto et al. 2010). Males in a subset of species (shown here for

*O. binodis* and *O. nigriventris*), however, convert this pupal outgrowth into a corresponding adult horn, which is then used as a weapon in male combat over access to females. In one highly unusual species (*O. sagittarius*), these sex-roles are reversed. See text for further description. (*Bottom panel*) Photic organs, or lanterns, of two firefly genera in the beetle family Lampyridae. Shown are (**a**) an adult *Photinus* firefly as well as close ups of *Photuris* larval (**b**), pupal (**c**), and *Photinus* adult (**d**) photic organs (note that larval/pupal lanterns are located on abdominal segment VIII (A8) in most lampyrids while the lanterns of adult males of both *Photinus* (shown) and *Photuris* (not shown) occupy A6-7). See text for further description

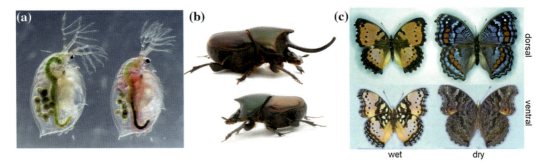

**Fig. 18.3** Three examples of developmental plasticity in which environmental conditions mediate extensive reorganization of development. **a** When the water flea *Daphnia magna* is confronted with poor oxygen concentrations, it increases haemoglobin concentration in the haemolymph by a factor of 15–20, colouring the body *red* (images by Dr. Shin-ichi Tokishita, Tokyo University of Pharmacy and Life Sciences). **b** Good or poor larval feeding conditions cause male horned beetles of many species (here *Onthophagus nigriventris*) to develop into distinct horned and hornless morphs, respectively, which in turn employ distinct fighting and sneaking reproductive behaviours (images by Alex Wild). **c** The Gaudy Commodore, *Precis octavia*, expresses alternative wing colours and pattern arrangements depending on season. Note that dorsal (**c**, *left*) and ventral (**c'**, *right*) wing surfaces adjust their development to season completely independent of each other, representing a striking example of the interactions between spatial modularity and developmental plasticity (images by Fred Nijhout)

and biophysical dependencies of developmental processes, for example, the temperature dependence of enzymatic reactions or the pH-dependent folding of proteins (Schlichting and Pigliucci 1995). On the other extreme are highly choreographed responses to environmental changes such as nutrition-dependent modification of growth and reproduction, physiological responses to temperature shock or oxygen deprivation, caste formation, seasonal migration, learning, and many more (Fig. 18.3). Here, plasticity represents a complex, evolved response that enables developing organisms to maintain high fitness in the face of environmental variability (Schlichting and Pigliucci 1998; Nijhout 1999b, 2003; West-Eberhard 2003). Lastly, plasticity also underlies many homoeostatic responses and canalization in development (Scharloo 1991), that is, processes intended to achieve phenotypic *constancy*, at least on specific levels of biological organization. From minimizing fluctuations in blood sugar levels in the face of nutritional variation to the maintenance of proper scaling relationships of body parts during growth, organisms flexibly adjust a vast array of plastic processes on some levels of biological organization to ensure phenotypic constancy on others (Moczek 2010). In the next section, we would like to highlight the means by which developmental plasticity in its various manifestations contributes to evolvability in general, and how it might have done so specifically during the diversification of certain arthropod lineages.

### 18.4.1 Contributions of Developmental Plasticity to Diversification and Innovation in Arthropods

Developmental plasticity is believed to contribute to organismal innovation and diversification through a diversity of mechanisms operating on a variety of levels of biological organization (Pfennig et al. 2010; Moczek et al. 2011). For instance, developmental plasticity is predicted to facilitate colonization of novel environments, thus increasing the likelihood of adaptive radiations and speciation events. While studies on fish and amphibians support this prediction (Pfennig and McGee 2010), no complementary studies have been conducted to date on any arthropod lineages.

Developmental plasticity is also predicted to facilitate diversification by providing additional targets, such as the developmental, genetic, or endocrine machinery underlying plastic responses, for evolutionary processes to act on. Here, much evidence exists from studies on diverse arthropods that illustrate that the nature of plastic responses can evolve, diversify, and mediate population divergences independent of trait means in different environments (e.g. lacewings: Tauber and Tauber 1972; beetles: Moczek and Nijhout 2002; pitcher-plant mosquitoes: Bradshaw et al. 2003; cabbage-white butterflies: Snell-Rood and Papaj 2009; water fleas: Scoville and Pfrender 2010).

Similarly, developmental plasticity is predicted to enhance modularity of development by providing reusable building blocks for the regulation of diverse development contexts. Several interesting examples exist in insects that suggest that much diversification, and in fact several key innovations, may have been facilitated through the repeated co-option of the same plastic regulatory processes into different contexts (reviewed in Nijhout 1994, 1999a, b, 2003). For instance, in holometabolous insects, the same endocrine machinery coordinates alternative reproductive decisions (whether to invest into growth or reproduction), alternative developmental decisions (moulting and metamorphosis), as well as decisions between alternative phenotypes (facultative diapause, host switch, caste, and morph expression).

Intriguingly, certain types of developmental plasticity may selectively enable the accumulation of genetic variation during variable environmental conditions, and the subsequent conditional release of genetic variation under periods of environmental stasis (reviewed in Snell-Rood et al. 2010). For instance, in cases in which gene expression is restricted to a subset of alternative phenotypes or environments, and individuals experience only one such environment during their lifetime, gene copies residing in non-expressing individuals are not screened by selection. Any mutations that may reside in such copies are predicted to accumulate in a population in proportion to the frequency of individuals experiencing the non-inducing environment (VanDyken and Wade 2010). Studies on male-specific gene expression in aphids (in which males are induced only every 10-20 generations; Brisson and Nuzhdin 2008) and maternal effect genes in *Drosophila* (Cruickshank and Wade 2008) support the prediction of mutations accumulating as a consequence of conditional gene expression. During periods of environmental stasis of inducing environments, the resulting accumulated variation could then be confronted with the full strength of selection, possibly enabling rapid evolutionary responses and adaptive divergences between populations. These predictions remain to be tested in natural populations (Snell-Rood et al. 2010).

Empirical support, especially from studies on insects, does exist for another form of developmental plasticity-mediated accumulation and release of genetic variation, namely under conditions of stress. Recall that developmental plasticity on some levels of biological organization often enables phenotypic constancy on others. Case in point is the facultative up-regulation of heat shock proteins in the face of temperature stress. Heath shock proteins act as chaperones and correct the 3-dimensional folding of proteins, which is increasingly prone to errors as temperatures become more stressful (Morimoto et al. 1997). In so doing, heat shock proteins may also act as buffers against genetic variants by corralling diverse genotypes to converge onto a single protein shape, that is, until the chaperoning capacity of heat shock proteins is exceeded, as might be the case during periods of prolonged stress or in response to sensitizing mutations. Laboratory studies on a diverse array of organisms between plants and fungi to animals, including insects, have highlighted the role of heat shock proteins and temperature stress as a means of accumulating and releasing selectable phenotypic diversity (Rutherford and Lindquist 1998; Queitsch et al. 2002; Cowen and Lindquist 2005; Suzuki and Nijhout 2006). In these studies, environmental stress resulted in a remarkable increase in the amount of selectable phenotypic variation, enabling rapid responses to artificial selection—including some reminiscent of naturally evolved phenotypes (Suzuki and Nijhout 2006). It is likely, though clearly in need of empirical confirmation, that many types of developmentally plastic processes other than heat shock protein induction similarly function as capacitors for dormant genetic variation that may be released during periods of stress. What is entirely unclear, however, are the roles such accumulation and release may play in natural populations and naturally evolved responses to environmental variation, representing one of the most exciting current frontiers at the interface of evolutionary- and ecological-developmental biology.

## 18.4.2 Developmental Plasticity, Evolvability, and the Differential Diversification of Arthropod Lineages

Are there reasons to believe that the contributions of developmental plasticity to evolvability highlighted above, which should be applicable to a wide range of organisms including arthropods, might have nevertheless disproportionately contributed to diversification and innovation of particular arthropod lineages? The answer is likely yes, though thorough comparisons akin to Yang's (2001) study introduced above are clearly needed to better understand this issue. For instance, it is very likely that the high levels of spatial and temporal modularity seen in certain arthropod taxa, such as the holometabolous insects, potentiated the degree to which developmental plasticity was able to facilitate subsequent diversification and innovation. For instance, developmental plasticity and spatial and temporal modularity frequently interact during insect development, enabling body-region and stage-specific diversification of conditional trait expression (see Fig. 18.3c for a spectacular example). Vivid examples of this can be seen during caste formation in social insects (e.g. Wheeler 1986, 1991; Emlen and Nijhout 2000) or the production of alternative male phenotypes (Emlen et al. 2005; Snell-Rood et al. 2011). In each case, facultative-, stage- and segment-specific modulation of development interact, allowing different body regions of the

same individual and stage to exhibit very different responses (from gene expression and growth allometries to pattern formation) to the same environmental changes (such as nutritional or seasonal conditions). This in turn has allowed taxa to diversify in the nature of body- and stage-specific responses, an evolutionary flexibility that likely contributed to the enormous diversity of social castes seen in the Hymenoptera or the diversification of alternative male morphologies observed in a wide range of insect orders (Emlen and Nijhout 2000).

## 18.5    Final Remarks

In this chapter, we posited that arthropod evolvability was differentially enabled in different lineages through spatial modularity, ontogenetic modularity, developmental plasticity, and the interactions among them. Combined, this allowed segments, appendages, and their developmental responses to environmental changes to diversify in a stage-specific manner, thereby elevating diversification rates and facilitating the evolution of complex novel traits. Given the persistence and continued diversification of many arthropod lineages into present times, there is no reason to believe that this process is somehow over. Instead, many opportunities exist, now perhaps more than ever, to examine the interplay between ecology and development in enabling and shaping arthropod evolution in nature.

**Acknowledgments** We thank the editors of this volume for the opportunity to contribute this chapter, as well as Lisa Nagy, Harald Parzer, and Alessandro Minelli for helpful comments on earlier drafts.

## References

Akam M (1989) Hox and HOM: homologous gene clusters in insects and vertebrates. Cell 57:347–349
Akam ME, Dawson I, Tear G (1988) Homeotic genes and the control of segment diversity. Development 104:123–133

Angelini DR, Kaufman TC (2005a) Comparative developmental genetics and the evolution of arthropod body plans. Annu Rev Genet 39:95–119

Angelini DR, Kaufman TC (2005b) Insect appendages and comparative ontogenetics. Dev Biol 286:57–77

Averof M, Patel NH (1997) Crustacean appendage evolution associated with changes in Hox gene expression. Nature 388:682–686

Bradshaw WE, Quebodeaux MC, Holzapfel CM (2003) Circadian rhythmicity and photoperiodism in the pitcher-plant mosquito: Adaptive response to the photic environment or correlated response to climatic adaptation? Am Nat 161:735–748

Branham MA, Wenzel JW (2003) The origin of photic behavior and the evolution of sexual communication in fireflies (Coleoptera: Lampyridae). Cladistics 19:1–22

Brisson JA, Nuzhdin SV (2008) Rarity of males in pea aphids results in mutational decay. Science 319:58

Brusca RC, Brusca GJ (2002) Invertebrates. Sinauer, Sunderland

Carroll SB, Gates J, Keys DN, Paddock SW, Panganiban GE, Selegue JE, Williams JA (1994) Pattern formation and eyespot determination in butterfly wings. Science 265:109–114

Chapman RF (1998) The insects: structure and function. Cambridge University Press, Cambridge

Cohen SM, Brönner G, Küttner F, Jürgens G, Jäckle H (1989) Distal-less encodes a homeodomain protein required for limb development in Drosophila. Nature 338:432–434

Conway Morris S, Peel JS, Higgins AK, Soper NJ, Davis NC (1987) A Burgess Shale-like fauna from the Lower Cambrian of North Greenland. Nature 326:181–183

Cowen LE, Lindquist S (2005) Hsp90 potentiates the rapid evolution of new traits: drug resistance in diverse fungi. Science 309:2185–2189

Cruickshank T, Wade MJ (2008) Microevolutionary support for a developmental hourglass: gene expression patterns shape sequence variation and divergence in Drosophila. Evol Dev 10:583–590

De Cock R, Matthysen E (1999) Aposematism and bioluminescence: experimental evidence from glow-worm larvae (Coleoptera: Lampyridae). Evol Ecol 13:619–639

Denlinger DL (2002) Regulation of diapause. Annu Rev Entomol 47:93–122

Emlen DJ, Hunt J, Simmons LW (2005) Evolution of sexual dimorphism and male dimorphism in the expression of beetle horns: phylogenetic evidence for modularity, evolutionary lability, and constraint. Am Nat 166:S42–S68

Emlen DJ, Nijhout HF (2000) The development and evolution of exaggerated morphologies in insects. Annu Rev Entomol 45:661–708

Force A, Lynch M, Pickett FB, Amores A, Yan YL, Postlethwait J (1999) Preservation of duplicate genes by complementary, degenerative mutations. Genetics 151:1531–1545

Galis F, Metz JAJ (2007) Evolutionary novelties: the making and breaking of pleiotropic constraints. Integr Comp Biol 47:409–419

Grimaldi DA, Engel MS (2005) Evolution of the insects. Cambridge University Press, New York

Gullan PJ, Cranston PS (2004) The insects: an outline of entomology, 3rd edn. Wiley, London

Halder G, Callaerts P, Gehring W (1995) Induction of ectopic eyes by targeted expression of the eyeless gene of Drosophila. Science 267:1788–1792

Harvey EN, Hall RT (1929) Will the adult firefly luminesce if its larval organs are entirely removed? Science 69:253–254

Heming BS (2003) Insect development and evolution. Cornell University Press, Ithaca

Hess WN (1922) Origin and development of the light-organs of Photuris pennsylvanica de Geer. J Morphol 36:244–277

Hou XG, Aldridge RJ, Bergström J, Siveter DJ, Feng XH (2004) The Cambrian fossils of Chengjiang, China. Blackwell Science, London

Hughes CL, Kaufman TC (2002a) Hox genes and the evolution of the arthropod body plan. Evol Dev 4:459–499

Hughes CL, Kaufman TC (2002b) Exploring the myriapod body plan: expression patterns of the ten Hox genes in a centipede. Development 129:1225–1238

Kazunori Y (2012) The treehopper's helmet is not homologous with wings (Hemiptera: Membracidae). Syst Entomol 37:2–6

Khadjeh S, Turetzek N, Pechmann M, Schwager EE, Wimmer EA, Damen WGM, Prpic NM (2012) Divergent role of the Hox gene Antennapedia in spiders is responsible for the convergent evolution of abdominal limb repression. Proc Natl Acad Sci USA 109:4921–4926

Kijimoto T, Andrews J, Moczek AP (2010) Programmed cell death shapes the expression of horns within and between species of horned beetles. Evol Dev 12:449–458

Koutsos AC, Blass C, Meister S, Schmidt S, MacCallum R, Soares MB, Collins FH, Benes V, Zdobnov E, Kafatos FC, Christophides GK (2007) Life cycle transcriptome of the malaria mosquito Anopheles gambiae and comparison with the fruitfly Drosophila melanogaster. Proc Natl Acad Sci USA 104:11304–11309

Lewis DL, DeCamillis M, Bennett RL (2000) Distinct roles of the homeotic genes Ubx and Abd-A in beetle embryonic abdominal appendage development. Proc Natl Acad Sci USA 97:4504–4509

Lewis EB (1978) A gene complex controlling segmentation in Drosophila. Nature 276:565–570

Lloyd JE (1965) Aggressive mimicry in Photuris: firefly femmes fatales. Science 149:653–654

Marinotti O, Calvo E, Nguyen QK, Dissanayake S, Ribeiro JMC, James AA (2006) Genome-wide analysis of gene expression in adult Anopheles gambiae. Insect Mol Biol 15:1–12

Mente E (2008) Reproductive biology of crustaceans: case studies of decapod crustaceans. Science Publishers, Enfield

Mikó I, Friedrich F, Yoder MJ, Hines HM, Deitz LL, Bertone MA, Seltmann KC, Wallace MS, Deans AR (2012) On dorsal prothoracic appendages in treehoppers (Hemiptera: Membracidae) and the nature of morphological evidence. PLoS ONE 7(1):e30137. doi:10.1371/journal.pone.0030137

Minelli A (2003) The development of animal form: ontogeny, morphology and evolution. Cambridge University Press, Cambridge

Minelli A (2009) Perspectives in animal phylogeny and evolution. Oxford University Press, Oxford

Minelli A, Brena C, Deflorian G, Maruzzo D, Fusco G (2006) From embryo to adult-beyond the conventional periodization of arthropod development. Dev Genes Evol 216:373–383

Moczek AP (2006) Pupal remodeling and the development and evolution of sexual dimorphism in horned beetles. Am Nat 168:711–729

Moczek AP (2010) Phenotypic plasticity and diversity in insects. In: Minelli A, Fusco G (eds) From polyphenism to complex metazoan life cycles, philosophical transactions of the royal society B vol 365. Royal Society of publishing, London, pp 593–603

Moczek AP, Cruickshank TE, Shelby JA (2006) When ontogeny reveals what phylogeny hides: gain and loss of horns during development and evolution of horned beetles. Evolution 60:2329–2341

Moczek AP, Nagy LM (2005) Diverse developmental mechanisms contribute to different levels of diversity in horned beetles. Evol Dev 7:175–185

Moczek AP, Nijhout HF (2002) Developmental mechanisms of threshold evolution in a polyphenic beetle. Evol Dev 4:252–264

Moczek AP, Rose DJ (2009) Differential recruitment of limb patterning genes during development and diversification of beetle horns. Proc Natl Acad Sci USA 106:8992–8997

Moczek AP, Sultan S, Foster S, Ledon-Rettig C, Dworkin I, Nijhout HF, Abouheif E, Pfennig D (2011) The role of developmental plasticity in evolutionary innovation. Proc R Soc B 278: 2705–2713

Morata G (2001) How Drosophila appendages develop. Nat Rev Mol Cell Biol 2:89–97

Morimoto RI, Kline MP, Bimston DN, Cotto JJ (1997) The heat-shock response: regulation and function of heat-shock proteins and molecular chaperones. Essays Biochem 32:17–29

Nijhout HF (1994) Insect hormones. Princeton University Press, Princeton

Nijhout HF (1999a) Hormonal control in larval development and evolution—insects. In: Hall BK, Wake MH (eds) The origin and evolution of larval forms. Academic, San Diego, pp 218–254

Nijhout HF (1999b) Control mechanisms of polyphenic development in insects. Bioscience 49:181–192

Nijhout HF (2003) Development and evolution of adaptive polyphenisms. Evol Dev 5:9–18

Nijhout HF (2013) Arthropod developmental endocrinology. In: Minelli A, Boxshall G, Fusco G (eds) Arthropod biology and evolution. Springer-Verlag Berlin Heidelberg

Ødegaard F (2000) How many species of arthropods? Erwin's estimate revised. Biol J Linn Soc 71:583–597

Ohno S (1970) Evolution by gene duplication. Springer, Berlin

Panganiban G, Irvine SM, Lowe C, Roehl H, Corley LS, Sherbon B, Grenier JK, Fallon JF, Kimble J, Walker M, Wray GA, Swalla BJ, Martindale MQ, Carroll SB (1997) The origins and evolution of animal appendages. Proc Natl Acad Sci USA 94:5162–5166

Pfennig DW, McGee M (2010) Resource polyphenism increases species richness: a test of the hypothesis. Phil Trans R Soc B 365:577–591

Pfennig D, Wund MA, Snell-Rood EC, Cruickshank T, Schlichting CD, Moczek AP (2010) Phenotypic plasticity's impacts on diversification and speciation. Trends Ecol Evol 25:459–467

Prud'homme B, Minervino C, Hocine M, Cande JD, Aouane A, Dufour HD, Kassner VA, Gompel N (2011) Body plan innovation in treehoppers through the evolution of an extra wing-like appendage. Nature 473:83–86

Queitsch C, Sangster TA, Lindquist S (2002) Hsp90 as a capacitor of phenotypic variation. Nature 417:618–624

Rabossi A, Acion L, Quesada-Allue LA (2000) Metamorphosis-associated proteolysis in Ceratitis capitata. Entomol Exp Appl 94:57–65

Regier JC, Shultz JW, Zwick A, Hussey A, Ball B, Wetzer R, Martin JW, Cunningham CW (2010) Arthropod relationships revealed by phylogenomic analysis of nuclear protein-coding sequences. Nature 463:1079–1082

Rutherford SL, Lindquist S (1998) Hsp90 as a capacitor for morphological evolution. Nature 396:336–342

Scharloo W (1991) Canalization: genetic and developmental aspects. Ann Rev Ecol Syst 22:65–93

Schlichting CD, Pigliucci M (1995) Gene regulation, quantitative genetics, and the evolution of reaction norms. Evol Ecol 9:154–168

Schlichting CD, Pigliucci M (1998) Phenotypic evolution: a reaction norm perspective. Sinauer, Sunderland

Scholz G (2004) On comparisons and causes in evolutionary developmental biology. In: Minelli A, Fusco G (eds) Evolving pathways: key themes in evolutionary developmental biology. Cambridge University Press, Cambridge, pp 144–159

Scoville A, Pfrender M (2010) Phenotypic plasticity facilitates recurrent rapid adaptation to introduced predators. Proc Natl Acad Sci USA 107:4260–4263

Snell-Rood EC, Cash A, Han MV, Kijimoto T, Andrews J, Moczek AP (2011) Developmental decoupling of alternative phenotypes: insights from the transcriptomes of horn-polyphenic beetles. Evolution 65:231–245

Snell-Rood EC, Papaj DR (2009) Patterns of phenotypic plasticity in common and rare environments: a study of host use and color learning in the cabbage white butterfly, *Pieris rapae*. Am Nat 173:615–631

Snell-Rood EC, VanDyken JD, Cruickshank TE, Wade MJ, Moczek AP (2010) Toward a population genetic framework of developmental evolution: costs, limits, and consequences of phenotypic plasticity. Bio Essays 32:71–81

Storch V, Welch U (1991) Systematische zoologie. Gustav Fischer Verlag, Stuttgart

Suzuki Y, Nijhout HF (2006) Evolution of a polyphenism by genetic accommodation. Science 311:650–652

Tauber MJ, Tauber CA (1972) Geographic variation in critical photoperiod and in diapause intensity of *Chrysopa carnea* (Neuroptera). J Ins Physiol 18:25–29

Truman JW, Riddiford LM (2002) Endocrine insights into the evolution of metamorphosis in insects. Annu Rev Entomol 47:467–500

Vachon G, Cohen B, Pfeifle C, McGuffin ME, Botas J, Cohen SM (1992) Homeotic genes of the Bithorax complex repress limb development in the abdomen of the *Drosophila* embryo through the target gene *Distal-less*. Cell 71:437–450

Valentine JW (2004) On the origin of phyla. University of Chicago Press, Chicago

Van Dyken JD, Wade MJ (2010) Quantifying the evolutionary consequences of conditional gene expression in time and space. Genetics 184:439–453

Warren RW, Nagy L, Selegue J, Gates J, Carroll S (1994) Evolution of homeotic gene-regulation and function in flies and butterflies. Nature 372:458–461

Wasik BR, Rose DJ, Moczek AP (2010) Beetle horns are regulated by the Hox gene, *Sex combs reduced*, in a species- and sex-specific manner. Evol Dev 12:353–362

West-Eberhard MJ (2003) Developmental plasticity and evolution. Oxford University Press, New York

Wheeler DE (1986) Developmental and physiological determinants of caste in social Hymenoptera: evolutionary implications. Am Nat 128:13–34

Wheeler DE (1991) Developmental basis of worker caste polymorphism in ants. Am Nat 138:1218–1238

Wheeler DE, Nijhout HF (2003) A perspective for understanding the modes of juvenile hormone action as a lipid signaling system. Bio Essays 25:994–1001

Whitman DW, Ananthakrishnan TN (2009) Phenotypic plasticity of insects: mechanisms and consequences. Science Publishers, Enfield

Yang AS (2001) Modularity, evolvability, and adaptive radiations: a comparison of the hemi- and holometabolous insects. Evol Dev 2:59–72

# Index

A

Abdomen
  definition, 215
  gene expression, 482
Abdominal appendages, 108
*Abdominal-A (abd-A)*, 260, 276, 481
  expression patterns, 214, *279*
*Abdominal-B (Abd-B)*, 260
  expression patterns, 214
*Acanthaeschna parvistigma*
  segmental mismatch, 209
*Acanthoscurria*, 252
Acanthosoma, 110
Acari (mites), 27, 96
  anamorphic development, 82
  fossil, 428
  germ disc, 79
  gut diverticula, 216
  heart, 350
  lack of regenerative potential, 152
  Lazarus appendages, 108
  limb regeneration, *153*, 155
  meiofauna, 420
  number of nymphal stages, 98
  olfactory glomeruli, 319
  parthenogenesis induced by bacteria, 453
  post-embryonic segment formation, 79
  pronymph, 112
  regeneration, *156*
  Rhynie fossil, 424
  segment number, *204*
  tagmosis, *205*
  total cleavage, 69
Accessory pulsatile organs, 346, 358, *367*, 369
  Crustacea, 363
  Hexapoda, 367
*Acentropus niveus*
  heartbeat rate, 349
Acercaria, 31
*Achaearanea tepidariorum*
  blastoderm, 65
  expression of *Twist*, 374

Achelata
  accessory olfactory lobes, 322
*Acheta*
  maxilla, 252
  ovipositor heart, 371
*Acheta domesticus*
  neurogenesis in the adult, 113
*Acleris minuta*, 102
*Acp65A*, 190
*Acraea encedon*
  behaviour affected by *Wolbachia*, 459
Acrididae
  lack of neurogenesis in the adult, 113
  structural changes during nymphal phase, 111
Acroceridae
  hypermetamorphosis, 115
*Acrolepiopsis assectella*, 464
Acron, 200, 228
Acrothoracica
  segment number, 205
  tagmosis, 205
Actin, 362
Actinotrichida
  anamorphosis, 94
  segment number, 199, 204
  tagmosis, 205
*Activator Protein-2*, 245
*Acyrtosiphon pisum* (pea aphid)
  association with lambdoid bacteriophage, 466
  genome, *43*, 49
  symbiosis with *Buchnera*, 443
  symbiosis with *Hamiltonella*, 448
  symbiosis with *Regiella*, 448
  symbiosis with *Serratia*, 448
Adephaga
  number of larval stages, 98
Adesmata
  sexual dimorphism in segment number, 203
Adherens junctions (AJ), *172*
*Adoxophyes orana*, *102*
Adult, 91, 95
  non-feeding, 97

Printed by Printforce, the Netherlands